Lecture Notes in Computer Science 16124

Advanced Research in Computing and Software Science

Subline of Lecture Notes in Computer Science

More information about this series at https://link.springer.com/bookseries/558

Henning Fernau · Philipp Kindermann
Editors

Graph-Theoretic Concepts in Computer Science

51st International Workshop, WG 2025
Otzenhausen, Germany, June 11–13, 2025
Revised Selected Papers

Springer

Editors
Henning Fernau
Universität Trier
Trier, Germany

Philipp Kindermann
Universität Trier
Trier, Germany

ISSN 0302-9743 ISSN 1611-3349 (electronic)
Lecture Notes in Computer Science
ISBN 978-3-032-11834-9 ISBN 978-3-032-11835-6 (eBook)
https://doi.org/10.1007/978-3-032-11835-6

This Springer imprint is published by the registered company Springer Nature Switzerland AG
The registered company address is: Gewerbestrasse 11, 6330 Cham, Switzerland

Preface

This volume contains the 32 papers presented at the 51st International Workshop on Graph-Theoretic Concepts in Computer Science (WG 2025). The conference was held in Otzenhausen, Germany from June 11–13, 2025, with a reception on the evening of June 10. About 45 participants from all over the world attended the conference in person.

The WG workshops have a long tradition going back to 1975: the previous workshops were held in Austria (twice), the Czech Republic (twice), France (three times), Germany (25 times), Greece, Israel, Italy, the Netherlands (five times), Norway, Poland, Slovakia, Slovenia, Spain, Switzerland (twice), Turkey, and the UK (twice). This was the 26th time the conference was held in Germany.

WG aims to merge theory and practice by demonstrating how concepts from graph theory can be applied to various areas in computer science, or by extracting new graph-theoretic problems from applications. The goal is to present emerging research results and to identify and explore directions of future research on diverse topics related to graph theory, including: design and analysis of sequential, parallel, randomized, and parameterized algorithms; distributed graph and network algorithms; structural graph theory with algorithmic or complexity applications; computational complexity of graph and network problems; graph grammars, graph rewriting systems and graph modeling; graph drawing and layouts; computational geometry; computational biology; graph mining; random graphs and models of the web and scale-free networks; and support of the above concepts by suitable implementations and applications. The conference is well balanced with respect to established researchers and junior scientists.

We received 105 submissions, fourteen of which were withdrawn before entering the (single-blind) review process. Submissions by the members of the Program Committee (PC) were not allowed. Each submission was reviewed by at least three PC members and/or subreviewers, and evaluated on its quality, originality, and relevance to the conference. After a careful discussion, the PC selected 32 papers for presentation at the conference and inclusion in the proceedings, which yields an acceptance ratio of 35%. Due to the high competition and a limited schedule, there were papers that could not be accepted although they deserved to be.

The program included two inspiring invited talks by Therese Biedl (University of Waterloo, Canada) on "From planar to 1-planar. or not?" and Mamadou Kanté (Université Clermont Auvergne, France) on "Some Tools from Formal Language Theory for Graph Theory".

The 5th WG Test of Time Award, given for a highly influential paper presented at a previous WG conference, was given to Derek G. Corneil, Stephan Olariu, and Lorna Stewart for their paper "Asteroidal Triple-Free Graphs" from WG 1993. An excellent third invited talk on this topic was given by Stephan Olariu (Old Dominion University, USA).

The Program Committee also selected two papers to receive the Best Paper Award and the Best Student Paper Award, respectively. These awards were sponsored by Springer.

During the conference, participants voted for a Best Presentation Award, which was handed out to two presenters that were tied for the most votes. The awardees are:

- Best Paper Award: Paweł Rzążewski and Bartosz Walczak. Polynomial-time recognition and maximum independent set in Burling graphs.
- Best Student Paper Award: Michael Lampis and Manolis Vasilakis. Structural Parameterizations for Induced and Acyclic Matching.
- Best Presentation Award: Simon D. Fink. The Peculiarities of Extending Queue Layouts.
- Best Presentation Award: Denise Graafsma. Playing Snake on a Graph.

The conference was dedicated to the memory of Hartmut Noltemeier, who passed away on May 10, 2025, at the age of 84. He was one of the co-founders of WG, and organized and chaired the conference three times: 2nd edition, Göttingen (1976); 6th edition, Bad Honnef (1980); and 11th edition, Schwanberg (1985).

We would like to thank all invited speakers for agreeing to give a talk at the conference, the Program Committee members who graciously gave their time and energy, and 149 external reviewers for their expertise.

We are very grateful to the Deutsche Forschungsgemeinschaft (DFG) for providing us with a grant that greatly helped us keeping conference fees at a reasonable price. We thank Springer for publishing the proceedings of WG 2025 in the ARCoSS subline of their LNCS series and for their financial support towards the best paper and the best student paper awards. Also, we are grateful to Easychair, which made the management work on WG really easy. Moreover, we like to thank the staff at Europäische Akademie Otzenhausen (EAO) for their friendliness and helpfulness that gave the conference a unique atmosphere in a modern environment embedded in a surroundings that stretch over millenia.

Finally, we thank the Steering Committee for giving us the opportunity to serve as program chairs of WG 2025 and for their continuous support.

July 2025

Henning Fernau
Philipp Kindermann

Organization

Steering Committee

Isolde Adler	University of Bamberg, Germany
Michael Bekos	University of Ioannina, Greece
Vida Dujmović	University of Ottawa, Canada
Fedor Fomin	University of Bergen, Norway
Michael Kaufmann	University of Tübingen Germany
Łukasz Kowalik	University of Warsaw, Poland
Daniel Král'	Masaryk University, Czech Republic
Jan Kratochvíl	Charles University, Czech Republic
Martin Milanič	University of Primorska, Slovenia
Haiko Müller	University of Leeds, UK
Daniel Paulusma (Chair)	Durham University, UK
Bernard Ries	University of Fribourg, Switzerland
Saket Saurabh	Institute of Mathematical Sciences, India
Kristina Vušković	University of Leeds, UK
Meirav Zehavi	Ben-Gurion University of the Israel, Negev

Program Committee

Faisal Abu-Khzam	Lebanese American University, Lebanon
Akanksha Agrawal	Indian Institute of Technology, Madras, India
Michael Bekos	University of Ioannina, Greece
Maike Buchin	Ruhr Universität Bochum, Germany
Katrin Casel	Humboldt University of Berlin, Germany
Maria Chudnovsky	Princeton University, USA
Giordano Da Lozzo	Roma Tre University, Italy
Henning Fernau (Chair)	Universität Trier, Germany
Celina Figueiredo	Universidade Federal do Rio de Janeiro, Brazil
Florent Foucaud	LIMOS — Université Clermont Auvergne, France
Henry Förster	University of Tübingen, Germany
Jan Goedgebeur	KU Leuven, Belgium
Siddharth Gupta	BITS Pilani, Goa Campus, India
Tesshu Hanaka	Kyushu University, Japan
Adriana Hansberg	Universidad Nacional Autónoma de México, Mexico
Pinar Heggernes	University of Bergen, Norway

Philipp Kindermann (Chair)	Universität Trier, Germany
Jonathan Klawitter	University of Auckland, New Zealand
Linda Kleist	Universität Potsdam, Germany
Boris Klemz	Universität Würzburg, Germany
Vadim Lozin	University of Warwick, UK
Irene Parada	Universitat Politècnica de Catalunya, Spain
Xingqin Qi	Shandong University, China
Alessandra Tappini	University of Perugia, Italy
Meirav Zehavi	Ben-Gurion University of the Negev, Israel

Additional Reviewers

Alecu, Bogdan
Almalki, Nada
Andreola, Giordano
Angelini, Patrizio
Antipov, Leonid
Antony, Dhanyamol
Araújo, Júlio
Atminas, Aistis
Basu Roy, Aniket
Benevides, Fabricio S.
Blažej, Václav
Bousquet, Nicolas
Brand, Timo
Cambie, Stijn
Cao, Zhulou
Capitelli, Nicolas
Caroppo, Susanna
Chakraborty, Dibyayan
Chakraborty, Dipayan
Chakraborty, Sankardeep
Chaudhary, Juhi
Chekan, Vera
Couto, Fernanda
da Silva, Murilo
D'Elia, Marco
Dailly, Antoine
Dallard, Clément
de Castro Mendes Gomes, Guilherme
de Lima, Paloma
Depian, Thomas
Di Giacomo, Emilio
van Dijk, Thomas C.
Dobler, Alexander
Dumas, Maël
Eagling-Vose, Tala
Ekim, Tinaz
Elbert Simões, Jefferson
Espuny Díaz, Alberto
Eto, Hiroshi
Fink, Simon D.
Frati, Fabrizio
Gahlawat, Harmender
Gerlach, Tim
Golovach, Petr
Gorsky, Maximilian
Groenland, Carla
Grosso, Fabrizio
Gutierrez, Marisa
Haase, Carolina
Hegemann, Tim
Hoang, Duc A.
Inamdar, Tanmay
Isenmann, Lucas
Iwamasa, Yuni
Jacob, Hugo
Jaffke, Lars
Jana, Satyabrata
Jooken, Jorik
Joret, Gwenaël
Kanesh, Lawqueen
Katsanou, Eleni
Kiefer, Sandra

Kißler, Wolf
Kiya, Hironori
Kluk, Kacper
Klute, Fabian
Knorr, Kristin
Kobayashi, Yasuaki
Kobayashi, Yusuke
Konstantinidis, Athanasios
Korpelainen, Nicholas
Krebs, Jonathan
Krenn, Daniel
Kryven, Myroslav
Kurita, Kazuhiro
Köhler, Noleen
Lauff, Robert
Legrand-Duchesne, Clément
Linial, Nathan
Lozano, Antoni
Maia, Ana Karolinna
Martin, Barnaby
Meiburg, Alexander
Montealegre, Pedro
Montecchiani, Fabrizio
Mouawad, Amer
Mulzer, Wolfgang
Nascimento, Julliano
Oijid, Nacim
Ophelders, Tim
Ortali, Giacomo
Pavlidi, Maria Eleni
Pedrosa, Lehilton L. C.
Perez, Anthony
Perz, Daniel
Piselli, Tommaso
Plätz, Lukas
Pournajafi, Pegah
Pyatkin, Artem
Rafiey, Arash
Raichel, Benjamin
Razgon, Igor
Rehs, Carolin
Reidl, Felix
Renders, Jarne
Rollin, Jonathan
Roshany Tabrizi, Aida
Rzążewski, Paweł
Sahu, Abhishek
Saitoh, Toshiki
Sato, Cristiane
Schulz, André
Sharma, Vidya Sagar
Siebertz, Sebastian
Sieper, Marie Diana
Sikora, Florian
Smith, Siani
Sommer, Frank
Spoerhase, Joachim
Storandt, Sabine
Štorgel, Kenny
Strippoli, Allegra
Stumpf, Peter
Subramanian, Aditya
Sun, Wuyang
Suomela, Jukka
Swiadek, Jan Erik
Takazawa, Kenjiro
Tey, Joaquín
Thiel, Lea
Thoumi, Sergio
Trotignon, Nicolas
Ueckerdt, Torsten
van der Hoog, Ivor
van Leeuwen, Erik Jan
van Dijk, Thomas C.
Vasilakis, Manolis
Vilmin, Simon
Vortmeier, Nils
Walczak, Bartosz
Wang, Fan
Weinberger, Alexandra
Wietheger, Simon
Wilhelm, Marcus
WĘgrzycki, Karol
Xu, Shou-Jun
Yu, Dongxiao
Zamaraev, Viktor
Zink, Johannes

The Long Tradition of WG

WG 1975	U. Pape – Berlin, Germany
WG 1976	H. Noltemeier – Göttingen, Germany
WG 1977	J. Mühlbacher – Linz, Austria
WG 1978	M. Nagl, H. J. Schneider – Burg Feuerstein, near Erlangen, Germany
WG 1979	U. Pape – Berlin, Germany
WG 1980	H. Noltemeier – Bad Honnef, Germany
WG 1981	J. Mühlbacher – Linz, Austria
WG 1982	H. J. Schneider, H. Göttler – Neuenkirchen, near Erlangen, Germany
WG 1983	M. Nagl, J. Perl – Haus Ohrbeck near Onasbrück, Germany
WG 1984	U. Pape – Berlin, Germany
WG 1985	H. Noltemeier – Schloß Schwanberg near Würzburg, Germany
WG 1986	G. Tinhofer, G. Schmidt – Stift Bernried near Munich, Germany
WG 1987	H. Göttler, H. J. Schneider – Kloster Banz near Bamberg, Germany
WG 1988	J. van Leeuwen – Amsterdam, The Netherlands
WG 1989	M. Nagl – Castle Rolduc, The Netherlands
WG 1990	R. H. Möhring – Johannesstift Berlin, Germany
WG 1991	G. Schmidt, R. Berghammer – Fischbachau near Munich, Germany
WG 1992	E. W. Mayr – Wilhelm-Kempf-Haus, Wiesbaden-Naurod, Germany
WG 1993	J. van Leeuwen – Utrecht, The Netherlands
WG 1994	G. Tinhofer, E. W. Mayr, G. Schmidt – Munich, Germany
WG 1995	M. Nagl – Haus Eich, Aachen, Germany
WG 1996	G. Ausiello, A. Marchetti-Spaccamela – Cadenabbia near Como, Italy
WG 1997	R. H. Möhring – Bildungszentrum am Müggelsee, Berlin, Germany
WG 1998	J. Hromkovič, O. Sýkora – Smolenice Castle, Slovakia
WG 1999	P. Widmayer – Monte Verità, Ascona, Switzerland
WG 2000	D. Wagner – Waldhaus Jakob, Konstanz, Germany
WG 2001	A. Brandstädt, Boltenhagen near Rostock, Germany

WG 2002	L. Kučera – Český Krumlov, Czech Republic
WG 2003	H. L. Bodlaender – Elspeet, The Netherlands
WG 2004	J. Hromkovič, M. Nagl – Bad Honnef, Germany
WG 2005	D. Kratsch – Île du Saulcy, Metz, France
WG 2006	F. V. Fomin – Sotra near Bergen, Norway
WG 2007	A. Brandstädt, D. Kratsch, H. Müller – Jena, Germany
WG 2008	H. Broersma, T. Erlebach – Durham, UK
WG 2009	C. Paul, M. Habib – Montpellier, France
WG 2010	D. M. Thilikos – Zarós, Crete, Greece
WG 2011	J. Kratochvíl – Teplá Monastery, West Bohemia, Czech Republic
WG 2012	M. C. Golumbic, G. Morgenstern, M. Stern, A. Levy – Israel
WG 2013	A. Brandstädt, K. Jansen, R. Reischuk – Lübeck, Germany
WG 2014	D. Kratsch, I. Todinca – Le Domaine de Chalès, Orléans, France
WG 2015	E. W. Mayr – Garching near Munich, Germany
WG 2016	P. Heggernes – Rumeli Hisarüstü, Istanbul, Turkey
WG 2017	H. L. Bodlaender, G. J. Woeginger – Eindhoven, The Netherlands
WG 2018	A. Brandstädt, E. Köhler, K. Meer – Cottbus, Germany
WG 2019	I. Sau, D. M. Thilikos –- Vall de Núria, Catalunya, Spain
WG 2020	I. Adler, H. Müller –- Leeds, UK (virtual)
WG 2021	Ł. Kowalik, M. Pilipczuk, P. Rzążewski –- Warsaw, Poland
WG 2022	M. A. Bekos, M. Kaufmann –- Tübingen, Germany
WG 2023	D. Paulusma, B. Ries – Fribourg, Switzerland
WG 2024	D. Kráľ, M. Milanič – Gozd Martuljek, Slovenia
WG 2025	H. Fernau, P. Kindermann – Otzenhausen, Germany

Contents

Boundaried Kernelization

Leonid Antipov(✉) and Stefan Kratsch

Humboldt-Universität zu Berlin, Berlin, Germany
leonid.antipov@hu-berlin.de, stefan.kratsch@hu-berlin.de

Abstract. The notion of a (polynomial) kernelization from parameterized complexity is a well-studied model for efficient preprocessing for hard computational problems. By now, it is quite well understood which parameterized problems do or (conditionally) do not admit a polynomial kernelization. Unfortunately, polynomial kernelizations seem to require strong restrictions on the *global structure* of inputs. To avoid this restriction, we propose a model for efficient *local preprocessing* that is aimed at local structure in inputs. Our notion, dubbed *boundaried kernelization*, is inspired by protrusions and protrusion replacement, which are tools in meta-kernelization [Bodlaender et al. J'ACM 2016]. Unlike previous work, we study the preprocessing of suitable boundaried graphs in their own right, in significantly more general settings, and aiming for polynomial rather than exponential bounds. We establish polynomial boundaried kernelizations for a number of problems, while *unconditionally* ruling out such results for others. We also show that boundaried kernelization can be a tool for regular kernelization by using it to obtain an improved kernelization for Vertex Cover parameterized by the vertex-deletion distance to a graph of bounded treedepth.

Keywords: local preprocessing · kernelization · parameterized complexity

1 Introduction

We introduce *boundaried kernelization* as a model for provably efficient *local preprocessing*, i.e., preprocessing that leverages local structure without having to explicitly consider the entire instance. This extends the notion of *kernelization* from *parameterized complexity* and is inspired by *protrusions* and *protrusion replacement*, which have been used, in particular, in meta theorems for kernelization (e.g., [9]). Unlike protrusions, which are subgraphs of constant treewidth that have a constant-size interface/boundary to the rest of the graph, we study different kinds of internal structure and we allow for larger boundaries. Further, we usually aim to reduce the size to at most polynomial in boundary size and optional further parameters, while for protrusion replacement it is sufficient to

The work of Leonid Antipov is funded by the Deutsche Forschungsgemeinschaft (DFG, German Research Foundation) Project number 526215872.

H. Fernau and P. Kindermann (Eds.): WG 2025, LNCS 16124, pp. 1–17, 2026.
https://doi.org/10.1007/978-3-032-11835-6_1

reduce to a size with any (usually at least exponential) dependence on treewidth and boundary size. Let us first give some context.

Parameterized complexity studies *parameterized problems*, which are classical problems augmented with one or more parameters such as solution size, dimension, or treewidth; often, parameters quantify structural properties. A central goal is to design algorithms that are fast when the chosen parameter value(s) are low, showing that the corresponding property or structure is algorithmically beneficial (called *fixed-parameter tractable* algorithms). This parameterized setting also allows for a robust notion of efficient preprocessing: A *kernelization* for a parameterized problem is an efficient algorithm that expects an instance (x, k), with parameter $k \in \mathbb{N}$, and returns an equivalent instance (x', k') of size upper bounded by some computable function $f(k)$. In particular, one is interested what parameterized problems admit *polynomial kernelizations* where the function f is polynomially bounded. Kernelization has been intensely studied (see, e.g., recent works [5,6,11,19,24,25,35,37,38,42,43,46] or the dedicated book [28]) and, by now, it is quite well understood which problems do or (conditionally) do not admit polynomial kernelizations. Moreover, there are both meta results (e.g., [9,27,29,44]), giving upper bounds for whole classes of problems, as well as dichotomy results (e.g., [11,18,39,40]), classifying all problems in some class into admitting or (conditionally) not admitting a polynomial kernelization.

We see *two main motivations* for studying local preprocessing: *First*, kernelizations usually rely on *reduction rules*, which usually apply to local configurations in the input. For example, this includes simple rules like *folding of degree-two vertices* for VERTEX COVER [14] and seeking *flowers*, i.e., cycles that pairwise intersect in the same vertex, for FEEDBACK VERTEX SET [8,45]. In other words, depending on the kind of reduction rule, the scope of the rule may vary, but beyond the required scope it usually does not matter what else is present in the instance.[1] Is there a sensible way of capturing local structure and giving local guarantees for the effectiveness of preprocessing? *Second*, most polynomial kernelizations in the literature are for parameterization by solution size or by the distance to some tractable special case for the problem. E.g., we are given a graph G and a modulator $X \subseteq V(G)$ such that $G - X$ belongs to a graph class where the target problem can be solved efficiently, with $|X|$ being the parameter. Intuitively, unless $\mathsf{P} = \mathsf{NP}$, efficient preprocessing can at best deal with tractable parts so the parameter must directly or indirectly allow to (at least) bound the size of the remainder of the instance, which may be structured arbitrarily. Together, this yields a fairly strict requirement for kernelization to be useful: Almost the entire instance must match some tractable special case for the problem in question. What if only parts of an instance fulfill such a restriction? Do we still get nontrivial guarantees for efficient preprocessing?

Our Work. Inspired by protrusions and protrusion replacement from meta kernelization [9], we use *boundaried graphs* to define *boundaried kernelization* (see

[1] A notably different case: Kernelizations obtained via cut-covering sets and other matroid-based techniques [41] use the entire instance to discover safe reduction steps.

Sect. 3 for formal definitions): A boundaried graph G_B is a graph G with the additional specification of a set $B \subseteq V(G)$ of boundary vertices. Informally, a boundaried kernelization for, e.g., the problem VERTEX COVER[fvs], i.e., VERTEX COVER parameterized by the size of a given feedback vertex set, is a polynomial-time algorithm that expects as input a boundaried graph G_B with forest-modulator X of size k, and returns a boundaried graph G'_B of size bounded by a function of parameter k and boundary size $|B|$ such that attaching the same graph H_B to either G_B or G'_B (by identification of boundary vertices, also called *gluing*) will yield graphs that are equivalent with respect to VERTEX COVER. In this way, local preprocessing for VERTEX COVER with respect to local structure as quantified by modulator X corresponds to a boundaried kernelization for VERTEX COVER[fvs]. Note that for $B = \emptyset$ and gluing the empty graph to G_B this includes the requirement that G and G' are equivalent for VERTEX COVER (and G' is of size bounded by a function of k), which essentially constitutes a regular kernelization for VERTEX COVER[fvs] (cf. Lemma 4). Accordingly, we mainly work with problems that admit a polynomial kernelization.

We establish polynomial size boundaried kernelizations for several problems that are known to admit regular polynomial kernelizations. This confirms the intuition that the local nature of reduction rules often allows us to work on the well-structured part only and ignore the rest of the instance.

Theorem 1. *The following parameterized problems admit a polynomial boundaried kernelization:* VERTEX COVER[vc], VERTEX COVER[fvs], FEEDBACK VERTEX SET[fvs], LONG CYCLE[vc], LONG PATH[vc], *and* HAMILTONIAN CYCLE[vc], HAMILTONIAN PATH[vc] *as well as* HAMILTONIAN CYCLE/PATH[$\#v, \deg(v) \neq 2$].

We complement this by proving *unconditionally* that some problems do not admit polynomial boundaried kernelizations despite admitting a regular polynomial kernelization. To round out the picture we also prove unconditionally that DOMINATING SET[vc], i.e., DOMINATING SET parameterized by the size of a given vertex cover, admits no polynomial boundaried kernelization, while it was already known to not admit a polynomial kernelization unless $\mathsf{NP} \subseteq \mathsf{coNP/poly}$.

Theorem 2. *The following parameterized problems do not admit a polynomial boundaried kernelization:* CLUSTER EDITING[cvd], CLUSTER EDITING[ce], MAXIMUM CUT[vc], TREE DELETION SET[vc], TREE DELETION SET[tds], LONG CYCLE[$\#v, \deg(v) \neq 2$], LONG PATH[$\#v, \deg(v) \neq 2$], DOMINATING SET[vc].

Finally, we show that, like protrusion replacement, boundaried kernelization can itself be useful for obtaining regular kernelization results. Concretely, we improve the size of a kernelization for VERTEX COVER[mod to td $\leq d$], i.e., VERTEX COVER parameterized by the size of a given deletion set/modulator to treedepth at most d, from $\mathcal{O}(k^{2^{\Theta(d^2)}})$ [12,31] to $\mathcal{O}(k^{2^{d-1}})$ vertices, which is a significant improvement and comes much closer to the known lower bound of $\mathcal{O}(k^{2^{d-2}+1})$ vertices [31].

Theorem 3. *For any constant $d \in \mathbb{N}_{\geq 1}$, the parameterized problem* VERTEX COVER[mod to td $\leq d$] *admits a polynomial kernelization with $\mathcal{O}(k^{2^{d-1}})$ vertices.*

Related Work. Arnborg et al. [3] used finite index of gluing-equivalence (defined in Sect. 3) and protrusion replacement for the construction of algorithms that solve the membership problem for MSO-definable graph classes with bounded treewidth. Likewise, Fellows and Langston [22] showed that under condition of finite index of gluing-equivalence, the obstruction set of a minor-closed graph class can be computed as soon as a treewidth bound for the former is known. Bodlaender and van Antwerpen-de Fluiter [7] extended the use to several decision, construction, and optimization problems on graphs of bounded treewidth.

As a tool for obtaining linear kernelizations for (CONNECTED) DOMINATING SET on graphs with a fixed excluded topological minor, Fomin et al. [26] introduce a notion dubbed "generalized protrusion:" This refers to a subgraph with constant-size boundary but, instead of having constant treewidth, it contains at most a constant number of vertices from the sought solution (presently, from the sought minimum (connected) dominating set). In a work on kernelization for INTEGER LINEAR PROGRAM FEASIBILITY, Jansen and Kratsch [34] consider shrinking protrusion-like subsystems of ILPs with a constant-size boundary that are either totally unimodular or have a Gaifman graph of bounded treewidth.

Other prior work has considered "dynamic sketching" [4] respectively "preprocessing under uncertainty" [20,21], mostly for tractable problems, where parts of the input are dynamic or unknown and the preprocessing needs to be consistent with any possible instantiation of the dynamic/unknown part, e.g., adjacency of vertices in dynamic part may change. Here the dynamic/unknown part is similar in spirit to the concept of a boundary, though the effects that either can have on the overall solution are very different.

Organization. Section 2 contains preliminaries regarding (graph) problems. In Sect. 3 we define boundaried graphs, introduce boundaried kernelization and obtain some first results. Several polynomial boundaried kernelization results are given in Sect. 4, while in Sect. 5 we unconditionally rule out existence of (polynomial) boundaried kernelization for several problems. In Sect. 6 we apply the notion of boundaried kernelization in order to show an improved polynomial kernelization for VERTEX COVER[mod to td $\leq d$]. We conclude in Sect. 7.

2 Preliminaries

Decision and Optimization Problems. We use common notation for decision problems $\Pi \subseteq \Sigma^*$ and *optimization problems* $\Pi\colon \Sigma^* \times \Sigma^* \to \mathbb{N} \cup \{\pm\infty\}$, as well as NP-*optimization problems.* For any optimization problem Π we define the decision problem Π_d such that for any instance $(x, k) \in \Sigma^* \times \mathbb{N}$ it holds that $(x, k) \in \Pi_d$ if and only if there exists a feasible solution with value at most k for minimization problems, resp., at least k for maximization problems.

Parameterized Complexity. A *parameterized problem* is a set $Q \subseteq \Sigma^* \times \mathbb{N}$. For a tuple $(x, k) \in \Sigma^* \times \mathbb{N}$, k is called the *parameter*. For an optimization problem Π, its *standard parameterization* is the decision variant Π_d with sought solution value k as the parameter, denoted by $\Pi[k]$. In contrast to this, for Π being some decision/optimization problem and ρ some minimization problem, we define the *structurally parameterized problem* $\Pi[\rho]$, for which it holds that $(x, s, k) \in \Pi[\rho]$ if and only if s is a feasible solution of value at most k for ρ on x, and (i) for Π being a decision problem, $x \in \Pi$; (ii) for Π being an optimization problem, $x \in \Pi_d$, i.e., $x = (x', \ell)$ where ℓ is the sought solution value for Π on x'.

Let Q be a parameterized problem and let f be a computable function. A *kernelization* for Q of *size* f is a polynomial-time algorithm which, given as input (x, k), outputs (x', k') such that $|x|, k' \leq f(k)$ and $(x, k) \in Q \Leftrightarrow (x', k') \in Q$. If f is a polynomial, the algorithm is called a *polynomial kernelization.* We will use the term *kernel* interchangeably with the term *kernelization.*

Graphs and Graph Problems. We use standard graph notation mostly following Diestel [16]. A *graph problem* is a problem for which every instance $x = (G, \hat{x})$ consists of a graph G and the remaining string $\hat{x}$. We say that Π is a *pure graph problem* if the remaining string $\hat{x}$ is empty, i.e., every instance is simply a graph G, and Π is isomorphism-invariant. For some graph class $\mathcal{C}$ and graph G, a $\mathcal{C}$-modulator (or vertex deletion set to $\mathcal{C}$) for G is a vertex set $X \subseteq V(G)$ such that $G - X \in \mathcal{C}$. We denote the problem of finding a minimum size modulator to $\mathcal{C}$ by MOD TO $\mathcal{C}$. All other problems discussed in this work are defined in detail in the full version of the paper [2].

The *treedepth* of a graph G, denoted $\mathrm{td}(G)$, is the minimum height d of a rooted forest F, called a *treedepth decomposition* of G, such that if $\{u, v\}$ is an edge in G, then u is an ancestor of v or vice versa. We denote by $\mathcal{C}^{\mathrm{td} \leq d}$ the class of graphs whose treedepth is at most d.

3 Boundaried Kernelization

Boundaried Graphs and Related Terms. We use boundaried graphs and gluing of such graphs in the usual way, e.g., as given by Bodlaender and van Antwerpen-de Fluiter [7]. We defer detailed definitions to the full version.

A (B-)*boundaried graph* consists of a graph G and a *boundary* $B \subseteq V(G)$, and is denoted as G_B. Given two boundaried graphs G_B and H_C, the *gluing* operation on those boundaried graphs, denoted as $G_B \oplus H_C$, results in a new (simple) graph, which is the disjoint union of G and H while identifying vertices from $B \cap C$. For a graph class $\mathcal{C}$, we denote by $\mathcal{C}_B$ the class of all $\mathcal{C}$-modulated B-boundaried graphs, i.e., all boundaried graphs G_B with $G - B \in \mathcal{C}$.

Let Π be a pure graph problem, let G_B and G'_B be boundaried graphs. We say that G_B and G'_B are *gluing equivalent with respect to* Π *and* B, denoted $G_B \equiv_{\Pi,B} G'_B$, if: (i) for Π being a decision problem, it holds for every H_B, that $G_B \oplus H_B \in \Pi \Leftrightarrow G'_B \oplus H_B \in \Pi$; (ii) for Π being an optimization problem, there exists some constant $\Delta \in \mathbb{Z}$ such that for every H_B it holds that $\mathrm{OPT}_\Pi(G_B \oplus$

$H_B) = \mathrm{OPT}_\Pi(G'_B \oplus H_B) + \Delta$. For some graph class $\mathcal{C}$ the equivalence relation $\equiv^{\mathcal{C}}_{\Pi,B}$ (resp. $\equiv^{\mathcal{C}_B}_{\Pi,B}$) is the restriction of $\equiv_{\Pi,B}$ to $\mathcal{C}$ (resp. $\mathcal{C}_B$). If Π, B and $\mathcal{C}$ or $\mathcal{C}_B$ are clear from context, we write $\equiv$.

We use *finite (integer) index* of decision/optimization problems in the usual way, i.e., that the number of equivalence classes of $\equiv$ needs to be bounded by some function of $|B|$. We also use index restricted to classes $\mathcal{C}$ and $\mathcal{C}_B$. Among problems with finite (integer) index, we highlight those having *single-exponential (integer) index*, which requires a constant c and polynomially bounded function f such that $\equiv_{\Pi,B}$ (resp. $\equiv^{\mathcal{C}}_{\Pi,B}$ and $\equiv^{\mathcal{C}_B}_{\Pi,B}$) has $\mathcal{O}(c^{f(|B|)})$ equivalence classes.

We conclude this part with two technical but simple lemmas.

Lemma 1. *Let Π be a pure graph problem, G and G' graphs, and B, C vertex sets of both $V(G)$ and $V(G')$ such that $B \subseteq C$. Then $G_C \equiv_{\Pi,C} G'_C$ implies $G_B \equiv_{\Pi,B} G'_B$, with the same offset Δ for these two equivalences, if Π is an optimization problem.*

Lemma 2. *Let Π be a pure graph problem, B some vertex set, and G^1, G^2, $\hat{G}^1$, $\hat{G}^2$ graphs, such that $G^1_B \equiv_{\Pi,B} \hat{G}^1_B$ and $G^2_B \equiv_{\Pi,B} \hat{G}^2_B$. Then it holds that $(G^1_B \oplus G^2_B)_B \equiv_{\Pi,B} (\hat{G}^1_B \oplus \hat{G}^2_B)_B$.*
If Π is an optimization problem and Δ_1, Δ_2 are the offset values for the equivalences between G^1_B and $\hat{G}^1_B$, resp. G^2_B and $\hat{G}^2_B$, when glued to any H_B, then the offset for the equivalence between $(G^1_B \oplus G^2_B)_B$ and $(\hat{G}^1_B \oplus \hat{G}^2_B)_B$ equals $\Delta_1 + \Delta_2$.

Boundaried Kernelization. We are now ready to formally introduce our notion of boundaried kernelization. Presently, we do this for the restricted setting of pure graph problems Π parameterized by a second pure graph minimization problem ρ. This enables us to prove a few general results on boundaried kernelizations while also keeping the required notation somewhat concise. Moreover, this encompasses all problems considered in this work. Possible extensions of the notion to other problem types are briefly discussed in the conclusion.

We remark that while we work with pure graph problems, the boundaried kernelization gets as input a *boundaried* graph G_B. This is consistent since the result of gluing G_B with any other boundaried graph H_B results in the *simple* graph $Q = G_B \oplus H_B$, which is then the actual input to the pure graph problem, for which the preprocessing was done on the known local part G_B. Also note that we distinguish between the encoding size $|s|$ of a problem and its value $\Pi(x, s)$ for optimization problem Π and instance x.

Unlike in regular kernelization, where, for Π being an optimization problem, a sought solution size k is given and can be changed by the kernel (adjusting to situations like safeness of including some vertex in the solution), this is not the case for boundaried kernelization, where only local structure of G_B is used, independently of what graph H_B could be glued to G_B and what the global solution size could be. For this reason, the boundaried kernelization needs to output its influence on the solution size in form of the solution size offset Δ, which does not need to be bounded. Note that this Δ is already part of the definition of gluing equivalence.

Definition 1 (Boundaried kernelization). *Let* Π *be a pure graph problem,* ρ *a pure graph minimization problem, and* f *a computable function. A boundaried kernelization of size* f *for* $\Pi[\rho]$ *is a polynomial-time algorithm that, given a boundaried graph* G_B *and* ρ*-solution* s*, outputs a* $\equiv_{\Pi,B}$*-equivalent boundaried graph* G'_B *with* ρ*-solution* s'*, such that* $|G'|, |s'|, \rho(G', s') \leq f(|B| + \rho(G, s))$*. Additionally, if* Π *is an optimization problem, the algorithm also needs to output an integer* Δ*, such that for all* H_B *it holds that* $\mathrm{OPT}_\Pi(G_B \oplus H_B) = \mathrm{OPT}_\Pi(G'_B \oplus H_B) + \Delta$*. If* f *is upper bounded by some polynomial function, the boundaried kernelization has polynomial size and is called a polynomial boundaried kernelization.*

Later we will show polynomial boundaried kernelizations for several problems parameterized by the size of a given modulator $|X|$ to some hereditary graph class $\mathcal{C}$. As the size of the output should be bounded polynomially in $|B| + |X|$, it suffices to bound/reduce the size of $R := V(G) \setminus (B \cup X)$. In order to simplify this, we show that it suffices to deal with the special case where B itself is a modulator to $\mathcal{C}$, i.e., where $G - B \in \mathcal{C}$.

Lemma 3. *Let* Π *be a pure graph problem,* $\mathcal{C}$ *be a hereditary graph class and* f *a computable function. If there exists a boundaried kernelization of size* $f(|B|)$ *for* $\Pi[$*mod to* $\mathcal{C}]$ *with the additional restriction that* B *coincides with the given and output solutions to* MOD TO $\mathcal{C}$*, then there also exists a boundaried kernelization of size* f *for* $\Pi[$*mod to* $\mathcal{C}]$ *without that restriction.*

The following lemma shows that, for parameterization by NP-minimization problem ρ, a boundaried kernelization of size f also implies a (regular) kernelization of size at most polynomial in f. In this sense, boundaried kernelizations are stronger than (regular) kernelizations, showing, in particular, that a polynomial kernelization is a prerequisite for admitting a polynomial boundaried kernelization. We note that most structural parameterizations are indeed based on NP-minimization problems, as one wishes to be able verify the (often provided) structure efficiently. (Parameterization by maximization problems is often unwieldy to deal with, e.g., because we need optimum solutions to get structural implications, but optimality is usually hard to verify. Nevertheless, parameterization by max leaf number comes to mind as a useful exception, cf. [23].)

Lemma 4. *Let* Π *and* ρ *be pure graph problems. Additionally, let* ρ *be an* NP*-minimization problem, and* Π *either a decision problem or an* NP*-optimization problem. If* $\Pi[\rho]$ *has a (polynomial) boundaried kernelization, it also has a (polynomial) kernelization.*

The next and final lemma of this section will be used to rule out (polynomial) boundaried kernelizations. In contrast to lower bounds for (regular) kernelization (cf. [10,15]), working with the index of the corresponding equivalence classes allows us to get unconditional lower bounds. In particular, it follows that if $\Pi[\text{mod to } \mathcal{C}]$ has a polynomial boundaried kernelization, then $\equiv^{\mathcal{C}_B}_{\Pi,B}$ and $\equiv^{\mathcal{C}}_{\Pi,B}$ have single-exponential index (and, conversely, larger index rules out such a boundaried kernelization).

Lemma 5. *Let* Π *be a pure graph problem,* ρ *a pure graph minimization problem and* f *a computable function such that* $\Pi[\rho]$ *admits a boundaried kernelization of size* f*. Further fix some vertex set* B *and graph class* $\mathcal{C}$ *such that for each* $G \in \mathcal{C}$ *with* $B \subseteq V(G)$ *exists some* ρ*-solution* s *with* $\rho(G, s) \leq g(|B|)$ *for some function* g*. Then the number of equivalence classes of* $\equiv^{\mathcal{C}}_{\Pi,B}$ *is upper bounded by* $\mathcal{O}(2^{f(|B|+g(|B|))^2})$.

4 Upper Bounds

In this section we describe a vertex-linear boundaried kernelization for the problem VERTEX COVER[vc] and outline main ideas of polynomial boundaried kernelizations for VERTEX COVER[fvs] and FEEDBACK VERTEX SET[fvs], which due to their closeness to their regular kernelization counterparts strengthen the hope for many reduction rules to be applicable and helpful for boundaried kernelization due to their rather local nature. Formal proofs for the latter two problems and the full descriptions for the remaining problems stated in Theorem 1 are given in the full version of the paper [2].

4.1 Vertex Cover[vc]

A classical tool in kernelization for the VERTEX COVER problem is the removal of so called *crowns*, which consist of an independent set I and its neighborhood H, such that exists an H-saturating matching between H and I, witnessing optimality of including H in the solution. This rule was introduced by Chen et al. [14] and explored further by, among others, Abu-Khzam et al. [1], and is known to lead to a kernel with linear number of vertices.

Lemma 6 ([1], Theorem 3). *If* G *is a graph with crown* (I, H)*, then there is a minimum vertex cover of* G*, containing all* H*-vertices and none from* I*.*

Observe that crowns in a boundaried graph G_B might become invalid after gluing to some boundaried graph F_B, e.g., I-vertices contained in B might become adjacent. For this reason we need to make sure that the intersection of I and B is empty. Furthermore we cannot simply remove vertices of B, as they will be re-introduced by gluing. Instead, we fix an H-saturating matching between I and H, and remove all other edges incident with H. As (I, H) remains a crown after that operation, we mark vertices in I and H to prevent multiple handling. Additionally, we remove any isolated vertices in $R := V(G) \setminus B$, as even after gluing, they will not be incident with any edge. Both reduction rules can be applied exhaustively in polynomial time by a modification of an algorithm given by Iwata et al. [32] for removing all crowns in a graph, as given in the appendix. Initialize $\tilde{I}, \tilde{H} = \emptyset$.

Reduction Rule 1. *If* $v \in R$ *is an isolated vertex, remove it.*

Reduction Rule 2. *If $G-(\tilde{I}\cup\tilde{H})$ contains a crown (I,H) with $I\subseteq R$ and an H-saturating matching M between H and I, then remove all edges incident with H but not contained in M, and add all vertices in I,H to $\tilde{I},\tilde{H}$, respectively.*

Proof (gluing safeness). Let (I,H) be a crown that satisfies the above conditions, let G'_B be the resulting graph after removing all edges incident with H, but not contained in M. We show that $\mathrm{OPT_{VC}}(G_B\oplus F_B)$ equals $\mathrm{OPT_{VC}}(G'_B\oplus F_B)$ for any B-boundaried graph F_B. First, note that (I,H) is also a crown in $G_B\oplus F_B$, as I is not part of the boundary and thus no vertex in I gets new neighbors, resulting in I still inducing an independent set in $G_B\oplus F_B$, H to be its neighborhood and the H-saturating matching M into I to persist. Clearly, removing all edges incident with H besides those in M does not change any of those points, and (I,H) is also a crown in $G'_B\oplus F_B$. Also note that, since the only difference between G and G' is about the edges of H, we have $(G_B\oplus F_B)-(I\cup H)=(G'_B\oplus F_B)-(I\cup H)$. Due to Lemma 6, we then get

$$\begin{aligned}\mathrm{OPT_{VC}}(G_B\oplus F_B)&=\mathrm{OPT_{VC}}((G_B\oplus F_B)-(I\cup H))+|H|\\&=\mathrm{OPT_{VC}}((G'_B\oplus F_B)-(I\cup H))+|H|\\&=\mathrm{OPT_{VC}}(G'_B\oplus F_B)\end{aligned}$$

□

Lemma 7 ([2]). *Rules 1 and 2 can be exhaustively applied in polynomial time.*

Such restrictive handling of crowns still suffices in order to bound the independent set number of $G[R]$, as stated in Lemma 9.

Lemma 8. *If there is a crown (I,H) in (G,B) with $I\subseteq R$ and $|I|>|H|$, then we can apply one among Rules 1 and 2.*

Proof. From the condition that (I,H) is a crown it follows that $H=N(I)$ and there exists an H-saturating matching M between H and I. If $E(H,I)\setminus M\neq\emptyset$, we apply Rule 2. Otherwise, we get $|I|>|H|=|M|=|E(H,I)|$ and thus at least one vertex in $I\subseteq R$ is isolated, so we apply Rule 1. □

Lemma 9. *If $\alpha(G[R])>\frac{1}{2}|V(G)|$, then we can apply one among Rules 1 and 2.*

Proof. Let Y be a maximum independent set of $G[R]$, i.e., such that $|Y|=\alpha(G[R])$. Let X be the neighborhood of Y, which is clearly a subset of $V(G)\setminus Y$. Note that since $|Y|=\alpha(G[R])>\frac{1}{2}|V(G)|$, it thus holds that $|X|\leq|V(G)|-|Y|<\frac{1}{2}|V(G)|<|Y|$.

If there exists an X-saturating matching into Y, then by Lemma 8 we can apply one of the reduction rules. Else we choose some maximal $\hat{X}\subseteq X$ such that $|\hat{X}|>|\hat{Y}|$ with $\hat{Y}=N(\hat{X})\cap Y$, which exists by Hall's Theorem. Let $X^*=X\setminus\hat{X}$ and $Y^*=Y\setminus\hat{Y}$. Note that even though X^* might have neighbors in $\hat{Y}=Y\setminus Y^*$, it holds that $N(Y^*)\subseteq X^*$, and by maximality of $\hat{X}$, also $N(Y^*)=X^*$. Further, as $Y^*\neq\emptyset$, if (Y^*,X^*) is not a crown, it can only be due to absence of an X^*-saturating matching into Y^*. Then, again by Hall's Theorem, there exists some $\bar{X}\subseteq X^*$ with $|\bar{X}|>|N(\bar{X})\cap Y^*|$, which contradicts maximality of $\hat{X}$. As a result, (Y^*,X^*) must be a crown with $|Y^*|=|Y|-|\hat{Y}|>|Y|-|\hat{X}|>|X|-|\hat{X}|=|X^*|$. Again, by Lemma 8 we can apply one of Rules 1 and 2. □

Theorem 4. *The parameterized problem* VERTEX COVER[vc] *admits a polynomial boundaried kernelization with at most* $2(|B| + k)$ *vertices.*

Proof. By Lemma 3 assume that we are given boundaried graph G_B with $G - B$ an independent set. Using Lemma 7 apply Rules 1 and 2 exhaustively in polynomial time, resulting in some boundaried graph G'_B. Observe that G' is a subgraph of G, implying that $R' := V(G') \setminus B$ is an independent set, leading to $|R'| = \alpha(G'[R']) \leq \frac{1}{2}|V(G')|$ by Lemma 9 and thus $|V(G')| = |B| + |R'| \leq 2|B|$. As a result, we can output G'_B with $G' - B$ an independent set and $\Delta = 0$. □

4.2 Sketches for Vertex Cover[fvs] and Feedback Vertex Set[fvs]

Vertex Cover[fvs]. The polynomial kernel for VERTEX COVER[fvs] by Bodlaender and Jansen [33] is one of the first kernelization results for a structurally parameterized problem. Main steps are to make sure that $G - X$ (with X a forest-modulator) has a perfect matching, to reduce the number of so called conflict structures in $G - X$, and to use that in order to bound the total size of R. We use the crown reduction rule from earlier and the resulting bound of $\alpha(G[R]) \leq 1/2|V(G)|$ for the first step, as it allows us to move vertices that are problematic for a perfect matching in $G[R]$ to B, with only a linear blow-up of the latter. Regarding the last two steps, we can mainly apply the same reduction rules as Bodlaender and Jansen. We only need to be careful to not delete boundary-vertices. Instead, we mark them by giving personal leaves and thus forcing them into a solution.

Feedback Vertex Set[fvs]. The state of the art vertex-quadratic kernelization for FEEDBACK VERTEX SET[k] was given by Thomassé [45]. It mainly works through handling vertices of low degree, as well as certain flowers, i.e., sets of cycles intersecting at exactly one vertex, which are shown to exist at vertices of high degree, and recognizing that this bounds the total number of vertices by a quadratic function on the sought solution size, unless certainly given a NO-instance, i.e. graph G and number k such that G has no feedback vertex set of size at most k. We have no sought solution size at hand, but we have a nice structure with $G[R]$ only consisting of trees as components. Using similar reduction rules to those given by Thomassé, we force the low-degree vertices of R (i.e., leaves and non-branching vertices) to be adjacent to B on one hand, and limit the maximum degree of the B-vertices by $\mathcal{O}(|B|)$ on the other hand. Together, this bounds the allowed number of vertices in R by $\mathcal{O}(|B|^2)$.

5 Lower Bounds

Complementing our polynomial boundaried kernels, Theorem 2 lists several problems that do not admit (polynomial) boundaried kernelization. We outline our constructions that imply infinite integer index for CLUSTER EDITING restricted to $\mathcal{C}^{\texttt{complete}}$ and more than single-exponential integer index for DOMINATING SET restricted to $\mathcal{C}_B^{\text{independent}}$. Observe, that graphs in these classes have cluster editing (resp. vertex cover) solution size 0 (resp. $|B|$), meaning in combination with Lemma 5 that CLUSTER EDITING[ce] does not admit boundaried kernelization, while DOMINATING SET[vc] does not admit polynomial boundaried kernelization. We refer to the full version of the paper [2] for the full proofs and the lower bounds for the remaining problems stated in Theorem 2.

Note that while for DOMINATING SET[vc] also exists a conditional exclusion of polynomial kernelization [17], the problem CLUSTER EDITING[k] and hence also CLUSTER EDITING[ce] admits a polynomial kernelization [13,30]. Further note that Jansen and Wulms [36] showed the problems VERTEX COVER and DOMINATING SET to not admit a single-exponential number of gluing-equivalence classes, even with the restriction to planar graphs with treewidth $|B| + \mathcal{O}(1)$. Together with Lemma 5 their result excludes polynomial boundaried kernelization for VERTEX COVER[tw + mod to planar] and DOMINATING SET[tw + mod to planar].

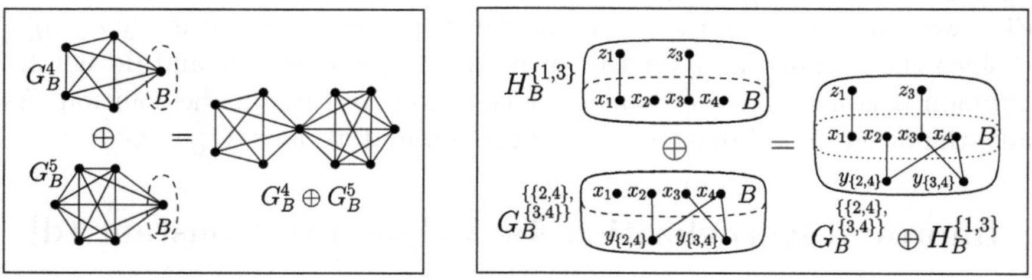

Fig. 1. Examples of graphs defined in proofs for Lemmas 10 (left) and 11 (right).

Lemma 10. CLUSTER EDITING *does not have finite integer index, even on the class of complete graphs.*

Proof (sketch, also see (Fig. 1)). We construct an infinite set of complete boundaried graphs $G_B^i, i \in \mathbb{N}$ such that for distinct $i, j \in \mathbb{N}$ it holds that $G_B^i \not\equiv G_B^j$. For that, let G_B^i be a complete graph on $i+1$ vertices with $B = \{x\}$. The graph G_B^0 consists only of vertex x. Note that for any $i, j \in \mathbb{N}$ with $i > j$ a minimum cluster editing solution for $G_B^i \oplus G_B^j$ is to remove the edges between x and the rest of G^j, and has thus size j. Now having fixed i and j for checking gluing-equivalence, this implies that $\mathrm{OPT}(G_B^i \oplus G_B^0) = 0 = \mathrm{OPT}(G_B^j \oplus G_B^0)$, forcing $\Delta = 0$ for the gluing-equivalence. However, such offset is not correct, since for $h > i, j$ it holds that $\mathrm{OPT}(G_B^i \oplus G_B^h) = i \neq j = \mathrm{OPT}(G_B^j \oplus G_B^h)$, resulting in $G_B^i \not\equiv G_B^j$. □

Lemma 11. DOMINATING SET *does not have single-exponential integer index, even on* $C_B^{independent}$.

Proof (sketch, also see (Fig. 1)). For some arbitrary, but fixed $q \in \mathbb{N}$, let $B = \{x_1, \ldots, x_q\}$. We give a set of $2^{\binom{q}{\lfloor q/2 \rfloor}}$-many pairwise not gluing-equivalent boundaried graphs. Since it holds that $\binom{q}{\lfloor q/2 \rfloor} \geq (\frac{q}{\lfloor q/2 \rfloor})^{\lfloor q/2 \rfloor} \geq 2^{\lfloor q/2 \rfloor} \in \omega(|B|^c)$ for any constant c, this excludes a single exponential integer index. Let $\mathcal{D}$ be a family of size-$\lfloor q/2 \rfloor$ subsets of $\{1, \ldots, q\}$. For each $\mathcal{I} \subseteq \mathcal{D}$ let $G_B^{\mathcal{I}}$ be a graph containing for each $I \in \mathcal{I}$ the vertex y_I which is adjacent to the B-vertices x_i with $i \in I$. To show that $G_B^{\mathcal{I}} \not\equiv G_B^{\mathcal{J}}$ for any $\mathcal{I}, \mathcal{J} \subseteq \mathcal{D}$, we construct graphs H_B^P with $P \subseteq \{1, \ldots, q\}$, containing for each $i \in P$ the vertex z_i which is adjacent to x_i. The idea is that after gluing some graph to H_B^P, we can assume the vertices x_i with $i \in P$ to be included in a minimum dominating set. Hence for fixed distinct $\mathcal{I}, \mathcal{J} \subseteq \mathcal{D}$ it holds that $\mathrm{OPT}(G_B^{\mathcal{I}} \oplus H_B^{[q]}) = q = \mathrm{OPT}(G_B^{\mathcal{J}} \oplus H_B^{[q]})$ and a hypothetical offset Δ for the gluing equivalence needs to equal zero. As $\mathcal{I} \neq \mathcal{J}$, without loss of generality, there exists some $I \in \mathcal{I} \setminus \mathcal{J}$. Note that $G_B^{\mathcal{I}} \oplus H_B^{[q] \setminus I}$ has a dominating set of size $\lceil q/2 \rceil + 1$: That solution contains y_I and every x_i with $i \in [q] \setminus I$. However, $G_B^{\mathcal{J}} \oplus H_B^{[q] \setminus I}$ does not have a solution of that size: As noted earlier, we can assume that an optimum solution contains x_i for every $i \in [q] \setminus I$; and since the remaining B-vertices form an independent set, and do not have a common neighbor, one needs at least two more vertices in the solution. As a result, the offset $\Delta = 0$ does not suit this case, and we have $G_B^{\mathcal{I}} \not\equiv G_B^{\mathcal{J}}$. □

6 Improved Kernelization for VC[mod to Constant td]

By applying the idea of boundaried kernelization we give an improved kernel for Vertex Cover[mod to td $\leq d$]. Bougeret and Sau [12] gave the first polynomial kernel for this problem on general graphs, which was then improved and generalized by Hols et al. [31]. The kernelization works in a recursive manner: If $d = 1$, then apply the vertex-linear kernel for Vertex Cover[vc]. Else, given a graph G and modulator X to constant treedepth $d > 1$, reduce the number of connected components in $G - X$ to at most $\mathcal{O}(|X|^{2^{d-2}+1})$, and put from each remaining component the root vertex of its treedepth decomposition to the modulator, reducing the task to computing a kernel for Vertex Cover[mod to td $\leq d-1$]. In total, this yields a kernel with $\mathcal{O}(|X|^{2^{\Theta(d^2)}})$ vertices. An issue with this approach is that the roots of all components are moved to the modulator simultaneously, leading to an unnecessarily large blow-up of the number of components in each recursive step. Instead, we take a depth-first approach: At each recursive step, after having reduced the number of components, we partition the set of remaining components into groups of size at most $|X|$, and recursively process each group individually as a boundaried graph, with X and the roots of the respective components forming the boundary and modulator to treedepth $d-1$. This way, we achieve a kernel with $\mathcal{O}(|X|^{2^{d-1}})$ vertices, which comes much closer to the best known lower bound of $\mathcal{O}(|X|^{2^{d-2}+1})$ vertices [31, Theorem 1.1]. The correctness of this approach follows from Theorem 4 and Lemma 12, a *gluing*-equivalence restatement for bounding the number of connected components in $G - X$.

Lemma 12 (Analogue of [31, Theorem 1.3]). *Let $\mathcal{C}$ be a hereditary graph class with minimal blocking set size b on which* Vertex Cover *can be solved in polynomial time. There is a polynomial-time algorithm that given boundaried graph $G_B \in \mathcal{C}_B$ returns a gluing-equivalent graph $G'_B \in \mathcal{C}_B$ with the respective offset Δ, such that $G' - B$ has at most $\mathcal{O}(|B|^b)$ connected components.*

7 Conclusion

We have introduced boundaried kernelization as a model for efficient local preprocessing, motivated by the (global) structural requirements for (regular) polynomial kernelization and inspired by protrusions and protrusion replacement from meta kernelization. We have showed a number of polynomial upper and (unconditional) lower bounds for boundaried kernelization. While finite (integer) index is a requirement for positive results, it is clearly not sufficient. Ditto for the existence of a (regular) polynomial kernelization.

Future work should naturally pursue which further problems with polynomial kernelization also admit polynomial boundaried kernelizations (for pure graph problems and beyond), and to establish unconditional lower bounds where this is infeasible. It is also interesting to perhaps identify some general properties of problems that admit a polynomial kernelization but no polynomial boundaried kernelization, like, e.g., Tree Deletion Set. To this end, one should lift boundaried kernelization to a more general setting by defining a gluing operator on the level of instances, i.e., not only for (pure) graph problems. General finite structures seem a good basis for this.

Finally, we showed that our polynomial boundaried kernelization for Vertex Cover[vc] can be used to get a polynomial kernelization for Vertex Cover[mod to td $\leq d$], and in fact got a significant improvement over the previous kernelization. Can this be further improved to match the lower bound via maximum size of minimal blocking sets [31]? Can one find more applications of polynomial boundaried kernelization for obtaining polynomial kernelizations for more complex parameterizations?

References

1. Abu-Khzam, F.N., Fellows, M.R., Langston, M.A., Suters, W.H.: Crown structures for vertex cover kernelization. Theory Comput. Syst. **41**(3), 411–430 (2007). https://doi.org/10.1007/S00224-007-1328-0
2. Antipov, L., Kratsch, S.: Boundaried Kernelization (2025). https://doi.org/10.48550/arXiv.2504.18476
3. Arnborg, S., Courcelle, B., Proskurowski, A., Seese, D.: An algebraic theory of graph reduction. J. ACM **40**(5), 1134–1164 (1993). https://doi.org/10.1145/174147.169807
4. Assadi, S., Khanna, S., Li, Y., Tannen, V.: Dynamic sketching for graph optimization problems with applications to cut-preserving sketches. In: Harsha, P., Ramalingam, G. (eds.) 35th IARCS Annual Conference on Foundation of Software Technology and Theoretical Computer Science, FSTTCS 2015, Bangalore, India, 16–18

December 2015. LIPIcs, vol. 45, pp. 52–68. Schloss Dagstuhl - Leibniz-Zentrum für Informatik (2015). 10.4230/LIPICS.FSTTCS.2015.52,
5. Bessy, S., Bougeret, M., Thilikos, D.M., Wiederrecht, S.: Kernelization for graph packing problems via rainbow matching. In: Bansal, N., Nagarajan, V. (eds.) Proceedings of the 2023 ACM-SIAM Symposium on Discrete Algorithms, SODA 2023, Florence, Italy, 22–25 January 2023, pp. 3654–3663. SIAM (2023). https://doi.org/10.1137/1.9781611977554.CH139
6. Bhyravarapu, S., Jana, S., Saurabh, S., Sharma, R.: Difference determines the degree: structural kernelizations of component order connectivity. In: Misra, N., Wahlström, M. (eds.) 18th International Symposium on Parameterized and Exact Computation, IPEC 2023, Amsterdam, The Netherlands, 6–8 September 2023. LIPIcs, vol. 285, pp. 5:1–5:14. Schloss Dagstuhl - Leibniz-Zentrum für Informatik (2023). https://doi.org/10.4230/LIPICS.IPEC.2023.5
7. Bodlaender, H.L., van Antwerpen-de Fluiter, B.: Reduction algorithms for graphs of small treewidth. Inf. Comput. **167**(2), 86–119 (2001). https://doi.org/10.1006/INCO.2000.2958
8. Bodlaender, H.L., van Dijk, T.C.: A cubic kernel for feedback vertex set and loop cutset. Theory Comput. Syst. **46**(3), 566–597 (2010). https://doi.org/10.1007/S00224-009-9234-2
9. Bodlaender, H.L., Fomin, F.V., Lokshtanov, D., Penninkx, E., Saurabh, S., Thilikos, D.M.: (Meta) kernelization. J. ACM **63**(5), 44:1–44:69 (2016). https://doi.org/10.1145/2973749
10. Bodlaender, H.L., Jansen, B.M.P., Kratsch, S.: Kernelization lower bounds by cross-composition. SIAM J. Disc. Math. **28**(1), 277–305 (2014). https://doi.org/10.1137/120880240
11. Bougeret, M., Jansen, B.M.P., Sau, I.: Kernelization dichotomies for hitting subgraphs under structural parameterizations. In: Bringmann, K., Grohe, M., Puppis, G., Svensson, O. (eds.) 51st International Colloquium on Automata, Languages, and Programming, ICALP 2024, Tallinn, Estonia, 8–12 July 2024. LIPIcs, vol. 297, pp. 33:1–33:20. Schloss Dagstuhl - Leibniz-Zentrum für Informatik (2024). 10.4230/LIPICS.ICALP.2024.33,
12. Bougeret, M., Sau, I.: How much does a treedepth modulator help to obtain polynomial kernels beyond sparse graphs? Algorithmica **81**(10), 4043–4068 (2019). https://doi.org/10.1007/S00453-018-0468-8
13. Cao, Y., Chen, J.: Cluster editing: kernelization based on edge cuts. Algorithmica **64**(1), 152–169 (2012). https://doi.org/10.1007/S00453-011-9595-1
14. Chen, J., Kanj, I.A., Jia, W.: Vertex cover: further observations and further improvements. J. Algor. **41**(2), 280–301 (2001). https://doi.org/10.1006/JAGM.2001.1186
15. Dell, H., van Melkebeek, D.: Satisfiability allows no nontrivial sparsification unless the polynomial-time hierarchy collapses. J. ACM **61**(4), 23:1–23:27 (2014). https://doi.org/10.1145/2629620
16. Diestel, R.: Graph Theory. GTM, vol. 173. Springer, Heidelberg (2017). https://doi.org/10.1007/978-3-662-53622-3
17. Dom, M., Lokshtanov, D., Saurabh, S.: Incompressibility through colors and IDs. In: Albers, S., Marchetti-Spaccamela, A., Matias, Y., Nikoletseas, S., Thomas, W. (eds.) ICALP 2009. LNCS, vol. 5555, pp. 378–389. Springer, Heidelberg (2009). https://doi.org/10.1007/978-3-642-02927-1_32
18. Donkers, H., Jansen, B.M.P.: A turing kernelization dichotomy for structural parameterizations of f-minor-free deletion. J. Comput. Syst. Sci. **119**, 164–182 (2021). https://doi.org/10.1016/J.JCSS.2021.02.005

19. Dumas, M., Perez, A.: An improved kernelization algorithm for trivially perfect editing. In: Misra, N., Wahlström, M. (eds.) 18th International Symposium on Parameterized and Exact Computation, IPEC 2023, Amsterdam, The Netherlands, 6–8 September 2023. LIPIcs, vol. 285, pp. 15:1–15:17. Schloss Dagstuhl - Leibniz-Zentrum für Informatik (2023). https://doi.org/10.4230/LIPICS.IPEC.2023.15
20. Fafianie, S., Hols, E.C., Kratsch, S., Quyen, V.A.: Preprocessing under uncertainty: Matroid intersection. In: Faliszewski, P., Muscholl, A., Niedermeier, R. (eds.) 41st International Symposium on Mathematical Foundations of Computer Science, MFCS 2016, Kraków, Poland, 22–26 August 2016. LIPIcs, vol. 58, pp. 35:1–35:14. Schloss Dagstuhl - Leibniz-Zentrum für Informatik (2016). https://doi.org/10.4230/LIPICS.MFCS.2016.35
21. Fafianie, S., Kratsch, S., Quyen, V.A.: Preprocessing under uncertainty. In: Ollinger, N., Vollmer, H. (eds.) 33rd Symposium on Theoretical Aspects of Computer Science, STACS 2016, Orléans, France, 17–20 February 2016. LIPIcs, vol. 47, pp. 33:1–33:13. Schloss Dagstuhl - Leibniz-Zentrum für Informatik (2016). DOI: https://doi.org/10.4230/LIPICS.STACS.2016.33
22. Fellows, M.R., Langston, M.A.: An analogue of the myhill-nerode theorem and its use in computing finite-basis characterizations (extended abstract). In: 30th Annual Symposium on Foundations of Computer Science, Research Triangle Park, North Carolina, USA, 30 October–1 November 1989, pp. 520–525. IEEE Computer Society (1989). https://doi.org/10.1109/SFCS.1989.63528
23. Fellows, M.R., Lokshtanov, D., Misra, N., Mnich, M., Rosamond, F.A., Saurabh, S.: The complexity ecology of parameters: an illustration using bounded max leaf number. Theory Comput. Syst. **45**(4), 822–848 (2009). https://doi.org/10.1007/S00224-009-9167-9
24. Fomin, F.V., Golovach, P.A., Inamdar, T., Saurabh, S., Zehavi, M.: Kernelization for spreading points. In: Gørtz, I.L., Farach-Colton, M., Puglisi, S.J., Herman, G. (eds.) 31st Annual European Symposium on Algorithms, ESA 2023, Amsterdam, The Netherlands, 4–6 September 2023. LIPIcs, vol. 274, pp. 48:1–48:16. Schloss Dagstuhl - Leibniz-Zentrum für Informatik (2023). https://doi.org/10.4230/LIPICS.ESA.2023.48
25. Fomin, F.V., Le, T., Lokshtanov, D., Saurabh, S., Thomassé, S., Zehavi, M.: Lossy kernelization for (implicit) hitting set problems. In: Gørtz, I.L., Farach-Colton, M., Puglisi, S.J., Herman, G. (eds.) 31st Annual European Symposium on Algorithms, ESA 2023, Amsterdam, The Netherlands, 4–6 September 2023. LIPIcs, vol. 274, pp. 49:1–49:14. Schloss Dagstuhl - Leibniz-Zentrum für Informatik (2023). https://doi.org/10.4230/LIPICS.ESA.2023.49
26. Fomin, F.V., Lokshtanov, D., Saurabh, S., Thilikos, D.M.: Kernels for (connected) dominating set on graphs with excluded topological minors. ACM Trans. Algor. **14**(1), 6:1–6:31 (2018). https://doi.org/10.1145/3155298
27. Fomin, F.V., Lokshtanov, D., Saurabh, S., Thilikos, D.M.: Bidimensionality and kernels. SIAM J. Comput. **49**(6), 1397–1422 (2020). https://doi.org/10.1137/16M1080264
28. Fomin, F.V., Lokshtanov, D., Saurabh, S., Zehavi, M.: Kernelization: theory of parameterized preprocessing. Cambridge University Press (2019). https://doi.org/10.1017/9781107415157
29. Ganian, R., Slivovsky, F., Szeider, S.: Meta-kernelization with structural parameters. J. Comput. Syst. Sci. **82**(2), 333–346 (2016). https://doi.org/10.1016/J.JCSS.2015.08.003

30. Gramm, J., Guo, J., Hüffner, F., Niedermeier, R.: Graph-modeled data clustering: exact algorithms for clique generation. Theory Comput. Syst. **38**(4), 373–392 (2005). https://doi.org/10.1007/S00224-004-1178-Y
31. Hols, E.C., Kratsch, S., Pieterse, A.: Elimination distances, blocking sets, and kernels for vertex cover. SIAM J. Disc. Math. **36**(3), 1955–1990 (2022). https://doi.org/10.1137/20M1335285
32. Iwata, Y., Oka, K., Yoshida, Y.: Linear-time FPT algorithms via network flow. In: Chekuri, C. (ed.) Proceedings of the Twenty-Fifth Annual ACM-SIAM Symposium on Discrete Algorithms, SODA 2014, Portland, Oregon, USA, 5–7 January 2014, pp. 1749–1761. SIAM (2014). https://doi.org/10.1137/1.9781611973402.127
33. Jansen, B.M.P., Bodlaender, H.L.: Vertex cover kernelization revisited - upper and lower bounds for a refined parameter. Theory Comput. Syst. **53**(2), 263–299 (2013). https://doi.org/10.1007/S00224-012-9393-4
34. Jansen, B.M.P., Kratsch, S.: A structural approach to kernels for ILPs: treewidth and total unimodularity. In: Bansal, N., Finocchi, I. (eds.) ESA 2015. LNCS, vol. 9294, pp. 779–791. Springer, Heidelberg (2015). https://doi.org/10.1007/978-3-662-48350-3_65
35. Jansen, B.M.P., van der Steenhoven, B.: Kernelization for counting problems on graphs: preserving the number of minimum solutions. In: Misra, N., Wahlström, M. (eds.) 18th International Symposium on Parameterized and Exact Computation, IPEC 2023, Amsterdam, The Netherlands, 6–8 September 2023. LIPIcs, vol. 285, pp. 27:1–27:15. Schloss Dagstuhl - Leibniz-Zentrum für Informatik (2023). https://doi.org/10.4230/LIPICS.IPEC.2023.27
36. Jansen, B.M.P., Wulms, J.J.H.M.: Lower bounds for protrusion replacement by counting equivalence classes. Disc. Appl. Math. **278**, 12–27 (2020). https://doi.org/10.1016/J.DAM.2019.02.024
37. Kammer, F., Sajenko, A.: Space-efficient graph kernelizations. In: Chen, X., Li, B. (eds.) Theory and Applications of Models of Computation - 18th Annual Conference, TAMC 2024, Hong Kong, China, 13–15 May 2024, Proceedings. Lecture Notes in Computer Science, vol. 14637, pp. 260–271. Springer, Heidelbegr (2024). https://doi.org/10.1007/978-981-97-2340-9_22
38. Kratsch, S., Kunz, P.: Approximate turing kernelization and lower bounds for domination problems. In: Misra, N., Wahlström, M. (eds.) 18th International Symposium on Parameterized and Exact Computation, IPEC 2023, Amsterdam, The Netherlands, 6–8 September 2023. LIPIcs, vol. 285, pp. 32:1–32:17. Schloss Dagstuhl - Leibniz-Zentrum für Informatik (2023). https://doi.org/10.4230/LIPICS.IPEC.2023.32
39. Kratsch, S., Marx, D., Wahlström, M.: Parameterized complexity and kernelizability of max ones and exact ones problems. ACM Trans. Comput. Theory **8**(1), 1:1–1:28 (2016). https://doi.org/10.1145/2858787
40. Kratsch, S., Wahlström, M.: Preprocessing of min ones problems: a dichotomy. In: Abramsky, S., Gavoille, C., Kirchner, C., Meyer auf der Heide, F., Spirakis, P.G. (eds.) ICALP 2010. LNCS, vol. 6198, pp. 653–665. Springer, Heidelberg (2010). https://doi.org/10.1007/978-3-642-14165-2_55
41. Kratsch, S., Wahlström, M.: Representative sets and irrelevant vertices: new tools for kernelization. J. ACM **67**(3), 16:1–16:50 (2020). https://doi.org/10.1145/3390887
42. Lokshtanov, D., Misra, P., Saurabh, S., Zehavi, M.: Kernelization of counting problems. In: Guruswami, V. (ed.) 15th Innovations in Theoretical Computer Science Conference, ITCS 2024, Berkeley, CA, USA, 30 January–2 February 2024. LIPIcs,

vol. 287, pp. 77:1–77:23. Schloss Dagstuhl - Leibniz-Zentrum für Informatik (2024). https://doi.org/10.4230/LIPICS.ITCS.2024.77
43. Sam, E., Bergougnoux, B., Golovach, P.A., Blaser, N.: Kernelization for finding lineal topologies (depth-first spanning trees) with many or few leaves. In: Fernau, H., Jansen, K. (eds.) Fundamentals of Computation Theory - 24th International Symposium, FCT 2023, Trier, Germany, 18–21 September 2023, Proceedings. Lecture Notes in Computer Science, vol. 14292, pp. 392–405. Springer, Heidelberg (2023). https://doi.org/10.1007/978-3-031-43587-4_28
44. Thilikos, D.M.: A retrospective on (meta) kernelization. In: Fomin, F.V., Kratsch, S., van Leeuwen, E.J. (eds.) Treewidth, Kernels, and Algorithms. LNCS, vol. 12160, pp. 222–246. Springer, Cham (2020). https://doi.org/10.1007/978-3-030-42071-0_16
45. Thomassé, S.: A $4k^2$ kernel for feedback vertex set. ACM Trans. Algor. **6**(2), 32:1–32:8 (2010). https://doi.org/10.1145/1721837.1721848
46. Xu, C., Dai, L., Liu, K.: Bloomfilter-based practical kernelization algorithms for minimum satisfiability. In: Luo, B., Cheng, L., Wu, Z., Li, H., Li, C. (eds.) Neural Information Processing - 30th International Conference, ICONIP 2023, Changsha, China, 20–23 November 2023, Proceedings, Part IX. Communications in Computer and Information Science, vol. 1963, pp. 38–47. Springer, Heidelberg (2023). https://doi.org/10.1007/978-981-99-8138-0_4

Unbent Collections of Orthogonal Drawings

Todor Antić[1], Giuseppe Liotta[2], Tomáš Masařík[3], Giacomo Ortali[2], Matthias Pfretzschner[4(✉)], Peter Stumpf[1,5], Alexander Wolff[6], and Johannes Zink[7]

[1] Charles University, Prague, Czech Republic
[2] Università degli Studi di Perugia, Perugia, Italy
[3] University of Warsaw, Warsaw, Poland
[4] Universität Passau, Passau, Germany
pfretzschner@fim.uni-passau.de
[5] Czech Technical University in Prague, Prague, Czech Republic
[6] Universität Würzburg, Würzburg, Germany
[7] Technische Universität München, Heilbronn, Germany

Abstract. Recently, there has been interest in representing single graphs by multiple drawings; for example, using graph stories, storyplans, or uncrossed collections. In this paper, we apply this idea to orthogonal graph drawing. Due to the orthogonal drawing style, we focus on 4-graphs, that is, graphs of maximum degree 4. We restrict ourselves to plane graphs, that is, planar graphs whose embedding is fixed. Our goal is to represent any plane 4-graph G by an *unbent collection*, that is, a collection of orthogonal drawings of G that adhere to the embedding of G and ensure that each edge of G is drawn without bends in at least one of the drawings. We investigate two objectives. First, we consider minimizing the number of drawings in an unbent collection. We prove that every plane 4-graph can be represented by a collection with at most three drawings, which is tight. We also give necessary and sufficient conditions for a graph to admit an unbent collection of size 2. Second, we consider minimizing the total number of bends over all drawings in an unbent collection. We show that this problem is NP-hard and give a 3-approximation algorithm. For the special case of plane triconnected cubic graphs, we show how to compute minimum-bend collections in linear time.

The full version of this paper is available on arXiv [2].

T. Antić—Supported by Czech Science Foundation grant no. 23-04949X and grant no. SVV-2023-260699.

G. Liotta—Research supported in part by MUR of Italy, PRIN Project no. 202 2TS4Y3N – EXPAND and PON Project ARS01_00540.

T. Masarík—Supported by the Polish National Science Centre SONATA-17 grant number 2021/43/D/ST6/03312.

M. Pfretzschner—Funded by the Deutsche Forschungsgemeinschaft (DFG, German Research Foundation) – grant 541433306.

P. Stumpf—Supported by Czech Science Foundation grant no. 23-04949X.

H. Fernau and P. Kindermann (Eds.): WG 2025, LNCS 16124, pp. 18–32, 2026.
https://doi.org/10.1007/978-3-032-11835-6_2

Keywords: Graph drawing · Orthogonal drawings · Bend minimization

1 Introduction

A way to determine the quality of a drawing of a graph is its *readability*, which, informally speaking, is the extent to which a drawing conveys the structure of the graph it represents. More formally, readability translates into a set of goals that a graph drawing algorithm must optimize. For example, in the context of orthogonal graph drawing, the typical objectives are crossing minimization and bend minimization. These objectives typically oppose each other: reducing the number of crossings may lead to adding bends along the edges, and vice versa.

Related Work. A recent research direction is to represent a graph by a collection or sequence of different drawings, each of which optimizes the readability of a particular subgraph, possibly at the expense of the rest. One contribution in this direction is the so-called *storyplan* approach, where a graph is represented by a sequence of drawings, called *frames*, each of which only shows a restricted (e.g., planar) subgraph. To see the whole (complicated) graph means to watch all (relatively simple) frames [5,8]. Another contribution are *graph stories*, where each frame contains the same number of vertices, vertices are shown in an interval of consecutive frames, and, for each vertex, the index of the first frame when it appears is given as input. As in a storyplan, it is required that the drawing of an edge and the position of a vertex is the same in every frame where the edge or the vertex appears. The task is to compute graph stories with specific properties (e.g., crossing-free frames [4]). Another approach is to visualize the graph by means of a collection of drawings, each of which includes all vertices and edges, but to optimize the drawings only with respect to specific subgraphs. Hliněný and Masařík [10] adopted this idea and introduced the problem of representing a non-planar graph G via multiple drawings such that G is fully represented in each drawing and each edge of G is not crossed in at least one of the drawings. They called such a collection an *uncrossed collection*. The *size* of a collection is the number of drawings it contains. Hliněný and Masařík studied two optimization problems. For a given graph G, the first problem called *the uncrossed number* is to find a minimum-size uncrossed collection of G. The second problem is to find an uncrossed collection that minimizes the total number of crossings over all drawings, for which they showed an NP-hardness and an FPT algorithm parameterized by the solution size. Recently, the authors of [3] determined the uncrossed number of complete and complete bipartite graphs exactly. They also gave a lower bound for the uncrossed number of general graphs in terms of vertex and edge numbers.

Our Contribution. We revisit the classical problem of computing a planar orthogonal layout with the minimum number of bends from the "collection perspective" of Hliněný and Masařík. Namely, we study the following problem: Given a plane

4-graph G (i.e., a planar graph with vertex degree at most 4 and with a fixed planar embedding), compute a set of planar orthogonal drawings of G such that every edge of G is straight (i.e., has no bend) in at least one drawing of the collection. Analogously to the terminology of Hliněný and Masařík, we call such a collection an *unbent collection*. We study two problems. For a given graph G, we want to find an unbent collection of minimum size, $\text{un}(G)$, and an unbent collection that minimizes the total number of bends, $\text{tbn}(G)$. Refer to Fig. 1.

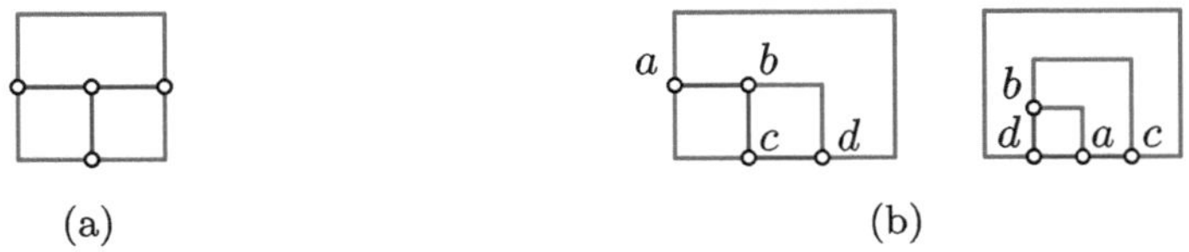

Fig. 1. (a) A minimum-bend orthogonal drawing of K_4. (b) An unbent collection for K_4 with two drawings showing that $\text{un}(K_4) = 2$ and $\text{tbn}(K_4) \leq 12$ (in fact, $\text{tbn}(K_4) = 12$); see Definition 1. Red edges are straight, whereas blue edges are bent. (Color figure online)

Specifically, we show the following. For every plane 4-graph, an unbent collection of size 3 always exists (this is tight) and can be found in almost-linear time; see Theorem 1. We also give a necessary and sufficient condition for graphs to have an unbent collection of size 2. The condition is based on a new concept of *balanced* 2-edge-coloring that we introduce in Sect. 2.

Then, we show that it is NP-hard to decide, given a plane 4-graph G and a positive integer k, whether G admits an unbent collection with at most k bends; see Sect. 3. Note that if the embedding of G is not fixed, it is even NP-hard to decide whether G admits a (single) drawing without bends [9]. Next, we give a 3-approximation algorithm for minimizing the number of bends in an unbent collection. It relies on Tamassia's flow network [15], combined with local modifications and the insight that every 4-graph has star arboricity at most 4.

For the special case of plane triconnected cubic graphs, we show how to compute minimum-bend collections in linear time; see Sect. 4. The collections that we compute have size 2, which is optimal. Our algorithm builds upon ideas of Nishizeki and Rahman [12,13], further refined by Didimo, Liotta, Ortali, and Patrignani [7].

We prove statements marked with a ($\star$) in the full version of this paper [2].

Preliminaries. As shorthand for the set $\{i, i+1, \ldots, j-1, j\}$, we use $[i, j]$, and for the set $[1, j]$, we use $[j]$. A *planar embedding*, or simply, an *embedding*, of a graph G is the collection of (counter-clockwise) circular orderings of incident edges around every vertex induced by a planar drawing of G. We will assume that all of our graphs are given with an embedding and a specified external face; such graphs are called *plane*. An *orthogonal drawing* of a graph is a drawing in which every edge is drawn as a sequence of vertical and horizontal line segments. Now, we can formally define the main concepts that we study in this paper.

Definition 1. *Let G be a plane 4-graph. A set of orthogonal drawings of G (all adhering to the embedding of G) is called an* unbent collection *of G if every edge of G is drawn without a bend (i.e., straight) in at least one drawing of the set. The* unbent number *of G, denoted $\mathrm{un}(G)$, is the size of a smallest unbent collection of G. The* unbent bend number *of G, denoted $\mathrm{tbn}(G)$, is the minimum total number of bends, taken over all unbent collections of G.*

Given a plane 4-graph G, Tamassia [15] defined a flow network N_G that has a node for each vertex and for each face of G, an arc from every face to each incident vertex, and two parallel arcs (one in each direction) for each edge e of G connecting the faces incident to e. A unit of flow in N_G represents a 90° angle at a vertex of G or a bend on an edge of G. For the sake of presentation, we invert the usual flow direction in this paper. Then, each vertex has a demand of 4, each inner face f has a demand of $4-\deg(f)$, and the external face f_{ext} has a demand of $-4-\deg(f_{\mathrm{ext}})$. In the case of negative demand (i.e., flow excess), we also speak of *supply* (of flow). Faces of degree greater than 4 and the external face supply flow, while vertices and faces of degree 3 demand flow. Flow in N_G has non-zero cost only for arcs between faces and vertices; otherwise, the cost is one for each unit of flow. The total cost equals the number of bends on the edges of G in the resulting *orthogonal representation* [15], which refines the given embedding by specifying angles and bends (but not the lengths of edges). Such a representation can then be turned into an orthogonal drawing in time linear in the number of vertices and bends [15].

Recently, a minimum-cost flow algorithm that runs in $O(m^{1+o(1)} \log U \log C)$ time was presented [6], where m is the number of edges, U is the maximum edge capacity, and C is the maximum absolute edge cost in the flow network. In N_G, $C \in O(1)$, $m \in O(n)$, and $U \in O(n)$, where $n = |V(G)|$. Since $\log n = n^{\log\log n/\log n} \in n^{o(1)}$, the algorithm runs in $O(n^{1+o(1)})$ time on input N_G.

2 Minimizing the Size of an Unbent Collection

In this section, we give, for any plane 4-graph G, an upper bound on $\mathrm{un}(G)$ in terms of the arboricity of G by slightly amending Tamassia's flow network [15].

Proposition 1. *If a plane 4-graph G can be partitioned into k forests, then $\mathrm{un}(G) \le k$. Given G and such a set of forests, an unbent collection of size k can be computed in $O(kn^{1+o(1)})$ time, where n is the number of vertices of G.*

Proof. For $i \in [k]$, let F_i be the i-th given forest. We construct Tamassia's flow network for G. We then make k copies of the network. In the i-th copy, for each edge $e \in F_i$, we delete the two arcs that connect the two nodes representing the incident faces of e. Since F_i contains no cycle, note that the nodes in the flow network representing the faces still form a strongly connected component, hence there always exists a solution. Solving these k flow networks individually produces k orthogonal drawings. Each edge e of G is drawn straight in at least one of these drawings, namely in the i-th drawing if e lies in F_i. For each

$i \in [k]$, constructing the i-th flow network and removing the links corresponding to edges in F_i can be done in linear time, and each flow network can be solved in $O(n^{1+o(1)})$ time [6]. □

Proposition 1 and Schnyder woods [14] yield an upper bound on $\text{un}(G)$.

Theorem 1. *Any plane 4-graph G admits an unbent collection of size at most 3. Such a collection can be computed in $O(n^{1+o(1)})$ time, where $n = |V(G)|$.*

Proof. Extend G to a plane triangulation G^+, and decompose $E(G^+)$ into three disjoint trees in linear time by Schnyder's algorithm [14]. Ignoring the edges from $G^+ \setminus G$ leaves three edge-disjoint forests, to which we apply Proposition 1. □

Theorem 1 tells us that for any planar graph G of maximum degree 4, we have $\text{un}(G) \leq 3$. We will later show that this bound is in fact tight, by constructing an infinite family of graphs with unbent number 3. Using Tamassia's flow network [15], we can decide efficiently whether $\text{un}(G) = 1$. Thus, the real challenge is to decide whether two drawings are sufficient or three are necessary. We give a partial answer to this question below.

Proposition 2. *Let G be a plane 4-graph. If, for every subset $V' \subseteq V(G)$, the induced subgraph $G[V']$ has at most $2|V'| - 2$ edges, then* $\text{un}(G) \leq 2$.

Proof. According to Nash-Williams' theorem [11], a graph H can be partitioned into k forests if and only if, for every $U' \subseteq V(H)$, the induced graph $H[U']$ has at most $k \cdot (|U'| - 1)$ edges. Applying Nash-Williams' theorem and Proposition 1 to G and $k = 2$ yields $\text{un}(G) \leq 2$. □

Testing the sufficient criterion presented in Proposition 2 naively takes exponential time. So, we next present a way to decompose G into relevant subgraphs.

Lemma 1 (⋆). *Let G be a plane 4-graph with n vertices and at most $2n - 2$ edges. In $O(n)$ time, we can construct a set $\mathcal{G}$ of plane 4-graphs such that (i) for each $G' \in \mathcal{G}$, $|E(G')| \geq 2|V(G')| - 1$, (ii) $\sum_{G' \in \mathcal{G}} |V(G')| \in O(n)$, and (iii) $\text{un}(G) \leq 2$ if and only if, for each $G' \in \mathcal{G}$, $\text{un}(G') \leq 2$.*

Proof (sketch). Essentially, we decompose G into induced subgraphs such that each such subgraph G' has at least $2|V(G')| - 1$ edges. (The other parts of G can be ignored because, due to Proposition 2, they admit an unbent collection of size 2 anyway.) Since G has maximum degree 4 and G' misses at most one edge to be 4-regular, G' is connected to $G \setminus G'$ with at most two edges. Hence, we find all 1- and 2-edge cuts in linear time, which decompose the graph into smaller graphs. In each smaller graph, we keep the edges of the cuts, which would end at leaves (i.e., vertices of degree 1). To still match $|E(G')| \geq 2|V(G')| - 1$ for each smaller graph G', we attach the gadget H shown in Fig. 2 to each such leaf. If $\text{un}(G') \leq 2$ holds for each G', we obtain an unbent collection of size 2 of G by merging the smaller graphs at the leaves of the corresponding cuts. Since each cut has size at most 2, we can keep one edge straight in each of the two drawings and add as many bends as necessary to the other edge. □

Lemma 1 tells us that we should focus on graphs with at least $2n-1$ edges (a 4-graph has at most $2n$ edges, so these are the "densest" plane 4-graphs) as the calculation of the unbent number of all graphs with lower density boils down to these graphs. It is natural to ask whether graphs with $\text{un}(G) = 3$ even exist, which we answer positively later. If G has at least $2n-1$ edges, then Proposition 1 cannot be applied to obtain $\text{un}(G) \leq 2$ as every 2-edge-coloring will contain a cycle in at least one color. So, we next characterize graphs of unbent number at most 2 via 2-edge-colorings with specific properties.

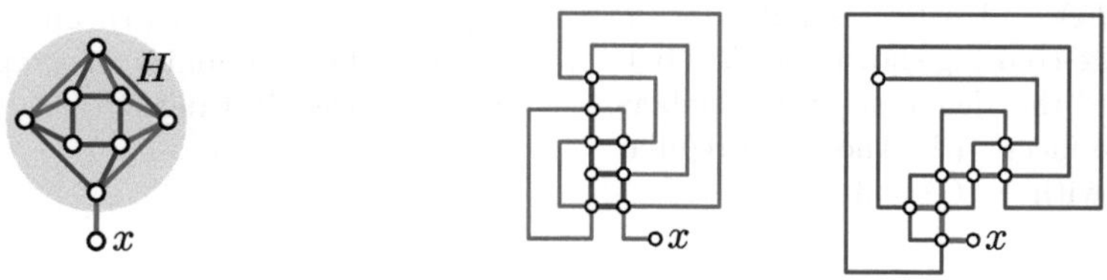

(a) (b)

Fig. 2. (a) The gadget H with $2n-1$ edges used to replace a degree-1 vertex in the proof of Lemma 1. (b) An unbent collection of size 2 for H.

We say that every vertex of G has $4-\deg(v)$ *free angles*. An *angle assignment* of G assigns each free angle of every vertex to one of its incident faces. Given an angle assignment, the *demand* of a face f of G is the number of angles it has been assigned plus $(4-\deg(f))$ (respectively $-4-\deg(f)$ if f is the external face). These demands of the faces correspond to a simplified version of Tamassia's flow network [15] where the flow of the nodes to the faces has already been assigned and all that still needs to be done is solving the flow network induced by the nodes representing the faces of G. A 2-edge-coloring of G is *balanced* if, for each of the two colors, there exists an angle assignment such that, for every cycle of that color, the demands of all faces it encloses sum up to 0. Using Tamassia's flow network [15], we can show that the graphs that admit an unbent collection of size at most 2 are exactly the graphs that admit a balanced 2-edge-coloring.

Proposition 3 ($\star$). *Let G be a plane 4-graph. Then* $\text{un}(G) \leq 2$ *if and only if G admits a balanced 2-edge-coloring.*

With this characterization, we can show that graphs with unbent number 3 exist. We define the *flower graph* with k petals, F_k, to be the graph obtained from two cycles of length k, $C = (v_1, \ldots, v_k)$ and $C' = (v'_1, \ldots, v'_k)$ by adding edges $\{v_i, v'_i\}$ and $\{v_i, v'_{i+1}\}$ for each $i \in [k]$. For an illustration, see Fig. 3. Note that F_3 is the octahedron.

The following Proposition 4 shows that K_5 minus an edge and the octahedron minus an edge have unbent number 3 if the large face is the external face; see Fig. 4a. Thus, plane 4-graphs that contain one of those two as an induced (plane) subgraph also have unbent number 3; see Fig. 4b. It also shows

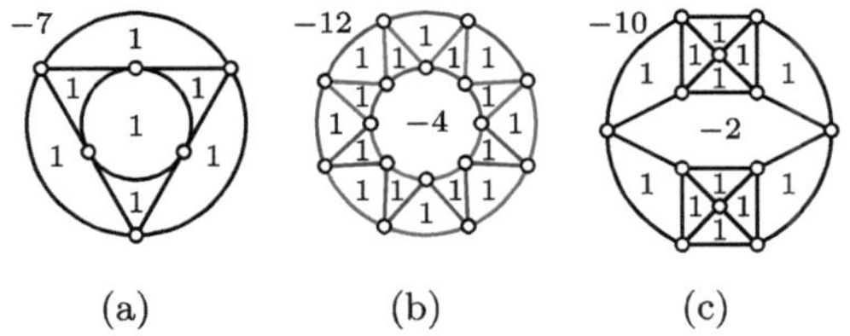

Fig. 3. (a), (b) Flower graphs F_3 and F_8 with 3 and 8 petals, respectively. Note that $\mathrm{un}(F_3) = 3$, whereas $\mathrm{un}(F_8) = 2$. The 2-edge-coloring shown for F_8 is balanced; see the demands (the numbers inside the faces). (d) Another 4-regular graph G with $\mathrm{un}(G) = 3$.

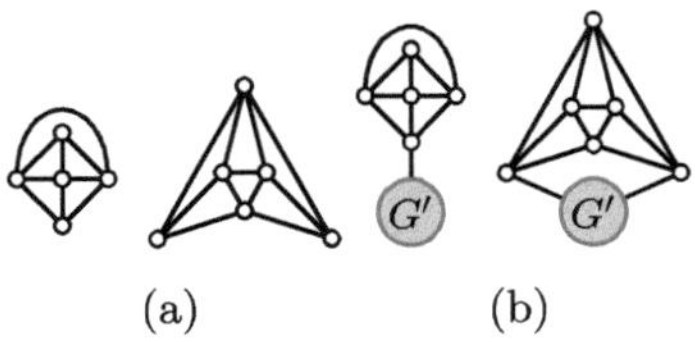

Fig. 4. (a) Two minimal graphs that require three drawings. (b) By extending the graphs from (a), one can construct infinite families of graphs that require three drawings.

$\mathrm{un}(F_k) = 3$ for $k \in [3, 7]$, and that the 4-regular graph in Fig. 3a has unbent number 3. We remark that $\mathrm{un}(F_k) = 2$ for $k \geq 8$; see Fig. 3b.

Proposition 4 (⋆). *Let G be a plane 4-graph with n vertices and m edges such that (i) $m = 2n - 1$ and every inner face is a triangle or (ii) $m = 2n$, all but the external face and exactly one inner face f are triangles and f shares an edge with the external face or f has at most seven edges. Then,* $\mathrm{un}(G) = 3$.

Proof (sketch). Suppose $\mathrm{un}(G) \leq 2$. Then, by Proposition 3, G admits a balanced 2-edge-coloring. Let us first assume condition (i). The edges of one color must induce a cycle since $m = 2n - 1$. This cycle, however, encloses only triangles and thus cannot be balanced. Hence, $\mathrm{un}(G) = 3$.

Now assume condition (ii). First suppose that one color induces two cycles c_1 and c_2 and let F_1 and F_2 be the sets of faces enclosed by c_1 and c_2, respectively. Because both cycles are balanced, both F_1 and F_2 contain f. The faces in the symmetric difference $F_1 \triangle F_2$ are only triangles that are enclosed by at least one monochromatic cycle, which cannot be balanced. Hence, the edges of each color induce a spanning graph with exactly one (balanced) cycle containing f. This is not possible if f shares an edge with the external face. Otherwise, let k be the number of edges of f. Then each edge e of f is incident to another triangle that is contained in the cycle of the color not assigned to e. This results in at least k triangles inside of both monochromatic cycles, while, for $k \leq 7$, there should be only $2(\deg(f) - 4) = 2k - 8 < k$ triangular faces to account for the demand of f. □

3 The Unbent Bend Number of Plane 4-Graphs

In this section, we first show that determining the unbent bend number is NP-complete. Afterwards, we present a 3-approximation algorithm for computing the unbent bend number of any graph that admits a crossing-free orthogonal drawing, that is, of any plane 4-graph.

As a new NP-complete version of the SAT problem, we introduce LevelPlanarMonotone2in4Sat, which is defined as follows. As in 2in4Sat, the input is a Boolean formula in conjective normal form where every clause consists of four literals, and the task is to decide whether an assignment of variables to Boolean values exists such that, in every clause, exactly two literals are true. Additionally, the variable–clause incidence graph is planar and admits a drawing where the variable vertices are placed on a set of parallel lines, called *levels*, and the clause vertices are placed between the levels such that no edge crosses a level. The levels divide the plane into rectangular strips. The clauses that are represented in the first, third, etc. strip contain only negated literals, while the clauses represented in the second, fourth, etc. strip contain only unnegated literals.

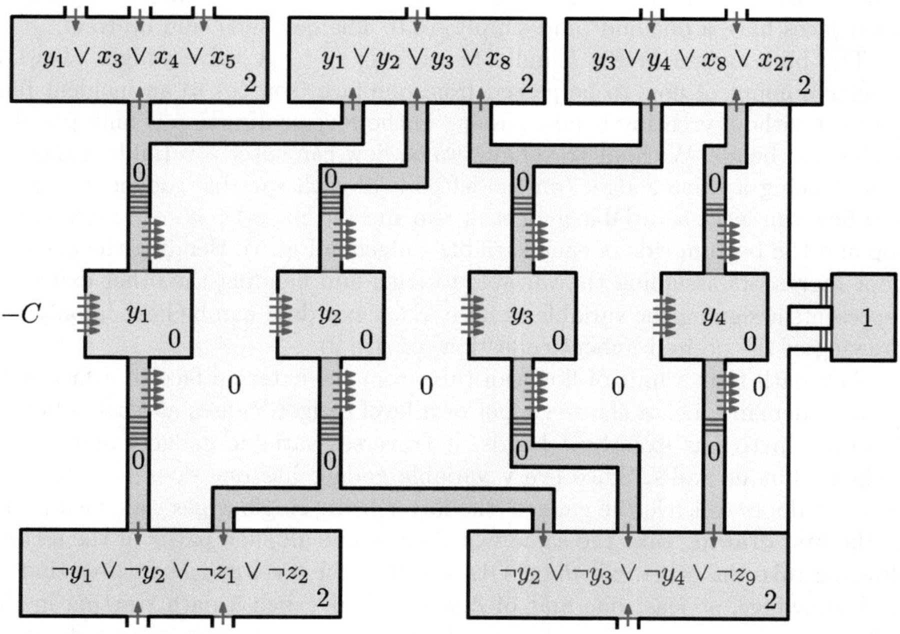

Fig. 5. NP-hardness: A layer of variable gadgets and the neighboring clause gadgets.

Lemma 2 (⋆). LevelPlanarMonotone2in4Sat *is NP-complete.*

This is the base problem for our NP-hardness reduction sketched next.

Theorem 2 (⋆). *Given a plane 4-graph G and an integer k, it is NP-complete to decide whether* $\mathrm{tbn}(G) \leq k$.

Proof (sketch). We sketch only the NP-hardness. Given a LEVELPLANARMONOTONE2IN4SAT instance Φ with n variables, m clauses, and ℓ levels, we construct a 4-plane graph G that follows the level structure of the variable–clause incidence graph (see Fig. 5 for one level) and has an unbent collection of size 2 with at most $k = 2nC$ bends, where $C = 2m + \ell$, if and only if Φ is satisfiable.

In G, there is a face for each variable vertex (*variable gadget*; blue in the middle row of Fig. 5), for each clause vertex (*clause gadget*; green and red in the top and bottom row of Fig. 5), and for each set of edges connecting a variable vertex to its clause vertices above and, separately, below (*tunnel gadget*; orange and thin in Fig. 5). Each such gadget is surrounded by a set of faces called *walls*, which make it costly for flow to enter the gadget unless it is through some *door*. (All walls have "thickness" $2n$, that is, each unit of flow traversing a wall bends at least $2n$ edges.) We construct each face with a specific demand. The clause gadgets have a demand of 2, the *level gadgets*, which are special faces existing once per level (light purple on the very right in Fig. 5), have a demand of 1, the external face has a supply of C (shown as a demand of $-C$ in Fig. 5), and all other faces have a demand (and supply) of 0. The demands sum up to C.

The blue arrows in Fig. 5 indicate *valve gadgets*. A valve gadget allows a specific amount of flow to be passed from one face (source) to an incident face (target) without creating bends; passing in the reverse direction is only possible with extra bends. Without traversing walls, flow can enter a variable gadget at a door being a valve gadget (on the left side of each variable gadget in Fig. 5), and flow can leave a variable gadget at two specific edges (*exit doors*; red on the top and the bottom side of each variable gadget in Fig. 5). Bending the one exit door represents assigning the variable to true, and bending the other exit door represents assigning the variable to false. Each exit door can be bent in only one drawing of the desired unbent collection (of size 2).

Any path that a unit of flow can take from the external face to a face with positive demand, i.e., a clause gadget or a level gadget, causes at least n bends. If such a path has exactly n bends, it traverses variable gadgets and tunnel gadgets, but no walls. Since every variable gadget has one door to enter and two exit doors to leave the gadget (the latter being single edges), all such paths in the first drawing take the same exit door, while all such paths in the second drawing take the other exit door. Observe that the level gadget assures that in both drawings, at least one unit of flow must take such a path starting in the external face and traversing all variable gadgets. The choices of exit doors at the variable gadgets correspond to an assignment of Boolean values to variables. Note that, in Fig. 5, there are additional red edges inside the tunnel gadgets. Their purpose is to make any path that a unit of flow can take from the external face to a clause gadget cost n (and not less). Moreover, they assure that the clause gadgets can only be reached via such a tunnel in one of the drawings.

Each clause gadget has four doors, each being a valve gadget allowing one unit of flow to enter. These doors are connected to tunnel gadgets, which, in turn, are connected to the variable gadgets of the four variables that occur in the clause. Two units of flow in a clause gadget need to come from two of these variables gadgets in the one drawing, while, in the other drawing, two units of flow come from the other two variables gadgets. Hence, an unbent collection exists if and only if the exit doors taken by the flow paths correspond to a satisfying truth assignment of Φ. □

Next, we describe a 3-approximation algorithm for the unbent bend number. We use the property that the star arboricity of a graph is bounded by its maximum degree. Algor and Alon [1] showed that every Δ-regular graph has star arboricity at most $\Delta/2 + O(\Delta^{2/3}(\log \Delta)^{1/3})$. We are interested only in the case $\Delta = 4$, and give the following simple explicit construction (stated for general Δ, although it is asymptotically worse than the bound of Algor and Alon).

Proposition 5 (⋆). *Every graph G of maximum degree Δ has star arboricity at most Δ. The corresponding partition of $E(G)$ into star forests can be computed in $O(\Delta \cdot |V(G)|)$ time.*

For every edge e of G, let f_e and g_e be the faces that are adjacent to e (possibly $f_e = g_e$). In Tamassia's flow network N_G, there are arcs (f_e, g_e) and (g_e, f_e) of unit cost between these faces. The amount of flow on these arcs corresponds to the number of bends that e makes into the appropriate directions. Without loss of generality, we can assume that at most one of the two arcs carries flow; let this arc be e^+. If none carries flow, the choice of e^+ is arbitrary.

Lemma 3. *Let G be a plane 4-graph, and let ϕ be a feasible flow in N_G. If v is a vertex of G that is incident to $d \geq 2$ edges $e_1, \dots, e_d$ in this order in the given embedding, then there exists a feasible flow ϕ' in N_G such that $\phi'(e_1^+) = 0$, for $i \in \{2, \dots, d\}$, it holds that $\phi'(e_i^+) \leq \phi(e_i^+) + \phi(e_1^+)$, and, for every arc $a \notin \{e_1^+, \dots, e_d^+\}$ of N_G, it holds that $\phi'(a) = \phi(a)$.*

Proof. Send the flow on e_1 around v in the other direction. This *elementary operation* increases the cost for the edges $e_2, \dots, e_d$ by at most $\phi(e_1^+)$. The resulting flow ϕ' is feasible and coincides with ϕ on all other arcs of N_G. □

Theorem 3. *There is an algorithm that computes, for any plane 4-graph G, an unbent collection with at most $3\,\mathrm{tbn}(G)$ bends. The algorithm runs in $O(n^{1+o(1)})$ time, where n is the number of vertices of G.*

Proof. We first compute a minimum-cost flow ϕ in N_G. This flow ϕ corresponds to a bend-minimal orthogonal representation R of G. Based on R, we compute an orthogonal drawing Γ of G [15]. Let b be the cost of ϕ (which corresponds to the number of bends in R and Γ). If $b = 0$, we return $\{\Gamma\}$ as an unbent collection (which is optimal), and we are done.

Otherwise, any optimal unbent collection has size at least 2 and, hence, $\mathrm{OPT} \geq 2b$. We use Proposition 5 to 4-color the edges of G in linear time such that each color induces a star forest. We now define an unbent collection of size 4. For each $i \in [4]$, we construct a drawing Γ_i where all edges of color i are straight. Each time, we start with R and proceed as follows. Let e be an edge of color i, and let $b_e = \phi(e)$. If $b_e = 0$, we do nothing. Otherwise we straighten e by

Lemma 3 with respect to the endpoint v of e that is *not* the center of the star that contains e. As a result, the contribution of e to the cost of ϕ reduces from b_e to 0 and, for each edge $e' \neq e$ that is incident to v, its contribution to the cost of ϕ increases by at most b_e. Since v has degree at most 4, the cost of the operation is bounded by $-e_b + 3b_e = 2b_e$. Performing all necessary elementary operations results in a feasible flow ϕ_i' whose cost is at most $\text{cost}(\phi) + \sum_{e \text{ of color } i} 2b_e$. As the edges of color i form a star forest and we never apply the elementary operation to the center of a star, the operation regarding one edge of color i does not increase the cost of any other edge of color i. Thus, using ϕ_i', we can compute an orthogonal representation R_i and an orthogonal drawing Γ_i where all edges of color i are straight. In total, the unbent collection $\{\Gamma_1, \ldots, \Gamma_4\}$ has at most $4\,\text{cost}(\phi) + \sum_{e \in E(G)} 2b_e = 4b + 2b = 6b$ bends, and the approximation factor of our algorithm is $\text{ALG} / \text{OPT} \leq 6b/(2b) = 3$. The algorithm for computing the minimum-cost flow [6] dominates the running time of our algorithm, which hence runs in $O(n^{1+o(1)})$ time. □

4 The Unbent Bend Number of Plane Triconnected Cubic Graphs

In this section we describe how to compute, for a given plane triconnected cubic graph, an unbent collection with the minimum number of bends. The collections that we compute consist of only two drawings, which is the minimum.

Let G be a plane triconnected cubic graph. Observe that computing an orthogonal representation of a graph with the minimum number of bends is equivalent to subdividing the edges of the graph the minimum number of times so that the graph admits an orthogonal representation without bends. Hence, our strategy will be to compute two different subdivisions of G such that each of the resulting graphs admits an orthogonal representation without bends and the total number of subdivision vertices is minimum. We call a subdivision vertex a *dummy*. Let f_{ext} denote the external face of G. Given a cycle C, a *leg* of C is an edge external to C and incident to a vertex of C. A *k-legged cycle* C of G is a cycle such that: (i) C has exactly k legs, and (ii) no two legs share a vertex. A k-legged cycle with $k \leq 3$ is *bad* if less than $4 - k$ dummies are placed along its edges. Otherwise, it is *good*. Observe that a 3-legged cycle (such as C_5 in Fig. 6) is good if we place a dummy along one of its edges, whereas the external cycle is 0-legged and hence it is good if we place four dummies along its edges (which we did in Fig. 6). We use the following lemma.

Lemma 4 [13]. *A plane cubic graph has an orthogonal representation without bends if and only if it does not contain any bad cycle. Given a plane cubic graph G without bad cycles, an orthogonal representation of G without bends can be computed in linear time.*

In Fig. 6a and 6b we show two ways to add dummies to a plane triconnected cubic graph G such that all cycles are good (and such that no edge of G receives

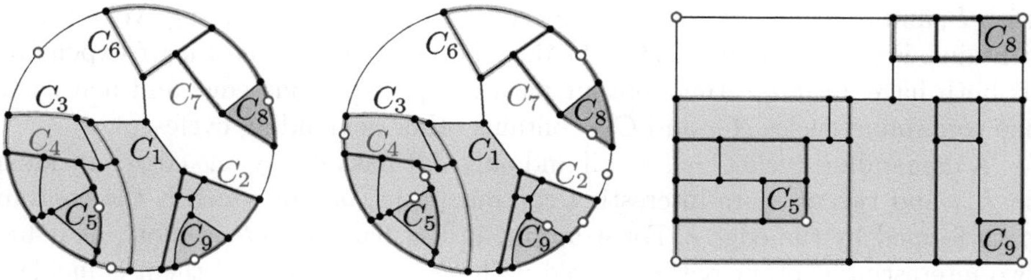

(a) G^1 (b) G^2 (c) orthogonal drawing of G^1

Fig. 6. (a–b) Unbent collection $\{G^1, G^2\}$ of a plane triconnected cubic graph G with 3-legged cycles $C_1, \ldots, C_9$ and dummies (blue circles). (c) Drawing of G^1 without bends.

dummies in both cases). In the resulting graphs G^1 and G^2, every 3-legged cycle contains a dummy and the external cycles contain four dummies. Due to this, both graphs admit orthogonal drawings without bends; see that of G^1 in Fig. 6c.

We say that the k paths of a k-legged cycle connecting the vertices incident to its legs are the *contour paths* of the cycle. A *leg face* of a k-legged cycle is a face incident to two of its legs. We now aim to lowerbound $\text{tbn}(G)$. To this end, we use specific 3-legged cycles of G that we call *demanding*. Informally, the demanding cycles form the smallest subset of 3-legged cycles with the property that, if we place a dummy on an edge of each of them, all 3-legged cycles in G will be good. A *non-demanding* cycle is a 3-legged cycle that is not demanding. For a formal definition of demanding cycles, see the full version [2]. The set of demanding cycles is unique. Let $D(G)$ be the largest set of demanding cycles of G that are pairwise vertex-disjoint.

For an example, see Fig. 6a, where $D(G) = \{C_3, C_5, C_8, C_9\}$. Observe that all these 3-legged cycles are good due to the placement of four dummy vertices, one for each cycle in $D(G)$. E.g., the dummy vertex on C_2 also lies on C_9 (since C_2 and C_9 share an edge). Hence, both these 3-legged cycles are good.

The demanding cycles and the set $D(G)$ are computed by using a coloring rule over a structure that describes the hierarchy of the 3-legged cycles of G. This structure and the coloring rule are described in the full version [2]. We now lowerbound $\text{tbn}(G)$ in terms of the numbers of certain types of 3-demanding cycles.

A cycle C *contains* another cycle C' if the edges of C' are edges of the subgraph of G induced by C and its interior vertices. Let C be a demanding cycle. A contour path P of C is *interesting* if every non-demanding cycle of G that contains C and shares some edges with C, contains P. For example, C_8 in Fig. 6a has two interesting contour paths, namely those sharing edges with C_6 and C_7. See also C_5 and C_9, which have only one interesting contour path each, sharing edges with C_4 and C_2, respectively.

A demanding cycle C is *expensive* if and only if C has exactly one interesting contour path P that contains a single edge, and there exists at least one non-demanding 3-legged cycle C' such that (i) C' contains C, (ii) C' shares edges with P, and (iii) C' does not share any edges with other demanding cycles that it contains. For example, in Fig. 6a, the demanding cycles C_5 and C_9 are expensive, as both have an interesting contour path composed of one edge and neither of the containing cycles (C_4 and C_2) contains other demanding cycles.

A demanding cycle C is *short* if and only if it has exactly one edge e incident to f_{ext} and two or more interesting contour paths, one of which is the contour path formed by the edge e. For example, in Fig. 6a, cycle C_8 is short, as it has two interesting contour paths (shared with C_6 and C_7), one of them is incident to f_{ext} and composed by one edge.

We now define three subsets of $D(G)$ and then use their cardinalities to lowerbound $\text{tbn}(G)$. Let $D_{\text{ext}}(G)$ be the set of demanding cycles that have f_{ext} as leg face, let $D_{\text{exp}}(G)$ be the set of expensive demanding cycles, and let $D_{\text{short}}(G)$ be the set of short demanding cycles.

Lemma 5 (⋆). *For any plane triconnected cubic graph G, it holds that*
$$\text{tbn}(G) \geq q(G) := 2|D(G)| + |D_{\text{exp}}(G)| + 8 - \min\{8, 2|D_{\text{ext}}(G)| - |D_{\text{short}}(G)|\}.$$

In Theorem 4 we prove that the lower bound of Lemma 5 is tight. Figure 6 shows an unbent collection $\mathcal{C} = \{G^1, G^2\}$ of plane triconnected cubic graph G with $D(G) = \{C_3, C_5, C_8.C_9\}$, $D_{\text{exp}}(G) = \{C_5, C_9\}$, $D_{\text{ext}}(G) = \{C_3, C_8, C_9\}$, and $D_{\text{short}}(G) = \{C_8\}$. Hence, $\mathcal{C}$ has $\text{tbn}(G) = q(G) = 2{\cdot}4{+}2{+}8{-}(2{\cdot}3{-}1) = 13$ dummies.

Theorem 4 (⋆). *There exists a linear-time algorithm that computes, for any plane triconnected cubic graph G, an unbent collection of G with* $\text{tbn}(G)$ *bends. The resulting collection consists of two drawings, which is optimal.*

Proof (sketch). We first create two copies of G, which we call G^1 and G^2. On these, we place $q(G)$ dummies in total. Below we prove that, then, none of the two resulting graphs has a bad cycle. Hence, by Lemma 4, they can be drawn without bends, in linear time. Replacing the dummies by bends yields an unbent collection with $q(G)$ bends. Finally, Lemma 5 implies the theorem.

We compute the sets $D(G)$, $D_{\text{exp}}(G)$, $D_{\text{ext}}(G)$, and $D_{\text{short}}(G)$. We first consider a demanding cycle C in $D_{\text{exp}}(G)$. In G^1, we place a dummy on the only edge belonging to an interesting contour path. In G^2, we place a dummy on another edge of C (not belonging to an interesting contour path), so that C is good in both G^1 and G^2. Still in G^2, we place a dummy on one of the two legs of C that is contained in all the non-demanding cycles that contain C and share edges with C (in the full version [2] we prove that such a leg always exists). Hence, for every cycle in $D_{\text{exp}}(G)$, we have placed three dummies; one in G^1 and two in G^2. As a result, all demanding cycles in $D_{\text{exp}}(G)$ and all non-demanding cycles sharing edges with them are good in G^1 and G^2.

We now consider the cycles in $D(G) \setminus D_{\text{exp}}(G)$. For each such cycle having two or more edges on an interesting contour path, we place a dummy on one such edge in G^1 and one in G^2. It remains to place dummies on (i) demanding (non-expensive) cycles that have one edge belonging to an interesting contour path and (ii) on the non-demanding cycles that share edges only with the demanding cycles in case (i). To this end, we use a data structure that we call *enriched genealogical tree*. With its help, we place the dummies that we need to place on such demanding cycles exclusively on interesting contour paths. As a result, all 3-legged cycles become good.

According to the placement strategy above, we have placed a total of three dummies for every cycle in $D_{\text{exp}}(G)$ and two dummies for every other demanding cycle. Hence, we have placed exactly $2|D(G) \setminus D_{\text{exp}}(G)| + 3|D_{\text{exp}}(G)| = 2|D(G)| + |D_{\text{exp}}(G)|$ dummies.

Since the external cycle is 0-legged, we have to ensure that it gets four dummies in G^1 and four (different ones) in G^2. We show that, for each cycle C in $D_{\text{ext}}(G) \setminus D_{\text{short}}(G)$, we can place two dummies incident to f_{ext} such that C and all the non-demanding cycles sharing edges with C become good cycles. (For the expensive cycles, we need three dummies, but we have already considered this extra cost before.) For the cycles in $D_{\text{short}}(G)$, we need three dummies, two of which incident to f_{ext}, to obtain the same results. Summing up, on the external cycle, we place $8 - \min\{8, 2|D_{\text{ext}}(G)| - |D_{\text{short}}(G)|\}$ dummies that we have not already considered in the term $2|D(G)| + |D_{\text{exp}}(G)|$. Hence, in total we have placed $2|D(G)| + |D_{\text{exp}}(G)| + 8 - \min\{8, 2|D_{\text{ext}}(G)| - |D_{\text{short}}(G)|\} = q(G)$ dummies. As a result, neither G^1 nor G^2 has any bad cycle. □

5 Open Questions and Further Research

We have introduced the concept of unbent collections. We propose here several open questions. Can we efficiently decide, for a given 4-plane graph G, whether $\text{un}(G) = 2$ or $\text{un}(G) = 3$? Does every plane 4-graph G admit an unbent collection of size $\text{un}(G)$ that minimizes $\text{tbn}(G)$? We have shown how to efficiently compute the unbent bend number of plane triconnected cubic graphs, but there are other natural classes to explore. For example, plane cubic graphs that are not triconnected, bipartite planar graphs, and series-parallel 4-graphs are good candidates as they admit unbent collections of size 2.

Lastly, we discuss a subtlety of Definition 1. Throughout the paper, we have required the drawings in the unbent collection to have the same embedding as that of the given plane 4-graph. This requirement stems from the NP-hardness of the bend minimization problem in the variable-embedding setting [9]. However, we suggest for future exploration a variant in which one does not have to adhere to the same embedding in all the drawings in the collection; we call a collection allowing for variable embedding *mutable*. Additionally, we propose a variant between the two definitions where we still allow only one fixed embedding which is however not given in the input; we call a corresponding collection *uniform*. Observe that the *mutable unbent bend number* is different from the unbent bend number even considering our introductory example of K_4. Indeed, the mutable collection can contain two copies of the left drawing in Fig. 1b. Hence, its mutable unbent bend number is at most 10 (as opposed to $\text{tbn}(K_4) = 12$).

Acknowledgments.. The authors thank the organizers and the other participants of the workshop HOMONOLO 2023 (Nová Louka, Czech Republic), where this work was started.

References

1. Algor, I., Alon, N.: The star arboricity of graphs. Disc. Math. **75**(1), 11–22 (1989). https://doi.org/10.1016/0012-365X(89)90073-3
2. Anti?, T., et al.: Unbent collections of orthogonal drawings (2025). https://arxiv.org/abs/2502.18390
3. Balko, M., Hliněný, P., Masařík, T., Orthaber, J., Vogtenhuber, B., Wagner, M.H.: On the uncrossed number of graphs. In: Felsner, S., Klein, K. (eds.) Graph Drawing & Network Vis. (GD). LIPIcs, vol. 320, pp. 18:1–18:13. Schloss Dagstuhl – Leibniz-Zentrum für Informatik (2024). https://doi.org/10.4230/LIPIcs.GD.2024.18
4. Battista, G.D., et al.: Small point-sets supporting graph stories. J. Graph Algor. Appl. **27**(8), 651–677 (2023). https://doi.org/10.7155/JGAA.00639
5. Binucci, C., et al.: On the complexity of the storyplan problem. J. Comput. Syst. Sci. **139**, 103466 (2024). https://doi.org/10.1016/J.JCSS.2023.103466
6. van den Brand, J., et al.: A deterministic almost-linear time algorithm for minimum-cost flow. In: Foundation of Computer Science (FOCS), pp. 503–514. IEEE (2023). https://doi.org/10.1109/FOCS57990.2023.00037
7. Didimo, W., Liotta, G., Ortali, G., Patrignani, M.: Optimal orthogonal drawings of planar 3-graphs in linear time. In: Chawla, S. (ed.) Discrete Algorithms (SODA), pp. 806–825 (2020). https://doi.org/10.1137/1.9781611975994.49
8. Fiala, J., Firman, O., Liotta, G., Wolff, A., Zink, J.: Outerplanar and forest storyplans. In: Fernau, H., Gaspers, S., Klasing, R. (eds.) Theory & Practice Comput. Sci. (SOFSEM). LNCS, vol. 14519, pp. 211–225. Springer, Heidelberg (2024). https://doi.org/10.1007/978-3-031-52113-3_15
9. Garg, A., Tamassia, R.: On the computational complexity of upward and rectilinear planarity testing. SIAM J. Comput. **31**(2), 601–625 (2001). https://doi.org/10.1137/S0097539794277123
10. Hliněný, P., Masařík, T.: Minimizing an uncrossed collection of drawings. In: Bekos, M., Chimani, M. (eds.) Graph Drawing & Network Vis. (GD). LNCS, vol. 14466, pp. 110–123. Springer, Heidelberg (2023). https://doi.org/10.1007/978-3-031-49272-3_8
11. Nash-Williams, C.S.J.A.: Edge-disjoint spanning trees of finite graphs. J. Lond. Math. Soc. **36**(1), 445–450 (1961). https://doi.org/10.1112/jlms/s1-36.1.445
12. Rahman, M.S., Nakano, S., Nishizeki, T.: A linear algorithm for bend-optimal orthogonal drawings of triconnected cubic plane graphs. J. Graph Algor. Appl. **3**(4), 31–62 (1999). https://doi.org/10.7155/JGAA.00017
13. Rahman, M.S., Nishizeki, T., Naznin, M.: Orthogonal drawings of plane graphs without bends. J. Graph Algor. Appl. **7**(4), 335–362 (2003). https://doi.org/10.7155/JGAA.00074
14. Schnyder, W.: Embedding planar graphs on the grid. In: Johnson, D.S. (ed.) Discrete Algorithms (SODA), pp. 138–148. SIAM (1990). http://dl.acm.org/citation.cfm?id=320176.320191
15. Tamassia, R.: On embedding a graph in the grid with the minimum number of bends. SIAM J. Comput. **16**(3), 421–444 (1987). https://doi.org/10.1137/0216030

Canadian Traveler Problems in Temporal Graphs

Thomas Bellitto[1], Johanne Cohen[2], Bruno Escoffier[1], Minh-Hang Nguyen[3], and Mikaël Rabie[3](✉)

[1] Sorbonne Université, CNRS, LIP6, 75005 Paris, France
{thomas.bellitto,bruno.escoffier}@lip6.fr
[2] LISN-CNRS, Université Paris-Saclay, Gif-sur-Yvette, France
Johanne.Cohen@lisn.fr
[3] CNRS, IRIF, Université Paris Cité, 75013 Paris, France
{minh-hang.nguyen,mikael.rabie}@irif.fr

Abstract. We study the Canadian Traveler Problem (CTP), where a traveler aims to travel from a vertex s to a vertex t in a network containing up to k potentially blocked edges. These blocked edges remain unknown until the traveler visits one of their endpoints. We investigate this problem in temporal graphs where edges are only accessible at specific times. We study classical variants of the shortest path problem in temporal graphs.

This paper formalizes the Canadian Traveler problem as a two-player positional game on graphs. In temporal graphs, the path can be viewed as a journey where the traveler takes trains from station s to station t. We consider two variants depending on how the traveler discovers a blocked edge (i.e., a canceled train). In the first variant, called *locally-informed*, the traveler learns which trains are canceled upon reaching a station. In the second variant, called *uninformed*, the traveler must wait until the departure time of the train to know whether it has been canceled or not. In the *uninformed* case, we provide a polynomial-time algorithm for each of the three shortest path variants. In the *locally-informed* case, we prove that finding a winning strategy for the traveler is PSPACE-complete. We also establish that the problem is polynomial-time solvable when $k = 1$, but becomes NP-hard for $k \geq 2$.

Additionally, we show that the standard (non-temporal) Canadian Traveler Problem is NP-hard when there are $k \geq 4$ blocked edges. To the best of our knowledge, this represents the first hardness result for CTP with a constant number of blocked edges.

Keywords: Shortest Path · Temporal graphs · Canadian traveler problem

1 Introduction

1.1 Motivation

A Canadian traveler planning to journey between cities may encounter unexpected obstacles, such as road closures due to snowfall. Consequently, choosing

H. Fernau and P. Kindermann (Eds.): WG 2025, LNCS 16124, pp. 33–47, 2026.
https://doi.org/10.1007/978-3-032-11835-6_3

the shortest path without considering potential roadblocks might not be the optimal strategy. For example, if the last road on the chosen route is blocked by snow, the traveler may be forced to return to the starting point and look for an alternative itinerary. Furthermore, the second-shortest path could be robust in the sense that for each planned road, there exists an alternative route that allows the traveler to adjust their journey in case the primary road becomes blocked. Therefore, a trip requires taking uncertainties into account.

The previous situation is modeled using static graphs. However, in many cases like train systems, temporal aspects of journeys must be taken into account (typically handled with temporal graphs). Here, potential blocks like train cancellations due to strikes may occur and should be considered in computing optimal strategies. In this context, when cancellation information is received becomes crucial for travel planning. Upon arriving at a train station, we might immediately be informed about which departing trains have been canceled. In such cases, we can plan an itinerary that lets us to return later to take a train that we know will not be canceled should our initial route be affected by a cancellation at another station. However, cancellation notifications may be received only at the last minute during certain strikes, just as the train is scheduled to depart. In such a scenario, it is essential to have an alternate route ready from the station we are at, allowing us to adjust our travel plans in real-time in response to the unexpected cancellation.

When the number of blocked roads (resp. canceled trains) is limited, the situation can be formalized as a two-player game: the first player, called *Traveler*, aims to reach a target destination within a specified time window while the second player, called *Blocker*, can block up to k segments. Each time Traveler arrives at a new intersection (resp. train station), Blocker reveals which adjacent roads are blocked (or which departing trains are canceled). Traveler must then dynamically adapt their itinerary to ensure they can reach their destination on time, regardless of the segments Blocker chooses to obstruct. Conversely, Blocker must strategically choose which roads (resp. trains) to block to prevent Traveler from reaching their destination within the time limit or even from reaching it at all.

The Canadian Traveler problem considers road maps (or static graphs) with a parameter k, meaning that up to k roads (or edges) may be blocked. In this work, we extend the Canadian Traveler problem to temporal graphs to better model train networks, where edges between vertices exist only during specific time intervals corresponding to train departure times. We explore several variants depending on how the problem is transposed to temporal graphs. Upon arriving at a train station, travelers may sometimes have advanced knowledge of which trains are unavailable, allowing them to adjust their plans accordingly. In other cases, train cancellations are revealed only at the last minute, necessitating real-time itinerary updates.

Temporal graphs introduce additional complexities beyond those found in static graphs in algorithmic problems. For example, finding a minimum set of edges that connects all the vertices of a graph is a well-known problem, as is finding a minimum spanning tree in the static case. However, temporal spanners raise many very challenging problems (see [5,6] for example). Given that the Canadian Traveler Problem is already PSPACE-complete on static graphs, investigating its temporal graph counterpart may initially seem hopeless. One of the reasons contributing to the difficulty of the Canadian Traveler Problem is the need for the traveler to remember which roads are blocked or passable while navigating the graph, as this information directly influences their next move. We refer the reader to Sect. 2.1 for a more detailed presentation of the Canadian Traveler Problem and an explanation of why remembering which edges are blocked is important. In contrast, edges appear and disappear over time in a temporal graph. Once an edge disappears, whether it was previously usable becomes irrelevant. This observation suggests that specific variants of our problem may be solvable more efficiently in temporal graphs than in static ones.

1.2 Related Work

Papadimitriou and Yannakakis introduced the Canadian Traveler Problem (CTP) in 1991 [14]. Their paper aims to find a path that minimizes the sum of the weight of its edges. The challenge arises from the fact that the traveler is given an interval of possible values for its weight for each edge. The exact weight of an edge is revealed only when the traveler reaches one of its adjacent vertices. If an edge turns out to be heavier than expected, the traveler can still choose an alternative route. The goal is to design a strategy that minimizes the worst possible ratio between the chosen and optimal routes (known as the competitive ratio). Blocking edges can be modeled by assigning them arbitrarily large weights.

From a complexity perspective, Papadimitriou and Yannakakis [14] showed that determining a strategy ensuring a specified competitive ratio is PSPACE-complete when the number of potentially blocked edges is not fixed. Later, Bar-Noy and Schieber [2] studied a problem closer to ours, where the weights of the edges are known in advance, but some edges can be blocked. They formulated the problem with a bounded number of blocked edges as a game between a traveler and an adversary who can block edges during the game. They show that the problem remains PSPACE-complete when the number of blocked edges k is a parameter, but it becomes polynomially solvable for a fixed k.

Despite the negative results, the CTP has inspired many further research studies. Several studies have investigated stochastic variants of the CTP, where edge weights are modeled as random variables with known probability distributions [13,15]. In particular, Nikolova and Karger [13] address this stochastic variant on directed acyclic and path-disjoint graphs, providing a polynomial-time algorithm to optimize the expected travel time.

The strategy for a traveler in CTP can be viewed as an *online algorithm* because the traveler lacks prior knowledge of which edges are blocked and only discovers them upon reaching their endpoints. In classical competitive analysis, many studies explicitly consider the impact of the number k of blocked edges on the analysis. In 2009, Westphal [19] proved that no deterministic online algorithm can achieve a competitive ratio smaller than $2k+1$. They also presented a simple online algorithm to reach this lower bound. Moreover, Westphal [19] showed that no randomized online algorithm can achieve a better than $(k+1)$-competitive ratio. Later, Bender and Westphal [4] showed that this bound is tight by constructing a randomized online algorithm for this case. Demaine, Huang, Liao, and Sadakane [7] introduced a randomized algorithm with polynomial time complexity that outperforms the best deterministic algorithms when there are at least two blockages. Their algorithm surpasses the lower bound of $2k+1$ by an $o(1)$ factor. Additionally, computing shortest paths in stochastic networks has been studied in [16].

Dynamic networks have been studied under various formalisms and terminologies. In particular, Temporal Graphs are graphs coupled with a discrete (often finite) time. At each time, a subset of edges is present (sometimes, with associated travel time). These graphs serve as models for real-world systems, such as transportation networks [10]. The concepts of paths and connectivity have been extensively studied; see, for example [1,11]. Questions typically revolve around the earliest arrival time, the latest departure time, or finding the shortest paths in terms of either travel time or distance (number of edges). In [9], the authors consider robust itineraries in the context of possible delays on edges. A stochastic approach considering a probability distribution over possible delays has been studied in [8].

2 Definitions and Contributions

2.1 Preliminaries

Before presenting our formal problem definitions, we first establish the foundational concepts of the classical Canadian Traveler Problem and temporal graphs.

Definition 1. *(Static Canadian Traveler Problem (CTP)) Given a graph G, a distance function on the edges $d : E \to \mathbb{N}^*$, two specific vertices s and t, an integer k, and a maximum travel time T, the game proceeds as follows between two players (Traveler and Blocker):*

- *(Blocker plays) When Traveler visits a vertex v for the first time, Blocker decides (and informs Traveler) which edges incident to v are blocked and which are not. The graph's total number of blocked edges cannot exceed k. If Traveler has already visited v before, then Blocker does not play, meaning the status of edges at v remains unchanged from the first visit.*
- *(Traveler plays) Traveler, currently at vertex v, chooses a non-blocked edge (v,u) to follow, spending time $d(v,u)$ to move from v to u.*

Traveler is initially at vertex s at time 0 and wins the game if they successfully reach vertex t within a total travel time at most T.

Note that this article sometimes uses multigraphs, for which the problem is defined similarly. We denote $d(e)$ as the distance or length of an edge e. According to this definition, we note that the status of an edge (blocked/non-blocked) remains unchanged throughout the game. Blocker makes an irrevocable decision about whether an edge is blocked or not the first time Traveler reaches one of its endpoints, and Blocker cannot change this status later. In particular, for all $(u, v) \in E$, if Traveler first visits u and later visits v, Blocker blocks (u, v) at the visit of v if and only if it was already blocked at the visit of u.

Let us now define temporal graphs.

Definition 2. *(Temporal graph) A* temporal graph $G = (V, E)$ *consists of a vertex set V and a set E of time edges. A* time edge (u, v, τ, d) *represents an edge (u, v) that exists at time $\tau \in \mathbb{N}$ and requires $d \in \mathbb{N} \setminus \{0\}$ units of time to be crossed. Note, as G is undirected, that $(u, v, \tau, d) = (v, u, \tau, d)$ is a single edge that can be described from both directions. By convention, if $d = 1$, we will omit it in the description, particularly in Sect. 4.*

For a time edge (u, v, τ, d), the parameter d represents the travel time required to traverse the edge and is also referred to as the length of the time edge. In the following, we will often use the term "edge" instead of "time edge" for simplicity when discussing temporal graphs. For a temporal graph $G = (V, E)$, a *temporal walk* is a sequence $(v_0, e_0, v_1, e_1, \ldots, e_{p-1}, v_p)$ with $v_i \in V$ and $e_i = (v_i, v_{i+1}, \tau_i, d_i) \in E$, such that $\tau_i + d_i \leq \tau_{i+1}$. In such a walk, Traveler arrives at v_{i+1} at time $\tau_i + d_i$ after traversing edge e_i, and leaves v_{i+1} at time τ_{i+1}. We assume that an edge can have multiple copies. In that case, we will explicitly specify how many copies we have in the graph.

2.2 Temporal CTP

In a temporal CTP, a traveler travels (i.e., follows temporal walks) through a temporal graph G. Some time edges may be blocked. For a given pair $u, v \in V^2$, each corresponding time edge must be blocked independently. This distinction is significant because if Blocker could block all time edges between two vertices simultaneously, the resulting problem would trivially generalize the static problem (which is the case of a temporal graph where the same edges are present at every time step). The complexity results follow immediately.

Traveler becomes aware of a blocked edge only upon reaching one of its endpoints. We define two variants of the problem:

- *Locally-informed model:* Traveler is made aware of all the blocked edges incident to u when they arrive at node u (including edges (u, v, τ, d) for τ greater than the arrival time).
- *Uninformed model*: Traveler is made aware of whether an edge (u, v, τ, d) is blocked only when located at node u (or at node v) at time τ.

Let us now formally define these variants. As mentioned in the introduction, CTP can be viewed as a two-player game.

Definition 3. *(Locally-informed game) Given a temporal graph G, two specific vertices s and t, an integer k, and two players named* Traveler *and* Blocker, *the game proceeds as follows:*

- *(Blocker plays) When Traveler visits for the first time vertex v, Blocker decides and tells Traveler about which edges incident to v are blocked and which are not. The total number of blocked edges in the graph cannot exceed k. If Traveler has previously visited v, then Blocker does not play.*
- *(Traveler plays) Traveler, currently at v at time τ, chooses a non-blocked edge (v, u, τ', d) to traverse (with $\tau' \geq \tau$), i.e., they move from v to u at time τ', arriving at u at time $\tau' + d$.*

Traveler is initially at vertex s at 0 *and aims to reach vertex t to win.*

We note that, here also, the set of blocked/non-blocked edges remains unchanged: once Blocker decides if edge (u, v, τ, d) is blocked (when Traveler reaches one of its endpoints for the first time), they cannot change this decision later.

Definition 4. *(Uninformed game) An* uninformed game *is defined similarly to a locally-informed game, with the following modification: Traveler must be at vertex v or u at time τ for Blocker to decide (and tell them) whether the edge (v, u, τ, d) is blocked or not.*

Note that this version gives more power to Blocker (as they reveal less information to Traveler). We introduce the different decision problems in those games:

Definition 5. *(Li-TCTP, U-TCTP) The* Locally-informed Temporal Canadian Traveler *Problem (resp.* Uninformed Temporal Canadian Traveler Problem*), Li-TCTP for short (resp. U-TCTP for short) asks whether Traveler has a winning strategy in the corresponding locally-informed (resp. uninformed) game.*

Problems Li-TCTP and U-TCTP are reachability problems. We can also consider their optimization versions. We define a (T_1, T_2)-*winning strategy* for Traveler as a strategy that guarantees reaching the destination t at time T_2 or earlier, while not departing from s before T_1 .

- *(Earliest Arrival)* We aim to find the smallest T_2 and a corresponding strategy such that Traveler has a $(0, T_2)$-winning strategy.
- *(Latest Departure)* We aim to find the largest T_1 (and a corresponding strategy) such that Traveler has a (T_1, ∞)-winning strategy.
- *(Shortest Path)* We want to find (T_1, T_2) with the smallest difference $T_2 - T_1$ and a corresponding strategy such that Traveler has a (T_1, T_2)-winning strategy.

Determining in polynomial time whether Traveler has a (T_1, T_2)-winning strategy or not allows us to solve all three optimization problems in polynomial time. Indeed, note that only $O(|E|)$ relevant values for T_1 and T_2 shall be considered: for T_1, the times τ for which there is an edge (s, u, τ, d), and for T_2 the times τ for which there is an edge $(u, v, \tau - d, d)$.

In the static version, reachability with k blocked edges is possible if and only if s and t are $k+1$-connected. If Blocker wins, the blocked edges necessarily form a cut between s and t, and vice versa. However, this approach fails in the temporal case. As time progresses, connectivity naturally decreases, even without blocking edges. It is not enough to ensure reachability from s to t with k blocked edges when the underlying graph of the temporal graph is $k+1$-connected.

To illustrate the difference between the locally-informed and uninformed models, we examine the example in Fig. 1, where all edges have length 1. First, observe that if there are $(k+1)$ parallel edges (u, v, τ, d) (where k is the maximal number of blocked edges), then at least one of them remains accessible, meaning that Traveler can always rely on taking one of these edges. We refer to an edge (u, v, τ, d) as a *forced edge* to mean that there are $(k+1)$ copies of it. Hence, Blocker cannot block a forced edge.

In U-TCTP (uninformed case), we can clearly observe that Traveler has no winning strategy. Traveler goes from s to v_0. At time 1, Traveler must either wait at v_0 or go to v_1. In the first case, Blocker blocks edge $(v_0, v_2, 2, 1)$, preventing further progress. In the other case, Blocker blocks all k parallel edges $(v_1, t, 2, 1)$, making it impossible to reach t. In both cases, Traveler is unable to win.

In Li-TCTP (locally informed case), Traveler has a winning strategy. Traveler goes to v_0. There, they know whether $(v_0, v_2, 2, 1)$ is blocked or not. If this edge is not blocked, Traveler takes it and reaches t (as $(v_2, t, 3, 1)$ is a forced edge). If $(v_0, v_2, 2, 1)$ is blocked, Traveler goes to v_1, and then to t: Blocker cannot block the k parallel edges $(v_1, t, 2, 1)$ as their remaining budget (corresponding to the remaining number of edges that can still be blocked) is $k-1$ (they have already blocked one edge).

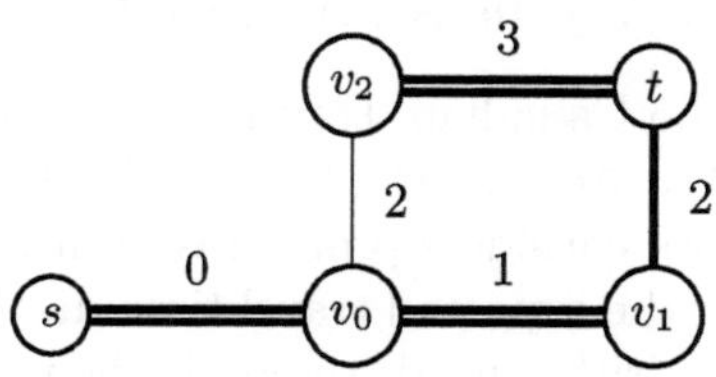

Fig. 1. Example of a temporal graph. Thick (resp. doubly thick) edges represent k parallel (resp. $(k+1)$-parallel) edges. The number of the edge corresponds to the time step in which it is present. All edges have length one.

This example highlights how the timing of when Traveler learns about blocked edges significantly impacts their ability to strategize effectively in temporal graphs.

2.3 Contributions

We focus on the computational complexity of solving both U-TCTP and Li-TCTP, highlighting a key difference between these problems. The intuition is that in the uninformed case, we can compute as we go since we never need to store knowledge (edges blocked at a previous time can be forgotten since they do not exist anymore). In contrast, in the locally-informed case, we need to store all the information we have. The following results confirm this intuition.

In Sect. 3, we show that U-TCTP can be solved in time polynomial in $n = |V|$, $m = |E|$ and k. This result, which also applies to the three optimization variants, is obtained by reducing our problem to a CTP in static *directed acyclic* graphs, a problem that we show how to solve efficiently. We leverage classical Dynamic Programming (DP) techniques to design a polynomial-time algorithm for U-TCTP, despite its static version being computationally hard.

We then tackle Li-TCTP in Sect. 4. We show that this problem, even in its reachability version, is PSPACE-hard. We further show that the case $k = 1$ can be solved in polynomial time.

In Sect. 5, we prove that Li-TCTP is NP-hard for $k = 2$. We also consider the static case (see Definition 1) and show that static CTP becomes NP-hard when there are $k = 4$ blocked edges. To the best of our knowledge, this result is the first hardness result for (static) CTP for a constant k.

We note that our hardness proofs (for the static and the locally-informed cases) rely on multigraphs, with edge multiplicity. This choice is made for clarity and ease of explanation.

The missing proofs, in addition to extra details such as how to transform multigraphs into simple graphs without affecting the complexity results can be found in [3].

3 Uninformed Temporal CTP

3.1 CTP for Directed Acyclic Static Graphs

This section focuses on the Canadian Traveler Decision Problem in a directed acyclic static graph (DAG). Acyclic graphs for CTP have already been studied in [13]. However, their work considers probabilistic distributions for arc weights and focuses on optimizing the expected travel time. In their model, each arc can be affected whereas our algorithm needs to ponder how many remaining arcs can be blocked. Our approach addresses a deterministic setting where the primary concern is the blocking of arcs. Specifically, we show that this problem in DAGs can be solved in polynomial time even when up to k arcs can be blocked (where k is part of the input).

Consider a directed acyclic graph (DAG) $G = (V, E)$ with source node s and target node t. A traveler wants to go from s to t when Blocker can block at most k arcs. Traveler can only discover if an arc (u, v) is blocked or not when they reach its head u.

Theorem 1. *The Canadian Traveler Decision Problem in a weighted DAG G can be solved in time $O(k^2(m + n) \log m)$.*

Briefly, each node v maintains a set of $k + 1$ values, denoted as $\pi_0, \ldots, \pi_k$. Informally, π_i represents the minimum amount of time τ such that Traveler can guarantee to go from v to t with time at most τ under the condition that Blocker can block i arcs between nodes v and t.

Since G is a DAG, we first compute a topological order on the vertices, of which we can assume that s is the first one and t is the last one. We then process the vertices in reverse topological order.

We first consider the case of a node v that has only t as an out-neighbor. Here, $\pi_i(v)$ is the maximum time required for Traveler to reach t from v when Blocker can block up to i arcs connecting v and t. Then, we have $\pi_0(v) = d(v,t)$ and $\pi_i(v) = \infty$ for $i \geq 1$.

Otherwise, let $N^+(v)$ be a set of all out-neighbours of v. The values $\pi_0, \ldots, \pi_k$ for each node are computed inductively as follow:

$$\pi_0(v) = \min_{u \in N^+(v)} (\pi_0(u) + d(v,u))$$

$$\pi_i(v) = \max \begin{cases} \min\limits_{u \in N^+(v)} (\pi_i(u) + d(v,u)) \\ \text{second } \min\limits_{u \in N^+(v)} (\pi_{i-1}(u) + d(v,u)) \\ \ldots \\ \text{i-th } \min\limits_{u \in N^+(v)} (\pi_1(u) + d(v,u)) \\ \text{(i+1)-th } \min\limits_{u \in N^+(v)} (\pi_0(u) + d(v,u)) \end{cases}$$

For each fixed index i, the values π_i for all vertices can be computed in time $O(k(m+n))$. Therefore, values $\pi_i(v)$, for $i = 0, \ldots, k$ and $v \in V$, can be computed in time $O(k^2(m+n)\log m)$. These previous arguments can be easily generalized to directed acyclic multigraphs. For example, in the case where v has only t as an out-neighbor, let h be the number of arcs from v to t. If $i \geq h$, then $\pi_i(v) = \infty$, otherwise $\pi_i(v)$ is the $(i+1)$-th minimum weight of an arc from v to t since Blocker can block i arcs with smallest weights.

We now consider the CTP for directed acyclic graphs as a two-player game.

Lemma 1. *For the Canadian Traveler Decision Problem in a weighted DAG G, Traveler has a winning strategy to reach t with an arrival time of at most T if and only if $\pi_k(s) \leq T$.*

3.2 CTP for Temporal Graph in the Uninformed Case

Theorem 2. *U-TCTP can be solved in polynomial time.*

Proof. Proving this theorem requires transforming the temporal CTP into a CTP on a static weighted DAG. Given a temporal graph $G = (V, E)$, we construct a weighted directed acyclic graph $G' = (V', E')$ corresponding to G. This construction is known as the static expansion of a temporal graph [12].

For each edge $(v, u, \tau, d) \in E$, we create two vertices (v, τ) and $(u, \tau + d)$ in V'. Additionally, we create one universal target vertex $\texttt{target} \in V'$.

- If $(v, u, \tau, d) \in E$, then $((v, \tau)(u, \tau + d)) \in E'$, and $d((v, \tau), (u, \tau + d)) = d$. We call $((v, \tau)(u, \tau + d))$ the corresponding edge of (v, u, τ, d).
- For all $v \in V$, $(v, \tau) \in V'$, and the minimum $\tau' > \tau$ such that $(v, \tau') \in V'$, we connect (v, τ) to (v, τ') by $k+1$ parallel edges, with $d((v, \tau), (v, \tau')) = \tau' - \tau$.
- For every $(t, \tau) \in V'$, connect node (t, τ) to `target` by $k + 1$ parallel edges, let $d((t, \tau), \texttt{target}) = 0$.

Since G is a temporal graph, G' is a directed acyclic graph. In G', the source node is $(s, 0)$ and the target node is `target`.

Then, Traveler has a winning strategy in U-TCTP on G if and only if there exists a winning strategy for Traveler in Canadian Traveler Decision Problem with reachability condition in the directed acyclic graph G'. □

Note that to solve U-TCTP for the three optimisation versions - Latest Departure, Early Arrival, and Shortest Time Traveling - we proceed as follows: For each $T_1 \leq T_2 \leq T$, where T is the maximum value of $\tau + d$ over all time edges (u, v, τ, d) (also known as the lifespan of the graph), we consider a subset of time edges in G, which are present between time T_1 and T_2. Then, we construct a corresponding directed acyclic graph $G'(T_1, T_2)$. As already noted previously, we can restrict to $O(|E|)$ relevant values for T_1 and T_2 (corresponding to times τ or $\tau + d$ of an edge u, v, τ, d).

4 Locally-Informed Temporal CTP

Theorem 3. *Li-TCTP is PSPACE-complete, even in instances where each edge has length* 1 *and is present at only one time step. This complexity applies to both directed and undirected temporal graphs.*

Proof. This problem constitutes a two-player game for which membership in PSPACE can be easily shown using standard techniques (Lemma 2.2 of [17]). Now, we prove completeness. As in [14], the reduction is from the quantified 3-satisfiability problem which is PSPACE-complete [18]:

Definition 6. *(Quantified 3-Sat):* **Given** *a set* $X = \{x_1, \ldots, x_n\}$ *of* n *variables* (n *even) and a quantified Boolean formula* $F = \exists x_1 \forall x_2 \exists x_3 \ldots \forall x_n : \phi(x_1, \ldots, x_n)$ *where* ϕ *is a propositional formula over* X *in 3-CNF (i.e., in conjunctive normal form with exactly three literals per clause).*
Question *: Is* F *true?*

Let $C_1, \ldots, C_m$ denote the clauses that make up ϕ: $\phi(x_1, \ldots, x_n) = C_1 \wedge C_2 \cdots \wedge C_m$, where for every $1 \leq i \leq m$, the clause C_i contains exactly three literals.

Given an instance $\mathcal{Q}$ of Quantified 3-Sat, we construct an instance $\mathcal{S}$ of a locally-informed game by creating a temporal graph $G = (V, E)$ as follows. First, Blocker can block $k = 2m + \frac{n}{2}$ edges.

In the temporal graph G, all edges have length 1. For clarity, we omit mentioning the length when referring to an edge. The vertex set of G contains:

- a start vertex s, a target vertex t, and a vertex w
- vertices v_i, x_i, $\overline{x_i}$, a_i, and z_i, for each integer i, $1 \leq i \leq n$, (one per variable)
- vertex b_i, for each even integer i, $1 \leq i \leq n+1$,
- vertex c_j, for each integer j, $1 \leq j \leq m$, (one per clause)
- two distinct vertices v_{n+1}, a_{n+1}.

Now, we describe how the vertices are connected (see Figs. 2 and 3 for illustrations). We will say that an edge (u, v, τ, d) is a *forced edge* to mean that there are $(k+1)$ copies of it. Hence, Blocker cannot block a forced edge. The set of edges can be divided into three parts. The first part corresponds to edges associated with variables of odd index, the second part corresponds to edges associated with variables of even index, and the third part corresponds to edges associated with clauses. We define $L = 7\frac{n}{2}$, a constant frequently used, corresponding to the time at which Traveler can reach v_{n+1}.

- for each variable x_i where i is odd, $1 \leq i \leq n$, we introduce an "existential" gadget. We put a diamond on 4 vertices $v_i, x_i, \overline{x_i}, v_{i+1}$, with 4 forced edges $(v_i, x_i, 7\lfloor\frac{i}{2}\rfloor+1)$, $(v_i, \overline{x_i}, 7\lfloor\frac{i}{2}\rfloor+1)$, $(x_i, v_{i+1}, 7\lfloor\frac{i}{2}\rfloor+2)$ and $(\overline{x_i}, v_{i+1}, 7\lfloor\frac{i}{2}\rfloor+2)$. Moreover, we add one forced edge $(v_i, a_i, 7\lfloor\frac{i}{2}\rfloor+1)$, and $k - \lfloor\frac{i}{2}\rfloor$ parallel edges $(a_i, t, L+5)$.
- for each variable x_i where i is even, $1 \leq i \leq n$, we introduce a "universal" gadget. We put a diamond on 4 vertices $v_i, x_i, \overline{x_i}, v_{i+1}$ with one edge $(v_i, x_i, 7\lfloor\frac{i-1}{2}\rfloor+3)$, one edge $(v_i, \overline{x_i}, 7\lfloor\frac{i-1}{2}\rfloor+6)$, 2 forced edges $(x_i, v_{i+1}, 7\lfloor\frac{i}{2}\rfloor)$ and $(\overline{x_i}, v_{i+1}, 7\lfloor\frac{i}{2}\rfloor)$. Moreover, there are 2 forced edges $(x_i, b_i, 7\lfloor\frac{i-1}{2}\rfloor+4)$ and $(b_i, v_i, 7\lfloor\frac{i-1}{2}\rfloor+5)$. Vertex b_i is linked to t with 2 copies of $(b_i, t, 7\lfloor\frac{i-1}{2}\rfloor+4)$. Finally, we add one forced edge $(v_i, a_i, 7\lfloor\frac{i-1}{2}\rfloor+3)$, and $k - \lfloor\frac{i}{2}\rfloor$ parallel edges $(a_i, t, L+5)$.
- for each i, $1 \leq i \leq n$, we have one edge $(\overline{x_i}, z_i, L+4)$, one edge $(x_i, z_i, L+4)$, and $2m$ parallel edges $(z_i, t, L+5)$.

We introduce the edges associated with the clauses (see Fig. 3). We have one forced edge $(v_{n+1}, w, L+1)$, one edge $(w, t, L+2)$, and for each clause $C_i = (\ell_1 \vee \ell_2 \vee \ell_3)$, $1 \leq i \leq m$:

- two parallel edges $(w, C_i, L+2)$;
- a forced edge $(C_i, \ell, L+3)$, for each $\ell \in \{\ell_1, \ell_2, \ell_3\}$.

Note that G is constructed in polynomial time.

The remaining part of the proof is given in the full version of this article [3]. We now give an intuition on how the reduction works. Intuitively, if the formula is true, Traveler follows the diamonds (vertex-gadgets) to reach v_{n+1}. When Traveler is at v_i for some even integer i, Blocker must block exactly one edge between $(v_i, x_i, 7\lfloor\frac{i-1}{2}\rfloor+3))$ or $(v_i, \overline{x_i}, 7\lfloor\frac{i-1}{2}\rfloor+6)$ to avoid losing. They choose the one they want corresponding to the universal variable $\forall x_i$. When i is odd, Traveler chooses whether to go to x_i or $\overline{x_i}$ according to the assignment they want to give to the existential variable $\exists x_i$.

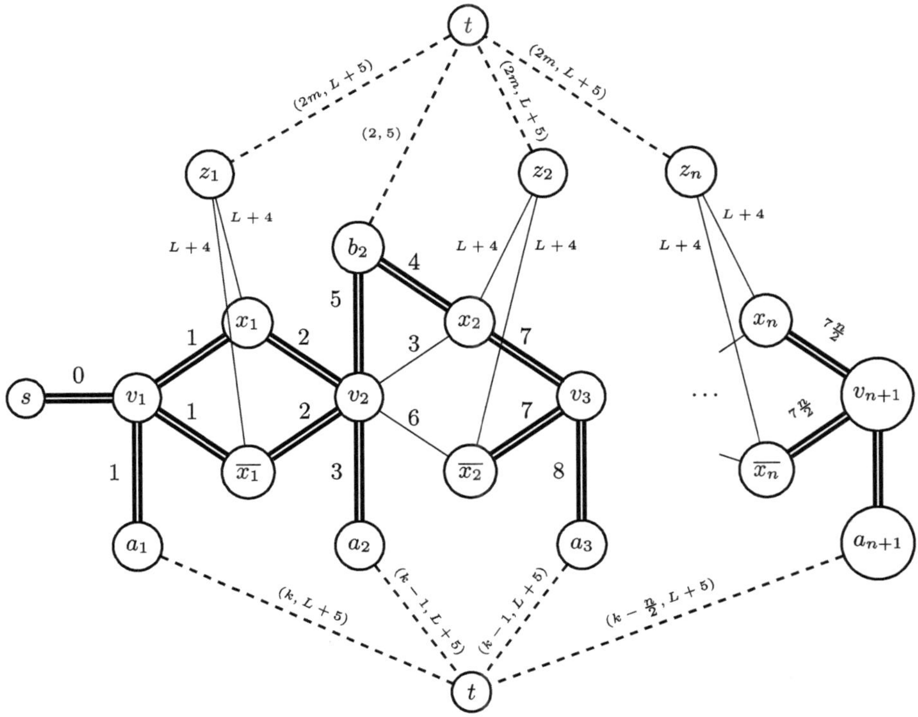

Fig. 2. A piece of graph G corresponding to the n variables. Double thick edges denote $(k+1)$-parallel edges (forced edges). The notation (α, τ) for the dashed edges indicates that there are α parallel edges appearing at time τ. For readability, there are several copies of t. As in the proof, $L = 7\frac{n}{2}$.

When Traveler reaches node x_i or $\overline{x_i}$, Blocker must not block the edge leading to z_i. Otherwise, they lose. Therefore, at v_{n+1}, Traveler has seen exactly $\frac{n}{2}$ blocked edges and has visited either x_i or $\overline{x_i}$ for all i (if Blocker had blocked more edges, Traveler would go to a_{n+1} and win).

Blocker then must block edge $(w, t, L+2)$, and cannot prevent Traveler from reaching a clause vertex C_j for some j. From C_j, Traveler can visit a vertex x_i or $\overline{x_i}$ that they already visited before (because the formula is satisfiable and they chose the assignments they could depending on the choices of Blocker accordingly). Consequently, from visited vertex x_i or $\overline{x_i}$, Traveler can go to vertex z_i and take one of the parallel edges $(z_i, t, L+5)$ to reach t and win.

If the formula is not satisfiable, thanks to the path to v_{n+1}, Blocker can force the assignment of universal variables in such a way that, regardless of Traveler's choices for existential variables, the formula remains unsatisfied. Blocker blocks edge (w, t). They can then force Traveler to visit a clause vertex C_j that is unsatisfied under the truth assignment, by using $2(m-1)$ edges to block access to the other clauses vertices. Traveler then must visit a vertex x_i or $\overline{x_i}$ that they

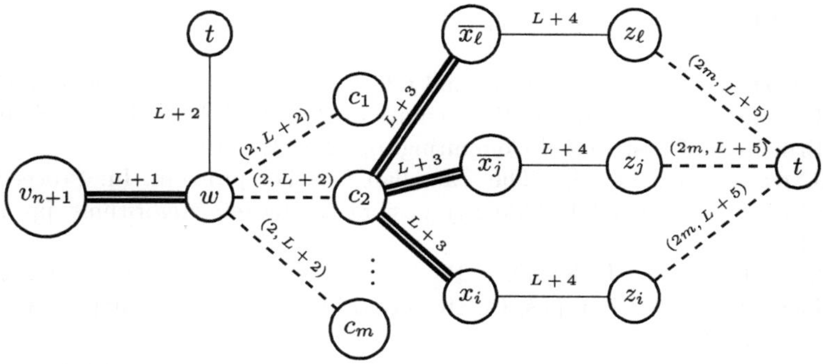

Fig. 3. A piece of graph G corresponding to the clauses. Double thick edges denote $(k+1)$-parallel edges (forced edges). The notation (α, τ) for the dashed edges indicates that there are α parallel edges appearing at time τ. For readability, there are several copies of the target vertex t. As in the proof, $L = 7\frac{n}{2}$. We assume that clause $c_2 = (x_i \vee \overline{x_j} \vee \overline{x_\ell})$.

have not visited before. This allows Blocker to block the edge from that node leading to z_i, preventing Traveler from reaching t. □

Since the problem is PSPACE-hard, we focus on the case where k is fixed. We show now that the problem is tractable when $k = 1$. Interestingly, in Sect. 5, we show that the problem becomes hard when $k = 2$.

Theorem 4. *The Locally-informed Temporal Canadian Traveler Problem is solvable in polynomial time when* $k = 1$.

5 NP-Hardness for a Constant Number of Blocked Edges for Static CTP and Li-TCTP

In this section, we present two theorems. The first states that the Canadian Traveler Problem is NP-complete in the static setting even for $k = 4$, using a reduction from 3-SAT. This reduction can then be adapted to show that Li-TCTP remains hard for $k = 2$.

Theorem 5. *k-CTP is NP-hard even for* $k = 4$.

Theorem 6. *Li-TCTP is NP-hard even for* $k = 2$.

An open question remains as to whether 2-CTP is also NP-hard, or if it can be solved in polynomial time.

References

1. Awerbuch, B., Even, S.: Efficient and reliable broadcast is achievable in an eventually connected network. In: Proceedings of the Third Annual ACM Symposium on Principles of Distributed Computing, pp. 278–281 (1984)
2. Bar-Noy, A., Schieber, B.: The Canadian traveller problem. In: Proceedings of the Second Annual ACM-SIAM Symposium on Discrete Algorithms, pp. 261–270 (1991)
3. Bellitto, T., Cohen, J., Escoffier, B., Nguyen, M., Rabie, M.: Canadian traveller problems in temporal graphs. CoRR arxiv:2407.16491 (2024). https://doi.org/10.48550/ARXIV.2407.16491
4. Bender, M., Westphal, S.: An optimal randomized online algorithm for the k-Canadian traveller problem on node-disjoint paths. J. Comb. Optim. **30**(1), 87–96 (2015)
5. Bilò, D., D'Angelo, G., Gualà, L., Leucci, S., Rossi, M.: Sparse temporal spanners with low stretch. In: Chechik, S., Navarro, G., Rotenberg, E., Herman, G. (eds.) 30th Annual European Symposium on Algorithms, ESA 2022, Berlin/Potsdam, Germany, 5–9 September 2022. LIPIcs, vol. 244, pp. 19:1–19:16. Schloss Dagstuhl - Leibniz-Zentrum für Informatik (2022). https://doi.org/10.4230/LIPICS.ESA.2022.19
6. Casteigts, A., Peters, J.G., Schoeters, J.: Temporal cliques admit sparse spanners. J. Comput. Syst. Sci. **121**, 1–17 (2021). https://doi.org/10.1016/J.JCSS.2021.04.004
7. Demaine, E.D., Huang, Y., Liao, C.S., Sadakane, K.: Approximating the Canadian traveller problem with online randomization. Algorithmica **83**, 1524–1543 (2021)
8. Dibbelt, J., Strasser, B., Wagner, D.: Delay-robust journeys in timetable networks with minimum expected arrival time. In: Funke, S., Mihalák, M. (eds.) 14th Workshop on Algorithmic Approaches for Transportation Modelling, Optimization, and Systems, ATMOS 2014, Wroclaw, Poland, 11 September 2014. OASIcs, vol. 42, pp. 1–14. Schloss Dagstuhl - Leibniz-Zentrum für Informatik (2014https://doi.org/10.4230/OASICS.ATMOS.2014.1
9. Füchsle, E., Molter, H., Niedermeier, R., Renken, M.: Delay-robust routes in temporal graphs. In: Berenbrink, P., Monmege, B. (eds.) 39th International Symposium on Theoretical Aspects of Computer Science, STACS 2022, Marseille, France (Virtual Conference), 15–18 March 2022. LIPIcs, vol. 219, pp. 30:1–30:15. Schloss Dagstuhl - Leibniz-Zentrum für Informatik (2022).https://doi.org/10.4230/LIPICS.STACS.2022.30
10. Holme, P., Saramäki, J.: Temporal networks. Phys. Rep. **519**(3), 97–125 (2012)
11. Kempe, D., Kleinberg, J.M., Kumar, A.: Connectivity and inference problems for temporal networks. J. Comput. Syst. Sci. **64**(4), 820–842 (2002). https://doi.org/10.1006/JCSS.2002.1829
12. Mertzios, G.B., Michail, O., Chatzigiannakis, I., Spirakis, P.G.: Temporal network optimization subject to connectivity constraints. In: Fomin, F.V., Freivalds, R., Kwiatkowska, M., Peleg, D. (eds.) ICALP 2013. LNCS, vol. 7966, pp. 657–668. Springer, Heidelberg (2013). https://doi.org/10.1007/978-3-642-39212-2_57
13. Nikolova, E., Karger, D.R.: Route planning under uncertainty: the Canadian traveller problem. In: AAAI, pp. 969–974 (2008)
14. Papadimitriou, C.H., Yannakakis, M.: Shortest paths without a map. Theor. Comput. Sci. **84**(1), 127–150 (1991)

15. Polychronopoulos, G.H., Tsitsiklis, J.N.: Stochastic shortest path problems with recourse. Netw. Int. J. **27**(2), 133–143 (1996)
16. Psaraftis, H.N., Tsitsiklis, J.N.: Dynamic shortest paths in acyclic networks with Markovian arc costs. Oper. Res. **41**(1), 91–101 (1993). https://doi.org/10.1287/OPRE.41.1.91
17. Schaefer, T.J.: On the complexity of some two-person perfect-information games. J. Comput. Syst. Sci. **16**(2), 185–225 (1978)
18. Stockmeyer, L.J., Meyer, A.R.: Word problems requiring exponential time (preliminary report). In: Proceedings of the Fifth Annual ACM Symposium on Theory of Computing, pp. 1–9 (1973)
19. Westphal, S.: A note on the k-Canadian traveller problem. Inf. Process. Lett. **106**(3), 87–89 (2008)

When Does FTP Become FPT?

Matthias Bentert[1], Fedor V. Fomin[1], Petr A. Golovach[1(✉)], and Laure Morelle[2]

[1] University of Bergen, Bergen, Norway
{fedor.fomin,petr.golovach}@uib.no
[2] LIRMM, Université de Montpellier, CNRS, Montpellier, France
laure.morelle@lirmm.fr

Abstract. In the problem FAULT-TOLERANT PATH (FTP), we are given an edge-weighted directed graph $G = (V, E)$, a subset $U \subseteq E$ of vulnerable edges, two vertices $s, t \in V$, and integers k and ℓ. The task is to decide whether there exists a subgraph H of G with total cost at most ℓ such that, after the removal of any k vulnerable edges, H still contains an s-t-path. We study whether FAULT-TOLERANT PATH is fixed-parameter tractable (FPT) and whether it admits a polynomial kernel under various parameterizations. Our choices of parameters include the number of vulnerable edges in the input graph, the number of safe (i.e., invulnerable) edges in the input graph, the budget ℓ, the minimum number of safe edges in any optimal solution, the minimum number of vulnerable edges in any optimal solution, the required redundancy k, and natural above- and below-guarantee parameterizations. We provide an almost complete description of the complexity landscape of FTP for these parameters.

Keywords: Graph connectivity · Non-uniform failure model · Parameterized complexity

1 Introduction

Fault-tolerant edge connectivity refers to the ability of a network to maintain connectivity even after some of its edges fail. Recently, there has been significant interest in a graph model where the edge set is partitioned into two distinct categories: safe edges S and vulnerable edges U [1,2,6,7]. Safe edges represent robust connections that are less likely to fail, while vulnerable edges are more prone to disruptions or failures. Our paper contributes to the expanding body of research on this intriguing model.

Adjiashvili et al. [2] introduced the following fault-tolerant generalization of Minimum-Cost Edge-Disjoint Paths (EDP), the classic network connectivity problem whose aim is to find edge-disjoint paths of minimum total cost between two given terminals. The input consists of an edge-weighted (directed) graph $G = (V, E)$, a subset $U \subseteq E$ of vulnerable edges, two vertices $s, t \in V$, and integers k and ℓ. The task is to decide whether there is a subgraph H of G of cost at most ℓ such that after removing any k vulnerable edges from H, there still remains an s-t-path. More formally, we study the following problem.

H. Fernau and P. Kindermann (Eds.): WG 2025, LNCS 16124, pp. 48–61, 2026.
https://doi.org/10.1007/978-3-032-11835-6_4

FAULT-TOLERANT PATH (FTP)

Input: A (directed) graph $G = (V, E = S \cup U)$ where S and U are disjoint sets of *safe* and *vulnerable* edges, respectively, an edge-weight function $w: E \to \mathbb{N}_{\geq 1}$, two vertices $s, t \in V$, and integers k and ℓ.

Question: Is there a set $K \subseteq E$ of edges of total cost (sum of weights) at most ℓ such that for each set $F \subseteq K \cap U$ with $|F| \leq k$, it holds that $G[K \setminus F]$ contains an s-t-path?

Note that when G only contains vulnerable edges, i.e., when $E = U$, then FTP is equivalent to deciding whether G contains $k+1$ edge-disjoint s-t-paths of total cost of at most ℓ. In other words, this is equivalent to determining whether a flow of $k+1$ units can be sent from s to t with a cost of at most ℓ—assuming all edges in the graph have unit capacity. While the minimum cost capacitated flow is solvable in polynomial time, Adjiashvili et al. showed that FTP is NP-complete [2]. They also provided a polynomial-time algorithm for FTP when $k = 1$. However, the complexity of the problem for any fixed $k \geq 2$ remains unknown. Additionally, they presented an $n^{\mathcal{O}(k)}$-time algorithm for directed acyclic graphs. Their work is the starting point for our research.

1.1 Our Results

We investigate the parameterized complexity of FAULT-TOLERANT PATH with respect to the following list of natural parameters. We start with the parameters k and ℓ. Next, we consider the numbers $|S|$ and $|U|$ of safe and vulnerable edges in the input graph as well as the minimum numbers p and q of safe and vulnerable edges, respectively, in a solution. Finally, we also investigate a natural "above-guarantee" and a "below-guarantee" parameter. First of all, any solution $K \subseteq E$ of edges contains an s-t-path, and thus we have a trivial no-instance if $\ell < \text{dist}(s,t)$, the distance from s to t in the weighted graph. Hence, we can assume that $\ell \geq \text{dist}(s,t)$, and this gives rise to the above-guarantee (that is, above $\text{dist}(s,t)$) parameter $a = \ell - \text{dist}(s,t)$. To define the last (below-guarantee) parameter, we need the following lemma, which directly follows from [2, Lemma 3]. We will use this lemma (sometimes implicitly) in several places.

Lemma 1. *An instance $I = (G = (V, S \cup U), s, t, k, \ell)$ of* FTP *is a yes-instance if and only if there are $k+1$ s-t-paths $P_1, P_2, \ldots, P_{k+1}$ with $P_i = (V_i, E_i)$ in G such that $E_i \cap E_j \subseteq S$ for all $i \neq j \in [k+1]$ and $\sum_{e \in \bigcup_{i \in [k+1]} E_i} w(e) \leq \ell$.*

Given an instance of FAULT-TOLERANT PATH, we define an instance of MINIMUM-COST FLOW as follows. We set the capacity of each safe edge $e \in S$ to be $k+1$ and the capacity of each vulnerable edge $e \in U$ to be one. Then, we ask for a minimum-cost $(k+1)$-flow from s to t. We denote the optimum cost of such a flow by C. We can assume, without loss of generality, that an optimal solution uses integer capacities on all edges and that such a flow solution corresponds

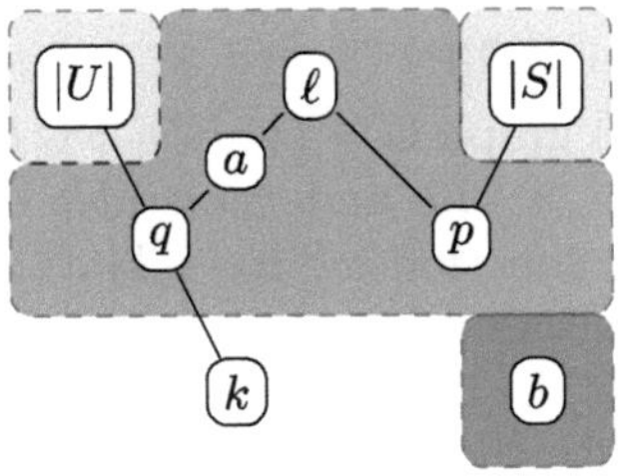

Fig. 1. Overview of our results. An edge between two parameters α and β, where β is above α, indicates that the value of β in any instance will always be at most a constant times larger than the value of α in that instance (or the problem can be solved trivially). Any hardness result for β immediately implies the same hardness result for α, and any positive result for α immediately implies the same positive result for β. The red box indicates that the parameter leads to para-NP-hardness (NP-hardness for constant parameter values). An orange box indicates that the problem is in XP and W[1]-hard and a yellow box indicates that the parameterized problem is fixed-parameter tractable, but does not admit a polynomial kernel. (Color figure online)

to $k+1$ paths in G that do not intersect in any vulnerable edges. By Lemma 1, there exists a solution of total cost at most C, that is, if $\ell \geq C$, then the instance I is a yes-instance. Note, however, that this is not an if and only if as the flow "pays" for each unit of flow that passes through a safe edge, whereas in Fault-Tolerant Path, we only "pay" once for each edge. This leads to the below-guarantee parameter $b = C - \ell$.

Relations Between Parameters. Before describing our result, let us discuss the relationships between the parameters (see Fig. 1), most of which are straightforward. First, the number p of safe edges in the solution is clearly upper-bounded by both the number of safe edges in the input graph and the total cost of edges in the solution (since each edge has a weight of at least one). Therefore, if $p > \ell$, the instance is a trivial no-instance. An analogous statement holds for vulnerable edges. The remaining parameter bounds are summarized in the following proposition. Here and further, the proofs of the claims labeled ($\star$) are omitted in this extended abstract and are available in the full version [5].

Proposition 1 ($\star$). *If the values of the parameters for an instance of* FTP *do not satisfy the inequality $k < q \leq 2a \leq 2\ell$, then the considered instance of* FTP *can be solved in polynomial time.*

Our results are depicted in Fig. 1.

As can be seen, we provide an almost complete description of the parameterized complexity landscape of FTP for our selection of natural parameters. The only remaining question is whether FTP is para-NP-hard or contained in XP when parameterized by k. Notice that FTP can be considered for both directed and undirected graphs, and there is no straightforward reduction for FTP on undirected graphs to the directed case (say, by replacing each undirected edge by a pair of opposite directed ones as for many other problems). We state all our algorithms for the directed case. However, they can be easily adapted for undirected graphs. Symmetrically, we show our algorithmic lower bounds for unweighted undirected graphs and the results also hold for directed graphs.

1.2 Related Work

The model of the classic s-t-PATH and s-t-FLOW problems with safe and vulnerable edges is due to Adjiashvili et al. [2], who studied the approximability of these problems. Subsequently, generalizations were also studied [4,7,9]. Closest to our work is the paper by Bentert et al. [6], who studied the parameterized complexity of computing a fault-tolerant spanning tree. They provide a fairly complete overview over the parameterized complexity landscape both in terms of natural (problem-specific) as well as structural parameters.

2 Preliminaries

In this section, we introduce the notation, definitions, and fundamental concepts used throughout the paper. For a positive integer $n \in \mathbb{N}$, we denote by $[n]$ the set $\{1, 2, \ldots, n\}$. A *graph* is a pair $G = (V, E)$, where V is a finite set of vertices and E is a set of edges. In an *undirected graph*, edges are unordered pairs of vertices, i.e., $E \subseteq \{\{u, v\} \mid u, v \in V, u \neq v\}$. In a *directed graph*, edges are ordered pairs of vertices, i.e., $E \subseteq \{(u, v) \mid u, v \in V, u \neq v\}$. Throughout the paper, we use n and m to denote the number of vertices and edges, respectively, if the considered graph is clear from the context. The underlying undirected graph of a directed graph $G = (V, E)$ is the graph where each directed edge is replaced by the corresponding undirected edge (and duplicates are eliminated). That is, the underlying undirected graph is $(V, \{\{u, v\} \mid (u, v) \in E \vee (v, u) \in E\})$. A *path* in a graph is a sequence $(v_1, v_2, \ldots, v_k)$ of vertices such that each consecutive pair is connected by an edge, that is, $\{v_i, v_{i+1}\} \in E$ (or $(v_i, v_{i+1}) \in E$ in the directed case). A *weighted graph* is a graph $G = (V, E)$ equipped with a weight function $w : E \to \mathbb{Q}_{\geq 0}$ that assigns a non-negative rational to each edge.

Parameterized Complexity. Parameterized complexity is a framework for analyzing computational problems based on two inputs: the main input size n and a parameter k. A problem is said to be *fixed-parameter tractable* (FPT) if it can be solved in $f(k) \cdot n^{\mathcal{O}(1)}$ time, where f is a computable function depending only on k. The class FPT contains all such problems.

The class XP consists of problems solvable in time $n^{f(k)}$ for some computable function f. A problem is W[1]-*hard* if it is at least as hard as the hardest problems in the class W[1], making it unlikely to be fixed-parameter tractable. A problem is *para-NP-hard* if it remains NP-hard even when the parameter k is a constant.

A *kernelization* algorithm reduces any instance (I, k) of a parameterized problem to an equivalent instance (I', k') in polynomial time, where the size of I' and k' depend only on k. If this size is bounded by a polynomial in k, the problem admits a *polynomial kernel.* It is known that a decidable problem is in FPT if and only if it admits a kernel of any size. Hence, problems admitting a polynomial kernel are a subclass of problems in FPT. To show that a problem does not admit a polynomial kernel (under standard complexity-theoretic assumptions), one can use polynomial parameter transformations. A *polynomial parameter transformation* is a polynomial-time reduction from one parameterized problem to another, such that the parameter in the new instance is bounded by a polynomial in the original parameter. If the original problem does not admit a polynomial parameter, then the problem reduced to also does not have a polynomial kernel. We refer to the textbook by Cygan et al. [10] for more information on kernelization and parameterized complexity.

Finally, we introduce three problems that are important later.

Hitting Set. Given a finite set $\mathcal{U}$ (the universe), a family $\mathcal{F}$ of subsets of $\mathcal{U}$, and an integer d, determine whether there exists a subset $H \subseteq \mathcal{U}$ of size at most d such that H intersects every set in $\mathcal{F}$ (i.e., $H \cap F \neq \emptyset$ for all $F \in \mathcal{F}$).

Steiner Tree. In the *Steiner Tree* problem, we are given an undirected edge-weighted graph $G = (V, E, w)$, a set of terminals $S \subseteq V$, and an integer d. The question is whether there is a tree $T = (V_T, E_T)$ in G that spans all vertices in S (i.e., $S \subseteq V_T$) with total cost $w(T) = \sum_{e \in E_T} w(e) \leq d$.

Biclique. A *biclique* in a bipartite graph $G = (U, V, E)$ is a complete bipartite subgraph $K_{a,b}$ where every vertex in a subset $A \subseteq U$ of size a is connected to every vertex in a subset $B \subseteq V$ of size b. Given a bipartite graph and an integer d, the *Biclique* problem asks whether G contains a $K_{d,d}$.

3 Algorithmic Results

In this section, we show our different FPT and XP algorithms. We mention that all algorithms can handle edge weights and directed graphs. We start with the two parameters related to the number of vulnerable edges (q and $|U|$). Therein, we will use the following problem, which we call DIRECTED STEINER LINKAGE and which seems to not have been studied in the literature before.

DIRECTED STEINER LINKAGE

Input: A directed graph $G = (V, A)$, an edge-weight function $w\colon A \to \mathbb{Q}_{\geq 0}$, two multisets S and T of vertices with $|S| = |T|$, and an integer ℓ.

Question: Is there a set $K \subseteq A$ of edges of total cost at most ℓ and a bijection $f\colon S \to T$ such that for each vertex $s \in S$, there is a directed path from s to $f(s) \in T$ in $G[K]$?

We suspect that DIRECTED STEINER LINKAGE is of independent interest. We start by showing that DIRECTED STEINER LINKAGE is fixed-parameter tractable when parameterized by the number $|S|$ of terminals in each of the two terminal sets. For that algorithm, we first state a helpful structural lemma regarding the structure of an optimal solution.

Lemma 2. *If an instance I is a yes-instance of* Directed Steiner Linkage, *then there exists a solution whose underlying undirected graph is a forest.*

Proof. Let $I = (G = (V, E), w, S, T, \ell)$ be a yes-instance of Directed Steiner Linkage. We represent a solution via a flow formulation. Note that Directed Steiner Linkage is equivalent to asking whether there is a set of edges of total cost at most ℓ such that in the induced graph (with arbitrarily large capacities), there is a feasible flow with one unit of supply in each vertex for each time it appears in S and one unit of demand in each vertex for each time it appears in T. Let K be a minimal set of edges of total cost at most ℓ such that there exists a feasible flow from S to T in $G[K]$ and let $C \subseteq K$ be a set of edges in K whose underlying undirected graph is a cycle. Let $\{v_0, v_1, \ldots, v_{k-1}\}$ be the vertices in C such that exactly one of $(v_{i-1}, v_{i \bmod k})$ and $(v_{i \bmod k}, v_{i-1})$ is contained in C for each $i \in [k]$. Note that if an edge is included in both directions, then reducing the flow along both edges by the same value yields another feasible flow solution. Hence, we can reduce the flow along at least one of the two edges to zero and get a solution in which one of the two edges is not used. We will next generalize this idea to arbitrary cycles.

Let d be the minimum flow send over any edge in C and let $e_d \in C$ be an edge over which exactly d flow was sent. We assume without loss of generality that $e_d = (v_{i-1}, v_{i \bmod k})$ for some $i \in [k]$. We say that for $j \in [k]$, edges of the form $(v_{j-1}, v_{j \bmod k})$ in C are *forward* edges and edges of the form $(v_{j \bmod k}, v_{j-1})$ are *backward* edges. We now reduce the flow on each forward edge by d and increase the flow of each backward edge by d. Each vertex incident to two backward edges or two forward edges in C now has precisely d less incoming and outgoing flow within C. For all other vertices, the total incoming flow and the total outgoing flow remains the same. Thus, for every vertex v_i for $i \in [k]$, the difference between the incoming and outgoing flow remains the same. This means that the change yields another feasible flow solution. We can now remove all edges with zero flow from the solution. As we remove at least the edge e_d, we found a subset of K which is a solution. Since this contradicts the minimality of K, this concludes the proof. □

We next show how to use the previous lemma to show that Directed Steiner Linkage parameterized by the number of terminals is in FPT.

Theorem 1. Directed Steiner Linkage *can be solved in $2^{\mathcal{O}(k \log k)} n^3$ time, where k is the number of terminals.*

Proof. Let $I = (G = (V, E), w, S, T, \ell)$ be an instance of Directed Steiner Linkage and let $k = |S|$. We first solve all-pairs shortest paths (APSP) in G in $\mathcal{O}(n^3)$ time using the standard FloydWarshall algorithm. Then we construct the auxiliary directed weighted graph H with the same set of vertices as G. For every pair (u, v) of distinct vertices $u, v \in V$, we construct an edge (u, v), and set its weight $w'(u, v) = \text{dist}(u, v)$, where $\text{dist}(u, v)$ is the distance in G between u and v, that is, the minimum cost of a directed u-v-path. We set $w'(u, v) = +\infty$ if such a path does not exist. We show the following claim.

Claim. The instances $I = (G, w, S, T, \ell)$ and $I' = (H, w', S, T, \ell)$ are equivalent. Furthermore, if (H, w', S, T, ℓ) is a yes-instance then there is a solution K' such that the underlying graph of $H[K']$ is a forest with at most $2k - 2$ internal vertices.

Proof (Proof of Claim). Suppose that I is a yes-instance. By Lemma 2, we can assume that it has a solution K whose underlying undirected graph is a forest. We assume that K is inclusion minimal. Then K induces a forest F in the underlying graph whose leaves are in $S \cup T$ (note that some terminals may be internal vertices). Since $|S| + |T| = 2k$, F has at most $2k - 2$ internal vertices of degree at least 3. Let X be the set of these vertices, and let $W = X \cup S \cup T$, and let $\mathcal{P}$ to be the set of paths in F with their end-points in W. Notice that each edge of K is in one of the paths of $\mathcal{P}$. We construct a subset K' of the set of edges of H with endpoints in W as follow. For every path $P = (v_1, \ldots, v_p) \in \mathcal{P}$, we observe that either P or its reversal $(v_p, \ldots, v_1)$ is a directed path in G because the internal vertices are not terminals. We assume without loss of generality that P is a directed path. Then, we add (v_1, v_p) to K'. Notice that $w'(v_1, v_p) = \text{dist}(v_1, v_p)$ is at most the cost of P. This immediately implies that the total cost of K' is at most ℓ. Also, because K is a solution to I, we have that K' is a solution to I', and as F is a forest, we have that underlying graph of $H[K']$ is a forest with at most $2k - 2$ internal vertices.

Assume now that I' is a yes-instance. Then I' has a solution K' of cost at most ℓ. For each $(u, v) \in K'$, G contains a u-v-path P_{uv} of total cost at most $w'(u, v) = \text{dist}(u, v)$. Consider the union K of the sets of edges of the paths P_{uv} taken over all edges of K'. By construction, the cost of K is at most ℓ. Furthermore, because we replace edges by directed paths, K is a solution to I. This proves the claim. ◇

We use this claim and solve Directed Steiner Linkage for I'. We are looking for a solution whose underlying undirected graph is a forest with at most $2k - 2$ internal vertices. We guess the structure of that forest, that is, we guess a forest F with at most $4k - 2$ vertices whose leaves are in $S \cup T$ together with a mapping of its vertices to the terminals. As shown by Takács [13], the number of (undirected) forests with at most $4k$ labeled vertices can be upper-bounded by $(4k)^{4k}$. Thus, we have at most $(4k)^{4k}$ guesses. Next, we turn F to a directed graph $\overrightarrow{F}$ by guessing the orientation of each edge of F in $\mathcal{O}(2^{4k})$ time, and then verify whether this orientation allows for a feasible S-T-flow in $(4k)^{1+o(1)}$ time [8]. Finally, we find a subgraph of H isomorphic to $\overrightarrow{F}$ respecting the mapping of terminals of minimum total cost. Because $\overrightarrow{F}$ has at most $4k - 2$ vertices, this can be done by the standard color coding algorithm of Alon, Yuster, and Zwick [3] in $2^{\mathcal{O}(k)} n^2 \log n$ time. If the cost of the solution is at most ℓ, we return the solution. Otherwise, we discard the current choice of $\overrightarrow{F}$. If we fail to find a solution for all possible choices of $\overrightarrow{F}$, we report that I' is a no-instance.

Since we have at most $(4k)^{4k} \cdot 2^{4k} \in 2^{\mathcal{O}(k \log k)}$ choices for $\overrightarrow{F}$, the overall running time is in $2^{\mathcal{O}(k \log k)} n^3$. This concludes the proof. □

With Theorem 1 at hand, we can show that FTP is FPT parameterized by $|U|$ and XP when parameterized by q.

Theorem 2. FTP *can be solved in* $2^{\mathcal{O}(|U|\log(|U|))}n^3$ *time and in* $2^{\mathcal{O}(q\log(q))}m^q n^3$ *time.*

Proof. We first guess the vulnerable edges of an optimal solution. This can be done in $\mathcal{O}(2^{|U|})$ and in $\mathcal{O}(m^q)$ time. Next, we reduce the remaining problem to DIRECTED STEINER LINKAGE. To this end, we delete all vulnerable edges from the input graph, define S to be the multiset of the heads of all guessed vulnerable edges and $k+1$ times the vertex s, and set T to be the multiset of all tails of guessed vulnerable edges and $k+1$ times the vertex t. Note that due to Proposition 1, we can assume that $|S| = |T| \leq q+k+1 \leq 2q \leq 2|U|$. Next, we use Theorem 1 to solve the constructed instance in $2^{\mathcal{O}(q\log(q))}n^3$ time. Note that a solution for DIRECTED STEINER LINKAGE (if such a solution exists) together with the guessed set of edges corresponds to a solution for FAULT-TOLERANT PATH. This is true since each guessed vulnerable edge in the solution takes one unit of demand out at its tail and adds one unit of supply back at its head. Note that this is equivalent to sending one unit of supply through the edge. Hence, any solution to DIRECTED STEINER LINKAGE together with the guessed vulnerable edges induces a graph in which $k+1$ units can flow from s to t (and there are potentially circles in the graph that circulate some amount of flow). Moreover, if the set of vulnerable edges is guessed correctly, then an optimal solution for DIRECTED STEINER LINKAGE together with the guessed set of edges also corresponds to an optimal solution for FTP (one in which no flow is circulated). To verify this, we show that any solution to FTP without the vulnerable edges is also a solution to DIRECTED STEINER LINKAGE. To this end, consider the set of vulnerable edges in an optimal solution. Lemma 1 says that the solution can be partitioned into $k+1$ paths from s to t that only intersect in safe edges. For each of these paths, consider the set of vulnerable edges in that path in the order they appear along the path. Between any two consecutive such edges, the safe edges in the solution connect the head of the first edge to the tail of the second (where we view s as the head of the first edge and t as the tail of the last edge). This gives both the bijection between the tails and heads of all vulnerable edges in the solution and the solution to connect the respective pairs. Thus, it is a solution to DIRECTED STEINER LINKAGE. This concludes the proof. □

As we will show later, FTP is W[1]-hard when parameterized by ℓ. As ℓ upper-bounds q, this also shows that we cannot hope to improve the algorithm for q to an FPT algorithm. We conclude this section by showing an algorithm that shows fixed-parameter tractability for $|S|$ and containment in XP for p.

Theorem 3. FTP *can be solved in* $2^{|S|} \cdot m^{1+o(1)}$ *time and in* $m^{p+1+o(1)}$ *time.*

Proof. First, we guess the set of safe edges in the solution. This can be done in $\mathcal{O}(2^{|S|})$ and in $\mathcal{O}(m^p)$ time. Let ℓ_1 be the sum of edge weights of all guessed safe edges. Next, we iterate over all edges and assign capacities and costs as follows. For safe edges that are guessed to be in the solution, we assign a capacity of $k+1$ and a cost of 0. For safe edges that are not guessed to be in the solution, we assign a capacity of 0 (which corresponds to removing the edge and hence the cost does not matter). For vulnerable edges, we assign a capacity of 1 and a cost equal to the weight of the edge. Note that the assignment of capacities and costs takes $\mathcal{O}(m)$ time. Finally, we compute a minimum-cost $k+1$-flow from s to t in the input graph with assigned capacities and costs. Note that such a flow corresponds to a solution of FTP by Lemma 1 where we only pay for the vulnerable edges (as all safe edges have a cost of 0). Hence, we return a solution only if the minimum cost of a $k+1$ flow plus ℓ_1 does not exceed ℓ. As shown in [8], a minimum-cost flow can be computed in $m^{1+o(1)}$ time. This concludes the proof. □

4 Hardness Results

In this section, we complement the results from the previous section. We start with a reduction from BICLIQUE that shows that FTP is W[1]-hard for the parameter ℓ even if $b = 1$. Hence, this also shows para-NP-hardness for b. Note that the problem is polynomial-time solvable for $b = 0$ and hence our result is tight with respect to b. All results in this section are for undirected and unweighted graphs.

Theorem 4. FTP *is W[1]-hard when parameterized by ℓ even when $b = 1$ and the input graph is undirected and unweighted. Moreover, assuming the ETH, the problem cannot be solved in $f(\ell) \cdot n^{g(b) \cdot o(\sqrt[4]{\ell})}$ time for any computable functions f and g.*

Proof. We reduce from BICLIQUE. This problem is known to be W[1]-hard when parameterized by d and cannot be solved in $f(d) \cdot n^{o(\sqrt{d})}$ time for any computable function f assuming the ETH [12]. Let $(G = (A \cup B, E), d)$ be an input instance. We assume without loss of generality that $|E| \geq 3d^2$.

To produce an equivalent instance of FAULT-TOLERANT PATH, we start with a copy of G, where all edges are vulnerable. Next, we add three vertices s, t, and v and add safe edges between v and each vertex in A and between t and each vertex in B. Moreover, we add a path P of $2d^2 - 2d - 1$ safe edges between s and v and d^2 paths between s and t of three vulnerable edges each. Let the set of edges in these latter paths be E_P. Finally, we set $k = d^2 - 1$ and $\ell = 3d^2 - 1$. The reduction is depicted in Fig. 2.

We next show that the two instances are equivalent. To this end, assume first that G contains a $K_{d,d}$ as a subgraph. We build a solution for FTP as follows. We pick all edges in the s-v-path, all edges in the supposed $K_{d,d}$, and all safe edges incident to the $2d$ endpoints. Note that this graph contains exactly $2d^2 - 2d - 1 + d^2 + 2d = \ell$ edges. Moreover, after removing any $k = d^2 - 1$ vulnerable edges, there still remains one edge $\{u, w\}$ with $u \in A$ and $w \in B$ in the solution graph. By construction, the s-v-path and the edges $\{v, u\}$ and $\{w, t\}$ are safe and contained in the solution. Thus, there remains an s-t-path after removing any k vulnerable edges.

For the reverse direction, assume that the constructed instance of FTP is a yes-instance. Then, there exists a subgraph consisting of at most $\ell = 3d^2 - 1$ edges such that after removing any $k = d^2 - 1$ vulnerable edges, there still remains an s-t-path. Note that this solution contains the entire s-v-path P as, otherwise, it has to contain all edges in E_P and therefore has size at least $\ell + 1$. Note that any s-t-path in the solution either contains at least one vertex of A and at least one vertex of B or is a path with three edges from E_P. In the second case, we can replace the edges of such a path in the solution by the edges of a path $Q = (v, x, y, t)$, where $x \in A$, $y \in B$, and $\{x, y\}$ is an edge of G that is not in the solution; such an edge exists because $|E| \geq 3d^2$. Since Q contains a single vulnerable edge and the edges of P are in the original solution, the obtained set of edges forms a solution. Thus, we can assume that any s-t-path in a solution contains a vertex from A and a vertex from B. Denote by K a solution of this type. Notice that it should contain at least $k+1 = d^2$ vulnerable edges between A and B. Also, because $\ell = 3d^2 - 1$ and P contains $2d^2 - 2d - 1$ edges, the total number of edges in K incident to the vertices $A \cup B$ is at most $d^2 + 2d$.

Denote by $X \subseteq A$ the set of vertices $x \in A$ such that $\{v, x\} \in K$, and let $Y \subseteq B$ be the set of vertices $y \in B$ such that $\{y, t\} \in K$. Let also $X' \subseteq A \setminus X$ and $Y' \subseteq B \setminus Y$ denote the subsets of vertices of A and B, respectively, incident to some edges in the solution but not adjacent to v or t. Notice that $|X| + |Y| \leq 2d$ because, otherwise, K would contain at most $2d^2 - 1$ vulnerable edges. See Fig. 3 for an illustration.

Denote by S_{XY} the set of vulnerable edges in K between X and Y. Furthermore, denote by $S_{XY'}$, $S_{X'Y}$, and $S_{X'Y'}$ the sets of edges in K between X and Y', X' and Y, and X' and Y', respectively. By this definition, S_{XY}, $S_{XY'}$, $S_{X'Y}$, and $S_{X'Y'}$ are disjoint, and their union is the set of vulnerable edges in K. Because K is a solution, we have that $|S_{XY}| + |S_{XY'}| \geq d^2$, $|S_{XY}| + |S_{X'Y}| \geq d^2$, and $|S_{XY}| + |S_{X'Y'}| \geq d^2$, as depicted in Fig. 3. Thus, K contains at least $3d^2 - 2|S_{XY}|$ vulnerable edges. Then

$$3d^2 - 2|S_{XY}| \leq d^2 + 2d - |X| - |Y| \tag{1}$$

because the total number of edges in K incident to the vertices $A \cup B$ is at most $d^2 + 2d$ and K contains $|X| + |Y|$ safe edges. Hence,

$$2d^2 - 2d \leq 2|S_{XY}| - |X| - |Y| \leq 2|X||Y| - |X| - |Y| \leq 2\Big(\frac{|X| + |Y|}{2}\Big)^2 - 2\frac{|X| + |Y|}{2}.$$

This implies that $|X| + |Y| \geq 2d$. We obtain that $|X| + |Y| = 2d$ and, moreover, $|X| = |Y| = d$ because $2|X||Y| - |X| - |Y| = 2|X||Y| - 2d$ achieves its maximum value for $|X| = |Y| = d$. Then, by Inequality 1, $|S_{XY}| \geq d^2$. As $|X| + |Y| = 2d$, K contains at most d^2 vulnerable edges. Therefore, $S_{XY'} = S_{X'Y} = S_{X'Y'} = \emptyset$ and $|S_{XY}| = d^2$. This proves that $G[X \cup Y]$ is a biclique $K_{d,d}$. This completes the proof of the equivalence of the instances.

To conclude the proof, observe that the reduction can be computed in polynomial time and that $b \leq 3d^2 - \ell = 1$ as the $3d^2$ edges in E_P ensure that the

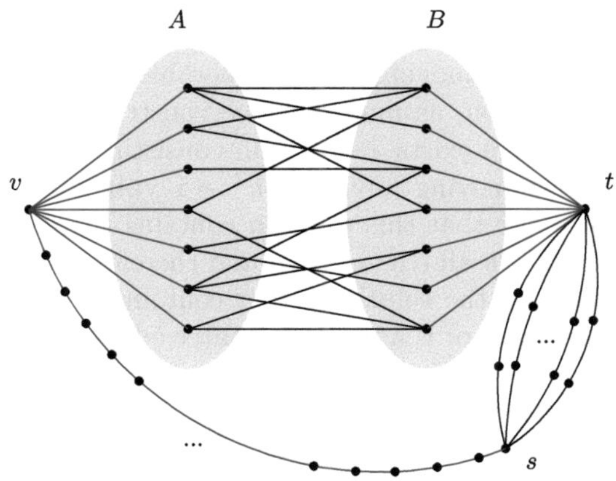

Fig. 2. The graph constructed in the proof of Theorem 4. Safe edges are in blue and vulnerable edges in black. (Color figure online)

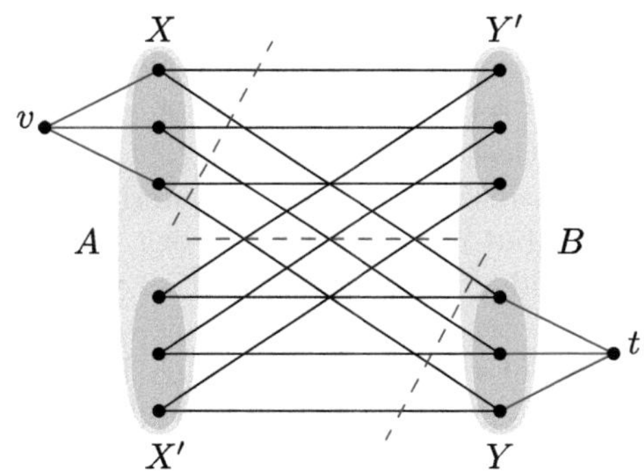

Fig. 3. The sets X, Y, X', and Y' constructed in Theorem 4. The three dashed lines corresponds to $S_{XY} \cup S_{XY'}$, $S_{XY} \cup S_{X'Y}$, and $S_{XY} \cup S_{X'Y'}$.

minimum-cost $(k+1)$-flow has cost at most $3d^2$. Thus, any algorithm that solves FTP in $f(\ell) \cdot n^{g(b) \cdot o(\sqrt[4]{\ell})}$ time for any computable functions f and g can be used to solve BICLIQUE in $f'(d) \cdot n^{o(\sqrt[4]{d^2})} = f'(d) \cdot n^{o(\sqrt{d})}$ time for some computable function f'. This refutes the ETH [12]. □

We next exclude polynomial kernels for $|S|$ and $|U|$.

Proposition 2 (⋆). FTP *does not admit a polynomial kernel when parameterized by the number $|U|$ of vulnerable edges in the input graph even if the input graph is undirected and unweighted.*

The basic idea behind the proof is a polynomial parameter transformation from STEINER TREE parameterized by the number of terminals. This problem is known to not admit a polynomial kernel [10]. We start with a copy of the input graph where all edges are safe edges. Next, we add a new vertex s and a

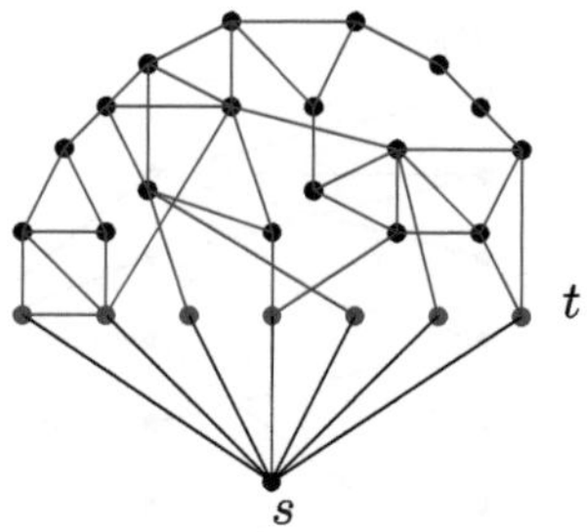

Fig. 4. An example of the construction behind the proof of Proposition 2. The blue (safe) edges are a copy of the input graph G and the terminals are represented in red. Black edges are vulnerable. (Color figure online)

vulnerable edge $\{s, u\}$ for each terminal $u \in T$. Finally, we arbitrarily pick one of the terminals in T and rename it to t, set $\ell = d + |T|$, and set $k = |T| - 1$. Figure 4 gives an example.

Note that any solution has to contain all $|T|$ vulnerable edges. Moreover, the remaining d safe edges have to connect all original terminals, that is, we are looking for a Steiner tree of size d.

Proposition 3 ($\star$). FTP *does not admit a polynomial kernel when parameterized by the number* $|S|$ *of safe edges in the input graph even if the input graph is undirected and unweighted.*

The claim is shown by constructing a polynomial parameter transformation from Hitting Set parameterized by the size of the universe which is known to not admit a polynomial kernel [11]. The main idea of the reduction is to build a bipartite graph where all elements in the universe are represented by a vertex on one side. On the other side, each set in the set family is represented by a vertex. There is a vulnerable edge between two vertices if the corresponding element is contained in the corresponding set. Moreover, each set vertex is connected to t by a vulnerable edge and each element vertex is connected to s via a safe edge. An example is depicted in Fig. 5. The value of k is set to $|\mathcal{F}|-1$ and $\ell = 2|\mathcal{F}| + k'$ where k' is the allowed solution size in the Hitting Set instance. Now a solution contains $2|\mathcal{F}|$ vulnerable edges and the k' additional safe edges encode a hitting set.

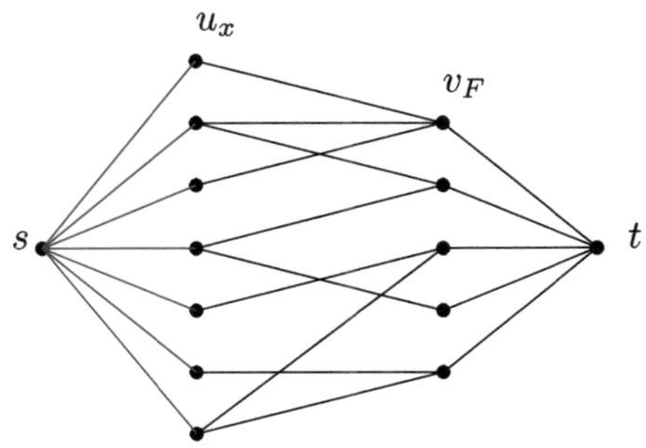

Fig. 5. The graph G in the proof of Proposition 3. Safe edges are represented in blue and vulnerable edges in black. (Color figure online)

5 Conclusion

We analyzed the parameterized complexity of FTP with respect to a collection of natural parameters. For these parameters, we gave an almost complete classification in terms of polynomial kernels, FPT, XP, and para-NP-hardness. Our results are summarized in Fig. 1.

The main open question remaining is the exact complexity for the parameter k (the required redundancy). Adjiashvili et al. [2] showed that the problem is polynomial-time solvable for $k = 1$. Theorem 4 shows that the problem is W[1]-hard. Whether the problem is in XP or para-NP-hard and even whether the problem is polynomial-time solvable for $k = 2$ remain open. Adjiashvili et al. [2] also showed that FTP is in XP when parameterized by k if the input graph is a DAG (a directed acyclic graph). Whether this can be improved to FPT or shown to be W[1]-hard is also an interesting open problem.

Finally, we mention that there is still a lot of room for investigating structural parameters like the treewidth of the input graph. Moreover, we believe that Directed Steiner Linkage is a problem of independent interest that deserves further investigation.

Acknowledgements. The research leading to these results has received funding from the Research Council of Norway via the project BWCA (grant no. 314528) and the Franco-Norwegian AURORA project (grant no. 349476), the European Research Council (ERC) under the European Union's Horizon 2020 research and innovation programme (grant agreement No. 819416), and the French ANR project ELIT (ANR-20-CE48-0008).

References

1. Adjiashvili, D., Hommelsheim, F., Mühlenthaler, M.: Flexible graph connectivity. Math. Program. **192**(1), 409–441 (2022)
2. Adjiashvili, D., Hommelsheim, F., Mühlenthaler, M., Schaudt, O.: Fault-tolerant edge-disjoint s-t paths – beyond uniform faults. In: Proceedings of the 18th Scandinavian Symposium and Workshops on Algorithm Theory (SWAT), pp. 5:1–5:19. Schloss Dagstuhl - Leibniz-Zentrum für Informatik (2022)
3. Alon, N., Yuster, R., Zwick, U.: Color-coding. J. ACM **42**(4), 844–856 (1995)
4. Bansal, I., Cheriyan, J., Grout, L., Ibrahimpur, S.: Improved approximation algorithms by generalizing the primal-dual method beyond uncrossable functions. In: Proceedings of the 50th International Colloquium on Automata, Languages, and Programming (ICALP), pp. 15:1–15:19. Schloss Dagstuhl–Leibniz-Zentrum für Informatik (2023)
5. Bentert, M., Fomin, F.V., Golovach, P.A., Morelle, L.: When does FTP become FPT? (2025). https://arxiv.org/abs/2506.17008
6. Bentert, M., Schestag, J., Sommer, F.: On the complexity of finding a sparse connected spanning subgraph in a non-uniform failure model. In: Proceedings of the 18th International Symposium on Parameterized and Exact Computation (IPEC), pp. 4:1–4:12. Schloss Dagstuhl - Leibniz-Zentrum für Informatik (2023)
7. Boyd, S.C., Cheriyan, J., Haddadan, A., Ibrahimpur, S.: Approximation algorithms for flexible graph connectivity. In: Proceedings of the 41st IARCS Annual Conference on Foundations of Software Technology and Theoretical Computer Science (FSTTCS), pp. 9:1–9:14. Schloss Dagstuhl–Leibniz-Zentrum für Informatik (2021)
8. van den Brand, J., et al.: A deterministic almost-linear time algorithm for minimum-cost flow. In: Proceedings of the 64th IEEE Annual Symposium on Foundations of Computer Science (FOCS), pp. 503–514. IEEE (2023)
9. Chekuri, C., Jain, R.: Approximation algorithms for network design in non-uniform fault models. In: Proceedings of the 50th International Colloquium on Automata, Languages, and Programming (ICALP), pp. 36:1–36:20. Schloss Dagstuhl–Leibniz-Zentrum für Informatik (2023)
10. Cygan, M., Fomin, F.V., Kowalik, Ł, Lokshtanov, D., Marx, D., Pilipczuk, M., Pilipczuk, M., Saurabh, S.: Lower bounds for kernelization. In: Parameterized Algorithms, pp. 523–555. Springer, Cham (2015). https://doi.org/10.1007/978-3-319-21275-3_15
11. Dom, M., Lokshtanov, D., Saurabh, S.: Incompressibility through colors and IDs. In: Albers, S., Marchetti-Spaccamela, A., Matias, Y., Nikoletseas, S., Thomas, W. (eds.) ICALP 2009. LNCS, vol. 5555, pp. 378–389. Springer, Heidelberg (2009). https://doi.org/10.1007/978-3-642-02927-1_32
12. Lin, B.: The parameterized complexity of the k-biclique problem. J. ACM **65**(5), 34:1–34:23 (2018)
13. Takács, L.: On Cayley's formula for counting forests. J. Comb. Theory, Ser. A **53**(2), 321–323 (1990)

Subgraph-Universal Planar Graphs for Trees

Helena Bergold[1], Vesna Iršič Chenoweth[2,3], Robert Lauff[4], Joachim Orthaber[5(✉)], Manfred Scheucher[4], and Alexandra Wesolek[4]

[1] Institute of Computer Science, Freie Universität Berlin, Berlin, Germany
helena.bergold@fu-berlin.de

[2] Faculty of Mathematics and Physics, University of Ljubljana, Ljubljana, Slovenia
vesna.irsic@fmf.uni-lj.si

[3] Institute of Mathematics, Physics and Mechanics, Ljubljana, Slovenia

[4] Institut für Mathematik, Technische Universität Berlin, Berlin, Germany
{lauff,scheucher,wesolek}@math.tu-berlin.de

[5] Institute of Algorithms and Theory, Graz University of Technology, Graz, Austria
orthaber@tugraz.at

Abstract. We show that there exists an outerplanar graph on $O(n^c)$ vertices for $c = \log_2(3 + \sqrt{10}) \approx 2.623$ that contains every tree on n vertices as a subgraph. This extends the line of research of Chung and Graham from 1983 who showed that there exist (non-planar) n-vertex graphs with $O(n \log n)$ edges that contain all trees on n vertices as subgraphs. Our result also stands in contrast to a result from Gol'dberg and Livshits from 1968 who showed that there exists a universal tree for n-vertex trees on $n^{O(\log(n))}$ vertices, and it was later shown that there is no polynomially sized universal tree. Furthermore, we investigate (outer)planar graphs containing all n-vertex (outer)planar graphs as subgraphs, determining exponential lower bounds in both cases. We can however construct a planar graph on $n^{O(\log(n))}$ vertices containing all n-vertex outerplanar graphs as subgraphs. Lastly, we determine the number of vertices needed in the worst case for a planar graph to contain three given trees as subgraphs to be on the order of $\frac{3}{2}n$, even if the three trees are caterpillars. This answers a question recently posed by Alecu et al. in 2024.

Keywords: Universal Graphs · Nash-Williams · Planar 3-Trees

H. Bergold was partially supported by the DFG Research Training Group 'Facets of Complexity' (DFG-GRK 2434). V. Iršič Chenoweth was supported by the Slovenian Research and Innovation Agency (ARIS) under the grants Z1-50003, P1-0297, N1-0218, N1-0285, and N1-0355, and by the European Union (ERC, KARST, 101071836). J. Orthaber was supported by the Austrian Science Fund (FWF) grant W1230. M. Scheucher was supported by the DFG Grant SCHE 2214/1-1. A. Wesolek was supported by the Deutsche Forschungsgemeinschaft (DFG, German Research Foundation) under Germany's Excellence Strategy The Berlin Mathematics Research Center MATH+ (EXC-2046/1, project ID: 390685689).

H. Fernau and P. Kindermann (Eds.): WG 2025, LNCS 16124, pp. 62–75, 2026.
https://doi.org/10.1007/978-3-032-11835-6_5

1 Introduction

The motivation of embedding multiple graphs on n vertices simultaneously in a (planar) graph has a long history, which originates in VLSI design. The aim is to find a universal circuit which can be used to create multiple smaller circuits. A circuit can be seen as a graph, and Valiant [21,22] used so-called universal graphs to construct universal circuits. A host graph H is called *(subgraph-)universal* for a family of graphs $\mathcal{G}$ if every graph $G \in \mathcal{G}$ appears as a subgraph of H. As trees are a central graph class both in graph theory and computer science, it is natural to search for (small) universal graphs containing all n-vertex trees. Already in 1978, Chung and Graham [6] showed that an n-vertex graph that is subgraph-universal for n-vertex trees and is edge-minimal has $\Theta(n \log n)$ edges. Comparing the number of vertices to edges, their construction is far from planar and in fact the graph contains cliques of size $\Omega(\log(n))$. Finding small universal graphs with more restricting criteria is much harder. For host graphs that are trees, the first result from Gol'dberg and Livshits [12] from 1968 shows that there exists a universal tree for n-vertex trees on $n^{O(\log(n))}$ vertices. Their construction was shown to be tight up to a polynomial factor in n by Chung, Graham and Coppersmith [7]. While Chung and Graham considered no restrictions on the host graph, Gol'dberg and Livshits considered the strongest restriction, that the host graph comes from the graph class it represents. However not much is known about host graphs which are close to the graph class they represent, for example, host graphs for trees which are outerplanar, which have small treewidth, or which have only small cliques.

Since host graphs for all n-vertex graphs from a class $\mathcal{G}$ that are in $\mathcal{G}$ themselves usually have many edges (and vertices), most of the research on universal graphs concentrated on putting no restrictions on the host graphs at all. In 1982, Babai et al. [2] started the systematic study of universal graphs for sparse graph classes. They gave a lower bound of $\Omega(\frac{n^2}{\log n})$ on the number of edges of a universal graph for the class of graphs with cn edges for $c > 1$, and they showed that asymptotically less than the trivial upper bound $\binom{n}{2}$ is needed. Regarding planar graphs, they showed that there is a universal (non-planar) graph with $O(n^{3/2})$ edges. After several improvements on the upper bound [2,5], recently Esperet et al. [11] constructed a universal (non-planar) graph for all n-vertex planar graphs on $n^{1+o(1)}$ edges and $O(n)$ vertices by improving a construction of Dujmović et al. [10] with $n^{1+o(1)}$ vertices.

Previous research on planar host graphs for planar graphs focused on universality for infinite graphs. Pach [19] showed that there is no countable planar graph containing all countable planar graphs, answering a question of Ulam. Recently, Huynh et al. [13] strengthened this result by showing that every countable graph that contains all countable planar graphs contains an infinite clique as a minor. When turning to finite graphs, a trivial upper bound on the number of vertices of a minimal universal planar graph for n-vertex planar graphs is the number of triangulations on n vertices (times n), which is $O((256/27)^n)$ as shown by Tutte [20]. We give the first exponential lower bound.

Our Contribution. We continue the line of research from [6] and [7] and study planar host graphs for trees. As our main result, we construct an outerplanar graph with polynomially many vertices in n that is subgraph-universal for all n-vertex trees. Since the construction is outerplanar, clearly the number of edges in it is of the same order as the number of vertices.

Theorem 1. *For every $n \in \mathbb{N}$, there exists an outerplanar graph on $O(n^c)$ vertices, $c = \log_2(3+\sqrt{10}) \approx 2.623$, that contains all n-vertex trees as subgraphs.*

Our result shows that imposing strong structural properties on the host graph does not necessarily blow up the number of its vertices (as is the case when restricting the host graph to be a tree). However, it seems that when the host graph is in the same graph class as the n-vertex graphs it represents, then the number of vertices it has is usually superpolynomial. We show that there are no subexponential sized planar graphs that contain all n-vertex planar graphs and the same is true for universal outerplanar graphs for outerplanar graphs.

Theorem 2. *Every universal planar graph for all n-vertex planar graphs has $\Omega(n^{-3/2}(27/4)^n)$ vertices.*

In addition, we can construct a universal planar graph with $n^{\theta(\log(n))}$ vertices which contains all n-vertex outerplanar graphs.

For practical purposes one might be interested in a universal graph for only a subset of n-vertex trees (or more generally from a graph class $\mathcal{G}$). We study the size of universal planar graphs for a family of three n-vertex trees. Nash-Williams' tree-covering theorem [18] implies that every connected planar graph contains three spanning trees that cover all edges. A natural question to ask is whether the converse is true, that is, whether any three trees on n vertices are contained in some planar graph on n vertices (as discussed by Alecu et al. [1]). We show that this is false for n large enough even when the three trees are caterpillars and the planar graph is allowed to have cn vertices for $c < \frac{3}{2}$.

Theorem 3. *There exist three caterpillars on n vertices each such that every minimal planar graph containing all three of them has $\frac{3}{2}n - O(n^{2/3})$ vertices.*

For the upper bound we can show in more generality that any three trees on n vertices are simultaneously contained in some planar graph on $\frac{3}{2}n$ vertices.

The paper is organized as follows: Theorem 1 is proven in Sect. 2. In Sect. 3 we discuss universal (outer)planar graphs containing all n-vertex (outer)planar graphs. The converse of Nash-Williams' tree-covering theorem is discussed in Sect. 4. And some directions for further research are given in Sect. 5.

2 A Universal Outerplanar Graph for Trees

The goal of this section is to construct, for every $n \in \mathbb{N}$, an outerplanar graph G on as few vertices as possible such that all n-vertex trees are contained in G

as subgraphs. We will see in the following that a number of vertices polynomial in n is sufficient for G.

Our approach is inspired by so-called stacked triangulations or planar 3-trees. While planar 3-trees are far from being outerplanar, our construction will be a subgraph of a planar 3-tree, where we delete sufficiently many edges (and vertices) such that it becomes outerplanar.

A first main idea is to find a vertex in a tree T that splits T into somewhat balanced subtrees. The following was probably first used by Jordan [15] and the respective vertex is nowadays known as the *centroid* of a tree [16].

Lemma 1 ([15]). *Every tree T on n vertices contains a vertex C such that all components of $T - C$ contain at most $\frac{n}{2}$ vertices.*

In the following, we will place C at a specified vertex of our construction and embed every subtree into the interior of a subconstruction. Recursively splitting each subtree again at a centroid, we can decompose any tree T into components of constant size with $\log_2(n)$ such steps.

However, after splitting T at C, in each subtree T' we have a *marked vertex* that needs to be adjacent to C. In the embedding, T' will be inside of some subconstruction and C will be mapped to one of its *outer vertices*. After each recursive step of splitting T at a centroid, we might have one additional marked vertex in some subtree. So after k steps a subtree might have k marked vertices that need to be adjacent to a different vertex each, which has been embedded at a previous step. To avoid this, we introduce an intermediate splitting step to keep the number of marked vertices below three in every subtree. The next lemma follows from the fact that trees are median graphs.

Lemma 2 ([14, Proposition 1.26]). *For every triple of vertices a, b, c in a tree T there exists a unique vertex M in T that lies on all three paths connecting a, b, and c in T.*

Using Lemma 2, we can split a subtree with three marked vertices at the *median* M of these three vertices. This ensures that after the splitting the number of marked vertices in each subtree is again at most two (where we also mark vertices adjacent to M). We now recursively define two outerplanar graphs G_d^1 and G_d^2. They will have 1 and 2, respectively, outer vertices o_1 (and o_2) and we call their remaining vertices *interior vertices*. The goal is that any tree on at most 2^d vertices and with at most 1 (or 2) marked vertices m_1 (and m_2) can be embedded on the interior vertices of G_d^1 (and G_d^2) such that m_1 is adjacent to o_1 (and m_2 to o_2). As the base case, G_0^1 is an edge consisting of one outer and one interior vertex and G_0^2 is a path of length two where the leaves are the outer vertices. Figure 1 gives an illustration of the recursive constructions.

G_d^1 consists of an edge between the outer vertex o_1 and a specified interior vertex p_1 and one copy of G_{d-1}^2 on top of that edge, that is, $o_1(G_{d-1}^2) = o_1(G_d^1)$ and $o_2(G_{d-1}^2) = p_1(G_d^1)$. In addition it contains five copies of G_{d-1}^1, four with their outer vertex $o_1(G_{d-1}^1)$ at $p_1(G_d^1)$ and the fifth with $o_1(G_{d-1}^1) = o_1(G_d^1)$. In G_d^2 there are two specified interior vertices p_1 and p_2 with edges between o_1

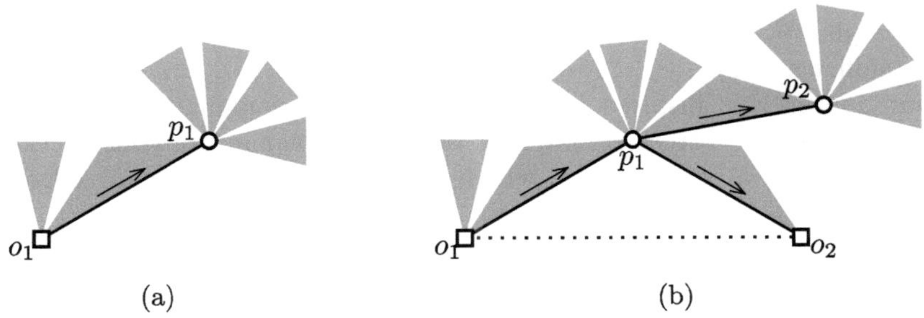

Fig. 1. The outerplanar graphs (a) G_d^1 and (b) G_d^2. Subgraphs G_{d-1}^1 and G_{d-1}^2 are represented by green and blue triangles, respectively. In G_{d-1}^2 the arrow points from its outer vertex o_1 to its outer vertex o_2.

and p_1, p_1 and p_2, and p_1 and o_2. Similar to before there is one copy of G_{d-1}^2 on top of each of these three edges. Furthermore, there are eight copies of G_{d-1}^1, for three of them their outer vertex $o_1(G_{d-1}^1)$ coincides with $p_1(G_d^2)$, for four of them it coincides with $p_2(G_d^2)$ and for one of them it coincides with $o_1(G_d^2)$.

An important part of the constructions is that $o_1(G_d^i)$, for $i \in \{1, 2\}$, coincides with o_1's of two smaller copies, G_{d-1}^1 and G_{d-1}^2. These copies are disjoint except for $o_1(G_d^i)$, that is, they are *internally disjoint*. Let $\mathcal{G}_d = \{G_d^1, G_d^2\}$.

Lemma 3. *Let ℓ and $d \geq d_1 \geq \ldots \geq d_k$ be nonnegative integers with*

$$\sum_{j=1}^{k} 2^{d_j} \leq 2^{d+\ell}. \tag{1}$$

Then 2^ℓ internally disjoint graphs in $\mathcal{G}_d$ with common outer vertex o_1 contain pairwise internally disjoint copies $G_{d_j}^{i_j}$ for $i_j \in \{1, 2\}$ and $j \in \{1, \ldots, k\}$ with outer vertex at o_1.

Proof. Every copy of G_d^i contains one copy of G_{d-1}^1 and G_{d-1}^2 each with the same outer vertex o_1 but internally disjoint. Hence, for every $m \in \{1, \ldots, d\}$, they contain exactly 2^m pairwise internally disjoint copies of graphs in $\mathcal{G}_{d-m}$ each, all having the same outer vertex o_1. Assigning a cost of 2^d to a graph in $\mathcal{G}_d$, this cost equally splits to a cost of 2^{d_j} for every subgraph $G_{d_j}^{i_j}$ with same outer vertex o_1. Further the 2^ℓ internally disjoint graphs in $\mathcal{G}_d$ give a budget of $2^{d+\ell}$ in total. Hence, within them we can greedily choose, for $j = 1, \ldots, k$, a copy of a graph in $\mathcal{G}_{d_j}$ with outer vertex o_1 and internally disjoint from the previous chosen copies. Indeed, by inequality (1), the cost of those graphs does not exceed the budget. And since we go over the d_j in order of descending size, no small graph will block a large one later. □

We will use this in the following way: Given a tree on $n \leq 2^d$ vertices, we split it along a centroid C. If $T_1, \ldots, T_k$ are the subtrees in $T - C$, consider $d_j = \lceil \log_2(|T_j|) \rceil$. Since C is a centroid, we have $d_j \leq d - 1$ for all $j \in \{1, \ldots, k\}$.

Hence, no matter how many subtrees there are, by Lemma 3 and due to rounding, we can fit them all into 4 copies of G^i_{d-1}. We are now ready to prove a slightly weaker version of Theorem 1.

Theorem 4. *Let T be a tree on $n \leq 2^d$ vertices for $d \in \mathbb{N}$, with up to two, not necessarily distinct, marked vertices m_1 and m_2. Then T can be embedded on the interior vertices of G^1_d such that m_1 is adjacent to o_1 of G^1_d and T can be embedded on the interior vertices of G^2_d such that m_1 and m_2 are adjacent to o_1 and o_2 of G^2_d, respectively.*

Proof. We use induction on the depth d, simultaneously for the outerplanar graphs G^1_d and G^2_d. For $d = 0$, the only tree on at most one vertex is mapped to the single interior vertex of G^1_0 or G^2_0, which is adjacent to all outer vertices.

For the inductive step, let C be a centroid of T, which exists by Lemma 1, and let $T_1, \ldots, T_k$ be the connected components of $T - C$, which are subtrees of T. Furthermore, let $d_j = \lceil \log_2(|T_j|) \rceil$ for $j \in \{1, \ldots, k\}$. Then d_j is the smallest integer such that $|T_j| \leq 2^{d_j}$ and we have $d_j \leq d - 1$, since C is a centroid. In each subtree T_j, we mark the vertex r_j adjacent to C. We distinguish three cases.

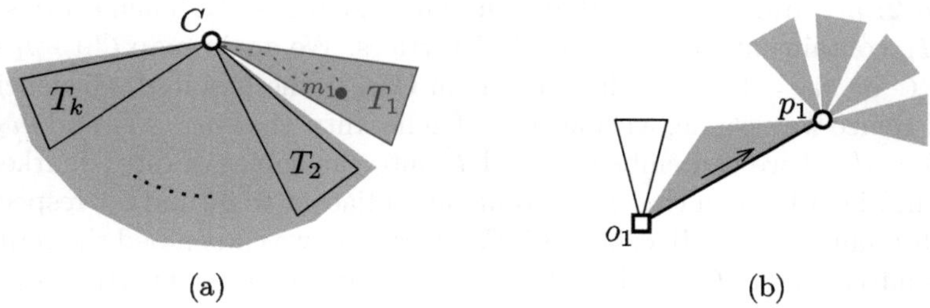

Fig. 2. (a) A splitting into subtrees with only one marked vertex m_1 in T. (b) Embedding the subtrees into G^1_d.

Case 1: T contains at most one marked vertex m_1. We map C to p_1 of G^i_d, for $i \in \{1, 2\}$. If $m_1 = C$ or there is no marked vertex, then the required adjacency is fulfilled since p_1 is adjacent to all outer vertices. Further the subtrees $T_1, \ldots, T_k$ have only one marked vertex. In both G^1_d and G^2_d there are at least four copies of G^i_{d-1} incident to p_1 with their outer vertex o_1 equal to p_1. We can apply Lemma 3 since the sizes of the subtrees are bounded by 2^{d-1} as C is a centroid and

$$\sum_{j=1}^{k} 2^{d_j} \leq \sum_{j=1}^{k} 2|T_j| \leq 2^{d+1}.$$

By the induction hypothesis, we can simultaneously embed the trees $T_1, \ldots, T_k$ into the interior of those copies of G^i_{d-1} such that $r_1, \ldots, r_k$ are all adjacent to C.

Otherwise, if $m_1 \neq C$, we assume without loss of generality that $m_1 \in T_1$. For $T_2, \ldots, T_k$ we do the same as before. Furthermore, by the induction hypothesis, we can embed T_1 into the copy of G^2_{d-1} such that r_1 is adjacent to $C = p_1$ and m_1 is adjacent to o_1, as claimed. Figure 2 illustrates the case for G^1_d, and as explained above, the case for G^2_d is similar.

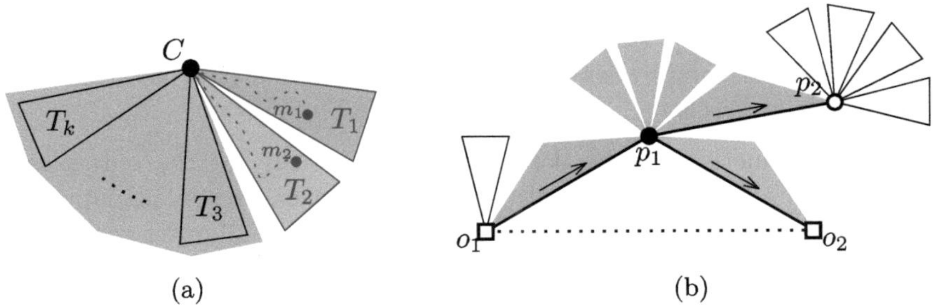

Fig. 3. (a) A splitting of T with centroid C such that m_1 and m_2 are in different subtrees. (b) Embedding the subtrees into G^2_d.

Case 2: m_1 and m_2 are not in the same subtree T_j. Then every subtree $T_1, \ldots, T_k$ contains at most two marked vertices. We again map C to p_1 of G^2_d. If $m_1 = C$ or $m_2 = C$, then similarly as in Case 1 the required adjacencies are fulfilled. Hence we assume, without loss of generality, that $m_1 \in T_1$ and $m_2 \in T_2$. As $d_1, d_2 \leq d - 1$ we can embed T_1 and T_2 into the copies of G^2_{d-1} marked blue and red in Fig. 3(b) such that m_1 and m_2 are adjacent to o_1 and o_2, respectively, and both r_1 and r_2 are adjacent to C. Furthermore, we can embed the remaining subtrees into copies of $G^i_{d_j}$ adjacent to C such that the marked vertices $r_3, \ldots, r_k$ are adjacent to C. This is again an application of Lemma 3 (Fig. 3).

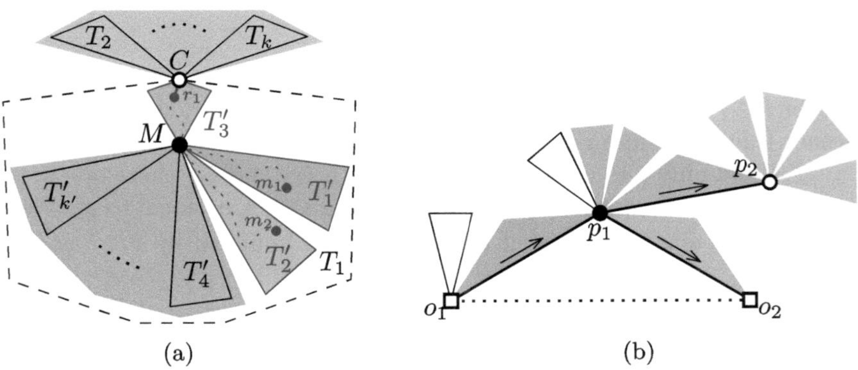

Fig. 4. (a) A splitting of T with a centroid C such that m_1 and m_2 are in the same subtree T_1, which is further split at the median M. (b) Embedding the subtrees into G^2_d.

Case 3: m_1 and m_2 are in the same subtree T_1. Then T_1 contains three marked vertices m_1, m_2, and r_1. Hence, using Lemma 2 and introducing the median M, we further split T_1 into subtrees T'_j, for $j = 1, \dots, k'$, and let $d'_j = \lceil \log_2(|T'_j|) \rceil$. This time we map M to p_1 and C to p_2 (Fig. 4). We embed the subtrees $T_2, \dots, T_k$ into the four copies of G^1_{d-1} incident to p_2, which is possible by Lemma 3. Further, we assume that m_1, m_2, and r_1 are contained in T'_1, T'_2, and T'_3, respectively. If any of them coincide with M, then the corresponding subtree is empty and the required adjacencies are fulfilled by $M = p_1$. In each subtree, we mark the vertex r'_j that is the neighbor of M in that subtree. Since $\sum 2^{d'_j} \leq 2^d$, we can embed the subtrees $T'_4, \dots, T'_{k'}$ into two copies of G^1_{d-1} incident to p_1 by Lemma 3 and the induction hypothesis. Further the following embeddings are ensured by the induction hypothesis:

- T'_1 is embedded into the copy of G^2_{d-1} between o_1 and p_1.
- T'_2 is embedded into the copy of G^2_{d-1} between p_1 and o_2.
- T'_3 is embedded into the copy of G^2_{d-1} between p_1 and p_2. □

Given a tree, we can mark any vertex, and then use Theorem 4. Since we can embed every tree on $n \leq 2^d$ vertices into G^i_d, this gives an outerplanar graph on $O(n^{\log_2(7)})$ vertices that contains all n-vertex trees as subgraphs. To reach the statement of Theorem 1, it remains to slightly reduce its number of vertices.

Theorem 1. *For every $n \in \mathbb{N}$, there exists an outerplanar graph on $O(n^c)$ vertices, $c = \log_2(3+\sqrt{10}) \approx 2.623$, that contains all n-vertex trees as subgraphs.*

Proof. We build on the constructions G^1_d and G^2_d from Theorem 4 to reach recursive constructions $\tilde{G}^1_d$ and $\tilde{G}^2_d$; see Fig. 5 for an illustration.

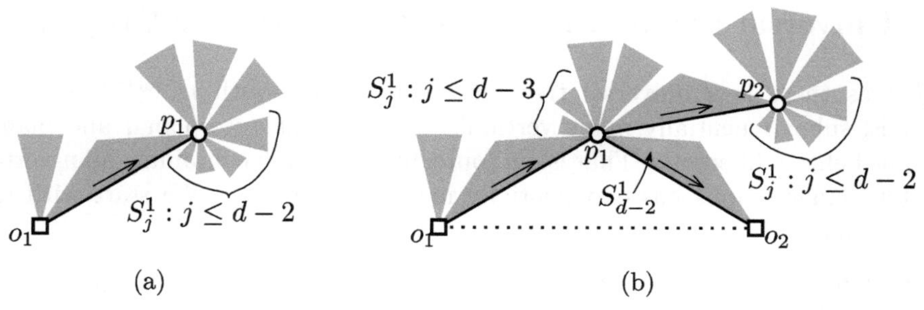

Fig. 5. The final recursive outerplanar constructions (a) $\tilde{G}^1_d$ and (b) $\tilde{G}^2_d$.

Consider now Lemma 3, which we use to embed multiple subtrees incident to one vertex in a tree on at most 2^d vertices. Observe that the sizes of the needed subconstructions actually sum up to at most $2^{d+1} - 1$, because at least the centroid C (or median M) is not part of any subtree. Therefore, with the same cost argument as in the proof of Lemma 3, we can replace one of the copies of G^1_{d-1} incident to p_1 and p_2, each, by one copy of G^1_j for every $j \leq d-2$. Furthermore, within G^2_d in the copy of G^2_{d-1} between p_1 and o_2 there is a copy of G^1_{d-2} incident to p_1 that is unused in all cases; see Fig. 5(b). Hence, we only need to add copies of G^1_j up to $j = d-3$ there. Implementing these changes and setting $a_i(d)$ to be the number of interior vertices of G^i_d, we end up with the system of recurrences

$$a_1(d) = 4a_1(d-1) + a_2(d-1) + 1 + \sum_{j=0}^{d-2} a_1(j)$$

$$a_2(d) = 6a_1(d-1) + 3a_2(d-1) + 2 + a_1(d-2) + 2\sum_{j=0}^{d-3} a_1(j).$$

Solving this system yields the asymptotic bound $O((3+\sqrt{10})^d) = O(n^{\log_2(3+\sqrt{10})})$ as claimed since $d = \log_2(n)$. To see that the constructions are outerplanar, observe that only G^2_d creates one triangle between o_1, o_2, and p_1 and that the remaining part of the construction can be placed outside of this triangle. □

Note that in our construction, every tree can be uniquely reconstructed from its vertex set when embedded following our proof. In particular, every edge of the tree is embedded to $\{o_1, o_2\}$ of some subgraph G^2_d. And the edge $\{o_1, o_2\}$ of some subgraph G^2_d is part of the tree if and only if the vertex p_1 of that G^2_d is not part of the tree.

Corollary 1. *Every tree on $n \leq 2^d$ vertices can be embedded into G^1_d such that it can be uniquely reconstructed by its vertex set.*

3 Universal Planar Graphs for (Outer)Planar Graphs

Whereas there is a universal tree for n-vertex trees with $n^{\theta(\log(n))}$ vertices, that is, subexponentially many vertices, we show that outerplanar and planar graphs behave differently. That is, an (outer)planar graph containing all n-vertex (outer)planar graphs has exponentially many vertices in n. We start with the planar case.

Theorem 2. *Every universal planar graph for all n-vertex planar graphs has $\Omega(n^{-3/2}(27/4)^n)$ vertices.*

Proof. Let $\mathcal{H}$ be the set of all n-vertex 4-connected triangulations and let G be a planar graph which is universal for $\mathcal{H}$. We consider a plane embedding D of G. Tutte [20] showed that the number $k_n = |\mathcal{H}|$ of n-vertex 4-connected triangulations is of the order of $n^{-5/2}(27/4)^n$. Consider two given 4-connected triangulations H_1 and H_2 from $\mathcal{H}$ in G (and their induced embeddings D_1 and D_2) that share vertices. Then either the vertices that they share appear on the outer face of both H_1 and H_2 in D_1 and D_2, or otherwise H_1 has to be contained in a bounded face of H_2 or vice versa (where the vertices of that face can be shared). Indeed, if H_2 has a vertex strictly inside and strictly outside a triangle of H_1 (or vice versa) then it has a cut set of size at most 3. This would be a contradiction to the 4-connectivity of H_2.

We say that $H_1 \leq H_2$ if H_1 is contained in a bounded face of H_2 in D. This gives us a partial order on the set $\mathcal{H}$ which we extend to a linear order $H_1, \ldots, H_{k_n}$ of $\mathcal{H}$. Note that removing all interior vertices of $H_1, \ldots, H_i$ in the embedding of G leaves a graph G' which is universal for $H_{i+1}, \ldots, H_{k_n}$. Hence

$$|V(G)| \geq (n-3) \cdot k_n = \Omega(n^{-3/2}(27/4)^n),$$

which finishes the proof. □

We turn now to outerplanar graphs. We call an outerplanar graph *path-like* if it is a maximal outerplanar graph and there is an embedding in which each triangle has an edge on the outer face. Note that the embedding with this property is unique up to isomorphism. Equivalently, the weak dual of a path-like outerplanar graph is a path. The *weak dual* of a plane graph is a graph whose vertices are the bounded faces of the original graph such that two bounded faces are connected by an edge in the weak dual if they share an edge in the original graph. Let $\mathcal{H}_n$ be the class of all path-like outerplanar graphs on n vertices.

Theorem 5. *Every universal outerplanar graph for all n-vertex outerplanar graphs has at least $\sqrt{2}^{n-5}$ vertices.*

Proof. An ear of a maximal outerplanar graph is a triangle with a vertex of degree 2. Note that each path-like outerplanar graph with $n \geq 4$ vertices has exactly two ears. If $n = 3$, $\mathcal{H}_n$ contains just a single triangle and for $n = 4, 5$ there is also a unique path-like outerplanar graph up to isomorphism (of the embedding). We give a lower bound on the pairs (H, Δ) where $H \in \mathcal{H}_n$ and Δ is an ear of H. Take the unique graph in $\mathcal{H}_5$ and Δ as one of its ears. We think of the outerplanar graph as rooted in Δ. We can recursively glue a triangle to an edge on the outer face of the other ear, that is, the ear that is not Δ. We have at each recursion the choice between two edges. This yields 2^{n-5} choices for $n \geq 6$ and each choice yields a different pair (H, Δ).

We can assume that our universal outerplanar graph G is maximal outerplanar, which means the number of triangles is $v(G) - 2$, where $v(G) = |V(G)|$. By the pigeonhole principle there exists a triangle Δ of G such that there are $2^{n-5}/(v(G) - 2)$ distinct path-like outerplanar subgraphs H for which Δ is an ear in H. Note that there is a bijection between path-like outerplanar graphs that have Δ as an ear and triangles at distance $n - 3$ to Δ (where the distance between triangles is measured as the graph distance in the weak dual, which is a tree for outerplanar graphs). But this means there are $2^{n-5}/(v(G) - 2)$ distinct triangles of distance exactly $n-3$ to Δ, hence $v(G)-2 \geq 2^{n-5}/(v(G)-2)$, which by rearranging proves the theorem. □

We however can construct a polynomially sized planar graph which is universal for $\mathcal{H}_n$ from which we can construct a subexponential sized planar graph for all outerplanar graphs. We give the proof in the full version of this paper [3].

Theorem 6. *There exists a universal planar graph on $n^{O(\log(n))}$ vertices for all n-vertex outerplanar graphs.*

4 Embedding Three Trees Into a Planar Graph

In this section we focus on the converse of Nash-Williams' tree-covering theorem: Given any three trees on n vertices we consider the smallest number of vertices of a planar graph containing them as subgraphs. In particular, we prove Theorem 3.

Recall that a *leaf* of a tree is a vertex of degree 1. A *caterpillar* is a tree such that the subgraph induced on all vertices of degree two or more is a path. This subgraph is called the *spine* of the caterpillar. Let $n, k \in \mathbb{N}$ and $k \mid n$. We define a *k-caterpillar on n vertices $C_{n,k}$* as a caterpillar with k vertices on the spine such that each of them is adjacent to $\frac{n}{k} - 1$ leaves. In particular this means that $|V(C_{n,k})| = n$ and that if $u \in V(C_{n,k})$ is not a leaf, then $|N(u)| \in \{\frac{n}{k}, \frac{n}{k} + 1\}$ for $k > 1$. Furthermore $C_{n,1}$ is a *star*, which is isomorphic to $K_{1,n-1}$.

We show now that there exist three caterpillars on n vertices each such that every planar graph containing all three of them as a subgraph has at least $\frac{3}{2}n - o(n)$ vertices.

Theorem 7. *Fix $1 < k < \ell$ and n large enough. Then every universal planar graph for $C_{n,1}$, $C_{n,k}$, and $C_{n,\ell}$ has size at least*

$$\left(\frac{3}{2} - \frac{1}{2k} - \frac{k}{2\ell}\right) n - 2(k+1) - \frac{\ell}{2}.$$

In particular, there are triples of caterpillars such that every universal planar graph for them has at least $\frac{3}{2}n - O(n^{2/3})$ vertices.

Proof. Let G be a universal planar graph for $C_{n,1}$, $C_{n,k}$, and $C_{n,\ell}$. We are going to count vertices in G that have to be different. Let x be the vertex in G corresponding to the spine vertex of the star $C_{n,1}$ and let L_x be its set of $n-1$ leaves in G. Next consider the spine of $C_{n,k}$. At least $k - 1$ of the corresponding vertices in G are different from x; pick $k-1$ of them if there are k. In $C_{n,k}$ those $k-1$ vertices form at most 2 connected components Y_1 and Y_2. Let L_y^1 and L_y^2 be the sets of vertices in G corresponding to the leaves of Y_1 and Y_2, respectively. Then $L_y := L_y^1 \cup L_y^2$ consists of $(k-1)(\frac{n}{k}-1)$ vertices. Similarly, considering the spine of $C_{n,\ell}$, there are at least $\ell - k$ corresponding vertices in G that are different from x, Y_1, and Y_2. In $C_{n,\ell}$ they form at most $k+1$ connected components $Z_1, \ldots, Z_{k+1}$ with respective leaf sets $L_z^1, \ldots, L_z^{k+1}$ in G. For $L_z := \bigcup_{j=1}^{k+1} L_z^j$ we have $|L_z| = (\ell - k)(\frac{n}{\ell} - 1)$.

Consider next the three sets L_x, L_y^i for $i \in \{1, 2\}$, and L_z^j for $j \in \{1, \ldots, k+1\}$ and assume they have at least 3 vertices a, b, and c in common in G. Then together with the corresponding spine sets x, Y_i, and Z_j, which are pairwise disjoint, they would form a $K_{3,3}$-minor in G, a contradiction to planarity. Hence, each of those triples has at most 2 vertices in common. In particular, summing up the vertices in L_x, L_y, and L_z we count at most $2 \cdot 2 \cdot (k+1)$ vertices three times in G. The remaining vertices are counted at most twice. Consequently G has at least

$$\frac{1}{2}\left((n-1)+(k-1)\left(\frac{n}{k}-1\right)+(\ell-k)\left(\frac{n}{\ell}-1\right)-4(k+1)\right)$$

vertices, which by rearranging yields the claimed bound. Furthermore, choosing $k \approx n^{1/3}$ and $\ell \approx n^{2/3}$ gives the asymptotic result, which finishes the proof. □

This means that the answer to Question 3 posed in [1] is negative, where it was asked whether for every triple of n-vertex caterpillars there is an n-vertex universal planar graph. Moreover, we can show that our asymptotic bound of $\frac{3}{2}n$ vertices is tight, which follows from the fact that trees have book thickness 1. We give the details in the full version of this paper [3].

Proposition 1. *For every triple of trees T_1, T_2 and T_3 on n vertices each, there exists a universal planar graph for $\{T_1, T_2, T_3\}$ on $\frac{3}{2}n$ vertices.*

Furthermore, we can show that already for $C_{48,1}$, $C_{48,2}$, and $C_{48,8}$ there is no planar graph on 48 vertices containing all three as subgraphs. In contrast to that, with computer assistance we found a universal planar graph on 22 vertices for all trees on 22 vertices. We give the details in the full version of this paper [3].

5 Concluding Remarks

In this paper we considered subgraph-universal planar graphs for triples of trees on n vertices and for all trees on n vertices. It would be interesting to investigate the minimum number of vertices of a subgraph-universal planar graph for all possible sets of k trees on n vertices in dependence of k. Studying universal graphs for a constant number of trees might help to understand universal graphs for all n-vertex trees. In particular, we wonder whether there is a universal planar graph for all n-vertex trees that has a quadratic number of vertices.

In some applications so-called induced-universal graphs are of interest, for example in the context of adjacency labeling schemes [17]. Although our construction is not induced-universal, considering Corollary 1 it comes close to being so. We wonder whether the number of vertices of an induced-universal planar graph for all n-vertex trees is also polynomial in n.

Although we have found three trees on n vertices that are not simultaneously contained in any planar graph on n vertices, one might ask how many non-isomorphic trees a planar graph on n vertices can contain. This is similar to [8], where different trees contained in a tree where studied.

Finally recall that trees are K_3-minor-free, outerplanar graphs are K_4-minor-free, and planar graphs are K_5-minor-free. Hence, a natural question extending our study is whether there is a polynomial size K_{t+1}-minor-free universal graph (or more general $K_{f(t)}$-minor-free for some function $f(t)$) for the class of K_t-minor-free graphs.

Acknowledgments. We would like to thank Louis Esperet and Cyril Gavoille for insightful remarks on the paper and Stefan Felsner for helpful discussions and pointing us towards reference [20].

References

1. Alecu, B., Lozin, V.V., Quiroz, D.A., Rabinovich, R., Razgon, I., Zamaraev, V.: The treewidth and pathwidth of graph unions. SIAM J. Disc. Math. **38**(1), 261–276 (2024). https://doi.org/10.1137/22M1524047
2. Babai, L., Chung, F.R.K., Erdős, P., Graham, R.L., Spencer, J.H.: On graphs which contain all sparse graphs. In: Theory and Practice of Combinatorics, North-Holland Mathematics Studies, vol. 60, pp. 21–26. North-Holland (1982).https://doi.org/10.1016/S0304-0208(08)73486-8
3. Bergold, H., Iršič, V., Lauff, R., Orthaber, J., Scheucher, M., Wesolek, A.: Subgraph-universal planar graphs for trees (2024). https://doi.org/10.48550/arXiv.2409.01678
4. Bonamy, M., Gavoille, C., Pilipczuk, M.: Shorter labeling schemes for planar graphs. In: Proceedings of the 31st Annual Symposium on Discrete Algorithms (SODA 2020), pp. 446–462 (2020). https://doi.org/10.1137/1.9781611975994.27
5. Bonamy, M., Gavoille, C., Pilipczuk, M.: Shorter labeling schemes for planar graphs. SIAM J. Disc. Math. **36**(3), 2082–2099 (2022). https://doi.org/10.1137/20M1330464. A conference version of this paper appeared at SODA 2020 [4]
6. Chung, F.R.K., Graham, R.L.: On universal graphs for spanning trees. J. Lond. Math. Soc. **s2-27**(2), 203–211 (1983). https://doi.org/10.1112/jlms/s2-27.2.203
7. Chung, F.R.K., Graham, R.L., Coppersmith, D.: On trees containing all small trees. In: Proceedings of the 4th International Conference on the Theory and Applications of Graphs (1980), pp. 265–272 (1981). https://mathweb.ucsd.edu/~fan/mypaps/fanpap/42trees.pdf
8. Czabarka, É., Székely, L.A., Wagner, S.: On the number of nonisomorphic subtrees of a tree. J. Graph Theory **87**(1), 89–95 (2018). https://doi.org/10.1002/jgt.22144
9. Dujmović, V., Esperet, L., Gavoille, C., Joret, G., Micek, P., Morin, P.: Adjacency labelling for planar graphs (and beyond). In: Proceedings of the 61st Annual Symposium on Foundations of Computer Science (FOCS 2020), pp. 577–588 (2020). https://doi.org/10.1109/FOCS46700.2020.00060
10. Dujmović, V., Esperet, L., Gavoille, C., Joret, G., Micek, P., Morin, P.: Adjacency labelling for planar graphs (and beyond). J. ACM **68**(6), 1–33 (2021). https://doi.org/10.1145/3477542. A conference version of this paper appeared at FOCS 2020 [9]
11. Esperet, L., Joret, G., Morin, P.: Sparse universal graphs for planarity. J. Lond. Math. Soc. **108**(4), 1333–1357 (2023). https://doi.org/10.1112/jlms.12781
12. Gol'dberg, M.K., Livshits, É.M.: On minimal universal trees. Math. Notes Acad. Sci. USSR **4**, 713–717 (1968). https://doi.org/10.1007/BF01116454
13. Huynh, T., Mohar, B., Šámal, R., Thomassen, C., Wood, D.R.: Universality in minor-closed graph classes (2021). https://doi.org/10.48550/arXiv.2109.00327
14. Imrich, W., Klavžar, S.: Product Graphs: Structure and Recognition. Wiley Series in Discrete Mathematics and Optimization, Wiley (2000). https://books.google.si/books?id=EOnuAAAAMAAJ
15. Jordan, C.: Sur les assemblages de lignes. J. für die reine und angewandte Mathematik **70**, 185–190 (1869). https://doi.org/10.1515/crll.1869.70.185

16. Kang, A.N., Ault, D.A.: Some properties of a centroid of a free tree. Inf. Process. Lett. **4**(1), 18–20 (1975). https://doi.org/10.1016/0020-0190(75)90055-1
17. Kannan, S., Naor, M., Rudich, S.: Implicit representation of graphs. SIAM J. Disc. Math. **5**(4), 596–603 (1992). https://doi.org/10.1137/0405049
18. Nash-Williams, C.S.A.: Decomposition of finite graphs into forests. J. Lond. Math. Soc. **s1-39**(1), 12 (1964). https://doi.org/10.1112/jlms/s1-39.1.12
19. Pach, J.: A problem of Ulam on planar graphs. Eur. J. Comb. **2**(4), 357–361 (1981). https://doi.org/10.1016/S0195-6698(81)80043-1
20. Tutte, W.T.: A census of planar triangulations. Can. J. Math. **14**, 21–38 (1962). https://doi.org/10.4153/CJM-1962-002-9
21. Valiant, L.G.: Universal circuits (preliminary report). In: Proceedings of the 8th Annual Symposium on the Theory of Computing (STOC 1976), pp. 196–203 (1976). https://doi.org/10.1145/800113.803649
22. Valiant, L.G.: Universality considerations in VLSI circuits. IEEE Trans. Comput. **C-30**(2), 135–140 (1981). https://doi.org/10.1109/TC.1981.6312176

Subset Feedback Vertex Set Parameterized by Multiway Cut is FPT

Sriram Bhyravarapu[1], Shashanka Kulamarva[2,3], Pritesh Kumar[2](✉), Shivesh K. Roy[2], and Saket Saurabh[2]

[1] Indian Institute of Technology Guwahati, Guwahati, India
sriramb@iitg.ac.in
[2] The Institute of Mathematical Sciences, HBNI, Chennai, India
{priteshk,shiveshkr,saket}@imsc.res.in
[3] Graduate School of Informatics, Kyoto University, Kyoto, Japan
kulamarva.shashanka.3k@kyoto-u.ac.jp

Abstract. In the Subset Feedback Vertex Set problem, we are given an undirected graph G, a subset of vertices $T \subseteq V(G)$ known as *terminals*, and a positive integer k. The objective is to determine whether there exists a set $H \subseteq V(G)$ of size at most k such that $G - H$ does not contain a cycle with a terminal. Subset Feedback Vertex Set generalizes the well-known Feedback Vertex Set problem when $T{:=}V(G)$; in the latter problem, the goal is to hit all cycles in the input graph, ensuring that $G - H$ is acyclic.

It was independently shown that Subset Feedback Vertex Set is fixed-parameter tractable when parameterized by the solution size k [Cygan, Pilipczuk, Pilipczuk, and Wojtaszczyk, SIDMA'13; Kawarabayashi, and Kobayashi, JCTB'12]. In this paper, we study Subset Feedback Vertex Set parameterized by a multiway cut $S \subseteq V(G)$ of the terminal set T, where S is a subset of vertices (potentially containing terminals) such that each connected component of $G - S$ contains at most one terminal. We show that Subset Feedback Vertex Set parameterized by a multiway cut is fixed-parameter tractable. The use of multiway cut as a parameter was recently studied by Jansen and Swennenhuis [ESA'24] in the context of the Steiner Tree problem.

1 Introduction

Feedback Vertex Set is one of the most fundamental problems in graph algorithms [1][2, Footnote 1]. It is one of the 21 NP-complete problems identified in Karp's seminal paper [3]. In this problem, we are given an undirected graph G and a positive integer k, and the goal is to find a set $H \subseteq V(G)$ of size at most k such that $G - H$ does not contain a cycle. The problem has applications in operating systems, database systems, constraint satisfaction problems, and Bayesian inference [4,5]; for example, one of its most important applications is deadlock recovery in operating systems [4]. The Feedback Vertex Set problem has also contributed significantly to the progress of parameterized algorithms [2,4,6–14].

H. Fernau and P. Kindermann (Eds.): WG 2025, LNCS 16124, pp. 76–90, 2026.
https://doi.org/10.1007/978-3-032-11835-6_6

A well-known generalization of Feedback Vertex Set is the Subset Feedback Vertex Set problem, where we are given an undirected graph G, a subset of vertices $T \subseteq V(G)$ called terminals, and a positive integer k. The goal is to find a set $H \subseteq V(G)$ of size at most k such that $G - H$ does not have a cycle that contains a vertex from T, henceforth called a T-*cycle*. It is easy to see that Subset Feedback Vertex Set generalizes Feedback Vertex Set when $T{:=}V(G)$. Moreover, it also generalizes the famous Multiway Cut problem [15]. In the Multiway Cut problem, we are given an undirected graph G, a subset of vertices $T \subseteq V(G)$ called terminals, and a positive integer k. The goal is to find a set $S \subseteq V(G)$ such that $G - S$ does not contain a t_i–t_j path for any $t_i \neq t_j \in T$.

It was posed as an open question by Kawarabayashi and Saurabh in 2009 whether Subset Feedback Vertex Set, parameterized by solution size k, is fixed-parameter tractable (FPT). Using tools from graph minors, Kawarabayashi and Kobayashi [16] obtained an algorithm for Subset Feedback Vertex Set with a running time of $f(k) \cdot \mathcal{O}(n^2 \cdot m)$, where $n = |V(G)|$ and $m = |E(G)|$ (see also [17, Theorem 1.5]). Independently, Cygan et al. [18] combined iterative compression [19, Chapter 4] with Menger's theorem [20] and Gallai's theorem [19, Theorem 9.2] to obtain an FPT algorithm for Subset Feedback Vertex Set running in time $2^{\mathcal{O}(k \cdot \log k)} \cdot n^{\mathcal{O}(1)}$.

Cygan et al. asked (in the conclusion of [18]) whether there exists an algorithm for Subset Feedback Vertex Set with a running time of $2^{\mathcal{O}(k)} \cdot n^{\mathcal{O}(1)}$. Iwata, Wahlström, and Yoshida provided the first single-exponential running time for Subset Feedback Vertex Set, achieving a time complexity of $4^k \cdot n^{\mathcal{O}(1)}$ [21]. It is known that the $2^{\mathcal{O}(k)}$ dependence on the parameter k is optimal under the Exponential Time Hypothesis (ETH) [22]. In the other direction, Lokshtanov, Ramanujan, and Saurabh showed that the dependence on the input size can be made linear [15].

In this paper, we continue the investigation of Subset Feedback Vertex Set parameterized by multiway cut of the terminal set T. Given a graph G and a set of terminals $T \subseteq V(G)$, a set $S \subseteq V(G)$ is called a *multiway cut* of T if $G - S$ does not contain a $t_i - t_j$ path for any $t_i \neq t_j \in T$. The problem is formally defined as follows.

Multiway Subset Feedback Vertex Set (M-SFVS) **Parameter:** $|S|$
Input: A graph G, a set of terminal vertices $T \subseteq V(G)$, a multiway cut $S \subseteq V(G)$ of the terminal set T, and a positive integer k.
Question: Does there exist a set $H \subseteq V(G)$ with $|H| \leq k$ such that no cycle in $G - H$ contains a vertex of T?

The main motivation for studying the above problem stems from the recent work of Jansen and Swennenhuis, who mainly studied the Steiner Tree problem parameterized by a multiway cut S (of the terminal vertices of Steiner Tree) and showed that it can be solved in time $2^{\mathcal{O}(|S| \log |S|)} \cdot n^{\mathcal{O}(1)}$ [23]. In their conclusion, they suggested that exploring "terminal-aware parameterizations" for other problems involving terminals would be an interesting direction for future research. Hence, our study of Subset Feedback Vertex Set parameterized by Multiway Cut can be seen as an effort to extend the investigation of terminal-aware parameterization.

Since we consider the deletable version of multiway cut (which was also studied by Jansen and Swennenhuis), i.e., the size of a minimum multiway cut is at most $|T| - 1$. Moreover, the size of a minimum subset feedback vertex set is at most $|T|$, because the terminal set forms a trivial solution.

However, the sizes of a minimum multiway cut and a minimum subset feedback vertex set can be unrelated. For example, consider a graph G consisting of $\ell > 0$ disjoint T-cycles, each containing exactly one terminal. In this case, the size of the minimum multiway cut is 0, whereas the size of a minimum subset feedback vertex set is ℓ. On the other hand, if G contains a cycle on vertices $v_1, v_2, \ldots, v_{4\ell}$ with $T{:=}\{v_i : i \text{ is odd}\}$, then the size of a minimum multiway cut is at least ℓ, whereas the size of a subset feedback vertex set is 1.

Thus, our parameterization of Multiway Subset Feedback Vertex Set is better in cases where the size of the multiway cut S is smaller compared to the solution size k. An example of an instance (G, T, S, k) satisfying this property is when the removal of the multiway cut S from G results in many connected components containing a T-cycle.

Our Result. The main result of our paper is the following theorem showing that Multiway Subset Feedback Vertex Set is fixed-parameter tractable.

Theorem 1. Multiway Subset Feedback Vertex Set *can be solved in time* $3^{\mathcal{O}(|S|^3)} \cdot n^{\mathcal{O}(1)}$.

Now, we discuss the main ideas behind the proof of Theorem 1. Given an instance (G, T, S, k) of Multiway Subset Feedback Vertex Set, we first bound the number of vertices in each connected component of $G - S$ as a function of $|S|$. To achieve this, we show that if there exists a connected component C of $G - S$ such that the number of vertices in C is more than $2^{|S|} \cdot 3^{|S|^2} \cdot (|S| + 1)$, then we can apply a reduction rule to reduce the size of C. Formally, we prove the following lemma in Sect. 3.

Lemma 1. *There exists an algorithm that, given an instance* (G, T, S, k) *of* Multiway Subset Feedback Vertex Set, *outputs an equivalent instance* (G^*, T, S, k) *such that the number of vertices in each connected component of* $G^* - S$ *is at most* $2^{|S|} \cdot 3^{|S|^2} \cdot (|S| + 1)$. *The running time of the algorithm is* $3^{\mathcal{O}(|S|^2)} \cdot n^{\mathcal{O}(1)}$.

Let $(G := G^*, T, S, k)$ be the reduced instance obtained after application of Lemma 1. Next, we guess a subset of vertices from the multiway cut S, say $S_H \subseteq S$, that belongs to a solution of Multiway Subset Feedback Vertex Set. What remains is to pick a set H of vertices from $G - S$ (that is disjoint from S) such that there is no T-cycle in $G - (S_H \cup H)$. Thus we arrive at the following disjoint version of the problem, which we call Multiway Disjoint SFVS (for convenience of notation, in the definition below we silently assume that $G{:=}G - S_H, T{:=} \mathrm{T} \setminus S_H, S{:=}S \setminus S_H$, and $k{:=}k - |S_H|$).

> MULTIWAY DISJOINT SFVS **Parameter:** $|S|$
> **Input:** An instance (G, T, S, k) of MULTIWAY SUBSET FEEDBACK VERTEX SET.
> **Question:** Does there exist a solution to (G, T, S, k) that is disjoint from S?

To obtain a solution to this problem, we design a dynamic programming based algorithm presented in Sect. 4. Formally, we prove the following lemma in Sect. 4.

Lemma 2. *There exists an algorithm that, given an instance* (G, T, S, k) *of* MULTIWAY DISJOINT SFVS, *where* $\mathcal{C}$ *is the set of connected components of* $G - S$, *solves* MULTIWAY DISJOINT SFVS *in time* $\mathcal{O}(3^{|S|^2} \cdot c^{|S|} \cdot n^{\mathcal{O}(1)})$, *where* $c = \max_{C \in \mathcal{C}} |V(C)|$ *is the size of the largest connected component in* $\mathcal{C}$.

Once we have the above two lemmas, a simple combination of them leads to a proof of Theorem 1, which we present in Sect. 5.

Organization. We begin with a brief overview of non-standard preliminaries, including crucial definitions, in Sect. 2. We prove Lemma 1 and Lemma 2 in Sects. 3 and 4, respectively. Finally, we prove Theorem 1 in Sect. 5. The basic preliminaries and proofs of statements marked with ($\bigstar$) are located in the full version.

2 Preliminaries

We use standard graph-theoretic notation, and for any undefined notation, we ask the reader to consult the book by Diestel [20]. The basic set of preliminaries for the paper is given in full version.

For a subset $T \subseteq V(G)$, we call a cycle C in G a *T-cycle* if $V(C) \cap T \neq \emptyset$. Similarly, we call a path P a *T-path* if $V(P) \cap T \neq \emptyset$. For a path $P = x_1, x_2, \cdots, x_q$ and two vertices x_i, x_j on the path P, we use $P[x_i, x_j]$ to denote the $x_i - x_j$ subpath of P. Furthermore, we use $P(x_i, x_j)$ to denote $x_{i+1} - x_{j-1}$ subpath of P. For $\ell \in \mathbb{N}$, we use $[\ell]$ to denote the set $\{1, 2, \cdots, \ell\}$.

A problem is said to be *fixed-parameter tractable* (FPT) with respect to a parameter k if it can be solved in time $f(k) \cdot n^{\mathcal{O}(1)}$, where f is a computable function that depends only on k. For further details on parameterized algorithms, we refer to the book by Cygan et al. [19].

Problem-Specific Preliminaries. An instance (G, T, S, k) of MULTIWAY SUBSET FEEDBACK VERTEX SET is called a YES-instance if there exists a set $H \subseteq V(G)$ with $|H| \leq k$ such that $G - H$ does not contain a T-cycle. Similarly, an instance (G, T, S, k) of MULTIWAY DISJOINT SFVS is a YES-instance if there exists a set $H \subseteq V(G) \setminus S$ with $|H| \leq k$ such that $G - H$ does not contain a T-cycle. Two instances (G, T, S, k) and (G', T', S', k') of MULTIWAY SUBSET FEEDBACK VERTEX SET are equivalent if (G, T, S, k) is a *YES*-instance if and only if (G', T', S', k') is a *YES*-instance.

In the reduction given in Sect. 3, we utilize the following known algorithm for the SUBSET FEEDBACK VERTEX SET problem, parameterized by the solution size k.

Theorem 2 ([21,24]). *There exists an algorithm for* SUBSET FEEDBACK VERTEX SET *that runs in time* $4^k \cdot n^{\mathcal{O}(1)}$. *Moreover, the algorithm also outputs a solution of size at most k if it is a YES-instance.*

Let G be a graph with terminal set $T \subseteq V(G)$. Let $S \subseteq V(G)$ be a multiway cut of the terminal set T, and let $\mathcal{C} = \{C_1, \cdots, C_\ell\}$ be the set of connected components of $G - S$. For an $i \in [\ell]$, we denote the graph induced by the vertex set $S \cup V(C_1) \cup \cdots \cup V(C_i)$ by G_i.

Next, we show that if C is a connected component of $G-S$ then any minimum solution of MULTIWAY SUBSET FEEDBACK VERTEX SET contains at most $|S|+1$ vertices from the set $S \cup V(C)$.

Proposition 1 (★). *Let* (G, T, S, k) *be an instance of* MULTIWAY SUBSET FEEDBACK VERTEX SET, *and let* $H \subseteq V(G)$ *be a minimum-sized solution of* (G, T, S, k). *Then, for any connected component* C *of* $G - S$, *it holds that* $|H \cap (S \cup V(C))| \leq |S| + 1$.

For a set $S \subseteq V(G)$, we define *configurations* that help us guess whether there exists a terminal path, a non-terminal path, or no path between each pair of vertices in S from the "outside". These configurations are represented by a function that assigns a value from the set $\{R, B, O\}$ to each pair of vertices in S, where R denotes a terminal path, B a non-terminal path, and O indicates no path.

Definition 1 (Configuration on S). *Given a set* $S \subseteq V(G)$, *a configuration on* S *is a function* $f\colon S \times S \to \{R, B, O\}$ *such that* $f(s_i, s_j) = f(s_j, s_i)$ *for all* $s_i, s_j \in S$. *We denote the set of all configuration functions by* $\mathcal{F}$.

We now define the notion of a configuration f induced by a graph G on a vertex set $S \subseteq V(G)$ as follows.

Definition 2 (Induced configuration). *Let* G *be a graph with a set of terminal vertices* $T \subseteq V(G)$ *such that there is no* T*-cycle in* G. *Let* $S \subseteq V(G)$ *be a subset of vertices. We say that* $f\colon S \times S \to \{R, B, O\}$ *is a configuration* induced *by* G *on* S *if the following conditions are satisfied for every pair of vertices* $s_i, s_j \in S$:

(a) $f(s_i, s_j) = R$ *if there exists a* s_i–s_j *path, say* P, *in* $G' := G - (S \setminus \{s_i, s_j\})$ *with* $(V(P) \setminus \{s_i, s_j\}) \cap T \neq \emptyset$,
(b) $f(s_i, s_j) = B$ *if Condition (a) is not satisfied for the pair* s_i, s_j *and there exists a* s_i–s_j *path in* G', *and*
(c) $f(s_i, s_j) = O$, *if neither Condition (a) nor Condition (b) is satisfied for the pair* s_i, s_j.

The following definition encodes the path information captured by a configuration $f\colon S \times S \to \{R, B, O\}$ using the notion of an auxiliary graph on the vertex set S.

Definition 3 (Construction of an auxiliary graph). *Given an instance (G, T, S, k) of* MULTIWAY SUBSET FEEDBACK VERTEX SET *with the set $S = \{s_1, \cdots, s_\ell\}$ and a configuration $f\colon S \times S \to \{R, B, O\}$, we construct an auxiliary graph $A_G(f, S)$ as follows. We begin with $A_G(f, S) := G[S]$ and $T_f := \emptyset$. Then, we add all the non-terminal path information to the auxiliary graph (if it is not already captured) as follows.*

(i) For $i, j \in [\ell]$ with $i \neq j$, if $f(s_i, s_j) = B$ and adding the edge $\{s_i, s_j\}$ in $A_G(f, S)$ does not form a cycle, we add edge $\{s_i, s_j\}$ in $A_G(f, S)$, and we call edge $\{s_i, s_j\}$ a B-edge.

Next, we capture all the terminal path information in the auxiliary graph.

(ii) For $i, j \in [\ell]$ with $i \neq j$, if $f(s_i, s_j) = R$. If there does not exist a s_i–s_j path in $A_G(f, S)$, then introduce a new vertex t_{ij} and add edges $\{s_i, t_{ij}\}$ and $\{t_{ij}, s_j\}$ in $A_G(f, S)$. We also add vertex t_{ij} in the set T_f. We call $s_i - t_{ij} - s_j$ as an auxiliary segment.

This completes the construction of $A_G(f, S)$ (observe that during the construction, we have not created any $(T \cup T_f)$-cycle in $A_G(f, S)$ if $G[S]$ is acyclic).

The auxiliary graph $A_G(f, S)$ serves as a compact representation of the connectivity among vertices in the set S for a given configuration function f. This abstraction allows us to analyze connectivity properties of any subgraph $\widetilde{G}$ (on the set S) that induces the same configuration f, by studying its corresponding auxiliary graph. The following lemma formalizes the relationship between the connectivity in such subgraphs and the corresponding auxiliary graph.

Lemma 3 (★). *Let (G, T, S, k) be an instance of* MULTIWAY SUBSET FEEDBACK VERTEX SET, *and let $\widetilde{G}$ be a subgraph of G that contains the terminal set $\widetilde{T} := T \cap V(\widetilde{G})$ and satisfies $S \subseteq V(\widetilde{G})$, with no $\widetilde{T}$-cycles in $\widetilde{G}$. Let f be the configuration induced by $\widetilde{G}$ on S, as defined in Definition 2. Let $A_G(f, S)$ be the auxiliary graph defined by Definition 3. Then for every pair of vertices $s_i, s_j \in S$, the following properties hold:*

1. *there exists an s_i–s_j path in $\widetilde{G}$ if and only if there is an s_i–s_j path in $A_G(f, S)$, and*
2. *there exists an s_i–s_j path in $\widetilde{G}$ that contains a vertex of $\widetilde{T}$ if and only if there is an s_i–s_j path in $A_G(f, S)$ that contains a vertex of $T_f \cup (T \cap S)$, where T_f is the set of auxiliary terminals returned by Definition 3.*

3 Bounding Size of connected components

In this section, we show that there exists an FPT algorithm, parameterized by the size of the multiway cut, which, given an instance (G, T, S, k) of Multiway Subset Feedback Vertex Set, outputs an equivalent instance in which each connected component of the graph obtained after removing the multiway cut is bounded. Formally, we prove Lemma 1.

Lemma 1. *There exists an algorithm that, given an instance* (G, T, S, k) *of* Multiway Subset Feedback Vertex Set, *outputs an equivalent instance* (G^*, T, S, k) *such that the number of vertices in each connected component of* $G^* - S$ *is at most* $2^{|S|} \cdot 3^{|S|^2} \cdot (|S| + 1)$. *The running time of the algorithm is* $3^{\mathcal{O}(|S|^2)} \cdot n^{\mathcal{O}(1)}$.

Proof. Given an instance (G, T, S, k) of Multiway Subset Feedback Vertex Set, we reduce the size of connected components of $G - S$ using a marking-scheme-based reduction procedure, which we discuss below.

Algorithm

1. While there exists a connected component C of $G - S$ such that $|V(C)| > 2^{|S|} \cdot 3^{|S|^2} \cdot (|S| + 1)$.
   (a) We begin with the marked set $M_c := S \cup (V(C) \cap T)$.
   (c) For every subset $X_i \subseteq S$ of size at most k, we obtain an instance $(G' := G - X_i, T' := T \setminus X_i, S' := S \setminus X_i, k' := k - |X_i|)$ of Multiway Subset Feedback Vertex Set.
      i. For every configuration $f_j \colon S' \times S' \to \{R, B, O\}$.
         - We obtain a graph $G_{c,i,j} := G[S' \cup V(C)] \cup A_{G'}(f_j, S')$, where $A_{G'}(f, S')$ is the auxiliary graph induced by configuration f_j on the set S' corresponding to the instance (G', T', S', k') according to Definition 3.
         - Let $T_{c,i,j} := (T \cap V(G_{c,i,j})) \cup T_f$, where T_f is the set of auxiliary terminals added during the construction of $A_{G'}(f_j, S')$.
         - By invoking Theorem 2, we compute the smallest nonnegative integer ℓ^* (using binary search) such that $(G_{c,i,j}, T_{c,i,j}, \ell^*)$ is a *YES*-instance of Subset Feedback Vertex Set.
         - Let $H_{c,i,j} \subseteq V(G_{c,i,j})$ be a solution for $(G_{c,i,j}, T_{c,i,j}, \ell^*)$.
         - Let $M_c := M_c \cup H_{c,i,j}$.

   (c) Let $U := V(C) \setminus M_c$ be the set of unmarked vertices of the connected component C.
   (d) Let $E_P := \emptyset$. For every pair of vertices u, $v \in M_c$ if there is a $u - v$ path say P_{uv} such that $V(P_{uv}(u, v)) \subseteq U$, i.e., all the internal vertices of P_{uv} belong to U, then we add an edge $\{u, v\}$ to E_P (note that we allow $u = v$ and hence we may get a self-loop).
   (e) We construct a reduced multi-graph $\widehat{G}$ with $V(\widehat{G}) := V(G) \setminus U$ and $E(\widehat{G}) := E(G - U) \cup E_P$. We update $G := \widehat{G}$.

2. Let $G^* := G$ be the reduced graph after termination of the while loop of Step 1. We return (G^*, T, S, k).

This concludes the description of the algorithm.

Analysis

We first show that after the above algorithm is applied, each connected component of $G^* - S$ has at most $2^{|S|} \cdot 3^{|S|^2} \cdot (|S| + 1)$ vertices. To see this, consider a fixed connected component C of $G - S$ with $V(C) > 2^{|S|} \cdot 3^{|S|^2} \cdot (|S| + 1)$. Since C satisfies the condition of the while loop of Step 1, it follows that for every guess $X_i \subseteq S$ of size at most k and for every configuration f_j, we computed the set $H_{c,i,j}$ and added it to the set M_c in Step 1(b)i.

Observe that the size of the set $H_{c,i,j}$ is bounded by $(|S|+1)$. This is because there exists a trivial solution of size at most $|S| + 1$ that selects all vertices in S and the terminal vertex of C (if it exists) for the Subset Feedback Vertex Set instance $(G_{c,i,j}, T_{c,i,j}, |S| + 1)$. Since the algorithm computes the minimum non-negative integer ℓ^* such that $(G_{c,i,j}, T_{c,i,j}, \ell^*)$ is a *YES*-instance, we have $\ell^* \leq |S| + 1$.

Since $|H_{c,i,j}| \leq |S| + 1$ for a fixed guess $X_i \subseteq S$ and for a fixed configuration f_j, the number of distinct guesses for X_i (of size at most k) is at most $2^{|S|}$, and the number of distinct valid configurations is at most $3^{|S|^2}$. Therefore, it follows that $|M_c \cap V(C)| \leq 2^{|S|} \cdot 3^{|S|^2} \cdot (|S| + 1)$. Consequently, for a connected component C whose size exceeds $2^{|S|} \cdot 3^{|S|^2} \cdot (|S|+1)$, we retain only the marked vertices $M_c \cap V(C)$, ensuring the desired bound on the connected components. It is easy to verify that the time taken for one iteration of the while loop is at most $2^{|S|} \cdot 3^{|S|^2} \cdot 4^{|S|} \cdot n^{\mathcal{O}(1)}$. Since the number of connected components of $G - S$ is at most n, and in each iteration of the while loop we bound the size of one component, the number of times the algorithm enters the while loop is also bounded by n. Hence, the overall running time of the algorithm is $3^{\mathcal{O}(|S|^2)} \cdot n^{\mathcal{O}(1)}$.

It remains to show that the instance (G^*, T, S, k) that we obtained is an equivalent instance. To establish this, we prove the following claim, which states that if we apply a single iteration of the while loop, we obtain an equivalent instance of Multiway Subset Feedback Vertex Set.

Claim 1. *Let (G, T, S, k) be an instance of* Multiway Subset Feedback Vertex Set *such that there exists a connected component C of $G - S$ with $|V(C)| > 2^{|S|} \cdot 3^{|S|^2} \cdot (|S| + 1)$ and the above algorithm chose the connected component C in the while loop of Step 1. Let M_c be the set of marked vertices computed by the algorithm in Step 1(b)i and $\widehat{G}$ be the reduced instance obtained by the algorithm in Step 1e. Then (G, T, S, k) is a* YES-*instance of* Multiway Subset Feedback Vertex Set *if and only if $(\widehat{G}, T, S, k)$ is a* YES-*instance.*

Proof. In the forward direction assume that (G, T, S, k) is a *YES*-instance of Multiway Subset Feedback Vertex Set. Let $H \subseteq V(G)$ be a solution of (G, T, S, k) such that $H \cap M_c$ is maximum. We first show that $H \subseteq V(\widehat{G})$. Towards this assume for a contradiction, i.e., there is a vertex $u \in U \cap H$

where $U = V(C) \setminus M_c$ is the set of unmarked vertices of the connected component C. Next we obtain a set $H^* \subseteq V(G)$ such that H^* is a solution of (G, T, S, k) and H^* contains strictly more vertices from the set M_c than the set H, yielding a contradiction to the maximality of $H \cap M_c$.

Towards this, let $G_H := G - (V(C) \cup H)$, $S_H := S \setminus H$, $X_H := S \cap H$, and $H_c := H \cap V(C)$. Since H is a solution of (G, T, S, k) and $G_H \subseteq G - H$, there does not exit a T-cycle in G_H. Moreover by definition of G_H and S_H, we have $S_H \subseteq V(G_H)$. Let $f_H : S_H \times S_H \to \{R, B, O\}$ be the configuration induced by G_H on the vertex set S_H. Since vertex $u \in U$ of the connected component C is unmarked, implying that for connected component C, $X_i := X_H$, and $f_j := f_H$, in Step 1(b)i we marked a smallest set $H_{c,i,j} \subseteq V(G_{c,i,j})$ such that $|H_{c,i,j}| \leq |H_c|$ and $H_{c,i,j} \subseteq M_c$. We obtain $H^* := (H \setminus H_c) \cup H_{c,i,j}$.

Next we show that H^* is a solution of (G, T, S, k). Assume for a contradiction that there is a T-cycle, say R, in $G - H^*$. First note that R can not be completely contained inside the connected components of $C - H^*$ (i.e., $V(R) \not\subseteq V(C) \setminus H^*$), since $H_{c,i,j} \subseteq H^*$ is a Subset Feedback Vertex Set of $(G_{c,i,j}, T_{c,i,j}, |H_{c,i,j}|)$ and $C \subseteq G_{c,i,j}$. Moreover R can not be completely contained in $G - (V(C) \cup H^*)$, because $(H \setminus H_c) \subseteq H^*$ and H is a Multiway Subset Feedback Vertex Set of (G, T, S, k).

Hence R contains at least one vertex from $V(C) \setminus H^*$ and at least one vertex from $V(G) \setminus (V(C) \cup H^*)$. Moreover, it is easy to note that $V(R) \cap S \neq \emptyset$, because a path from a vertex of $V(C) \setminus H^*$ to a vertex of $V(G) \setminus (V(C) \cup H^*)$ has to pass through S.

Next, we show that there is a $T_{c,i,j}$-cycle in $G_{c,i,j} - H_{c,i,j}$, yielding a contradiction to the fact that $H_{c,i,j}$ is a Subset Feedback Vertex Set of $(G_{c,i,j}, T_{c,i,j}, |H_{c,i,j}|)$. Towards this, we decompose the T-cycle R of $G - H^*$ as follows:

$$R = s_1 \cdot R_1 \cdot s_2 \cdot R_2 \cdot s_3 \cdot \ldots \cdot s_{w-1} \cdot R_{w-1} \cdot s_w = s_1$$

where:

- for each $i \in [w]$, vertex $s_i \in S$,
- for each $\ell \in [w - 1]$, the subpath R_ℓ is contained in $G - S$ (note that some of R_ℓ may be empty).

Let $\mathcal{R}_o := \{R_i : R_i$ is contained in a connected component apart formC$\}$. Observe that for each $R_i \in \mathcal{R}_o$, since there is a s_i–s_{i+1} path $s_i \cdot R_i \cdot s_{i+1}$ in $G - (V(C) \cup S \setminus \{s_i, s_{i+1}\})$ such that all the internal vertices $V(R_i)$ belong to $G - (C \cup S \cup H)$, as $H^* \setminus V(C) = H \setminus V(C)$.

Since f_H is the configuration induced by $G_H := G - (V(C) \cup H)$, for each $R_i \in \mathcal{R}o$, the path $s_i \cdot R_i \cdot si + 1$ witnesses that $f(s_i, s_{i+1}) = B/R$ depending on whether there exist a terminal on the path R_i. Hence due to Definiton 3, during the construction of the auxiliary graph $A_G(f_H, S_H)$, we ensure that $A_G(f_H, S_H)$ either contain $\{s_i, s_{i+1}\}$ as a B-edge (when $f(s_i, s_{i+1}) = B$) or an suxiliary segment segment $s_i - t_{i,i+1} - s_{i+1}$. Hence the replacemnt of all such $R_i \in \mathcal{R}o$

yields a $T_{c,i,j}$-cycle in $Gc, i, j - H_{c,i,j}$, a contradiction to the fact that $H_{c,i,j}$ is a SUBSET FEEDBACK VERTEX SET of $(G_{c,i,j}, T_{c,i,j}, |H_{c,i,j}|)$.

Hence from now on we assume that $H \subseteq V(\widehat{G})$. Next, we show that H is also a solution of $(\widehat{G}, T, S, k)$. Towards this, assume for a contradiction that H is not a solution and hence there exists a T-cycle, say R, in $\widehat{G} - H$.

Note that, by the construction of the graph $\widehat{G}$ in Step 1e and the fact that R is a T-cycle in $\widehat{G} - H$, the cycle R must contain at least one edge from the set E_P. If R does not contain any edge from E_P, then we would have $R \subseteq G - H$, which contradicts the assumption that H is a solution to (G, T, S, k). Consider R be the smallest such cycle.

Consider replacing each edge $\{u, v\} \in E_P \cap E(R)$ with a corresponding $u - v$ path $P_{u,v}$ in $G[U]$. We claim that for any two distinct edges $\{u, v\}, \{x, y\} \in E_P$ in R, the vertex sets of their corresponding paths are disjoint, i.e., $V(P_{u,v}) \cap V(P_{x,y}) = \emptyset$. Suppose for contradiction that $V(P_{u,v}) \cap V(P_{x,y}) \neq \emptyset$. Then, there exist paths between the pairs (u, x), (u, y), (v, x), and (v, y) in $G[U]$. By the construction in Step 1d, this implies the existence of the edges $\{u, x\}$, $\{u, y\}$, $\{v, x\}$, $\{v, y\}$ belong to E_P. Adding any of these edges to R would yield a shorter T-cycle containing at least one edge from E_P, contradicting the minimality of R.

Replacing each edge $\{u, v\} \in E_P \cap E(R)$ with the corresponding path $P_{u,v}$, we obtain a new cycle W with the following vertex and edge sets:

$$V(W) = V(R) \cup \bigcup_{\{u,v\} \in E_P \cap E(R)} V(P_{u,v}),$$

$$E(W) = (E(R) \setminus E_P) \cup \bigcup_{\{u,v\} \in E_P \cap E(R)} E(P_{u,v}).$$

Since R was a T-cycle and $V(R) \subseteq V(W)$, it follows that W is also a T-cycle. We can obtain a $T-$cycle in $G - H$, a contradiction to the fact that H is a solution of (G, T, S, k).

In the reverse direction assume that $(\widehat{G}, T, S, k)$ is a *YES*-instance and let $H \subseteq V(\widehat{G})$ be a solution. We claim that H is also a solution of (G, T, S, k). Towards this assume for a contradiction that H is not a solution of (G, T, S, k), implying that there is a T-cycle R in $G - H$. Observe that the cycle R must contain at least one vertex from the unmarked set U. Moreover, since there is no terminal in the set U, cycle R also contains at least one vertex outside U. Let $u, v \in V(C) \cap V(M_c)$ such that the subpath $R(u, v)$ is contained inside $G[U]$. Note that by construction of the set E_P and graph $\widehat{G}$, we have $\{u, v\} \in E(\widehat{G})$. Hence replacing all such occurrence of all such subpaths $R(u, v)$ by the edge $\{u, v\}$ results in a T-cycle $\hat{R}$ in $\widehat{G} - H$, a contradiction to the fact that H is a solution of $(\widehat{G}, T, S, k)$. This concludes the proof of the claim and the correctness of the algorithm. □

This concludes the proof of Lemma 1. □

4 Dynamic Programming on a connected component

In this section we prove Lemma 2. Recall that S is a multiway cut of G. First, we look at some definitions required to solve the problem.

Definition 4 (Compatible Configurations). *Given two configurations f_a and f_b on S, we say f_a is* compatible *with f_b if the following conditions hold:*

- $\mathcal{R}_{f_a} \subseteq \mathcal{R}_{f_b}$, *where* $\mathcal{R}_{f_b} = \{\{u,v\} \mid u,v \in S$ *and* $f_b(u,v) = R\}$ *and* $\mathcal{R}_{f_a} = \{\{u,v\} \mid u,v \in S$ *and* $f_a(u,v) = R\}$, *and*
- $\mathcal{B}_{f_a} \subseteq \mathcal{B}_{f_b}$, *where* $\mathcal{B}_{f_b} = \{\{u,v\} \mid u,v \in S$ *and* $f_b(u,v) = B\}$ *and* $\mathcal{B}_{f_a} = \{\{u,v\} \mid u,v \in S$ *and* $f_a(u,v) = B\}$.

Definition 5. *Given a component $C \in \mathcal{C}$, two configurations $f_a, f_b \in \mathcal{F}$ on S, and a set $X \subseteq C$, we say (f_a, X)* respects *f_b if the following conditions hold:*

- *f_a is compatible with f_b,*
- $|X| \leq |S| + 1$,
- *$G[V(A_G(f_a,S)) \cup (C \setminus X)]$ induces f_b, where $A_G(f_a,S)$ is the auxiliary graph constructed given the configuration f_a (see Definition 3), and*
- *$G[V(A_G(f_a,S)) \cup (C \setminus X)]$ does not contain a T-cycle.*

Let Z be a subgraph of G that does not contain a T-cycle, and $S \subseteq V(G)$ be a set of vertices of Z. Let $f \in \mathcal{F}$ be a configuration on S such that Z induces f and $A_G(f,S)$ is the corresponding auxiliary graph obtained according to Definition 3. Now we show attaching a graph Q to the subgraph Z has the same property as attaching Q to $A_G(f,S)$ in terms of the solution size of the problem. We state the result formally.

Lemma 4 (★). *Let (G,T,S,k) be an instance of* Multiway Disjoint SFVS *and $\mathcal{C} = \{C_1, C_2, \ldots, C_\ell\}$ be the connected components of $G - S$. Let $\hat{\mathcal{C}} \subseteq \mathcal{C}$ and G_1 be the subgraph induced by $S \cup V(\hat{\mathcal{C}})$. Let $Q = G - V(\hat{\mathcal{C}})$ and Z be a subgraph of G_1 such that Z does not contain a T-cycle.*

Let f be the induced configuration *for the graph Z and $A_G(f,S)$ be the auxiliary graph constructed according to Definition 3. Let T_f be the set of newly introduced terminal vertices in $A_G(f,S)$. Define $T_Z = T \cap V(Z)$ and $T_Q = T \cap V(Q)$.*

A set H_Q is a solution to the instance $(Z \cup Q, T_Z \cup T_Q, S, k)$ of Multiway Disjoint SFVS *if and only if H_Q is also a solution to the instance $(Q \cup A_G(f,S), T_Z \cup T_Q \cup T_f, S \cup T_f, k)$ of* Multiway Disjoint SFVS.

Lemma 2. *There exists an algorithm that, given an instance (G,T,S,k) of* Multiway Disjoint SFVS, *where $\mathcal{C}$ is the set of connected components of $G - S$, solves* Multiway Disjoint SFVS *in time $\mathcal{O}(3^{|S|^2} \cdot c^{|S|} \cdot n^{\mathcal{O}(1)})$, where $c = \max_{C \in \mathcal{C}} |V(C)|$ is the size of the largest connected component in $\mathcal{C}$.*

Proof. We give a dynamic programming based algorithm to show the lemma. Let $\mathcal{C} = \{C_1, C_2, \ldots, C_\ell\}$ denote the set of components of $G - S$. Let G_i be the graph induced by the vertex set $S \cup \{C_1, C_2, \ldots, C_i\}$. The components of $\mathcal{C}$ are processed sequentially starting from C_1.

For a given component C_i and a configuration f_b on S, let $h[C_i, f_b]$ represent the minimum size of a set $D \subseteq V(G_i - S)$ such that the following conditions are satisfied:

1. there is no T-cycle in $G_i - D$,
2. the graph $G_i - D$ induces the configuration f_b on S (see Definition 2).

Base Case: We consider the graph induced on S. Let C_0 be an empty graph. We initialize $h[C_0, f_a] = 0$ and $h[C_0, f'_a] = +\infty$ for all $f'_a \in \mathcal{F} \setminus \{f_a\}$. Here f_a is the configuration on S such that $A_G(f, S) = G[S]$ (refer to the Definition 3 for the auxiliary graph).

Recursive Step: Consider the component C_i, where $i \geq 1$. For each configuration $f_b \in \mathcal{F}$, the entry $h[C_i, f_b]$ is computed as follows.

$$h[C_i, f_b] = \begin{cases} \min\limits_{f_a \in \mathcal{F}, X \subseteq C_i} \Big(h[C_{i-1}, f_a] + |X|\Big), & \text{if } (f_a, X) \text{ respects } f_b; \\ +\infty, & \text{otherwise.} \end{cases}$$

Answer: Finally, after processing all components $C_1, C_2, \ldots, C_\ell$, the solution to the problem is obtained by looking at the entries in the table corresponding to the component C_ℓ. Let $c = \min_{f_b \in \mathcal{F}} h[C_\ell, f_b]$. If $c \leq k$, then we conclude that there is a solution of size k for the given instance of Multiway Disjoint SFVS. Otherwise, we conclude that the given instance is a no-instance of Multiway Disjoint SFVS.

To prove the correctness of the above algorithm, we have the following claim. Its proof is deferred to the full version due to space constraints.

Claim (★). *The recurrence step is correct.*

We now present the running time of the algorithm. Since the number of configurations is $3^{|S|^2}$ and the number of components is at most n, the number of entries in a table is $3^{|S|^2} n$. To compute the value of $h[\cdot]$, we go over all the subsets of the component of size at most $|S| + 1$ and check whether (f_a, X) respects f_b which can be checked in polynomial time. Since the number of subsets of a component C is at most $|C|^{|S|+1}$, we get the desired running time. □

5 Proof of Theorem 1

Theorem 1. Multiway Subset Feedback Vertex Set *can be solved in time* $3^{\mathcal{O}(|S|^3)} \cdot n^{\mathcal{O}(1)}$.

Proof. Given an instance (G, T, S, k) of Multiway Subset Feedback Vertex Set, we first apply Lemma 1, which generates a reduced instance (G^*, T, S, k) of Multiway Subset Feedback Vertex Set. In this instance, the number of vertices in each of the connected components of $G^* - S$ is bounded by $(|S| + 1) \cdot 2^{|S|} \cdot 3^{|S|^2}$.

Now given the instance (G^*, T, S, k) of Multiway Subset Feedback Vertex Set we guess the set of vertices from S that belong to a solution of Multiway Subset Feedback Vertex Set. Let us focus on a set $S_H \subseteq S$ that belongs to the solution. Notice that if there is a solution H for $(G^* - S_H, T \setminus S_H, S \setminus S_H, k - |S_H|)$ of Multiway Subset Feedback Vertex Set whose intersection with S is empty then $H \cup S_H$ is a solution for (G^*, T, S, k) of Multiway Subset Feedback Vertex Set. To incorporate this guess, we remove S_H from G^*, update T to $T \setminus S_H$, update S to $S \setminus S_H$, and reduce k by $|S_H|$, thereby obtaining a new instance (G^*, T, S, k) of Multiway Disjoint SFVS. We then apply Lemma 2 to this instance. If, for any choice of S_H, Lemma 2 returns a solution of size at most k, we conclude that the original instance is a YES-instance of Multiway Subset Feedback Vertex Set.

The correctness of the algorithm follows from Lemma 1 and Lemma 2. Due to space constraints, the running time analysis of the algorithm is presented in the full version. □

6 Conclusion

In this paper, we looked at the terminal-aware parameterization for the Subset Feedback Vertex Set problem. This parameterization was recently studied by Jansen and Swennenhuis [23]. We show that Multiway Subset Feedback Vertex Set admits a fixed-parameter algorithm with running time $3^{\mathcal{O}(|S|^3)} \cdot n^{\mathcal{O}(1)}$. Our algorithm is based on a marking-scheme-based reduction rule and then the application of dynamic programming on the reduced instance. Because of the guesses on the number of different configurations, our algorithm inherently has the $3^{|S|^2}$ factor hidden. This raises the question of whether we can design an algorithm for Multiway Subset Feedback Vertex Set with running time $2^{\mathcal{O}(|S| \log |S|)} \cdot n^{\mathcal{O}(1)}$ or even $c^{\mathcal{O}(|S|)} \cdot n^{\mathcal{O}(1)}$ for a fixed constant $c > 0$. Note that such an algorithm exists for the Subset Feedback Vertex Set problem when parameterized by the solution size k.

Another direction for future work lies in further studying terminal-aware parameterization for other problems that rely on terminals.

References

1. Festa, P., Pardalos, P.M., Resende, M.G.C.: Feedback set problems. In: Floudas, C.A., Pardalos, P.M. (eds.) Encyclopedia of Optimization, Second Edition, pp. 1005–1016. Springer, Heidelberg (2009)

2. Cao, Y.: A naive algorithm for feedback vertex set. In: Seidel, R. (ed.) 1st Symposium on Simplicity in Algorithms, SOSA 2018, New Orleans, LA, USA, 7–10 January 2018, vol. 61 of OASIcs, pp. 1:1–1:9. Schloss Dagstuhl - Leibniz-Zentrum für Informatik (2018)
3. Karp, R.M.: Reducibility among combinatorial problems. In: Miller, R.E., Thatcher, J.W. (eds.) Proceedings of a symposium on the Complexity of Computer Computations, held March 20-22, 1972, at the IBM Thomas J. Watson Research Center, Yorktown Heights, New York, USA, The IBM Research Symposia Series, pp. 85–103. Plenum Press, New York (1972)
4. Chen, J., Fomin, F.V., Liu, Y., Songjian, L., Villanger, Y.: Improved algorithms for feedback vertex set problems. J. Comput. Syst. Sci. **74**(7), 1188–1198 (2008)
5. Bar-Yehuda, R., Geiger, D., Naor, J., Roth, R.M.: Approximation algorithms for the feedback vertex set problem with applications to constraint satisfaction and bayesian inference. SIAM J. Comput. **27**(4), 942–959 (1998)
6. Guo, J., Gramm, J., Hüffner, F., Niedermeier, R., Wernicke, S.: Compression-based fixed-parameter algorithms for feedback vertex set and edge bipartization. J. Comput. Syst. Sci. **72**(8), 1386–1396 (2006)
7. Raman, V., Saurabh, S., Subramanian, C.R.: Faster fixed parameter tractable algorithms for finding feedback vertex sets. ACM Trans. Algor. **2**(3), 403–415 (2006)
8. Bodlaender, H.L., van Dijk, T.C.: A cubic kernel for feedback vertex set and loop cutset. Theory Comput. Syst. **46**(3), 566–597 (2010)
9. Agrawal, A., Lokshtanov, D., Misra, P., Saurabh, S., Zehavi, M.: Feedback vertex set inspired kernel for chordal vertex deletion. ACM Trans. Algor. **15**(1), 11:1–11:28 (2019)
10. Thomassé, S.: A $4k^2$ kernel for feedback vertex set. ACM Trans. Algor. **6**(2), 32:1–32:8 (2010)
11. Galby, E., Khazaliya, L., Mc Inerney, F., Sharma, R., Tale, P.: Metric dimension parameterized by feedback vertex set and other structural parameters. SIAM J. Disc. Math. **37**(4), 2241–2264 (2023)
12. Jansen, B.M.P., Bodlaender, H.L.: Vertex cover kernelization revisited - upper and lower bounds for a refined parameter. Theory Comput. Syst. **53**(2), 263–299 (2013)
13. Cygan, M., Nederlof, J., Pilipczuk, M., Pilipczuk, M., van Rooij, J.M.M., Wojtaszczyk, J.O.: Solving connectivity problems parameterized by treewidth in single exponential time. ACM Trans. Algor. **18**(2), 17:1–17:31 (2022)
14. Li, J., Nederlof, J.: Detecting feedback vertex sets of size k in $O^\star$ (2.7^k) time. ACM Trans. Algorithms **18**(4), 34:1–34:26 (2022)
15. Lokshtanov, D., Ramanujan, M.S., Saurabh, S.: Linear time parameterized algorithms for subset feedback vertex set. ACM Trans. Algor. **14**(1), 7:1–7:37 (2018)
16. Kawarabayashi, K., Kobayashi, Y.: Fixed-parameter tractability for the subset feedback set problem and the S-cycle packing problem. J. Comb. Theory B **102**(4), 1020–1034 (2012)
17. Kakimura, N., Kawarabayashi, K., Kobayashi, Y.: Erdös-Pósa property and its algorithmic applications: parity constraints, subset feedback set, and subset packing. In: Rabani, Y. (ed.) Proceedings of the Twenty-Third Annual ACM-SIAM Symposium on Discrete Algorithms, SODA 2012, Kyoto, Japan, 17–19 January 2012, pp. 1726–1736. SIAM (2012)
18. Cygan, M., Pilipczuk, M., Pilipczuk, M., Wojtaszczyk, J.O.: Subset feedback vertex set is fixed-parameter tractable. SIAM J. Disc. Math. **27**(1), 290–309 (2013)
19. Cygan, M., et al.: Parameterized Algorithms. Springer, Marcin Pilipczuk (2015)
20. Diestel, R.: Graph Theory, 5th edn, vol. 173 of Graduate texts in mathematics. Springer, Heidelberg (2017)

21. Iwata, Y., Wahlström, M., Yoshida, Y.: Half-integrality, LP-branching, and FPT algorithms. SIAM J. Comput. **45**(4), 1377–1411 (2016)
22. Impagliazzo, R., Paturi, R., Zane, F.: Which problems have strongly exponential complexity? J. Comput. Syst. Sci. **63**(4), 512–530 (2001)
23. Jansen, B.M.P., Swennenhuis, C.M.F.: Steiner tree parameterized by multiway cut and even less. In: Chan, T.M., Fischer, J., Iacono, J., Herman, G. (eds.) 32nd Annual European Symposium on Algorithms, ESA 2024, Royal Holloway, London, United Kingdom, 2–4 September 2024, vol. 308 of LIPIcs, pp. 76:1–76:16. Schloss Dagstuhl - Leibniz-Zentrum für Informatik (2024)
24. Cygan, M., Pilipczuk, M., Pilipczuk, M.: On group feedback vertex set parameterized by the size of the cutset. Algorithmica **74**(2), 630–642 (2016)

Chromatic Index Under Parameterized Settings

Sriram Bhyravarapu[1], Soumen Mandal[3(✉)], Ashutosh Rai[3], Saket Saurabh[2,4], and Shaily Verma[5]

[1] Indian Institute of Technology Guwahati, Guwahati, India
sriramb@iitg.ac.in

[2] The Institute of Mathematical Sciences, HBNI, Chennai, India
saket@imsc.res.in

[3] Indian Institute of Technology Delhi, Delhi, India
{soumen.mandal,ashutosh.rai}@maths.iitd.ac.in

[4] University of Bergen, Bergen, Norway

[5] Hasso Plattner Institute, Potsdam, Germany
Shaily.Verma@hpi.de

Abstract. Edge coloring is one of the most fundamental graph problems. A *proper edge coloring* of a graph is an assignment of colors to the edges of the graph such that no two adjacent edges get the same color and the minimum number of colors required for a proper edge coloring of a graph is called as *chromatic index* of the graph. Vizing's theorem says that the chromatic index of a graph is either Δ or $\Delta + 1$, where Δ is the maximum degree of the graph. Given a graph G, the Chromatic Index problem asks if the chromatic index of G is Δ. This problem is known to be NP-complete even for 3-regular graphs, which makes this problem unlikely to be in FPT (unless P = NP) when parameterized by $\Delta(G)$ or by the clique number of the graph. In this paper, we initiate the study of the parameterized complexity of the problem.

We design an FPT algorithm using dynamic programming on a given (nice) tree decomposition of the graph with respect to the treewidth parameter. This algorithm matches the running time of the best known algorithm for the problem in partial k-trees (graphs with treewidth at most k) given by X. Zhou et.al. [J. Algorithms, 1996], which uses a parallel algorithm for Chromatic Index as a subroutine. This result follows that the Chromatic Index problem is FPT for chordal graphs, when parameterized by the clique number of the graph. Moreover, we show that the problem does not admit a polynomial kernel when parameterized by the *treewidth* parameter unless PH collapses. On the positive side, we design a quadratic kernel parameterized by the size of a given *feedback vertex set*. We also design a polynomial kernel for the problem with respect to the parameters cluster vertex deletion number and clique number.

Keywords: Chromatic index · Treewidth · Kernelization · Feedback vertex set

H. Fernau and P. Kindermann (Eds.): WG 2025, LNCS 16124, pp. 91–104, 2026.
https://doi.org/10.1007/978-3-032-11835-6_7

1 Introduction

Edge coloring is a fundamental problem in graph theory that has many applications in various scheduling tasks [1,9,22]. An edge coloring of a graph is an assignment of colors to the edges such that no two adjacent edges have the same color. The minimum number of colors required to color a graph G is known as the *chromatic index* of the graph and is denoted by $\chi'(G)$. A graph G is called *k-edge colorable* if there exists an edge coloring of G using k colors. Note that edge coloring of any graph G with maximum degree $\Delta(G)$ needs at least $\Delta(G)$ colors, i.e., $\chi'(G) \geq \Delta(G)$. In 1964, Vizing [20] proved that any graph G is $(\Delta+1)$-edge colorable. It implies that the chromatic index of a graph G is equal to $\Delta(G)$ or $\Delta(G)+1$. A graph G is said to be in *class 1*, if $\chi'(G) = \Delta$ and in *class 2*, otherwise. The problem of deciding if a graph is in class 1, known as CHROMATIC INDEX, is NP-complete even for cubic graphs (graphs with maximum degree 3) [14].

CHROMATIC INDEX
Input: A graph $G = (V, E)$.
Question: Is $\chi'(G) = \Delta(G)$?

CHROMATIC INDEX is one such problem that is widely popular and actively studied, but still, we do not know much about the problem compared to its counterpart, vertex coloring. In 2022, Figueiredo et al. [12] revised the summary table (consisting of 11 problems) given in Garey and Johnson's famous "Guide to the Theory of NP-completeness" [13]. The article points out that CHROMATIC INDEX is by far the hardest problem from the table, with the largest number of open entries remaining open even for classes considered "easy" as, for instance, cographs or split graphs. Despite a long list of papers [2,3,7,8,10,19], we still do not know the complete classification for split graphs. The complexity status of the problem is still open in chordal graphs or its important subgraph classes like interval graphs, where only partial results are known [6].

It is well known that bipartite graphs are in class 1 [16]. The planar graphs with the maximum degree of at least 7 are in class 1 [18,21]. On the other hand, there are some planar graphs known with maximum degree in $\{2, 3, 4, 5\}$ that are in class 2. The status of planar graphs with maximum degree 6 is still unknown.

In 1990, Hans. Bodlaender [4] study the CHROMATIC INDEX problem in partial k-trees, which are graphs with treewidth bounded by k. The paper gives an algorithm for the problem with running time $\mathcal{O}(n(\Delta+1)^{2^{(k+1)}})$, which is an FPT algorithm with the treewidth and maximum degree of the graph. Later in 1996, Zhou et al. [23] give an algorithm for the problem in partial k-trees with running time $\mathcal{O}((6k)^{k(k+1)/2}kn)$, which is an FPT algorithm parameterized by treewidth alone. In order to design this algorithm, they proved the following important lemma.

Lemma 1. ([23]). *If a partial k-tree graph H has the maximum degree $\Delta(H) \geq 2k$, then $\chi'(H) = \Delta(H)$.*

Our Contribution. In this paper, we study the CHROMATIC INDEX problem under parameterized settings. In particular, we prove the following results:

- In this paper, we improve on Bodleander's algorithm [4] and design an algorithm for CHROMATIC INDEX problem in graphs of treewidth at most k that runs in time $\Delta^{\mathcal{O}(k^2+\Delta k)} \cdot n^{\mathcal{O}(1)}$. Because of Lemma 1, if we consider $\Delta(G) < 2k$, the running time of our algorithm matches with the running time of the algorithm of Zhou et al. [23]. However, our algorithm might be useful to improve the known algorithms for different variants of edge coloring problems like *strong edge coloring* or *distance-edge coloring* problem on partial k-trees [15,17], which runs in double exponential-time in Δ and k similar to Bodlaender's algorithm. Furthermore, this result also implies that for chordal graphs, the CHROMATIC INDEX problem, when parameterized by clique number, is fixed-parameter tractable.
  Recall that the CHROMATIC INDEX problem is NP-hard even for 3-regular graphs [14]. This implies that the CHROMATIC INDEX problem is para-NP-hard (does not admit an XP-algorithm) when parameterized by the maximum degree, $\Delta(G)$, or the clique number $\omega(G)$. Additionally, the complexity status of the problem is still unknown for chordal graphs. All this makes the above implication worth mentioning.
- Next, we initiate the kernelization study of CHROMATIC INDEX problem and show that there does not exist a polynomial kernel for CHROMATIC INDEX when parameterized by the treewidth of the graph. To prove this, we use the cross-composition technique.
- On the positive side, we design a quadratic kernel for the problem when parameterized by the *feedback vertex set number* of the graph, where feedback vertex set number is the minimum size of any vertex set whose removal from the graph results in a forest.
- We also design a polynomial kernel for the CHROMATIC INDEX problem when parameterized by the cluster vertex deletion number and the clique number of the given graph. Note that we cannot expect a polynomial kernel by the clique number alone as CHROMATIC INDEX parameterized by clique number is para-NP-hard. However, designing a (polynomial) kernel with respect to the cluster vertex deletion number alone is an interesting open question.

Organization. First, we discuss some basic definitions, notations, and results we need in Sect. 2. In Sect. 3, we study the CHROMATIC INDEX problem parameterized by the treewidth of the graph, design an alternate FPT algorithm, and show that the problem does not admit a polynomial kernel. In Sect. 4, we design a quadratic kernel for the problem parameterized by feedback vertex set number. Furthermore, in Sect. 5, we design a polynomial kernel for the problem when parameterized by the combined parameters, cluster vertex deletion number, and clique number of the graph.

2 Preliminaries

Throughout this paper, we only consider simple graphs. For a graph $G = (V, E)$, we denote the set of vertices by $V(G)$ and the set of edges by $E(G)$. We denote $|V(G)|$ by n, where the graph is clear from the context. For a set $S \subseteq V(G)$, the subgraph of G *induced* by S is denoted by $G[S]$ and defined as the subgraph of G with the vertex set S and the edge set $\{uv \in E(G) : u, v \in S\}$. The subgraph obtained after deleting S (and the edges incident to the vertices in S) is denoted as $G - S$. For any graph $G = (V, E)$ and $v \in V$, a vertex $u \in V$ is said to be a neighbour of v if $vu \in E(G)$. The open neighbourhood of v is denoted by $N(v)$ and it is defined as $N(v) = \{u \in V : vu \in E(G)\}$. We denote the degree of a vertex v by $deg(v)$ where $deg(v) = |N(v)|$. We denote the maximum degree of a vertex in G by $\Delta(G)$. Sometimes we denote the maximum degree as Δ only when the graph is clear from the context. The *clique number* of a graph is the size of the largest clique in the graph and it is denoted by ω. A graph G is called *cluster graph* if each component of G is a clique. A set of vertices M is called *cluster vertex deletion set* if $G - M$ is a cluster graph and the minimum size of any cluster vertex deletion set is called *cluster vertex deletion (cvd) number.* A set of vertices F is said to be a *feedback vertex set* of a graph G if $G - F$ is acyclic. The minimum size of any feedback vertex set is called *feedback vertex set (fvs) number* of the graph. A graph G is said to be *chordal graph* if for every cycle of four or more vertices of G there is an edge that connects two nonadjacent vertices. For any set S, we denote its power set by $\mathscr{P}(S)$.

Definition 1 (Tree decomposition [11]**).** *A tree decomposition of a graph G is a pair $\mathcal{T} = \left(T, \{X_t\}_{t \in V(T)}\right)$, where T is a tree where every node t is assigned a vertex subset $X_t \subseteq V(G)$, called a bag, such that the following three conditions hold:*

- $\bigcup_{t \in V(T)} X_t = V(G)$. *In other words, every vertex of G is in at least one bag,*
- *For every $uv \in E(G)$, there exists a node t of T such that bag X_t contains both u and v, and*
- *For every $u \in V(G)$, the set $T_u = \{t \in V(T) : u \in X_t\}$, i.e., the set of nodes whose corresponding bags contain u, induces a connected sub-tree of T.*

The width of tree decomposition $\mathcal{T} = \left(T, \{X_t\}_{t \in V(T)}\right)$ equals $\max_{t \in V(T)} |X_t| - 1$, that is, the maximum size of its bag minus 1. The treewidth of a graph G, denoted by $\text{tw}(G)$, is the minimum possible width of a tree decomposition of G.

To compute an edge coloring of a graph G, we consider a special tree decomposition of G called *nice tree decomposition.* In a nice tree decomposition, we consider T to be a rooted tree with root r, and other vertices are identified as a leaf node, vertex introduce node, forget node, or join node. Here is a formal definition of a nice tree decomposition:

Definition 2 (Nice tree decomposition). *A rooted tree decomposition* $\left(T, \{X_t\}_{t \in V(T)}\right)$ *is* nice *if the following conditions are satisfied:*

- $X_r = \emptyset$ *and* $X_\ell = \emptyset$ *for every leaf* ℓ *of* T. *In other words, all the leaves as well as the root contain empty bags.*
- *Every non-leaf node of* T *is of one of the following three types:*
  1. *Introduce node: a node* t *with exactly one child* t' *such that* $X_t = X_{t'} \cup \{v\}$ *for some vertex* $v \notin X_{t'}$*; we say that* v *is introduced at* t.
  2. *Forget node: a node* t *with exactly one child* t' *such that* $X_t = X_{t'} \setminus \{w\}$ *for some vertex* $w \in X_{t'}$*; we say that* w *is forgotten at* t.
  3. *Join node: a node* t *with two children* t_1, t_2 *such that* $X_t = X_{t_1} = X_{t_2}$.

3 Parameterized by Treewidth

In this section, we present an FPT algorithm for Chromatic Index parameterized by treewidth.

Theorem 1 *For a given graph* G *and a tree decomposition of width* k, Chromatic Index *can be solved in time* $2^{\mathcal{O}(k^2 \log k)} n^{\mathcal{O}(1)}$.

We design a dynamic programming based algorithm on the given nice tree decomposition $\mathcal{T} = (T, \{X_t\}_{t \in V(T)})$ of the input graph G of width k. We may also assume that no bag is empty and that the leaf bag has exactly one vertex.

For each node $t \in V(T)$, let X_t represent a bag, which is a set of vertices, in a nice tree decomposition of G. We define two functions f_t and g_t at the bag X_t where $f_t : E(G[X_t]) \rightarrow \{1, 2, \ldots, \Delta\}$, and $g_t : X_t \rightarrow \mathcal{P}\{1, 2, \ldots, \Delta\}$. In some sense, the function f_t captures the coloring of the edges in the bag X_t, and the function g_t captures the set of colors used for the edges incident to each vertex in the subtree rooted at the bag X_t. We denote the graph induced by the vertices in the bags corresponding to the node t and all of its descendants by G_t.

We now define the table entry. The entry $T[t, f_t, g_t]$ indicates the existence of a proper edge coloring of $E(G_t)$ satisfying the functions f_t and g_t. We set $T[t, f_t, g_t]$ =TRUE, if there is a proper edge coloring $h : E(G_t) \rightarrow [\Delta]$ of the edges of the subtree rooted at the node t such that

- the set of colors used for the edges incident to a vertex $u \in X_t$ is $g_t(u)$, i.e., for each $u \in X_t$, $g_t(u) = \{h(uv) \mid v \in V(G_t) \text{ and } uv \in E(G_t)\}$, and
- the coloring of the edges of $G[X_t]$ is given by f_t, i.e., $f_t = \{h(uv) \mid u, v \in V(X_t) \text{ and } uv \in E(G[X_t])\}$.

Otherwise, we set $T[t, f_t, g_t]$ =FALSE. We say f_t *respects* g_t if for each edge $uv \in E(X_t)$, we have $f_t(uv) \in g_t(u)$ and $f_t(uv) \in g_t(v)$. Otherwise, we say f_t does not respect g_t.

We now show how to compute the table at each node in a nice tree decomposition.

- **Leaf node.**
  Let t be a leaf node such that $X_t = \{v\}$. We assign the entry $T[t, f_t, g_t]$ to TRUE if and only if $g_t(v) = \emptyset$ and f_t is a null function.
- **Introduce node.**
  Let t be an introduce node with child t' such that $X_t = X_{t'} \cup \{v\}$. For each f_t and g_t, we assign $T[t, f_t, g_t]$ to be FALSE if f_t is not a proper edge coloring of $G[X_t]$ or f_t does not respect g_t. Otherwise, the entry $T[t, f_t, g_t]$ is computed as follows:
  $$T[t, f_t, g_t] = \bigvee_{(f_{t'}, g_{t'}) \text{ is i-good}} T[t', f_{t'}, g_{t'}]$$
  We say a tuple $(f_{t'}, g_{t'})$ is i-good if the following are satisfied:
  - $f_{t'}(e) = f_t(e)$ for all edges e in $G[X_{t'}]$, and
  - for all $u \in X_{t'}$ if $uv \in E(G[X_t])$ then $g_t(u) = g_{t'}(u) \cup \{f_t(uv)\}$ and $f_t(uv) \notin g_{t'}(u)$, otherwise $g_t(u) = g_{t'}(u)$.

  We go over each of the entries $T[t', f_{t'}, g_{t'}]$ at the node t' where $(f_{t'}, g_{t'})$ is i-good and check if $T[t', f_{t'}, g_{t'}] =$ TRUE. If such an entry exists, we set $T[t, f_t, g_t]$ to TRUE and FALSE otherwise.
- **Forget node.**
  Suppose t is a forget node with child node t' such that $X_t = X_{t'} \setminus \{w\}$ for some $w \in X_{t'}$. For each f_t and g_t, we assign $T[t, f_t, g_t]$ to be FALSE if f_t is not a proper edge coloring of $G[X_t]$ or f_t does not respect g_t. Otherwise, the entry $T[t, f_t, g_t]$ is computed as follows:
  $$T[t, f_t, g_t] = \bigvee_{(f_{t'}, g_{t'}) \text{ is f-good}} T[t', f_{t'}, g_{t'}].$$
  We say a tuple $(f_{t'}, g_{t'})$ is f-good if $f_t = f_{t'}|_{X_{t'} \setminus \{w\}}$ and $g_t = g_{t'}|_{X_{t'} \setminus \{w\}}$. We go over each of the entries $T[t', f_{t'}, g_{t'}]$ at the node t' where $(f_{t'}, g_{t'})$ is f-good and check if $T[t', f_{t'}, g_{t'}] =$ TRUE. If such an entry exists, we set $T[t, f_t, g_t]$ to TRUE and FALSE otherwise.
- **Join node.**
  Let t be a join node with two children t_1 and t_2 such that $X_t = X_{t_1} = X_{t_2}$. For each f_t and g_t, we assign $T[t, f_t, g_t]$ to be FALSE if f_t is not a proper edge coloring of $G[X_t]$ or f_t does not respect g_t. Otherwise, the entry $T[t, f_t, g_t]$ is computed as follows:
  $$T[t, f_t, g_t] = \bigvee_{(f_{t_1}, g_{t_1}), (f_{t_2}, g_{t_2}) \text{ compatible with } (f_t, g_t)} (T[t_1, f_{t_1}, g_{t_1}] \wedge T[t_2, f_{t_2}, g_{t_2}]).$$
  We say two pairs (f_{t_1}, g_{t_1}) and (f_{t_2}, g_{t_2}) are compatible with (f_t, g_t) if the following holds:
  - $f_t(e) = f_{t_1}(e) = f_{t_2}(e)$, $\forall e \in E(G[X_t])$.
  - $g_t(v) = g_{t_1}(v) \cup g_{t_2}(v)$, $\forall v \in X_t$, and
  - $g_{t_1}(v) \cap g_{t_2}(v) = \{f_t(uv) : u \in N(v) \cap X_t\}$, $\forall v \in X_t$.

  For each entry $T[t_1, f_{t_1}, g_{t_1}]$ we go over each entry $T[t_2, f_{t_2}, g_{t_2}]$ and check if (f_{t_1}, g_{t_1}) and (f_{t_2}, g_{t_2}) are compatible with (f_t, g_t), and $T[t_1, f_{t_1}, g_{t_1}] = T[t_2, f_{t_2}, g_{t_2}] =$ TRUE. If such an entry exists, we set $T[t, f_t, g_t]$ to TRUE and FALSE otherwise.

Lemma 2. *For a given graph G and a tree decomposition of width k,* CHROMATIC INDEX *can be solved in time $\Delta^{\frac{3k(k-1)}{2}} \cdot 2^{3\Delta k} \cdot n^{\mathcal{O}(1)}$.*

Proof. For a given tree decomposition $\mathcal{T} = (T, \{X_t\}_{t \in V(T)})$ of width k for graph G, we can obtain a nice tree decomposition for G of width k in polynomial time. Thus we assume that $\mathcal{T} = (T, \{X_t\}_{t \in V(T)})$ is a nice tree decomposition of G of width k. Then by the recursion described above, we compute $T[t, f_t, g_t]$ correctly for every possible $f_t : E[X_t] \to \{1, 2, \ldots, \Delta\}$ and every $g_t : X_t \to \mathscr{P}(\{1, 2, \ldots, \Delta\})$ in each bag X_t. Here, $T[t, f_t, g_t] = \mathsf{TRUE}$ means that there is a proper edge coloring for graph G_t that satisfies f_t and g_t. We first start with a leaf node and then go up in the tree T. Let r be the root of T, then we know that $X_r = \emptyset$ and $G_r = G$. So, at the root node, if $T[\emptyset, \emptyset, \emptyset] = \mathsf{TRUE}$, then we say that the input graph G has a proper edge coloring with Δ colors, that is, we say that G is in class 1. Otherwise, we say that G belongs to class 2.

In each bag X_t, we compute $T[t, f_t, g_t]$ for all possible f_t and g_t. Note that there can be at most $\Delta^{\frac{k(k-1)}{2}}$ many such f_t and $2^{\Delta k}$ many g_t. In an introduce node of the tree T, to compute an entry $T[t, f_t, g_t]$, we spend $\Delta^{k(k-1)/2} \cdot 2^{\Delta k}$. So, in total, we need $\Delta^{k(k-1)} \cdot 2^{2\Delta k}$ time in an introduced node of the tree. In a very similar way, we need $\Delta^{k(k-1)} \cdot 2^{2\Delta k}$ time in each of the forget nodes of the tree. For each join node, to compute an entry $T[t, f_t, g_t]$, we need $\Delta^{k(k-1)} \cdot 2^{2\Delta k}$ time. Overall in a join node, we require $\Delta^{\frac{3k(k-1)}{2}} \cdot 2^{3\Delta k}$ time. There are $n^{\mathcal{O}(1)}$ nodes in T. So in total our dynamic programming runs in $\Delta^{\frac{3k(k-1)}{2}} \cdot 2^{3\Delta k} \cdot n^{\mathcal{O}(1)}$ time. □

Proof (Proof of Theorem 1). Since the class of graphs with treewidth at most k is equivalent to that of partial k-trees, from Lemma 1, we can say that $\chi'(G) = \Delta(G)$ when $\Delta(G) \geq 2k$. Now, if $\Delta(G) < 2k$, then Theorem 2 runs in $2^{\mathcal{O}(k^2 logk)} \cdot n^{\mathcal{O}(1)}$ time. Since the maximum degree of a graph can be found in polynomial time, we can decide whether $\Delta(G) \geq 2k$ in polynomial time. This proves the Theorem 1. □

Note that chordal graphs admits a clique-tree decomposition, that is a tree decomposition where each bag is a clique. Also, the treewidth of a chordal graph is at most the clique number of the graph. Therefore, Theorem 1 implies the following result.

Corollary 1. *Let G be a chordal graph and ω be the clique number of G. Then* CHROMATIC INDEX *can be solved in time $2^{\mathcal{O}(\omega^2 \log \omega)} n^{\mathcal{O}(1)}$.*

3.1 No Polynomial Kernel

In this subsection, we prove that CHROMATIC INDEX does not admit a polynomial kernel. To prove this, first we need the following definitions and result.

Definition 3 (Polynomial equivalence relation [5]**).** *An equivalence relation R on Σ^* is called a polynomial equivalence relation if the following two conditions hold:*

1. *There is an algorithm that given two strings $x, y \in \Sigma^*$ decides whether x and y belong to the same equivalence class in $(|x| + |y|)^{O(1)}$ time.*
2. *For any finite set $S \subseteq \Sigma^*$ the equivalence relation R partitions the elements of S into at most $(\max_{x \in S} |x|)^{O(1)}$ classes.*

Definition 4 (Cross Composition [5]**).** *Let $L \subseteq \Sigma^*$ be a set and let $Q \subseteq \Sigma^* \times N$ be a parameterized problem. We say that L cross-composes into Q if there is a polynomial equivalence relation R and an algorithm which, given t strings $x_1, x_2, \ldots, x_t$ belonging to the same equivalence class of R, computes an instance $(x^*, k^*) \in \Sigma^* \times N$ in time polynomial in $\sum_{i=1}^{t} |x_i|$ such that:*

1. *$(x^*, k^*) \in Q \iff x_i \in L$ for some $1 \leq i \leq t$,*
2. *k^* is bounded by a polynomial in $\max_{i=1}^{t} |x_i| + \log t$.*

Theorem 2 ([5])**.** *If some problem L is* NP*-hard under Karp reductions and L cross-composes into the parameterized problem Q then there is no polynomial kernel for Q unless* NP $\subseteq$ coNP*/poly.*

We will show a cross-composition algorithm for Chromatic Index problem parameterized by treewidth. It leads to the following result, by Theorem 2.

Theorem 3. Chromatic Index *parameterized by treewidth does not admit a polynomial kernel, unless* NP $\subseteq$ coNP*/poly.*

Proof. We define the polynomial equivalence relation R such that two instances (G_1, k_1) and (G_2, k_2) of Chromatic Index belong to R if and only if the number of vertices in G_1 is equal to the number of vertices in G_2, the treewidth of both instances is the same ($k_1 = k_2$), and the maximum degree in G_1 and G_2 is the same ($\Delta(G_1) = \Delta(G_2)$). Thus we assume that we are given with t instances of Chromatic Index from R. Let the instances be $\{(G_i, k) \mid i \in [t]\}$. Let G' be the disjoint union of the graphs $G_1, G_2, \ldots, G_t$. We show that (G', k) is a Yes-instance of Chromatic Index if and only if each of the instances (G_i, k), for all $i \in [t]$, is a Yes-instance of Chromatic Index.

We consider the Chromatic Index problem on graphs with maximum degree Δ which is NP-hard as 3-edge colorability of 3-regular graphs is NP-hard [14]. For cross-composition algorithm, let the sequence $(G_1, G_2, \ldots, G_t)$ be the given input, where G_i for all $i \in [t]$, is a graph with maximum degree Δ, we output (G', k), where G' is the disjoint union of the t graphs in the input and k is the treewidth of graph G'. Next, we only need to prove that the instance (G', k) is a Yes instance if and only if each G_i is Δ-edge colorable, where $i \in [t]$.

For the forward direction, if (G', k) is a Yes-instance. Let f be a proper edge coloring of G' using $\Delta_{G'}$ colors and k be the treewidth of the graph. Note that $\Delta(G') = \Delta$. Note that the coloring f restricted on G_i is a proper edge coloring using Δ colors, for each $i \in [t]$. For the backward direction, let each G_i be Δ-edge colorable. It follows that the graph G' is also Δ-edge colorable.

Hence, the result follows by Theorem 2. □

4 Kernel Parameterized by fvs Number

In this section, we show a quadratic kernel for Chromatic Index parameterized by the feedback vertex set number of the input graph. Note that the Chromatic Index problem parameterized by the feedback vertex set number takes the input (G, k), where G is the input graph and k is the fvs number of G and it asks whether $\chi'(G) = \Delta(G)$. We prove the following result.

Theorem 4. Chromatic Index *admits a kernel with $\mathcal{O}(k^2)$ vertices and edges, where k is the feedback vertex set number.*

To design a kernel, we use Lemma 1 to bound the maximum degree as a function of the size of feedback vertex set. It is known that $\mathsf{tw}(G) \leq \mathsf{fvs}(G) + 1$, for any graph G where $\mathsf{tw}(G)$ and $\mathsf{fvs}(G)$ represents the treewidth and feedback vertex set number of the graph G. Since the class of graphs with treewidth at most k is equivalent to that of partial k-trees, from Lemma 1, we get the following.

Corollary 2. *Let G be a graph. If $\Delta(G) \geq 2(1 + \mathsf{fvs}(G))$, then $\chi'(G) = \Delta(G)$.*

To obtain the desired kernel, we design some reduction rules and it is easy to observe that the first reduction rule is safe.

Reduction Rule 1. *If u is an isolated vertex in G, remove the vertex u.*

Reduction Rule 2. *For any $uv \in E(G)$, if $deg_G(u) + deg_G(v) \leq \Delta(G) + 1$ and none of u,v is a unique maximum degree vertex, then delete the edge uv from G.*

Lemma 3. *Reduction Rule 2 is safe.*

Proof. To prove the safeness of Reduction Rule 2, we need to show that (G, k) is a `Yes` instance if and only if $(G - uv, k)$ is a `Yes` instance where $G - uv$ is obtained after applying Reduction Rule 2. Since neither u nor v is a unique maximum degree vertex, the maximum degrees of G and $G - uv$ are the same. Let us take, $\Delta = \Delta(G - uv) = \Delta(G)$.

The forward direction is easy. If we can properly color the edges of G with Δ colors, then the same coloring restricted to graph $G - uv$ is a proper edge coloring with at most Δ colors as $G - uv$ is a proper subgraph of G.

For the reverse direction, let $G' = G - uv$ and f be an edge coloring of the graph G' with Δ colors. Since $deg_G(u) + deg_G(v) \leq \Delta + 1$ and uv is an edge, $deg_{G'}(u) + 1 + deg_{G'}(v) + 1 \leq \Delta + 1$. It implies that at most $\Delta - 1$ edges are incident to the vertices u and v in G'. Therefore, there exists some color c that is not used to color any edge incident on u or v in the coloring f of G'. By assigning the color c to the edge uv, we can extend the proper edge coloring of G' to a proper edge coloring of G with Δ colors. Thus, (G, k) is also a `Yes` instance. □

Reduction Rule 3. *For any $uv \in E(G)$, if $deg_G(u) + deg_G(v) \leq \Delta(G) + 1$ and u is the unique maximum degree vertex, then return* `Yes`.

Lemma 4. *Reduction Rule 3 is safe.*

Proof. Since u is the unique maximum degree vertex then $d_{G-uv}(u) = \Delta(G - uv) = \Delta(G) - 1$. Also, since $deg_G(u) + deg_G(v) \leq \Delta(G) + 1$, we must have $d(v) = 1$ in G. We know that $\chi'(G - uv) \leq \Delta(G - uv) + 1 = \Delta(G)$. Now if we color the edges of $G - uv$ properly with $\Delta(G)$ colors, then there must be a color c from the set of $\Delta(G)$ colors which has not been used to color any edge incident to u in $G - uv$. Thus if we give an edge coloring f of G that uses same proper edge coloring of $G - uv$ with $\Delta(G)$ colors and we assign $f(uv) = c$, then f will be a proper edge coloring for G that uses same number of colors (i.e., $\Delta(G)$) as the proper edge coloring of $G - uv$. □

Notice that for any edge $uv \in E(G)$, if $deg_G(u) + deg_G(v) \leq \Delta(G) + 1$, then one of Reduction Rule 2 or Reduction Rule 3 would be applicable. Let F be a feedback vertex set of G of size k. Next, we show that after an exhaustive application of Reduction Rule 2 and Reduction Rule 3, the number of leaves and vertices with degree at least three in graph $G - F$ gets bounded.

Lemma 5. *Let G be a graph and F be a feedback vertex set of size k in G. Let L be the set of leaves and $S_{\geq 3}$ be the set of vertices with degree at least 3 in $G - F$. If Reduction Rule 2 and Reduction Rule 3 are not applicable to G and $\Delta(G) < 2(k+1)$, then the following statements are true:*

(i) $|L| = \mathcal{O}(k^2)$
(ii) $|S_{\geq 3}| = \mathcal{O}(k^2)$

Proof. Let it be the case that Reduction Rule 2 and Reduction Rule 3 are not applicable on G. To prove the statement (i), first, we claim that there is no pendent vertex (a vertex of degree one) in G. Assume that there is a pendant vertex u in G, and v is the only neighbor of u. It implies that $deg_G(u) + deg_G(v) \leq \Delta(G) + 1$ and uv is an edge of G, which contradicts the fact that Reduction Rule 2 and Reduction Rule 3 are not applicable to G. Therefore, every leaf vertex in $G - F$ definitely has some neighbor in F. Since $\Delta(G) < 2(k+1)$, there are at most $2k(k+1) = \mathcal{O}(k^2)$ neighbours of F in $G - F$. It follows that the number of leaves in $G - F$ is at most $\mathcal{O}(k^2)$, that is, $|L| = \mathcal{O}(k^2)$. It proves the statement (i).

Let $S_{=2}$ be the set of vertices of degree two in $G - F$. We know that $\sum_{v \in V(G-F)} deg(v) = 2|E(G-F)|$, by the Handshaking lemma. Since $G - F$ is a forest, we have $|L| + 2|S_{=2}| + 3|S_{\geq 3}| \leq 2(|L| + |S_{=2}| + |S_{\geq 3}| - 1)$, that is, $|S_{\geq 3}| \leq |L| - 2$. Therefore, $|S_{\geq 3}| = \mathcal{O}(k^2)$ from the statement (i). It completes our proof. □

Now, the only goal remaining, to bound the size of graph $G - F$, is to bound the size of set $S_{=2}$, which is the set of degree two vertices in graph $G - F$. To achieve this, we define the notion of *degree-2 path* and consider the set $S_{=2}$ as the union of degree-2 paths. Therefore, we bound the number of degree 2-paths and the total number of vertices in these degree-2 paths.

Definition 5 (Degree-2 path). *A path $P = (u, v_1, v_2, v_3, .., v_t, v)$ is said to be a degree-2 path if $d(v_i) = 2$, $\forall i \in [t]$. P is said to be a maximal degree-2 path if $d(u) \neq 2$ and $d(v) \neq 2$.*

Lemma 6. *If Reduction Rule 2 and Reduction Rule 3 are not applicable to G and $\Delta(G) < 2(k+1)$, then the following statements are true:*

(i) The number of maximal degree-2 paths in $G - F$ is $\mathcal{O}(k^2)$,
(ii) The total number of vertices and edges that appear in all maximal degree-2 paths in $G - F$ is $\mathcal{O}(k^2)$.

Proof. If we consider each maximal degree-2 path of $G - F$ as an edge, then $G - F$ will still remain a tree, say T, with no vertices of degree two. Now all the vertices of T either belong to L or $S_{\geq 3}$. Since Reduction Rule 2 and Reduction Rule 3 are not applicable on G and $\Delta(G) \leq 2k + 1$, the size of L and the size of $S_{\geq 3}$ is bounded by $\mathcal{O}(k^2)$ by Lemma 5. Thus, the number of edges of T and the number of maximal degree-2 paths in $G - F$ is $\mathcal{O}(k^2)$. It proves the statement (i).

To prove the statement (ii), we claim that for any edge e in a maximal degree-2 path of $G - F$, at least one of the endpoints of e is either incident to a terminal vertex of that path (a vertex of degree one or at least three in $G-F$) or incident to some neighbor of F. Suppose it is not true, then the sum of degrees of endpoints of e is exactly 4, which is not more than $\Delta(G) + 1$ as $\Delta(G) \geq 3$ (else, we can solve the problem in polynomial time). It follows that either Reduction Rule 2 or Reduction Rule 3 is applicable to the edge e, which is a contradiction. Since there are at most $\mathcal{O}(k^2)$ maximal degree-2 paths in $G - F$, in total, there will be at most $\mathcal{O}(k^2)$ terminal vertices for all the maximal degree-2 paths in $G - F$. Also, recall that F has at most $\mathcal{O}(k^2)$ neighbours in $G - F$. Therefore, the total number of edges (and vertices) that appear in any maximal degree-2 paths in $G - F$ is $\mathcal{O}(k^2)$. It completes the proof. □

We are now ready to prove Theorem 4.

Proof (Proof of Theorem 4). First, if $\Delta(G) \geq 2(k+1)$, we return that the given instance is a `Yes` instance, by Corollary 2. Otherwise, note that the total number of edges completely inside $G[F]$ and the edges between F and $G - F$ are at most $\mathcal{O}(k^2)$, as $|F| = k$ and $\Delta(G) < 2(k+1)$. Then, we will apply the reduction rules to G exhaustively. Due to Reduction Rule 2 and Reduction Rule 3, the number of edges in G can only be reduced. Let G' be the reduced graph when none of the reduction rules are applicable anymore.

Note that the total number of edges completely inside $G'[F]$ and the edges between F and $G' - F$ remain at most $\mathcal{O}(k^2)$. Next, we bound the number of edges in $G' - F$. Due to Reduction Rule 1, the graph will have no isolated vertices. Since Reduction Rule 2 and Reduction Rule 3 do not apply to G', the number of leaves and the number of vertices with degree more than two in $G' - F$ is $\mathcal{O}(k^2)$, by Lemma 5. Moreover, by Lemma 6, the number of maximal degree-2 paths and the total number of vertices over all maximal degree-2 path in $G' - F$ is $\mathcal{O}(k^2)$. Thus, the total number of vertices in $G' - F$ is $\mathcal{O}(k^2)$. Since

$G' - F$ is a forest, the total number of edges in $G' - F$ is $\mathcal{O}(k^2)$. Therefore, the total number of edges in G' is $\mathcal{O}(k^2)$. Hence, we get a kernel for Chromatic Index with $\mathcal{O}(k^2)$ vertices and edges, where k is the size of feedback vertex set. □

5 Kernel Parameterized by cvd Number and Clique Number

In this section, we give a polynomial kernel for Chromatic Index parameterized by cluster vertex deletion number and clique number. We state the result formally below.

Theorem 5. *Let G be a graph with clique number, ω and s be the cluster vertex deletion number of G. Then,* Chromatic Index *admits a kernel with $\mathcal{O}(s^2\omega + s\omega^3)$ edges.*

Let $G = (V, E)$ be a graph and M be a cluster vertex deletion set of size s and ω be the clique number which is at least the size of a maximum-sized clique in $G[V \setminus M]$. Clearly, $\mathsf{tw}(G) \leq s + \omega$ and thus if $\Delta(G) \geq 2(s + \omega)$, the graph G is $\Delta(G)$-colorable due to Corollary 2. So we will consider the case that $\Delta(G) < 2(s + \omega)$. Now we will prove the above theorem using Reduction Rule 2 and Reduction Rule 3 in the following two lemmas.

Lemma 7. *If Reduction Rule 2 and Reduction Rule 3 are not applicable to the graph G and $2\omega \leq \Delta(G) < 2(s + \omega)$, then G has at most $\mathcal{O}(s^2\omega + s\omega)$ edges.*

Proof. Let M be a cluster vertex deletion set of size s of G. Since $\Delta(G) < 2(s + \omega)$, the number of edges between M and $V \setminus M$ are at most $2s(s + \omega)$. On the other hand, for every edge $uv \in E(G - M)$, at least one of u or v must have a neighbor in M, otherwise, one of Reduction Rule 2 or Reduction Rule 3 would have been applied. Also, for every $w \in V \setminus M$, $\deg_{G-M}(w) \leq \omega - 1$, as the size of a largest clique in $G - M$ is at most ω. So for every edge e with one end point in M and the other endpoint in $V \setminus M$, there can be at most ω edges in $G - M$ that are adjacent to e. Hence, the total number of edges in G is at most $s^2 + 2s(s + \omega) + 2s(s + \omega)\omega$, which sums up to $\mathcal{O}(s^2\omega + s\omega^2)$. □

Lemma 8. *If Reduction Rule 2 and Reduction Rule 3 are not applicable to the graph G and $\Delta(G) < 2\omega$, then G has at most $\mathcal{O}(s^2 + s\omega^3)$ edges.*

Proof. Let M be a cluster vertex deletion set of size s of G. If there is any isolated clique in G as a result of application of Reduction Rule 2 and Reduction Rule 3, we solve the problem on that clique independently and remove it from the graph. As a consequence of the above step, each clique in $G - M$ has a vertex that is adjacent to a vertex in M. Since $\Delta(G) < 2\omega$, the number of edges with exactly one endpoint in M is at most $2s\omega$ and thus there are at most $2s\omega$ cliques in $G - M$. Hence, the total number of edges in G will be at most $s^2 + 2s\omega + 2s\omega \cdot \omega^2$ which is $\mathcal{O}(s^2 + s\omega^3)$. □

For a given input graph G with cluster vertex deletion number s, we apply our Reduction Rule 2 and Reduction Rule 3 as long as they are applicable. Once the reduction rules are no longer applicable, we get the desired kernel from Lemmas 7 and 8. Hence, Theorem 5 follows.

References

1. Aggarwal, G., Motwani, R., Shah, D., Zhu, A.: Switch scheduling via randomized edge coloring. In: 2003 Proceedings of the 44th Annual IEEE Symposium on Foundations of Computer Science, pp. 502–512. IEEE (2003)
2. de Almeida, S.M., de Mello, C.P., Morgana, A.: On the classification problem for split graphs. J. Braz. Comput. Soc. **18**, 95–101 (2012)
3. de Almeida, S.M., de Mello, C.P., Morgana, A.: Edge-coloring of split graphs. Ars Comb. **119**, 363–375 (2015)
4. Bodlaender, H.L.: Polynomial algorithms for graph isomorphism and chromatic index on partial k-trees. J. Algorithms **11**(4), 631–643 (1990)
5. Bodlaender, H.L., Jansen, B.M., Kratsch, S.: Cross-composition: a new technique for kernelization lower bounds. In: 28th International Symposium on Theoretical Aspects of Computer Science (STACS 2011). Schloss-Dagstuhl-Leibniz Zentrum für Informatik (2011)
6. Bojarshinov, V.: Edge and total coloring of interval graphs. Discret. Appl. Math. **114**(1–3), 23–28 (2001)
7. Campos, C.N., de Figueiredo, C., Machado, R., de Mello, C.P.: The total chromatic number of split-indifference graphs. Discret. Math. **312**(17), 2690–2693 (2012)
8. Chen, B., Fu, H., Ko, M., et al.: Total chromatic number and chromatic index of split graphs. J. Combin. Math. Combin. Comput. **17**, 137–146 (1995)
9. Cheng, M., Yin, L.: Transmission scheduling in sensor networks via directed edge coloring. In: 2007 IEEE International Conference on Communications, pp. 3710–3715. IEEE (2007)
10. Couto, F., Ferraz, D.A., Klein, S.: New results on edge-coloring and total-coloring of split graphs. Discret. Appl. Math. **360**, 297–306 (2025)
11. Cygan, M., et al.: Parameterized Algorithms. Springer, Cham (2015). https://doi.org/10.1007/978-3-319-21275-3
12. de Figueiredo, C.M., de Melo, A.A., Sasaki, D., Silva, A.: Revising Johnson's table for the 21st century. Discret. Appl. Math. **323**, 184–200 (2022)
13. Garey, M.R., Johnson, D.S.: Computers and Intractability: A Guide to the Theory of NP-Completeness. W. H. Freeman and Co. (1979)
14. Holyer, I.: The NP-completeness of edge-coloring. SIAM J. Comput. **10**(4), 718–720 (1981)
15. Ito, T., Kato, A., Zhou, X., Nishizeki, T.: Algorithms for finding distance-edge-colorings of graphs. In: Wang, L. (ed.) COCOON 2005. LNCS, vol. 3595, pp. 798–807. Springer, Heidelberg (2005). https://doi.org/10.1007/11533719_81
16. König, D.: Über graphen und ihre anwendung auf determinantentheorie und mengenlehre. Math. Ann. **77**(4), 453–465 (1916)
17. Salavatipour, M.R.: A polynomial time algorithm for strong edge coloring of partial k-trees. Discret. Appl. Math. **143**(1–3), 285–291 (2004)
18. Sanders, D.P., Zhao, Y.: Planar graphs of maximum degree seven are class I. J. Combin. Theory Ser. B **83**(2), 201–212 (2001)

19. da Soledade Gonzaga, L.G., de Almeida, S.M., de Sousa Cruz, J.B., et al.: The chromatic index of split-interval graphs. Procedia Comput. Sci. **195**, 325–333 (2021)
20. Vizing, V.G.: On an estimate of the chromatic class of a p-graph. Diskret Analiz. **3**, 25–30 (1964)
21. Vizing, V.G.: Critical graphs with given chromatic class (in Russian). Metody Discret. Analiz. **5**, 9–17 (1965)
22. Williamson, D.P., et al.: Short shop schedules. Oper. Res. **45**(2), 288–294 (1997)
23. Zhou, X., Nakano, S., Nishizeki, T.: Edge-coloring partial k-trees. J. Algorithms **21**(3), 598–617 (1996). https://doi.org/10.1006/jagm.1996.0061

Concurrency Constrained Scheduling with Tree-Like Constraints

Hans L. Bodlaender[1](✉), Danny Hermelin[2], and Erik Jan van Leeuwen[1]

[1] Department of Information and Computing Sciences, Utrecht University, Utrecht, The Netherlands
h.l.bodlaender@uu.nl

[2] Department of Industrial Engineering and Management, Ben-Gurion University of the Negev, Beer-Sheva, Israel

Abstract. This paper investigates concurrency-constrained scheduling problems, where the objective is to construct a schedule for a set of jobs subject to concurrency restrictions. Formally, we are given a conflict graph G defined over a set of n jobs, where an edge between two jobs in G indicates that these jobs cannot be executed concurrently. Each job may have distinct attributes, such as processing time, due date, weight, and release time. The goal is to determine a schedule that optimizes a specified scheduling criterion while adhering to all concurrency constraints. This framework offers a versatile model for analyzing resource allocation problems where processes compete for shared resources, such as access to shared memory. From a theoretical perspective, it encompasses several classical graph coloring problems, including Chromatic Number, Sum Coloring, and Interval Chromatic Number.

Given that even the simplest concurrency-constrained scheduling problems are NP-hard for general conflict graphs, this study focuses on conflict graphs with bounded treewidth. Our results establish a dichotomy: Some problems in this setting can be solved in FPT time, while others are shown to be XALP-complete for treewidth as parameter. Along the way, we generalize several previously known results on coloring problems for bounded treewidth graphs. Several of the FPT algorithms are based on the insight that completion times are bounded by the Grundy number of the conflict graph—the fact that this number is bounded by the product of treewidth and the logarithm of the number of vertices then leads to the FPT time bound.

Keywords: treewidth · scheduling · parameterized complexity · concurrency constraints · FPT · XALP

1 Introduction

In offline concurrency constrained scheduling, we have a set of n jobs $\{1, \ldots, n\}$ for non-preemptive processing. We have an unlimited number (n is

H. Fernau and P. Kindermann (Eds.): WG 2025, LNCS 16124, pp. 105–120, 2026.
https://doi.org/10.1007/978-3-032-11835-6_8

enough) of identical machines. In the most basic setting, each job $j \in \{1, \ldots, n\}$ requires $p_j = 1$ (unit) *processing time* on any of the given machines. However, some pairs of jobs cannot be processed concurrently at the same time. These *concurrency constraints* are given by a *conflict graph* $G = (\{1, \ldots, n\}, E)$, where two jobs i and j cannot be processed concurrently if and only if $\{i, j\} \in E$. Bar-Noy *et al.* [5,6] observed that concurrency constrained scheduling occurs in resource allocation problems with constraints imposed by the competition of processes on conflicting resource requirements. Similar situations occur in other settings, e.g. almost all modern database systems implement some sort of mutual exclusion scheme. A *schedule* is a function $C : \{1, \ldots, n\} \to \mathbb{N}$ specifying the completion time of each job. We write $C_j = C(j)$ following standard scheduling notation. A schedule C is said to be *feasible* if $C(i) \neq C(j)$ for every $\{i, j\} \in E$. A feasible schedule thus is exactly a *proper coloring* of G, where no two adjacent vertices get the same color. In concurrency constrained scheduling we aim to find a proper coloring of G that optimises some scheduling criterion (as discussed below).

Following the three-field notation of Graham *et al.* for scheduling problems [27], we denote the concurrency constrained setting with unit processing times by $P_n \mid conc, p_j = 1 \mid *$. The first field denotes the machine model being used, in our case n identical machines that run in parallel. The second field denotes constraints on the problem, where 'conc' denotes that concurrency constrains (via a conflict graph) are present, while $p_j = 1$ signifies that all jobs require a unit amount of processing time on any of the n machines. Additional constraints and restrictions can be added to this field, as will be discussed later on. Finally, the last field '$*$' denotes the objective function that we optimise.

Two notable and well-studied concurrency constrained scheduling problems are obtained by considering the *makespan* $C_{\max} = \max_j C_j$ and the *total completion time* $\sum C_j$ objective functions. In the former, we wish to minimize the time in which the last job completes in our schedule. This directly translates to the classical Chromatic Number problem, as the minimum number of colors $\chi(G)$ necessary to properly color G is precisely $C_{\max}$ in a schedule with minimum makespan. In the latter, we wish to minimize the sum of completion times in our schedule, which is the same as minimizing the average completion time in the schedule. This directly translates to the Chromatic Sum problem [39,40,51], a widely-studied graph coloring problem (see e.g. [5,7,16,17,25,29,34,35,39,40,43,44,46,49,50,52]) which has been originally studied in the context of VLSI design [51]. In our setting, these two problems are denoted by $P_n \mid conc, p_j = 1 \mid C_{\max}$ and $P_n \mid conc, p_j = 1 \mid \sum C_j$ respectively.

Besides unit processing times, we may also consider that each job j has an integer processing time $p_j \geq 1$ (which it requires on any machine). The time intervals in which two conflicting jobs i and j are processed must be disjoint. Thus, we now require a feasible schedule to have:

- $C_j - p_j \geq 0$ for all $j \in \{1, \ldots, n\}$ (no jobs can start before time 0), *and*
- $[C_i - p_i, C_i) \cap [C_j - p_j, C_j) = \emptyset$ for all $\{i, j\} \in E$ (all concurrency constraints are met).

We denote this setting by $P_n \mid conc \mid *$. Note that in the case of arbitrary processing times, some proper colorings of G no longer correspond to feasible schedules. However, every feasible schedule is still a proper coloring.

The $P_n \mid conc \mid C_{\max}$ problem was previously studied in the literature under the names INTERVAL CHROMATIC NUMBER [26,38] or NON-PREEMPTIVE MULTICHROMATIC NUMBER [6]. When restricting the conflict graph G to be an interval graph we get the classical DYNAMIC STORAGE ALLOCATION problem [24] (also known as the SHIP BUILDING problem [26]), and when G is a line graph we get the most basic case of the SCHEDULING FILE TRANSFERS problem [18]. The $P_n \mid conc \mid \sum C_j$ problem was previously studied in the literature under the name NON-PREEMPTIVE MULTICHROMATIC SUM [6].

Apart from $C_{\max}$ and $\sum C_j$, there are several other natural criteria in the context of concurrency constrained scheduling. First, we can assume that each job j has a due date d_j associated with it, where if it completes after time d_j there is a penalty incurred to our schedule. Two such criteria that are widely considered in scheduling literature (see *e.g.* [48]) are the *maximum lateness* criterion and the *total tardiness* criterion. In the former, we wish to minimize the maximum lateness $L_{\max} = \max_j L_j$ in our schedule, where the *lateness* of a job j is defined by $L_j = C_j - d_j$. In the latter, we wish to minimize the total tardiness $\sum_j T_j$ of our schedule, where the *tardiness* of a job j is given by $T_j = \max\{0, C_j - d_j\}$. Note that $C_{\max}$ is a special case of $L_{\max}$ where each job has due date $d_j = 0$, and in this special case $\sum T_j$ becomes $\sum C_j$ as well. Weighted variants, where each job j has an additional integer weight w_j, exist for all criteria above as well.

Second, we can consider a release time $r_j \geq 0$ for each job $j \in \{1, \dots, n\}$. Note that in the above, all jobs are released at time $r_j = 0$. Release times are different than due dates in that release times are hard constraints that must be satisfied, where we require any feasible schedule to have $C_j - p_j \geq r_j$ for all jobs j. We assume throughout the paper that all job parameters p_j, d_j, w_j, and r_j are given in unary, and so are bounded by the total input length.

Bounded Treewidth Conflict Graphs: As the most basic cases of concurrency constrained scheduling, CHROMATIC NUMBER and CHROMATIC SUM, are already NP-hard [37,40], it makes sense to study cases where the conflict graph has some restricted structure. We focus on graphs of bounded treewidth and pathwidth. CHROMATIC NUMBER is well known to be solvable in FPT time parameterized by treewidth, as there exists a $O^*(q^{tw+1})$-time[1] algorithm for determining whether a graph G can be properly colored with q colors given a tree decomposition of width tw for G [19,53]. Moreover, this running time cannot substantially be improved assuming the Strong Exponential Time Hypothesis (SETH) [41]. Since the chromatic number of a graph is at most its treewidth, this yields an algorithm with running time $O^*(tw^{tw+1})$ for $P_n \mid conc, p_j = 1 \mid C_{\max}$. The CHROMATIC SUM (*i.e.* $P_n \mid conc, p_j = 1 \mid \sum C_j$) problem has also been previously studied in the context of bounded treewidth graphs. Specifically, Jansen [35] proved that an optimal coloring for bounded treewidth graphs uses $O(tw \log n)$ colors, and used

[1] Here and elsewhere in the paper we use $O^*()$ to suppress polynomial factors.

the $O^*(q^{tw+1})$-time algorithm mentioned above to solve a slightly more generalised version of the $P_n \mid conc, p_j = 1 \mid \sum C_j$ problem in $O^*((tw \cdot \log n)^{tw+1})$ time (which is also FPT time, albeit a bit slower than $O^*(tw^{tw+1})$). Halldórsson and Kortsarz showed that the more general $P_n \mid conc, p_j \leq p \mid C_{\max}$ and $P_n \mid conc, p_j \leq p \mid \sum C_j$ problems, where all processing times are bounded by some fixed integer p, can be solved in $O^*((p \cdot tw \cdot \log n)^{tw+1})$ time [29].

XNLP/XALP Problems: XNLP, introduced by [11,22], is the class of all problems solvable in FPT non-deterministic time using only $f(k) \log n$ space. Several natural problems problems parameterized by pathwidth turned out to be complete for this class [10], and in particular problems that require storing $n^{\Omega(f(pw))}$ space (for a non-decreasing f that grows to infinity) in their standard path decomposition dynamic programming algorithm. The class XALP was introduced by [10] as analogue to XNLP. Similar to XNLP, XALP (formally defined in Sect. 2) captures problems whose dynamic programming algorithms on tree decompositions require $n^{\Omega(f(tw))}$ time and space for similar f (see e.g., [10,14,15]).

Our Results

We investigate in detail the complexity of concurrency-constrained scheduling problems parameterized by the treewidth (resp. pathwidth) of the conflict graph. We provide a complete classification of all problems discussed above into two categories: those that are solvable in FPT time and those that are XALP-complete (resp. XNLP-complete). See the overview in Table 1.

Table 1. Complexity results for unweighted concurrency constrained scheduling problems, with respectively treewidth and pathwidth as parameter. The columns represent the objective function, while the rows correspond to constraints on release times and processing times. The running times presented ignore polynomial factors in n.

	$C_{\max}$	$\sum C_j$	$L_{\max}$	$\sum T_j$
$r_j = 0, p_j = 1$	$tw^{O(tw)}$ [53] $pw^{O(pw)}$ [53]	$(tw \cdot \log n)^{O(tw)}$ [35] $pw^{O(pw)}$ [35]	$(tw \cdot \log n)^{O(tw)}$ $pw^{O(pw)}$	$(tw \cdot \log n)^{O(tw)}$ $pw^{O(pw)}$
$r_j \geq 0, p_j = 1$	$(tw \cdot \log n)^{O(tw)}$ $pw^{O(pw)}$	XALP-complete XNLP-complete	XALP-complete XNLP-complete	XALP-complete XNLP-complete
$r_j = 0, p_j \geq 1$	XALP-complete XNLP-complete	XALP-complete XNLP-complete	XALP-complete XNLP-complete	XALP-complete XNLP-complete

On the positive side, we show that for the case of unit processing times and no release times (*i.e.* $p_j = 1$ and $r_j = 0$ for all $j \in \{1, \ldots, n\}$), all problems we consider are FPT. In particular, we generalize the result of Jansen [35] and show that for any so called *regular scheduling criterion* [48], *i.e.* any scheduling criterion which is non-decreasing in the completion time of the jobs, there exists an optimal schedule C with makespan $C_{\max} \leq tw(G) \cdot \log n + 1$. All scheduling criteria discussed above, as well as many others, are regular. We also show that this bound increases by a factor of $O(p)$ for the case where the processing time of each job is at most p, generalizing the result of Halldórsson and Kortsarz [29].

Some of our FPT algorithms are based on the insight that completion times are bounded by the Grundy number of the conflict graph. Telle and Proskurowski [53] showed that the Grundy number of a graph with treewidth tw is $O(tw \cdot \log n)$. If k bounds the maximum completion time, then an algorithm with running time $k^{O(tw)} \cdot n^{O(1)}$ and the Grundy number bound implies the problem is FPT, as $O(tw \cdot \log n)^{O(tw)} \cdot n^{O(1)}$ is of the form $f(tw) \cdot n^{O(1)}$ (cf. [19, p. 74, Answer 3.18]).

On the negative side, we show that once we allow jobs to have unbounded processing times or we introduce release times, then only $P_n \mid conc, r_j \geq 0, p_j = 1 \mid C_{\max}$ remains FPT; all other problems become XALP-complete. Our hardness results here are obtained by developing reductions from the XALP-complete Precoloring Extension problem. We note that XALP-completeness results have two important implications. First, these problems are in XP, i.e., have an algorithm with running time of the form $O(|I|^{f(k)})$, where $|I|$ is the size of the instance. Second, a conjecture by Pilipczuk and Wrochna [47] implies that XALP-hard problems are unlikely to have an XP-algorithm with space usage $f(k) \cdot |I|^{O(1)}$.

We also consider the weighted variants of our problems, where each job j has an integer weight w_j specifying its significance. In this context, we write $WC_{\max} = \max_j w_j C_j$ and $WL_{\max} = \max_j w_j L_j$. Again, when $p_j = 1$ and $r_j = 0$ for all jobs j, all problems are FPT, although with a slightly slower running time of $O^*((tw \cdot \log n)^{tw+1}))$ for $P_n \mid conc, p_j = 1 \mid WC_{\max}$ in comparison to the unweighted case. We also show that $P_n \mid conc, p_j = 1, r_j \geq 0 \mid WC_{\max}$ is XALP-complete, as opposed to the unweighted $P_n \mid conc, p_j = 1, r_j \geq 0 \mid C_{\max}$ problem which is FPT. This hardness result again follows via a reduction from Precoloring Extension parameterized by treewidth.

For pathwidth, the running time of the algorithms decreases, as the Grundy number of a graph of pathwidth pw is $O(pw)$ [8,21]. For the hardness results, since Precoloring Extension is XNLP-complete parameterized by pathwidth [11], we note that we can reuse the above hardness constructions.

2 Preliminaries

We use $G = (\{1, \dots, n\}, E)$ to denote the conflict graph of our concurrency constrained scheduling problems. Standard graph-theoretic notation is found in [20]. We assume readers are familiar with the concepts of treewidth and pathwidth.

XALP-Completeness: The class XALP is the class of all parameterized problems that can be solved in $f(k) \cdot n^{O(1)}$ time and $f(k) \cdot \log n$ space on a non-deterministic Turing Machine with access to an auxiliary stack (with only top element lookup allowed) [10]. XALP is closed by reductions using at most $f(k) \log n$ space and

FPT time. These conditions are implied by using at most $f(k)+O(\log n)$ space. Reductions respecting the latter condition are *parameterized logspace* reductions.

We use the PRECOLORING EXTENSION problem frequently in our reductions. In PRECOLORING EXTENSION, we are given a graph $H = (V, F)$, and integer k, and a proper coloring $\chi : V_0 \to \{1, \ldots, k\}$ defined only on a subset of vertices $V_0 \subseteq V$. Our goal is to determine whether we can extend χ to the set $V \setminus V_0$ of remaining vertices. That is, is there a proper coloring $\chi' : V \to \{1, \ldots, k\}$ with $\chi'(v) = \chi(v)$ for all $v \in V_0$. It is XALP-complete for parameter treewidth [10] and XNLP-complete for parameter pathwidth [11].

Tree Decomposition Dynamic Programming: We show that one can compute an optimal schedule C for any concurrency constrained scheduling problem considered in this paper in $O^*(k^{tw+1})$ time, assuming there exists such a schedule with $C_{\max} \leq k$. We prove this specifically for the two most general problems we consider in the paper, the $P_n \mid conc, r_j \geq 0 \mid WL_{\max}$ and $P_n \mid conc, r_j \geq 0 \mid \sum w_j T_j$ problems. The proof uses an algorithm very similar to the $O^*(k^{tw+1})$-time algorithm for properly coloring graphs of treewidth tw with k colors [19,53].

Theorem 1 (See Full Version). *The $P_n \mid conc, r_j \geq 0 \mid WL_{\max}$ and $P_n \mid conc, r_j \geq 0 \mid \sum w_j T_j$ problems are both solvable in $O^*(k^{tw+1})$-time, where k is a bound on the makespan of some optimal schedule.*

Observe that Theorem 1 above implies that both $P_n \mid conc, r_j \geq 0 \mid WL_{\max}$ and $P_n \mid conc, r_j \geq 0 \mid \sum w_j T_j$ are solvable in $|I|^{O(tw)}$-time, where $|I|$ denotes the size of the input, as we assume that all processing times are given in unary. Hence, the total processing time of all jobs, and the makespan of any schedule, is always polynomial in $|I|$. The dynamic programming algorithm of Theorem 1 can even be turned into an XALP-membership proof:

Corollary 1 (See Appendix). *$P_n \mid conc, r_j \geq 0 \mid WL_{\max}$ and $P_n \mid conc, r_j \geq 0 \mid \sum w_j T_j$ are both in XALP with treewidth as parameter.*

3 Bounded Processing Times

In the following section we focus on cases where job processing times are either unit, or relatively small, *i.e.* $P_n \mid conc, p_j = 1 \mid *$ and $P_n \mid conc, p_j \leq p \mid *$ for some fixed $p \in O(1)$. Previous work has shown that for the $\sum C_j$ objective, there always exists an optimal schedule C with makespan at most $tw(G) \cdot \log n$ for the unit processing time case [35], and $tw(G) \cdot p \log n$ for the $p_j \leq p$ case [29]. In what follows we generalize and unify these results into our framework. For this, we first introduce the notion of minimal feasible schedules.

Definition 1. *A feasible schedule C is said to be* minimal *for a set of jobs $\{1, \ldots, n\}$ if decreasing the completion time of any job by any amount of time causes C to become infeasible.*

We begin by considering the case of unit processing times. Note that this case is equivalent to the case of $p_j = p$ for all jobs j, since one can simply multiply the completion times in any feasible schedule for a $P_n \mid conc, p_j = 1 \mid *$ instance by p. We will heavily rely on the notion of a *Grundy number* of a graph: A *greedy coloring* of a graph $G = (V, E)$ is a proper coloring $C : \{1, \ldots, n\} \to \mathbb{N}$ such that for every vertex j and any color $c \in \{1, \ldots, C(j) - 1\}$ there is some vertex i with $\{i, j\} \in E$ and $C(i) = c$. The *value* of a greedy coloring C of $G = (V, E)$ is $C_{\max} = \max_{v \in V} C(v)$. The *Grundy number* of $\Gamma(G)$ of a graph G is the maximum value of any greedy coloring of G. From [53], we have:

Lemma 1 **([53]).** $\Gamma(G) \leq tw(G) \cdot \log n + 1$ *for any graph* G *on* n *vertices.*

Lemma 2 (See full version). *A feasible minimal schedule of any* $P_n \mid conc, p_j = 1 \mid *$ *instance corresponds to a greedy coloring of the conflict graph* G.

Recall that a scheduling criterion is *regular* if it is non-decreasing in the completion time of the jobs; i.e., decreasing the completion time of any job does not increase the value of the schedule. We thus combine Lemma 1 and 2 to get:

Corollary 2. *Any* $P_n \mid conc, p_j = 1 \mid *$ *instance where* $*$ *is regular has an optimal schedule* C *with* $C_{\max} \leq tw \cdot \log n + 1$.

In the full version, we discuss whether the bound of Corollary 2 is tight for the different objectives.

Note that all scheduling criteria discussed in this paper are regular. Thus, by using the algorithm of Theorem 1, we get an algorithm for all these criteria in the unit processing-times case. We state this for the two most general criteria:

Theorem 2. *The* $P_n \mid conc, p_j = 1 \mid WL_{\max}$ *and* $P_n \mid conc, p_j = 1 \mid \sum w_j T_j$ *problems are both solvable in* $O^*((tw \cdot \log n + 1)^{tw+1})$*-time.*

We next consider the case where the processing times of jobs may vary, but are still rather small. We denote this case by $P_n \mid conc, p_j \leq p \mid *$. We extend the bound of Corollary 2 to this case as well, exploiting the connection between treedepth and treewidth, although at the price of being slightly less precise.

Lemma 3 (See full version). *There exists a feasible minimal schedule* C *for any* $P_n \mid conc, p_j \leq p \mid *$ *instance such that* $C_{\max} = O(tw(G) \cdot p \log n)$.

Theorem 3. *The* $P_n \mid conc, p_j \leq p \mid WL_{\max}$ *and* $P_n \mid conc, p_j \leq p \mid \sum w_j T_j$ *problems are both solvable in* $O^*((tw \cdot p \cdot \log n)^{O(tw)})$*-time.*

4 Unbounded Processing Times

We next consider the case where the processing times of jobs are unbounded, *i.e.* problems of the form $P_n \mid conc \mid *$. We show that already for the two most basic objective functions of $C_{\max}$ and $\sum C_j$, the corresponding scheduling with concurrency constrained problem becomes XALP-complete parameterized by the treewidth of the conflict graph. As $P_n \mid conc \mid C_{\max}$ and $P_n \mid conc \mid \sum C_j$ are special cases of any $P_n \mid conc \mid *$ problem considered in this paper, this will complete the picture for the second row in Table 1. For both these problems, we show XALP-hardness by a reduction from PRECOLORING EXTENSION.

4.1 The Makespan Objective

Let (H, k, χ) be a given instance of PRECOLORING EXTENSION with $H = (V, F)$, $V = \{1, \ldots, n\}$, k an integer, and $\chi : V_0 \to \{1, \ldots, k\}$ a given precoloring of H defined on a subset of vertices $V_0 \subseteq V$. We may assume that H has at least one edge. The main idea of our reduction is as follows: We construct an instance of $P_n \mid conc \mid C_{\max}$ by initially starting with the set of n jobs $\{1, \ldots, n\}$ corresponding to the vertices of H, each with processing time $p_j = 1$, and a conflict graph G which is identical to H. We then add auxiliary jobs to each job $j \in V_0$ whose goal is to ensure that we get $C_j = \chi(j)$ in any feasible schedule C with makespan at most k. In this way, feasible schedules of makespan at most k correspond to valid extensions of χ when restricted to $\{1, \ldots, n\}$.

Our construction proceeds as follows: We start with $G = H$, and define the processing time of each job $j \in \{1, \ldots, n\}$ to be $p_j = 1$. We then construct four jobs a, a', b, and b' such that:

- $p_a = p_b = 1$ and $p_{a'} = p_{b'} = k - 1$.
- $\{a, a'\}$, $\{b, b'\}$, and $\{a, b\}$ are edges in G.

The purpose of a and b is to serve as delimiter jobs, as is formalized in the following easy to verify lemma:

Lemma 4. *If C is a feasible schedule for $\{a, a', b, b'\}$ with $C_{\max} \leq k$ then $\{C_a, C_b\} = \{1, k\}$.*

Next add at most two additional jobs per each precolored $j \in V_0$. Let $j \in V_0$, and suppose that $\chi(j) \in \{2, \ldots, k-1\}$. Create two jobs $j(1)$ and $j(2)$ such that:

- $p_{j(1)} = \chi(j) - 1$ and $p_{j(2)} = k - \chi(j)$.
- $\{j, j(1)\}$, $\{j, j(2)\}$, and $\{j(1), j(2)\}$ are edges in G.
- $\{a, j\}$, $\{b, j\}$ $\{a, j(2)\}$ and $\{b, j(1)\}$ are edges in G.

For $j \in V_0$ with $\chi(j) = 1$, we create only job $j(2)$ above, and we do not add the edge $\{a, j\}$ to G. For $j \in V_0$ with $\chi(j) = k$, we create only job $j(1)$ above, and we do not add the edge $\{b, j\}$ to G. This completes our construction.

Lemma 5 (See full version). *$tw(G) \leq tw(H) + 2$ and $pw(G) \leq pw(H) + 4$.*

Lemma 6. *If there is a solution $\chi' : \{1, \ldots, n\} \to \{1, \ldots, k\}$ to (H, k, χ), then there is a feasible schedule C for the constructed $P_n \mid conc \mid C_{\max}$ instance with $C_{\max} \leq k$.*

Proof. It is easy to verify that the following schedule C is feasible and has makespan at most k: set $C_j = \chi'(j)$ for all $j \in \{1, \ldots, n\}$; $C_a = 1$, $C_b = k$, $C_{a'} = k$, and $C_{b'} = k - 1$; $C_{j(1)} = \chi'(j) - 1$ for all $j \in V_0$ with $\chi'(j) > 1$; $C_{j(2)} = k$ for all $j \in V_0$ with $\chi'(j) < k$. □

Lemma 7 (See full version). *If there is a feasible schedule C for the constructed $P_n \mid conc \mid C_{\max}$ instance with $C_{\max} \leq k$, then there is a solution $\chi' : \{1, \ldots, n\} \to \{1, \ldots, k\}$ to (H, k, χ).*

The reduction can be carried out in polynomial time and an additional $O(\log n)$ working space. Correctness of the reduction is given by Lemma 6 and Lemma 7. Moreover, by Lemma 5 we have $tw(G) = O(tw(H))$. Thus, in total, we have described a parameterized logspace reduction from the XALP-hard PRECOLORING EXTENSION problem to $P_n \mid conc \mid C_{\max}$. Since XALP-membership follows from Corollary 1, we obtain the following theorem:

Theorem 4. *$P_n \mid conc \mid C_{\max}$, parameterized by treewidth, is XALP-complete.*

4.2 Total Completion Time

Let (H, k, χ) be a given instance of PRECOLORING EXTENSION with $H = (\{1, \ldots, n\}, F)$, k an integer, and $\chi : V_0 \to \{1, \ldots, k\}$ a given precoloring of H defined on a subset of vertices $V_0 \subseteq V$. We may assume that H has at least one edge, and thus $n \geq 2$, or the instance is trivial. On a high level the main idea of our reduction is similar to the one for $P_n \mid conc \mid \sum C_{\max}$: We initially start with H as our conflict graph for the set of jobs $\{1, \ldots, n\}$. We then add auxiliary jobs to each job $j \in \{1, \ldots, n\}$ whose goal is to ensure that in an optimal schedule C, jobs $j \in V_0$ complete precisely at time $\chi(j)$, while jobs in $\{1, \ldots, n\} \setminus V_0$ complete anywhere in $\{1, \ldots, k\}$. However, the technical details here are a bit more involved in comparison to the $P_n \mid conc \mid \sum C_{\max}$ problem.

We first set $X = nk + 1$, and $G = H$ with $p_j = 1$ for all $j \in \{1, \ldots, n\}$. We then construct a set of *upper bound jobs* for each $j \in \{1, \ldots, n\}$. Define $u_j = \chi(j)$ for each $j \in V_0$, and $u_j = k$ for each $j \in \{1, \ldots, n\} \setminus V_0$. Then u_j is an upper bound on the color assigned to vertex j in a solution χ' for (H, k, χ); that is, $\chi'(j) \leq u_j$ for any such χ'. For each $j \in \{1, \ldots, n\}$, we construct (see Fig. 1):

- jobs $j(1), \ldots, j(X)$, each having a processing time of $X - u_j$. These are the *primary* upper bound jobs of j. Add the edges $\{j, j(1)\}, \ldots, \{j, j(X)\}$ to G.
- jobs $j(i, 1), \ldots, j(i, X)$ for each $i \in \{1, \ldots, X\}$, each with processing time u_j. Add all edges $\{j(i), j(i, 1)\}, \ldots, \{j(i), j(i, X)\}$, for each $i \in \{1, \ldots, X\}$, to G. We collectively call all these jobs the *secondary* upper bound jobs of j. Note that the secondary upper bound jobs of j are not in conflict with j itself.

Next, we construct *lower bound jobs* for each $j \in \{1, \ldots, n\}$. Define $\ell_j = \chi(j)$ for each $j \in V_0$, and $\ell_j = 1$ for each $j \in \{1, \ldots, n\} \setminus V_0$. Then ℓ_j is a lower bound on the color assigned to vertex j in a solution χ' for (H, k, χ); that is, $\chi'(j) \geq \ell_j$ for any such χ'. For each $j \in V_0$ with $\chi(j) > 1$, we construct (see Fig. 1):

- jobs $j^*(1), \ldots, j^*(X)$, each of which has processing time of $\ell_j - 1$ and is in conflict only with j. We add all edges $\{j, j^*(1)\}, \ldots, \{j, j^*(X)\}$ to G.

This completes our construction of the jobs for the constructed $P_n \mid conc \mid \sum C_j$ instance. Note that $tw(G) = tw(H)$, as one only needs to append nodes of size 2 to a tree decomposition of H to obtain a tree decomposition of G. Similarly, $pw(G) \leq pw(H) + 2$, as one only needs to add copies of existing nodes with two additional vertices. Finally, we set the required total completion time of all jobs to be $K = \sum_{j=1}^{n} \Big[(1 + u_j) \cdot X^2 + (\ell_j - 1) \cdot X\Big] + nk$.

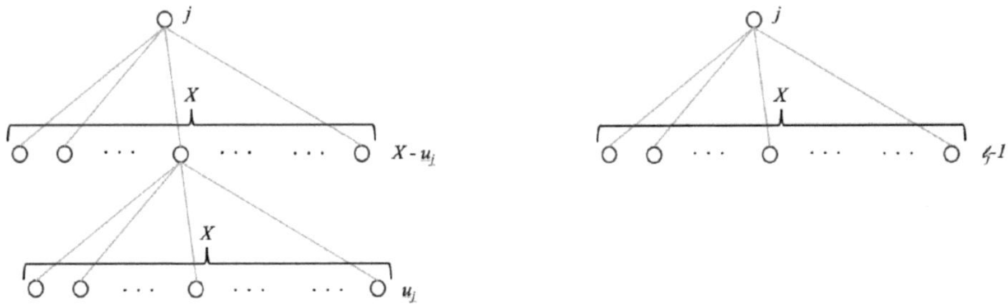

Fig. 1. The conflict graph of the upper bound (left) and lower bound jobs (right) of some job j. The processing times are depicted to the right of each job.

Lemma 8. *If there is a solution $\chi' : \{1, \ldots, n\} \to \{1, \ldots, k\}$ to (H, k, χ), then there is a feasible schedule C for the constructed $P_n \mid conc \mid \sum C_j$ instance with total completion time at most K.*

Proof. Define a schedule C as follows:

- $C(j) = \chi'(j)$ for all $j \in \{1, \ldots, n\}$.
- $C(j(i)) = X$ for all $j \in \{1, \ldots, n\}$ and $i \in \{1, \ldots, X\}$.
- $C(j(i, i_0)) = u_j$ for all $j \in \{1, \ldots, n\}$ and $i, i_0 \in \{1, \ldots, X\}$.
- $C(j^*(i)) = \ell_j - 1$ for all $j \in V_0$ such that $\chi(j) > 1$ and $i \in \{1, \ldots, X\}$.

Then C is feasible, as χ' is a proper coloring on H with $\ell_j < \chi'(j) \leq u_j$ for all $j \in \{1, \ldots, n\}$. Moreover, for every $j \in \{1, \ldots, n\}$, the total completion time of all upper bound jobs of j is $X^2 \cdot u_j + X \cdot X = (1 + u_j) \cdot X^2$, and the total completion time of all lower bound jobs of j is $(\ell_j - 1) \cdot X$. As $\sum_{j=1}^{n} C(j) = \sum_{j=1}^{n} \chi'(j) \leq nk$, it follows that the total completion time of all jobs in C is at most K. □

We now prove the converse direction, where the idea is that if $C(j) < \ell_j$ or $C(j) > u_j$, then the total completion time of the auxiliary jobs is too large.

Lemma 9. *Let C be a minimal feasible schedule for the constructed $P_n \mid conc \mid \sum C_j$ instance. Consider some $j \in \{1, \ldots, n\}$, and let J denote the set of upper bound jobs of j. Then the total completion time of all jobs in J in C is at least $(u_j + 1) \cdot X^2$. Moreover, if $C(j) \in \{u_j + 1, \ldots, X\}$, then the total completion time of all jobs in J is at least $(u_j + 1) \cdot X^2 + X$.*

Proof (Sketch). We prove the first part; the second follows (see the full version). Let $j(i)$ be a primary upper bound job for $i \in \{1, \ldots, X\}$. If $C(j(i)) \geq X$, then by minimality of the schedule, all secondary upper bound jobs $j(i, 1), \ldots, j(i, X)$ complete at the same time u_j. Hence, $C(j(i)) + \sum_{i_0=1}^{X} C(j(i, i_0)) \geq X(u_j + 1)$. If $C(j(i)) > X$, then this sum is strictly larger than $X(u_j + 1)$. If $C(j(i)) < X$, then all secondary upper bound jobs $j(i, 1), \ldots, j(i, X)$ conflict with $j(i)$ in the schedule and, by minimality of the schedule, complete at the same time $C(j(i, 1)) = \cdots = C(j(i, X)) \geq X$. Hence, since $n \geq 2$, $C(j(i)) + \sum_{i_0=1}^{X} C(j(i, i_0)) \geq (X - u_j) + X \cdot X > X \cdot X > X \cdot (u_j + 2) = X(u_j + 1) + X$. Then the total completion time of all jobs in J is at least $(u_j + 1) \cdot X^2$. □

Lemma 10. *Let C be a minimal feasible schedule for the constructed $P_n \mid conc \mid \sum C_j$ instance. Consider some $j \in V_0$ with $\chi(j) > 1$, and let J denote the set of lower bound jobs of j. Then the total completion time of all jobs in J in C is at least $(\ell_j - 1) \cdot X$. Moreover, if $C(j) < \ell_j$, then the total completion time of all jobs in J is at least $\ell_j \cdot X$.*

Proof. The total completion time of J is at least $(\ell_j - 1) \cdot X$ because this is the total processing time of J. If $C(j) < \ell_j$, then the completion time of any job in J is at least ℓ_j, as all jobs in J are conflicting with j and have processing time $\ell_j - 1$. Since $|J| = X$, the total completion time of J is at least $\ell_j \cdot X$. □

Combining Lemma 9 with Lemma 10 we obtain the following:

Lemma 11. *If there is a minimal feasible schedule C for the constructed $P_n \mid conc \mid \sum C_j$ instance with $\sum C_j \leq K$, then there is a solution $\chi' : \{1, \dots, n\} \to \{1, \dots, k\}$ for (H, k, χ).*

Proof. Let C be a minimal feasible schedule for the constructed $P_n \mid conc \mid \sum C_j$ instance with total completion time at most K. By Lemma 9 and Lemma 10, the total completion time of all jobs other than $\{1, \dots, n\}$ in C is at least $K' = \sum_{j=1}^{n} \left[(1 + u_j) \cdot X^2 + (\ell_j - 1) \cdot X\right]$. Since $K - K' = nk < X$, we have that $C(j) < X$. Therefore, $C(j) \leq u_j$ for all $j \in \{1, \dots, n\}$ by Lemma 9. Also, $C(j) \geq \ell_j$ for all $j \in V_0$ by Lemma 10. It follows that the restriction of C onto $\{1, \dots, n\}$ is a proper extension of the coloring χ onto $\{1, \dots, n\}$. □

The reduction above can be carried out in polynomial time and an additional $O(\log n)$ working space. Moreover, $tw(G) = tw(H)$. Correctness of the reduction is given by Lemma 8 and Lemma 11. This yields a parameterized logspace reduction from the XALP-hard PRECOLORING EXTENSION to $P_n \mid conc \mid \sum C_j$. Membership in XALP follows again from Corollary 1, and so:

Theorem 5. *$P_n \mid conc \mid \sum C_j$ parameterized by treewidth is XALP-complete.*

5 Non-zero Release Times

We now consider the case where each job j has an integer release time $r_j \geq 0$ and in a feasible schedule C we require that $C(j) \geq r_j + p_j$ for all jobs j. We show that the complexity of the concurrency scheduling problems considered above substantially differs from the $r_j = 0$ case. As all problems are XALP-complete for arbitrary processing times, we only consider the unit processing time case.

We begin with makespan objective. By a simple reduction to the $P_n \mid conc, p_j = 1 \mid L_{\max}$ problem, by reversing the order of schedules, we obtain:

Theorem 6 (See full version). *$P_n \mid conc, p_j = 1, r_j \geq 0 \mid C_{\max}$ is solvable in $O^*((tw \cdot \log n + 1)^{tw+1})$-time.*

We next consider the $L_{\max}$ and $WC_{\max}$ objectives. For both of these objectives we previously obtained FPT algorithms when all jobs are released at time 0. However, when this is not true, the problems become XALP-complete:

Theorem 7 (See full version). *$P_n \mid conc, p_j = 1, r_j \geq 0 \mid L_{\max}$ and $P_n \mid conc, p_j = 1, r_j \geq 0 \mid WC_{\max}$ are XALP-complete parameterized by treewidth.*

Finally, we consider the $\sum C_j$ objective.

Theorem 8. *$P_n \mid conc, p_j = 1, r_j \geq 0 \mid \sum C_j$ is XALP-complete parameterized by treewidth.*

Proof (Sketch). The full proof is in the full version. Reduce from PRECOLORING EXTENSION. Let $u_j = \chi(j)$ for each $j \in V_0$, and $u_j = k$ for each $j \in \{1, \ldots, n\} \setminus V_0$. Let $\ell_j = \chi(j) - 1$ for each $j \in V_0$, and $\ell_j = 0$ for each $j \in \{1, \ldots, n\} \setminus V_0$. Then $\chi' : \{1, \ldots, k\}$ is a solution for (H, k, C) if and only if it is a proper coloring of H with $\ell_j < \chi'(j) \leq u_j$ for all $j \in \{1, \ldots, n\}$. Let $X = nk + 1$.

Create n jobs $\{1, \ldots, n\}$, corresponding to the vertices of H, and a conflict graph $G = H$. The release time of each job $j \in \{1, \ldots, n\}$ is $r_j = \ell_j$, to ensure the lower bounds are met. Add, for each $j \in \{1, \ldots, n\}$, a set of upper bound jobs. For $i \in \{u_j, \ldots, X-2\}$, add jobs $j(i,1), \ldots, j(i,X)$ released at time i. Add the edges $\{j, j(i, i_0)\}$ to G for each $j \in \{1, \ldots, n\}$, $i \in \{u_j, \ldots, X\}$, and $i_0 \in \{1, \ldots, X\}$. The required total completion time of all jobs is $K = nk + X \cdot \sum_{j=1}^{n} \sum_{i=u_j}^{X-2} (i+1)$. If in a minimal feasible schedule C, $C(j) > u_j$ for a job j, then the upper bound jobs of type $i = C(j) - 1$ complete after time $i + 1$. This adds X to the total completion time, prohibited by the bound K. Thus, $C(j) \leq u_j$. □

6 Discussion

We may also consider further aspects of scheduling problems. First, if we consider preemption, then $p_j \geq 1$ but the times at which a job is processed is not necessarily a contiguous interval. The problem $P_n \mid conc, pmtn \mid C_{\max}$ is known as PREEMPTIVE MULTICOLORING, MULTICHROMATIC NUMBER, WEIGHTED COLORING, or MINIMUM INTEGER WEIGHTED COLORING [3,4,28,31–33,45,54]. We can reduce PREEMPTIVE MULTICOLORING to COLORING by replacing each job j by p_j jobs that are pairwise in conflict. That is, we replace each vertex by p_j (true) twins. Note that this step can increase the pathwidth and treewidth of the conflict graph by a multiplicative factor of $\max p_j$. Thus, we can transfer positive results for the non-preemptive case with $p_j = 1$ to positive results for the preemptive case with $p_j \geq 1$, when taking $\max p_j$ as additional parameter. It is interesting to investigate what happens when the p_j are not parameter bounded. We may also consider $P_n \mid conc, pmtn \mid \sum C_j$, known as PREEMPTIVE MULTICHROMATIC SUM [6,29].

Second, we can consider a bounded number m of machines. If $p_j = 1$, then $P_m \mid conc \mid C_{\max}$ is equivalent to the well-studied m-BOUNDED COLORING problem and has also been called MUTUAL EXCLUSION SCHEDULING and χ_m-COLORING; see e.g. [1,2,9,12,13,30,36,42]. In particular, for fixed m, this problem is fixed-parameter tractable parameterized by the treewidth of the conflict graph [36], while for variable m, it is W[1]-hard [23] and in XP [9] parameterized by the treewidth of the conflict graph plus the makespan. This problem is also strongly related to the EQUITABLE COLORING problem, where the goal is to color a graph such that the size of any two color classes differs by at most 1.

Acknowledgments. The first author wants to thank Meike Hatzel, Pascal Gollin, Matjaz Krnc, and Martin Milanic for discussions on CHROMATIC SUM and related problems.

References

1. Alon, N.: A note on the decomposition of graphs into isomorphic matchings. Acta Math. Hungar. **42**, 221–223 (1983). https://doi.org/10.1007/BF01956769
2. Baker, B.S., Coffman, E.G., Jr.: Mutual exclusion scheduling. Theor. Comput. Sci. **162**(2), 225–243 (1996). https://doi.org/10.1016/0304-3975(96)00031-X
3. Balas, E., Xue, J.: Minimum weighted coloring of triangulated graphs, with application to maximum weight vertex packing and clique finding in arbitrary graphs. SIAM J. Comput. **20**(2), 209–221 (1991). https://doi.org/10.1137/0220012
4. Balas, E., Xue, J.: Addendum: Minimum weighted coloring of triangulated graphs, with application to maximum weight vertex packing and clique finding in arbitrary graphs. SIAM J. Comput. **21**(5), 1000 (1992). https://doi.org/10.1137/0221058
5. Bar-Noy, A., Bellare, M., Halldórsson, M.M., Shachnai, H., Tamir, T.: On chromatic sums and distributed resource allocation. Inf. Comput. **140**(2), 183–202 (1998). https://doi.org/10.1006/INCO.1997.2677
6. Bar-Noy, A., Halldórsson, M.M., Kortsarz, G., Salman, R., Shachnai, H.: Sum multicoloring of graphs. J. Algor. **37**(2), 422–450 (2000). https://doi.org/10.1006/JAGM.2000.1106
7. Bar-Noy, A., Kortsarz, G.: Minimum color sum of bipartite graphs. J. Algor. **28**(2), 339–365 (1998). https://doi.org/10.1006/JAGM.1998.0938
8. Belmonte, R., Kim, E.J., Lampis, M., Mitsou, V., Otachi, Y.: Grundy distinguishes treewidth from pathwidth. SIAM J. Disc. Math. **36**(3), 1761–1787 (2022). https://doi.org/10.1137/20M1385779
9. Bodlaender, H.L., Fomin, F.V.: Equitable colorings of bounded treewidth graphs. Theor. Comput. Sci. **349**(1), 22–30 (2005). https://doi.org/10.1016/J.TCS.2005.09.027
10. Bodlaender, H.L., Groenland, C., Jacob, H., Pilipczuk, M., Pilipczuk, M.: On the complexity of problems on tree-structured graphs. In: Dell, H., Nederlof, J. (eds.) 17th International Symposium on Parameterized and Exact Computation, IPEC 2022, vol. 249 of LIPIcs, pp. 6:1–6:17. Schloss Dagstuhl - Leibniz-Zentrum für Informatik (2022). https://doi.org/10.4230/LIPICS.IPEC.2022.6
11. Bodlaender, H.L., Groenland, C., Nederlof, J., Swennenhuis, C.: Parameterized problems complete for nondeterministic FPT time and logarithmic space. Inf. Comput. **300**, 105195 (2024). https://doi.org/10.1016/j.ic.2024.105195

12. Bodlaender, H.L., Jansen, K.: On the complexity of scheduling incompatible jobs with unit-times. In: Borzyszkowski, A.M., Sokołowski, S. (eds.) MFCS 1993. LNCS, vol. 711, pp. 291–300. Springer, Heidelberg (1993). https://doi.org/10.1007/3-540-57182-5_21
13. Bodlaender, H.L., Jansen, K.: Restrictions of graph partition problems. Part I. Theor. Comput. Sci. **148**(1), 93–109 (1995). https://doi.org/10.1016/0304-3975(95)00057-4
14. Bodlaender, H.L., Mannens, I., Oostveen, J.J., Pandey, S., van Leeuwen, E.J.: The parameterised complexity of integer multicommodity flow. In: Misra, N., Wahlström, M. (eds.) Proceedings of the 18th International Symposium on Parameterized and Exact Computation, IPEC 2023, vol. 285 of LIPIcs, pp. 6:1–6:19. Schloss Dagstuhl - Leibniz-Zentrum für Informatik (2023). https://doi.org/10.4230/LIPICS.IPEC.2023.6
15. Bodlaender, H.L., Szilágyi, K.: XNLP-hardness of parameterized problems on planar graphs. In: Král, D., Milanic, M. (eds.) Proceedings of the 50th International Workshop on Graph-Theoretic Concepts in Computer Science, WG 2024, vol. 14760 of Lecture Notes in Computer Science, pp. 107–120. Springer, Heidelberg (2024). https://doi.org/10.1007/978-3-031-75409-8_8
16. Bonomo, F., Valencia-Pabon, M.: Minimum sum coloring of p_4-sparse graphs. Electron. Notes Disc. Math. **35**, 293–298 (2009). https://doi.org/10.1016/J.ENDM.2009.11.048
17. Borodin, A., Ivan, I., Ye, Y., Zimny, B.: On sum coloring and sum multi-coloring for restricted families of graphs. Theor. Comput. Sci. **418**, 1–13 (2012). https://doi.org/10.1016/J.TCS.2011.11.010
18. Coffman, E.G., Jr., Garey, M.R., Johnson, D.S., LaPaugh, A.S.: Scheduling file transfers. SIAM J. Comput. **14**(3), 744–780 (1985). https://doi.org/10.1137/0214054
19. Cygan, M., et al.: Parameterized Algorithms. Springer, Cham (2015). https://doi.org/10.1007/978-3-319-21275-3
20. Diestel, R.: Graph Theory, 4th edn, vol. 173 of Graduate Texts in Mathematics. Springer, Heidelberg (2012)
21. Dujmovic, V., Joret, G., Wood, D.R.: An improved bound for first-fit on posets without two long incomparable chains. SIAM J. Disc. Math. **26**(3), 1068–1075 (2012). https://doi.org/10.1137/110855806
22. Elberfeld, M., Stockhusen, C., Tantau, T.: On the space and circuit complexity of parameterized problems: classes and completeness. Algorithmica **71**(3), 661–701 (2014). https://doi.org/10.1007/s00453-014-9944-y
23. Fellows, M.R., et al.: On the complexity of some colorful problems parameterized by treewidth. Inf. Comput. **209**(2), 143–153 (2011). https://doi.org/10.1016/J.IC.2010.11.026
24. Garey, M.R., Johnson, D.S.: Computers and Intractability: A Guide to the Theory of NP-Completeness. W. H. Freeman, New York (1979)
25. Gavril, F.: Algorithms for minimum coloring, maximum clique, minimum covering by cliques, and maximum independent set of a chordal graph. SIAM J. Comput. **1**(2), 180–187 (1972). https://doi.org/10.1137/0201013
26. Golumbic, M.C.: Algorithmic Graph Theory and Perfect Graphs. Academic Press, New York (1980)
27. Graham, R., Lawler, E., Lenstra, J., Kan, A.: Optimization and approximation in deterministic sequencing and scheduling: a survey. Ann. Disc. Math. **3**, 287–326 (1979). https://doi.org/10.1016/S0167-5060(08)70356-X

28. Grötschel, M., Lovász, L., Schrijver, A.: Geometric Algorithms and Combinatorial Optimization, volume 2 of Algorithms and Combinatorics. Springer, Heidelberg (1988). https://doi.org/10.1007/978-3-642-97881-4
29. Halldórsson, M.M., Kortsarz, G.: Tools for multicoloring with applications to planar graphs and partial k-trees. J. Algor. **42**(2), 334–366 (2002). https://doi.org/10.1006/JAGM.2001.1210
30. Hansen, P., Hertz, A., Kuplinsky, J.: Bounded vertex colorings of graphs. Disc. Math. **111**(1–3), 305–312 (1993). https://doi.org/10.1016/0012-365X(93)90165-P
31. Hoàng, C.T.: Efficient algorithms for minimum weighted colouring of some classes of perfect graphs. Disc. Appl. Math. **55**(2), 133–143 (1994). https://doi.org/10.1016/0166-218X(94)90004-3
32. Ito, T., Nishizeki, T., Zhou, X.: Algorithms for the multicolorings of partial k-trees. In: Ibarra, O.H., Zhang, L. (eds.) COCOON 2002. LNCS, vol. 2387, pp. 430–439. Springer, Heidelberg (2002). https://doi.org/10.1007/3-540-45655-4_46
33. Ito, T., Nishizeki, T., Zhou, X.: Algorithms for multicolorings of partial k-trees. IEICE Trans. Inf. Syst. **E86-D**(2), 191–200 (2003). https://search.ieice.org/bin/summary.php?id=e86-d_2_191
34. Jansen, K.: Complexity results for the optimum cost chromatic partition problem. Manuscript (1997)
35. Jansen, K.: The optimum cost chromatic partition problem. In: Bongiovanni, G., Bovet, D.P., Di Battista, G. (eds.) CIAC 1997. LNCS, vol. 1203, pp. 25–36. Springer, Heidelberg (1997). https://doi.org/10.1007/3-540-62592-5_58
36. Kaller, D., Gupta, A., Shermer, T.: The χt-coloring problem. In: Mayr, E.W., Puech, C. (eds.) STACS 1995. LNCS, vol. 900, pp. 409–420. Springer, Heidelberg (1995). https://doi.org/10.1007/3-540-59042-0_92
37. Karp, R.M.: Reducibility among combinatorial problems. In: Miller, R.E., Thatcher, J.W. (eds.) Proceedings of a symposium on the Complexity of Computer Computations, at the IBM Thomas J. Watson Research Center, Yorktown Heights, New York, USA, The IBM Research Symposia Series, 20–22 March 1972, pp. 85–103. Plenum Press, New York (1972). https://doi.org/10.1007/978-1-4684-2001-2_9
38. Kubale, M.: Interval vertex-coloring of a graph with forbidden colors. Disc. Math. **74**(1), 125–136 (1989). https://doi.org/10.1016/0012-365X(89)90204-5
39. Kubicka, E.M.: The chromatic sum of a graph: history and recent developments. Int. J. Math. Math. Sci. **2004**(30), 1563–1573 (2004). https://doi.org/10.1155/S0161171204306216
40. Kubicka, E.M., Schwenk, A.J.: An introduction to chromatic sums. In: Riehl, A.M. (ed.) Computer Trends in the 1990s - Proceedings of the 1989 ACM 17th Annual Computer Science Conference, Louisville, Kentucky, USA, 21–23 February 1989, pp. 39–45. ACM (1989). https://doi.org/10.1145/75427.75430
41. Lokshtanov, D., Marx, D., Saurabh, S.: Known algorithms on graphs of bounded treewidth are probably optimal. ACM Trans. Algor. **14**(2), 13:1–13:30 (2018). https://doi.org/10.1145/3170442
42. Lonc, Z.: On complexity of some chain and antichain partition problems. In: Schmidt, G., Berghammer, R. (eds.) WG 1991. LNCS, vol. 570, pp. 97–104. Springer, Heidelberg (1992). https://doi.org/10.1007/3-540-55121-2_9
43. Malafiejski, M., Giaro, K., Janczewski, R., Kubale, M.: Sum coloring of bipartite graphs with bounded degree. Algorithmica **40**(4), 235–244 (2004). https://doi.org/10.1007/S00453-004-1111-4

44. Marx, D.: A short proof of the NP-completeness of minimum sum interval coloring. Oper. Res. Lett. **33**(4), 382–384 (2005). https://doi.org/10.1016/J.ORL.2004.07.006
45. McDiarmid, C., Reed, B.A.: Channel assignment and weighted coloring. Networks **36**(2), 114–117 (2000). 10.1002/1097-0037(200009)36:2<114::AID-NET6>3.0.CO;2-G
46. Nicoloso, S., Sarrafzadeh, M., Song, X.: On the sum coloring problem on interval graphs. Algorithmica **23**(2), 109–126 (1999). https://doi.org/10.1007/PL00009252
47. Pilipczuk, M., Wrochna, M.: On space efficiency of algorithms working on structural decompositions of graphs. ACM Trans. Comput. Theory **9**(4), 18:1–18:36 (2018). https://doi.org/10.1145/3154856
48. Pinedo, M.: Scheduling: Theory, Algorithms, and Systems, 5th edn. Springer, Heidelberg (2016)
49. Salavatipour, M.: On sum coloring of graphs. Master's thesis, University of Toronto (2000). https://hdl.handle.net/1807/15002
50. Salavatipour, M.R.: On sum coloring of graphs. Disc. Appl. Math. **127**(3), 477–488 (2003). https://doi.org/10.1016/S0166-218X(02)00249-4
51. Supowit, K.: Finding a maximum planar subset of a set of nets in a channel. IEEE Trans. Comput. Aided Des. Integr. Circuits Syst. **6**(1), 93–94 (1987). https://doi.org/10.1109/TCAD.1987.1270250
52. Szkaliczki, T.: Routing with minimum wire length in the dogleg-free manhattan model is NP-complete. SIAM J. Comput. **29**(1), 274–287 (1999). https://doi.org/10.1137/S0097539796303123
53. Telle, J.A., Proskurowski, A.: Algorithms for vertex partitioning problems on partial k-trees. SIAM J. Disc. Math. **10**(4), 529–550 (1997). https://doi.org/10.1137/S0895480194275825
54. Zhou, X., Nishizeki, T.: Multicolorings of series-parallel graphs. Algorithmica **38**(2), 271–297 (2004). https://doi.org/10.1007/S00453-003-1060-3

Non-crossing H-Graphs: A Generalization of Proper Interval Graphs Admitting FPT Algorithms

Flavia Bonomo-Braberman[1,2], Nick Brettell[3], Andrea Munaro[4], and Daniël Paulusma[5(✉)]

[1] Departamento de Computación, FCEN, Universidad de Buenos Aires, Buenos Aires, Argentina
[2] ICC CONICET/UBA, Buenos Aires, Argentina
fbonomo@dc.uba.ar
[3] School of Mathematics and Statistics, Victoria University of Wellington, Wellington, New Zealand
nick.brettell@vuw.ac.nz
[4] Department of Mathematical, Physical and Computer Sciences, University of Parma, Parma, Italy
andrea.munaro@unipr.it
[5] Department of Computer Science, Durham University, Durham, UK
daniel.paulusma@durham.ac.uk

Abstract. We prove new parameterized complexity results for the FO Model Checking problem and in particular for Independent Set, for two recently introduced subclasses of H-graphs, namely proper H-graphs and non-crossing H-graphs. It is known that proper H-graphs, and thus H-graphs, may have unbounded twin-width. However, we prove that for every multigraph H, non-crossing H-graphs have bounded proper mixed-thinness, and thus bounded twin-width. Consequently, we can apply a well-known result of Bonnet, Kim, Thomassé, and Watrigant (2021) to find that the FO Model Checking problem is in FPT for non-crossing H-graphs when parameterized by $\|H\| + \ell$, where $\|H\|$ is the size of H and ℓ is the size of a formula. In particular, this implies that Independent Set is in FPT for non-crossing H-graphs when parameterized by $\|H\|+k$, where k is the solution size. In contrast, Independent Set for general H-graphs is known to be $\mathsf{W}[1]$-hard when parameterized by $\|H\| + k$. We strengthen the latter result by proving that Independent Set is $\mathsf{W}[1]$-hard even on proper H-graphs when parameterized by $\|H\| + k$. In this way, we solve, subject to $\mathsf{W}[1] \neq \mathsf{FPT}$, an open problem of Chaplick (2023), who asked whether there exist problems that can be solved faster for non-crossing H-graphs than for proper H-graphs.

Keywords: H-graphs · FO Model Checking · parameterized complexity · proper mixed-thinness · twin-width

1 Introduction

The FO Model Checking problem is to decide whether a given graph G satisfies a given first-order formula ϕ. As such, the FO Model Checking problem captures

H. Fernau and P. Kindermann (Eds.): WG 2025, LNCS 16124, pp. 121–134, 2026.
https://doi.org/10.1007/978-3-032-11835-6_9

a large class of graph problems, including Clique, Dominating Set and Independent Set. We prove new parameterized complexity results for FO Model Checking by restricting the input to certain classes of intersection graphs and proving boundedness of proper mixed-thinness. Showing the latter is the focus of our paper, as it implies boundedness of twin-width, which enables us to apply a well-known result of Bonnet et al. [4]; we explain this in more detail below.

We consider the framework of intersection graphs, because this framework captures many well-known graph classes and plays a central role in algorithmic graph theory. An *intersection graph* $G_{\mathcal{S}}$ of a family $\mathcal{S}$ of subsets of a set X has a unique vertex s_i for each $S_i \in \mathcal{S}$ and there is an edge between two distinct vertices s_i and s_j if and only if $S_i \cap S_j \neq \varnothing$. Two very popular classes of intersection graphs are those of *interval graphs* and *circular-arc graphs*. We obtain these classes by letting $\mathcal{S}$ be a family of intervals of the real line or arcs of the circle, respectively. The reason that both graph classes are so widely known is that they have various "good" algorithmic properties, often leading to fast algorithms.

It is natural to try to extend a graph class $\mathcal{G}$ into a larger graph class that preserves the desirable algorithmic properties of $\mathcal{G}$. Applying this approach to interval graphs and circular-arc graphs has led to the notion of H-graphs, introduced in 1992 by Bíró, Hujter and Tuza [2]. In order to define H-graphs, we first need to introduce some terminology. The *subdivision* of an edge uv in a graph replaces uv by a new vertex w and edges uw and wv. A graph H' is a *subdivision* of a graph H if we can modify H into H' by a sequence of edge subdivisions. A vertex subset of a graph is *connected* if it induces a connected subgraph (the empty set is also connected). A *multigraph* may have parallel edges (edges between the same pair of vertices) and self-loops (edges uu). A *simple graph* is a graph with no parallel edges and no self-loops.

Definition 1. *For a multigraph H, a (simple) graph G is an H-*graph *if G is the intersection graph of a family $\mathcal{S}$ of connected vertex subsets of a subdivision H' of H; we say that $\mathcal{S}$ is an H-*representation *of G.*

We refer to Fig. 1 for an example of an H-graph G. We also note that the P_2-graphs are exactly the interval graphs, and that the C_2-graphs are exactly the circular-arc graphs; here, P_r is the r-vertex path and C_s is the s-vertex cycle, where C_2 is the 2-vertex cycle, consisting of two vertices with two parallel edges. As such, the size $\|H\| = |E(H)|$ (counting multiplicities) of H is a natural parameter for "measuring" how far an H-graph is from being interval or circular-arc. Hence, H-graphs form a *parameterized* generalization of interval graphs and circular-arc graphs. This led Chaplick et al. [11] to initiate an algorithmic study of H-graphs. They showed that fundamental problems, such as Independent Set and Dominating Set, are all in XP on H-graphs, when parameterized by $\|H\|$, and also that Clique is para-NP-hard on H-graphs when parameterized by $\|H\|$ (see [12] for background on parameterized complexity). Fomin, Golovach and Raymond [15] complemented these results by showing that Independent Set and Dominating Set on H-graphs are W[1]-hard when parameterized by $\|H\| + k$, where k denotes the size of the independent (resp. dominating) set.

a proper and non-crossing | b non-crossing but not proper | c proper but not non-crossing | d neither proper nor non-crossing

Fig. 1. Four different representations of the claw ($K_{1,3}$) as a claw-graph.

We refer to [7,11,20,27] for NP-completeness and polynomial-time results for recognizing H-graphs, depending on the structure of H.

We obtain the classes of *proper* interval graphs and *proper* circular-arc graphs by requiring the corresponding family $\mathcal{S}$ of intervals of the real line or arcs of the circle, respectively, to be *proper*; here, a family $\mathcal{S}$ is *proper* if no set in $\mathcal{S}$ properly contains another. Proper interval graphs and proper circular-arc graphs have certain structural properties that interval graphs and circular-arc graphs do not have, and that can be exploited algorithmically. In particular, FO Model Checking is well known to be FPT on proper interval graphs [16] and even on proper circular-arc graphs [17] when parameterized by the length of the formula. As explained in [17], such a result does not hold for interval graphs. This is due to the fact that Induced Subgraph Isomorphism (does a given graph G contain a given graph H as an induced subgraph?) is W[1]-hard on interval graphs when parameterized by $|V(H)|$ [23].

The underlying reason a problem is polynomial-time solvable on some graph class $\mathcal{G}$ might be the fact that $\mathcal{G}$ has *bounded* width for some width parameter p, that is, there exists a constant c such that $p(G) \leq c$ for every graph $G \in \mathcal{G}$. Bonnet et al. [4] proved that FO Model Checking is FPT when parameterized by twin-width plus the length of the formula. As it turns out, proper circular-arc graphs, and thus proper interval graphs, have bounded twin-width [19], whereas interval graphs, and thus circular-arc graphs, and more generally H-graphs, have unbounded twin-width [3]. This raises the question whether it is possible to generalize proper interval graphs and proper circular-arc graphs to subclasses of H-graphs whose twin-width is at most $f(\|H\|)$ for some function f that only depends on $\|H\|$. Due to [4], the latter would imply that FO Model Checking is FPT when parameterized by $\|H\|$ plus the length of the formula.

Recently, two new rich subclasses of H-graphs were introduced. First, Chaplick et al. [9] defined proper H-graphs (see also Fig. 1).

Definition 2. *For a multigraph H, a (simple) graph G is a* proper *H-graph if G is the intersection graph of a proper family $\mathcal{S}$ of connected vertex subsets of a subdivision H' of H; we say that $\mathcal{S}$ is a* proper *H-representation of G.*

Note that the proper P_2-graphs are exactly the proper interval graphs, and that the proper C_2-graphs are exactly the proper circular-arc graphs. We refer to [10,13,20] for NP-completeness and polynomial-time results for recognizing proper H-graphs, depending on H. Chaplick et al. [9] showed the existence of polynomial kernels for Hamiltonian-type problems on proper H-graphs. However, as a possible drawback, Jedelský [19] proved that already proper $K_{1,3}$-graphs may have unbounded twin-width; here, $K_{1,3}$ denotes the *claw*, i.e., the 4-vertex star.

Second, Chaplick [8] defined non-crossing H-graphs, whose definition requires an extra notion. Namely, a family $\mathcal{S}$ of connected vertex subsets of some graph is *non-crossing* if, for every pair A, B in $\mathcal{S}$, both $A \setminus B$ and $B \setminus A$ are connected. See also Fig. 1 and see [21] for a different type of non-crossing intersection graphs that includes disk graphs.

Definition 3. *For a multigraph H, a (simple) graph G is a* non-crossing H-graph *if G is the intersection graph of a non-crossing family $\mathcal{S}$ of connected vertex subsets of a subdivision H' of H; we say that $\mathcal{S}$ is a* non-crossing H-representation *of G.*

The classes of non-crossing P_2-graphs and proper P_2-graphs coincide [8]. However, proper H-graphs and non-crossing H-graphs form incomparable graph classes in general. Even though every proper C_2-graph is non-crossing, the claw is already an example of a non-crossing C_2-graph that is not proper. On the other hand, we can show that there exists a proper claw-graph, namely the 4-fan, that is not a non-crossing T-graph for any tree T (proof omitted).

Chaplick [8] gave structural properties of non-crossing H-graphs and variants to obtain linear-time algorithms not only for recognizing certain non-crossing H-graphs, but also, for example, for deciding if a chordal graph is claw-free. Given the structure of non-crossing H-graphs, Chaplick [8] asked the following:

Can some problems be solved faster for non-crossing H-graphs than for proper H-graphs?

We aim to answer this question with respect to the FO Model Checking problem, and we will therefore also focus on:

Do non-crossing H-graphs have twin-width at most $f(\|H\|)$ for some function f that only depends on $\|H\|$?

Our Results. We provide positive answers to both questions. To answer the second one, we sketch in Sect. 3 a proof of boundedness of a less powerful width parameter, namely proper mixed-thinness (see Sect. 2 for its definition). Balabán, Hlinený and Jedelský [1] introduced the notion of proper mixed-thinness and, as we discuss below, they also proved that it is a less powerful width parameter than twin-width.

Theorem 1. *For a multigraph H, every non-crossing H-graph G has proper mixed-thinness at most $t = 2^{\|H\|}(4^{\|H\|}-1)(\|H\|^2+1)+\|H\|$; moreover, a proper t-mixed-thin representation can be obtained in polynomial time from a non-crossing H-representation of G.*

Table 1. Complexity of INDEPENDENT SET on H-free graphs, where k denotes the size of a solution. It is still open whether INDEPENDENT SET is in FPT for non-crossing H-graphs when parameterized by $\|H\|$ only.

Parameter	Graph class	Complexity	Reference
$\|H\|$	H-graphs	XP	[11]
	proper H-graphs	XP	[11] (from H-graphs)
	non-crossing H-graphs	XP	[11] (from H-graphs)
$\|H\| + k$	H-graphs	W[1]-hard	[15] (strengthened by Theorem 2)
	proper H-graphs	W[1]-hard	Theorem 2
	non-crossing H-graphs	FPT	Corollary 1

In general, proper H-graphs may have unbounded proper mixed-thinness, because they may have unbounded twin-width [19]. However, we prove Theorem 1 by adapting a proof of a corresponding but incomparable result of Balabán, Hlinený and Jedelský [1], which shows that some proper H-graphs *have* bounded proper mixed-thinness, namely the class of proper H-graphs that have a proper H-representation $\mathcal{S}$, in which each set induces a path in the corresponding subdivision H' of H. We state their result as Theorem 3 in Sect. 3.

Balabán, Hlinený and Jedelský [1] proved that for every integer $t \geq 1$, the twin-width of a graph G with proper mixed-thinness t is at most $9t$, and moreover a $9t$-contraction sequence for G can be computed in polynomial time from a proper t-mixed-thin representation of G. Recall also that Bonnet et al. [4] proved that deciding a first-order formula ϕ for an n-vertex graph G with a given d-contraction sequence can be done in $f(d, |\phi|) \cdot O(n)$ time for some function f only depending on d and $|\phi|$. Combining these two results with Theorem 1 yields:

Corollary 1. *The FO Model Checking problem is in* FPT *for non-crossing H-graphs when parameterized by $\|H\| + \ell$, where ℓ is the size of a formula, provided that a non-crossing H-representation of the input graph G is given.*

Note that Corollary 1 applies to classic problems, such as INDEPENDENT SET, CLIQUE and DOMINATING SET, which are all special cases of the FO Model Checking problem. Hence, in particular, we find that for a multigraph H, it is possible to check in $g(\|H\|, k) \cdot O(n)$ time if an n-vertex graph G with a given non-crossing H-representation has an independent set of size at least k, where g is a function that only depends on $\|H\|$ and k.

The fact that proper H-graphs may have unbounded twin-width does not necessarily excludes FPT algorithms. However, our next result, proven in Sect. 4, provides hard evidence. Together with the above result it shows that there exist problems that can be solved faster for non-crossing H-graphs than for proper H-graphs (assuming W[1] $\neq$ FPT). Hence, we also provide a positive answer to our first research question, originally posed by Chaplick [8].

Theorem 2. Independent Set *is* W[1]*-hard for proper* H*-graphs when parameterized by* $\|H\| + k$*, even if a proper* H*-representation of the input graph* G *is given.*

Theorem 2 strengthens the aforementioned W[1]-hardness result of Fomin, Golovach and Raymond [15] for general H-graphs, when parameterized by $\|H\| + k$; see also the summary in Table 1.

From our results we conclude that, perhaps somewhat surprisingly, non-crossing H-graphs are indeed the "correct" generalization of proper interval graphs in the H-graph setting, at least from the viewpoint of parameterized complexity. Nevertheless, there are still many open structural and algorithmic problems resulting from our paper, which we list in Sect. 5.

2 Preliminaries

***H*-Graph Representations.** For a multigraph H, let G be an H-graph with H-representation $\mathcal{S}$. Thus, $\mathcal{S}$ is a family of connected vertex subsets of some subdivision H' of H, in which case we call H' the *framework*. Moreover, for every vertex $u \in V(G)$, there exists a unique vertex subset $S_u \subseteq V(H')$, which we call the *representative* of u in $\mathcal{S}$. By definition, for every two distinct vertices u and v in G it holds that u and v are adjacent if and only if $S_u \cap S_v \neq \varnothing$. Now, let x and y be two adjacent vertices of H. To obtain H', the edge xy may be replaced by a path $xz_1 \cdots z_r y$ for some $r \geq 1$. In that case, we say that $z_1, \ldots, z_r$ are the *internal vertices* of the edge xy.

Width Parameters. A graph $G = (V, E)$ is *k-thin* if there is an ordering $v_1, \ldots, v_n$ of V and a partition of V into k classes $(V^1, \ldots, V^k)$ such that for every triple (r, s, t) with $r < s < t$ the following holds: if v_r and v_s belong to the same class V^i for some $i \in \{1, \ldots, k\}$ and moreover $v_r v_t \in E$, then $v_s v_t \in E$. In this case, the ordering $v_1, \ldots, v_n$ and the partition $(V^1, \ldots, V^k)$ are said to be *consistent* and form a *k-thin representation* of G. The *thinness* $\text{thin}(G)$ of G, introduced by Mannino et al. [22], is the minimum integer k such that G is k-thin. Interval graphs are known to be equivalent to 1-thin graphs [24,25], and a vertex ordering consistent with the partition into a single class is called an *interval order* for G.

A graph $G = (V, E)$ is *proper k-thin* if there is an ordering $v_1, \ldots, v_n$ of V and a partition of V into k classes $(V^1, \ldots, V^k)$ such that both $v_1, \ldots, v_n$ and $v_n, \ldots, v_1$ are consistent with the partition. In this case, the partition and the ordering $v_1, \ldots, v_n$ are said to be *strongly consistent*, and form a *proper k-thin representation* of G. The *proper thinness* $\text{pthin}(G)$ of G, introduced by Bonomo and de Estrada [5], is the minimum k such that G is proper k-thin and is denoted by $\text{pthin}(G)$. Proper interval graphs are known to be equivalent to proper 1-thin graphs [26], and a vertex ordering strongly consistent with the partition into a single class is called a *proper interval order* for G.

We now define the parameters mixed-thinness and proper mixed-thinness. These parameters were introduced by Balabán, Hlinený and Jedelský [1] and form a lower bound for thinness and proper thinness, respectively. Let $G = (V, E)$ be a graph and let $k > 0$ be an integer. For disjoint $W, W' \subseteq V$, we let $G[W, W']$ be the bipartite subgraph of G induced by the edges with one end-vertex in W and the other in W'. For convenience, we also let $G[W, W] = G[W]$. For two linear orders $<$ and $<'$ on the same set, we say that $<$ and $<'$ are *aligned* if they are the same or one is the inverse of the other. Now, G is *k-mixed-thin* if there exists a partition $(V^1, \dots, V^k)$ of V and, for each $1 \le i \le j \le k$, a linear order $<_{ij}$ on $V^i \cup V^j$ and a choice of $E_{i,j} \in \{E(G[V^i, V^j]), E(\overline{G}[V^i, V^j])\}$ such that, for every $1 \le i \le j \le k$, the following hold:

(AL) the restriction of $<_{ij}$ to V^i (resp. to V^j) is aligned with $<_{ii}$ (resp. $<_{jj}$);
(CO) for every triple u, v, w such that ($\{u, v\} \subseteq V^i$ and $w \in V^j$) or ($\{u, v\} \subseteq V^j$ and $w \in V^i$), we have that if $u <_{ij} v <_{ij} w$ and $uw \in E_{i,j}$, then $vw \in E_{i,j}$.

A k-mixed-thin graph G is *proper* if, in addition to (AL) and (CO), the following holds:

(SC) for every triple u, v, w such that ($\{v, w\} \subseteq V^i$ and $u \in V^j$) or ($\{v, w\} \subseteq V^j$ and $u \in V^i$), we have that if $u <_{ij} v <_{ij} w$ and $uw \in E_{i,j}$, then $uv \in E_{i,j}$.

A (proper) k-mixed-thin graph G is *inversion-free* if (AL) is replaced with the following:

(IN) the restriction of $<_{ij}$ to V^i (resp. to V^j) is equal to $<_{ii}$ (resp. $<_{jj}$).

We can represent the edge sets choice by a matrix $R \in \{-1, 1\}^{k \times k}$ such that, for every $1 \le i \le j \le k$, $R_{ij} = 1$ if $E_{i,j} = E(G[V^i, V^j])$ and $R_{ij} = -1$ if $E_{i,j} = E(\overline{G}[V^i, V^j])$. We then call the 3-tuple $((V^1, \dots, V^k), \{<_{i,j} : 1 \le i \le j \le k\}, R)$ a *(proper) k-mixed-thin representation* of G.

3 A Sketch of the Proof of Theorem 1

In this section we sketch the proof of Theorem 1. As mentioned, we prove Theorem 1 by adapting a proof of a result of Balabán, Hlinený and Jedelský [1] using a different approach (as we explain in detail later).

Theorem 3 (Balabán, Hlinený and Jedelský [1]**).** *Let H be a simple connected graph*[1]*, and let k be the sum of the number of paths and the number of cycles in H. Let G be a proper H-graph, admitting a proper H-representation where each vertex set induces a path in the subdivision of H. Then G is proper $||H||^2k$-mixed-thin, and a proper $||H||^2k$-mixed-thin representation of G can be computed in polynomial time from such an H-representation of G.*

[1] We remark that the proof of Theorem 3 in [1] in fact works for every multigraph H.

Note that Theorem 1 and Theorem 3 address incomparable graph classes. We will need an auxiliary result, Lemma 1, which states that a proper circular-arc graph is inversion-free proper 2-mixed-thin, and describes how to obtain a representation. The key idea behind Lemma 1 appears in [1] (in the proof of Theorem 3.7, see cases 3 and 6).

Before stating Lemma 1, we need one more notion. Let G be a proper circular-arc graph. We now describe a natural way to obtain a partition (W, W') of $V(G)$, and total orders $<_W$ and $<_{W'}$ on W and W' respectively, from a representation of G. So let $\mathcal{S}$ be a C_2-representation of G, with framework on vertex set $\{x_1, x_2, \ldots, x_p\}$, where $x_1, x_2, \ldots, x_p, x_1$ is a cycle. Note that we do not require that the representation $\mathcal{S}$ is proper (in particular, it might be non-crossing but not proper). Each vertex $s_i \in V(G)$ has a representative in $\mathcal{S}$ that we denote by S_i. Let W be the subset of vertices of G given by $\{s_i : S_i \text{ contains } x_1\}$, and let $W' = V(G) \setminus W$. For each $s_i \in W'$, we have $S_i = \{x_i, \ldots, x_j\}$ for $1 < i \leq j < p+1$; we obtain an order $<_{W'}$ on $s_i \in W'$ by first ordering on i, and then ordering on j, where we break ties arbitrarily. For each $s_i \in W$, we have either $S_i = \{x_i, \ldots, x_p, x_1, \ldots, x_j\}$ with $j \leq i$, in which case we let $i' = i$; or $S_i = \{x_1, \ldots, x_j\}$ for $j > 1$, in which case we let $i' = p + 1$; we obtain an order $<_W$ on $s_i \in W$ by first ordering on i', and then ordering on j, where we break ties arbitrarily. We call $((W, W'), <_W, <_{W'})$ a *natural bipartition ordering of G from $\mathcal{S}$*.

Lemma 1. *Let G be a proper circular-arc graph. Then G is inversion-free proper 2-mixed-thin. Moreover, for a natural bipartition ordering $((W, W'), <_W, <_{W'})$ of G, there is a total ordering $<$ of $V(G)$ that extends both $<_W$ and $<_{W'}$ such that $((W, W'), \{<_W, <, <_{W'}\}, \left(\begin{smallmatrix} 1 & -1 \\ -1 & 1 \end{smallmatrix}\right))$ is a proper 2-mixed-thin representation of G that satisfies (IN), (CO), and (SC).*

We are finally ready to sketch the proof of Theorem 1, which we restate below for convenience. The proof is inspired by the proof of Theorem 3 by Balabán, Hlinený and Jedelský [1]. Their approach was the following. Suppose G is the intersection graph of a proper family of paths in a subdivision H' of H. First, partition the vertices of G according to the vertices and (subdivided) edges of H involved in their representatives. We can assume, by possibly choosing a "finer" subdivision H', that no end of a representative in H' is a vertex of H, i.e., that the ends of all paths "lie inside" the subdivided edges of H. Then, refine the partition according to the (at most two) edges of H whose subdivisions contain the endpoints of the representative path. Let us say that these are the "special" edges of H for the class. Finally, we can order each class in a strongly consistent way, and show that for every pair of classes, the respective orders can be extended to a strongly consistent common total order (by possibly complementing the edges between the classes). This last part is heavily based on the fact that the family of paths is proper, which allows a vertex order within each class to be defined using the order of the corresponding paths endpoints on each of the (at most two) special edges of H for that class.

In our proof, we start in a similar way: based on a non-crossing H-representation of a graph G with framework H', we first partition the vertices of G

according to the vertices and (subdivided) edges of H involved in their representatives. Note that, unlike for the endpoints of the paths, there can be more than two degree-1 vertices of representatives, so it is not straightforward how to define an order with the desired properties. Our approach is to fix an "ambassador" of each class, and prove that, thanks to the non-crossing property, for every other vertex in the class, its representative and the representative of the ambassador differ on at most two (subdivided) edges of H. We refine the partition classes according to edges, and prove that, again thanks to the non-crossing property, we are able to order these refined classes in a strongly consistent way, and such that, for every pair of classes, the respective orders can be extended to a strongly consistent common total order (after possibly complementing the edges between the classes). For this last part (as well as for Lemma 1), we simplify our arguments by using a bipartite pattern characterization of 2-thin graphs from [6]. This characterization allows us to show, given two classes and a proper interval order of each of them, the existence of a linear extension of both the orders satisfying (CO) and (SC).

Theorem 1. *For a multigraph H, every non-crossing H-graph G has proper mixed-thinness at most $t = 2^{\|H\|}(4^{\|H\|}-1)(\|H\|^2+1)+\|H\|$; moreover, a proper t-mixed-thin representation can be obtained in polynomial time from a non-crossing H-representation of G.*

Proof (Sketch). Suppose G has a non-crossing H-representation $\mathcal{X}$ on framework H', where each $v \in V(G)$ has a representative X_v in $\mathcal{X}$ (so X_v is a connected vertex subset of H'). For each vertex $v \in V(G)$, let $E_v^1 \subseteq E(H)$ be the set of edges e of H such that both endpoints of e and all the internal vertices of e in H' belong to X_v; let $E_v^2 \subseteq E(H)$ be the set of edges e of H such that both endpoints of e belong to X_v but not all the internal vertices of e in H' belong to X_v; let $E_v^3 \subseteq E(H)$ be the set of edges e of H such that X_v contains exactly one endpoint of e (and possibly some internal vertices of e in H'); let $E_v^4 \subseteq E(H)$ be the set of edges e of H such that none of the endpoints of e belongs to X_v but at least one internal vertex of e in H' belongs to X_v; finally, let $F_v^3 \subseteq E(H) \times V$ be the set of pairs (e, x) such that $e \in E_v^3$ and x is the endpoint of e that belongs to X_v. We call $P_v = (E_v^1, E_v^2, F_v^3, E_v^4)$ the *profile* of v. See Fig. 2 for an example.

Note that, since every X_v is a connected vertex subset of H', either $E_v^4 = \varnothing$ or we have $|E_v^4| = 1$ and $E_v^1 = E_v^2 = E_v^3 = F_v^3 = \varnothing$. There are m possible profiles for which $|E^4| = 1$. When $E^4 = \varnothing$, there are $4^m - 1$ ways of distributing the edges of H into E^1, E^2, E^3 and $E(H) \setminus (E^1 \cup E^2 \cup E^3)$ in such a way that $E^1 \cup E^2 \cup E^3$ is non-empty; for each such way, there are at most $2^{|E^3|} \leq 2^m$ possibilities for choosing the endpoint of each edge in F^3. Therefore, there are at most $(4^m - 1)2^m + m$ different profiles. Consider now the partition $\mathcal{P} = (V[P_u] : u \in V(G)) = (V_1, V_2, \ldots, V_k)$ of $V(G)$, where $V[P_u] = \{v \in V(G) : P_v = P_u\}$.

Claim 1. Let v and w be distinct vertices in V_i, for some $i \in \{1, 2, \ldots, k\}$. Then $X_v \setminus X_w$ is either empty or a non-empty connected subset of the internal vertices

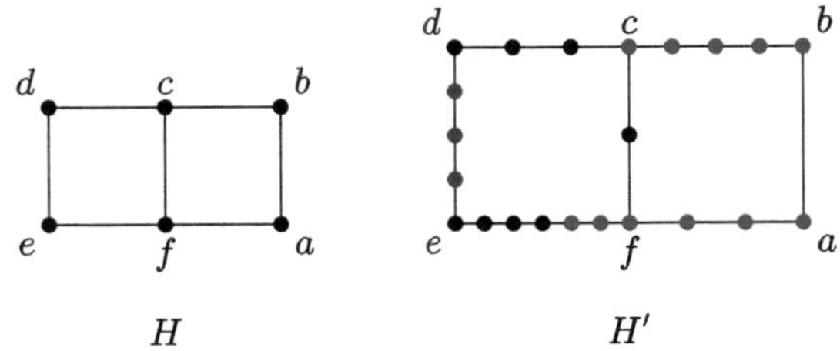

Fig. 2. A graph H and a subdivision H' of H. The profiles of vertices v (red) and w (blue), as defined in the proof of Theorem 1, are: $E_v^1 = \{ab, bc, af\}, E_v^2 = \{cf\}, E_v^3 = \{cd, ef\}, F_v^3 = \{(cd, c), (ef, f)\}$, $E_v^4 = \varnothing$; $E_w^1 = E_w^2 = E_w^3 = F_w^3 = \varnothing$, $E_w^4 = \{de\}$. (Color figure online)

of some edge e_1 of $E(H)$, and $X_w \setminus X_v$ is either empty or a non-empty connected subset of the internal vertices of some edge e_2 of $E(H)$, where neither e_1 nor e_2 belongs to $E_v^1 = E_w^1$.

We prove the claim by using the fact that the representation is non-crossing, and we use it to further refine the partition classes of $\mathcal{P}$ corresponding to profiles where $E^4 = \varnothing$. Let $i \in \{1, \dots, k\}$ be such that V_i is such a partition class in $\mathcal{P}$. We fix a representative v_i in V_i and let $V_i' = \{w \in V(G) : X_w = X_{v_i}\}$. In particular, $v_i \in V_i'$. For $e \in E(H)$, we define $W_i(e)$ as the set of vertices $w \in V_i$ such that at least one of $X_{v_i} \setminus X_w$ and $X_w \setminus X_{v_i}$ is non-empty and both are subsets of the internal vertices of e. For $e_1, e_2 \in E(H)$ with $e_1 \neq e_2$, we define $Z_i(e_1, e_2)$ as the set of vertices $w \in V_i$ such that $X_{v_i} \setminus X_w$ is a non-empty subset of the internal vertices of e_1 and $X_w \setminus X_{v_i}$ is a non-empty subset of the internal vertices of e_2. Now, by the claim above, there is a partition of V_i given by $\mathcal{V}_i = \{V_i'\} \cup \{W_i(e) : e \in E(H) \text{ and } W_i(e) \neq \varnothing\} \cup \{Z_i(e_1, e_2) : e_1, e_2 \in E(H), e_1 \neq e_2 \text{ and } Z_i(e_1, e_2) \neq \varnothing\}$. We refine the partition $\mathcal{P}$ to a partition $\mathcal{P}'$ of $V(G)$ by replacing each class $V_i = V[P_u]$ of $\mathcal{P}$ for which $E_u^4 = \varnothing$ with $\mathcal{V}_i$. Observe that the partition $\mathcal{P}'$ has at most $((4^m - 1)2^m)(m(m-1) + m + 1) + m = 2^m(4^m - 1)(m^2 + 1) + m$ partition classes.

For each pair W, W' of partition classes in $\mathcal{P}'$, we then explain how to choose a linear order on $W \cup W'$ and an appropriate $E_{W,W'}$ satisfying (AL), (CO) and (SC). This technical part of the proof is done by a non-trivial case analysis, and makes use of Lemma 1 and the characterization lemma of [6]. □

4 The Proof of Theorem 2

In this section we prove our W[1]-hardness result for Independent Set on proper H-graphs parameterized by $\|H\| + k$. To put our result in perspective, we start with a small remark.

Remark 1. It is not true that for every H-graph G there exists a function f such that G is a proper H'-graph for some graph H' with $\|H'\| = f(\|H\|)$, even in the case $H = P_2$. Indeed, as the claw is not a proper interval graph, every proper H'-representation of a claw has at least one vertex subset containing an original vertex of H'. It is then enough to consider the P_2-graph (interval graph) that is the disjoint union of an arbitrarily large number of claws. ◇

The following result is crucial for the proof.

Proposition 1. *Given an H-graph G together with an H-representation of G, it is possible to find, in time polynomial in $|V(G)| + |V(H)|$, a parameter k_1 such that $k_1 \leq \alpha(G)$, a graph H' with $|V(H')| = 2|V(H)| + 4k_1$, and a proper H'-representation of G.*

Proof. Without loss of generality, we will consider representations in which no two vertices are represented by exactly the same set. Otherwise, we can keep only one such vertex, say v, and add afterwards the necessary copies of v, all of them being represented by the new representative set of v.

Let $\mathcal{S}$ be an H-representation of G with framework H'', where the representative of each $v \in V(G)$ is S_v. If we remove from G the vertices whose representatives contain a vertex of H, the remaining vertices are exactly those whose representatives are intervals of the subdivision of an edge of H not containing any vertex of H. For each of these edges of H, we traverse the edge in some arbitrary direction and compute a maximum set of pairwise disjoint intervals using the following greedy algorithm: Start with the empty set and iteratively add, among those intervals that do not intersect the current set, the one that ends first, breaking ties with the one that starts last. Observe that, by construction, every unpicked interval contains at least one of the endpoints of some picked interval. The union I over the edges of H of these sets of pairwise disjoint intervals clearly corresponds to an independent set of G and so, if $|I| = k_1$, then $k_1 \leq \alpha(G)$.

Let now W be the union of $V(H)$ and the $2k_1$ vertices of H'' corresponding to the endpoints of the intervals in I. Then, each representative of a vertex of G in the original representation $\mathcal{S}$ contains at least one vertex of W. Moreover, if in the representation $\mathcal{S}$ the representative of the vertex v is properly contained in that of v', then both contain a common vertex in W (one of the vertices in W of the representative of v).

We now construct H' and a proper H'-representation of G. First, take H and subdivide its edges by adding precisely the $2k_1$ vertices corresponding to the endpoints of the intervals in I. That is, we initially have that $V(H') = W$ and H'' is a subdivision of H'. Now, for each vertex $w \in W$, we consider the set of vertices of G whose representatives contain w. The inclusion of representatives defines a partial order on this set, and we can order them by extending such partial order. In H' we add a leaf w' attached to w, and extend the representatives of the vertices of G that contain w onto the edge ww' (via a subdivision of ww') by making them end up in reverse order to that of inclusion; in this way we avoid representatives properly included in others among the vertices whose representatives contain w. Clearly, $|V(H')| = 2|V(H)| + 4k_1$. Moreover, as observed earlier, if the representative of a vertex v is properly included in that of v' in the original representation $\mathcal{S}$, then both contain a common vertex in W, and so the new representation of G as an H'-graph is indeed proper. (Recall that we assumed that no two vertices are originally represented by the same set.) □

We are now ready to prove Theorem 2, which we restate below.

Theorem 2. Independent Set *is* W[1]*-hard for proper H-graphs when parameterized by $\|H\| + k$, even if a proper H-representation of the input graph G is given.*

Proof. We provide a parameterized reduction from Independent Set on H-graphs, which is W[1]-hard parameterized by $\|H\| + k$, even if a representation of G as an H-graph is given [15]. Let (G, k) be an instance of this problem, where G is given together with an H-representation and $k \in \mathbb{N}$. Recall that the question is whether G has an independent set of size at least k.

By Theorem 1, we find, in time polynomial in $|V(G)| + |V(H)|$, a parameter k_1 such that $k_1 \leq \alpha(G)$, a graph H' with $|V(H')| = 2|V(H)| + 4k_1$, and a proper H'-representation of G. If $k_1 \geq k$, then (G, k) is a yes-instance. Otherwise, we have a representation of G as a proper H'-graph with $|V(H')| < 2\,V(H)| + 4k$, and we ask whether G has an independent set of size at least k. The conclusion follows from the fact that $|E(H')| + k_1 = (|E(H)| + 2\,V(H)| + 4k_1) + k_1 < |E(H)| + 2\,V(H)| + 5k \leq g(|E(H)| + k)$, for some function g. □

5 Conclusions

We showed that for every multigraph H, the FO Model Checking problem is in FPT for non-crossing H-graphs when parameterized by $\|H\| + \ell$, where ℓ is the size of a formula. We did this by proving that for every multigraph H, the class of non-crossing H-graphs has bounded proper mixed-thinness and thus bounded twin-width. Moreover, we strengthened a result of Fomin, Golovach and Raymond [15] by proving that a special case of the FO Model Checking problem, namely Independent Set, is W[1]-hard even on proper H-graphs when parameterized by $\|H\| + k$, where k is the solution size. This answered, subject to W[1] $\neq$ FPT, a recent question of Chaplick [8].

We finish our paper with a number of open problems resulting from our work. First of all, it would be good to increase our insights in recognizing non-crossing H-graphs. Recall that recognizing H-graphs or proper H-graphs is NP-complete for certain graphs H. However, we do not know any graph H for which recognizing non-crossing H-graphs is NP-complete and whether boundedness of proper mixed-thinness would be of help.

Open Problem 1. Is recognizing non-crossing H-graphs NP-complete for some H?

Note that, unless P = PSPACE, FO Model Checking is not in FPT for non-crossing H-graphs when parameterized by $\|H\|$ only, as it is PSPACE-complete on any class of structures that contains at least one structure with at least two elements (see, e.g., [14]). However, we leave as an open problem to determine whether INDEPENDENT SET and CLIQUE are in FPT for non-crossing H-graphs even when parameterized by $\|H\|$ only. Recall that Chaplick et al. [11] proved that CLIQUE is para-NP-hard on H-graphs when parameterized by $\|H\|$, while the problem is also still open for proper H-graphs.

Open Problem 2. Is INDEPENDENT SET in FPT for non-crossing H-graphs when parameterized by $\|H\|$ only?

Open Problem 3. Determine the complexity of CLIQUE for proper H-graphs and non-crossing H-graphs when parameterized by $\|H\|$.

We also wonder whether there are other problems expressible in first-order logic that exhibit the same behaviour as INDEPENDENT SET on proper H-graphs. Note that DOMINATING SET is captured by the FO Model Checking framework. Hence, it is FPT for non-crossing H-graphs when parameterized by $\|H\| + k$ due to Corollary 1. However, our technique for proving W[1]-hardness for INDEPENDENT SET does not work for DOMINATING SET.

Open Problem 4. Is DOMINATING SET W[1]-hard for proper H-graphs when parameterized by $\|H\| + k$?

Finally, the good algorithmic properties of non-crossing H-graphs might hold beyond FO Model Checking, and FEEDBACK VERTEX SET seems a good candidate to consider. This is because FEEDBACK VERTEX SET, restricted to H-graphs, is in XP [11,15] but W[1]-hard, when parameterized by $\|H\|$ [18].

Open Problem 5. Is FEEDBACK VERTEX SET in FPT for proper H-graphs and non-crossing H-graphs when parameterized by $\|H\|$?

Acknowledgments. Bonomo was supported by UBACyT (20020220300079BA and 20020190100126BA), CONICET (PIP 11220200100084CO), and ANPCyT (PICT-2021-I-A-00755); Brettell by the New Zealand Marsden Fund; Paulusma by the Leverhulme Trust (RPG-2016-258). Munaro is a member of the Gruppo Nazionale Calcolo Scientifico-Istituto Nazionale di Alta Matematica (GNCS-INdAM).

References

1. Balabán, J., Hliněný, P., Jedelský, J.: Twin-width and transductions of proper k-mixed-thin graphs. Disc. Math. **347**, 113876 (2024)
2. Biró, M., Hujter, M., Tuza, Z.: Precoloring extension. I. Interval graphs. Disc. Math. **100**, 267–279 (1992)
3. Bonnet, É., Geniet, C., Kim, E.J., Thomassé, S., Watrigant, R.: Twin-width II: small classes. Comb. Theory **2**, #10 (2022)
4. Bonnet, E., Kim, E.J., Thomassé, S., Watrigant, R.: Twin-width I: tractable FO model checking. J. ACM **69**, 3:1–3:46 (2021)
5. Bonomo, F., de Estrada, D.: On the thinness and proper thinness of a graph. Disc. Appl. Math. **261**, 78–92 (2019)

6. Bonomo-Braberman, F., Brito, G.A.: Intersection models and forbidden pattern characterizations for 2-thin and proper 2-thin graphs. Disc. Appl. Math. **339**, 53–77 (2023)
7. Çagirici, D.A., et al.: Recognizing H-graphs - beyond circular-arc graphs. In: Proceedings of MFCS 2023, LIPIcs, vol. 272, pp. 8:1–8:14 (2023)
8. Chaplick, S.: Intersection graphs of non-crossing paths. Disc. Math. **346**, 113498 (2023)
9. Chaplick, S., Fomin, F.V., Golovach, P.A., Knop, D., Zeman, P.: Kernelization of graph hamiltonicity: proper H-graphs. SIAM J. Disc. Math. **35**, 840–892 (2021)
10. Chaplick, S., Golovach, P.A., Hartmann, T.A., Knop, D.: Recognizing proper tree-graphs. In: Proceedings of IPEC 2020, LIPIcs, vol. 180, pp. 8:1–8:15 (2020)
11. Chaplick, S., Töpfer, M., Voborník, J., Zeman, P.: On H-topological intersection graphs. Algorithmica **83**, 3281–3318 (2021)
12. Cygan, M., et al.: Parameterized algorithms. Springer, Cham (2015). https://doi.org/10.1007/978-3-319-21275-3
13. Deng, X., Hell, P., Huang, J.: Linear-time representation algorithms for proper circular-arc graphs and proper interval graphs. SIAM J. Comput. **25**, 390–403 (1996)
14. Flum, J., Grohe, M.: Parameterized Complexity Theory. Springer, Heidelberg (2006). https://doi.org/10.1007/3-540-29953-X
15. Fomin, F.V., Golovach, P.A., Raymond, J.: On the tractability of optimization problems on H-graphs. Algorithmica **82**, 2432–2473 (2020)
16. Ganian, R., Hliněný, P., Král, D., Obdrzálek, J., Schwartz, J., Teska, J.: FO model checking of interval graphs. Log. Methods Comput. Sci. **11**(4) (2015)
17. Hliněný, P., Pokrývka, F., Roy, B.: FO model checking on geometric graphs. Comput. Geom. **78**, 1–19 (2019)
18. Jaffke, L., Kwon, O., Telle, J.A.: Mim-width II. The feedback vertex set problem. Algorithmica **82**, 118–145 (2020)
19. Jedelský, J.: Classes of bounded and unbounded twin-width. Bachelor's thesis, Masaryk University (2021)
20. Klavík, P., Kratochvíl, J., Otachi, Y., Saitoh, T.: Extending partial representations of subclasses of chordal graphs. Theor. Comput. Sci. **576**, 85–101 (2015)
21. Kratochvíl, J.: Intersection graphs of noncrossing arc-connected sets in the plane. In: North, S. (ed.) GD 1996. LNCS, vol. 1190, pp. 257–270. Springer, Heidelberg (1997). https://doi.org/10.1007/3-540-62495-3_53
22. Mannino, C., Oriolo, G., Ricci, F., Chandran, S.: The stable set problem and the thinness of a graph. Oper. Res. Lett. **35**, 1–9 (2007)
23. Marx, D., Schlotter, I.: Cleaning interval graphs. Algorithmica **65**(2), 275–316 (2013)
24. Olariu, S.: An optimal greedy heuristic to color interval graphs. Inf. Process. Lett. **37**, 21–25 (1991)
25. Ramalingam, G., Pandu Rangan, C.: A unified approach to domination problems on interval graphs. Inf. Process. Lett. **27**, 271–274 (1988)
26. Roberts, F.S.: Indifference graphs. In: Harary, F. (ed.) Proof Techniques in Graph Theory, pp. 139–146. Academic Press (1969)
27. Tucker, A.: An efficient test for circular-arc graphs. SIAM J. Comput. **9**, 1–24 (1980)

Excluding an Induced Wheel Minor in Graphs Without Large Induced Stars

Mujin Choi[1,2](✉), Claire Hilaire[3], Martin Milanič[3], and Sebastian Wiederrecht[4]

[1] Department of Mathematical Sciences, KAIST, Daejeon, Korea
mujinchoi@kaist.ac.kr
[2] Discrete Mathematics Group, Institute for Basic Science, Daejeon, South Korea
[3] FAMNIT and IAM, University of Primorska, Koper, Slovenia
[4] School of Computing, KAIST, Daejeon, Korea

Abstract. We study a conjecture due to Dallard, Krnc, Kwon, Milanič, Munaro, Štorgel, and Wiederrecht stating that for any positive integer d and any planar graph H, the class of all $K_{1,d}$-free graphs without H as an induced minor has bounded tree-independence number. A k-wheel is the graph obtained from a cycle of length k by adding a vertex adjacent to all vertices of the cycle. We show that the conjecture of Dallard et al. is true when H is a k-wheel for any $k \geq 3$. Our proof uses a generalization of the concept of brambles to tree-independence number. As a consequence of our main result, several important NP-hard problems such as Maximum Independent Set are tractable on $K_{1,d}$-free graphs without large induced wheel minors. Moreover, for fixed d and k, we provide a polynomial-time algorithm that, given a $K_{1,d}$-free graph G as input, finds an induced minor model of a k-wheel in G if one exists.

Keywords: Induced minor · Wheel · Tree-independence number · Maximum Independent Set

1 Introduction

Treewidth is one of the most widely studied graph parameters based on so called *tree-decompositions*[1]. The popularity of treewidth in theoretical computer science is due to its powerful properties for the design of (parameterized) algorithms. As a prominent example of these properties, Courcelle's theorem (see, e.g., [11]) states that any decision problem expressible in *counting monadic second-order logic*, written CMSO_2 (see [12] for the definition), can be solved in linear time when the input is restricted to graphs whose treewidth is bounded by some constant. On the structural graph theory side, treewidth plays a central role in the Graph Minor Theory by Robertson and Seymour (see, e.g., [29,30]).

[1] We postpone the definition of tree-decompositions and treewidth to Sect. 2.

H. Fernau and P. Kindermann (Eds.): WG 2025, LNCS 16124, pp. 135–148, 2026.
https://doi.org/10.1007/978-3-032-11835-6_10

Unfortunately, graph classes of bounded treewidth are rather restricted. This is because treewidth is a *minor-monotone* parameter and thus graphs of bounded treewidth are forced to be sparse: for every graph G of treewidth at most k we have that $|E(G)| \leq k|V(G)|$. Hence, bounding the treewidth naturally fails to consider several important but very simple graph classes. There have been various attempts to deal with this problem by defining new width parameters that can be used to understand dense graph classes more thoroughly. Examples of such parameters are clique-width [13] (and the related rank-width [26]) but also more recent parameters such as twin-width [5]. Another possible approach to the design of new width parameters is to change the way the width of a tree-decomposition is evaluated to account for dense but still naturally tree-like graphs such as chordal graphs. In this context, Yolov [33], and independently, Dallard, Milanič, and Štorgel [15] defined *tree-independence number*, denoted[2] by $\alpha\text{-}\mathsf{tw}$. This new parameter measures the independence number of each bag in a tree-decomposition instead of its size. While graphs of treewidth at most 1 are exactly the forests, and thus treewidth can be considered to measure how close a given graph is to being a forest, graphs of tree-independence number at most 1 are precisely the chordal graphs; hence, tree-independence number can be understood as a measure for how close a given graph is to being chordal.

Tree-independence number has gotten a lot of attention recently, for both structural and algorithmic reasons. It appears to be a natural parameter for the study of *induced-minor-closed* graph classes [9,15,21]. We say that a graph H is an *induced minor* of a graph G if H can be obtained from G by vertex deletions and edge contractions. The induced minor relation has been under consideration for quite some time for both its structural properties and algorithmic implications (see [4,19,21,23,25]). In particular, tree-independence number is monotone under the induced minor relation (see [15]).

From the algorithmic point of view, in [33] (see also [15]), the author showed that the well-known Maximum Weight Independent Set (MWIS) problem can be solved in polynomial time in graph classes with bounded tree-independence number. In fact, in [16], Dallard, Fomin, Golovach, Korhonen, and Milanič showed that tree-independence number is the most general parameter among similar tree-decomposition based parameters that implies tractability of MWIS (unless $\mathsf{P} = \mathsf{NP}$). Bounded tree-independence number implies poylnomial-time solvability of a number of algorithmic problems other than MWIS (see [15,24,33]). In particular, Lima, Milanič, Muršič, Okrasa, Rzążewski, and Štorgel [24] proved an *algorithmic metatheorem*, giving a generic algorithm that finds a maximum-weight induced subgraph of bounded clique number satisfying a fixed property definable in CMSO_2. As we will see in Sect. 6.1, this metatheorem allows us to efficiently solve H-Induced Minor Containment in graphs of bounded tree-independence number (a similar result was recently obtained by Bousquet, Dallard, Dumas, Hilaire, Milanič, Perez, and Trotignon [6]). Note that H-Induced Minor Containment is known to be NP-complete for some graphs H (see [19]), even if H is a tree (see [23]).

[2] The authors of [15] denoted the tree-independence number by $\mathsf{tree}\text{-}\alpha$.

In this context, and from a more structural point of view, it is interesting to characterize graph classes with bounded tree-independence number, in particular in terms of excluded induced minors. Recently, Dallard et al. [17]conjectured that for every positive integer d and every planar graph H, every graph that excludes $K_{1,d}$ as an induced subgraph and H as an induced minor has bounded tree-independence number. Here, $K_{1,d}$ denotes the *star* with d leaves, i.e. the complete bipartite graph where one side has exactly one vertex and the other side has d vertices. Alternatively, $K_{1,d}$-free[3] graphs are exactly those where the maximum size of an independent set in the neighbourhood of any vertex is at most $d-1$. The conjecture of Dallard et al. can be seen as an induced minor version of the celebrated Grid Theorem by Robertson and Seymour [29]. Moreover, the conjecture generalises a related theorem by Korhonen [22], stating that for every positive integer d, every graph with maximum degree at most d that excludes some fixed planar graph as an induced minor has bounded treewidth.

1.1 Our Results

We prove a first result that shows that for every integer $k \geq 4$, there exists a 3-connected planar graph H on k vertices such that excluding H as an induced minor in $K_{1,d}$-free graphs yields bounded tree-independence number.

Let $\ell \geq 3$ be an integer. The *ℓ-wheel*, denoted by W_ℓ, is the graph obtained from the cycle on ℓ vertices by introducing one additional vertex adjacent to all vertices of the cycle. Before this work, it was known that graphs excluding W_3 or W_4 as an induced minor have bounded tree-independence number [14]. In the same paper it was proven that for any $\ell \geq 5$, excluding W_ℓ alone as an induced minor is not enough to bound the tree-independence number.

The main result of this paper is, that in $K_{1,d}$-free graphs, excluding W_ℓ suffices to bound the tree-independence number for all integers $d \geq 1$ and $\ell \geq 3$.

Theorem 1. *There exists a function $f_1(d,\ell) \in \mathcal{O}\big(d\ell(\ell^{10} + 2^{(\max\{d+2,\ell\})^5})\big)$ such that for all integers $d \geq 1$ and $\ell \geq 3$ and every $K_{1,d}$-free graph G, either G contains W_ℓ as an induced minor, or $\alpha\text{-}\mathsf{tw}(G) \leq f_1(d,\ell)$.*

Our proofs are constructive, in the sense that they can be transformed into an algorithm that provides, given a $K_{1,d}$-free graph G and an integer ℓ as input, either an induced minor model of W_ℓ in G, or a tree-decomposition where every bag has independence number at most $f_1(d,\ell)$ in time $|V(G)|^{g(d,\ell)}$, where g is a non-decreasing computable function. As a consequence, we obtain an algorithm that finds an induced W_ℓ-minor, if there is one, in a $K_{1,d}$-free graph in polynomial time (see Theorem 9). In particular, this result identifies islands of tractability of the H-Induced Minor Containment problem even if the tree-independence number of the input graph is unbounded.

Due to space restrictions, proofs of some of the results are omitted and can be found in [7].

[3] Given a graph H, a graph G is *H-free* if it does not contain H as an induced subgraph.

1.2 Our Method: A New Tool for the Study of Tree-Independence Number

One strength of treewidth is that it has many useful dual notions, such as brambles, well-linked sets, and tangles. These have all been key concepts in proofs for the Grid Theorem [10,28,31]. As a key tool towards our main result, we provide similar dual notions for the tree-independence number. A similar approach was originally studied by Isolde Adler in her doctoral dissertation [1], for the more general notions of f-hypertree-width and the f-bramble number, where f is any width function on a graph. In this paper, we focus on the independence number to establish the following approximate duality between tree-independence number and a relevant notion for brambles. We postpone the precise definitions to Sect. 3. (For readers familiar with the notion of brambles: a *strong bramble* is a bramble where every pair of elements has a non-empty intersection, and the *α-order* of a bramble $\mathcal{B}$ is the smallest independence number of a cover for $\mathcal{B}$.)

Theorem 2. *Let G be a graph. Then, the following statements hold for every positive integer k.*

1. *If G has a strong bramble of α-order k, then $\alpha\text{-tw}(G) \geq k$.*
2. *If $\alpha\text{-tw}(G) \geq 4k - 2$, then G has a strong bramble of α-order at least k.*

In order to explain our first application of this notion of brambles, we need to recall a result of Seymour [32]. He proved that for every integer $k \geq 4$, if a graph G excludes all induced cycles of length at least k, then G admits a tree-decomposition in which the subgraph induced by each bag has a dominating path with at most $k - 3$ vertices. If in addition G is $K_{1,d}$-free for some $d \geq 2$, the above property implies that the tree-independence number of G is at most $(d-1)(k-3)$. Using Theorem 2, this implies the following.

Corollary 1. *Let $d \geq 2$ and $k \geq 4$ be integers and G be a $K_{1,d}$-free graph with a strong bramble $\mathcal{B}$ of α-order at least dk. Then there exists an induced cycle C of length at least k in G.*

We strengthen Theorem 1 to tighten the relation between the induced cycle and the bramble as follows.

Theorem 3. *Let $d \geq 1$ and $k \geq 2$ be integers and G be a $K_{1,d}$-free graph with a strong bramble $\mathcal{B}$ of α-order at least dk. Then there exists an induced cycle C of length at least k in G such that $N[C] \cap B \neq \emptyset$ for all $B \in \mathcal{B}$.*

We believe that understanding the structural properties of strong brambles of large α-order is key for a possible resolution of the conjecture of Dallard et al. In our case, the long induced cycle we are guaranteed to find in relation to any strong bramble of large α-order due to Corollary 3 directly allows us to construct a large induced wheel minor in any $K_{1,d}$-free graph of large tree-independence number.

1.3 Related Work

The *Grid Theorem* of Robertson and Seymour [29] can be understood as a min-max relation between the treewidth of a graph and its largest grid minor. Recently, considerable attempts have been made to understand treewidth through a similar min-max relation using induced subgraphs (see, for example, [3]). Replacing treewidth with tree-independence number often allows for more streamlined arguments for the study of forbidden induced objects [9,15]. We believe that the existence of canonical obstructions such as the brambles discussed in this paper will be crucial in further streamlining arguments about tree-independence number and forbidden induced substructures.

For graphs G of bounded degree, i.e. a proper subclass of graphs of bounded clique number, by the arguments above there is only a qualitative difference between[4] $\mathsf{tw}(G)$ and $\alpha\text{-}\mathsf{tw}(G)$. Hence, the study of both parameters essentially coincides within classes of bounded degree. Korhonen [22] proved that for a graph with bounded maximum degree, if its treewidth is sufficiently large, the graph must contain a large grid as an induced minor.

Theorem 4 (Korhonen [22]). *There exists a function $f_4(d,k) \in \mathcal{O}(k^{10} + 2^{d^5})$ such that every graph G with maximum degree at most d and $\mathsf{tw}(G) \geq f_4(d,k)$ contains the $(k \times k)$-grid as an induced minor.*

Notice that graphs of maximum degree at most $d-1$ are exactly those that do not contain $K_{1,d}$ as a *subgraph*. A natural generalization of this property to the world of induced subgraphs would be to exclude $K_{1,d}$ as an *induced* subgraph. In other words, for each vertex in a $K_{1,d}$-free graph, its neighbourhood has bounded independence number. Thus, we can consider $K_{1,d}$-free graphs as those of bounded "induced" maximum degree. It is apparent that the equivalence of treewidth and tree-independence number can no longer hold: the class of complete graphs is contained in the class of $K_{1,2}$-free graphs, has unbounded treewidth but tree-independence number bounded by 1. Building on this intuition, in [17], Dallard et al. conjectured an analogue of Theorem 4 for $K_{1,d}$-free graphs.

Conjecture 1 (Dallard et al. [17]*).* There exists a function $f(d,k)$ such that every $K_{1,d}$-free graph G with $\alpha\text{-}\mathsf{tw}(G) \geq f(d,k)$ contains the $(k \times k)$-grid as an induced minor.

In the same paper, Dallard et al. provide partial evidence towards their conjecture. They prove two results: the first is that every $\{K_{1,d}, P_k\}$-free graph G satisfies $\alpha\text{-}\mathsf{tw}(G) \leq (d-1)(k-2)$ and the second is that every $K_{1,d}$-free graph that excludes a k-times subdivided claw[5] as an induced minor has tree-independence number bounded by some function depending only on d and k (see also [8] for a more general result). Recently, Chudnovsky, Hajebi, and Trotignon [9] proved

[4] We denote the treewidth of a graph G by $\mathsf{tw}(G)$.

[5] For $k \in \mathbb{N}$, a *k-times subdivided claw* is a graph obtained from $K_{1,3}$ by subdividing every edge k times.

that excluding so-called *thetas* and *prisms* from $K_{1,d}$-free graphs also results in classes of bounded tree-independence number. Furthermore, Ahn, Gollin, Huynh, and Kwon [2] proved that there is a function f such that for every $k \in \mathbb{N}$, every $K_{1,d}$-free graph G either contains k pairwise disjoint and pairwise non-adjacent induced cycles, or there is a set $X \subseteq V(G)$ such that $\alpha(G[X]) \leq f(k)$ and $G - X$ is chordal, implying that $\alpha\text{-}\mathsf{tw}(G) \leq f(k) + 1$.

The results above are the first that provide evidence towards Conjecture 1 by showing that the exclusion of small planar patterns forces bounded tree-independence number in $K_{1,d}$-free graphs. The wheels W_ℓ, $\ell \geq 3$, while containing long induced paths and cycles, are in general incomparable with respect to the induced minor relation with both the thetas and prisms of Chudnovsky et al. and the subdivided claws of Dallard et al. In this regard, our main result, Theorem 1, presents a novel structure whose exclusion forces small tree-independence number in $K_{1,d}$-free graphs. Moreover, wheels mark the first infinite class of 3-*connected* planar graphs whose exclusion as induced minors bounds the tree-independence number of $K_{1,d}$-free graphs. With grids being (almost) 3-connected, this can be considered to be a major milestone towards a (hopefully positive) resolution of Conjecture 1.

2 Preliminaries

Let $a \in \mathbb{Z}$. We denote the set $\{1, \ldots, a\}$ by $[a]$. All graphs we consider in this paper are finite, simple, and undirected. For a graph G, let $V(G)$ denote the vertex set of G, and let $E(G)$ denote the edge set of G.

Let G and H be graphs. We say that H is an *induced subgraph* of G if H can be obtained from G by a sequence of vertex deletions. For graphs G and H, we say that G is H-*free* if H is not an induced subgraph of G. If H is a subgraph of G we write $G - H$ for $G - V(H)$. Let G and H be graphs and $e = xy \in E(G)$ be an edge. We say that H is obtained from G by *contracting* e if there exists a vertex $u \in V(H)$ such that $H - u = G - x - y$ and $N_H(u) = N_G(x) \cup N_G(y)$. We say that H is an *induced minor* of G if H can be obtained from G by a sequence of vertex deletions and edge contractions.

Let G be a graph and $X \subseteq V(G)$. We denote by $N_G(X)$ the set $\{v \in V(G) \setminus X \colon vx \in E(G) \text{ for some } x \in X\}$. We call $N_G(X)$ the *neighbourhood* of X in G. Moreover, $N_G[X] := N_G(X) \cup X$ is called the *closed neighbourhood* of X. By a slight abuse of notation, if H is an induced subgraph of G, we write $N_G[H]$ (respectively $N_G(H)$) for $N_G[V(H)]$ (respectively $N_G(V(H))$). In case G is understood from the context, we write N instead of N_G and, for $X \subseteq V(G)$, we write $\alpha(X)$ to denote the value $\alpha(G[X])$ for the maximum size of an independent set in the subgraph of G induced by X.

A *tree-decomposition* for a graph G is a tuple $\mathcal{T} = (T, \beta)$ where T is a tree and $\beta \colon V(T) \to 2^{V(G)}$ (here $2^{V(G)}$ denotes the power set of the vertex set of G), such that: (1.) $\bigcup_{t \in V(T)} \beta(t) = V(G)$, (2.) for every $e \in E(G)$ there exists some $t \in V(T)$ such that $e \subseteq \beta(t)$, and (3.) for every vertex $v \in V(G)$ the set $\{t \in V(T) \colon v \in \beta(t)\}$ induces a subtree of T.

The *width* of $\mathcal{T}$ is defined as the value $\max_{t \in V(\mathcal{T})} |\beta(t)| - 1$ and the *treewidth* of G, denoted by $\mathsf{tw}(G)$, is the minimum width over all tree-decompositions for G. The *α-width* of $\mathcal{T}$ is defined as the value $\alpha\text{-}\mathsf{width}(\mathcal{T}) = \max_{t \in V(\mathcal{T})} \alpha(\beta(t))$ and the *tree-independence number* of G, denoted by $\alpha\text{-}\mathsf{tw}(G)$, is the minimum α-width over all tree-decompositions for G.

3 Bramble Duality for Tree-Independence Number

Our goal is to prove an approximate duality between tree-independence number and a suitable notion of brambles that allows us to easily show lower bounds and provides us with additional information on the structure of graphs with large tree-independence number. The notion of brambles we introduce here is slightly more restrictive than the usual definition (see, e.g., [18]), hence we add the attribute *strong*. We make this choice here because it allows slightly simpler arguments while not using Menger's Theorem. For a proof of bramble/treewidth duality see, e.g., [18, Theorem 12.4.3].

Let G be a graph. We say that a set $X \subseteq V(G)$ is *connected* if $G[X]$ is connected.

A *strong bramble* in G is a collection $\mathcal{B} = \{B_1, \ldots, B_n\}$ of connected sets $B_i \subseteq V(G)$ such that $B_i \cap B_j \neq \emptyset$ for all $i, j \in [n]$.

A *cover* for a strong bramble $\mathcal{B} = \{B_1, \ldots, B_n\}$ is a set $X \subseteq V(G)$ such that $B_i \cap X \neq \emptyset$ for all $i \in [n]$. The *α-order* of $\mathcal{B}$ is defined as $\min\{\alpha(X) : X \subseteq V(G) \text{ is a cover for } \mathcal{B}\}$.

Our proof for Theorem 2 takes a slight detour through the concept of balanced separators and highly linked sets, i.e. those kinds of sets that do not have such separators. Our proofs for the results in this section follow along very well-established strategies (see, e.g. [1]).

Let G be a graph and $X \subseteq V(G)$ be a set of vertices. We say that a set $S \subseteq V(G)$ is an *α-balanced separator* for X if every component C of $G - S$ satisfies $\alpha(V(C) \cap X) \leq \frac{\alpha(X)}{2}$.

Now let $k \geq 1$ be an integer and G be a graph. A set $X \subseteq V(G)$ is *k-α-linked* if there does not exist an α-balanced separator S for X with $\alpha(S) \leq k$.

Theorem 5. *Let $k \geq 1$ be an integer. If a graph G does not contain a k-α-linked set, then $\alpha\text{-}\mathsf{tw}(G) \leq 2k + 1$.*

The proof of Theorem 2 now consists of two steps. First, we show that the existence of a strong bramble of large α-order forces large tree-independence number. Second, assume that a given graph G has large tree-independence number, namely, at least $4k-2$. By Theorem 5 this implies that G contains a $(2k-2)$-α-linked set for some k. Finally, notice that for each $S \subseteq V(G)$ with independence number at most $k-1$, $G-S$ has a unique component J_S whose intersection with S has a large independent set. The collection $\mathcal{B}$ of all of those components J_S can then be observed to be the strong bramble of α-order at least k, as desired.

4 Dominating Paths and Cycles for Strong Brambles

In this section we show that there is always an induced path and, in most cases, an induced cycle such that for each of them, their closed neighbourhood dominates a given strong bramble $\mathcal{B}$. The proof comes in two steps. First we show that one can, in essence, mimic the Gyárfás path argument (see [20]) to inductively grow a path P until all members of $\mathcal{B}$ intersect the closed neighbourhood of P. In a second step we show that P can be chosen in such a way that it can be closed to an induced cycle. This cycle will later be used as a starting point to find a desired wheel graph as an induced minor. Furthermore, if we assume that the graph is $K_{1,d}$-free and the strong bramble $\mathcal{B}$ has large α-order, then we can guarantee that the path or cycle must be long. We then show that this implies Corollary 3.

Lemma 1. *Let $\mathcal{B}$ be a strong bramble in a graph G. Then, there is an induced path P in G such that $N[P] \cap B \neq \emptyset$ for all $B \in \mathcal{B}$.*

Theorem 6. *Let $\mathcal{B}$ be a strong bramble in a graph G. Then, G contains either a vertex v such that $N[v] \cap B \neq \emptyset$ for all $B \in \mathcal{B}$, or an induced cycle C such that $N[C] \cap B \neq \emptyset$ for all $B \in \mathcal{B}$.*

Proof. By Lemma 1, there exists an induced path P such that $N[P] \cap B \neq \emptyset$ for all $B \in \mathcal{B}$. We may assume that P is shortest among such paths, and let s and t be the endpoints of P. If $s = t$, the first case holds, thus suppose that $s \neq t$.

By the minimality of P, for each endpoint x of P there exists a $B_x \in \mathcal{B}$ such that $N[P - x] \cap B_x = \emptyset$. Note that $B_s \neq B_t$, since otherwise $N[P - s] \cap B_s = \emptyset$, implying $N[s] \cap B_s \neq \emptyset$, which is in contradiction with $N[P - t] \cap B_s = N[P - t] \cap B_t = \emptyset$. Recall that $B_s \cup B_t$ induces a connected subgraph of G and the only neighbours of $B_s \cup B_t$ in P are s and t. Therefore, there exists a shortest path R from s to t such that R has length at least 2 and all the internal vertices of R are in $B_s \cup B_t$. Note that R is an induced path, unless P has length 1. Furthermore, no internal vertex of R has a neighbour in P other than s and t from the choice of B_s and B_t. Hence, $C{:=}P \cup R$ is an induced cycle and $N[C] \cap B \supseteq N[P] \cap B \neq \emptyset$ for all $B \in \mathcal{B}$. □

Now we analyse the length of the cycle under the condition that G is $K_{1,d}$-free and $\mathcal{B}$ is a strong bramble of α-order at least dk (as in Corollary 3), where $d \geq 1$ and $k \geq 2$. By Lemma 6, there exists an induced cycle C in G such that $N[C]$ is a cover of $\mathcal{B}$. Then $dk \leq \alpha(N[C]) \leq d \cdot |V(C)|$. This proves Corollary 3.

5 Excluding an Induced Wheel Minor

We now proceed with a sketch for the proof of our main theorem, i.e. Theorem 1. This proof follows along two steps. First we prove a technical auxiliary lemma, and then we use the lemma together with Theorem 2 to obtain a contradiction to the assumption that there exists a $K_{1,d}$-free graph without an induced W_ℓ-minor and whose tree-independence number exceeds the bound claimed in our main theorem.

Roughly speaking, our auxiliary lemma says that in a $K_{1,d}$-free graph G without an induced wheel minor, given a long induced cycle C, there exists a tree-decomposition (T, β) of G whose internal bags all have small independence number and whose leaves correspond to the components of $G - N[C]$. Here we only give the main ideas of the proof, omitting some of the details.

Lemma 2. *There exists a function $f_2(d, \ell) \in \mathcal{O}\big(d\ell(\ell^{10} + 2^{(\max\{d+2,\ell\})^5})\big)$ such that for every choice of integers $d \geq 1$ and $\ell \geq 3$ and every $K_{1,d}$-free graph G containing an induced cycle C of length at least ℓ, one of the following statements holds:*

1. *G contains W_ℓ as an induced minor, or*
2. *G has a tree-decomposition (T, β) such that there exists a partition of $V(T)$ into two sets F_1 and F_2 such that*
   - *every vertex in F_2 is a leaf of T,*
   - *$\alpha(\beta(t)) \leq f_2(d, \ell)$ for all $t \in F_1$,*
   - *$\alpha(\beta(t_1) \cap \beta(t_2)) \leq f_2(d, \ell)$ for all $t_1t_2 \in E(T)$,*
   - *there is a bijection ϕ between the components of $G - N[C]$ and F_2 such that for each component J of $G - N[C]$ it holds that $N[J] = \beta(\phi(J))$, and*
   - *for every $t \in F_2$ with neighbour $t' \in F_1$ we have that $\beta(t) \cap \beta(t') = N(\phi^{-1}(t))$.*

Proof (with some details omitted). Suppose that G does not contain W_ℓ as an induced minor. Let G_1 be the induced minor of G obtained by contracting each component K of $G - C$ into a single vertex v_K. Then we have $\Delta(G_1) \leq \Delta := \max\{d + 2, \ell - 1\}$. If $\mathsf{tw}(G_1) \geq f_4(\Delta, \ell)$, we get that G_1, and therefore G, contains the $(\ell \times \ell)$-grid as an induced minor by Theorem 4. This implies that G contains W_ℓ as an induced minor, so we may assume that $\mathsf{tw}(G_1) < f_4(\Delta, \ell)$. Let (T', β_1) be a tree-decomposition of G_1 of width at most $f_4(\Delta, \ell) - 1$.

Now let G_2 be obtained from G by contracting every component J of $G - N[C]$ into a vertex u_J. We can now construct a tree-decomposition (T', β_2) of G_2 from (T', β_1) by replacing every $v_K \in V(G_1)$ in each bag $\beta_1(t)$ with the union of all vertices $u_J \in V(G_2)$ such that $J \subseteq K$, together with all vertices in $N(C) \cap V(K)$.

Next, let (T', β_3) be the tree-decomposition of $G[N[C]]$ obtained from (T', β_2) by removing all vertices u_J as above from its bags where J is a component of $G - N[C]$. It can then be verified that, for every $t \in V(T')$, $\alpha(G[\beta_3(t)]) \leq d(\ell - 1) f_4(\Delta, \ell)$.

Finally, we construct a tree-decomposition (T, β) of G as follows. Let T be a tree obtained from T' by introducing a new vertex t_J adjacent to some vertex $t \in V(T')$ such that $N(J) \subseteq \beta_3(t)$ for every component J of $G - N[C]$. For every $t \in V(T) \cap V(T')$ we set $\beta(t) := \beta_3(t)$ and for every J as above we set $\beta(t_J) := N[J]$. Then (T, β) is a tree-decomposition of G satisfying the conditions of the second outcome of our assertion by setting $F_1 := V(T')$, $F_2 := V(T) \setminus F_1$ and $f_2(d, \ell) := d(\ell - 1) f_4(\Delta, \ell)$. □

From here, towards our main theorem, we assume that G has large tree-independence number but no large induced wheel minor. Thus, by Theorem 2, G has a strong bramble $\mathcal{B}$ of large α-order. This means, according to Theorem 3, that there is a long induced cycle C in G whose closed neighbourhood covers $\mathcal{B}$.

Using Lemma 2, we construct a tree-decomposition (T, β) of G such that the intersection of any two adjacent bags has small independence number; moreover, every bag, except possibly for the leaf bags that contain a component of $G - N[C]$, has small independence number. Finally, we employ a strategy that is very similar to the proof of Theorem 2, orienting T as follows. For each edge t_1t_2, since $\mathcal{B}$ has large α-order, there exists $B \in \mathcal{B}$ that does not intersect $\beta(t_1) \cap \beta(t_2)$. Since all the elements of $\mathcal{B}$ are connected and pairwise intersect, all the elements not intersecting $\beta(t_1) \cap \beta(t_2)$ are included in the union of the bags of one of the two trees of $T - t_1t_2$: we orient t_1t_2 toward that subtree. This orientation results in a sink t that cannot be a leaf, and thus $\alpha(\beta(t))$ is small. However, now $\beta(t)$ is a cover for $\mathcal{B}$ that has large α-order, so we reach a contradiction.

6 Detecting an Induced Wheel Minor in $K_{1,d}$-Free Graphs

Let H be an arbitrary graph. We are interested in the following problem:
H-Induced Minor Containment
Input: A graph G.
Question: Does G contain H as an induced minor?

Indeed, we are interested in the search version, that is, if the answer to H-Induced Minor Containment is Yes, then we want to find an *induced minor model* of H in G.

Let H and G be graphs. An *induced minor model* of H in G is a collection $\{X_v\}_{v \in V(H)}$ of pairwise vertex disjoint sets of vertices of G such that X_v is connected for all $v \in V(H)$, and for all distinct $u, v \in V(H)$ there exists an edge $xy \in E(G)$ with $x \in X_u$ and $y \in X_v$ if and only if $uv \in E(H)$.

In this section we discuss the algorithmic implication of our main theorem, i.e. Theorem 1, for W_ℓ-Induced Minor Containment in $K_{1,d}$-free graphs.

6.1 Induced Minor Containment in Graphs of Small Tree-Independence Number

In this subsection we provide a short argument for the following theorem. A version of this theorem was recently announced by Bousquet et al. [6].

Theorem 7. *There exists a function $g_7 \colon \mathbb{N}^2 \to \mathbb{N}$ such that for every positive integer k and every graph H there is an algorithm that solves the problem H-Induced Minor Containment in time $|V(G)|^{g_7(|V(H)|,k)}$ on every graph G with α-$\mathsf{tw}(G) \leq k$. Moreover, the algorithm returns an induced minor model of H if one exists.*

Given two positive integers a and b, we denote by $\mathsf{Ramsey}(a,b)$ the smallest positive integer such that every graph on at least $\mathsf{Ramsey}(a,b)$ vertices contains either a clique of size a or an independent set of size b. A classic theorem of Ramsey [27] says that $\mathsf{Ramsey}(a,b)$ always exists.

Theorem 8 (Lima et al. [24]). *For every k,r and every CMSO_2 formula ψ there exists a positive integer $f_8(k,r,\psi)$ such that the following holds: Let G be a graph given together with a tree-decomposition (T,β) of independence number at most k, and let $w\colon V(G) \to \mathbb{Q}^+$ be a weight function. Then, there exists an algorithm that, given G and (T,β) as input, finds a maximum-weight set $Y \subseteq V(G)$ such that $G[Y] \models \psi$ and $\omega(G[Y]) \leq r$, or concludes correctly that no such set Y exists. Moreover, this algorithm runs in time $f_8(k,r,\psi) \cdot |V(G)|^{\mathcal{O}(\mathsf{Ramsey}(k+1,r+1))} \cdot |V(T)|$.*

We say that an induced minor model $\{X_v\}_{v\in V(H)}$ in a graph G is *small* if $\omega(G[\bigcup_{v\in V(H)} X_v]) \leq |V(H)|^2$.

It is relatively easy to see that one can write a formula ψ_H in CMSO_2 that expresses, for a given graph G' with vertex set Y if there exists an induced minor model $\{X_v\}_{v\in V(H)}$ of H in G' with $Y = \bigcup_{v\in V(H)} X_v$. Moreover, the size of this formula only depends on H. Combining this fact with Theorem 8 and the algorithm of Dallard et al. for approximating the tree-independence number [16] implies, that all we have to do in order to prove Theorem 7 is to show that for all graphs H and G, if G contains H as an induced minor, then G has a small induced minor model of H.

Lemma 3. *Let G and H be graphs. Then G contains H as an induced minor if and only if G contains a small induced minor model of H.*

Proof. Clearly, if G has a small induced minor model of H, then G must contain H as an induced minor.

Now for the converse let us assume that G contains H as an induced minor and, hence, G has an induced minor model $\{X_v\}_{v\in V(H)}$ of H. We select $\{X_v\}_{v\in V(H)}$ that minimizes the number of vertices in $\bigcup_{v\in V(H)} X_v$.

If $\{X_v\}_{v\in V(H)}$ is small, we are done, so we may assume that it is not small. Let $G'{:=}G[\bigcup_{v\in V(H)} X_v]$. By the assumption, we have that $\omega(G') > |V(H)|^2$. Then there exists $v \in V(H)$ such that $\omega(G[X_v]) > |V(H)|$ and we may select $Y \subseteq X_v$ to be a maximum clique in $G[X_v]$. Now let (T,r) be a BFS-tree[6] in $G[X_v]$ rooted at an arbitrary vertex of Y. It follows from $\omega(G[X_v]) > |V(H)|$ that (T,r) has at least $|V(H)|$ many leaves. We may select for each $u \in N(v)$ some leaf t_u of (T,r) such that X_u has a neighbour in the unique r-t_u-path in T. Then $Z{:=}\{t_u\colon u \in N(H)\}$ contains at most $|V(H)|-1$ vertices. Let $t \in V(T)$ be an arbitrary leaf of (T,r) that does not belong to Z. It follows that $G'-t$ still contains H as an induced minor and thus G has an induced minor model of H on $|V(G')|-1$ vertices, contradicting our choice of $\{X_v\}_{v\in V(H)}$. □

[6] We denote by (T,r) a *rooted tree* where T is a tree and $r \in V(T)$ is its *root*.

Note that the bound $|V(H)|^2$ is chosen to optimise the simplicity of the argument; in [6] Bousquet et al. achieve stronger bounds with a more in-depth analysis.

Proof of Theorem 7. From the discussion above, Lemma 3, and Theorem 8 we obtain an algorithm that either finds an induced subgraph G' of G that has a small induced minor model of H spanning its vertex set, or correctly determines that G does not contain H as an induced minor. Notice that G' itself has bounded tree-independence number since it is an induced subgraph of G. Since its clique number is also bounded this implies that G' has bounded treewidth. Therefore, we may now apply Courcelle's Theorem [11] for graphs of bounded treewidth to find an induced minor model of H in G'. □

6.2 Efficiently Finding an Induced Wheel Minor

While W_ℓ-Induced Minor Containment in general graphs is polynomial-time solvable for $\ell \in \{3, 4\}$ (see [6]), the complexity of the problem is open for $\ell \geq 5$. We show that the problem can be solved in polynomial time in any class of graphs excluding some fixed induced star.

Theorem 9. *Let d, ℓ be positive integers with $\ell \geq 3$. Then, there exists a polynomial-time algorithm that, given a $K_{1,d}$-free graph G as input either decides that G does not contain W_ℓ as an induced minor or finds an induced minor model of W_ℓ in G.*

Proof (sketch). We show that it is possible to reduce the problem to the H-Induced Minor Containment problem in graphs of bounded tree-independence number. Then the theorem will follow from Theorem 7.

Let $\mathcal{A}$ be the approximation algorithm for tree-independence number of Dallard et al. [16]. Given a graph G and an integer k, $\mathcal{A}$ determines either that *(i)* $\alpha\text{-tw}(G) > k$ or *(ii)* finds a tree-decomposition of independence number in $\mathcal{O}(k^3)$ in time $|V(G)|^{\mathcal{O}(k^3)}$. Let $k:=f_1(d, \ell) + 1$, where f_1 is the function given by Theorem 1. Now, if $\mathcal{A}$ returns the tree-decomposition from outcome *(ii)*, then using the algorithm from Theorem 7 yields the desired result. Hence, we may assume that $\mathcal{A}$ determines that $\alpha\text{-tw}(G) > k$. Now let $v \in V(G)$ be an arbitrary vertex. It follows that $\alpha\text{-tw}(G - v) \geq k$. Therefore, by our choice of k and Theorem 1, we know that $G - v$ must contain W_ℓ as an induced minor and we may iterate the process with $G - v$ instead of G. Clearly after at most $|V(G)|$ steps, $\mathcal{A}$ must return a tree-decomposition of bounded independence number. This completes the proof. □

Acknowledgement. Mujin Choi was supported by the Institute for Basic Science (IBS-R029-C1). Claire Hilaire was partially supported by the Slovenian Research and Innovation Agency (J1-4008 and N1-0370). Martin Milanič was partially supported the Slovenian Research and Innovation Agency (I0-0035, research program P1-0285 and research projects J1-3003, J1-4008, J1-4084, J1-60012, and N1-0370) and by the research program CogniCom (0013103) at the University of Primorska. Sebastian Wiederrecht was partially supported by the Institute for Basic Science (IBS-R029-C1).

References

1. Adler, I.: Width functions for hypertree decompositions. Freiburg im Breisgau: Univ. Freiburg, Fakultät für Mathematik und Physik (Dissertation) (2006). d-nb.info/979896851
2. Ahn, J., Gollin, J.P., Huynh, T., Kwon, O.: A coarse Erdős-Pósa theorem. In: Azar, Y., Panigrahi, D. (eds.) Proceedings of the 2025 Annual ACM-SIAM Symposium on Discrete Algorithms, SODA 2025, New Orleans, LA, USA, 12–15 January 2025, pp. 3363–3381. SIAM (2025). https://doi.org/10.1137/1.9781611978322.109
3. Alecu, B., Chudnovsky, M., Hajebi, S., Spirkl, S.: Induced subgraphs and tree decompositions XIII. Basic obstructions in $\mathcal{H}$-free graphs for finite $\mathcal{H}$ (2023). https://arxiv.org/abs/2311.05066
4. Błasiok, J., Kamiński, M., Raymond, J.F., Trunck, T.: Induced minors and well-quasi-ordering. J. Comb. Theory Ser. B **134**, 110–142 (2019). https://doi.org/10.1016/j.jctb.2018.05.005
5. Bonnet, E., Kim, E.J., Thomassé, S., Watrigant, R.: Twin-width I: tractable FO model checking. J. ACM **69**(1) (2021). https://doi.org/10.1145/3486655
6. Bousquet, N., et al.: Induced minor models. I. Structural properties and algorithmic consequences (2025). https://arxiv.org/abs/2402.08332
7. Choi, M., Hilaire, C., Milanič, M., Wiederrecht, S.: Excluding an induced wheel minor in graphs without large induced stars (2025). https://arxiv.org/abs/2506.08829
8. Chudnovsky, M., Codsi, J., Lokshtanov, D., Milanič, M., Sivashankar, V.: Tree independence number V. Walls and claws (2025). https://arxiv.org/abs/2501.14658
9. Chudnovsky, M., Hajebi, S., Trotignon, N.: Tree independence number III. Thetas, prisms and stars (2024). https://arxiv.org/abs/2406.13053
10. Chuzhoy, J., Tan, Z.: Towards tight(er) bounds for the excluded grid theorem. J. Comb. Theory Ser. B **146**, 219–265 (2021). https://doi.org/10.1016/j.jctb.2020.09.010
11. Courcelle, B.: The monadic second-order logic of graphs. I. Recognizable sets of finite graphs. Inf. Comput. **85**(1), 12–75 (1990). https://doi.org/10.1016/0890-5401(90)90043-H
12. Courcelle, B., Engelfriet, J.: Graph Structure and Monadic Second-Order Logic: A Language-Theoretic Approach, vol. 138. Cambridge University Press, Cambridge (2012)
13. Courcelle, B., Olariu, S.: Upper bounds to the clique width of graphs. Disc. Appl. Math. **101**(1), 77–114 (2000). https://doi.org/10.1016/S0166-218X(99)00184-5
14. Dallard, C., Milanič, M., Štorgel, K.: Treewidth versus clique number. III. Tree-independence number of graphs with a forbidden structure. J. Comb. Theory Ser. B **167**, 338–391 (2024). https://doi.org/10.1016/j.jctb.2024.03.005
15. Dallard, C., Milanič, M., Štorgel, K.: Treewidth versus clique number. II. Tree-independence number. J. Comb. Theory Ser. B **164**, 404–442 (2024). https://doi.org/10.1016/j.jctb.2023.10.006
16. Dallard, C., Fomin, F.V., Golovach, P.A., Korhonen, T., Milanič, M.: Computing tree decompositions with small independence number. In: Bringmann, K., Grohe, M., Puppis, G., Svensson, O. (eds.) 51st International Colloquium on Automata, Languages, and Programming, ICALP 2024, Tallinn, Estonia, 8–12 July 2024. LIPIcs, vol. 297, pp. 51:1–51:18. Schloss Dagstuhl - Leibniz-Zentrum für Informatik (2024). https://doi.org/10.4230/LIPICS.ICALP.2024.51

17. Dallard, C., et al.: Treewidth versus clique number. IV. Tree-independence number of graphs excluding an induced star (2024). https://arxiv.org/abs/2402.11222
18. Diestel, R.: Graph Theory. Graduate Texts in Mathematics, vol. 173, 6th edn. Springer, Heidelberg (2025). https://doi.org/10.1007/978-3-662-70107-2
19. Fellows, M.R., Kratochvíl, J., Middendorf, M., Pfeiffer, F.: The complexity of induced minors and related problems. Algorithmica **13**(3), 266–282 (1995). https://doi.org/10.1007/BF01190507
20. Gyárfás, A.: Problems from the world surrounding perfect graphs. Zastos. Mat. **19**(3–4), 413–441 (1987). https://doi.org/10.4064/am-19-3-4-413-441
21. Hilaire, C., Milanič, M., Trotignon, N., Vasić, Đ.: Treewidth versus clique number: induced minors (2024). https://arxiv.org/abs/2410.17979
22. Korhonen, T.: Grid induced minor theorem for graphs of small degree. J. Comb. Theory Ser. B **160**, 206–214 (2023). https://doi.org/10.1016/j.jctb.2023.01.002
23. Korhonen, T., Lokshtanov, D.: Induced-minor-free graphs: separator theorem, subexponential algorithms, and improved hardness of recognition. In: Woodruff, D.P. (ed.) Proceedings of the 2024 ACM-SIAM Symposium on Discrete Algorithms, SODA 2024, Alexandria, VA, USA, 7–10 January 2024, pp. 5249–5275. SIAM (2024). https://doi.org/10.1137/1.9781611977912.188
24. Lima, P.T., Milanič, M., Muršič, P., Okrasa, K., Rzążewski, P., Štorgel, K.: Tree decompositions meet induced matchings: Beyond max weight independent set. In: Chan, T.M., Fischer, J., Iacono, J., Herman, G. (eds.) 32nd Annual European Symposium on Algorithms, ESA 2024, Royal Holloway, London, United Kingdom, 2–4 September 2024. LIPIcs, vol. 308, pp. 85:1–85:17. Schloss Dagstuhl - Leibniz-Zentrum für Informatik (2024). https://doi.org/10.4230/LIPICS.ESA.2024.85
25. Matoušek, J., Nešetřil, J., Thomas, R.: On polynomial time decidability of induced-minor-closed classes. Comment. Math. Univ. Carolin. **29**(4), 703–710 (1988). http://eudml.org/doc/17683
26. Oum, S., Seymour, P.: Approximating clique-width and branch-width. J. Comb. Theory Ser. B **96**(4), 514–528 (2006). https://doi.org/10.1016/j.jctb.2005.10.006
27. Ramsey, F.P.: On a problem of formal logic. Proc. London Math. Soc. (2) **30**(4), 264–286 (1929). https://doi.org/10.1112/plms/s2-30.1.264
28. Reed, B., Wood, D.: Polynomial treewidth forces a large grid-like-minor. Eur. J. Comb. **33**(3), 374–379 (2012). https://doi.org/10.1016/j.ejc.2011.09.004
29. Robertson, N., Seymour, P.: Graph minors. V. Excluding a planar graph. J. Comb. Theory Ser. B **41**, 92–114 (1986). https://doi.org/10.1016/0095-8956(86)90030-4
30. Robertson, N., Seymour, P.: Graph minors. IV. Tree-width and well-quasi-ordering. J. Comb. Theory Ser. B **48**(2), 227–254 (1990). https://doi.org/10.1016/0095-8956(90)90120-O
31. Robertson, N., Seymour, P., Thomas, R.: Quickly excluding a planar graph. J. Comb. Theory Ser. B **62**(2), 323–348 (1994). https://doi.org/10.1006/jctb.1994.1073
32. Seymour, P.: Tree-chromatic number. J. Comb. Theory Ser. B **116**, 229–237 (2016). https://doi.org/10.1016/j.jctb.2015.08.002
33. Yolov, N.: Minor-matching hypertree width. In: Proceedings of the Twenty-Ninth Annual ACM-SIAM Symposium on Discrete Algorithms, pp. 219–233. SIAM (2018). https://doi.org/10.1137/1.9781611975031.16

A Simple Quadratic Kernel for Token Jumping on Surfaces

Daniel W. Cranston[1], Moritz Mühlenthaler[2], and Benjamin Peyrille[2(✉)]

[1] Department of Computer Science, Virginia Commonwealth University, Richmond, USA

[2] Grenoble INP, Université Grenoble Alpes, Grenoble, France
{moritz.muhlenthaler,benjamin.peyrille}@grenoble-inp.fr

Abstract. The problem Token Jumping asks whether, given a graph G and two independent sets of *tokens* I and J of G, we can transform I into J by changing the position of a single token in each step and having an independent set of tokens throughout. We show that there is a polynomial-time algorithm that, given an instance of Token Jumping, computes an equivalent instance of size $O((g+k)^2)$, where g is the genus of the input graph and k is the size of the independent sets. Our algorithm is very simple and does not require any information about the genus of the input graph.

Keywords: independent set · reconfiguration · kernelization · fixed-parameter algorithm · graphs on surfaces

1 Introduction

We consider an independent set of a graph to be a set of *tokens* placed on pairwise non-adjacent vertices. From an independent set we may obtain another one by letting a token "jump" to some vertex that has no token on it and no token in its neighborhood. We say that two independent sets are *TJ-equivalent*, if one can be obtained from the other by a sequence of jump-operations. The problem Token Jumping asks whether two given independent sets of a graph are TJ-equivalent. Token Jumping can be thought of as a motion planning problem on a graph, where the tokens correspond to moving agents. It is a variant of the *independent set reconfiguration* problem, which has attracted considerable attention in the last ten years, in particular in the context of parameterized complexity [9,11,16].

Token Jumping is known to be PSPACE-complete even on subcubic planar graphs of bounded bandwidth [19]. Furthermore, the problem has been shown to be XL-complete when parameterized by the number k of tokens [6]. This means that, under standard complexity assumptions, Token Jumping is not fixed-parameter tractable (FPT), that is, it does not admit an $h(k) \cdot \mathsf{poly}(n)$-time algorithm, where h is some computable function. Under certain conditions however, some of which we list below, Token Jumping has been shown to be

H. Fernau and P. Kindermann (Eds.): WG 2025, LNCS 16124, pp. 149–161, 2026.
https://doi.org/10.1007/978-3-032-11835-6_11

FPT by the existence of a kernelization algorithm. Such an algorithm reduces in polynomial time an instance of the problem to an equivalent instance of size $f(k)$, where f is some computable function.

Notably, there are several kernelization results for sparse graphs: TOKEN JUMPING parameterized by k admits a linear kernel ($f = O(k)$) on graphs of bounded degree [4] and a polynomial kernel ($f = \mathsf{poly}(k)$) on graphs of bounded degeneracy [14]. Furthermore, Ito, Kamiński, and Ono showed that TOKEN JUMPING admits a kernelization algorithm on planar graphs and, more generally, on $K_{3,t}$-free graphs [13], where $K_{r,s}$ denotes the complete bipartite graph with r and s vertices on the two sides, respectively. Finally, Bousquet, Mary, and Parreau gave a polynomial kernel for $K_{t,t}$-free graphs [8], which implies a polynomial kernel for graphs embeddable on any fixed surface. However, when the problem is parameterized by the size k of the independent sets and the genus g of the surface (rather than forbidding $K_{t,t}$, where t is fixed), their kernel is not polynomial in g and k. Our main result is the following theorem.

Theorem 1. *TOKEN JUMPING parameterized by the size k of the independent sets and the genus g of the input graph admits a kernel of size $O((g+k)^2)$.*

To the best of our knowledge problems parameterized by the the genus of the input graph are rarely considered, so Theorem 1 could open up new possibilities for future research. Notice that in previous work, kernelization algorithms were proposed for certain classes of graphs, for instance planar graphs [13] and $K_{t,t}$-free graphs [8]. In contrast, our kernelization algorithm works on *any* input graph and in particular, it does not require any information about the genus.

The general approach is inspired by that in [13]. Consider an instance of TOKEN JUMPING given by a graph G and two independent sets I and J of G, each of size k. For $Y \subseteq I \cup J$, let $C_Y \subseteq G-(I \cup J)$ be the set of vertices that are adjacent to precisely Y in $I \cup J$. To obtain the kernel of Theorem 1, we partition the vertices of G not in I and J into three classes: those that have at most one, exactly two, or at least three neighbors in $I \cup J$.

$$\mathcal{C}_1 := \bigcup_{Y \subseteq I \cup J, |Y| \leq 1} C_Y \qquad \mathcal{C}_2 := \bigcup_{Y \subseteq I \cup J, |Y| = 2} C_Y \qquad \mathcal{C}_3 := \bigcup_{Y \subseteq I \cup J, |Y| \geq 3} C_Y$$

Figure 1 presents an example of the classes $\mathcal{C}_1$, $\mathcal{C}_2$ and $\mathcal{C}_3$. The tokens of I are represented by small circles on vertices of G while the tokens of J are represented by larger hatched vertices.

To obtain a kernel of size $O((g+k)^2)$, we show that each of the three vertex classes can be reduced to size $O((g+k)^2)$. This significantly improves on the results of [13] for planar graphs and [8] for bounded-genus graphs. For a comparison of the kernel sizes obtained for these graph classes, please see table 1.

In contrast to our analysis, Ito, Kamiński, and Ono use a bound on the Ramsey numbers to show that for planar graphs, if the subgraph induced by the neighbors of a pair $Y \in \mathcal{C}_2$ is larger than some $f(k)$, then it can be replaced by an independent set of size k [13] (the bounds on $\mathcal{C}_1$ and $\mathcal{C}_3$ are straight-forward). Using this approach, they obtain a kernel of size $O(2^{6k})$ for planar graphs. For

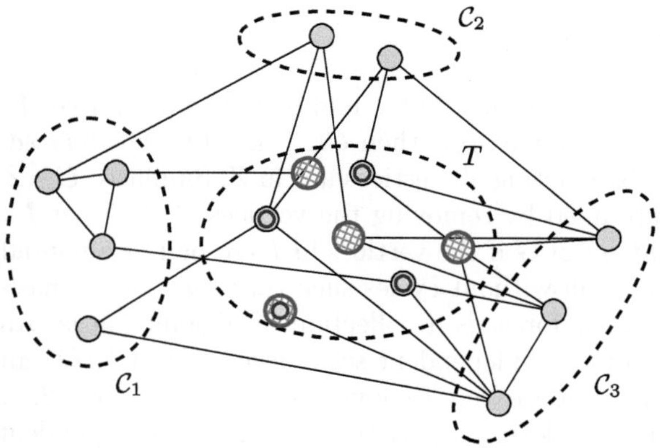

Fig. 1. Illustration of the classes $\mathcal{C}_1$, $\mathcal{C}_2$ and $\mathcal{C}_3$ for a given subset T of the tokens.

Table 1. Kernel sizes for different classes of graphs for TOKEN JUMPING

Class	Kernel size	For genus g
$K_{3,t}$-free [13]	Ramsey$((2t+1)k, t+3)$	Ramsey$((8g+7)k, 4g+6)$
$K_{t,t}$-free [8]	$O(f(t) \cdot k^{t \cdot 3^t})$	$O(h(g) \cdot k^{(4g+3) \cdot 3^{4g+3}})$
d-degenerate [14]	$O(d! \cdot k^{2d+1})$	$O(\sqrt{g}! \cdot k^{2\sqrt{g}+1})$
any graph (Thm. 1)	$O((g+k)^2)$	-

$K_{3,t}$-free graphs they do not give an explicit bound, but the reduced instance seems to be of size at least $2^{2k+(k+t+1)^{t+2}}$. Bousquet, Mary, and Parreau use a theorem of Kövári, Sós, and Turán to obtain a kernel for $K_{t,t}$-free graphs of size $O(f(t) \cdot k^{t \cdot 3^t})$ [8]. Since we do not require any arguments from extremal graph/set theory, our constants are reasonably small.

A direct consequence of Theorem 1 is the following result, which improves on the polynomial kernel given in [8].

Corollary 1. *TOKEN JUMPING parameterized by the size k of the independent sets admits a kernel of size $O(k^2)$ on graphs of bounded genus.*

The same algorithm, with an even simpler analysis, gives a sub-quadratic kernel for graphs with no $K_{2,3}$-subgraph; notably, this class includes all outerplanar graphs.

An interesting open question is whether a linear dependence on k can be obtained in the setting of Theorem 1 and Corollary 1, or on planar graphs (see [9, Question 1]). This is known to be the case for instance for graphs of bounded degree [4].

2 Preliminaries

All graphs in this article are finite, undirected, and simple. Let $G = (V, E)$ be a graph. If Z is a vertex set, then $G[Z]$ denotes the *induced subgraph* of G obtained from G by deleting all vertices not in Z. Similarly, $G - Z$ is the induced subgraph of G obtained by removing the vertices of Z. A set $I \subseteq V$ of vertices is an *independent set* of G if the vertices in I are pairwise non-adjacent in G. A linear forest is a graph without cycles such that each vertex has degree at most two. That is, a linear forest is a collection of disjoint paths. Any linear forest on n vertices admits an independent set of size at least $\lceil n/2 \rceil$ and we may find one in linear time by iteratively removing a vertex of degree 2. A proper vertex coloring of G with k colors is a partition of V into k independent sets.

Two independent sets I and J of G are *TJ-equivalent* if there exists a sequence $I_1, I_2, \ldots, I_\ell$ of independent sets of G, such that $I = I_1$ and $J = I_\ell$ and for $1 \leq i < \ell$, we have $|I_i \setminus I_{i+1}| = |I_{i+1} \setminus I_i| = 1$. Intuitively, at each step i, the independent set I_{i+1} is obtained from I_i by changing the position of a single token. Given a graph G together with two independent sets I and J of G, each of size k, the problem Token Jumping asks whether I and J are TJ-equivalent.

The (minimum, orientable) *genus* of a graph G is the minimum number g such that the graph G admits a crossing-free drawing on an orientable surface Σ of genus g. Any such drawing specifies a cyclic ordering of the edges incident to a vertex, and hence of its neighbors. Notice that a graph of genus 0 is *planar*, that is, it admits a crossing-free drawing on the sphere. The *Heawood number* $H(g)$, defined as $H(g) = \lfloor (7 + \sqrt{1 + 48g})/2 \rfloor$, is the minimum number of colors that are sufficient to color any graph of genus g. Heawood's 1890 proof of sufficiency [10] did not handle the planar case. The planar case was later solved by Appel and Haken and is well known as the four color theorem [1–3]. For more information about graphs on surfaces, please consult [15].

A decision problem parameterized by some parameter k is *fixed-parameter tractable (FPT)* if there exists an algorithm that decides an instance of size n with parameter $k \in \mathbb{N}$ in time $f(k) \cdot \mathsf{poly}(n)$, where $f : \mathbb{N} \to \mathbb{N}$ is some computable function. A *kernelization algorithm* is a polynomial-time algorithm that takes as input an instance of size n with parameter k and outputs an equivalent instance (called *kernel*) of size $s(k)$, where $s : \mathbb{N} \to \mathbb{N}$ is some computable function. By *equivalent* we mean that the kernel is a Yes-instance if and only if the input instance is a Yes-instance.

3 Kernelization Algorithm

Let us consider an instance of Token Jumping given by a graph $G = (V, E)$ and two independent sets I and J of G, each of size k. In the following, we describe our kernelization algorithm that, given such an instance, will yield an equivalent instance of Token Jumping of size $O((g + k)^2)$, where g is the genus of G. The analysis of the algorithm is deferred to the next section.

Recall that $\mathcal{C}_1$ (resp., $\mathcal{C}_2$, $\mathcal{C}_3$) is the set of vertices of $V \setminus (I \cup J)$ that have at most one (resp., exactly two, at least three) neighbors in $I \cup J$. We first show that either $|\mathcal{C}_1| < H(g) \cdot k$ or else the algorithm can report YES (Lemma 1). Notice that the algorithm does not need any information about the genus g of the input graph for this decision. We next prove that $|\mathcal{C}_3| = O(g^2 + gk + k)$ (Lemma 2). The bounds on the sizes of $\mathcal{C}_1$ and $\mathcal{C}_3$ follow, respectively, from the formulas of Heawood and Euler. Finally, we prove that we can replace $\mathcal{C}_2$ by a vertex subset of size $O((g+k)^2)$. This is the most involved part, which we describe in the following.

Recall that for $Y \subseteq X = I \cup J$, we denote by C_Y the set of vertices in $V \setminus X$ whose neighbors in X are precisely the vertices Y, that is, $C_Y := \{v \in V \setminus X \mid N(v) \cap X = Y\}$. Let $\mathcal{P}$ be the set of pairs u, v such that $u, v \in X = I \cup J$ and $C_{\{u,v\}} \neq \emptyset$. For each pair $Y \in \mathcal{P}$, we will execute Algorithm 1 on C_Y in order to find an independent set T_Y of size $2k$ if one exists and replace C_Y by T_Y. If $|C_Y|$ is large enough, we will show we are guaranteed to find such a T_Y.

Algorithm 1: Filtering algorithm

Input: Instance (G, I, J) of TOKEN JUMPING, a pair $Y \in \mathcal{P}$.
Output: A linear forest Z_Y contained in C_Y such that no vertex $v \in V - (C_Y \cup Y)$ is adjacent to more than 2 vertices of Z_Y.

1 $Z := C_Y$
2 **for** $v \in V(G) - (C_Y \cup Y)$ **do**
3 | **if** *v has at least 3 neighbors in C_Y* **then** $Z \leftarrow Z - N(v)$
4 **end**
5 **for** $w \in Z$ **do**
6 | **if** *w has degree at least 3 in $G[Z]$* **then** $Z \leftarrow Z - w$
7 **end**
8 Remove arbitrarily one vertex from each cycle in $G[Z]$
9 **return** Z

The main idea behind Algorithm 1 is that whenever C_Y is large enough $(\Omega(g) + 4k)$, we can replace C_Y by an independent set T_Y (of size $2k$), such that any vertex outside of $C_Y \cup Y$ is adjacent to at most two vertices of T_Y. This adjacency condition, enforced in Lines 3 and 6, is key to the analysis. It ensures that if Y contains at most one token, then there exists at least one token that can move to T_Y.

Algorithm 1 exploits the structure of C_Y as follows. On any given surface, a crossing-free drawing of G places the vertices of C_Y into disjoint planar *zones* (an example of zones on a torus is provided in Fig. 2), such that the vertices inside these zones, the *inner vertices*, are adjacent to at most 2 vertices of C_Y and form a linear forest. The number of vertices on the border of these zones, the *outer vertices*, is bounded by a linear function of the genus of G. Thus, if C_Y is large enough, the number of inner vertices is also large (at least $4k$), and we can find an independent set of size $2k$ among these inner vertices.

Recall that we are not given a drawing of G on any surface, so our algorithm cannot determine which vertices are inner and which are outer with perfect accuracy. But using the structure of C_Y, we show that we can extract a large linear forest using most of the inner vertices by the conditions in Lines 3, 6 and 8.

4 Proof of Theorem 1

In this section we prove our main result, Theorem 1. In the following, let (G, I, J) be an instance of TOKEN JUMPING, where $G = (V, E)$ is a graph and I and J are independent sets of G, each of size k. Let g be the genus of the graph G and let $X = I \cup J$. Notice that $|X| \leq 2k$.

Lemma 1. *If $|\mathcal{C}_1| \geq H(g) \cdot k$, then (G, I, J) is a YES-instance.*

Proof. Assume that $|\mathcal{C}_1| \geq H(g) \cdot k$. Since G admits a proper vertex coloring with $H(g)$ colors, $G[\mathcal{C}_1]$ contains an independent set I_m of size k. We will now prove we can move all tokens from I to I_m. Initially, all tokens are on $I \subseteq X$. Consider a vertex v of I_m. Since I_m is an independent set, v is adjacent to no other vertices of I_m. Furthermore, since $v \in \mathcal{C}_1$, vertex v is adjacent to at most one vertex of X. Thus, if v is adjacent to a token, then that token must be on X and we can move it directly to v. Otherwise, we can move any token directly from X to v. Similarly, we can move all tokens from J to I_m. □

We remark that the independent set I_m in the proof of Lemma 1 can be computed in polynomial time, if it exists: If G is planar, which can be tested in linear time [12], then we can find I_m using the 4-coloring algorithm of Robertson, Sanders, Seymour and Thomas [18]. Alternatively, we may use the fact that planar graphs are 5-degenerate in order to greedily color G with 6 colors by recursively coloring vertices of minimum degree, which is much simpler but gives slightly worse constants. If G is not planar, then by Euler's formula ([15], Theorem 8.3.1 due to Heawood), the graph G is $(H(g) - 1)$-degenerate and we can thus color it greedily with $H(g)$ colors as described above.

From now on we assume that $|\mathcal{C}_1| \leq H(g) \cdot k$. We bound the size of $\mathcal{C}_3$ using Euler's formula.

Lemma 2. $|\mathcal{C}_3| \leq 16g^2 + 8g(2k - 1) + 8k$.

Proof. We will bound the number of sets $Y \subseteq X$ such that $|Y| \geq 3$ and $C_Y \neq \emptyset$. Consider a drawing of G on a surface of genus g. We want to construct from G a graph G' such that we can map injectively to a face of G' each $Y \subset X$ such that $|Y| \geq 3$ and $C_Y \neq \emptyset$.

We start with the subgraph of G induced by the vertices X. For each $Y \subseteq X$ such that $|Y| \geq 3$ and $C_Y \neq \emptyset$, we will arbitrarily add one vertex v_Y from C_Y and only keep the edges from v_Y to Y.

For each such Y, the drawing of G induces a cyclic ordering $\{v_1, \cdots, v_t\}$ of the neighbors of v_Y. We now want to create a contractible cycle with the neighbors of v_Y but we must not create any edge crossing. First, we add edge $v_i v_{i+1}$, following the path v_i, v_Y, v_{i+1}, for each $i \in \{1, \cdots, t-1\}$; similarly, we add edge $v_1 v_t$ following the path v_1, v_Y, v_t. Following those paths ensures all edges can be drawn without creating crossings. Furthermore, this guarantees the cycle is contractible and its interior only contains v_Y. We then remove the vertex v_Y and eliminate any parallel edges; in the process, we create a unique face f_Y that we associate to the set Y. Let G' be the resulting graph.

Let G'' be a triangulation of G'. Then $2|E(G'')| = \sum_{v \in V(G'')} d(v) = 3|F(G'')|$ and thus $|E(G'')| = 3|F(G'')|/2$. Let g' be the genus of G'' and observe that $g' \leq g$. By Euler's formula,

$$2 - 2g \leq 2 - 2g' = |V(G'')| - |E(G'')| + |F(G'')| = |V(G'')| - |F(G'')|/2 .$$

Thus, $|F(G')| \leq |F(G'')| \leq 2|V(G'')| + 4(g-1) \leq 2|X| + 4(g-1)$.

Hence, there are at most $2|X| + 4(g-1)$ sets $Y \subseteq X$ such that $|Y| \geq 3$ and $C_Y \neq \emptyset$. By [7,17], if a graph has genus g, it contains no $K_{3,m}$ as a subgraph when $m \geq 4g+3$. So we obtain

$$\begin{aligned}
|\bigcup_{Y \subseteq X,\ |Y| \geq 3} C_Y| &\leq (4g+2)(2|X| + 4(g-1)) \\
&= 16g(g-1) + 8g|X| + 8g + 4|X| - 8 \\
&\leq 16g^2 - 8g + 16gk + 8k \\
&= 16g^2 + 8g(2k-1) + 8k.
\end{aligned}$$

□

We now reduce the size of $\mathcal{C}_2$. First, like in the proof of 2, we show that the number of pairs in $\mathcal{P}$ is small.

Lemma 3. $|\mathcal{P}| \leq 3|X| + 6(g-1)$.

Proof. Similar to the proof of Lemma 2, consider a drawing of G on a surface of genus g. As before, we draw a graph G' such that G' is obtained by restricting G to the vertices of $\bigcup_{Y \in \mathcal{P}} Y$ and one vertex v_Y in C_Y for each $Y \in \mathcal{P}$. For each $\{u, v\} \in \mathcal{P}$, we add an edge u, v (if it does not already exist) following the path u, v_Y, v and then remove the vertex v_Y. In order to apply Euler's formula, let G'' be a triangulation of G'. The number of edges of G' corresponds to the number of elements in $\mathcal{P}$; that is, $|\mathcal{P}| = |E(G')|$. Thus, the lemma follows from the inequality below. Since $g(G'') \leq g$ and $3|F(G'')| = 2|E(G'')|$, we have

$$\begin{aligned}
|E(G'')| &\leq |V(G'')| + |F(G'')| - (2-2g) \\
&= 3\,V(G'')| - 3(2-2g) \\
&\leq 3|X| + 6(g-1).
\end{aligned}$$

□

For a given pair $Y \in \mathcal{P}$, the size of C_Y can be arbitrarily large. However, if C_Y is large enough, then we will show that we can find a vertex set T_Y in C_Y of size $O(k)$ such that replacing C_Y by T_Y creates an equivalent instance (G', I, J), where G' is a subgraph of G. To prove the existence of such a vertex set, we need some topological arguments.

A 2-cell embedding of a graph on a surface S is a drawing of the graph on S such that every face is homeomorphic to an open disk. Let $Y = \{u, v\} \in \mathcal{P}$ and consider the connected component G_Y of G containing all vertices reachable from Y. That component admits a 2-cell embedding on an orientable surface of genus at most g and a corresponding crossing-free drawing D_Y (see [15, Section 3.2]).

Consider the restriction $D_Y[C_Y \cup Y]$ of D_Y to the drawing of the subgraph $G_Y[C_Y \cup Y]$. The drawing $D_Y[C_Y \cup Y]$ partitions the paths $u - w - v$, where $w \in C_Y$, into homotopy classes[1]. Let $\mathcal{C}$ be any such homotopy class and let C be the set of midpoints of the paths in $\mathcal{C}$. Since D_Y corresponds to a 2-cell embedding, the restriction of D_Y to the paths in $\mathcal{C}$ is homeomorphic to a planar drawing of $\mathcal{C}$. Therefore, there exists an outer face f_0 in the restriction of D_Y to $\mathcal{C}$. If more than 2 vertices in C are contained in the boundary of f_0, say w, w', and w'', then remark we obtain an embedding of $K_{3,3}$ in the plane. First, u and v are both adjacent to w, w' and w'' by definition. Then, we can place a vertex inside f_0 and connect it to w, w', and w'', giving us an embedding of $K_{3,3}$ on a planar surface, which is impossible. Therefore, at most two vertices of C touch f_0. We call those vertices *outer* and all other vertices of C *inner*. We denote by $N^p(\{u, v\})$ the set of inner vertices corresponding to the pair u, v. By iteratively removing outer vertices and their corresponding paths in $\mathcal{C}$ from the drawing (breaking ties arbitrarily), we obtain a linear order $w_1, w_2, \ldots, w_\ell$ on the midpoints C.

By the discussion above, the homotopy classes of the paths $u-w-v$ partition the drawing D_Y restricted to the paths $u-w-v$ into (not necessarily connected) regions of the surface that we call *zones*. By the following lemma, the number of homotopy classes, and therefore the number of zones, is bounded by a function of g (see [15, Proposition 4.2.7] and the discussion after its proof). Let $f_M(g) := \max\{1, 4g\}$.

Lemma 4 (see [15, Proposition 4.2.7]). *Let G be a connected graph embedded on an orientable surface of genus g. If $P_1, \cdots, P_k$ are u-v paths of G ($u \neq v$) such that no two are homotopy-equivalent with respect to the embedding, then $k \leq f_M(G)$.*

Figure 2 shows a drawing of a set of u-v paths on the torus having 8 outer vertices and 4 zones. Clearly, there are no edges between vertices in two different zones of the same pair u, v in $\mathcal{P}$.

We now reduce the size of $\mathcal{C}_2$ to $O(g^2 + gk + k^2)$. Fix $Y \in \mathcal{P}$ such that $|C_Y| \geq 2f_M(g) + 4k$. Notice that the number of inner vertices of Y is at least $4k$, because the total number of outer vertices of Y (over all zones) is at most $2f_M(g)$. Note that a vertex $v \notin X$ is either inside a zone of Y, or else outside all its zones. If v is outside, then v is adjacent to none of the inner vertices of C_Y. But if v is inside a zone of Y, then v is adjacent to at most two vertices of C_Y.

First, we will find in polynomial time an independent subset T_Y of C_Y of size $2k$, such that each vertex $v \notin (C_Y \cup Y)$ is adjacent to at most two elements

[1] Two curves are *homotopy-equivalent* if there exists a continuous transformation between them, see [15].

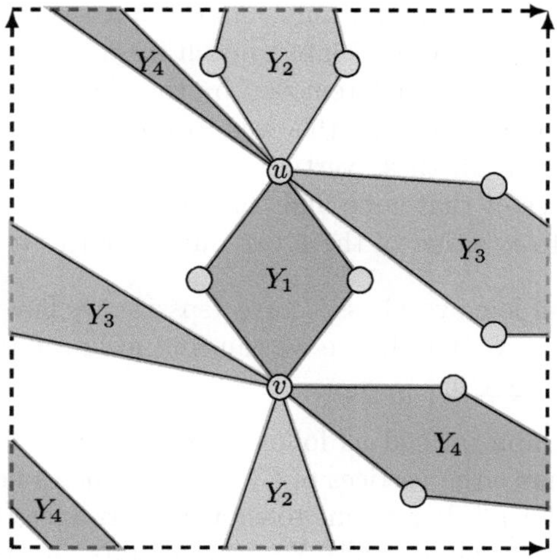

Fig. 2. Four zones of a pair $Y = \{u, v\}$ on a torus ($g = 1$). Notice that the torus is obtained by identifying opposing sides of the rectangle.

of T_Y. Then, we will show that restricting C_Y to T_Y produces an equivalent instance.

We first show that T_Y can be computed in polynomial time.

Lemma 5. *Let Z_Y be the output of Algorithm 1 for a pair $Y \in \mathcal{P}$. Then $G[Z_Y]$ is a linear forest such that any vertex $v \notin (C_Y \cup Y)$ has at most 2 neighbors in Z_Y.*

Proof. First, we show $G[Z_Y]$ is a linear forest. Assume for a contradiction that, just before reaching Line 8, there exists a vertex $w \in Z_Y$ such that $d_{G[Z_Y]}(w) \geq 3$, then w would have been removed by Line 6. It follows that when reaching Line 8, the graph $G[Z_Y]$ is of maximum degree 2. The algorithm then removes one vertex from each cycle of $G[Z_Y]$, so the vertex set Z_Y returned by Algorithm 1 induces a linear forest.

Finally, Line 3 ensures that every vertex $v \notin (C_Y \cup Y)$ has at most 2 neighbors in Z_Y. □

Lemma 6. *Let Z_Y be the output of Algorithm 1 for a pair $Y \in \mathcal{P}$. We have $|Z_Y| \geq |N^p(Y)|$.*

Proof. We will first show Line 3 only removes outer vertices of C_Y. Let v be a vertex such that $v \notin (C_Y \cup Y)$. If v is inside some zone of C_Y, then v can have at most two neighbors in C_Y (these are the predecessor and successor of v in the linear order constructed after Lemma 3). If v is outside of any zone of C_Y, then v cannot be adjacent to any inner vertex of C_Y (because the two outer vertices of any zone of C_Y, together with the two vertices in Y, induce a cycle that separates v from the inner vertices of that zone). So Line 3 can only remove outer vertices of C_Y. Because an inner vertex can only be adjacent to two vertices of C_Y, Line 6 can only remove outer vertices. To conclude, it remains to show that we can associate to each inner vertex removed in Line 8 a unique outer vertex that is not removed by the algorithm. If an inner vertex w of a zone of Y is removed, then this was to break a cycle that contained all vertices in the zone; so all these vertices were in Z_Y. Therefore, Z_Y contains the two outer vertices of that zone and they appear in only one cycle. We can therefore associate to w either of these two outer vertices of the zone. □

By Lemmas 5 and lem:zyspshasspscorrectspssize, we have that Z_Y is a linear forest of size at least $|N^p(Y)|$, hence we can find in linear time an independent set T_Y of size $2k \leq \left\lceil \frac{|N^p(Y)|}{2} \right\rceil$ in Z_Y.

Let G' be the graph formed as follows: for each $Y \in \mathcal{P}$ such that $|C_Y| \geq 2f_M(g) + 4k$, we remove the vertices of $C_Y \setminus T_Y$. In the same way we defined $\mathcal{C}_2$ for G, we define $\mathcal{C}_2'$ for G'. It remains to show the instance (G', I, J) is indeed a kernel of (G, I, J) of size $O((g+k)^2)$. We start by proving the size of our kernel.

Lemma 7. $|V(G')| = O((g+k)^2)$.

Proof. The size of $\mathcal{C'}_2$ in G' can be bounded directly as each pair Y either satisfied $|C_Y| \leq 2f_M(g) + 4k$ or was restricted using Algorithm 1 to a subset T_Y of size $2k$.

$$
\begin{aligned}
|V(G')| &= |X| + |\mathcal{C}_1| + |\mathcal{C}_3| + |\mathcal{C}_2'| \\
&\leq 2k + H(g)k + (16g^2 + 8g(2k-1) + 8k) + (|\mathcal{P}|(2f_M(g) + 4k)) \\
&\leq 16g^2 + 8g(2k-1) + H(g)k + 10k + \\
&\quad ((3|X| + 6(g-1))(2f_M(g) + 4k)) \\
&\leq 16g^2 + 8g(2k-1) + H(g)k + 10k + \\
&\quad 12((k + (g-1))(f_M(g) + 2k)) \\
&= 16g^2 + 16gk - 8g + H(g)k + 10k + \\
&\quad 12(f_M(g)k + 2k^2 + f_M(g)g + 2gk - f(g) - 2k) \\
&= 16g^2 + 12f_M(g)g + 40gk + \\
&\quad 12f_M(g)k - 8g - 12f_M(g) + H(g)k - 14k + 24k^2
\end{aligned}
$$

If $g = 0$, then $|V(G')| \leq 24k^2 + 2k$.
If $g \geq 1$, then $|V(G')| \leq 64g^2 + 88gk + H(g)k - 56g - 14k + 24k^2$. □

To conclude our proof, we show that our new instance is equivalent to the original one.

Theorem 2. *The instances (G, I, J) and (G', I, J) of* Token Jumping *are equivalent.*

Proof. Since G' is an induced subgraph of G, if (G', I, J) is a Yes-instance, then also (G, I, J) a is Yes-instance. It remains to show that if (G, I, J) is a Yes-instance, then also (G', I, J) is a Yes-instance.

Assume that (G, I, J) is a YES-instance and let σ be a sequence of token jumps certifying this, that is, a sequence of token jumps transforming I into J. Let $\mathcal{Y}$ be the set of pairs $Y \in \mathcal{P}$ such that C_Y was reduced to T_Y in the construction of G'. If σ only uses vertices of G', then we are done. Suppose that σ uses a vertex from G that is not in G'. Such a vertex must be in $C_Y - T_Y$ for some $Y \in \mathcal{Y}$. Let Y_1 correspond to the first pair of tokens such that σ moves a token into $C_{Y_1} - T_{Y_1}$. Similarly, let Y_2 be the last pair of tokens such that σ removes a token from $C_{Y_2} - T_{Y_2}$. We will show that we can move all tokens of I to T_{Y_1} and all tokens of J to T_{Y_2}. We will then move all tokens from T_{Y_1} to T_{Y_2} to obtain a transformation from I to J in G', concluding our proof.

We follow the token jump sequence σ from the beginning to the first time that a token of I jumps to a vertex in $C_{Y_1} - T_{Y_1}$. Remark that when this jump is completed, there are no more tokens of I on Y_1. If the token came from Y_1, we can move it to a vertex of T_{Y_1} and then move all other tokens to T_{Y_1}, since every token not on Y_1 is adjacent to at most two vertices of T_{Y_1} and $|T_{Y_1}| = 2k$. Otherwise, we can instead move directly all tokens to T_{Y_1} for the same reason. Let I' be the resulting independent set and notice that (G', I, I') is a YES-instance.

Reversing the sequence σ and exchanging I and J, we may apply the same reasoning to move the tokens of J to $|T_{Y_2}|$. Let J' be the resulting independent set and notice that (G', J, J') is a YES-instance. It remains to show that (G', I', J') is a YES-instance. We first move the tokens from $I' \subseteq T_{Y_1}$ to T_{Y_2} in G' as follows. Since each token on T_{Y_1} is adjacent to at most 2 vertices of T_{Y_2} and $|T_{Y_2}| = 2k$, each token of I' has at least two valid destinations on T_{Y_2}. Finally, since T_{Y_2} is an independent set, we may move the tokens to J'. This means that (G, I', J') is a YES-instance, which concludes the proof. □

By combining Lemma 7 and Theorem 2 we obtain Theorem 1.

5 A Kernel for Token Jumping on Graphs with no $K_{2,3}$-Subgraph

TOKEN JUMPING is PSPACE-complete on graphs with no $K_{2,3}$-induced-subgraph by [5, Theorem 2], however the complexity of TOKEN JUMPING is not known for graphs with no $K_{2,3}$-subgraph and even outerplanar graphs.

In this section we show that the algorithm from Sect. 3 can be applied directly to obtain a sub-quadratic kernel for TJ on graphs with no $K_{2,3}$-subgraph, which notably include outerplanar graphs. The analysis will use the same principles as Sect. 4 but will be slightly different for $\mathcal{C}_3$ and significantly simpler for $\mathcal{C}_2$. We will use the same notation.

Theorem 3. *For input graphs with no $K_{2,3}$-subgraph, TOKEN JUMPING parameterized by the size k of the independent sets and the genus g of the graph admits a kernel of size $O(g + \sqrt{g}k + k)$.*

Proof. From Lemma 1, we know that if $|\mathcal{C}_1| \geq H(g) \cdot k$, we have a YES-instance. We therefore assume that $|\mathcal{C}_1| < H(g) \cdot k$.

We now show that $|\mathcal{C}_2| \leq 12k + 12g$. By Lemma 3, $|\mathcal{P}| \leq 6k + 6g$. However there are no $K_{2,3}$ subgraphs, so for any $Y \in \mathcal{P}$, we have $|C_Y| \leq 2$. Hence $|\mathcal{C}_2| \leq 12k + 12g$.

Finally, we show that $|\mathcal{C}_3| \leq 4k + 4g$. The beginning of the proof for Lemma 2 still holds: there are at most $2|X| + 4(g-1)$ sets $Y \subseteq X$ such that $|Y| \geq 3$ and $C_Y \neq \emptyset$. However, since there is no $K_{2,3}$ subgraph, we have that $|C_Y| = 1$ for any such Y.

Combining the bounds above we obtain

$$\begin{aligned} |V(G)| &= |X| + |\mathcal{C}_1| + |\mathcal{C}_2| + |\mathcal{C}_3| \\ &\leq 2k + H(g)k + (12k + 12g) + (4k + 4g) \\ &\leq 18g + H(g)k + 18k. \end{aligned}$$ □

Observe that all that is required for the kernelization algorithm is to compute the size of $|\mathcal{C}_1|$, which can be done in time $O(k \cdot |E(G)|)$. Noticing that any outerplanar graph is 3-colorable, we may assume that $|\mathcal{C}_1| < 3k$, which gives a linear kernel for Token Jumping on outerplanar graphs (Token Jumping on outerplanar graphs is not known to be PSPACE-hard).

Corollary 2. *Token Jumping parameterized by the size k of the independent sets admits a kernel of size $21k$ on outerplanar graphs.*

References

1. Appel, K., Haken, W.: Every planar map is four colorable. I. Discharging. Illinois J. Math. **21**(3), 429–490 (1977). http://projecteuclid.org.proxy.library.vcu.edu/euclid.ijm/1256049011
2. Appel, K., Haken, W., Koch, J.: Every planar map is four colorable. II. Reducibility. Illinois J. Math. **21**(3), 491–567 (1977). http://projecteuclid.org.proxy.library.vcu.edu/euclid.ijm/1256049012
3. Appel, K.I., Haken, W.: Every planar map is four colorable, vol. 98. American Mathematical Soc. (1989)
4. Bartier, V., Bousquet, N., Dallard, C., Lomer, K., Mouawad, A.E.: On girth and the parameterized complexity of token sliding and token jumping. Algorithmica **83**(9), 2914–2951 (2021). https://doi.org/10.1007/S00453-021-00848-1
5. Bartier, V., Bousquet, N., Mühlenthaler, M.: Independent set reconfiguration in h-free graphs. In: International Workshop on Graph-Theoretic Concepts in Computer Science, pp. 35–49. Springer (2024)
6. Bodlaender, H.L., Groenland, C., Swennenhuis, C.M.F.: Parameterized complexities of dominating and independent set reconfiguration (2023). https://arxiv.org/abs/2106.15907
7. Bouchet, A.: Orientable and nonorientable genus of the complete bipartite graph. J. Comb. Theory Ser. B **24**(1), 24–33 (1978). https://doi.org/10.1016/0095-8956(78)90073-4
8. Bousquet, N., Mary, A., Parreau, A.: Token jumping in minor-closed classes. In: International Symposium on Fundamentals of Computation Theory, pp. 136–149. Springer (2017)

9. Bousquet, N., Mouawad, A.E., Nishimura, N., Siebertz, S.: A survey on the parameterized complexity of reconfiguration problems. Comput. Sci. Rev. **53**, 100663 (2024). https://doi.org/10.1016/J.COSREV.2024.100663
10. Heawood, P.J.: Map colour theorem. Quart. J. Math. **24**, 332–333 (1890)
11. van den Heuvel, J.: The complexity of change. In: Blackburn, S.R., Gerke, S., Wildon, M. (eds.) Surveys in Combinatorics 2013, London Mathematical Society Lecture Note Series, vol. 409, pp. 127–160. Cambridge University Press (2013). https://doi.org/10.1017/CBO9781139506748.005
12. Hopcroft, J., Tarjan, R.: Efficient planarity testing. J. ACM (JACM) **21**(4), 549–568 (1974)
13. Ito, T., Kaminski, M.J., Ono, H.: Fixed-parameter tractability of token jumping on planar graphs. In: Algorithms and Computation – 25th International Symposium, ISAAC. Lecture Notes in Computer Science, vol. 8889, pp. 208–219. Springer (2014). https://doi.org/10.1007/978-3-319-13075-0_17
14. Lokshtanov, D., Mouawad, A.E., Panolan, F., Ramanujan, M.S., Saurabh, S.: Reconfiguration on sparse graphs. J. Comput. Syst. Sci. **95**, 122–131 (2018). https://doi.org/10.1016/J.JCSS.2018.02.004
15. Mohar, B., Thomassen, C.: Graphs on surfaces. Johns Hopkins University Press (2001)
16. Nishimura, N.: Introduction to reconfiguration. Algorithms **11**(4), 52 (2018). https://doi.org/10.3390/A11040052
17. von Ringel, G.: Das Geschlecht des vollständigen paaren Graphen. Abh. Math. Semin. Univ. Hambg. **28**, 139–150 (1965)
18. Robertson, N., Sanders, D.P., Seymour, P., Thomas, R.: Efficiently four-coloring planar graphs. In: Proceedings of the twenty-eighth annual ACM symposium on Theory of computing, pp. 571–575 (1996)
19. van der Zanden, T.C.: Parameterized complexity of graph constraint logic. In: 10th International Symposium on Parameterized and Exact Computation, IPEC. LIPIcs, vol. 43, pp. 282–293. Schloss Dagstuhl - Leibniz-Zentrum für Informatik (2015). https://doi.org/10.4230/LIPICS.IPEC.2015.282

Bounding Width on Graph Classes of Constant Diameter

Konrad K. Dabrowski[1], Tala Eagling-Vose[2], Noleen Köhler[3(✉)], Sebastian Ordyniak[3], and Daniël Paulusma[2]

[1] School of Computing, Newcastle University, Newcastle, UK
konrad.dabrowski@newcastle.ac.uk
[2] Department of Computer Science, Durham University, Durham, UK
{tala.j.eagling-vose,daniel.paulusma}@durham.ac.uk
[3] School of Computer Science, University of Leeds, Leeds, UK
sordyniak@gmail.com, n.koehler@leeds.ac.uk

Abstract. We determine if the width of a graph class $\mathcal{G}$ changes from unbounded to bounded if we consider only those graphs from $\mathcal{G}$ whose diameter is bounded. As parameters we consider treedepth, pathwidth, treewidth and clique-width, and as graph classes we consider classes defined by forbidding some specific graph F as a minor, induced subgraph or subgraph, respectively. Our main focus is on treedepth for F-subgraph-free graphs of diameter at most d for some fixed integer d. We give classifications of boundedness of treedepth for $d \in \{4, 5, \ldots\}$ and partial classifications for $d = 2$ and $d = 3$.

1 Introduction

Graph width parameters play a prominent role in modern graph theory. One of the reasons is that large sets of NP-complete graph problems may become polynomial-time solvable on graph classes on which some width parameter is bounded (by a constant). For example, the celebrated meta-theorem of Courcelle [4] states that every problem definable in MSO_2 is polynomial-time solvable on graph classes of bounded treewidth. Another well-known meta-theorem, due to Courcelle, Makowsky and Rotics [5], states that every problem definable in MSO_1 is polynomial-time solvable on graph classes of bounded clique-width. The logic MSO_1 is more restrictive than MSO_2. However, any graph class of bounded treewidth has bounded clique-width, whereas the reverse statement does not hold. That is, clique-width is more *powerful* than treewidth.

Due to the above algorithmic implications and also out of a graph-structural interest, there exist many papers in the literature that research whether certain graph classes have bounded width. The framework of graph containment opens the way for a more systematic approach. For example, the recent treewidth dichotomy of Lozin and Razgon [24] determines exactly for which finite sets $\mathcal{F}$, the class of $\mathcal{F}$-free graphs has bounded treewidth; here, *$\mathcal{F}$-free* means not containing any graph from $\mathcal{F}$ as an *induced* subgraph. Hickingbotham [20] showed

H. Fernau and P. Kindermann (Eds.): WG 2025, LNCS 16124, pp. 162–176, 2026.
https://doi.org/10.1007/978-3-032-11835-6_12

that in the treewidth dichotomy we may replace "treewidth" by "pathwidth". In contrast, there still exist some pairs (H_1, H_2) for which boundedness of clique-width is open for (H_1, H_2)-free graphs; see the survey [8] for details.

Our Focus. A width parameter p that is unbounded on a graph class $\mathcal{G}$ may be bounded on a subclass $\mathcal{G}'$ of $\mathcal{G}$. Ideally, we would like $\mathcal{G}'$ to be as large as possible to optimally benefit from the algorithmic benefits if p is bounded. We consider the *diameter* of the input graph and ask:

For a graph class $\mathcal{G}$ of unbounded width, what is the largest d such that the graphs in $\mathcal{G}$ of diameter at most d have bounded width?

This is a natural question, as there exist numerous NP-complete problems that stay NP-complete even on graphs of diameter $d = 2$; see, e.g. the diameter study for k-Colouring [11,27], in particular for $\mathcal{F}$-free graphs in [11,22,26,27]. There are even problems that are NP-complete *only if* $d = 2$ (e.g. Disconnected Cut [25]). Answering the above question will have a wide range of algorithmic consequences, in particular due to meta-theorems, as we discussed above [5,6].

Our Approach. We work within the framework of graph containment and thus focus on graph classes defined by some set $\mathcal{F}$ of forbidden graphs. To get a handle on these, we restrict ourselves to the case when $\mathcal{F}$ consists of a single graph F. To answer our research question, we selected some classical graph containment relations and width parameters. We forbid F as an induced subgraph, subgraph or minor. We say that a graph G is *F-free*, *F-subgraph-free* or *F-minor-free* if it does not contain F as an induced subgraph, subgraph or minor, respectively. Note that F-minor-free graphs are F-subgraph-free and F-subgraph-free graphs are F-free. As width parameters, we will consider pathwidth pw, treewidth tw, clique-width cw and also treedepth td (which has algorithmic applications for many problems where treewidth cannot be used; see e.g. [16,18,19,21,23]). We write $p \triangleright q$ if p is more powerful than q. It is well known [2,6] that cw$\triangleright$tw$\triangleright$pw$\triangleright$td.

Known Results. We first describe, in Table 1, the situation without a diameter bound. For $r \geq 1$, the graph P_r is the r-vertex path. The set $\mathcal{S}$ consists of all graphs, every component of which is a path or *subdivided claw* (cubic tree with exactly one vertex of degree 3). The set $\overline{\mathcal{S}}$ consists of all graphs that are subgraphs of any *subdivided star* (any tree with exactly one vertex of degree at least 3).

Table 1 also includes known results on diameter-width. Eppstein [14] defined a graph class $\mathcal{G}$ to have the *diameter-treewidth* property if the treewidth of every graph in $\mathcal{G}$ is bounded by a function of the diameter of G. For graph classes closed under taking subgraphs, this notion coincides with bounded local treewidth, a crucial notion in bidimensionality theory. We define the properties of diameter-clique-width, diameter-pathwidth and diameter-treedepth analogously. A graph G is *apex* $\mathcal{G}$ for a graph class $\mathcal{G}$ if $G - v \in \mathcal{G}$ for some vertex $v \in V(G)$. So, for instance, the class of *apex linear forests* consists of all graphs that become *linear forests* (disjoint unions of paths) after removing at most one vertex.

The $(k \times k)$-wall, is the k-by-k hexagonal grid. We first note that by adding a dominating vertex to a wall we obtain a graph of diameter $d = 2$ whose clique-width can be arbitrarily large. This graph only contains apex planar graphs as minors. Hence, the result of Eppstein [14] implies that a class of F-minor-free graphs of diameter 2 has bounded clique-width if and only if F is apex planar. As this is not true for $d = 1$ (just take $F = K_6$), we say that $d = 2$ is *tight* for diameter-clique-width for minors. By adding a dominating vertex to a full binary tree, we find that $d = 2$ is also tight for diameter-pathwidth for minors.

Table 1. Overview of known and new results. Entries without brackets classify the graphs F such that the width of F-free graphs is bounded. For example, a class of F-minor free graphs has bounded clique-width if and only if F is a planar graph. Entries within brackets classify the graphs F such that the class of F-free graphs has the diameter-width property. We use the notation $F \subseteq_i G$ to indicate that F is an induced subgraph of G. Unreferenced results indicate a trivial/folklore result. A "?" indicates an open case. Results in bold/blue are new results proven in this paper.

	minor		induced subgraph		subgraph	
cw	planar [9]	**(apex planar)**	$\subseteq_i P_4$ [6]	$(\subseteq_i \mathbf{P_4})$	$\mathcal{S}$ [3]	(?)
tw	planar [30]	(apex planar [14])	$\subseteq_i P_2$	$(\subseteq_i \mathbf{P_2})$	$\mathcal{S}$ [29]	(?)
pw	forest [1]	(apex forest [12])	$\subseteq_i P_2$	$(\subseteq_i \mathbf{P_2})$	$\mathcal{S}$ [29]	(?)
td	linear forest [28]	**(apex linear forest)**	$\subseteq_i P_2$	$(\subseteq_i \mathbf{P_2})$	linear forest [28]	$(\overline{\mathcal{S}})$

Our Results. Table 1 also contains several new results, and all the new and known results together already show a clear impact of bounding the diameter. For each of the new results in Table 1 we show that diameter $d = 2$ is tight (see preprint [7]) except for one result. Namely, to classify the diameter-treedepth property, we prove the following two results in Sect. 3, the second of which shows that $d = 5$ is tight. In our proofs, we will exploit a result of Galvin, Rival and Sands [17], who proved that a graph of large treedepth must either contain a large complete bipartite graph as a subgraph or a large induced path; see Sect. 2 for a more detailed discussion of this result and its consequences.

Theorem 1 (Classification for diameter $d \geq 5$). *Let $d \geq 5$. For a graph F, the class of F-subgraph-free graphs of diameter at most d has bounded treedepth if and only if F is a subgraph of a subdivided star.*

Theorem 2 (Classification for diameter 4). *For a graph F, the class of F-subgraph-free graphs of diameter at most 4 has bounded treedepth if and only if F is a subgraph of a subdivided star or H_2^ℓ for some $\ell \in \mathbb{N}$ (see also Fig. 1).*

Theorems 1 and 2 show that there is a considerable scope for improvement by considering graphs of diameter 2 and 3, in line with our research question. This is unlike the other cases in Table 1, for which $d = 2$ is always tight.

We were not able to give complete classifications for treedepth under the subgraph relation for $d = 2$ and $d = 3$. However, by a deeper exploration of the result of Galvin, Rival and Sands [17] and using a result on polarity graphs, due to Erdős, Rényii and Sós [15], we were able to prove a variety of results summarized in the state-of-the-art theorem below. We again refer to Fig. 1 for an explanation of the various graphs that we define in these statements.

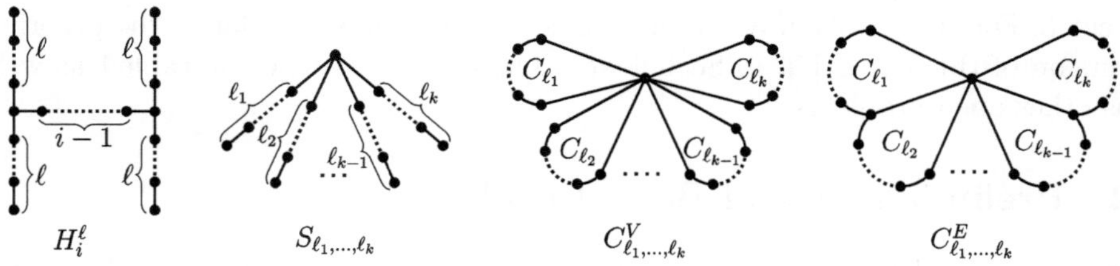

Fig. 1. The subdivided "H"-graph H_i^ℓ, the subdivided star $S_{\ell_1,\ldots,\ell_k}$, the V-type graph $C^V_{\ell_1,\ldots,\ell_k}$ (set of cycles sharing one common vertex) and the E-type graph $C^V_{\ell_1,\ldots,\ell_k}$ (set of consecutive cycles sharing an edge). We write $C^V_{\ell_1,\ldots,\ell_k} = C^V_{k\times[\ell_1]}$ and $C^E_{\ell_1,\ldots,\ell_k} = C^E_{k\times[\ell_1]}$ if $\ell_1 = \cdots = \ell_k$, and $C^V_{\ell_1,\ldots,\ell_k} = C^V_{i\times[\ell_1],k-i\times[\ell_k]}$ if $\ell_1 = \cdots = \ell_i$, $\ell_{i+1} = \cdots = \ell_k$.

Theorem 3 (Partial classification for diameters 2 and 3). *For a graph F and $d \in \{2,3\}$, the class of F-subgraph-free graphs of diameter at most d has:*

(i) bounded treedepth if:
1. *$d = 3$ and F is an acyclic apex linear forest (Theorem 12);*
2. *$d = 3$ and F is C_8;*
3. *$d = 2$ and F is a bipartite, C_4-subgraph-free apex linear forest that contains exactly one cycle;*
4. *$d = 2$ and F is $C^{\mathrm{V}}_{2\ell_1,2\ell_2}$ for some $\ell_1, \ell_2 \geq 3$;*
5. *$d = 2$ and F is a subgraph of $C^{\mathrm{V}}_{k\times[2\ell]}$ for some $\ell \geq 3$ and $k \geq 1$ or*
6. *$d = 2$ and F is $C^{\mathrm{E}}_{2\ell_1,2\ell_2}$ for some $\ell_1, \ell_2 \geq 3$;*

(ii) unbounded treedepth if:
1. *$d = 2$ and F is not bipartite (Observation 7);*
2. *$d = 2$ and F is bipartite, but is not an apex linear forest (Observation 7);*
3. *$d = 2$ and F is a supergraph of C_4 (Theorem 9 [15]);*
4. *$d = 2$ and F is a supergraph of $C^{\mathrm{V}}_{12\times[6],12\times[8]}$;*
5. *$d = 2$ and F is a supergraph of $C^{\mathrm{E}}_{k\times[2\ell]}$ for some $\ell \geq 3$ and $k = 2(2\ell - 3)$ (Theorem 13);*
6. *$d = 3$ and F is a supergraph of C_6 or*
7. *$d = 3$ and F is a supergraph of either $C^{\mathrm{V}}_{4\ell_1,4\ell_2}$, $C^{\mathrm{V}}_{2\ell,2\ell}$, $C^{\mathrm{E}}_{4\ell_1,4\ell_2}$ or $C^{\mathrm{E}}_{2\ell,2\ell}$ for $\ell_1, \ell_2 \geq 2$ and $\ell \geq 4$.*

From Theorem 3 we note a jump from boundedness for diameter $d = 2$ to unboundedness for diameter $d = 3$ if $F = C_6$. Moreover, there is a difference for $d = 2$ when we forbid a graph F whose cycles share a unique vertex or an edge with a common end-vertex, i.e. a *V-type* graph $C^{\mathrm{V}}_{k\times[2\ell]}$ or *E-type* graph $C^{\mathrm{E}}_{k\times[2\ell]}$, respectively. The treedepth also becomes unbounded even for $d = 2$ if the cycles of a V-type graph or E-type graph have only two different lengths.

In Sect. 2, we show Observation 7 and how to derive Theorem 9 from a result on polarity graphs [15]. We prove Theorem 12 in Sect. 4 and Theorem 13 in Sect. 5. For the proofs of all other results in Theorem 3 we refer to the preprint version of this paper [7]. In Sect. 6 we discuss the open cases in Table 1 as well as other open problems.

2 Preliminaries and Basic Results

We only consider finite, simple, undirected graphs $G = (V(G), E(G))$. Let G be a graph. We denote the *neighbourhood* of a vertex $v \in V(G)$ by $N_G(v) = \{u \in V(G) \mid uv \in E(G)\}$ (we may also just write $N(v)$). For a subset $S \subseteq V(G)$, we write $N_G(S) = \bigcup_{v\in S} N_G(v)$. A graph H is a *subgraph* of G if H can be obtained from G by a sequence of vertex deletions and edge deletions, whereas H is an *induced subgraph* of G if H can be obtained from G by a sequence of vertex deletions. For a vertex set $S \subseteq V(G)$, we write $G[S]$ to denote the subgraph of G *induced by* S, that is, the graph obtained from G after deleting the vertices not in S. The *contraction* of an edge $e = uv$ in G replaces u and v by a new vertex w that is adjacent to every vertex in $(N_G(u) \cup N_G(v)) \setminus \{u, v\}$ (without creating parallel edges). We let G/e denote the graph obtained from G after contracting e. A graph H is a *minor* of G if H can be obtained from G by a sequence of edge deletions, edge contractions and vertex deletions. For a set of graphs $\mathcal{F}$, we say that G is *$\mathcal{F}$-subgraph free*, *$\mathcal{F}$-free*, or *$\mathcal{F}$-minor free* if G does not contain any graph in $\mathcal{F}$ as a subgraph, induced subgraph or minor, respectively.

We may refer to a path P with vertices $u_0, \ldots, u_l$ and edges $u_{i-1}u_i$ for $1 \leq i \leq l$ by the sequence $(u_0, u_1, u_2, \ldots, u_l)$. The *length* of P is its number of edges l. The *distance* $d_G(u, v)$ between two vertices u and v of a graph G is the length of a shortest path from u to v. For two graphs G and H, we let $G \bowtie H$ denote the graph obtained from the disjoint union of G and H after adding all edges between the vertices in $V(G)$ and the vertices in $V(H)$.

An *elimination forest* of a graph G is a rooted forest T such that $V(G) = V(T)$ and for every $uv \in E(G)$ both u and v are on the same root-to-leaf path of T. The *treedepth* $\mathrm{td}(G)$ is the minimum height of an elimination forest of G.

Fact 4 ([28]) *For a graph G with a longest path of length ℓ,* $\log(\ell) \leq \mathrm{td}(G) \leq \ell$.

The following theorem, combined with Fact 4, has a useful consequence.

Theorem 5 ([17]) *For all $r, s, \ell \in \mathbb{N}$, there is a number $c(r, s, \ell)$ such that every $K_{r,s}$-subgraph-free graph with a path of length $c(r, s, \ell)$ has an induced P_ℓ.*

Corollary 6. *For all $r, s, \ell \in \mathbb{N}$, there is a number $c(r, s, \ell)$ such that every $K_{r,s}$-subgraph-free graph of treedepth $c(r, s, \ell)$ has an induced P_ℓ.*

By Fact 4, complete bipartite graphs and graphs $P_n \bowtie K_1$ have unbounded treedepth. As both classes consist of graphs of diameter at most 2, we obtain:

Observation 7 *Let $d \geq 2$. For a graph F, the class of F-subgraph-free graphs with diameter at most d has unbounded treedepth if there is no integer n such that F is a subgraph of both $K_{n,n}$ and of $P_n \bowtie K_1$.*

Every subgraph of $P_n \bowtie K_1$ can have at most one component which is not a path. Moreover, in a graph of large enough treedepth, we can find any number of disjoint paths as subgraphs. Hence, Observation 7 implies the following:

Observation 8 *For any graph F, there is a component C of F such that the class of F-subgraph-free graphs with diameter d has bounded treedepth if and only if the class of C-subgraph-free graphs with diameter d has bounded treedepth.*

Therefore, from now on we consider all forbidden subgraphs F to be connected.

Erdős, Rényii and Sós [15] showed how a family of C_4-subgraph-free graphs with diameter 2 can be constructed from a polarity of a projective plane. Making the observation that this family has unbounded minimum degree and so also unbounded treewidth we obtain the following.

Theorem 9 **([15]).** *The class of C_4-subgraph-free graphs of diameter 2 has unbounded treedepth.*

We also use polarity graphs to show that the class of C_6-subgraph-free graphs of diameter 3 has unbounded treedepth (see preprint [7]).

3 Proofs of Theorems 1 and 2

We first note that the 1-subdivisions of complete bipartite graphs have diameter 4 and unbounded treedepth alongside the graphs P_n^{1001}, i.e. for $b \in \{0,1\}^*$, where P_n^b is the graph obtained from P_n after adding a new vertex u and making u adjacent to the i-th vertex of P_n if the ($i \bmod |b|$)-th bit of b is 1. Likewise, the 2-subdivisions of complete graphs have diameter 5 and unbounded treedepth.

If F is a subgraph of the 1-subdivision of a complete bipartite graph and some P_n^{1001}, then F is a subgraph of H_2^ℓ or $S_{\ell_1,\ldots,\ell_k}$ for some $\ell \geq 1$ and $\ell_1, \ldots, \ell_k \geq 1$. If F is also a subgraph of the 2-subdivision of a complete graph, this leaves only $S_{\ell_1,\ldots,\ell_k}$ for some $\ell \geq 1$. Hence, Theorems 1 and 2 follow from the following two lemmas. Forbidding $S_{\ell_1,\ldots,\ell_k}$ in the first lemma bounds the treedepth of a graph of any diameter d by $c((k\ell)/2+1, (k\ell)/2, \ell(k^d+1))$ where $\ell = \max\{\ell_1, \ldots, \ell_k\}$ and c is the function from Theorem 5. We did not try to optimize this function.

Lemma 10. *Let $d \geq 1$. For all $\ell_1, \ldots, \ell_k \in \mathbb{N}$, the class of $S_{\ell_1,\ldots,\ell_k}$-subgraph-free graphs of diameter at most d has bounded treedepth.*

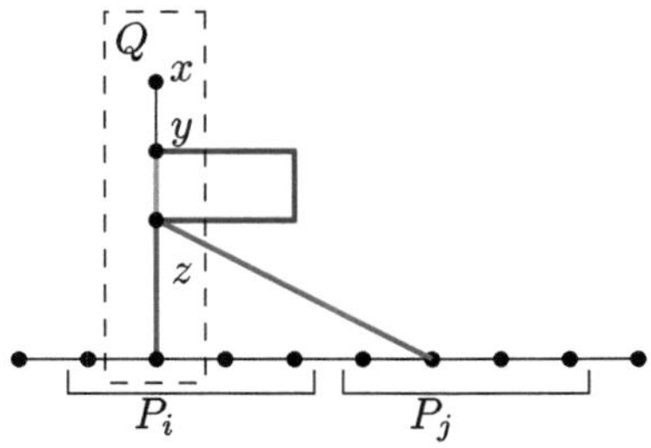

Fig. 2. Path Q. Paths $Q^1, Q^2, \overline{Q}^1$ and $\overline{Q}^2$ are orange, red, blue, and green, resp. (Color figure online)

Proof. Let G be some $S_{\ell_1,\ldots,\ell_k}$-subgraph-free graph of diameter at most d and $\ell = \max\{\ell_1, \ldots, \ell_k\}$. We claim $td(G) < c((k\ell)/2+1, (k\ell)/2, \ell(k^d+1))$. Suppose for contradiction $td(G) \geq c((k\ell)/2+1, (k\ell)/2, 2\ell(k^d+1))$. As $K_{(k\ell)/2+1,(k\ell)/2}$ contains $S_{\ell_1,\ldots,\ell_k}$ as a subgraph, G cannot contain a large complete bipartite graph as a subgraph. From Corollary 6, G contains some induced path P of length $2\ell(k^d+1)$. We may assume that $k \geq 2$ by Fact 4.

Pick disjoint subpaths $P_1, \ldots, P_{k^d+1}$ of P of length 2ℓ. In the following we prove inductively that for every $x \in V(G)$ and every $\delta \in [d]$ there is a path of length at most δ from x to some vertex on P_i for at most k^δ different $i \in [1, k^d+1]$. For $\delta = d$ this yields a contradiction to G having diameter at most d. First observe that every vertex $x \in V(G)$ can have a neighbour in at most $k-1$ distinct subpaths from $\{P_1, \ldots, P_{k^d+1}\}$ else G contains $S_{\ell_1,\ldots,\ell_k}$. Considering that x might be on P, we obtain that there is a path of length at most 1 from x to some vertex on P_i for at most k different $i \in [k^d+1]$.

Now, let $x \in V(G)$ and $\mathcal{I} \subseteq [k^d+1]$ be all indices such that there is a path of length at most $\delta+1$ from x to some vertex on P_i. Let $\mathcal{Q}$ be a set consisting a single minimum length path Q from x to any vertex q on P_i for every $i \in \mathcal{I}$. Towards a contradiction, assume that $|\mathcal{Q}| > k^{\delta+1}$. As there are paths of length at most δ from any $v \in V(G)$ to some $q \in P_i$ for at most k^δ different $i \in [k^d+1]$ by assumption, there is a subset $\mathcal{Q}_0 \subseteq \mathcal{Q}$ of paths that have length $\delta+1$ with $|\mathcal{Q}_0| \geq (k-1)k^\delta + 1$.

Claim 10.1. *For any path $Q \in \mathcal{Q}_0$ there are at most $k^\delta - 1$ other paths $\overline{Q} \in \mathcal{Q}_0$ that intersect Q in more than one vertex.*

Proof of Claim: Let $i \in [k^d+1]$ be the index such that the last vertex of $Q \in \mathcal{Q}$ is on P_i. Let y be the second vertex of Q. Let $j \in [k^d+1]$, $j \neq i$ be any index such that the path $\overline{Q} \in \mathcal{Q}$ from x to some vertex in P_j intersects Q in some vertex $z \neq x$. We argue that there is a path from y to some vertex on P_j of length δ. Let Q^1 be the subpath of Q from y to z, $\overline{Q}^1$ be the subpath of $\overline{Q}$ from y to z and $\overline{Q}^2$ subpath of $\overline{Q}$ from z to some vertex on P_j (see Fig. 2 for an illustration). As Q is a shortest path from x to any vertex on P_i of length $\delta+1$ and $\overline{Q}$ is a shortest paths from x to any vertex on P_j of length $\delta+1$, Q^1 and

$\overline{Q}^1$ have the same length. Hence, the concatenation of Q^1 and $\overline{Q}^2$ yields a path of length δ from y to some vertex on P_j. Since, additionally to i, there can be paths of length δ to some vertex on P_j for at most $k^\delta - 1$ different $j \in [k^d + 1]$, $j \neq i$, the claim follows. $\diamond$

We now iteratively choose paths $Q_1, \ldots, Q_m$ in such a way that Q_i is any path from the set $\mathcal{Q}_i \subseteq \mathcal{Q}_0$ of paths that intersect each Q_j, $j \in [i-1]$ in x only. Here m is the minimum integer such that $\mathcal{Q}_{m+1} = \emptyset$. By Claim 10.1 we know that $|\mathcal{Q}_i| \geq |\mathcal{Q}_{i-1}| - k^\delta = (k-i)k^\delta + 1$. Hence, $|\mathcal{Q}_k| \geq 1$ which implies that there are paths $Q_1, \ldots, Q_k$ pairwise only intersecting in vertex x. Therefore, G contains $S_{\ell_1,\ldots,\ell_k}$ as a subgraph, a contradiction. $\square$

For the proof of the next lemma we refer to [7].

Lemma 11. *For any $\ell \in \mathbb{N}$, the class of H_2^ℓ-subgraph-free graphs of diameter at most 4 has bounded treedepth.*

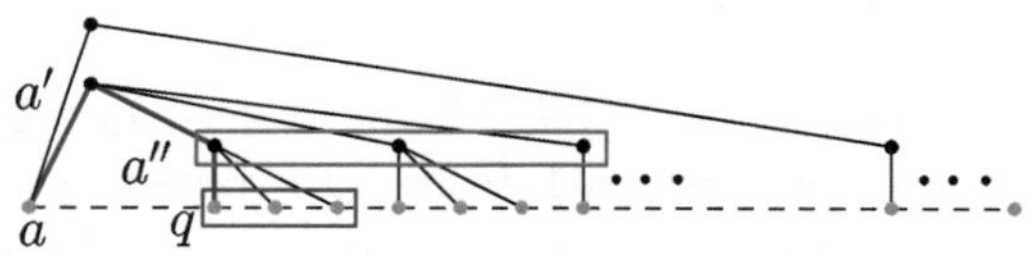

Fig. 3. The dashed line is P with vertices of A in orange. Some vertices from $X(A, a)$ with their respective shortest paths are drawn, with the path (a, a', a'', q) in blue. (Color figure online)

4 Proof of Theorem 12

Theorem 12. *For every tree F, the class of F-subgraph-free graphs of diameter at most 3 has bounded treedepth if and only if F is an acyclic apex linear forest.*

We give a short overview of the proof of Theorem 12 before proceeding with the proof. The forward direction essentially follows from Observation 7. For the backwards direction, aiming for a proof by contradiction, we assume that for some acyclic apex linear forest F, the class of F-subgraph-free graphs of diameter at most 3 has unbounded treedepth. Since F is contained as a subgraph in $K_{n,n}$ for some sufficiently large n, we use Corollary 6 to obtain a graph G in the class containing a long induced path P. We will proceed by showing that it follows that G contains, as an induced subgraph, the disjoint union of γ copies of $S_{k\times[\ell]}$ for some large number γ. Let $\mathcal{S}$ denote this set of γ copies of $S_{k\times[\ell]}$ and for each $S \in \mathcal{S}$ let $X_1(S)$ denote the set of vertices at distance one from the centre of S. We claim that, if γ is sufficiently large, there exists some $S \in \mathcal{S}$ such that there is a bijective mapping from the vertices in $X_1(S)$ to k distinct stars in $\mathcal{S} \setminus \{S\}$, such that there are disjoint paths between those vertices in $X_1(S)$ and their corresponding stars which contain no edges from S. Finally, we show that the construction results in a copy of F in G, a contradiction.

Proof. A graph is an apex linear forest if, and only if it is a subgraph of $P_n \bowtie K_1$ for some integer n. This together with Observation 7 shows the forward direction.

For the backward direction, let F be an acyclic apex linear forest. There exists some k and ℓ such that F is a subgraph of the graph obtained from the disjoint union of a vertex v and k paths of length $\ell - 1$ after making v adjacent to exactly one vertex on every path. We claim the class $\mathcal{C}$ of F- subgraph-free graphs of diameter at most 3 have bounded treedepth.

Suppose for a contradiction that $\mathcal{C}$ has unbounded treedepth, it must contain some graph G such that $\text{td}(G) \geq c((k\ell+1)/2, (k\ell+1)/2, \gamma(\ell+1)(5k)^7)$, where γ is some function of k and the function c is given by Corollary 6. As F is contained in $K_{(k\ell+1)/2,(k\ell+1)/2}$, G has no large complete bipartite subgraph. Corollary 6 implies that G contains some induced path P with length at least $\gamma(\ell+1)(5k)^7$. We further, let $P^{\ell+1} = \{p_i : i \mod (\ell+1) = 0\} \setminus p_0$, where p_i is the i-th vertex of P (starting from $i = 0$).

Claim 12.1. *If $|P^{\ell+1}| > \gamma(5k)^7$, then G contains the disjoint union of γ copies of $S_{k\times[\ell]}$ as an induced subgraph.*

Proof of Claim: Consider some $A \subseteq P^{\ell+1}$ and $a \in A$. Let $X(A, a)$ contain all vertices that lie on some shortest path from a to some $a' \in A$ in G. We claim that if $|A| \geq (k'-1)^3+1$, then there is some vertex $x \in X(A, a)$ with k' disjoint shortest paths to vertices in A for every $k' \geq k$. Note that every vertex in G is adjacent to at most $k-1$ vertices in A; else F is a subgraph of G; see the purple vertices in Fig. 3. Consider some $x \in X(A, a)$ together with all shortest paths of length at most 2 from x to some vertex in A. These paths can contain at most $k'-1$ distinct intermediate vertices, else there are k' disjoint paths from x to vertices in A. Each intermediate vertex (green in Fig. 3) has a path of length at most 1 to at most $k'-1$ vertices in A, so there are at most $(k'-1)^2$ paths of length at most 2 from x to some vertex in A. As G has diameter 3, a has a path of length at most 3 to every vertex in A. Let (a, a', a'', q) be such a path. There are at most $(k'-1)^2$ shortest paths of length 2 from a' to vertices in A. At most $(k'-1)^2$ shortest paths from a to some vertex in A contain either a' or a'', so at least $(k'-1)^3+1-(k'-1)^2$ paths (among all shortest paths from a to some vertex in A) are disjoint from this a to q path. As $|A| \geq (k'-1)^3+1$, there are at least k' disjoint shortest paths from a to vertices in A.

Let $x \in X(A, a)$ have k' disjoint shortest paths to vertices in A, these paths together with sections of P describe a $S_{k'\times[\ell]}$ subgraph, which we call S. Let $X_1(S)$ and $X_2(S)$ be the sets of vertices of S containing all vertices at distance 1 and 2 from x in S, respectively. We refer to edges in $G[S]$ but not in S as cross edges. Given all vertices with distance at least 3 from x in S are in P and P is an induced path, all cross edges have some endpoint in $\{x\} \cup X_1(S) \cup X_2(S)$.

Claim 12.2. *There is a set of at most $4k-3$ branches of S whose removal from S leaves no cross edges with an endpoint in $X_1(S) \cup \{x\}$.*

Proof of Claim: If a cross edge is incident to x, then the other endpoint is in P as $N(x) \cap X_2(S) = \emptyset$. If x has some cross edge to $2k$ different branches, then x

has k neighbours in P each with pairwise distance at least $\ell - 1$ along P, which would imply that F is a subgraph of G. Therefore, after removing at most $2k-1$ branches, all cross edges have some endpoint in $X_1(S)$ or $X_2(S)$.

The graph G_{X_1} has one vertex for every branch of S and an edge (b, b') between two distinct branches b and b' if a vertex $x' \in X_1(S)$ laying on b is adjacent to some vertex in b'. If G_{X_1} has a matching of size k, then G has F with centre x as a subgraph. Matching edges indicate k vertices in $X_1(S)$ with a pair of disjoint paths of length ℓ, the first via its respective branch in S and the second via the branch given by the matching. As G_{X_1} has a maximum matching of size at most $k-1$, it has a vertex cover R of size at most $2(k-1)$. By deleting the branches indicated by R, no cross edge is adjacent to $X_1(S)$ either. ◇

If some $x'' \in X_2(S)$ has cross edges to $5k-3$ branches, let $S_{x''}$ be that $S_{(5k-3)\times[\ell]}$ with centre x'' resulting from these cross edges. As $X_2(S_{x''}) \subseteq P$ all cross edges of $S_{x''}$ have an endpoint in $\{x''\} \cup X_1(S_{x''})$. From Claim 12.2, removing at most $4k-3$ branches of $S_{x''}$ gives some induced $S_{k\times[\ell]}$. If each vertex in $X_2(S)$ has a cross edge to at most $5k-4$ other branches and S has $k(5k-4)+5k-4$ branches at least k of these have no cross edges. That is, if $k' \geq (5k-4)(k+1)$, then there is some induced $S_{k\times[\ell]}$.

If there are $\gamma k(2k+1)$ disjoint subsets of $P^{\ell+1}$ of size $(5k-4)^3(k+1)^3$ (recall that if $|A| \geq (k'-1)^3+1$, then G has a $S_{k'\times[\ell]}$ as a subgraph) there must be $\gamma k(2k+1)$ induced stars, $S_1, \cdots, S_{\gamma(2k+1)}$, as described above. It remains to show that at least γ among those stars are pairwise disjoint and have no edges between each other. Recall that every S_i contains at most $2k+1$ vertices that are external to the path P. No vertex can be contained in or adjacent to k of these stars as otherwise G contains F as a subgraph. Therefore, since we are given $\gamma k(2k+1)$ stars, there must be at least γ that are pairwise disjoint and have no edges between each other. ◇

Let $\mathcal{S} = \{S_1, \cdots, S_\gamma\}$ be the set of pairwise distinct, non-adjacent copies of $S_{k\times[\ell]}$ in G obtained from Claim 12.1. For every $S \in \mathcal{S}$, let x_S be the centre of S. For $b \in \{1,2\}$, let $X_b(S)$ be the set of all vertices in S of distance b from x_S in S.

Let $T \subseteq \mathcal{S}$ with $|T| \geq 2k$. For $S \in T$, if there is some bijective mapping between k vertices in $X_1(S)$ and k stars in T, such that there are disjoint paths between these k vertices in $X_1(S)$ and their respective star in T containing no edges from S, then G contains F. For any pair $S, S' \in T$, the shortest path between any pair $u \in X_1(S)$ and $v \in X_1(S')$ has length at most 3 and has at most two vertices not in $V(S) \cup V(S')$. Hence, there is a set $Z(S,T)$ of $2(k-1)$ vertices such that for every $x' \in X_1(S)$ with a neighbour $x'' \in X_2(S)$, the shortest path from x' to each vertex in $X_1(S')$ contains some vertex in $Z(S,T) \cup \{x_S, x''\}$.

For $\rho \leq k$, let $\Gamma_\rho \subseteq \mathcal{S}$ be such that for every $S \in \Gamma_\rho$ there are ρ vertex disjoint shortest paths containing no edges from S each from a vertex in $X_1(S)$ to its own star in $\mathcal{S} \setminus \Gamma_\rho$, i.e. there is a one-to-one mapping between the ρ vertices and the ρ stars. See Fig. 4. Note that $\Gamma_0 = \mathcal{S}$, we claim that that if $|\Gamma_\rho| \geq 2^{\rho+2}k^{\rho+2}$, then $|\Gamma_{\rho+1}| \geq \frac{|\Gamma_\rho|}{2^{\rho+1}k^{\rho+1}} - (k-1)^2$. This implies there is a constant γ which is a function of k such that if $|\mathcal{S}| = \gamma$, then $|\Gamma_k| \geq 1$. If $|\Gamma_k| \geq 1$ then G contains F as a

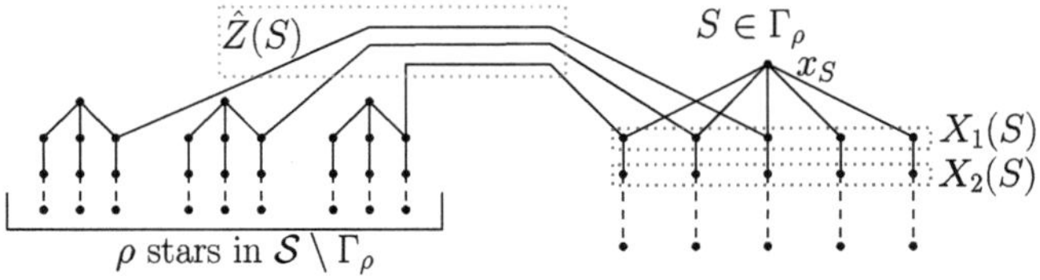

Fig. 4. Some $S \in \Gamma_\rho$. Boxes in green indicate $X_1(S)$ and $X_2(S)$ and in orange $\hat{Z}(S)$.

subgraph, a contradiction. Therefore, for the proof of this theorem, it remains only to show that $|\Gamma_{\rho+1}| \geq \frac{|\Gamma_\rho|}{2^{\rho+1}k^{\rho+1}} - (k-1)^2$ for every $\rho \in \{0, \ldots, k-1\}$.

For $S \in \Gamma_\rho$, we let $\hat{Z}(S)$ be the set of all vertices that lie on the ρ disjoint paths from $X_1(S)$ to other stars and we denote by x'_S an arbitrary vertex in $X_1(S) \setminus \hat{Z}(S)$; note that x'_S always exists because $k > \rho$. Let x''_S denote the vertex in $N(x'_S) \cap X_2(S)$. For every $S' \in \Gamma_\rho$, the shortest path from x'_S to $x'_{S'}$ must contain some vertex from $Z(S, \Gamma_\rho) \cup \{x_S, x''_S\}$. As $|Z(S, \Gamma_\rho) \cup \{x_S, x''_S\}| \leq 2k$, there exists some $z_S(\Gamma_\rho) \in Z(S, \Gamma_\rho) \cup \{x_S, x''_S\}$ that lies on at least $\frac{|\Gamma_\rho|-1}{2k}$ of these paths. Such a $z_S(T)$ can also be defined for every $T \subseteq \Gamma_\rho$ with size at least $2k$.

Claim 12.3. *For every $r \geq k^2$, if $|\Gamma_\rho| \geq r2^{\rho+1}k^{\rho+1}$, then there is some $\bar{S} \in \Gamma_\rho$ and $\Gamma'_\rho \subseteq \Gamma_\rho$ of size r such that for some $z \notin \bigcup_{S' \in \Gamma'_\rho} \hat{Z}(S')$, z lies on the shortest path from $x'_{\bar{S}}$ to $x'_{S'}$ for every $S' \in \Gamma'_\rho$.*

Proof of Claim: Say, for contradiction, such set and vertex does not exist. We claim there exists a set $Q_{\rho+1} \subseteq \Gamma_\rho$ of size at least r and $\rho+1$ stars $A = \{S_0, \ldots, S_\rho\} \subseteq \Gamma_\rho \setminus Q_{\rho+1}$ such that the following holds. For every $S \in A$ there is a distinct vertex z, such that for every $S' \in Q_{\rho+1}$, $z \in \hat{Z}(S')$, $N(z) \cap V(S') = \emptyset$, and z lies on the shortest path from x'_S to $x_{S'}$: a contradiction as for any $S \in \Gamma_\rho$ there are at most ρ vertices $\hat{z} \in \hat{Z}(S)$ with $\hat{z}$ not adjacent to S.

Consider some arbitrary $S_0 \in \Gamma_\rho$ and its respective $z_{S_0}(\Gamma_\rho)$. Let $T_0 \subseteq \Gamma_\rho$ be those $S \in \Gamma_\rho$ such that $z_{S_0}(\Gamma_\rho)$ lies on the shortest path from x'_{S_0} to x'_S. By assumption there are at most $r-1$ different $S \in T_0$ such that $z_{S_0}(\Gamma_\rho) \notin \hat{Z}(S)$. Given $z_{S_0}(\Gamma_\rho)$ has some neighbour in at most $k-1$ different stars in $\mathcal{S}$, there must exists some $Q_0 \subseteq T_0$ with size at least $|T_0|-(k-1)-r \geq \frac{|\Gamma_\rho|-1}{2k}-(k-1)-r$ such that $z_{S_0}(S) \in \hat{Z}(S)$ and $N(z_{S_0}(S)) \cap V(S) = \emptyset$ for every $S \in Q_0$.

Assume there exists some $Q_\delta \subseteq \Gamma_\rho$ and stars $A_\delta = \{S_0, \ldots, S_\delta\} \subseteq \Gamma_\rho \setminus Q_{\delta-1}$. For every $S \in A_\delta$ there is some distinct vertex z such that for every $S' \in Q_\delta$, $z \in \hat{Z}(S')$; $N(z) \cap V(S') = \emptyset$; and z lies on the shortest path from x'_S to $x_{S'}$.

Consider some arbitrary $S_{\delta+1} \in Q_\delta$ with its respective $z_{S_{\delta+1}}(Q_\delta)$, let $T_{\delta+1} \subseteq Q_\delta$ be those $S \in Q_\delta$ such that $z_{S_{\delta+1}}(S_\delta)$ lies on the shortest path from $x'_{S_{\delta+1}}$ to x'_S. If there are at most r stars $S \in T_{\delta+1}$ such that $z_{S_{\delta+1}}(Q_\delta) \notin \hat{Z}(S)$, then there is some $Q_\delta \subseteq T_\delta$ with size at least $|T_\delta| - (k-1) - r \geq \frac{|Q_\delta|-k(r-1)}{2k}$ such that $z_{S_{\delta+1}}(Q_\delta) \in \hat{Z}(S)$ and $N(z_{S_{\delta+1}}(Q_\delta)) \cap V(S) = \emptyset$ for every $S \in Q_\delta$. If

$|\Gamma_\rho| = 2^\rho k^\rho + (r-1)(\sum_{i=1}^{\rho} 2^i k^i) \leq r 2^{\rho+1} k^{\rho+1}$ (for sufficiently large r and ρ), then $|Q_{\rho+1}| \geq r$. Let $\Gamma'_\rho = Q_{\rho+1}$ which concludes the proof of this claim. ◇

Let $S \in \Gamma_\rho$, $\Gamma'_\rho \subseteq \Gamma_\rho$ and z be those obtained from Claim 12.3. Let $\mathcal{P}$ be the set of shortest paths containing z from x'_S to $x'_{S'}$ for every $S' \in \Gamma'_\rho$. There are at most $k-1$ intermediate vertices q on the paths in $\mathcal{P}$ with $q \in \hat{Z}(S')$ for some $S' \in \Gamma'_\rho$. Each such vertex q is adjacent to some star, so q can lie on at most $k-1$ of these paths from z to some $S' \in \Gamma'_\rho$. Hence, there is some $\Gamma_{\rho+1} \subseteq \Gamma'_\rho$ of size at least $|\Gamma'_\rho| - (k-1)^2$ such that for each $S' \in \Gamma_{\rho+1}$, the shortest path from z to $x'_{S'}$ has no vertex from $\bigcup_{S'' \in \Gamma'_\rho} \hat{Z}(S'')$, thus concluding our proof. □

5 The Proof of Theorem 13

Theorem 13. *For any $\ell \geq 3$, there exists some $k \leq 2(2\ell-3)$ such that the class of $C^{\mathrm{E}}_{k\times[2\ell]}$-subgraph-free graphs of diameter at most 2 has unbounded treedepth.*

Proof. We set $k = 2(2\ell-3)$. For all $n \in \mathbb{N}$ we construct a $C^{\mathrm{E}}_{k\times[2\ell]}$-subgraph-free graph G_n with diameter at most 2 and treedepth at least $\log(n)$ by taking $2\ell-3$ vertices $x_0, \ldots, x_{2\ell-4}$ and a path $P = (p_0, \ldots, p_n)$. Consider the binary string $R = 1(10)^{\ell-2}$ consisting of 1 followed by $\ell-2$ repetitions of the string 10. We add the edge $p_i x_j$ if the $i - j \mod (2\ell-3)$th bit of R is 1. See Fig. 5. By Fact 4, $\mathrm{td}(G_n) \geq \log(n)$.

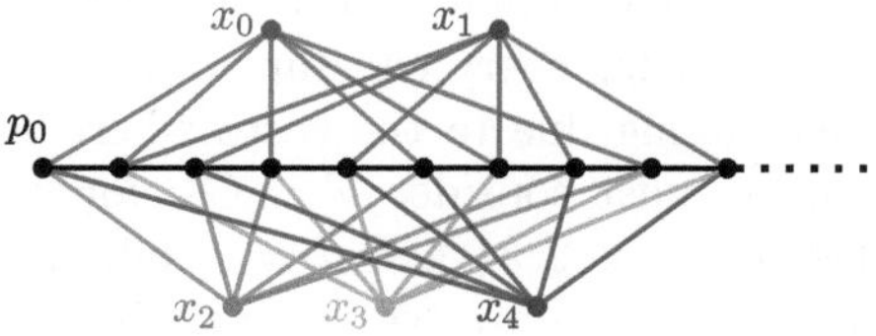

Fig. 5. The construction from Theorem 13 for $\ell = 4$ for which the pattern is $R = 11010$.

Next we argue that G_n has diameter 2 by showing that any pair of vertices has distance at most 2. Each x_i is either adjacent to p_{j-1} or p_j for every $j \in [n]$. Next consider p_i and p_j and observe that p_i is both adjacent to $x_{i \bmod (2\ell-3)}$ and $x_{i+1 \bmod (2\ell-3)}$. Furthermore, p_j must be either adjacent to $x_{i \bmod (2\ell-3)}$ or $x_{i+1 \bmod (2\ell-3)}$. Finally, consider x_i and x_j. Note that x_i is adjacent to both p_i and p_{i+1} and x_j must be adjacent to either p_i or p_{i+1}.

Now, for a contradiction, assume G_n contains $C^{\mathrm{E}}_{k\times[2\ell]}$ as a subgraph and $x \in V(G_n)$ is the vertex common to all cycles. As each vertex on P has degree $\ell-1$, we have $x \in \{x_0, \ldots, x_{2\ell-4}\}$. Any $C_{2\ell}$ in $G[\{x\} \cup V(P)]$ is of the form $(x, p_i, \ldots, p_{2\ell-2}, x)$. As R is of length $2\ell-3$, there is either one or two indices $j \in \{i, \ldots, 2\ell-3\}$ such that x is adjacent to both p_j and p_{j+1}. If there is only one such index j, x cannot be adjacent to both p_i and $p_{i+2\ell-2}$. Hence, there are two. But then there is only one index $j \in \{2\ell-2, \ldots, 4\ell-5\}$ such that x is adjacent to both p_j and p_{j+1}. Hence, x cannot be adjacent to $p_{i+4\ell-4}$ and thus not both $(x, p_i, \ldots, p_{i+2\ell-2}, x)$ and $(x, p_{i+2\ell-2}, \ldots, p_{i+4\ell-4}, x)$ are subgraphs of G_n. Therefore, at least every second of the k cycles of the subgraph isomorphic to $C^{\mathrm{E}}_{k\times[2\ell]}$ contains some $y \in \{x_0, \ldots, x_{2\ell-3}\}\setminus\{x\}$. As x is not adjacent to any vertex in $\{x_0, \ldots, x_{2\ell-3}\}\setminus\{x\}$ we get that each of the vertices $y \in \{x_0, \ldots, x_{2\ell-3}\}\setminus\{x\}$ can be in at most one cycle, a contradiction as we may only obtain $2(2\ell-4)+1 < k$ cycles. □

6 Conclusions

We showed that bounding the diameter has a significant impact on the boundedness of width of a graph class, in particular for the subgraph relation and treedepth. We pose some open problems. First, recall that Table 1 still contains three open cases: which classes of F-subgraph-free graphs have the diameter-width property for clique-width, treewidth or pathwidth? Towards solving these questions, we can show there are graphs F such that F-subgraph-free graphs of diameter 2 have bounded pathwidth but unbounded treedepth; and also that there are graphs F such that F-subgraph-free graphs of diameter 2 have bounded clique-width but unbounded treewidth (see the preprint of this paper [7] for proofs).

Second, we note that Demaine and Hajiaghayi [10] proved the diameter-treewidth property with even a linear (and optimal) diameter bound. We leave it as an open problem whether our diameter bound in Theorem 1 can be optimized.

Third, recall that we classified boundedness of treedepth of F-subgraph-free graphs of diameter at most d for every constant $d \geq 4$. Completing the classification for $d = 2$ is challenging due to the cases where F is of V-type C^V_x or E-type C^E_x. For $d = 3$, we must consider $F = C_{2r}$ for $r \geq 2$. We showed the treedepth is unbounded for $r \in \{2, 3\}$ and $d = 2$, but bounded for $r = 4$ and $d = 3$. Our proof for the latter is an involved case analysis, which seems not easy to extend to larger values of r. But by use of a computer [13] we can also prove boundedness for $r \in \{5, \ldots, 12\}$ and $d = 3$. We therefore conjecture:

Conjecture 14. For every $r \geq 4$, the class of C_{2r}-subgraph-free graphs of diameter at most 3 has bounded treedepth.

If a stronger version of Conjecture 14 involving all graphs with one cycle (analogous to [7, Theorem 29]) is true, just like the following conjecture (which is supported by all our results so far), then we will obtain a complete classification for $d = 3$.

Conjecture 15. For a graph F with at least two cycles, the class $\mathcal{C}$ of F-subgraph-free graphs of diameter at most 3 has unbounded treedepth.

Acknowledgements. We would like to thank Bogdan Alecu for pointing out Theorem 5. The fifth author was supported by the Engineering and Physical Sciences Research Council [EP/X01357X/1].

References

1. Bienstock, D., Robertson, N., Seymour, P.D., Thomas, R.: Quickly excluding a forest. J. Comb. Theory Ser. B **52**(2), 274–283 (1991)
2. Bodlaender, H.L., Gilbert, J.R., Hafsteinsson, H., Kloks, T.: Approximating treewidth, pathwidth, frontsize, and shortest elimination tree. J. Algorithms **18**(2), 238–255 (1995)
3. Boliac, R., Lozin, V.: On the clique-width of graphs in hereditary classes. In: Bose, P., Morin, P. (eds.) ISAAC 2002. LNCS, vol. 2518, pp. 44–54. Springer, Heidelberg (2002). https://doi.org/10.1007/3-540-36136-7_5
4. Courcelle, B.: The monadic second-order logic of graphs. I. Recognizable sets of finite graphs. Inf. Comput. **85**(1), 12–75 (1990)
5. Courcelle, B., Makowsky, J.A., Rotics, U.: Linear time solvable optimization problems on graphs of bounded clique-width. Theory Comput. Syst. **33**(2), 125–150 (2000)
6. Courcelle, B., Olariu, S.: Upper bounds to the clique width of graphs. Discret. Appl. Math. **101**(1), 77–114 (2000)
7. Dabrowski, K.K., Eagling-Vose, T., Köhler, N., Ordyniak, S., Paulusma, D.: Bounding width on graph classes of constant diameter. arXiv preprint arXiv:2505.19926 (2025)
8. Dabrowski, K.K., Johnson, M., Paulusma, D.: Clique-width for hereditary graph classes. London Math. Soc. Lecture Note Ser. **456**, 1–56 (2019)
9. Dabrowski, K.K., Paulusma, D.: Clique-width of graph classes defined by two forbidden induced subgraphs. Comput. J. **59**(5), 650–666 (2016)
10. Demaine, E.D., Hajiaghayi, M.: Equivalence of local treewidth and linear local treewidth and its algorithmic applications. Proc. SODA **2004**, 840–849 (2004)
11. Dębski, M., Piecyk, M., Rzążewski, P.: Faster 3-coloring of small-diameter graphs. SIAM J. Discret. Math. **36**(3), 2205–2224 (2022)
12. Dujmović, V., Eppstein, D., Joret, G., Morin, P., Wood, D.R.: Minor-closed graph classes with bounded layered pathwidth. SIAM J. Discret. Math. **34**(3), 1693–1709 (2020)
13. Eagling-Vose, T.: (2025). https://github.com/tevose/treedepth-diameter-3
14. Eppstein, D.: Diameter and treewidth in minor-closed graph families. Algorithmica **27**(3), 275–291 (2000)
15. Erdős, P., Rényi, A., Sós, V.T.: On a problem of graph theory. Stud. Sci. Math. Hung. **1**, 215–235 (1966)
16. Gajarský, J., Hlinený, P.: Faster deciding MSO properties of trees of fixed height, and some consequences. In: Proceedings of the FSTTCS 2012, Leibniz International Proceedings in Informatics (LIPIcs), vol. 18, pp. 112–123 (2012)
17. Galvin, F., Rival, I., Sands, B.: A Ramsey-type theorem for traceable graphs. J. Comb. Theory Ser. B **33**(1), 7–16 (1982)
18. Ganian, R., Ordyniak, S.: The complexity landscape of decompositional parameters for ILP. Artif. Intell. **257**, 61–71 (2018)
19. Gutin, G., Jones, M., Wahlström, M.: The mixed Chinese postman problem parameterized by pathwidth and treedepth. SIAM J. Discret. Math. **30**(4), 2177–2205 (2016)
20. Hickingbotham, R.: Induced subgraphs and path decompositions. Electron. J. Comb. **30**(2), P2.37 (2023)
21. Iwata, Y., Ogasawara, T., Ohsaka, N.: On the power of tree-depth for fully polynomial FPT algorithms. In: Proceedings of the STACS 2018, Leibniz International Proceedings in Informatics (LIPIcs), vol. 96, pp. 41:1–41:14 (2018)

22. Klimosová, T., Sahlot, V.: 3-coloring C_4 or C_3-free diameter two graphs. In: Proceedings of the WADS 2023, Lecture Notes in Computer Science, vol. 14079, pp. 547–560 (2023)
23. Koutecký, M., Levin, A., Onn, S.: A parameterized strongly polynomial algorithm for block structured integer programs. In: Proceedings of the ICALP 2018, Leibniz International Proceedings in Informatics (LIPIcs), vol. 107, pp. 85:1–85:14 (2018)
24. Lozin, V.V., Razgon, I.: Tree-width dichotomy. Eur. J. Comb. **103**, 103517 (2022)
25. Martin, B., Paulusma, D.: The computational complexity of disconnected cut and $2K_2$-partition. J. Comb. Theory Ser. B **111**, 17–37 (2015)
26. Martin, B., Paulusma, D., Smith, S.: Colouring graphs of bounded diameter in the absence of small cycles. Discret. Appl. Math. **314**, 150–161 (2022)
27. Mertzios, G.B., Spirakis, P.G.: Algorithms and almost tight results for 3-Colorability of small diameter graphs. Algorithmica **74**(1), 385–414 (2016)
28. Nešetřil, J., de Mendez, P.O.: Sparsity. AC, vol. 28. Springer, Heidelberg (2012). https://doi.org/10.1007/978-3-642-27875-4
29. Robertson, N., Seymour, P.D.: Graph minors. III. Planar tree-width. J. Comb. Theory Ser. B **36**(1), 49–64 (1984)
30. Robertson, N., Seymour, P.D.: Graph minors. V. Excluding a planar graph. J. Comb. Theory Ser. B **41**(1), 92–114 (1986)

The Peculiarities of Extending Queue Layouts

Thomas Depian(✉), Simon D. Fink, Robert Ganian, and Martin Nöllenburg

Algorithms and Complexity Group, TU Wien, Vienna, Austria
{tdepian,sfink,rganian,noellenburg}@ac.tuwien.ac.at

Abstract. We consider the problem of computing ℓ-page queue layouts, which are linear arrangements of vertices accompanied with an assignment of the edges to pages from one to ℓ that avoid the nesting of edges on any of the pages. Inspired by previous work in the extension of stack layouts, here we consider the setting of extending a partial ℓ-page queue layout into a complete one and primarily analyze the problem through the refined lens of parameterized complexity. We obtain novel algorithms and lower bounds which provide a detailed picture of the problem's complexity under various measures of incompleteness, and identify surprising distinctions between queue and stack layouts in the extension setting.

Keywords: Queue layouts · Parameterized complexity · Linear layouts · Extension problems

1 Introduction

The computation of fundamental representations or visualizations of graphs is a core research topic in several fields including theoretical computer science, computational geometry and graph drawing. While some representations can be computed in polynomial time (e.g., planar drawings of planar graphs), for many others computing a valid representation from the input graph alone is intractable (and, in particular, NP-hard when phrased as a decision question).

Nevertheless, it is entirely plausible—and often natural—that a fixed partial representation (such as a partial drawing) is already available, and our task is merely to complete this into a full representation. This perspective has recently led to the systematic study of graph representations through the perspective of so-called *extension problems*. A classical result in this line of work is the linear-time algorithm for extending partial planar drawings [1]; however, when the representations are NP-hard to compute "from scratch" one cannot expect polynomial-time algorithms in the strictly more general extension setting. That is why work on extension problems has predominantly been carried out through the lens of parameterized complexity [4–6,12,14,20,21,23], where one can use

Full proofs of statements marked with ★ can be found in the full version [13].

H. Fernau and P. Kindermann (Eds.): WG 2025, LNCS 16124, pp. 177–191, 2026.
https://doi.org/10.1007/978-3-032-11835-6_13

parameters to capture how much of the input graph still needs to be added to the partial representation.

In this article we present the next step in the aforementioned systematic investigation of extension problems, specifically targeting the notion of *queue layouts*. An ℓ-page queue layout of a graph G is a linear arrangement of $V(G)$ accompanied with an assignment of $E(G)$ to pages from $[\ell]$ such that the preimage of each page contains no pair of nesting edges (see Sect. 2 and Fig. 1). Queue layouts have been extensively studied from a variety of contexts, including their relationship to planarity [3,17], classical as well as parameterized complexity [7,8,24] and connections to other graph parameters [16,27]. However, unlike for the closely related notion of stack layouts (which differ by excluding crossings instead of nestings) [9,10,12], to the best of our knowledge up to now nothing was known about the extension problem for queue layouts:

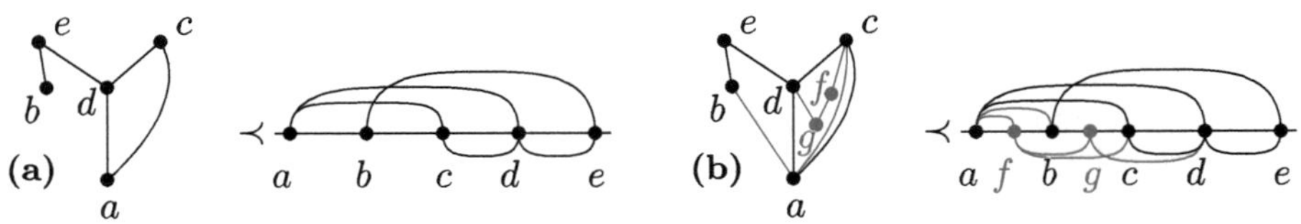

Fig. 1. **(a)** Graph H with a two-page queue layout $\langle \prec_H, \sigma_H \rangle$. Observe the twist on the upper page between the edges ac and be. **(b)** An extension of H and $\langle \prec_H, \sigma_H \rangle$ by new vertices and edges in blue. (Color figure online)

> QUEUE LAYOUT EXTENSION (QLE)
> **Given** An integer $\ell \geq 1$, a graph G, a subgraph H of G, and an ℓ-page queue layout $\langle \prec_H, \sigma_H \rangle$ of H.
> **Question** Does there exist an ℓ-page queue layout $\langle \prec_G, \sigma_G \rangle$ of G that extends $\langle \prec_H, \sigma_H \rangle$?

QLE is easily seen to be NP-hard already when $\ell = 1$, as it generalizes the NP-hard problem of computing a 1-page queue layout from scratch [24]—indeed, this occurs when $V(H) = \emptyset$. An interesting special case occurs when $V(H) = V(G)$ and $E(H) = \emptyset$, i.e., H contains all vertices of G but none of its edges: in this case, QLE coincides with the well-studied problem of computing an ℓ-page queue layout with a fixed linear arrangement of the vertices, which is known to be polynomial-time solvable [24]. Since the case where $E(H) = E(G)$ is trivial, this raises the question of whether QLE is polynomial-time solvable whenever $V(G) = V(H)$. We begin our investigation by building upon a known relation between queue layouts with fixed spine order and colorings of specifically constructed permutation graphs [18,19] to establish our first minor contribution:

Theorem 1 (★). *When restricted to the case where* $V(G) = V(H)$, QLE *is* (1) NP-*complete and* (2) *in* XP *when parameterized by the number* ℓ *of pages.*

By slightly adapting the ideas used for extending stack layouts [12, Theorem 3.2], we obtain our second minor contribution: fixed-parameter tractability for the special case above w.r.t. the number of missing edges.

Theorem 2 (★). *When restricted to the case where* $V(G) = V(H)$, QLE *when only* m_{add} *edges are missing from* $\langle \prec_H, \sigma_H \rangle$ *is fixed-parameter tractable when parameterized by* m_{add}.

Moving on to the general case where both edges and vertices may be missing[1] from H, we begin by providing our third and final "minor" contribution:

Theorem 3 (★). QLE *is in* XP *when parameterized by the number* κ *of missing elements.*

Similarly to Theorem 2, the proof of the above result follows by adapting the ideas from an analogous algorithm in the setting of stack layouts [12, Theorem 5.1]. Essentially, while Theorems 1 to 3 are important pieces in our overall understanding of QLE, in terms of technical contributions they merely set the stage for our three main results. The first of these establishes a corresponding lower bound to Theorem 3:

Theorem 4 (★). QLE *is* W[1]*-hard when parameterized by the number* κ *of missing elements.*

Theorem 4 also has a matching result in the stack layout setting [12, Theorem 6.4]—however, the gadgets used to establish the previous lower bound seem impossible to directly translate to the queue layout setting, and circumventing this issue by redesigning the reduction required overcoming several technical difficulties that were not present in the stack layout setting.

Since Theorem 4 excludes fixed-parameter algorithms w.r.t. the number of missing elements alone, the natural question is whether one can at least obtain such an algorithm if the parameter also includes the number ℓ of pages. This directly corresponds to the main open question highlighted in the previous work on stack layout extension [12, Section 8]. As our second main contribution, for QLE we answer this in the affirmative:

Theorem 5. QLE *is fixed-parameter tractable w.r.t. the number* ℓ *of pages plus the number* κ *of missing elements.*

The proof of Theorem 5 relies on a combination of branching and structural insights into queue layouts that allow us to reduce the problem to solving a set of 2-SAT instances. One high-level take-away from this result is that even if both vertices and edges are missing from H, QLE seems to behave differently from stack layout extension—in particular, the approach used to establish Theorem 5 does not seem applicable to the latter. As our final result, we provide "hard" evidence for this intuition by establishing the polynomial-time tractability of the special case where only two vertices and their incident edges are missing:

[1] We say that an element, i.e., a vertex or edge, is *missing* if it is in G but not in H.

Theorem 6 (★). *When restricted to the case where H is obtained by deleting two vertices from G,* QLE *can be solved in polynomial time.*

The reason Theorem 6 is noteworthy is that it directly contrasts the known NP-hardness of the corresponding restriction in the stack layout setting [12, Theorem 4.2]. The proof is also non-trivial: the difficulty is that many missing edges may need to be assigned to a large number of available pages. We resolve this by developing a set of reduction rules based on novel insights into the behavior of queue layouts. Table 1 summarizes our results.

Table 1. Overview of our results. SLE is the problem of extending stack layouts and VEDD = 2 indicates two missing vertices. Results for QLE are shown in this paper.

Problem	$V(H) = V(G)$	κ	$\kappa + \ell$	VEDD = 2
SLE	NPc, FPT(m_{add})	XP, W[1]-hard	Open	NPc
	Theorems 1 [26] & 3.2 [12]	& 6.4 [12]		Theorem 4.2 [12]
QLE	NPc, FPT(m_{add})	XP, W[1]-hard	FPT	Polynomial time
	Theorems 1 & 2	Theorems 3 & 4	Theorem 5	Theorem 6

2 Preliminaries

We assume familiarity with standard notions from graph theory [15] and parameterized complexity [11]. Unless stated otherwise, all graphs G are simple and undirected with vertex set $V(G)$ and edge set $E(G)$. We let $[\ell]$ denote the set $\{1, \ldots, \ell\}$ for $\ell \geq 1$ and $G[X]$ the subgraph of G induced by X for $X \subseteq V(G)$.

Given two edges u_1v_1 and u_2v_2 and a linear order $\prec$ of $V(G)$ with $u_1 \prec u_2 \prec v_2 \prec v_1$, we say that u_1v_1 *nests* u_2v_2 or, conversely, that u_2v_2 *is nested* by u_1v_1. In addition, the edges are in a *nesting relation*, indicated as $u_1v_1 \Cap_{\prec_G} u_2v_2$. We omit $\prec_G$ if it is clear from context and observe that only edges with pairwise distinct endpoints can be in a nesting relation. An *ℓ-page queue layout* consists of a linear order $\prec_G$ of $V(G)$ (the *spine order*) and a function $\sigma_G \colon E(G) \to [\ell]$ (the *page assignment*) that assigns each edge $uv \in E(G)$ to a *page* $\sigma_G(uv)$ such that for any two edges u_1v_1 and u_2v_2 with $\sigma_G(u_1v_1) = \sigma_G(u_2v_2)$ we do not have $u_1v_1 \Cap u_2v_2$. If $u_1 \prec u_2 \prec v_1 \prec v_2$, then we say that u_1v_1 and u_2v_2 form a *twist* in $\langle \prec_G, \sigma_G \rangle$; see also Fig. 1a. We write $\langle \prec, \sigma \rangle$ instead of $\langle \prec_G, \sigma_G \rangle$ if G is clear from context. Let $u, v \in V(G)$ be two vertices and $\langle \prec, \sigma \rangle$ a queue layout of G. A vertex $v \in V(G)$ *sees* a vertex $u \in V(G)$ on a page $p \in [\ell]$ if $\langle \prec, \sigma' \rangle$, where $\sigma'(uv) = p$ and $\sigma'(e) = \sigma(e)$ otherwise, is a valid queue layout of $G \cup \{uv\}$. Note that the visibility relation is symmetric. Furthermore, v is *left* of u if $v \prec u$ and *right* of u if $u \prec v$. An *interval* $[u, v]$, with $u \prec v$, is a segment on the spine such that there is no $w \in V(G)$ with $u \prec w \prec v$. A queue layout $\langle \prec_G, \sigma_G \rangle$ *extends* a queue layout $\langle \prec_H, \sigma_H \rangle$ of a subgraph $H \subseteq G$ of G if $\sigma_H \subseteq \sigma_G$ and

$\prec_H \subseteq \prec_G$, i.e., $\sigma_H(e) = \sigma_G(e)$ for all $e \in E(H)$ and $(u \prec_H v) \Leftrightarrow (u \prec_G v)$ for all $u, v \in V(H)$.

Let $\mathcal{I}$ be an instance of QLE. We use $|\mathcal{I}|$ as a shorthand for $|V(G)|+|E(G)|+\ell$ and refer to $\langle \prec_G, \sigma_G \rangle$, if it exists, as a *solution* of $\mathcal{I}$. For the remainder of this paper, we use $V_{\text{add}} := V(G) \setminus V(H)$ and $E_{\text{add}} := E(G) \setminus E(H)$ to speak about the set of *new* vertices and edges and denote with n_{add} and m_{add} their cardinality, respectively. Vertices and edges of H are called *old*. We use *missing* as a synonym for *new*. Furthermore, $E_{\text{add}}^H := \{e = uv \in E_{\text{add}} \mid u, v \in V(H)\}$ is the set of new edges e with only old endpoints u and v.

3 QLE With Only Missing Edges

For our first result, we will use a known equivalence between finding a certain queue layout and a coloring problem in a restricted graph class, combined with existing complexity results on the latter problem. Dujmović and Wood already note that "the problem of assigning edges to queues in a fixed vertex ordering is equivalent to colouring a permutation graph" [18] and attribute this result to a 1941 publication by Dushnik and Miller [19]. In the full version [13], we define the term permutation graph and also make this insight more formal and accessible in modern terms by independently showing this relationship. Here, it suffices to note that we not only have equivalence of both problems, but also that additional constraints on a permutation graph k-coloring can equivalently be expressed as additional constraints on a k-page-assignment with a fixed vertex order that yields a valid queue layout. Especially, if we want a certain k-page queue layout, for example one that observes some predefined pages for certain edges, this is equivalent to requiring certain colors for some vertices in the coloring, i.e., an extension of a precoloring. Using known results from permutation graph coloring, that precoloring extension is NP-complete [25] while list-coloring for up to k colors is in XP parameterized by k [22], we thus obtain the following theorem.

Theorem 1 (★). *When restricted to the case where $V(G) = V(H)$, QLE is (1) NP-complete and (2) in XP when parameterized by the number ℓ of pages.*

In light of Theorem 1, we now turn our attention to the parameterized complexity of QLE when only edges are missing and show that this problem is fixed-parameter tractable in their number. This matches the complexity of extending stack layouts in the same setting [12, Theorem 3.2]. In particular, we can use the same proof strategy to obtain the result: Remove all new edges e that have at least m_{add} different pages p such that there is no old edge e' with $\sigma(e') = p$ and $e \Cap e'$, branch to determine the page assignment of the remaining edges, and, finally, check each of the $\mathcal{O}({m_{\text{add}}}^{m_{\text{add}}})$ different branches in linear time if it results in a valid queue layout. Nevertheless, for reasons of completeness, we provide the full details of the following theorem in the appendix.

Theorem 2 (★). *When restricted to the case where $V(G) = V(H)$, QLE when only m_{add} edges are missing from $\langle \prec_H, \sigma_H \rangle$ is fixed-parameter tractable when parameterized by m_{add}.*

4 QLE With Missing Vertices and Edges

While Theorem 2 shows tractability of QLE when parameterized by the number of missing edges, it only applies to the highly restrictive setting where H already contains all vertices of G. However, in general also vertices are missing from $\langle \prec_H, \sigma_H \rangle$ and our task is to find a spine position for them. We now turn our attention to the complexity of QLE when both vertices and edges are missing and recall that the problem of deciding whether a graph has a 1-page queue layout is already NP-complete [24]. Thus, we investigate the complexity of QLE when parameterized by the number $\kappa := n_{\text{add}} + m_{\text{add}}$ of missing elements. We show that under this parameter, extending queue and stack layouts admit the same complexity-theoretic behavior: being XP-tractable but W[1]-hard [12].

4.1 QLE Parameterized by the Number of New Elements is in XP

We first establish XP-membership of QLE when parameterized by κ by using an algorithm analogous to the one for extending stack layouts [12, Theorem 5.1]: We first branch to determine the spine position of the new vertices. This gives us $\mathcal{O}(|\mathcal{I}|^{n_{\text{add}}})$ different branches, each corresponding to a different spine order $\prec_G$. The fixed spine order allows us to employ our FPT-algorithm from Theorem 2, resulting in the following theorem.

Theorem 3 (★). QLE *is in* XP *when parameterized by the number* κ *of missing elements.*

Theorem 3 raises the question whether we can get fixed parameter tractability under κ. In what follows, we answer this question negatively and show W[1]-hardness of QLE when parameterized by the number of missing elements.

4.2 QLE Parameterized by the Number of New Elements is Hard

We now show W[1]-hardness of QLE when parameterized by κ. Our reduction uses the idea and framework of the reduction for establishing W[1]-hardness of the problem of extending stack layouts under this parameter [12, Theorem 6.4]. However, the substantially different notion of visibility makes the construction of new gadgets inevitable. We reduce from MULTI-COLORED CLIQUE (MCC): Given a graph G_C, an integer $k \geq 1$, and a partition of $V(G_C)$ into k independent color sets, i.e., $V(G_C) = V_1 \,\dot{\cup}\, \dots \,\dot{\cup}\, V_k$, does there exist a colorful k-clique $\mathcal{C}$ in G_C? MCC is well-known to be W[1]-hard parameterized by k [11].

For the remainder of this section, we let $(G_C, k, (V_1, \dots, V_k))$ be an instance of MCC with $N = |V(G_C)|$ vertices and $M = |E(G_C)|$ edges. We use Greek letters to denote the color of a vertex to separate it from its index, i.e., we let v_α^i denote the ith vertex of color α. Based on the instance of MCC, we construct an instance $(\ell, G, H, \langle \prec_H, \sigma_H \rangle)$ of QLE. To facilitate description and presentation of our reduction, we allow multi-edges in H. We show in the full version [13] how we can redistribute the multi-edges over multiple auxiliary vertices while preserving the correctness of the reduction.

Overview of the Reduction. At a high level, the constructed instance of QLE contains k new vertices, one for each color. For every vertex of G_C, we create an interval on the spine $\prec_H$. We represent membership in $\mathcal{C}$ through the placement of new vertices in the intervals corresponding to vertices in $\mathcal{C}$. To guarantee that every solution to McC corresponds to a solution to QLE and vice versa, we have to ensure that we select exactly one vertex for each color $\alpha \in [k]$, and that they form a clique. For the former requirement, we use a gadget that forces one of the new vertices to be placed in an interval of color α. For the latter property, we let the new vertices form a k-clique in G and introduce for each edge e of G_C a further gadget. This gadget contains a twist that restricts the placement of the endpoints of edges assigned to this page. In particular, it forces them to lie in the intervals corresponding to the endpoints of e, i.e., we only select pairwise adjacent vertices. With the above intuition in mind, we now describe our reduction in more detail.

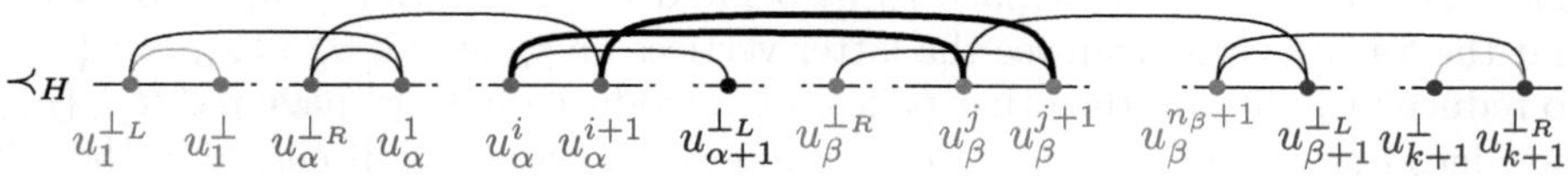

Fig. 2. Visualization of the spine order $\prec_H$ and the edge gadget for the edge $e = v_\alpha^i v_\beta^j \in E(G_C)$ on page p_e. The bold edges form the twist on p_e that we use to synchronize the placement of the vertices and the gray edges are part of the fixation gadget.

Description of the Reduction. The spine order $\prec_H$ forms the core of our reduction. To construct it, we introduce for each of the N *original* vertices $v_\alpha^i \in V(G_C)$ a *copy* $u_\alpha^i \in V(H)$. Furthermore, we introduce for each color α four dummy vertices $u_\alpha^{\perp_L}$, $u_\alpha^{\perp}$, $u_\alpha^{\perp_R}$, and $u_\alpha^{n_\alpha+1}$, with $n_\alpha := |V_\alpha|$, and overall three further dummy vertices $u_{k+1}^{\perp_L}$, $u_{k+1}^{\perp}$, and $u_{k+1}^{\perp_R}$. Figure 2 visualizes the spine order $\prec_H$, which is the transitive closure of the following orders: For every $\alpha \in [k+1]$, we set $u_\alpha^{\perp_L} \prec u_\alpha^{\perp} \prec u_\alpha^{\perp_R}$. Furthermore, for every $\alpha \in [k]$, we set $u_\alpha^{\perp_R} \prec u_\alpha^1$ and $u_\alpha^{n_\alpha+1} \prec u_{\alpha+1}^{\perp_L}$, and for every $j \in [n_\alpha]$ we set $u_\alpha^j \prec u_\alpha^{j+1}$.

Observe that for each original vertex $v_\alpha^i \in V(G_C)$ we have the interval $[u_\alpha^i, u_\alpha^{i+1}]$ on the spine. We say that the interval $[u_\alpha^i, u_\alpha^{i+1}]$ *corresponds* to v_α^i and is of color α. We address $[u_\alpha^i, u_\alpha^{i+1}]$ with $\curlyvee(v_\alpha^i)$. The graph $G[V_{\text{add}}]$ forms a k-clique representing $\mathcal{C}$ in G on the new vertices $V_{\text{add}} = \{x_1, \ldots, x_k\}$.

To ensure that the selected vertices form a clique in G_C, we introduce for each edge $e = v_\alpha^i v_\beta^j \in E(G_C)$, $\alpha < \beta$, a so-called *edge gadget*; see Fig. 2. This gadget consists of a page p_e on which we place a twist between $\curlyvee(v_\alpha^i)$ and $\curlyvee(v_\beta^j)$. We observe that we introduce one page for each edge of G_C. The aforementioned twist ensures that if x_α is placed in $\curlyvee(v_\alpha^i)$, it can see x_β only if the latter vertex is placed in $\curlyvee(v_\beta^j)$. To ensure that the new edges on page p_e can only be used to connect x_α and x_β if they are placed in the right intervals, we furthermore introduce edges that block visibility for intervals of color γ with $\gamma < \alpha$ (Fig. 2

left), $\alpha \leq \gamma \leq \beta$ (middle), and $\beta < \gamma$ (right). More formally, we show in the full version [13] that the edge gadget has the following properties.

Property 1. Let $\langle \prec, \sigma \rangle$ be a solution to an instance of QLE with $e = v_\alpha^i v_\beta^j \in E(G_C)$ and $x_\alpha, x_\beta \in V_{\text{add}}$ with $\alpha < \beta$. If $u_\alpha^1 \prec x_\alpha \prec u_\alpha^{n_\alpha+1}$, $u_\beta^1 \prec x_\beta \prec u_\beta^{n_\beta+1}$, and $\sigma(x_\alpha x_\beta) = p_e$, then x_α is placed in $\Upsilon(v_\alpha^i)$ and x_β in $\Upsilon(v_\beta^j)$.

Property 2. For every color $\alpha \in [k]$ and for the page $p \in [\ell]$ of every edge gadget, the vertex $u_\alpha^\perp$ is not visible on p for any vertex v with $u_1^{\perp L} \prec v \prec u_\alpha^{\perp L}$ or $u_\alpha^{n_\alpha+1} \prec v \prec u_{k+1}^{\perp R}$.

To ensure that $\mathcal{C}$ is colorful, we must force the vertex x_α to be placed in an interval of color α for all $\alpha \in [k]$. We introduce for this purpose the *fixation gadget.* It consists of $2k$ new edges connecting x_α with $u_\alpha^\perp$ and $u_{\alpha+1}^\perp$ for all $\alpha \in [k]$. These edges are accompanied by the edges $u_\alpha^\perp u_\alpha^1$, $u_\alpha^1 u_\alpha^{n_\alpha+1}$, and $u_\alpha^{n_\alpha+1} u_{\alpha+1}^\perp$. We place these edges on a new page p_d associated to the fixation gadget to ensure that the former vertex can see the latter vertices on p_d only if $u_\alpha^1 \prec x_\alpha \prec u_\alpha^{n_\alpha+1}$. To reduce visibility on the other pages, we introduce on every page $p \in [\ell] \setminus \{p_d\}$ the edges $u_1^{\perp L} u_1^\perp$ and $u_{k+1}^\perp u_{k+1}^{\perp R}$, drawn in gray in Fig. 2. As discussed in the full version [13], the gadget has the following property.

Property 3. Let $\langle \prec, \sigma \rangle$ be a solution to an instance of QLE that contains a fixation gadget on k vertices and has Property 2 for every page $p \in [\ell] \setminus \{p_d\}$. For every color $\alpha \in [k]$ we have $u_\alpha^1 \prec x_\alpha \prec u_\alpha^{n_\alpha+1}$.

Correctness of the Reduction. The above (informal) description of our construction should only convey its main idea. We provide in the full version [13] a precise description of the gadgets and show a small illustrative example in Fig. 3.

Observe that even for $u_\alpha^1 \prec x_\alpha \prec u_\alpha^{n_\alpha+1}$, the edges of the edge gadget prevent visibility to $u_\alpha^\perp$ and $u_{\alpha+1}^\perp$. Thus, the edges $x_\alpha u_\alpha^\perp$ and $x_\alpha u_{\alpha+1}^\perp$ can only be placed on the dedicated page for the fixation gadget. Furthermore, for every new edge $x_\alpha x_\beta$ in $G[V_{\text{add}}]$ we have $\alpha \neq \beta$ and since $\sigma(u_\alpha^\perp u_\alpha^1) = \sigma(u_\alpha^{n_\alpha+1} u_{\alpha+1}^\perp) = p_d$, we conclude that these are the only edges on p_d. More formally, we obtain:

Observation 1. *Let $\langle \prec, \sigma \rangle$ be a solution to our created instance of* QLE. *For every $\alpha \in [k]$ we have $\sigma(x_\alpha u_\alpha^\perp) = \sigma(x_\alpha u_{\alpha+1}^\perp) = p_d$. Furthermore, for every edge $e \in E_{add}$ with $\sigma(e) = p_d$ we have $e \in \{x_\alpha u_\alpha^\perp, x_\alpha u_{\alpha+1}^\perp \mid \alpha \in [k]\}$.*

Furthermore, observe that the page p_e for an edge $e = v_\alpha^i v_\beta^j \in E(G_C)$ can only be used for new edges that have their endpoints placed in intervals of color α and β. If at least one endpoint is placed in an interval of a different color γ, we create, depending on the relation of α, β, and γ, always a nesting relation with at least one edge on p_e. The following observation summarizes this property.

Observation 2. *Let $\langle \prec, \sigma \rangle$ be a solution to our constructed instance of* QLE. *For every new edge $x_\gamma x_\delta \in E(G[V_{add}])$ assigned to the page $p_e \in [\ell] \setminus \{p_d\}$ with $e = v_\alpha^i v_\beta^j \in E(G_C)$, $\alpha < \beta$, and $\gamma < \delta$, we have $\alpha = \gamma$ and $\beta = \delta$.*

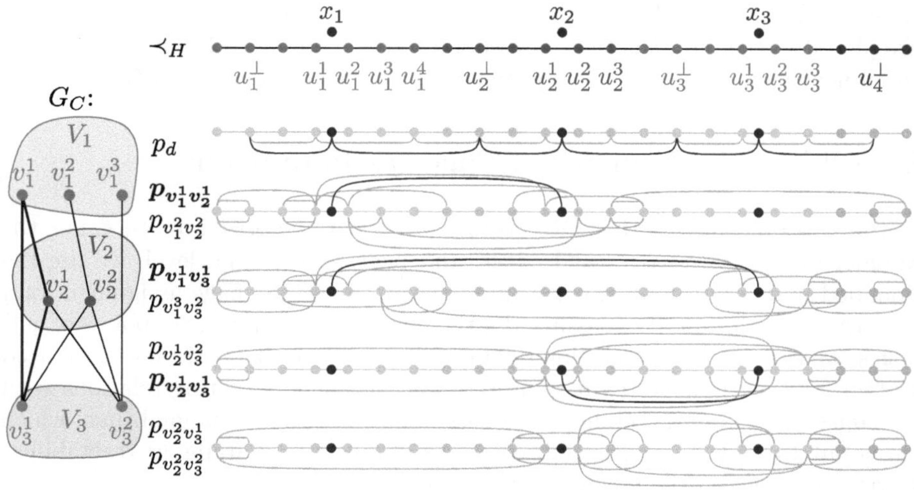

Fig. 3. An example of the reduction. We indicate the solution to McC in red (left) and the corresponding solution to the instance $(\ell, G, H, \langle \prec_H, \sigma_H \rangle)$ of QLE in saturated colors (right). To increase readability, we omit the labels for the vertices $u_\alpha^{\perp L}$ and $u_\alpha^{\perp R}$. (Color figure online)

We now have all the ingredients to show the main theorem of this section.

Theorem 4 (★). QLE *is* W[1]*-hard when parameterized by the number* κ *of missing elements.*

Proof Sketch. Let $(G_C, k, (V_1, \dots, V_k))$ be an instance of McC and $\mathcal{I}$ the instance of QLE obtained by our reduction. A careful analysis of $\mathcal{I}$ reveals that we have $|\mathcal{I}| = \mathcal{O}(|G_C| + k^2)$ and $\kappa = \mathcal{O}(k^2)$. In particular, $n_{\text{add}} = k$ and $m_{\text{add}} = \binom{k}{2} + 2k$. We now proceed with discussing correctness of our reduction.

For the forward direction, let $\mathcal{C}$ be a colorful k-clique. To construct a solution of $\mathcal{I}$, we consider every vertex $v_\alpha^i \in \mathcal{C}$ and place x_α in $\Upsilon(v_\alpha^i)$. As $\mathcal{C}$ is colorful, this placement is well-defined and yields the spine order $\prec_G$. For the page assignment σ_G, we set $\sigma_G(x_\alpha u_\alpha^\perp) = \sigma_G(x_\alpha u_{\alpha+1}^\perp) = p_d$ for every $\alpha \in [k]$. As this resembles the necessary page assignment from Observation 1, we do not introduce a nesting relation on page p_d. Furthermore, for each new edge $x_\alpha x_\beta \in E(G[V_{\text{add}}])$, we consider the intervals $\Upsilon(v_\alpha^i)$ and $\Upsilon(v_\beta^j)$ into which x_α and x_β are placed. As $\mathcal{C}$ is a clique, there exists the edge $e = v_\alpha^i v_\beta^j \in E(G_C)$ and consequently the page p_e in $\mathcal{I}$ that can accommodate $x_\alpha x_\beta$. This shows that $\langle \prec_G, \sigma_G \rangle$ is a solution to $\mathcal{I}$.

For the backward direction, let $\langle \prec_G, \sigma_G \rangle$ be a solution to $\mathcal{I}$. Thanks to Properties 2 and 3, we have $u_\alpha^1 \prec_G x_\alpha \prec_G u_\alpha^{n_\alpha+1}$ for each new vertex x_α. As each interval between u_α^1 and $u_\alpha^{n_\alpha+1}$ corresponds to a vertex of G_C, x_α is placed in $\Upsilon(v_\alpha^i)$ for some $v_\alpha^i \in V(G_C)$. We add v_α^i to our solution $\mathcal{C}$ of the instance of McC and observe that it is colorful. To see that $\mathcal{C}$ is a clique in G_C, we consider the page p for a new edge $x_\alpha x_\beta \in E(G[V_{\text{add}}])$. Observation 1 ensures $p = p_e$ for some $e \in E(G_C)$ and Property 1 and Observation 2 guarantee that x_α and x_β are placed in the intervals that correspond to the endpoints of e. Extending this to all new edges $E(G[V_{\text{add}}])$, we conclude that $\mathcal{C}$ is also a clique in G_C. □

5 Adding the Number of Pages as Parameter

In this section, we show that QLE is FPT when parameterized by $\kappa + \ell$, i.e., the number of missing vertices and edges, and pages. On a high level, we use the bounded number of missing edges and pages to guess the page assignment of a hypothetical solution. By guessing also the relative order among endpoints of new edges, we can check for a nesting relation between new edges. However, we need to find for each new vertex a spine position that avoids nesting relations with old edges. In Lemma 1, we show that this can be done in polynomial time, which turns out to be the essential ingredient to obtain fixed-parameter tractability. Lemma 1 builds on the following observation:

Observation 3. *Let $\langle \prec, \sigma \rangle$ be a queue layout of G and $u_1v_1, u_2v_2 \in E(G)$ two edges of G with $u_1 \prec v_1$, $u_2 \prec v_2$, and $\sigma(u_1v_1) = \sigma(u_2v_2)$. We have $u_1 \prec u_2 \Leftrightarrow v_1 \prec v_2$.*

Observation 3 allows us to encode the (non-existence of a) nesting relation as a 2-SAT expression. By introducing one variable $x_{u,v}$ for each pair of vertices $u, v \in V(G)$, we can express Observation 3 as $x_{u_1,u_2} \Leftrightarrow x_{v_1,v_2}$. Hence, we can describe an ℓ-page queue layout with a fixed page assignment as a 2-SAT formula plus anti-symmetry ($x_{u,v} \Leftrightarrow \neg x_{v,u}$) and transitivity ($x_{u,v} \wedge x_{v,w} \Rightarrow x_{u,w}$) constraints. While the latter constraints require in general clauses on three literals, the total order among old and new vertices, respectively, fixes for each three vertices $u, v, w \in V(G)$ the relative order among at least two of them. Thus, we know the truth value of $x_{u,v}$, $x_{v,w}$, or $x_{u,w}$, which allows us to also incorporate transitivity constraints in our 2-SAT formula. As the satisfiability of a 2-SAT formula can be checked in linear time [2], we obtain:

Lemma 1 (★). *Given an instance $\mathcal{I} = (\ell, G, H, \langle \prec_H, \sigma_H \rangle)$ of QLE, a page assignment σ_G for all edges, and a total order $\prec_{E_{\text{add}}}$ in which the endpoints of new edges will appear on the spine. In $\mathcal{O}(|\mathcal{I}|^2)$ time, we can check if $\mathcal{I}$ admits a solution $\langle \prec_G, \sigma_G \rangle$ where $\prec_G$ extends $\prec_{E_{\text{add}}}$ or report that no such solution exists.*

Finally, we observe that there are $\mathcal{O}(\ell^{m_{\text{add}}})$ different page assignments σ_G and $\mathcal{O}(n_{\text{add}}! \cdot {m_{\text{add}}}^{n_{\text{add}}})$ potential orders among endpoints of new edges. Since we apply Lemma 1 in each of the $\mathcal{O}(\ell^{m_{\text{add}}} \cdot n_{\text{add}}! \cdot {m_{\text{add}}}^{n_{\text{add}}})$ different branches, we obtain the following theorem.

Theorem 5. QLE *is fixed-parameter tractable w.r.t. the number ℓ of pages plus the number κ of missing elements.*

6 QLE With Two Missing Vertices

In this section, we show that QLE is polynomial-time solvable for the case where the missing part consists of only two vertices of arbitrary degree. Recall that the problem of extending stack layouts is already NP-hard in this setting [12].

Let $\mathcal{I}$ be an instance of QLE with $V_{\text{add}} = \{u, v\}$. First, we branch to determine the placement of u and v on the spine. In each of the $\mathcal{O}(|\mathcal{I}|^2)$ branches, the spine order $\prec_G$ is fixed, and we only need to assign the new edges to pages. However, in contrast to Theorem 3, the number of new edges is unbounded. Nevertheless, we can avoid the intractability result of Theorem 1 by using the fact that all missing edges are incident to u or v, which allows us to observe useful properties of the visibility relation on certain pages.

For the remainder of this section, we assume a fixed spine order $\prec_G$ of G where, w.l.o.g., $u \prec_G v$. Furthermore, we can branch to determine the page assignment $\sigma(uv)$ of uv, if it exists, and check in linear time whether uv is in a nesting relation with an old edge on $\sigma(uv)$. Hence, we consider the edge uv for the upcoming discussion as an old edge.

We define for every edge $e \in E_{\text{add}}$ the sets $P(e)$ and $\Cap(e)$ of *admissible* pages and *conflicting* new edges, respectively. More formally, $P(e)$ contains all pages $p \in [\ell]$ such that $\langle \prec_G, \sigma'_H \rangle$ is a queue layout of the graph G' with $V(G') = V(G)$ and $E(G') = E(H) \cup \{e\}$ where $\sigma'_H(e) = p$ and $\sigma'_H(e') = \sigma_H(e')$ for all $e' \in E(H)$. Furthermore, $\Cap(e)$ contains all new edges $e' \in E_{\text{add}}$ that are in a nesting relation with e w.r.t. $\prec_G$. We now make a simple observation about the relation among the sets $P(e)$ and $\Cap(e)$ and the existence of a solution and its page assignment.

Observation 4. *Let $e \in E_{add}$ be a new edge. If $P(e) = \emptyset$, then $\mathcal{I}$ is a negative instance. Furthermore, if $\mathcal{I}$ is a positive instance and there exists a page $p \in P(e) \setminus \bigcup_{e' \in \Cap(e)} P(e')$, then there exists a solution $\langle \prec_G, \sigma \rangle$ with $\sigma(e) = p$.*

Observation 4 allows us to restrict our attention to parts of the instance where for every missing edge e each admissible page p is also admissible for some other new edge $e' \in \Cap(e)$. Note that this also implies $\Cap(e) \neq \emptyset$. Next, we show a relation among the admissible pages for different new edges.

Lemma 2 (★). *Let $e_1 = vx$, $e_2 = uy$, and $e_3 = uz$ be three new edges such that $u \prec x \prec y \prec z$ and $v \prec y$. For any page $p \in P(e_1) \cap P(e_3)$ it holds $p \in P(e_2)$.*

Proof Sketch. For the sake of a contradiction, assume $p \notin P(e_2)$. This means there is an old edge $e = ab$ on p that is in a nesting relation with e_2. If e is nested by e_2, it is also nested by e_3; see Fig. 4a. Thus, we have $p \notin P(e_3)$, which is a contradiction. Similarly, if e nests e_2, then it must also nest e_1; see Fig. 4b. As this implies $p \notin P(e_1)$, we again arrive at a contradiction. As both cases lead to a contradiction, we conclude that $p \in P(e_2)$ must hold. □

A solution *uses* a spine order $\prec^*_G$ if we have $\langle \prec^*_G, \sigma_G \rangle$. We now use Lemma 2 to show that for two missing vertices with a fixed placement on the spine we can safely remove some edges with at least two admissible pages:

Fig. 4. If the page p is admissible for e_1 and e_3, it is also admissible for e_2: No edge e (on page p) can be in a nesting relation with only e_2.

Lemma 3 (★). *Let* $\mathcal{I} = (\ell, G, H, \langle \prec_H, \sigma_H \rangle)$ *be an instance of* QLE *with* $V_{add} = \{u, v\}$ *and let* $\prec_H \subseteq \prec_G$ *be a spine order with* $u \prec_G v$ *that contains an edge* $e = vx \in E_{add}$ *with* $u \prec_G x$ *and* $|P(e)| \geq 2$. $\mathcal{I}$ *admits a solution that uses* $\prec_G$ *if and only if* $\mathcal{I}' = (\ell, G' = G \setminus \{e\}, H, \langle \prec_H, \sigma_H \rangle)$ *admits one that uses* $\prec_G$.

Proof Sketch. Since deleting an edge can never invalidate an existing solution, we focus on the more involved direction of showing that we can re-insert e into a solution $\langle \prec_G, \sigma_{G'} \rangle$ to $\mathcal{I}'$. As the spine order $\prec_G$ is fixed, we can list all new edges $\Cap(e) = \{uw_1, uw_2, \dots, uw_k\}$ that are in a nesting relation with e. Note that they must be incident to u. Thus, we can sort them by their single old endpoint, i.e., $w_1 \prec_G \dots \prec_G w_k$. If there exists a page $p \in P(e)$ such that for all $e' \in \Cap(e)$ we have $\sigma_{G'}(e') \neq p$, we can set $\sigma_G(e) = p$. Otherwise, we process the w_i, $i \in k$ in decreasing order from $i = k - 1$ to $i = 1$. If $\sigma_{G'}(e_{i+1}) \neq \sigma_{G'}(e_i)$, we change the page assignment of e_i to match the one from e_{i+1}. By Lemma 2, we know that e_i cannot be in a nesting relation with an old edge on page $\sigma_{G'}(e_{i+1})$. Furthermore, e_i cannot be in a nesting relation with a new edge e' on page $\sigma_{G'}(e_{i+1})$ as then $\sigma_{G'}(e_{i+1})$ would also be in a nesting relation with e' as Fig. 5a underlines. Thus, setting $\sigma_{G'}(e_i) = \sigma_{G'}(e_{i+1})$ is safe. Once we have eventually processed the last edge e_1, there exists at least one page $p \in P(e)$ such that no edge $e' \in \Cap(e)$ is still assigned to p. We can now safely set $\sigma_G(e) = p$. □

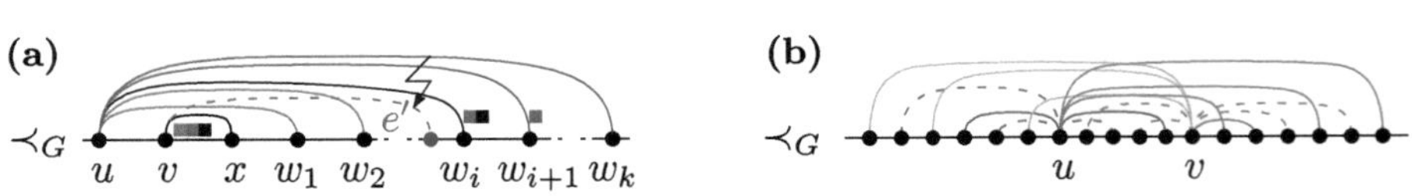

Fig. 5. (a) Visualization of the reduction rule from Lemma 3. Colors indicate the page assignment and squares the admissible pages for some of new edges. The dashed edge e' cannot exist. **(b)** The obtained instance after applying the reduction rule. Solid, colored edges have only one admissible page, dashed, colored edges got removed, and the page assignment of the gray edges depends on the one of the solid, colored edges. These edges are thus retained.

We now have all tools at hand to show our final result:

Theorem 6 (★). *When restricted to the case where* H *is obtained by deleting two vertices from* G, QLE *can be solved in polynomial time.*

Proof Sketch. We branch to determine the spine order $\prec_G$ and the page assignment for the edge uv. For each of the $\mathcal{O}(|\mathcal{I}|^2 \cdot \ell)$ branches, we check if the edge uv is in a nesting relation with an old edge. If not, we compute the sets $P(\cdot)$ and $⋒(\cdot)$. Using Observation 4, we remove edges e with $\left(P(e) \setminus \bigcup_{e' \in ⋒(e)} P(e')\right) \neq \emptyset$. Furthermore, we remove all edges $vx \in E_{\text{add}}$ with $u \prec_G x$ and $|P(vx)| \geq 2$. Lemma 3 ensures that this operation is safe. Symmetrically, we remove all edges $uy \in E_{\text{add}}$ with $y \prec_G v$ and $|P(uy)| \geq 2$. Safeness is guaranteed by a reduction rule that is symmetric to Lemma 3. We provide the formal statement in the full version [13, Lemma 21]. Each new edge e now either has $|P(e)| = 1$ or is of the form ux' with $v \prec_G x'$ or vy' with $y' \prec u$; see Fig. 5b for an example. Observe that these edges ux' and vy' are not in a nesting relation. We assign edges with a single admissible page to the respective page. Afterwards, we assign the remaining new edges e to a page $\sigma(e) = p$ such that for all new edges $e' \in ⋒(e)$ with $|P(e')| = 1$ we have $\sigma(e') \neq p$. If this is not possible, the instance has no solution due to Lemma 3 and [13, Lemma 21]. Otherwise, the same lemmas in concert with Observation 4 allow us to complete the obtained page assignment to a solution. Overall, this takes $\mathcal{O}(m_{\text{add}} \cdot |\mathcal{I}|)$ time per branch. □

7 Concluding Remarks

This paper provides a detailed analysis of the (parameterized) complexity of the problem of extending queue layouts. While we can draw for some results parallels to the related extension problem for stack layouts, our investigation also uncovers surprising differences between the two problems.

We see generalizing Theorem 6 to k (high-degree) missing vertices as the main challenge for future work. While the equivalence between queue layouts and colorings of permutation graphs immediately yields membership in XP when parameterized by n_{add} and ℓ, Theorem 6 suggests that we can maintain membership in XP even after dropping the number of pages ℓ from our parametrization. Recall that we required H to be a subgraph of G. A generalization of the studied problem relaxes this requirement, allowing the spine order for some vertices and the page assignment for some edges of G to be specified as part of the input. We see this as an interesting direction for future investigations. Finally, a natural next step is to extend our results to other linear layout drawing styles.

Acknowledgments. All authors acknowledge support from the Vienna Science and Technology Fund (WWTF) [10.47379/ICT22029]. Robert Ganian and Thomas Depian furthermore acknowledge support from the Austrian Science Fund (FWF) [10.55776/Y1329]. We want to thank P. Bachmann, T. Brand, G. Gutowski, M. Pfretzschner, I. Rutter, P. Stumpf, S. Wolf, and J. Zink who at the HOMONOLO 2024 workshop helped us uncover and independently prove the relationship to permutation graphs used in Sect. 3. We also want to thank the organizers for making this enjoyable event possible.

References

1. Angelini, P., et al.: Testing planarity of partially embedded graphs. ACM Trans. Algorithms **11**(4), 32:1–32:42 (2015). https://doi.org/10.1145/2629341
2. Aspvall, B., Plass, M.F., Tarjan, R.E.: A linear-time algorithm for testing the truth of certain quantified Boolean formulas. Inf. Process. Lett. **8**(3), 121–123 (1979). https://doi.org/10.1016/0020-0190(79)90002-4
3. Bekos, M.A.: Planar graphs of bounded degree have bounded queue number. SIAM J. Comput. **48**(5), 1487–1502 (2019). https://doi.org/10.1137/19M125340X
4. Bhore, S., Da Lozzo, G., Montecchiani, F., Nöllenburg, M.: On the upward book thickness problem: combinatorial and complexity results. In: Purchase, H.C., Rutter, I. (eds.) Proceedings of the 29th International Symposium on Graph Drawing and Network Visualization (GD'21). Lecture Notes in Computer Science, vol. 12868, pp. 242–256. Springer (2021). https://doi.org/10.1007/978-3-030-92931-2_18
5. Bhore, S., Da Lozzo, G., Montecchiani, F., Nöllenburg, M.: On the upward book thickness problem: combinatorial and complexity results. Eur. J. Comb. **110**, 103662 (2023). https://doi.org/10.1016/j.ejc.2022.103662
6. Bhore, S., Ganian, R., Khazaliya, L., Montecchiani, F., Nöllenburg, M.: Extending orthogonal planar graph drawings is fixed-parameter tractable. In: Chambers, E.W., Gudmundsson, J. (eds.) Proceedings of the 39th International Symposium on Computational Geometry (SoCG'23). LIPIcs, vol. 258, pp. 18:1–18:16. Schloss Dagstuhl - Leibniz-Zentrum für Informatik (2023). https://doi.org/10.4230/LIPICS.SOCG.2023.18
7. Bhore, S., Ganian, R., Montecchiani, F., Nöllenburg, M.: Parameterized algorithms for queue layouts. J. Graph Algorithms Appl. **26**(3), 335–352 (2022). https://doi.org/10.7155/JGAA.00597
8. Bhore, S., Ganian, R., Montecchiani, F., Nöllenburg, M.: Parameterized Algorithms for Queue Layouts. In: Auber, D., Valtr, P. (eds.) Proceedings of the 28th International Symposium on Graph Drawing and Network Visualization (GD'20). Lecture Notes in Computer Science, vol. 12590, pp. 40–54. Springer (2020). https://doi.org/10.1007/978-3-030-68766-3_4
9. Bilski, T.: Embedding graphs in books: a survey. IEE Proc. E (Comput. Digital Techniques) **139**(2), 134 (1992). https://doi.org/10.1049/ip-e.1992.0021
10. Chung, F.R.K., Leighton, F.T., Rosenberg, A.L.: Embedding graphs in books: a layout problem with applications to VLSI design. SIAM J. Algebraic Discret. Methods **8**(1), 33–58 (1987). https://doi.org/10.1137/0608002
11. Parameterized Algorithms. Springer, Cham (2015). https://doi.org/10.1007/978-3-319-21275-3
12. Depian, T., Fink, S.D., Ganian, R., Nöllenburg, M.: The Parameterized Complexity Of Extending Stack Layouts. In: Felsner, S., Klein, K. (eds.) Proceedings of the 32nd International Symposium on Graph Drawing and Network Visualization (GD'24). LIPIcs, vol. 320, pp. 12:1–12:17. Schloss Dagstuhl – Leibniz-Zentrum für Informatik (2024). https://doi.org/10.4230/LIPICS.GD.2024.12
13. Depian, T., Fink, S.D., Ganian, R., Nöllenburg, M.: The Peculiarities of extending queue layouts. https://doi.org/10.48550/ARXIV.2506.05156
14. Depian, T., Fink, S.D., Firbas, A., Ganian, R., Nöllenburg, M.: Pathways to tractability for geometric thickness. In: Kralovic, R., Kårková, V. (eds.) Proceedings of the 50th Conference on Current Trends in Theory and Practice of Computer Science (SOFSEM'25). Lecture Notes in Computer Science, vol. 15538, pp. 209–224. Springer (2025). https://doi.org/10.1007/978-3-031-82670-2_16

15. Diestel, R.: Graph Theory, 4th Edition, Graduate Texts in Mathematics, vol. 173. Springer (2012)
16. Dujmovic, V., Eppstein, D., Hickingbotham, R., Morin, P., Wood, D.R.: Stack-Number is not bounded by queue-number. Combinatorica **42**(2), 151–164 (2022). https://doi.org/10.1007/S00493-021-4585-7
17. Dujmovic, V., et al.: Planar graphs have bounded queue-number. J. ACM **67**(4), 22:1–22:38 (2020). https://doi.org/10.1145/3385731
18. Dujmović, V., Wood, D.R.: On Linear layouts of graphs. Discrete Math. Theoretical Comput. Sci. **6**(2) (2004). https://doi.org/10.46298/dmtcs.317
19. Dushnik, B., Miller, E.W.: Partially ordered sets. Am. J. Math. **63**(3), 600 (1941). https://doi.org/10.2307/2371374
20. Eiben, E., Ganian, R., Hamm, T., Klute, F., Nöllenburg, M.: Extending partial 1-planar drawings. In: Czumaj, A., Dawar, A., Merelli, E. (eds.) Proceedings of the 47th International Colloquium on Automata, Languages and Programming (ICALP'20). LIPIcs, vol. 168, pp. 43:1–43:19. Schloss Dagstuhl - Leibniz-Zentrum für Informatik (2020). https://doi.org/10.4230/LIPICS.ICALP.2020.43
21. Eiben, E., Ganian, R., Hamm, T., Klute, F., Nöllenburg, M.: Extending Nearly Complete 1-Planar Drawings in Polynomial Time. In: Esparza, J., Král', D. (eds.) Proceedings of the 45th Mathematical Foundations of Computer Science (MFCS'20). LIPIcs, vol. 170, pp. 31:1–31:16. Schloss Dagstuhl – Leibniz-Zentrum für Informatik (2020). https://doi.org/10.4230/LIPICS.MFCS.2020.31
22. Enright, J., Stewart, L., Tardos, G.: On list coloring and list homomorphism of permutation and interval graphs. SIAM J. Discret. Math. **28**(4), 1675–1685 (2014). https://doi.org/10.1137/13090465x
23. Ganian, R., Hamm, T., Klute, F., Parada, I., Vogtenhuber, B.: Crossing-Optimal extension of simple drawings. In: Bansal, N., Merelli, E., Worrell, J. (eds.) Proceedings of the 48th International Colloquium on Automata, Languages and Programming (ICALP'21). LIPIcs, vol. 198, pp. 72:1–72:17. Schloss Dagstuhl – Leibniz-Zentrum für Informatik (2021). https://doi.org/10.4230/LIPICS.ICALP.2021.72
24. Heath, L.S., Rosenberg, A.L.: Laying out graphs using queues. SIAM J. Comput. **21**(5), 927–958 (1992). https://doi.org/10.1137/0221055
25. Jansen, K.: The optimum cost chromatic partition problem. In: Bongiovanni, G.C., Bovet, D.P., Battista, G.D. (eds.) Proceedings of the 3rd International Conference on Algorithms and Complexity (CIAC'97). Lecture Notes in Computer Science, vol. 1203, pp. 25–36. Springer (1997). https://doi.org/10.1007/3-540-62592-5_58
26. Unger, W.: On the k-Colouring of circle-graphs. In: Cori, R., Wirsing, M. (eds.) Proceedings of the 5th Symposium on Theoretical Aspects of Computer Science (STACS'88). Lecture Notes in Computer Science, vol. 294, pp. 61–72. Springer (1988). https://doi.org/10.1007/BFB0035832
27. Wiechert, V.: On the queue-number of graphs with bounded tree-width. Electron. J. Comb. **24**(1), 1 (2017). https://doi.org/10.37236/6429

Reductions in Local Certification

Louis Esperet[1(✉)] and Sébastien Zeitoun[2]

[1] Univ. Grenoble Alpes, CNRS, Laboratoire G-SCOP, Grenoble, France
louis.esperet@grenoble-inp.fr
[2] CNRS, INSA Lyon, UCBL, LIRIS, UMR5205, 69622 Villeurbanne, France
sebastien.zeitoun@univ-lyon1.fr

Abstract. Local certification is a topic originating from distributed computing, where a prover tries to convince the vertices of a graph G that G satisfies some property $\mathcal{P}$. To convince the vertices, the prover gives a small piece of information, called certificate, to each vertex, and the vertices then decide whether the property $\mathcal{P}$ is satisfied by just looking at their certificate and the certificates of their neighbors. When studying a property $\mathcal{P}$ in the perspective of local certification, the aim is to find the optimal size of the certificates needed to certify $\mathcal{P}$, which can be viewed as a measure of the local complexity of $\mathcal{P}$.

A certification scheme is considered to be efficient if the size of the certificates is polylogarithmic in the number of vertices. While there have been a number of meta-theorems providing efficient certification schemes for general graph classes, the proofs of the lower bounds on the size of the certificates are usually very problem-dependent.

In this work, we introduce a notion of hardness reduction in local certification, and show that we can transfer a lower bound on the certificates for a property $\mathcal{P}$ to a lower bound for another property $\mathcal{P}'$, via a (local) hardness reduction from $\mathcal{P}$ to $\mathcal{P}'$. We then give a number of applications in which we obtain polynomial lower bounds for many classical properties using such reductions.

Keywords: Local certification · Proof labeling scheme · Hardness reduction

1 Introduction

The main focus of the paper is on *local certification*, which is a field at the intersection of distributed computing, graph theory, and complexity studying how the vertices of a graph G can be convinced that G satisfies some given property $\mathcal{P}$, even if each vertex of G only has a very local view of the graph (for instance, if it can only interact with its neighbors). More precisely, a *proof labeling scheme* for a property $\mathcal{P}$ in a graph G works as follows:

(1) a prover assigns (short) certificates to the vertices of the graph G;
(2) each vertex v checks its *local view* (its certificate and the certificates of its neighbors),

H. Fernau and P. Kindermann (Eds.): WG 2025, LNCS 16124, pp. 192–204, 2026.
https://doi.org/10.1007/978-3-032-11835-6_14

(3) based only on its local view, each vertex must decide whether G satisfies the given property $\mathcal{P}$.

Note that there is no requirement on the computational power of the prover and the vertices. If the graph indeed satisfies $\mathcal{P}$, all vertices must accept the instance. If the graph does not satisfy $\mathcal{P}$, then for any possible assignment of certificates, at least one vertex must reject the instance. For $n \in \mathbb{N}$, the *local complexity* of $\mathcal{P}$ is the minimum $f(n)$ such that for every n-vertex graph G, the property $\mathcal{P}$ can be locally certified in G with certificates of at most $f(n)$ bits per vertex. We can also study more generally the local complexity of a property $\mathcal{P}$ within a graph class $\mathcal{C}$, by which we mean that the minimum above is taken over all n-vertex graphs from $\mathcal{C}$. Note that every property has local complexity $O(n^2)$: it suffices to write the adjacency matrix of the graph in the certificate of each vertex. By doing so, each vertex knows the whole graph and is able to decide whether the property is satisfied or not [11]. On the other hand, a counting argument shows that some properties involving symmetries have local complexity $\Omega(n^2)$. The original work of Göös and Suomela [11] identified three natural ranges of local complexity for graph classes:

- $\Theta(1)$: this includes k-colorability for fixed k, and in particular bipartiteness;
- $\Theta(\log n)$: this includes non-bipartiteness, acyclicity, planarity [6] and more generally bounded genus [2,7];
- $\Theta(\text{poly}(n))$: this includes non-3-colorability [11], unit-distance graphs, unit-disk graphs, and 1-planar graphs [1], as well as problems involving symmetries [11].

Many of the results mentioned above (either lower bounds or upper bounds) are proved using rather ad-hoc arguments that do not seem to be applicable to a large number of graph properties at once. On the other hand, a number of positive results (upper bounds on the local complexity) have been obtained via so-called "meta-theorems", that is theorems applicable to a wide range of problems and graph classes. It was proved that any graph class of bounded treewidth which is expressible in monadic second order logic has local complexity $O(\log^2 n)$ [9]. This implies in particular that for any fixed k, the class of graphs of treewidth at most k has local complexity $O(\log^2 n)$. Similar meta-theorems involving graph properties expressible in first-order logic were proved for graphs of bounded treedepth [5] and graphs of bounded cliquewidth [8]. Another type of meta-theorem was obtained in [12]: any property that can be decided in linear time has an *interactive* proof labeling scheme with compact certificates (in this setting several rounds of interactions are allowed between the prover and the vertices). However, no general purpose tool has been designed for proving lower bounds on the local complexity of a large variety of problems.

Proof labeling schemes in graphs are considered to be efficient if their complexity is polylogarithmic in the size of the graph. The class of all properties whose local complexity is polylogarithmic in the size of the graph can be seen as a distributed analogue of the class NP. However, a major feature of NP has so far been missing in the toolbox of local certification: the possibility of doing hardness reductions. The purpose of this paper is to introduce a framework allowing to use classical NP-hardness reductions to prove automatic lower bounds on the local complexity of graph properties. Our techniques apply whenever the reduction is sufficiently local. In particular, using the result of Göös and Suomela [11] stating that non-3-colorability has local complexity $\Omega(n^2/\log n)$, we prove that a number of classical coNP-hard problems have polynomial local complexity. The table in Fig. 1 sums up our lower bounds.

Property $\mathcal{P}$	Class of graphs $\mathcal{C}$	Lower bound for $\mathcal{P}$ in $\mathcal{C}$	Theorem
Non-3-colorability	Planar graphs with $\Delta \leqslant 4$	$\Omega\left(\frac{\sqrt{n}}{\log^2 n}\right)$	3
Non-k-colorability ($k \geqslant 4$)	General graphs	$\Omega\left(\frac{n^2}{\log n}\right)$	4
Non-k-colorability ($k \geqslant 3$)	Graphs with $\Delta \leqslant k + \lceil\sqrt{k}\rceil - 1$	$\Omega\left(\frac{n}{\log n}\right)$	12
Domatic number at most 2	Graphs with $\Delta \leqslant 8$	$\Omega\left(\frac{n}{\log n}\right)$	6
Non-existence of a cubic subgraph	Graphs with $\Delta \leqslant 7$	$\Omega\left(\frac{n}{\log n}\right)$	7
Non-existence of a partition in k acyclic subgraphs ($k \geqslant 3$)	General graphs	$\Omega\left(\frac{n}{\log n}\right)$	8
Existence of a monochromatic triangle in every 2-edge-coloring	Graphs with $\Delta \leqslant 40$	$\Omega\left(\frac{n}{\log n}\right)$	9
Non-existence of a Hamiltonian cycle	Graphs with $\Delta \leqslant 4$	$\Omega\left(\frac{\sqrt{n}}{\log n}\right)$	10
Chromatic index equal to $\Delta + 1$	Cubic graphs	$\Omega\left(\frac{n}{\log n}\right)$	11

Fig. 1. Summary of our lower bounds. In this table, Δ denotes the maximum degree.

Note that most of the lower bounds in the table of Fig. 1 hold in a bounded-degree graph class (so in general graphs too). However, in a bounded-degree graph class, every property has local complexity $O(n \log n)$. Indeed, $O(n \log n)$ bits are sufficient to encode all the edges of a bounded-degree graph (using adjacency lists), and we can just write the description of the whole graph in the certificate of every vertex. Thus, most of the results above in bounded-degree classes are optimal (within a polylogarithmic factor). The two exceptions are the local complexity of non-3-colorability in planar graphs of maximum degree 4, and the local complexity of non-hamiltonicity in graphs of maximum degree 4. We believe that in both cases the local complexity should be linear.

As alluded to above, in the distributed world "efficient" usually means logarithmic or polylogarithmic. From this point of view, our results show that all these problems have very high local complexity. On the other hand, it turns out that the complement problems all have very low local complexity (either constant or logarithmic).

Organization of the Paper. We start with some preliminary definitions on graphs and local certification in Section 2. In Sect. 3, we mention the lower bound on the local complexity of non-3-colorability by Göös and Suomela [11], and deduce lowers bounds in the more specific cases of graphs with maximum degree at most 4, and planar graphs. These lower bounds will be crucial in the remainder of the paper, as all problems we consider will be reduced to one of these problems (directly or indirectly). In Sect. 4, we then present as a motivating example a local reduction between non-3-colorability and non-k-colorability, and discuss the main features we expect of a local reduction. This motivates the general definition of a local reduction between problems that we present in Section 5, together with a number of applications that give polynomial lower bounds for several coNP-hard problems. We conclude with some final remarks in Sect. 6.

2 Preliminaries

2.1 Graphs

In this paper, graphs are assumed to be simple, loopless, undirected, and connected. The *distance* between two vertices u and v in a graph G, denoted by $d_G(u, v)$ is the minimum length of a path between u and v (that is, its number of edges). The *neighborhood* of a vertex v in a graph G, denoted by $N_G(v)$ (or $N(v)$ if G is clear from the context), is the set of vertices at distance exactly 1 from v. The *closed neighborhood* of v, denoted by $N_G[v] := \{v\} \cup N_G(v)$, is the set of vertices at distance at most 1 from v. For a set S of vertices of G, we define $N_G[S] := \bigcup_{v \in S} N_G[v]$. Finally, for every $k \geqslant 1$, we define $N_G^k[v]$ as the set of vertices at distance at most k from v (in particular, we have $N_G^1[v] = N_G[v]$).

2.2 Local Certification

In all the n-vertex graphs G that we consider, the vertices of G are assumed to be assigned distinct (but otherwise arbitrary) identifiers $(\mathrm{id}(v))_{v \in V(G)}$ from a set $\{1, \ldots, \mathrm{poly}(n)\}$. This is a standard assumption in local certification, which is necessary if we want to be able to distinguish a graph from its universal cover (and thus prove any meaningful result). When we refer to a subgraph H of a graph G, we implicitly refer to the corresponding labeled subgraph of G. Note that the identifiers of each of the vertices of G can be stored using $O(\log n)$ bits.

We follow the terminology introduced by Göös and Suomela [11].

Proofs and Provers. A *proof* for a graph G is a function $P : V(G) \to \{0, 1\}^*$ (as G is a labeled graph, the proof P is allowed to depend on the identifiers of the vertices of G). The binary words $P(v)$ are called *certificates*. The *size* of P is the maximum size of a certificate $P(v)$, for $v \in V(G)$. A *prover* for a graph class $\mathcal{G}$ is a function that maps every $G \in \mathcal{G}$ to a proof for G.

Local Verifiers. A *verifier* $\mathcal{A}$ is a function that takes a graph G, a proof P for G, and a vertex $v \in V(G)$ as inputs, and outputs an element of $\{0, 1\}$. We say that v *accepts* the instance if $\mathcal{A}(G, P, v) = 1$ and that v *rejects* the instance if $\mathcal{A}(G, P, v) = 0$. In the definition we do not require that $\mathcal{A}$ is efficiently computable or even decidable (but in practice it will almost always be the case).

Consider a graph G, a proof P for G, and a vertex $v \in V(G)$. We denote by $G[v]$ the subgraph of G induced by $N[v]$, the closed neighborhood of v, and similarly we denote by $P[v]$ the restriction of P to $N[v]$.

A verifier $\mathcal{A}$ is *local* if for any $v \in G$, the output of v only depends on its identifier and $P[v]$.

We note that there is a small difference here with the original setting of Göös and Suomela [11], where the verifier was allowed to depend on $G[v]$ as well (so vertices have the knowledge of the subgraph induced by their neighborhood). Lower bounds in the setting of Göös and Suomela imply lower bounds in our setting, but the converse does not hold in general. However, in the case of bounded degree graphs (which is the setting of most of our results) the two models are very close, as encoding the subgraphs induced by each neighborhood can be done at an extra cost of $O(\log n)$ bits per vertex.

Since the lower bounds of Göös and Suomela [11] on non-3-colorability hold in their stronger setting, all the lower bounds we deduce from it in this paper also hold in their setting. But since our main goal was to design a general framework for reductions in local certification, we have preferred to work with the weaker setting, which applies more broadly (see [4] for an overview of the different models of local certification).

A second difference between the setting of this paper and that of Göös and Suomela [11] is that vertices only look at distance at their neighbors 1 before accepting or rejecting the instance, while in [11] they can look at their neighborhood at distance r, for some constant r called the *local horizon*. The lower bounds of Göös and Suomela hold in this more general setting, and so do the result in our paper (most of the time, without any modification to the proofs). However, there are quite a few places where having an arbitrary local horizon r adds a layer of technicality on top of some already quite technical statements and proofs. So, for the sake of readability we have chosen to only state and prove our results in this paper in the case where the local horizon $r = 1$, which already contains most of the difficulty of the problems we consider.

Proof Labeling Schemes. A *proof labeling scheme* for a graph class $\mathcal{G}$ is a prover-verifier pair $(\mathcal{P}, \mathcal{A})$ where $\mathcal{A}$ is local, with the following properties.

Completeness: If $G \in \mathcal{G}$, then $P := \mathcal{P}(G)$ is a proof for G such that for any vertex $v \in V(G)$, $\mathcal{A}(G, P, v) = 1$.

Soundness: If $G \notin \mathcal{G}$, then for every proof P' for G, there exists a vertex $v \in V(G)$ such that $\mathcal{A}(G, P', v) = 0$.

In other words, upon looking at its closed neighborhood (labeled by the identifiers and certificates), the local verifier of each vertex of a graph $G \in \mathcal{G}$

accepts the instance, while if $G \notin \mathcal{G}$, for every possible choice of certificates, the local verifier of at least one vertex rejects the instance.

The *complexity* of the proof labeling scheme is the maximum size of a proof $P = \mathcal{P}(G)$ for an n-vertex graph $G \in \mathcal{G}$, and the *local complexity* of $\mathcal{G}$ is the minimum complexity of a proof labeling scheme for $\mathcal{G}$. If we say that the complexity is $O(f(n))$, for some function f, the $O(\cdot)$ notation refers to $n \to \infty$. See [4,11] for more details on proof labeling schemes and local certification in general.

We implicitly identify every graph property $\mathcal{P}$, for instance 3-colorability, or non-3-colorability, with the class of all graphs that satisfy $\mathcal{P}$.

3 Non-3-Colorability

The proofs omitted in this section are available in the full version of the paper [3]. We will use the following result, proved by Göös and Suomela [11]:

Theorem 1 ([11]). *Non-3-colorability has local complexity $\Omega(n^2/\log n)$.*

Our first result is the following variant of Theorem 1, which will be crucial in our subsequent lower bounds.

Theorem 2. *Non-3-colorability of graphs of maximum degree 4 has local complexity $\Omega(n/\log n)$.*

Using a classical uncrossing technique we also obtain the following variant for planar graphs (of bounded degree).

Theorem 3. *Non-3-colorability of planar graphs of maximum degree 4 has local complexity $\Omega(\sqrt{n}/\log^2 n)$.*

4 Non-k-Colorability

In this section we explain how to obtain a $\Omega(n^2/\log n)$ lower bound on the local complexity of non-k-colorability, for every $k > 3$, using the results of the previous section. This will also be a direct consequence of our main reduction result, Theorem 5, but since the statement of this theorem is quite technical we prefer to illustrate the main ideas of the proof with this simple example before actually stating the theorem.

So let us prove that at least $\Omega(n^2/\log n)$ bits are necessary to locally certify non-k-colorability for every $k \geqslant 4$. Assume that there is a proof labeling scheme of complexity $f(n)$ for this problem, and consider a graph G for which we wish to certify that G is not 3-colorable. The main idea will be to simulate locally the construction of a graph G', so that G' is k-colorable if and only if G is 3-colorable, use the local certificates of non-k-colorability of G' and assign them to the vertices of G, in such a way that G can also simulate the verification of the certificates of G'.

First Attempt. We start by defining the graph G' as the graph obtained from G by adding $k-3$ universal vertices (such a graph G' is k-colorable if and only if G is 3-colorable).

Given a proof labeling scheme for non-k-colorability in G', our goal is to assign to each vertex $u \in V(G)$ the list of certificates of a subset C_u of vertices of G', together with their identifiers, in such a way that u can simulate the verification procedure of a subset V_u of vertices of G'. In this case, our choice for C_u and V_u is the following: for any $u \in V(G)$, we set $C_u = V(G')$ (that is, we assign to each vertex of G the list of the certificates of all the vertices of G' together with their identifier), and V_u consists of u together with the $k-3$ universal vertices of G'. This allows each vertex $u \in V(G)$ to simulate the verification process of the corresponding vertex in G', since its local view in G' consists of its neighbors in G, together with the universal vertices. Each vertex $u \in V(G)$ also simulates the verification of the $k-3$ universal vertices (whose local view is the whole graph). Now each vertex $u \in V(G)$ accepts the instance if and only if it would have accepted the instance in G' and each of the universal vertices would also have accepted the instance in G'. The size of the certificates is at most n times the size of the certificates for non-k-colorability, so by Theorem 1 this shows that non-k-colorability requires certificates of size $\frac{1}{n} \cdot \Omega(n^2/\log n) = \Omega(n/\log n)$.

The resulting bound of $\Omega(n/\log n)$ is much weaker than the bound of $\Omega(n^2/\log n)$ for non-3-colorability. The main reason is that adding universal vertices is very far from being a local operation, and this should be avoided as much as possible. Let us now consider a second (more local) way of defining the graph G'.

Second Attempt. The new graph G' is obtained from the disjoint union of two graphs G_0 and G_1 defined as follows. The graph G_0 is an isomorphic copy of G. The graph G_1 is obtained from the graph G by replacing each vertex u by a clique of size k with vertices numbered from 1 to k. We denote this clique by K_u, and for any $1 \leqslant i \leqslant k$, the vertex numbered i in K_u is denoted by $K_u[i]$. Finally, each edge uv of G is replaced by an antimatching between K_u and K_v in G_1 (that is, for any $i \neq j$ we add an edge between $K_u[i]$ and $K_v[j]$ in G_1). The edges between G_0 and G_1 are the following: for each $u \in V(G_0)$, we put an edge between u and $K_u[i]$ for every $i \in \{4, \ldots, k\}$. We can define an identifier for each vertex of G' describing whether it belongs to G_0, or to G_1 (and its number from $\{1, \ldots, k\}$ in the latter the case), together with the identifier of the corresponding vertex in G. All this information can be encoded on $O(\log n)$ bits.

It is straightforward to show that G is 3-colorable if and only if G' is k-colorable: indeed, the antimatchings enforce that, in any k-coloring of G', for each $i \in \{1, \ldots, k\}$ the vertices from $\{K_v[i] : v \in V(G)\}$ all have the same color, and thus there are only three possible remaining colors for the vertices of G_0, which is an isomorphic copy of G.

For every $u \in V(G)$, we define $V_u := C_u$ as being the union of the copy of u in G_0 and the clique K_u in G_1 (as above, C_u is the list of vertices of G' whose certificate will be given to u and V_u is the set of vertices of G' whose verification process will be simulated by u). Note that these lists of certificates are enough for each vertex u of $V(G)$ to simulate the verification process of V_u, as for any neighbor w of a vertex $v \in V_u$ in G', the certificate of w in G' will be contained in the certificate assigned to some neighbor of u in G.

Finally, let us compute the size of certificates in the resulting proof labeling scheme for non-3-colorability, assuming the local complexity of non-k-colorability is $f(n)$. If G has n vertices, then G' has $(k+1)n$ vertices. So the size of the certificate assigned to each vertex of G is $(k+1)\big(O(\log n) + f\big((k+1)n\big)\big)$. As k is constant, it follows from Theorem 1 that non-k-colorability requires certificates of size $\Omega(n^2/\log n)$, as desired. Note also that, if G has maximum degree 4, then G' has maximum degree $5k-4$. Thus, it follows from Theorem 2 that non-k-colorability requires certificates of size $\Omega(n/\log n)$ in graphs of maximum degree $5k-4$.

We have thus proved the following extension of Theorems 1 and 2.

Theorem 4. *For any fixed $k \geqslant 3$, non-k-colorability has local complexity at least $\Omega(n^2/\log n)$ in general graphs, and at least $\Omega(n/\log n)$ in graphs of maximum degree $5k-4$.*

Remark 1. In Theorem 12, we will prove that non-k-colorability has local complexity at least $\Omega(n/\log n)$ even in graphs of maximum degree $k + \lceil\sqrt{k}\rceil - 1$.

In the next section, we identify the key components of the proof above and explain how it can be used to define a general class of local reductions between graph problems. In particular we formalize the necessary and sufficient conditions on the sets V_u and C_u allowing the local simulation of the verification process, and show how the relative sizes of these two sets influence the relation between the local complexities of the two problems.

5 Reductions from $\mathcal{P}$ to $\mathcal{P}'$

Let $\mathcal{C}$, $\mathcal{C}'$ be two graph classes, and $\mathcal{P}$, $\mathcal{P}'$ be two graph properties on $\mathcal{C}$, $\mathcal{C}'$ respectively. Let $\alpha, \beta : \mathbb{N} \to \mathbb{N}$, with $\beta(n) \geqslant n$ for every integer n.

We say that there exists a *local reduction* from $\mathcal{P}$ to $\mathcal{P}'$ with local expansion α and global expansion β if there exists a function $f_{\mathcal{P},\mathcal{P}'}$ which takes as input a graph $G \in \mathcal{C}$ with n vertices and unique identifiers in $\{1, \ldots, n\}$[1], and outputs a graph (with unique identifiers) $G' = f_{\mathcal{P},\mathcal{P}'}(G) \in \mathcal{C}'$ having at most $\beta(n)$ vertices, such that the following properties are satisfied for all G:

(R1) the identifiers of G' are written on $O(\log n)$ bits,
(R2) G satisfies $\mathcal{P}$ if and only if G' satisfies $\mathcal{P}'$,

[1] We emphasize that the range of identifiers is more restricted than in the definition of local certification, where they were assumed to be in $\{1, \ldots, n^c\}$, for some $c \geqslant 1$. This is just to make the applications of our framework easier (but we will show that we do not lose any level of generality).

(R3) for every vertex u of G, there exist two sets $C_u, V_u \subseteq V(G')$, such that, for every $u \in V(G)$, the following properties are satisfied:
  (a) $V_u \subseteq C_u$,
  (b) $\bigcup_{v \in V(G)} V_v = V(G')$,
  (c) $|C_u| \leqslant \alpha(n)$,
  (d) for every $t \in V(G')$, the subgraph of G induced by $\{v \in V(G) \mid t \in C_v\}$ is connected,
  (e) for every $t \in V_u$, $N_{G'}[t]$ is included in $\bigcup_{v \in N_G[u]} C_v$,
  (f) V_u and its neighborhood (resp. C_u) only depend on the identifiers of the vertices in $N_G[u]$. In other words: if H is another n-vertex graph from $\mathcal{C}$ with unique identifiers in $\{1, \dots, n\}$, and if the subgraph with identifiers formed by u and its adjacent edges is the same in G and H, then the sets V_u and the subgraphs with identifiers formed by the vertices in V_u and their adjacent edges in G' and $H' = f_{\mathcal{P},\mathcal{P}'}(H)$ are the same (resp. the sets C_u are the same in G' and H').

Let us elaborate on the different parts of the definition of a local reduction, in view of the results from the previous section. Condition (R1) says that there is no significant blow-up in the size of G' compared to that of G (G' has size polynomial in that of G). It also states implicitly that there is a way to describe the identifiers of G' as a function of the identifiers of G. Condition (R2) is the underlying principle of a reduction. In condition (R3), the set V_u again stands for the subsets of vertices of G' whose verification in G' will be simulated by the vertex u of G, and C_u corresponds to the set of vertices of G' whose certificates will be given to u. It is reasonable to require that u is given the certificates of the vertices it is supposed to simulate (that is $V_u \subseteq C_u$, according to (R3a)). It is not strictly necessary, as u could also find the certificates of some vertices of V_u in the certificates assigned to its neighbors in G, but assuming $V_u \subseteq C_u$ makes some of our arguments simpler. Condition (R3b) simply says that every vertex of G' has to be simulated by at least one vertex of G. Condition (R3d) is necessary to make sure that if the certificate of t in G' is given to several vertices of G, then all these vertices are given the same certificate. Condition (R3e) is simply saying that in order to simulate the verification process of every vertex of V_u in G', u has all the required certificates in its neighborhood in G. Finally, condition (R3f) makes sure u can reconstruct the local views of all vertices of V_u in G', only based on its local view in G.

The next result shows that whenever we have a local reduction from a problem $\mathcal{P}$ to a problem $\mathcal{P}'$, any local certification scheme for $\mathcal{P}'$ can be turned into a local certification scheme for $\mathcal{P}$.

Theorem 5. *Assume that:*

- *there exists a certification scheme for $\mathcal{P}'$ with certificates of size $s(n)$, where $s : \mathbb{N} \to \mathbb{N}$ is an increasing function satisfying $s(n) = \Omega(\log n)$, and*
- *there exists a local reduction from $\mathcal{P}$ to $\mathcal{P}'$ with local expansion α and global expansion β.*

Then there exists a certification scheme for $\mathcal{P}$ with certificates of size $O(\alpha(n) \cdot s(\beta(n)))$.

We obtain the following immediate corollary (which is the result that we will use in practice to prove lower bounds).

Corollary 1. *Let $\mathcal{P}$ be a graph property (on some graph class $\mathcal{C}$) and let $k \geqslant 3$.*

1. *If there exists a local reduction from k-colorability (in general graphs) to $\mathcal{P}$ with local expansion $O(n^{\delta})$ for some $0 \leqslant \delta < 2$, and global expansion $O(n^{\gamma})$, then $\overline{\mathcal{P}}$ has local complexity $\Omega\big(n^{(2-\delta)/\gamma}/\log n\big)$. In particular, since $\gamma \leqslant \delta+1$, $\overline{\mathcal{P}}$ has local complexity $\Omega\big(n^{\frac{3}{\delta+1}-1}/\log n\big)$.*
2. *If there exists a local reduction from k-colorability in graphs of maximum degree $k + \lceil\sqrt{k}\rceil - 1$ to $\mathcal{P}$ with local expansion $O(n^{\delta})$ for some $0 \leqslant \delta < 1$, and global expansion $O(n^{\gamma})$, then $\overline{\mathcal{P}}$ has local complexity $\Omega\big(n^{(1-\delta)/\gamma}/\log n\big)$. In particular, since $\gamma \leqslant \delta + 1$, $\overline{\mathcal{P}}$ has local complexity $\Omega\big(n^{\frac{2}{\delta+1}-1}/\log n\big)$.*

Our local reduction framework, namely Theorem 5 and Corollary 1, can be combined with the lower bounds for non-3-colorability from Sect. 3, allowing us to provide optimal (or close to optimal) lower bounds on the local complexity of a number of coNP-hard problems.

A *dominating set* in a graph G is a subset S of vertices such that every vertex of G is in S or is adjacent to a vertex of S. The *domatic number* of a graph G is the maximum integer k such that the vertex set of G can be partitioned into k dominating sets.

Theorem 6. *The property of having domatic number at most two has local complexity $\Omega(n/\log n)$, even in graphs of maximum degree at most 8.*

Proof. Let $\mathcal{P}$ be the property of having domatic number at least 3. We will show that there exists a local reduction from 3-colorability in graphs of maximum degree 4 to $\mathcal{P}$ with local expansion $O(1)$, and the result will follow from Corollary 1. Let G be a graph of maximum degree at most 4, with n vertices having unique identifiers in $\{1, \ldots, n\}$. The construction of the graph $G' = f_{\text{3-col},\mathcal{P}}(G)$ is the following. For each edge $uv \in E(G)$, we add a new vertex w connected only to u and v. The identifier of each vertex in G' indicates if it is a vertex of G (and which one), or a vertex added for an edge of G (and which edge). This information can be encoded on $O(\log n)$ bits. For each vertex u of G, we define $C_u := V_u$ as the set containing u and all the vertices corresponding to an edge uv in G (see Fig. 2).

Let us show that G admits a proper 3-coloring if and only if the domatic number of G' is at least 3. Assume that G is 3-colorable. Then, we create a partition of the vertices of G' in the following way: we keep the partition into color classes, and for each new vertex w corresponding to an edge uv of G, we assign it to the part to which neither u nor v belongs to. Each part of the partition is a dominating set: indeed, each vertex of G' belongs to some triangle formed by two original adjacent vertices of G and the new vertex corresponding to the

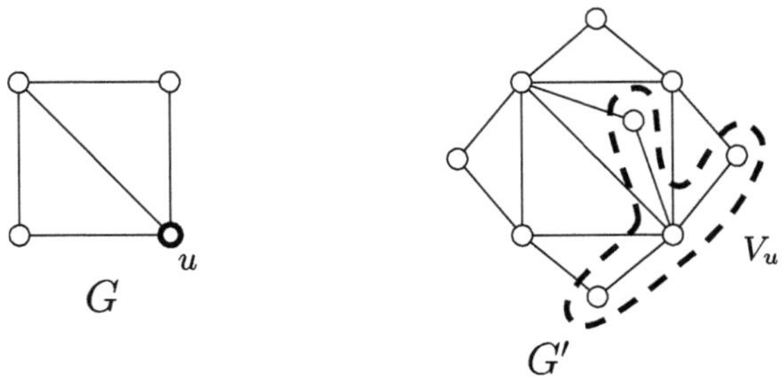

Fig. 2. Some graphs G and $G' = f_{3\text{-col},\mathcal{P}}(G)$.

edge, and by construction, each of the three parts of the partition dominates this triangle. Conversely, assume that the vertices of G' can be partitioned into at least 3 dominating sets. Then, we obtain a proper 3-coloring of G by restricting this partition to the original vertices of G: indeed, let uv be an edge of G, and w be the vertex of G' corresponding to this edge. Since there are only three vertices in $N[w]$ (which are u, v and w), these 3 vertices are in different parts of the partition and the partition consists of exactly 3 dominating sets (otherwise, some part does not dominate w). In particular, u and v are in different parts, and the corresponding 3-coloring of G is indeed proper.

Finally, let us show that the conditions of the definition of local reduction are satisfied. We have already seen that (R1) and (R2) both hold. Properties (R3a), (R3b), (R3e) and (R3f) simply follow from the definition of V_u and C_u. For (R3c), since G has maximum degree at most 4, for all $u \in V(G)$ we have $|C_u| \leqslant 5$ and thus $|C_u| = O(1)$. For (R3d), if $C_u \cap C_v \neq \emptyset$, then u and v are neighbors. So all conditions of a local reduction are verified, and Corollary 1 gives us a lower bound of order $\Omega(n/\log n)$ on the local complexity of the problem. Since G has maximum degree at most 4, $G' = f_{3\text{-col},\mathcal{P}}(G)$ has maximum degree at most 8. So this lower bound holds even for graphs of maximum degree at most 8. □

We prove the following results in the full version of the paper [3].

Theorem 7. *The local complexity of the property of not containing a cubic subgraph is $\Omega(n/\log n)$, even in graphs of maximum degree at most 7.*

Theorem 8. *For any fixed $k \geqslant 3$, the property that the vertex set of a given graph G cannot be partitioned into k induced forests has local complexity $\Omega\left(\frac{n}{\log n}\right)$.*

Theorem 9. *The property of having no 2-edge-coloring without monochromatic triangles has local complexity $\Omega(n/\log n)$, even in graphs of maximum degree 40.*

The problems mentioned above are described in the book by Garey and Johnson [10], but the reference for the NP-hardness proof for each of them is either an unpublished result or a personal communication. Thus, we present here some reductions that we found ourselves, which are maybe the same as the ones cited by Garey and Johnson [10], but to our knowledge, these reductions do not appear in any prior publication. Note that, for the problem of determining whether there exists a cubic subgraph, some reductions were already published (see e.g. [13,14]), but to our knowledge they are all different from the one we present here (we do not use these previous reductions here, because they are not sufficiently local).

While every coNP-hard problem has a polynomial reduction from non-3-colorability, these reductions are often non-local, and sometimes reductions are much easier to perform from problems such as 3-SAT (which is not a purely graph theoretic problem). We obtain a variant of Theorem 5 in the specific context of local reduction from 3-SAT. As an application, we prove the following two results in the full version of the paper [3].

Theorem 10. *The property of not having a Hamiltonian cycle has local complexity $\Omega(\sqrt{n}/\log n)$, even in graphs of maximum degree at most 4.*

Theorem 11. *The property of having chromatic index $\Delta+1$ (where Δ denotes the maximum degree) has local complexity $\Omega(n/\log n)$, even in cubic graphs.*

Finally we introduce another variant of our reduction framework which is specifically designed for bounded degree graphs, and captures problems that were not captured originally. We prove the following results in the full version of the paper [3].

Theorem 12. *Let $k \geqslant 3$. In graphs of maximum degree $k + \lceil\sqrt{k}\rceil - 1$, non-k-colorability has local complexity $\Omega(n/\log n)$.*

Theorem 13. *Let k be a sufficiently large integer. In graphs of maximum degree $k + \lceil\sqrt{k}\rceil - 3$, non-k-colorability has local complexity $O(\log n)$.*

6 Final Remarks

Since the polynomial lower bounds obtained in this paper are all based on NP-hardness reductions, all the applications presented so far relate to coNP-hard problems. It might be tempting to conjecture that all such problems have at least polynomial local complexity. In the full version of the paper [3], we observe that this is not the case, by giving a coNP-hard problem with logarithmic complexity.

We emphasize that our reduction framework also applies to problems that are not coNP-hard. Instead of starting from non-3-colorability, it suffices to start with a problem that has high local complexity, but which is easy globally (i.e. solvable in polynomial time). In the full version of the paper [3], we give examples of reductions that apply to problems that are not coNP-hard.

Acknowledgement. The authors are partially supported by the French ANR Project TWIN-WIDTH (ANR-21-CE48-0014-01), and ENEDISC (ANR-24-CE48-7768-01) by LabEx PERSYVAL-lab (ANR-11-LABX-0025).

Disclosure of Interests. The authors have no competing interests to declare that are relevant to the content of this article.

References

1. Defrain, O., Esperet, L., Lagoutte, A., Morin, P., Raymond, J.: Local certification of geometric graph classes. In: Královic, R., Kucera, A. (eds.) 49th International Symposium on Mathematical Foundations of Computer Science, MFCS 2024, 26–30 August 2024, Bratislava, Slovakia. LIPIcs, vol. 306, pp. 48:1–48:14. Schloss Dagstuhl - Leibniz-Zentrum für Informatik (2024). https://doi.org/10.4230/LIPICS.MFCS.2024.48
2. Esperet, L., Lévêque, B.: Local certification of graphs on surfaces. Theoret. Comput. Sci. **909**, 68–75 (2022). https://doi.org/10.1016/j.tcs.2022.01.023
3. Esperet, L., Zeitoun, S.: Reductions in local certification (2025). https://arxiv.org/abs/2502.01551
4. Feuilloley, L.: Introduction to local certification. Discret. Math. Theor. Comput. Sci. **23**(3) (2021). https://doi.org/10.46298/DMTCS.6280
5. Feuilloley, L., Bousquet, N., Pierron, T.: What can be certified compactly? Compact local certification of MSO properties in tree-like graphs. In: Milani, A., Woelfel, P. (eds.) PODC 2022: ACM Symposium on Principles of Distributed Computing, Salerno, Italy, 25–29 July 2022, pp. 131–140. ACM (2022). https://doi.org/10.1145/3519270.3538416
6. Feuilloley, L., Fraigniaud, P., Montealegre, P., Rapaport, I., Rémila, É., Todinca, I.: Compact distributed certification of planar graphs. Algorithmica **83**(7), 2215–2244 (2021). https://doi.org/10.1007/s00453-021-00823-w
7. Feuilloley, L., Fraigniaud, P., Montealegre, P., Rapaport, I., Rémila, É., Todinca, I.: Local certification of graphs with bounded genus. Discret. Appl. Math. **325**, 9–36 (2023). https://doi.org/10.1016/j.dam.2022.10.004
8. Fraigniaud, P., Mazoit, F., Montealegre, P., Rapaport, I., Todinca, I.: Distributed certification for classes of dense graphs. In: Oshman, R. (ed.) 37th International Symposium on Distributed Computing, DISC 2023, 10–12 October 2023, L'Aquila, Italy. LIPIcs, vol. 281, pp. 20:1–20:17. Schloss Dagstuhl - Leibniz-Zentrum für Informatik (2023). https://doi.org/10.4230/LIPICS.DISC.2023.20
9. Fraigniaud, P., Montealegre, P., Rapaport, I., Todinca, I.: A meta-theorem for distributed certification. In: Parter, M. (ed.) SIROCCO 2022. LNCS, vol. 13298, pp. 116–134. Springer, Cham (2022). https://doi.org/10.1007/978-3-031-09993-9_7
10. Garey, M.R., Johnson, D.S., Stockmeyer, L.J.: Some simplified NP-complete graph problems. Theor. Comput. Sci. **1**(3), 237–267 (1976). https://doi.org/10.1016/0304-3975(76)90059-1
11. Göös, M., Suomela, J.: Locally checkable proofs in distributed computing. Theory Comput. **12**(1), 1–33 (2016). https://doi.org/10.4086/toc.2016.v012a019
12. Naor, M., Parter, M., Yogev, E.: The power of distributed verifiers in interactive proofs. In: Chawla, S. (ed.) Proceedings of the 2020 ACM-SIAM Symposium on Discrete Algorithms, SODA 2020, Salt Lake City, UT, USA, 5–8 January 2020, pp. 1096–115. SIAM (2020). https://doi.org/10.1137/1.9781611975994.67
13. Stewart, I.A.: Deciding whether a planar graph has a cubic subgraph is NP-complete. Discret. Math. **126**(1–3), 349–357 (1994). https://doi.org/10.1016/0012-365X(94)90277-1
14. Stewart, I.A.: On locating cubic subgraphs in bounded-degree connected bipartite graphs. Discret. Math. **163**(1–3), 319–324 (1997). https://doi.org/10.1016/0012-365X(95)00324-P

Segment Intersection Representations, Level Planarity and Constrained Ordering Problems

Simon D. Fink[1], Matthias Pfretzschner[2(✉)], and Peter Stumpf[3,4]

[1] Algorithms and Complexity Group, Technische Universität Wien, Vienna, Austria
sfink@ac.tuwien.ac.at

[2] Faculty of Computer Science and Mathematics, University of Passau, Passau, Germany
pfretzschner@fim.uni-passau.de

[3] Department of Theoretical Computer Science, Faculty of Information Technology, Czech Technical University in Prague, Prague, Czech Republic
stumpf@kam.mff.cuni.cz

[4] Department of Applied Mathematics, Charles University Prague, Prague, Czech Republic

Abstract. In the Segment Intersection Graph Representation Problem, we want to represent the vertices of a graph as straight-line segments in the plane such that two segments cross if and only if there is an edge between the corresponding vertices. This problem is NP-hard (even $\exists\mathbb{R}$-complete [21]) in the general case [15] and remains so if we restrict the segments to be axis-aligned, i.e., horizontal and vertical [14]. A long standing open question for the latter variant is its complexity when the order of segments along one axis (say the vertical order of horizontal segments) is already given [14,16].

We resolve this question by giving efficient quartic-time solutions using two very different approaches that are interesting on their own. First, using a graph-drawing perspective, we relate the problem to a variant of the well-known Level Planarity problem, where vertices have to lie on pre-assigned horizontal levels. In our case, each level also carries consecutivity constraints on its vertices; this Level Planarity variant is known to have a quadratic solution if all edges connect adjacent levels.

Second, we use an entirely combinatorial approach and show that both problems can equivalently be formulated as a linear ordering problem subject to certain consecutivity constraints. While the complexity of such problems varies greatly, we show that in this case the constraints are well-structured in a way that allows a direct quadratic solution. Thus, we obtain three different-but-equivalent perspectives on this problem: the initial geometric one, one from planar graph drawing and a purely combinatorial one.

Keywords: segment intersection representation · grid intersection graph · level planarity

H. Fernau and P. Kindermann (Eds.): WG 2025, LNCS 16124, pp. 205–218, 2026.
https://doi.org/10.1007/978-3-032-11835-6_15

1 Introduction

In a seminal work from 1989, Kratochvíl and Matoušek initiated the systematic study of intersection graphs of segments forming the class SEG [15]. A graph is in SEG if its vertices can be represented by straight line segments in the plane that intersect if and only if the corresponding vertices are adjacent. They not only showed that testing membership in SEG is NP-hard, but, by showing that a representation of a graph in SEG might require coordinates with exponential precision [15], also laid the foundation for the study of a new complexity class denoted as $\exists\mathbb{R}$ [21]. They studied further variants of SEG (see Fig. 1 of [15]), in one direction further allowing the segments to have a bounded number of bends [17,18], and in the other direction restricting the segments to a bounded number of slopes [14]. In the latter setting, even the case where the graph is bipartite and only two directions, say horizontal and vertical, are allowed[1] turned out NP-complete [14].

Kratochvíl and Neŝetril observed that further prescribing the orders of horizontal and vertical segments, separately along both axes, finally turned the problem tractable [14,16]. They did so through two different equivalent reformulations of the problem. First note that, in the case without overlaps, we are given a bipartite graph $G = (H \cup V, E)$ together with orders σ_H and σ_V of H and V, respectively; see Fig. 1 for an example instance. In their first formulation, they used a 2-layer drawing of G with H and V respectively on two parallel lines, using σ_H as order on the one layer and σ_V on the other; see Fig. 1a and [14, Fig. 19]. They observed that an instance is positive if and only if it avoided a certain configuration of edges in this drawing, which they called "volkswagen". Alternatively, following an approach of Hartman, Newman and Ziv [11], the yes-instances can also be classified via their adjacency matrix, which shall not contain a configuration called "cross", see Fig. 1c and [11, Proposition 3.1].

With the complexity of recognizing segment intersection graphs thus rather well established, Kratochvíl and Neŝetril posed the following open problem [14, 16]: what if only one of the two orders (e.g. the vertical order of horizontal segments) is fixed? In the above equivalent formulations, this is equivalent to either allowing to permute the vertices along one of the two parallel lines, or to permute the matrix along one axis (e.g. permuting its columns). Formally, this can be stated as follows; see also Fig. 2.

Problem Semi-Fixed HV-Segment Intersection Graph Recognition
Input A bipartite graph $G = (H \cup V, E)$ and a linear order σ_H of H.
Question Is G the intersection graph of horizontal segments H and vertical segments V, where the horizontal segments have the fixed vertical order σ_H?

[1] In their work, this setting is denoted as PURE-2-DIR, while the not-necessarily bipartite variant that allows segment overlaps is called 2-DIR. Other authors also called the bipartite setting Grid Intersection Graph (GIG) [11] and showed its equivalence to having a boxicity of 2 [4].

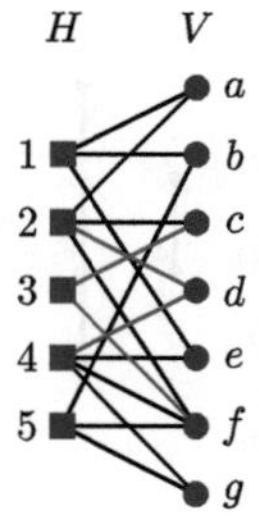

(a)

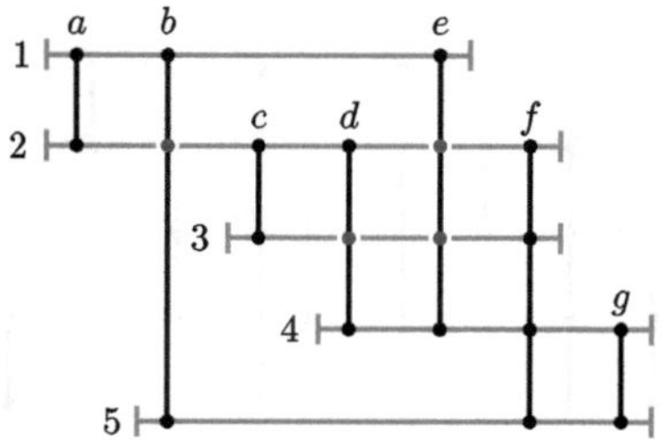

(b)

$$\begin{array}{c} \\ 1 \\ 2 \\ 3 \\ 4 \\ 5 \end{array} \begin{array}{c} a \ b \ c \ d \ e \ f \ g \\ \begin{pmatrix} 1 & 1 & & & 1 & & \\ 1 & & 1 & 1 & & 1 & \\ & & 1 & 0 & & 1 & \\ & & & 1 & 1 & 1 & 1 \\ & 1 & & & & 1 & 1 \end{pmatrix} \end{array}$$

(c)

Fig. 1. (a) A 2-layer drawing of a bipartite graph $G = (H \cup V, E)$. Highlighted in red is one "volkswagen" configuration present due to the orders used for H and V as well as the missing $3d$ edge. **(b)** An invalid HV-segment representation using the same vertex orders, with unwanted intersections highlighted in red. **(c)** The adjacency matrix of G, with an unwanted "cross" configuration centered on the absent $3d$ edge highlighted in orange. Empty cells correspond to a value of 0.

In this work, we will finally resolve this question after roughly 30 years, using two further entirely different, yet equivalent perspectives on the problem. Akin to the visual "volkswagen" classification, our first perspective will be formulated in terms of Graph Drawing using a problem called LEVEL PLANARITY. In Sect. 2, we will show that SEMI-FIXED HV-SEGMENT INTERSECTION GRAPH RECOGNITION (or SF-HV-SEG for short) can be reduced to a variant of LEVEL PLANARITY, thus also allowing us to use its known quadratic-time solution. In Sect. 3 we will conversely show that we can also reduce the problems in the other direction, showing their equivalence up to (in)efficiencies in representing problem instances. We will turn to our second perspective in Sect. 4, where we give an entirely combinatorial formulation of SF-HV-SEG. We again show equivalence for this formulation and also show it can directly be solved in quadratic time without the need for further reductions.

2 Semi-fixed HV-Segment Intersection Graphs

In this section, we show that SF-HV-SEG can be solved in quartic time using a reduction to some variant of the LEVEL PLANARITY problem. For LEVEL PLANARITY, we are given a graph together with prescribed y-coordinates for its vertices (also called *levels*), and seek a crossing-free drawing that respects these y-coordinates and also has all edges drawn y-monotone. This problem has been thoroughly researched in the field of Graph Drawing, e.g. in terms of efficient and also simple solutions [8–10,12,13] and also its relationship to other planar graph drawing problems [3,22]. We call the tuple (G, ℓ) consisting of a graph $G = (V, E)$ and a *leveling function* $\ell : V \to \mathbb{N}$ a *level graph*. In this paper, we will assume all edges to be directed from the lower to the higher level. A level graph (and especially its leveling function) is called *proper* if all its edges have their endpoints on adjacent levels, i.e., if we have $\ell(v) = \ell(u) + 1$ for all $uv \in E$.

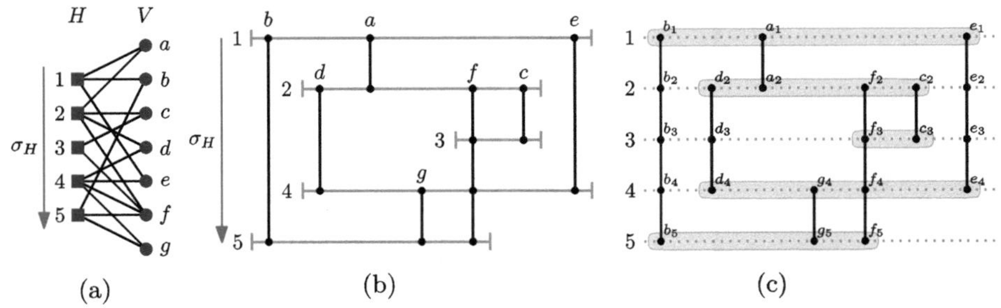

Fig. 2. **(a)** The SF-HV-SEG instance from Fig. 1a, only fixing the order σ_H. **(b)** A corresponding HV-segment representation with a different order of V than in Fig. 1b, showing that it is a yes-instance. **(c)** An equivalent instance of PROPER $\mathcal{T}$-LEVEL PLANARITY. The consecutive sets of every level are highlighted in orange.

Similarly, a path $v_1, v_2, \ldots, v_k$ is *level-monotone* if $\ell(v_{i+1}) = \ell(v_i) + 1$ for all $i \in [k-1]$ with $[k-1] :- \{1, \ldots, k-1\}$.

In the context of our work, we will use a further constrained variant that also allows us to require sets of vertices on the same level to be *consecutive*, i.e., ordered such that no other vertices of that level lie between them. For now, we require the sets of consecutive vertices to form a laminar family, i.e., any two sets are either disjoint or one contains the other. This problem is then known to have a quadratic-time solution if the instance is proper, while it is NP-complete on non-proper instances [2,3]. Formally, the former, tractable case of the problem is defined as follows.

Problem PROPER $\mathcal{T}$-LEVEL PLANARITY
Input A graph $G = (V, E)$, a proper leveling function $\ell : V \to \mathbb{N}$, and for each level i a laminar family T_i of consecutivity constraints on $\ell^{-1}(i)^2$.
Question Is there a planar drawing of G where each vertex $v \in V$ has y-coordinate $\ell(v)$, all edges are drawn y-monotone, and the horizontal order of vertices on each level i satisfies the consecutivity constraints of T_i?

To now reduce SF-HV-SEG to this problem, we create a level for each horizontal segment along the given order; see Fig. 2. For each vertical segment, we create an edge that spans from the first to the last level of horizontal segments that shall be intersected. We subdivide the edges to make the instance proper. The crucial insight to SF-HV-SEG is now that all vertical segments $\mathcal{I}(i)$ that intersect a given horizontal segment h_i need to be consecutive on the horizontal line corresponding to h_i. This means that any other vertical segment that spans, but does not intersect h_i, needs to cross level i before or after $\mathcal{I}(i)$, but not within. This can easily be modeled using the per-level consecutivity constraints of PROPER $\mathcal{T}$-LEVEL PLANARITY.

[3] Using the containment-relationship of laminar families allows us to equivalently represent T_i as a tree of linear size, thus obtaining the original formulation of the problem [3].

Theorem 1. *There exists a quadratic-time reduction from SF-HV-SEG to* PROPER $\mathcal{T}$-LEVEL PLANARITY *where the input graph is the disjoint union of level-monotone paths.*

Proof. Let $((H \cup V, E), \sigma_H)$ be an instance of SF-HV-SEG. For $k := |H|$, let $\ell : H \to [k]$ be the bijection that maps every horizontal segment in H to its corresponding position in the linear order defined by σ_H. We call $\ell(h)$ the *level* of $h \in H$.

For every vertical segment $v \in V$, we let the interval $L(v) = [\min\{\ell(h) : vh \in E\}, \max\{\ell(h) : vh \in E\}] \subseteq [k]$ denote the *lifespan* of v, that is the interval spanned by the lowest and highest levels of vertices adjacent to v. For every $i \in [k]$ and its corresponding horizontal segment $h = \ell^{-1}(i)$, the set $\mathcal{A}(i) := \{v \in V : i \in L(v)\}$ denotes the *active segments* of level i and the set $\mathcal{I}(i) = \{v \in V : vh \in E\} \subseteq \mathcal{A}(i)$ describes the *intersecting segments* of level i.

We construct an equivalent PROPER $\mathcal{T}$-LEVEL PLANARITY instance $(G, \gamma, (T_1, \ldots, T_k))$ with k levels as follows. The horizontal segments H correspond bijectively to the levels of (G, γ) as defined by ℓ. Every vertical segment $v \in V$ is represented by a level-monotone path in $G = (V, E)$ that contains a vertex on every level that v is active in, i.e., $V := \{v_i : i \in [k], v \in \mathcal{A}(i)\}$ with $\gamma(v_i) := i$ and $E := \{v_i v_{i+1} : i \in [k-1], v \in \mathcal{A}(i) \cap \mathcal{A}(i+1)\}$. Finally, on every level $i \in [k]$, we constrain the vertices corresponding to the intersecting segments $\mathcal{I}(i)$ to be consecutive on that level, that is, we set $T_i := \{\{v_i : v \in \mathcal{I}(i)\}\}$. In particular, observe that we obtain only a single consecutivity constraint per level.

Note that the resulting level graph (G, γ) is proper and, in the worst case, the number of vertices in G is quadratic in $|H \cup V|$. We show that $((H \cup V, E), \sigma_H)$ is a yes-instance of SF-HV-SEG if and only if $(G, \gamma, (T_1, \ldots, T_k))$ is a yes-instance of PROPER $\mathcal{T}$-LEVEL PLANARITY.

First assume that $(G, \gamma, (T_1, \ldots, T_k))$ is a yes-instance of PROPER $\mathcal{T}$-LEVEL PLANARITY and let Γ be a corresponding drawing of G. Since G is the disjoint union of monotone paths and Γ is planar, we can alter Γ such that each path is represented by a straight vertical line, without changing the order of vertices on any level. We obtain an HV-segment representation of $((H \cup V, E), \sigma_H)$ by replacing each of these paths with a vertical segment, and subsequently adding on every level i a horizontal segment that intersects the vertices of T_i. Since these vertices are consecutive on level i in Γ, the horizontal segment intersects precisely the vertical segments it is adjacent to in $(H \cup V, E)$.

Conversely, assume that $((H \cup V, E), \sigma_H)$ is a yes-instance of SF-HV-SEG and assume that a corresponding HV-segment representation Δ is given. We obtain a level-planar drawing Γ of (G, γ) by simply replacing every vertical segment with a path that contains a vertex on every level where it is active. Now assume that, for some level $i \in [k]$, the single consecutivity constraint of T_i is not satisfied in Γ. This implies that the intersecting segments $\mathcal{I}(i)$ of level i are not consecutive with respect to all active segments $\mathcal{A}(i)$ of level i in Δ. Therefore, a vertical segment $v \in \mathcal{A}(i) \setminus \mathcal{I}(i)$ lies between two vertical segments of $\mathcal{I}(i)$ on level i. Since the horizontal segment $\ell^{-1}(i)$ intersects exactly the segments in $\mathcal{I}(i)$ and $v \notin \mathcal{I}(i)$, i cannot be part of the lifespan of v. But this is a contradiction, since $v \in \mathcal{A}(i)$. Therefore, $(G, \gamma, (T_1, \dots, T_k))$ is a yes-instance of PROPER $\mathcal{T}$-LEVEL PLANARITY. □

Since PROPER $\mathcal{T}$-LEVEL PLANARITY admits a quadratic-time algorithm [2,3] and our reduction increases the size to be at most quadratic, we can thereby solve the initial instance in quartic time overall.

Corollary 1. *SF-HV-SEG can be solved in $O(n^4)$ time.*

3 Relation to Level-Planarity Variants

In this section, we show that SF-HV-SEG is quadratic-time equivalent to PROPER $\mathcal{T}$-LEVEL PLANARITY. Since SF-HV-SEG reduces to PROPER $\mathcal{T}$-LEVEL PLANARITY in quadratic time by Theorem 1, it remains to show the other direction. To this end, we first show that every instance of PROPER $\mathcal{T}$-LEVEL PLANARITY can be reduced in linear time to an equivalent instance where the graph is a matching. The edges of this matching then correspond to the vertical segments of the equivalent SF-HV-SEG instance, while each consecutivity constraint results in a horizontal segment. Since each level of the PROPER $\mathcal{T}$-LEVEL PLANARITY instance may have a linear number of consecutivity constraints, this yields a quadratic-time reduction.

Lemma 1. *Given an instance of PROPER $\mathcal{T}$-LEVEL PLANARITY, an equivalent instance in which the graph is a matching can be computed in linear time.*

Proof. We assume without loss of generality that the input graph contains no isolated vertices. Given an instance of PROPER $\mathcal{T}$-LEVEL PLANARITY, first "explode" every vertex v, i.e., replace v with $\deg(v)$ degree-1 vertices on the same level, adjacent to the original neighbors of v. Subsequently, add an additional consecutivity constraint to the level of v that ensures that the new degree-1 vertices that stem from v are consecutive. Due to this consecutivity, a corresponding drawing of the original graph can be obtained by simply merging the new degree-1 vertices back into a single vertex. Conversely, to obtain a drawing of the new instance, perform the explosions for each vertex and order the new degree-1 vertices on the same level according to the order of edges around the original vertex.

Observe that the resulting family of consecutivity constraints is still laminar and since every vertex now has degree 1, the graph is a matching. Moreover, the number of new vertices and the combined size of the new consecutivity constraints each correspond to the sum of vertex-degrees in the original instance. Therefore, the reduction is linear. □

Lemma 2. *There exists a quadratic-time reduction from PROPER $\mathcal{T}$-LEVEL PLANARITY to SF-HV-SEG.*

Proof. Due to Lemma 1, we can assume that the input graph is a matching. We first show that we can compute an equivalent instance of PROPER $\mathcal{T}$-LEVEL PLANARITY with only one consecutivity constraint per level. For each level i, let T_i denote the corresponding laminar family with $|T_i|$ consecutivity constraints. We split i into $|T_i|$ levels $i^1, \ldots, i^{|T_i|}$ and redistribute its constraints as follows. Replace each vertex v on ℓ by a path of $|T_i|$ vertices $v^1, \ldots, v^{|T_i|}$ on levels $i^1, \ldots, i^{|T_i|}$, respectively. Since the input graph is a matching, v has exactly one neighbor. If this neighbor lies on a lower level, we assign the corresponding edge to v^1, otherwise we assign it to $v^{|T_i|}$. Observe that due to these parallel paths, the vertices on all $|T_i|$ levels have the same order in a level planar drawing. This allows us to distribute T_i over the levels $i^1, \ldots, i^{|T_i|}$, leaving one constraint per level. As each level i of the input instance may have a linear number of constraints, this yields a quadratic-time reduction. Observe that the resulting instance is the disjoint union of level-monotone paths and each level has exactly one constraint.

Now consider an instance $I = (G, \ell, (T_1, \ldots, T_k))$ of PROPER $\mathcal{T}$-LEVEL PLANARITY where G consists of disjoint level-monotone paths and that has exactly one consecutivity constraint per level, i.e., $|T_i| = 1$ for all $i \in [k]$. We construct an equivalent instance $I' = ((H \cup V, E), \sigma_H)$ of SF-HV-SEG as follows. The (vertices represented by) vertical segments V correspond bijectively to the level-monotone paths of G, while the (vertices represented by) horizontal segments H correspond bijectively to the levels of I. The order σ_H orders the horizontal segments according to the order of their corresponding levels. For each level i, the horizontal segment corresponding to level i is adjacent to the vertical segments of V whose corresponding paths in I have a vertex in T_i; see Fig. 2.

Note that, as in the proof of Theorem 1, the level-monotone paths of I correspond bijectively to the vertical segments in I', while the one-per-level consecutivity constraints correspond to the horizontal segments. The equivalence of the two instances can thus be shown analogously as done in the proof of Theorem 1. □

This now allows us to reduce between SF-HV-SEG and PROPER $\mathcal{T}$-LEVEL PLANARITY with a quadratic factor that is mostly due to differences in how efficiently the two problems represent their input instances. That is, the latter one represents multiple consecutivity constraints per level more efficiently than the former, using the representation of a laminar family of sets as a tree. In contrast, it also on each level needs to make explicit those elements that shall lie outside of the consecutive subset, which is implicit in the former representation.

4 Relation to Constrained Ordering Problems

The reduction in Theorem 1 already hints that SF-HV-SEG is essentially a constrained ordering problem: we seek a linear order of the vertical segments, such that, on each of the fixed horizontal layers, the active segments satisfy a prescribed consecutivity constraint. Finding a linear order of a ground set X

subject to some constraints that require a subset A to appear consecutively, that is, uninterrupted with regards to some other set B, is indeed a common combinatorial problem. This problem is in general NP-complete for arbitrary A and B, due to a straight-forward reduction from the well-known BETWEENNESS problem [19]. In contrast it is, e.g., linear-time solvable if we restrict B to be the complement of A with regards to X (using solely the PQ-tree data structure described further below [7]). In this section, we want to show that SF-HV-SEG and PROPER $\mathcal{T}$-LEVEL PLANARITY can equivalently be seen as a relaxed variant of this latter setting, where all given (A, B) consecutivity constraints can be ordered such that every element of X appears in a consecutive subsequence of constraints.

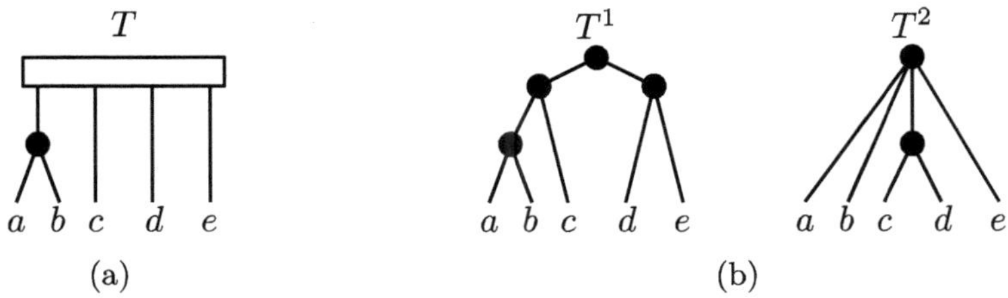

Fig. 3. **(a)** A PQ-tree T with $L(T) = \{a, b, c, d, e\}$ and consecutivity constraints $\{a, b\}, \{a, b, c\}, \{c, d\}$, and $\{d, e\}$. P-nodes and Q-nodes are represented by black circles and white rectangles, respectively. **(b)** Two PQ-trees T^1 and T^2 with the laminar consecutivity constraints $\{a, b\}, \{a, b, c\}, \{d, e\}$ and $\{c, d\}$, respectively.

Recall that a laminar family of consecutivity constraints can succinctly be represented as a tree with the constrained elements as leaves. If we now drop the restriction that the constraints form a laminar family, i.e., allow them to have arbitrary intersections, we can still represent them efficiently using a data structure called PQ-tree [7]. A *PQ-tree* T is a rooted ordered tree that represents linear orders of its leaves $L(T)$, subject to a set of constraints that require certain leaves to be consecutive; see Fig. 3a for an example. Each of its inner nodes is either a *P-node* or a *Q-node*, where the former allow their children to be permuted arbitrarily, while the latter fix the order of their children up to reversal. The linear orders of $L(T)$ that can be obtained in this way are the orders that are *represented* by T. Conversely, we say that a linear order σ of a set $X \supseteq L(T)$ *satisfies* T if σ restricted to $L(T)$ is represented by T. For a given (not necessarily laminar) family of consecutivity constraints on a ground set X, the PQ-tree T with $L(T) = X$ that represents exactly those orders that satisfy all constraints can be found in linear time [7]. If there exists no order of $L(T)$ that satisfies all consecutivity constraints, then T is the *null tree*. We now use this data structure to formulate the following constrained ordering problem and, in the later parts of this section, show it to be the combinatorial analogue of the problems from the previous sections.

Problem SEQUENTIAL PQ-ORDERING
Input A ground set X and a sequence $T_1, \ldots, T_k$ of PQ-trees with $X = L(T_1) \cup \cdots \cup L(T_k)$ such that, for every $x \in X$, the PQ-trees that contain x are consecutive in the sequence $T_1, \ldots, T_k$.
Question Is there a linear order of X that satisfies all PQ-trees $T_1, \ldots, T_k$?

We remark that, when we drop the requirement that each element of the ground set must appear in a consecutive subset of the PQ-tree sequence, the problem becomes NP-complete.[3] This can again easily be proven via reduction from the BETWEENNESS problem by representing each betweenness constraint individually by a PQ-tree consisting of a single Q-node. However, the following lemma shows that thanks to the additional requirement in SEQUENTIAL PQ-ORDERING, it suffices to ensure that common elements are ordered consistently for PQ-trees that are adjacent in the sequence.

Lemma 3. *Let $\mathcal{I} = (X, (T_1, \ldots, T_k))$ be an instance of* SEQUENTIAL PQ-ORDERING. *Then $\mathcal{I}$ is a yes-instance if and only if there exist linear orders $\sigma_1, \ldots, \sigma_k$ of $L(T_1), \ldots, L(T_k)$ such that*

1. *for every $i \in [k]$, σ_i satisfies T_i and*
2. *for every $i \in [k-1]$ and every $x, y \in L(T_i) \cap L(T_{i+1})$ it is $x \leq_{\sigma_i} y \Leftrightarrow x \leq_{\sigma_{i+1}} y$*

Proof. It is clear that both properties hold if $\mathcal{I}$ is a yes-instance. Conversely, assume there exist linear orders $\sigma_1, \ldots, \sigma_k$ of $L(T_1), \ldots, L(T_k)$ that satisfy both properties. We show that X admits a linear order σ that extends $\sigma_1, \ldots, \sigma_k$ by induction over the number of PQ-trees. The statement clearly holds for $k = 1$. Assume $k > 1$ and that the statement holds for $\mathcal{I}' = (X', (T_1, \ldots, T_{k-1}))$ with $X' = L(T_1) \cup \cdots \cup L(T_{k-1})$, i.e., there is a linear order σ' of X' that extends $\sigma_1, \ldots, \sigma_{k-1}$. We construct a linear order σ of X that extends $\sigma_1, \ldots, \sigma_k$ as follows. By Property 2, σ_{k-1} and σ_k (and thus σ' and σ_k) induce the same linear order τ on $X' \cap L(T_k) = L(T_{k-1}) \cap L(T_k)$. Consider $\sigma' \cup \sigma_k$ as the edges of a directed graph D on vertex set X, i.e., for two distinct elements $a, b \in X$, D contains the directed edge (a, b) if and only if $a \leq_{\sigma'} b$ or $a \leq_{\sigma_k} b$. Observe that D is acyclic since σ' and σ_k are linear orders that induce the same linear order τ on their common elements. Hence, we can choose any topological ordering of the graph as σ. By construction σ extends $\sigma_1, \ldots, \sigma_k$ and thus satisfies $T_1, \ldots, T_k$. This means that $\mathcal{I}$ is a yes-instance of SEQUENTIAL PQ-ORDERING. □

We now will use a known similar characterization for proper level graphs to establish the equivalence of SEQUENTIAL PQ-ORDERING and PROPER $\mathcal{T}$-LEVEL PLANARITY. It was initially given in [20] and states that a straight-line drawing is level planar if and only if the endpoints of every pair of edges between adjacent levels are ordered consistently.

[3] In a sense, this is analogous to dropping the "proper" part of PROPER $\mathcal{T}$-LEVEL PLANARITY, where the then-allowed long edges do not appear in the trees of levels they span.

Lemma 4 ([20]). *A proper level graph (G, ℓ) on k levels is level planar if and only if there exists linear orders $\sigma_1, \dots, \sigma_k$ of the vertices on individual levels $\ell^{-1}(1), \dots, \ell^{-1}(k)$, respectively, that satisfy the following property. For every pair $(a,b), (u,v)$ of disjoint edges with $i = \ell(a) = \ell(u) = \ell(b) - 1 = \ell(v) - 1$, we have $a <_{\sigma_i} u \Leftrightarrow b <_{\sigma_{i+1}} v$.*

Using Lemmas 3 and 4, we can now show that SEQUENTIAL PQ-ORDERING and PROPER $\mathcal{T}$-LEVEL PLANARITY are linear-time equivalent.

Lemma 5. *There exists a linear-time reduction from SEQUENTIAL PQ-ORDERING to PROPER $\mathcal{T}$-LEVEL PLANARITY.*

Proof. Recall that PROPER $\mathcal{T}$-LEVEL PLANARITY requires a laminar set of consecutivity constraints on each level. We therefore first show how we can translate each PQ-tree into an equivalent set of PQ-trees that represent laminar consecutivity constraints.

Each P-node corresponds to a constraint that requires consecutivity of the leaves in its subtree. Similarly, a Q-node can be formulated as multiple consecutivity constraints: for each pair of adjacent children of a Q-node, the union of the leaves of their subtrees must be consecutive [6]. This allows us to translate each PQ-tree T into two equivalent PQ-trees T^1 and T^2 that each represent laminar families of consecutivity constraints as follows. Assign all constraints that stem from P-nodes to T^1, and distribute the constraints that stem from pairs of children of Q-nodes alternatingly between T^1 and T^2; see Fig. 3b for an example. Alternating between the two trees ensures that two constraints in the same tree do not intersect if one is not contained in the other; T^1 and T^2 thus contain no Q-nodes. The PQ-trees T^1 and T^2 together contain precisely the consecutivity constraints of T and can be computed in linear time. Given an instance of SEQUENTIAL PQ-ORDERING, we can therefore compute in linear time an equivalent instance $\mathcal{I} = (X, (T_1, \dots, T_k))$ where each T_i represents a laminar set of consecutivity constraints.

We now construct an equivalent instance of PROPER $\mathcal{T}$-LEVEL PLANARITY as follows. We obtain the level graph $(G = (V, E), \ell)$ by introducing a level for each $i \in [k]$, which corresponds to the PQ-tree T_i. Each level i contains a vertex for every element of $L(T_i)$, i.e., $V := \{x_i : i \in [k], x \in L(T_i)\}$ and $\ell(x_i) = i$. Moreover, we connect vertices on adjacent levels that correspond to the same element, i.e., $E := \{x_i x_{i+1} : i \in [k-1], x \in L(T_i) \cap L(T_{i+1})\}$. Finally, we annotate each level i with the laminar family T_i' of consecutivity constraints that is obtained from T_i by replacing every leaf $x \in L(T_i)$ with the corresponding vertex x_i, resulting in the instance $\mathcal{I}' = (G, \ell, (T_1', \dots, T_k'))$.

For each level $i \in [k]$, the linear order of the vertices on level i corresponds to the linear order σ_i that a linear order σ of X induces for the vertices corresponding to $L(T_i)$. Note that T_i and T_i' enforce the same restrictions on these orders. Since the construction ensures that vertices on adjacent levels that correspond to the same element of X are adjacent in G, the vertices are ordered consistently between adjacent levels by Lemma 4, which corresponds to Property 2 of Lemma 3. Therefore, $\mathcal{I}$ and $\mathcal{I}'$ are equivalent. □

Lemma 6. *There exists a linear-time reduction from* PROPER $\mathcal{T}$-LEVEL PLANARITY *to* SEQUENTIAL PQ-ORDERING.

Proof. Let $\mathcal{I} = (G, \ell, (C_1, \ldots, C_k))$ be an instance of PROPER $\mathcal{T}$-LEVEL PLANARITY. Recall that we can assume that each laminar family C_i is represented by a tree with leaves $\ell^{-1}(i)$. Due to Lemma 1, we can additionally assume that G is a matching. We construct an equivalent instance $\mathcal{I}' = (X, (T_1, \ldots, T_k))$ of SEQUENTIAL PQ-ORDERING as follows. The elements of X correspond bijectively to the edges of G. For each $i \in [k]$, the PQ-tree T_i represents the consecutivity constraints obtained from C_i by replacing each vertex with the unique edge it is incident to. As before, the linear order of the vertices on each level i in a solution to $\mathcal{I}$ corresponds to a satisfying order of the corresponding PQ-tree T_i in $\mathcal{I}'$. For the converse direction, we also need to ensure level planarity, that is the characterization of Lemma 4. As each edge of $\mathcal{I}$ between levels i and $i+1$ is represented by a unique element of X in $\mathcal{I}'$, this is ensured by Property 2 of Lemma 3. □

This shows that both problems are linear-time equivalent. The initial quadratic-time algorithm for solving PROPER $\mathcal{T}$-LEVEL PLANARITY uses a chain of linear-time reductions via the problems CONNECTED SEFE-2 and 2-fixed SIMULTANEOUS PQ-ORDERING [2,3], where for the latter an explicit quadratic solution exists [5]. More recently, a direct reduction from PROPER $\mathcal{T}$-LEVEL PLANARITY to 2-fixed SIMULTANEOUS PQ-ORDERING was given, which also generalizes to other surfaces such as the torus [1, Theorem 3 ff.]. For the sake of completeness, we want to first shortly introduce this problem and then give a brief, but explicit reduction of our SEQUENTIAL PQ-ORDERING to the quadratic-time solvable 2-fixed instances of SIMULTANEOUS PQ-ORDERING.

The problem SIMULTANEOUS PQ-ORDERING, introduced by Bläsius and Rutter [5], takes as input a DAG whose vertices are PQ-trees and where every arc with source T_a and target T_b injectively maps $L(T_b)$ to $L(T_a)$. The question is whether there exist linear orders for the leaves of all PQ-trees such that for every arc, the chosen order for the target induces a suborder of the chosen order for the source. While this problem is NP-complete in the general case, Bläsius and Rutter showed that the problem can be solved in quadratic time if the instance is what they call *2-fixed* [5]. In the following, we show that we can model an instance of PROPER $\mathcal{T}$-LEVEL PLANARITY as an equivalent instance of SIMULTANEOUS PQ-ORDERING where the DAG has maximum degree 2, which also implies 2-fixedness [5].

Theorem 2. *There exists a linear-time reduction from* SEQUENTIAL PQ-ORDERING *to 2-fixed* SIMULTANEOUS PQ-ORDERING. *The resulting instance can be solved in quadratic time.*

Proof. Let $\mathcal{I} = (X, (T_1, \ldots, T_k))$ be an instance of SEQUENTIAL PQ-ORDERING. The DAG in the equivalent instance of SIMULTANEOUS PQ-ORDERING simply consists of the PQ-trees $T_1, \ldots, T_k$, together with a set of $k-1$ further PQ-trees that ensure that, between each pair of consecutive PQ-trees in $T_1, \ldots, T_k$, their common elements are ordered consistently. More specifically, for each $i \in [k-1]$, we add a PQ-tree S_i consisting of a single inner P-node with $|L(T_i) \cap L(T_{i+1})|$ leaves as children and add arcs from S_i to T_i and T_{i+1} that map each pair of common elements in T_i and T_{i+1} to a distinct leaf of S_i. Thus, the DAG is a path with edges of alternating orientation, and the vertices thus alternate between being sinks $(T_1, \dots, T_k)$ and sources $(S_1, \dots, S_{k-1})$. The equivalence of the instances is immediate by Lemma 3. Moreover, since every vertex of the DAG has degree at most 2, the instance is 2-fixed and can be solved in quadratic time [5]. □

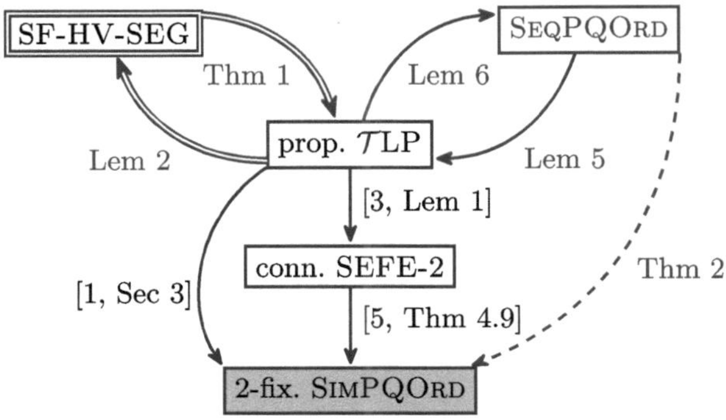

Fig. 4. Relation of the problems we consider, with new problems and relations shown in blue. Dashed, single- and double-line arrows represent trivial, linear- and quadratic-time reductions, respectively. Double-border boxes can be solved (possibly via reductions) in quartic time, single-border boxes in quadratic time. Shaded boxes have an explicit solution not using reductions.

5 Conclusion

In this paper, we resolve the remaining open question from the 1989 seminal works by Kratochvíl and Matoušek on segment intersection graphs with restricted orders [14–16]. When posing the question, the authors already suggested two reformulations of the problem in terms of forbidden volkswagen patterns in 2-layer drawings or forbidden cross patterns in the adjacency matrix[4]. Still, the problem remained open for 30 years until another different perspective finally allowed us to show tractability using the constrained planar graph drawing problem PROPER $\mathcal{T}$-LEVEL PLANARITY. Interestingly, PROPER $\mathcal{T}$-LEVEL PLANARITY itself is also a reformulation of another problem called k-ARY TANGLEGRAMS [3,23], which arose in computational biology to visualize evolutionary

[4] Another reformulation in terms of the chromatic number of an auxiliary graph was attempted [16], but turned out to be insufficient [14].

histories of species. In addition to this very visual solution, we also give another equivalent characterization as a constrained ordering problem that allows a direct solution without involved reductions. Figure 4 gives a graphical overview over the reductions and equivalences thus shown in this work. Through the links established in our paper, we now know that we can view the same tractable problem through a large variety of different-yet-equivalent formulations, some of which arose naturally and independently, and taking different perspectives from either a graph drawing or entirely combinatorial standpoint. It will be interesting to see whether further problems turn out to join this order of equivalent formulations.

Acknowledgements. We want to thank Jan Kratochvíl for introducing us to the "Volkswagen-Problem" at HOMONOLO'22 and, together with Jiří Fiala, also for organizing this very enjoyable workshop. We also want to thank Marie Diana Sieper and Felix Klesen who joined us at the workshop for developing early approaches towards the SF-HV-SEG problem.

Funding Information. Simon D. Fink was funded by the Vienna Science and Technology Fund (WWTF) [10.47379/ICT22029]. Matthias Pfretzschner was funded by the Deutsche Forschungsgemeinschaft (DFG, German Research Foundation) – 541433306. Peter Stumpf was funded by Czech Science Foundation grant no. 23-04949X.

References

1. Angelini, P., Da Lozzo, G., Di Battista, G., Frati, F., Patrignani, M., Rutter, I.: Beyond level planarity: cyclic, torus, and simultaneous level planarity. Theoret. Comput. Sci. **804**, 161–170 (2020). https://doi.org/10.1016/J.TCS.2019.11.024
2. Angelini, P., Da Lozzo, G., Di Battista, G., Frati, F., Roselli, V.: The importance of being proper. In: Duncan, C., Symvonis, A. (eds.) GD 2014. LNCS, vol. 8871, pp. 246–258. Springer, Heidelberg (2014). https://doi.org/10.1007/978-3-662-45803-7_21
3. Angelini, P., Da Lozzo, G., Di Battista, G., Frati, F., Roselli, V.: The importance of being proper: (in clustered-level planarity and t-level planarity). Theor. Comput. Sci. **571**, 1–9 (2015). https://doi.org/10.1016/J.TCS.2014.12.019
4. Bellantoni, S., Ben-Arroyo Hartman, I., Przytycka, T., Whitesides, S.: Grid intersection graphs and boxicity. Discret. Math. **114**(1), 41–49 (1993). https://doi.org/10.1016/0012-365X(93)90354-V
5. Bläsius, T., Rutter, I.: Simultaneous PQ-ordering with applications to constrained embedding problems. ACM Trans. Algorithms **12**(2), 16:1–16:46 (2016). https://doi.org/10.1145/2738054
6. Booth, K.S.: PQ-tree algorithms. Ph.D. thesis, University of California, Berkeley (1975)
7. Booth, K.S., Lueker, G.S.: Testing for the consecutive ones property, interval graphs, and graph planarity using PQ-tree algorithms. J. Comput. Syst. Sci. **13**(3), 335–379 (1976). https://doi.org/10.1016/S0022-0000(76)80045-1
8. Brückner, G., Rutter, I., Stumpf, P.: Level planarity: transitivity vs. even crossings. In: Biedl, T., Kerren, A. (eds.) GD 2018. LNCS, vol. 11282, pp. 39–52. Springer, Cham (2018). https://doi.org/10.1007/978-3-030-04414-5_3

9. Brückner, G., Rutter, I., Stumpf, P.: Level-planarity: transitivity vs. even crossings. Electron. J. Combin. **29**(4) (2022). https://doi.org/10.37236/10814
10. Fulek, R., Pelsmajer, M.J., Schaefer, M., Štefankovič, D.: Hanani-Tutte, monotone drawings, and level-planarity. In: Pach, J. (ed.) Thirty Essays on Geometric Graph Theory, pp. 263–287. Springer, New York (2013). https://doi.org/10.1007/978-1-4614-0110-0_14
11. Hartman, I.B.A., Newman, I., Ziv, R.: On grid intersection graphs. Discret. Math. **87**(1), 41–52 (1991). https://doi.org/10.1016/0012-365X(91)90069-E
12. Jünger, M., Leipert, S.: Level planar embedding in linear time. J. Graph Algorithms Appl. **6**(1), 67–113 (2002). https://doi.org/10.7155/jgaa.00045
13. Jünger, M., Leipert, S., Mutzel, P.: Level planarity testing in linear time. In: Whitesides, S.H. (ed.) GD 1998. LNCS, vol. 1547, pp. 224–237. Springer, Heidelberg (1998). https://doi.org/10.1007/3-540-37623-2_17
14. Kratochvíl, J.: A special planar satisfiability problem and a consequence of its np-completeness. Discret. Appl. Math. **52**(3), 233–252 (1994). https://doi.org/10.1016/0166-218X(94)90143-0
15. Kratochvíl, J., Matoušek, J.: Intersection graphs of segments. J. Combin. Theory Ser. B **62**(2), 289–315 (1994). https://doi.org/10.1006/jctb.1994.1071
16. Kratochvíl, J., Nesetril, J.: Problems proposed at the problem session of the prachatice conference on graph theory. In: Nesetril, J., Fiedler, M. (eds.) Proceeding of the Fourth Czechoslovakian Symposium on Combinatorics, Graphs and Complexity. Annals of Discrete Mathematics, vol. 51, pp. 375–384. Elsevier (1992). https://doi.org/10.1016/S0167-5060(08)70659-9
17. Kratochvíl, J.: String graphs. II. Recognizing string graphs is NP-hard. J. Combin. Theory Ser. B **52**(1), 67–78 (1991). https://doi.org/10.1016/0095-8956(91)90091-W
18. Kratochvíl, J., Matoušek, J.: NP-hardness results for intersection graphs. Commentationes Math. Univ. Carolinae **030**(4), 761–773 (1989). https://eudml.org/doc/17790
19. Opatrny, J.: Total ordering problem. SIAM J. Comput. **8**(1), 111–114 (1979). https://doi.org/10.1137/0208008
20. Randerath, B., et al.: A satisfiability formulation of problems on level graphs. Electron. Notes Discrete Math. **9**, 269–277 (2001). https://doi.org/10.1016/S1571-0653(04)00327-0
21. Schaefer, M.: Complexity of some geometric and topological problems. In: Eppstein, D., Gansner, E.R. (eds.) GD 2009. LNCS, vol. 5849, pp. 334–344. Springer, Heidelberg (2010). https://doi.org/10.1007/978-3-642-11805-0_32
22. Schaefer, M.: Toward a theory of planarity: Hanani-Tutte and planarity variants. J. Graph Algorithms Appl. **17**(4), 367–440 (2013). https://doi.org/10.7155/jgaa.00298
23. Wotzlaw, A., Speckenmeyer, E., Porschen, S.: Generalized k-ary tanglegrams on level graphs: a satisfiability-based approach and its evaluation. Discret. Appl. Math. **160**(16–17), 2349–2363 (2012). https://doi.org/10.1016/J.DAM.2012.05.028

On the Time Complexity of Finding a Well-Spread Perfect Matching in Bridgeless Cubic Graphs

Babak Ghanbari(✉) and Robert Šámal

Computer Science Institute, Charles University, Prague, Czech Republic
{babak,samal}@iuuk.mff.cuni.cz

Abstract. We present an algorithm for finding a perfect matching in a 3-edge-connected cubic graph that intersects every 3-edge cut in exactly one edge. Specifically, we propose an algorithm with a time complexity of $O(n \log^4 n)$, which significantly improves upon the previously known $O(n^3)$-time algorithms for the same problem. The technique we use for the improvement is efficient use of the cactus model of 3-edge cuts. As an application, we use our algorithm to compute embeddings of 3-edge-connected cubic graphs with limited number of singular edges (i.e., edges that are twice in the boundary of one face) in $O(n \log^4 n)$ time; this application contributes to the study of the well-known Cycle Double Cover conjecture.

Keywords: Algorithm · Perfect matching · Cut representation · Embedding

1 Introduction

It is well known that every bridgeless cubic graph admits a perfect matching [Pet91]. Using Edmonds' blossom algorithm, a maximum matching can be found in polynomial time. Gabow [Gab90] demonstrated that the weighted matching problem on general graphs can be solved in time $O(n(m + n \log n))$.

Consequently, the minimum weight perfect matching problem for bridgeless cubic graphs can be solved in $O(n^2 \log n)$ time. Diks and Stańczyk [DS10] proposed an improved algorithm for finding perfect matchings in bridgeless cubic graphs with time complexity of $O(n \log^2 n)$.

In this paper we study the complexity of finding a perfect matching M in a bridgeless cubic graph such that in every 3-edge cut M contains exactly one edge. By parity, M can contain one or three edges in a 3-edge cut; so our condition says no 3-edge cut is contained in M. We shortly express this by saying M is *well-spread.*

B. Ghanbari and R. Šámal—Supported by the project GAUK182623 of the Charles University Grant Agency and by grant 25-16627S of the Czech Science Foundation.

H. Fernau and P. Kindermann (Eds.): WG 2025, LNCS 16124, pp. 219–229, 2026.
https://doi.org/10.1007/978-3-032-11835-6_16

Kaiser et al. [KKN06] showed that a well-spread perfect matching exists in every bridgeless cubic graph (and went on to use this to find how much of the graph can be covered by two, three, etc. perfect matchings). Their proof uses Edmonds' perfect matching polytope (vector $(1/3, 1/3, \dots)$ is a fractional perfect matching, thus it is a convex combination of perfect matchings – each of them is well-spread). This, however, doesn't yield an efficient algorithm.

In [BIT13], Boyd et al. focus on developing algorithms for finding 2-factors in bridgeless cubic graphs that cover specific edge-cuts, which brings these 2-factors closer to Hamiltonian cycles. They provide an efficient algorithm that finds a minimum-weight 2-factor that covers all 3-edge cuts in weighted bridgeless cubic graphs (in contrast with finding a Hamiltonian cycle, which is computationally hard). They also provide both a polyhedral description of such 2-factors and of their complements – well-spread perfect matchings. This is the first known polynomial-time algorithm for this problem, with a time complexity of $O(n^3)$, where n is the number of vertices. In this work, we improve this result for 3-edge-connected cubic graphs using Algorithm 1 with time complexity $O(n \log^4 n)$.

Boyd et al. find a "peripheral 3-edge-cut" (a cut such that one side of it is internally 4-edge-connected) and then use recursion. We save computation by using the cactus model, also called a tree of cuts. In this tree it is easy to find the peripheral cut and also to update the tree when we contract a part of the cut. This leads to much improved time complexity, although only for 3-edge-connected graphs.

We then proceed to apply this result to the study of the well-known Cycle Double Cover (or CDC) Conjecture [Sze73, Sey79]. In the language of graph embedding, the CDC conjecture is equivalent to every bridgeless cubic graph having a surface embedding with no singular edges (i.e., with no edge that is on the boundary of one face twice). As an application of our work, for a 3-edge-connected cubic graph we can find embeddings with a bounded number of singular edges in time $O(n \log^4 n)$.

The structure of the paper is as follows. In the next section, we introduce the necessary definitions and concepts. In Sect. 3, we discuss the cactus model, a key component in deriving our results. Section 4 presents our main result, an efficient algorithm to find the well-spread perfect matching. Finally, in Sect. 5, we apply our algorithm to get results about the well-known CDC conjecture.

2 Preliminaries

Let $G = (V, E)$ be a graph with vertex set V and edge set E. A graph in which all vertices have degree 3 is called *cubic*. A *cycle* is a connected 2-regular graph. A *bridge* in a graph G is an edge whose removal increases the number of components of G. Equivalently, a bridge is an edge that is not contained in any cycles of G. A graph is *bridgeless* if it contains no bridge. An *edge cut* in a graph is a set of edges whose removal increases the number of connected components of the graph. An edge cut C in G is called a *non-trivial cut* if every component of $G - C$ has at least two vertices. Otherwise it is called *trivial.* A *k-edge-cut* in a graph is an edge cut that contains exactly k edges. A connected graph is *k-edge-connected* if it remains connected whenever fewer than k edges are removed. A graph is *cyclically k-edge-connected*, if at least k edges must be removed to disconnect it into two components such that each component contains a cycle. We say that a subset $F \subset E$ *covers* an edge-cut D if $F \cap D \neq \emptyset$. For a subset $S \subseteq E$, G/S is the graph obtained from G by contracting all the edges in S. Note that we keep multiple edges and only remove loops in such contraction.

Petersen's theorem [Pet91] states that every bridgeless cubic graph contains a perfect matching. Schönberger [Sch34] proved the following strengthened form of Petersen's theorem.

Theorem 1 ([Sch34]). *Let $G = (V, E)$ be a bridgeless cubic graph with specified edge $e^* \in E$. Then there exists a perfect matching of G that contains e^*.*

A perfect matching containing a specified edge e^* in a bridgeless cubic graph can be found in $O(n \log^4 n)$ time [BBDL01].

A graph G is *embedded* in a surface S if the vertices of G are distinct elements of S and every edge of G is a simple arc connecting in S the two vertices which it joins in G, such that its interior is disjoint from other edges and vertices. An *embedding* of a graph G in S is an isomorphism of G with a graph G' embedded in S.

Let G be a graph that is cellularly embedded in a surface S, that is, every face is homeomorphic to an open disk. Let $\pi = \{\pi_v | v \in V(G)\}$ where π_v is the cyclic permutation of the edges incident with the vertex v such that $\pi_v(e)$ is the successor of e in the clockwise ordering around v. The cyclic permutation π_v is called the *local rotation* at v, and the set π is the *rotation system* of the given embedding of G in S.

Let G be a connected multigraph. A *combinatorial embedding* of G is a pair (π, λ) where $\pi = \{\pi_v | v \in V(G)\}$ is a rotation system, and λ is a signature mapping which assigns to each edge $e \in E(G)$ a sign $\lambda(e) \in \{-1, 1\}$. If e is an edge incident with $v \in V(G)$, then the cyclic sequence $e, \pi_v(e), \pi_v^2(e), \ldots$ is called the *π-clockwise ordering* around v (or the *local rotation* at v). Given an embedding (π, λ) of G we say that G is *(π, λ)-embedded.* It is known that the combinatorial embedding uniquely determines a cellular embedding to some surface, up to homeomorphism [MT01, Theorems 3.2.4 and 3.3.1].

A *closed walk* is a sequence $(v_0, e_0, v_1, e_1, \ldots, e_{n-1}, v_n)$ where $v_0 = v_n$ and for every i, $e_i = \{v_i, v_{i+1}\}$. We say that a collection of closed walks $C_1, \ldots, C_n$ forms a *partial circuit double cover* (or *partial CDC*) if each edge is covered at most once by one of C_i's or exactly twice by two different C_i and C_j, and for every vertex v and edges e and f where $e \cap f = \{v\}$, there exists at most one closed walk C_i such that $\{e, f\} \subseteq E(C_i)$.

3 The Cactus Model

Here we follow the notations and definition of [NI08]. A general way to represent a subset of cuts within a graph G involves constructing a cactus representation (T, φ). In this representation, T is a graph, and φ is a function mapping vertices from $V(G)$ to $V(T)$. The cactus representation (or the *cactus model*) was introduced in [DKL76]. It was further utilized in [Gab91,Din93]. This model plays a crucial role in understanding both edge connectivity and graph rigidity problems.

Definition 1. *Let G be a graph. A pair (T, φ) with $\varphi : V(G) \to V(T)$ is called a cactus representation (or cactus model) for the graph G if it meets the following criteria:*

1. *For an arbitrary minimum cut $\{S, V(T) - S\} \in C(T)$, the cut $\{X, \bar{X}\}$ in G defined by $X = \{u \in V(G) \mid \varphi(u) \in S\}$, and $\bar{X} = \{u \in V(G) \mid \varphi(u) \in V(T) - S\}$ is a minimum cut in G.*
2. *Conversely, for every minimum cut $\{X, \bar{X}\}$ in G, there exists a minimum cut $\{S, V(T) - S\} \in C(T)$ such that $X = \{u \in V(G) \mid \varphi(u) \in S\}$, and $\bar{X} = \{u \in V(G) \mid \varphi(u) \in V(T) - S\}$.*

In this context, we use *vertex* when referring to elements of $V(G)$ and *node* for elements of $V(T)$. The set $V(T)$ may include a node x that does not correspond to any vertex $v \in V(G)$ with $\varphi(v) = x$; such a node is referred to as an *empty node*. We use $C(T)$ for the set of all minimum cuts in T.

It was shown in [DKL76] that every G has a cactus representation using a *cactus graph* – a multigraph where every edge is in exactly one cycle. A special case of such graph is a tree with every edge doubled. It is known that when the connectivity is odd, the cactus representation is of this special type. In this case we will just call it a tree of cuts (or cactus tree). For a 3-edge-connected cubic graph G, the vertices of the graph G correspond to the leaf nodes of the cactus tree T and all the interior nodes of T are empty nodes (see Fig. 1). We will also use the following estimate for the size of T.

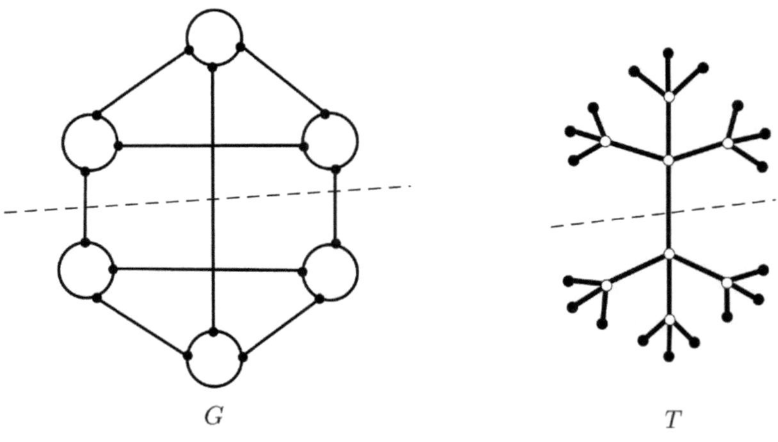

Fig. 1. A 3-edge-connected cubic graph G and its cactus tree T. Small white circles represent empty vertices. Dashed lines represent a non-trivial 3-edge cut in G and its equivalent edge in T.

Lemma 1 ([Din93]). *Let G be a 3-edge-connected cubic graph with n vertices and let T be its cactus tree. Then*

$$|E(T)| \leq 2n - 3.$$

4 Well-Spread Perfect Matchings in Bridgeless Cubic Graphs

In this section, we present an algorithm that finds a perfect matching in a 3-edge-connected cubic graph that intersects all 3-edge cuts in exactly one edge and is substantially faster than the algorithm introduced by Boyd et al. [BIT13]. We first explain the underlying technique. Let G be a graph with a 3-edge cut $\{A, \bar{A}\}$; that is $|\delta(A)| = 3$. We let $G_1 = G/A$ (all vertices in A are contracted to a single vertex a) and $G_2 = G/\bar{A}$ (with a new vertex $\bar{a}$). A well-spread matching M in G contains exactly one edge of $\delta(A)$, thus it gives us a perfect matching M_1 in G_1 and M_2 in G_2. As edges out of a in G_1 and out of $\bar{a}$ in G_2 are in natural correspondence with edges of $\delta(A)$, we will identify them and say that M_1 and M_2 *agree on* $\delta(A)$. For our algorithm the following is crucial.

Theorem 2. *Let G, A, G_1 and G_2 be as above. Given a well-spread perfect matching M_1 in G_1 and M_2 in G_2 that agree on the cut $\delta(A)$, the union $M_1 \cup M_2$ is a well-spread perfect matching in G.*

Proof. The fact that $M_1 \cup M_2$ is a perfect matching is clear as these matchings agree on $\delta(A)$. The fact that M is well-spread follows from the fact that 3-edge cuts in a 3-edge-connected graph do not cross, see [BIT13] for details. □

Boyd et al. [BIT13] use the above theorem together with repeated finding of a "peripheral 3-edge-cut" $\{A, \bar{A}\}$: one where G/A is internally 4-edge-connected. We refine their approach by utilizing (and updating) the model for all 3-edge cuts that will allow us to find peripheral cuts quickly. Also, we will use the model of the cuts to quickly decompose the graph G in G/A and $G/\bar{A}$ and then combine them back.

We will also use the following information about structure of cuts in G_1 and G_2.

Theorem 3. *Let G, A, G_1, G_2, A and $\bar{A}$ be as above. Let (φ, T) be the cactus tree for G. Let $e = x_1x_2$ be the edge of T corresponding to A; that is the components of $T - e$ are T'_1, T'_2, $x_i \in V(T'_i)$ and $A = \varphi^{-1}(V(T'_2))$. Put $T_i = T'_i + e$ for $i = 1, 2$. Define φ_1 as a restriction of φ to $\bar{A}$ and put $\varphi_1(a) = x_1$. We define φ_2 symmetrically.*

Then (φ_i, T_i) is the cactus tree representation of 3-edge cuts in G_i (for $i = 1, 2$).

The proof follows the same idea as that of Theorem 2: minimum odd edge-cuts do not cross. We omit the details in this extended abstract.

In our application we have a 3-regular 3-edge-connected graph, so all vertices are mapped to leaf nodes of the cactus tree and nontrivial 3-cuts correspond to edges that are not incident with a leaf. In particular, when the cactus tree is a star there are no nontrivial 3-cuts and we may simply find any perfect matching. In a sense, the purpose of our algorithm is to efficiently combine matchings coming from these "special cases" of internally 4-edge-connected graphs.

To this end, we pick a root in the cut model T. For a vertex x, let e_x be the edge connecting x to its parent. We let C_x be the three edges in the cut corresponding to e_x. We let A_x be the vertices of G that are mapped by φ to descendants of x (including x), so that $C_x = \delta(A_x)$.

We do not use this definition directly, instead we recursively update A_x, C_x while traversing the tree. We can compute A_x as a union of A_y for all children y of x. We can compute C_x by going over C_y for all children y of x, excluding edges that lie within A_x (between A_y and $A_{y'}$ for two children of x). To do this efficiently we utilize the Union-Find algorithm.

In a typical step of the decomposition part of the algorithm, we use Theorem 3, with the new graphs being G_x (which we store for later) and updated version of G. To illustrate the algorithm, in Fig. 2 we show one step including the updated trees. We see that T_x is a star – which means G_x is internally 4-edge-connected. Updated T simply removes children of x. However, we actually do not need to update T and compute T_x in the algorithm.

In a typical step of the assembling part of the algorithm, we use Theorem 2. For graph G we have our well-spread matching from the recursion. For the other graph, G_x, we compute it easily, as the graph is internally 4-edge-connected, so any perfect matching will do. Figure 3 illustrates this part of the algorithm.

Theorem 4. *The output of Algorithm 1 is a well-spread perfect matching.*

Proof. Follows from Theorem 2 and 3 and the discussion above. □

4.1 Complexity Analysis

The construction of the tree of cuts (the cactus model) can be done in time $O(n \log n)$ [Gab91]. By Lemma 1 we have $|V(T)| = O(n)$. Next, the algorithm decomposes the graph G along 3-edge cuts. Here the main work is keeping track of the sets A_x and C_x, updating the graph G and G_x. For this we use the standard Union-Find algorithm plus a constant amount of work for changing three edges of the cut C_x at each step. Thus the total time required for this loop is $O(n\alpha(n))$.

In the assembly step, we repeatedly look for a perfect matching in an internally 4-edge-connected cubic graph G_x containing a given edge. By [BBDL01], this can be done in time $O(|G_x| \log^4 |G_x|)$. (The initial step where x is the root is faster, as we don't need to specify an edge, but this bound works as well.) The remaining work (combining the matchings) needs only a constant time per edge, so linear in total.

Algorithm 1

Require: A 3-edge-connected cubic graph G.
Ensure: A well-spread perfect matching M.

```
Find φ : G → T using [Gab91].
Choose any node of T as root.
▷ Decomposing G along 3-edge-cuts ◁
for all x ∈ V(T) in post-order do
    if x is the root then                                  ▷ no proper 3-edge-cut remains
        Break out of the for loop.
    else if x is a leaf then                               ▷ Dealing with a trivial cut
        A_x ← φ^{-1}(x)                                    ▷ Single vertex
        C_x ← δ(A_x)                                       ▷ three edges incident to it
    else
        A_x ← ⋃{A_y | y a child of x}                      ▷ Use Union-Find
        C_x ← {u, v} ∈ ⋃{C_y | y a child of x} s.t. only one of u, v is in A_x
        G_x ← G/Ā_x                                        ▷ contract Ā_x to a single vertex ā_x
        put a link to C_x, G_x to x
        G ← G/A_x                                          ▷ contract A_x to a single vertex a_x
        φ ← φ/Ā_x ∪ {(a_x, x)}
▷ Assembling the perfect matching ◁
for all x ∈ V(T) in reverse pre-order do
    if x is a root of T then
        Find any perfect matching M in G.  ▷ G is now internally 4-edge-connected
    else if x is not a leaf then
        ▷ First we find which edge of C_x does M use. ◁
        b ← vertex s.t. a_x b ∈ M
        a ← vertex s.t. ab ∈ C_x
        Find a perfect matching M_x in G_x using the edge aā_x. ▷ G_x is internally 4-edge-connected
        M ← M ∪ M_x
return M
```

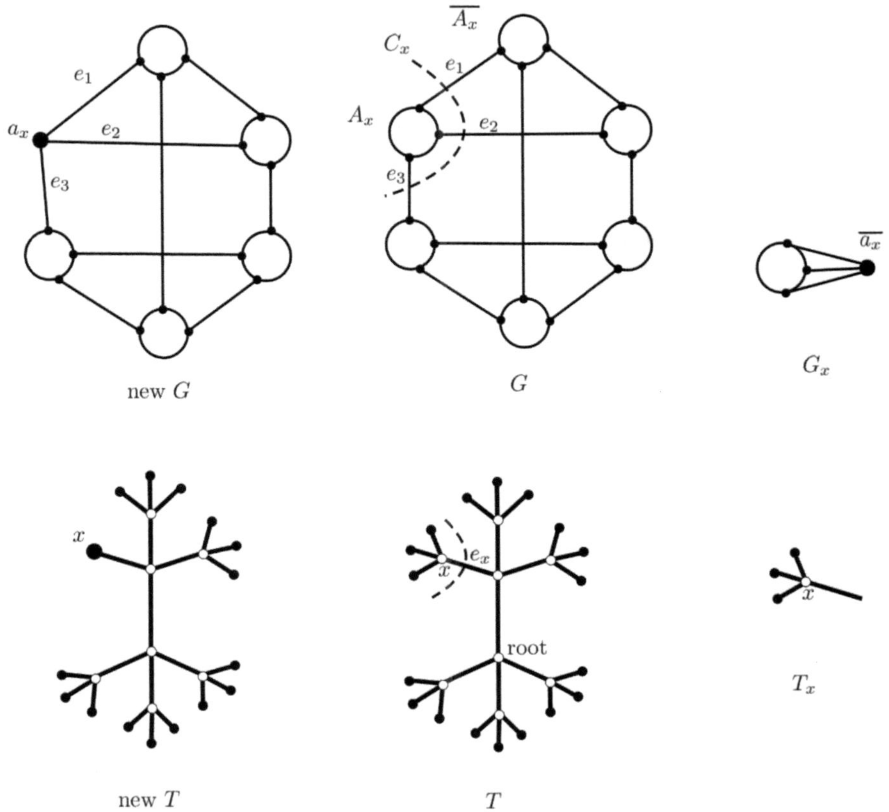

Fig. 2. A typical step in the decomposition part of the algorithm. We show a concrete 3-edge-connected cubic graph G and its cactus tree T, together with the updated version of G and T and the internally 4-edge-connected graph G_x with the star cactus tree T_x.

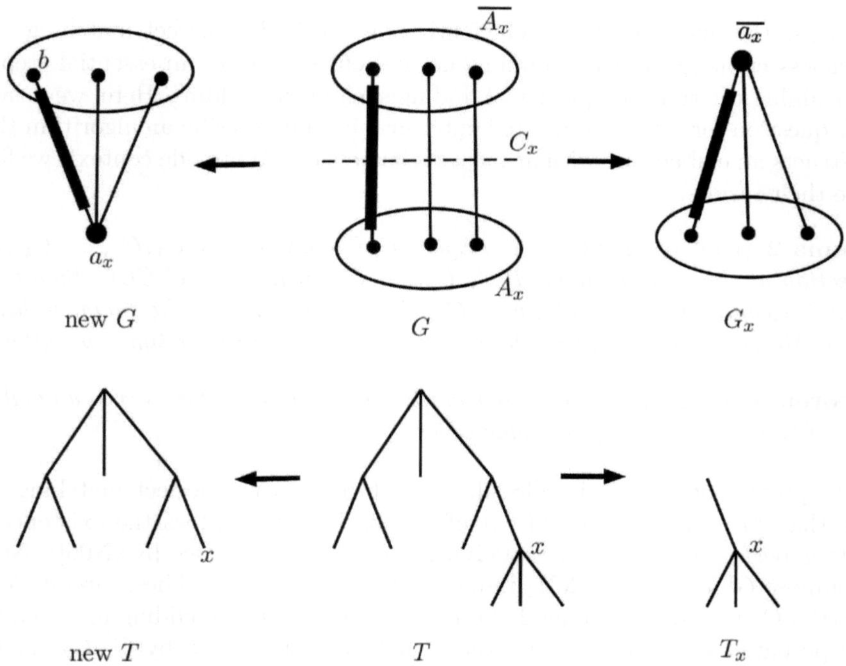

Fig. 3. A typical step in the assembly part of the algorithm: we combine a perfect matching M in "new G" and M_x in G_x to get a matching in the original graph G.

Note that $|G_x|$ is the degree of x in T; let d_x denote this quantity. We know that $\sum_x d_x = 2|E(T)| = O(n)$. The time complexity is thus

$$O(n \log n) + O(n\alpha(n)) + O(n) + \sum_{x \in V(T)} d_x \log^4 d_x.$$

So the leading term in the time complexity is the final sum. As the function $f(d) = d \log^4 d$ is convex, the sum will be maximal when one d_x is as large as possible (namely, $O(n)$) and the other as small as possible (namely, 1) – otherwise we can increase the sum of $f(d_x)$, while keeping sum of d_x constant. For this extreme case we get our bound for time complexity of the algorithm, $O(n \log^4 n)$.

Theorem 5. *Algorithm 1 finds a well-spread perfect matching M in a 3-edge-connected cubic graph G in time $O(n \log^4 n)$.*

5 Application

In [GŠ24], Ghanbari and Šamal establish an upper bound of $\frac{n}{10}$ on the number of singular edges in an embedding of a bridgeless cubic graph on a surface. They also raise the question of how efficiently one can find a perfect matching in a bridgeless cubic graph that contains no odd cut of size 3—an essential step in determining the time complexity of finding such an embedding. Here, we answer their question for 3-edge connected cubic graphs and describe an algorithm that constructs an embedding with at most $\frac{n}{10}$ bad edges. To provide context, we first state their results.

Lemma 2 ([GŠ24]). *Let G be a bridgeless cubic graph, and $C_1, C_2, \ldots, C_t$ be a collection of closed walks in G. If $C_1, C_2, \ldots, C_t$ form a partial CDC, then there is an embedding (π, λ) of G where $C_1, \ldots, C_t$ are some of the facial walks of (π, λ). Moreover, such an embedding can be found by a linear time algorithm.*

Theorem 6 ([GŠ24]). *Let G be a bridgeless cubic graph. There exists an embedding of G with at most $\frac{n}{10}$ singular edges.*

To prove Theorem 6, in [GŠ24], we first construct a perfect matching M_1 such that M_1 contains no odd cut of size 3. This guarantees the existence of another perfect matching M_2 satisfying $|M_1 \cap M_2| \leq \frac{n}{10}$ (see [KKN06]). Next, decompose $G \backslash M_1$ and $M_1 \Delta M_2$ into circuits $C_1, C_2, \ldots, C_t$. These circuits form a partial CDC and by Lemma 2 can be extended to an embedding in which the only potential singular edges are those that are not covered by $C_1, C_2, \ldots, C_t$, that is, edges in $M_1 \cap M_2$. See [GŠ24] for the details.

Suppose that the cubic graph G is 3-edge-connected. To find M_1, we employ Algorithm 1, which finds a perfect matching in time $O(n \log^4 n)$. Consequently, determining the overall time complexity reduces to finding the perfect matching M_2. We utilize Diks and Stańczyk's algorithm [DS10] to find M_2 in time $O(n \log^2 n)$. Thus, the embedding guaranteed by Theorem 6 can be found in total time

$$O(n \log^2 n) + O(n \log^4 n) = O(n \log^4 n).$$

6 Conclusion

We studied a new approach to finding a perfect matching that includes all 3-edge cuts in a 3-edge-connected cubic graph, utilizing the cactus representation of the graph. We referred to such a perfect matching as *well-spread*. In Sect. 4, we presented an algorithm that finds a well-spread perfect matching in a 3-edge-connected cubic graph in time $O(n \log^4 n)$.

In general, the best-known algorithm for finding a well-spread perfect matching in a bridgeless cubic graph has a time complexity of $O(n^3)$ [BIT13]. In the future, it would be interesting to determine whether the cactus representation can be leveraged to develop a faster algorithm for finding such a well-spread perfect matching. However, the issue is that representing 3-edge-cuts in a graph that may contain 2-edge-cuts is more complicated.

References

[BBDL01] Biedl, T.C., Bose, P., Demaine, E.D., Lubiw, A.: Efficient algorithms for Petersen's matching theorem. J. Algorithms **38**(1), 110–134 (2001)

[BIT13] Boyd, S., Iwata, S., Takazawa, K.: Finding 2-factors closer to TSP tours in cubic graphs. SIAM J. Discret. Math. **27**(2), 918–939 (2013)

[Din93] Dinitz, E.: The 3-edge-components and a structural description of all 3-edge-cuts in a graph. In: Mayr, E.W. (ed.) WG 1992. LNCS, vol. 657, pp. 145–157. Springer, Heidelberg (1993). https://doi.org/10.1007/3-540-56402-0_44

[DKL76] Dinitz, E., Karzanov, A.V., Lomonosov, M.V.: On the structure of the system of minimum edge cuts of a graph. Issledovaniya Diskretnoi Optim. 290–306 (1976)

[DS10] Diks, K., Stanczyk, P.: Perfect matching for biconnected cubic graphs in $O(n \log^2 n)$ time. In: van Leeuwen, J., Muscholl, A., Peleg, D., Pokorný, J., Rumpe, B. (eds.) SOFSEM 2010. LNCS, vol. 5901, pp. 321–333. Springer, Heidelberg (2010). https://doi.org/10.1007/978-3-642-11266-9_27

[Gab90] Gabow, H.N.: Data structures for weighted matching and nearest common ancestors with linking. In: Proceedings of the First Annual ACM-SIAM Symposium on Discrete Algorithms, 22–24 January 1990, San Francisco, California, USA, pp. 434–443 (1990)

[Gab91] Gabow, H.N.: Applications of a poset representation to edge connectivity and graph rigidity. In: 32nd Annual Symposium on Foundations of Computer Science, San Juan, Puerto Rico, 1–4 October 1991, pp. 812–821 (1991)

[GŠ24] Ghanbari, B., Šámal, R.: Approximate cycle double cover. In: Rescigno, A.A., Vaccaro, U. (eds.) IWOCA 2024. LNCS, vol. 14764, pp. 421–432. Springer, Cham (2024). https://doi.org/10.1007/978-3-031-63021-7_32

[KKN06] Kaiser, T., Král, D., Norine, S.: Unions of perfect matchings in cubic graphs. In: Klazar, M., Kratochvíl, J., Loebl, M., Matoušek, J., Valtr, P., Thomas, R. (eds.) Topics in Discrete Mathematics. Algorithms and Combinatorics, vol. 26, pp. 225–230. Springer, Heidelberg (2006). https://doi.org/10.1007/3-540-33700-8_14

[MT01] Mohar, B., Thomassen, C.: Graphs on Surfaces. Johns Hopkins University Press (2001)

[NI08] Nagamochi, H., Ibaraki, T.: Algorithmic Aspects of Graph Connectivity. Cambridge University Press (2008)

[Pet91] Petersen, J.: Die Theorie der regulären graphs. Acta Math. **15**(1), 193–220 (1891)

[Sch34] Schönberger, T.: Ein Beweis des Petersenschen Graphensatzes. Acta Litt. Sci. Szeged **7**, 51–57 (1934)

[Sey79] Seymour, P.D.: Sums of circuits. Graph Theory Relat. Top. **1**, 341–355 (1979)

[Sze73] Szekeres, G.: Polyhedral decompositions of cubic graphs. Bull. Aust. Math. Soc. **8**(3), 367–387 (1973)

Playing Snake on a Graph

Denise Graafsma(✉), Bodo Manthey, and Alexander Skopalik

Mathematics of Operations Research, University of Twente,
Enschede, The Netherlands
d.f.graafsma@utwente.nl

Abstract. Snake is a classic computer game, which has been around for decades. Based on this game, we study the game of Snake on arbitrary undirected graphs. A snake forms a simple path that has to move to an apple while avoiding colliding with itself. When the snake reaches the apple, it grows longer, and a new apple appears. A graph on which the snake has a strategy to keep eating apples until it covers all the vertices of the graph is called snake-winnable.

We prove that determining whether a graph is snake-winnable is NP-hard, even when restricted to grid graphs. We fully characterize snake-winnable graphs for odd-sized bipartite graphs and graphs with vertex-connectivity 1. While Hamiltonian graphs are always snake-winnable, we show that non-Hamiltonian snake-winnable graphs have a girth of at most 6 and that this bound is tight.

1 Introduction

In the game of Snake, players control a snake and have to guide it to apples that appear on the screen. With each apple consumed, the snake grows longer. The challenge is to grow the snake as long as possible while avoiding collisions with its own body. Based on this game, we study the game of Snake on a graph.

1.1 Related Work

Over the years, many puzzles and games have been studied through a mathematical lens. Demaine and Hearn provide an extensive overview of the area of combinatorial games [6], and Kendall et al. published a survey of NP-complete puzzles [10]. A number of combinatorial games are played on a graph and involve two players with different roles. In pursuit-evasion games, such as Cops and Robbers [1–3,11], one plays the pursuer, while the other tries to avoid getting caught. In Maker-Breaker games [4,7,8], the players alternately claim previously unclaimed edges. The Maker tries to possess a winning set, while the Breaker wants to prevent this. An example of a Maker-Breaker game is the Hamiltonicity game, where the Maker tries to possess a set of edges that form a Hamiltonian cycle on a complete graph. For these types of games, one of the most important

Full version available at http://arxiv.org/abs/2506.21281.

H. Fernau and P. Kindermann (Eds.): WG 2025, LNCS 16124, pp. 230–243, 2026.
https://doi.org/10.1007/978-3-032-11835-6_17

questions is: given a graph, does one of the two players have a winning strategy. Nowakowski and Winkler gave a characterization for the graphs with a winning cop strategy [11]. For the Hamiltonicity game, Chvátal and Erdős proved that for large enough complete graphs, the Maker has a winning strategy [4]. While the game of Snake on a graph can neither be described as a pursuit-evasion game, nor as a Maker-Breaker game, it does have two players with different roles: the snake who wants to grow as long as possible and the apple placer who tries to prevent this.

The work by De Biasi and Ophelders [5] on the Nibbler food collection problem is inspired by a different variant of Snake known as Nibbler. This problem asks the following: given a graph, the food locations, growth rate, and starting position of the snake, can the snake collect all the food? The growth rate indicates how much the snake grows each time it eats a piece of food. The Nibbler food collection problem is NP-hard, even when restricted to solid grid graphs. Furthermore, if the growth rate is at least 2, then it is also NP-hard on rectangular grid graphs.

1.2 Our Contributions

While the work by De Biasi and Ophelders is primarily inspired by the game Nibbler, the version of Snake we consider is closer to the Nokia version. Instead of having all the apples on the graph from the beginning, only one apple is present at a time. A new apple is placed only when the previous one is consumed, meaning the snake does not know the locations of future apples. Hence, it will have to adjust its strategy according to where the next apple appears. In contrast to the game described by De Biasi and Ophelders, which is framed as a motion planning problem, our game can be viewed as a two-player game where the apple placer acts as an adversary to the snake. We also generalize the game to be played on any connected graph, rather than strictly adhering to the original game of Snake and only considering grid graphs.

For Snake on a graph, we show that determining whether the snake has a winning strategy is NP-hard, even when restricted to grid graphs (Sect. 3). We also prove that the snake can never win on non-Hamiltonian graphs with a girth greater than 6 (Sect. 4). Due to space constraints, some proofs have been omitted and can be found in the full version.

2 The Game of Snake on a Graph

In this section, we formally define the game of Snake on a graph and introduce some notation. The game is played on a simple graph $G = (V, E)$ that is connected and has $|V| \geq 3$. From now on, we assume all our graphs have this property.

2.1 Snake Position

During the game, the snake always occupies an ordered set of vertices, which must form a simple path. We define the *length* of the snake as the number of vertices on this path. Since the length of a path is defined as the number of edges it contains, the length of the snake is always one more than the length of the path it forms.

Let ℓ be the current length of the snake. We denote the position of the snake by $S = (s_1, \ldots, s_\ell)$. We refer to s_1 as the head, and s_ℓ as the tail of the snake. By $\overline{S}$, we denote all vertices in V that are not on S. We refer to $\overline{S}$ as the *unoccupied set.*

In some cases, we index the snake's position by time to better describe the snake's movement. By $S^t = (s_1^t, \ldots, s_\ell^t)$, we denote the position of the snake at time t.

2.2 Snake Movement

Let $S^t = (s_1^t, \ldots, s_\ell^t)$ be the current position of the snake. For a vertex $v \in V$, the neighbor set $N(v)$ denotes the set of vertices in V that are adjacent to v. The head of the snake must move to a vertex in $N(s_1^t)$. Suppose the head moves from s_1^t to some vertex $v \in N(s_1^t)$. Then the next position of the snake becomes $S^{t+1} = (v, s_1^t, \ldots, s_{\ell-1}^t)$. In other words, we add v to the beginning of the path and remove s_ℓ^t from the end.

By our rules, S^{t+1} should still form a simple path. It follows that the head can move to any adjacent vertex that is either unoccupied or the current tail vertex, as depicted in Fig. 1(a). More formally, the head must move to some vertex in $N(s_1^t) \cap \left(\overline{S^t} \cup \{s_\ell^t\}\right)$. In Fig. 1(b), we can see that if the snake moves to an occupied vertex that is not the tail, then the snake will no longer form a simple path. Hence, we forbid this type of movement.

An exception to the rules is made when the snake has length $\ell \leq 2$. For these shorter lengths, we do not allow the head to move to the tail vertex, since this would allow the snake to turn around, occupying the same edge twice.

2.3 Eating an Apple

At the start of the game, the location of the first apple can be any vertex. Let a be the first apple location, then the snake automatically starts on a with $S^0 = (a)$. The next apple is placed on some vertex $a' \neq a$. The game continues as follows. At all times, there is exactly one apple on the graph. The snake eats the apple by moving its head to this vertex, at which point a new apple is immediately placed on one of the unoccupied vertices. We sometimes refer to the apple location as the apple itself.

Let $S^t = (s_1^t, \ldots, s_\ell^t)$ be the current snake position and a be the current apple location. Suppose the head moves from s_1^t to a. Then the snake eats the apple on a and the next snake position will be $S^{t+1} = (a, s_1^t, \ldots, s_\ell^t)$. So different to a

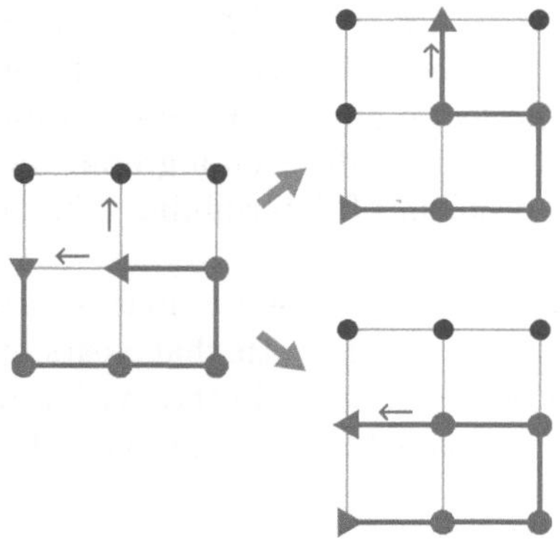

(a) The snake can move to an unoccupied vertex or the tail vertex.

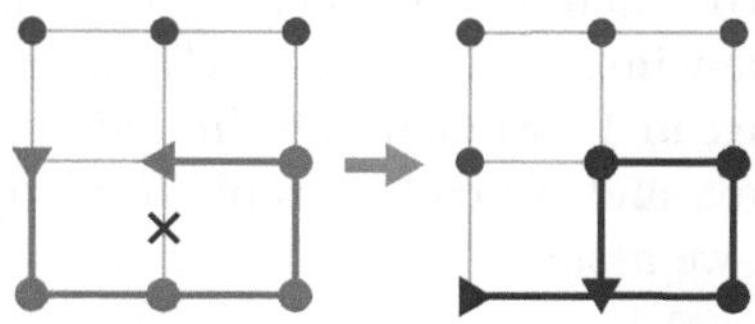

(b) After an illegal move, the snake no longer forms a simple path.

Fig. 1. The rules for the snake's movement are demonstrated on a grid graph.

"normal" movement, s_ℓ^t is not removed from the path and the snake grows one vertex longer.

Observation 2.1. *Let a be the location of the apple at time t, then $a \notin S^t$.*

2.4 Winning and Losing Conditions

If the snake manages to reach length $|V|$, then we say that the snake wins. Note that when this happens, the snake will occupy all the vertices. Hence, there are no more vertices an apple could be placed on.

If there is no vertex the snake can move to, then the snake loses. To be more precise, if $\ell < |V|$, the snake is in the position $S = (s_1, \ldots, s_\ell)$, and

$$N(s_1) \cap \left(\overline{S} \cup s_\ell\right) = \emptyset,$$

then the snake loses.

We want to avoid strategies where the snake can keep moving around without making progress. Imagine, for example, a scenario where the snake will lose if it eats the apple, but it can keep moving in cycles, postponing its loss forever. To this end, the snake also loses if it repeats a previous position. More precisely, if, for the current position S^t, there is some $t' < t$ such that $S^{t'} = S^t$, then the snake loses. Note that the snake can never repeat a position it was in at a shorter length. Furthermore, any strategy that repeats a position can be reformulated as one where the position is not repeated: we simply remove the set of moves between the two identical positions. Because of this, we will not be very careful with this rule when formulating winning snake strategies. If we find a winning strategy that violates it, we know there exists a winning strategy that does adhere to this rule.

Our aim is to determine on which graphs the snake can always win, regardless of the apple placement. To this end, we view it as a two-player game: one player controls the snake, and the other places the apples. While the *snake* tries to grow to length $|V|$, the *apple placer* tries to prevent the snake from doing so. We usually assume that both players play perfectly and always use a winning strategy if possible. When the snake has a winning strategy on a graph, we call the graph *snake-winnable*. We refer to the problem of determining whether a graph is snake-winnable as the *snake problem*.

To conclude our overview of the game, we make a few observations regarding snake-winnable graphs. A *Hamiltonian path* is a simple path that visits every vertex in V. Similarly, a *Hamiltonian cycle* is a simple cycle that visits every vertex, and we call a graph *Hamiltonian* if it has a Hamiltonian cycle. Recall that we assume that $|V| \geq 3$.

Observation 2.2. *If G does not contain a Hamiltonian path, then G is not snake-winnable.*

Observation 2.3. *If G is Hamiltonian, then G is snake-winnable.*

Observation 2.4. *Let $G = (X \cup Y, E)$ be a bipartite graph. If $||X| - |Y|| > 1$, then G is not snake-winnable.*

Observation 2.5. *For a graph $G = (V, E)$, if there is some $v \in V$ with degree $d(v) = 1$, then G is not snake-winnable.*

3 The Complexity of Snake on a Graph

In this section, we show that the snake problem is NP-hard, even when restricted to grid graphs. To do this, we first characterize the odd-sized bipartite graphs that are snake-winnable. We then use this characterization to formulate a reduction from the Hamiltonian cycle problem on grid graphs to the snake problem on odd-sized grid graphs.

Definition 3.1. *A* ***rectangular grid graph*** *$G = (V, E)$ is a graph that has an embedding with $V = [m] \times [n]$ with $m, n \in \mathbb{N}$. For any $u, v \in V$ we have $uv \in E$ if and only if $\|u - v\| = 1$.*

Definition 3.2. *A graph $G = (V, E)$ is a* ***grid graph*** *if and only if it is a vertex-induced subgraph of some rectangular grid graph.*

Definition 3.3. *The* ***theta graph*** *$\Theta(p, q, r)$ is constructed by taking two vertices u and v and connecting them by three internally disjoint paths of lengths p, q and r. If at least one of p, q or r is 0, then $u = v$.*

Lemma 3.4. *If a graph $G = (V, E)$ has $\Theta(|V| - 3, 2, 2)$ as a spanning subgraph, then it is snake-winnable.*

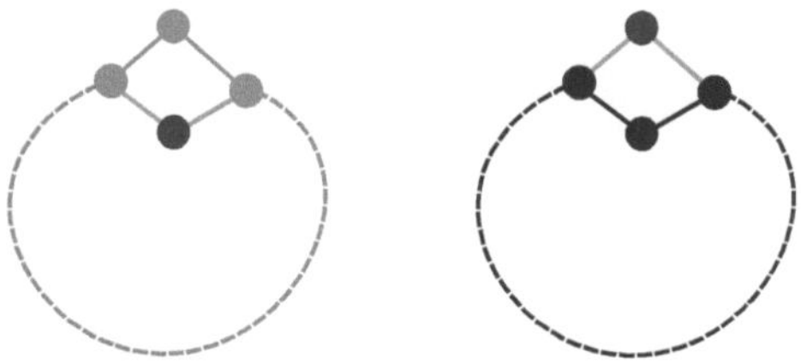

Fig. 2. The snake always moves along one of the two cycles of length $|V| - 1$ in the $\Theta(|V| - 3, 2, 2)$ subgraph.

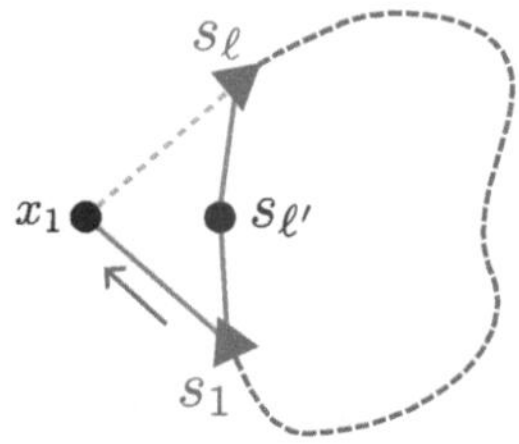

Fig. 3. The snake is on a cycle of length $\ell + 1$ right before it eats the second to last apple.

Proof. The snake can employ the following strategy, which is depicted in Fig. 2. Until it has length $|V| - 1$, the snake always moves along one of the two cycles that consist of the path of length $|V| - 3$, and one of the paths of length 2. This means that there is always only one vertex that is not on this cycle, namely the vertex on the remaining path of length 2. If the apple is placed on this vertex, then the snake switches to the other cycle that does contain this vertex. □

Theorem 3.5. *Let $G = (V, E)$ be a bipartite graph with partition $V = X \cup Y$ and $|V|$ odd. Then G is snake-winnable if and only if $\Theta(|V| - 3, 2, 2)$ is a spanning subgraph of G.*

Proof (Proof sketch.). If $\Theta(|V| - 3, 2, 2)$ is a spanning subgraph of G, then the snake can use the strategy from Lemma 3.4 to win.

It remains to show that if there is no $\Theta(|V| - 3, 2, 2)$ spanning subgraph, then the apple placer has a winning strategy. We give a brief overview of this strategy, the full proof can be found in the full version.

By cleverly placing the apples, the apple placer can guarantee the snake is on a cycle of length $\ell + 1$ right before it eats the second to last apple. This is depicted in Fig. 3. After eating this apple, both the head and the remaining unoccupied vertex are adjacent to the previous head position. Hence, they must be in the same part and cannot be adjacent. It follows that the head can only move to the tail. However, by doing so, the snake would create a $\Theta(|V| - 3, 2, 2)$ spanning subgraph. Thus, there is no vertex the head can move to after eating the second to last apple, and thus the snake will lose. □

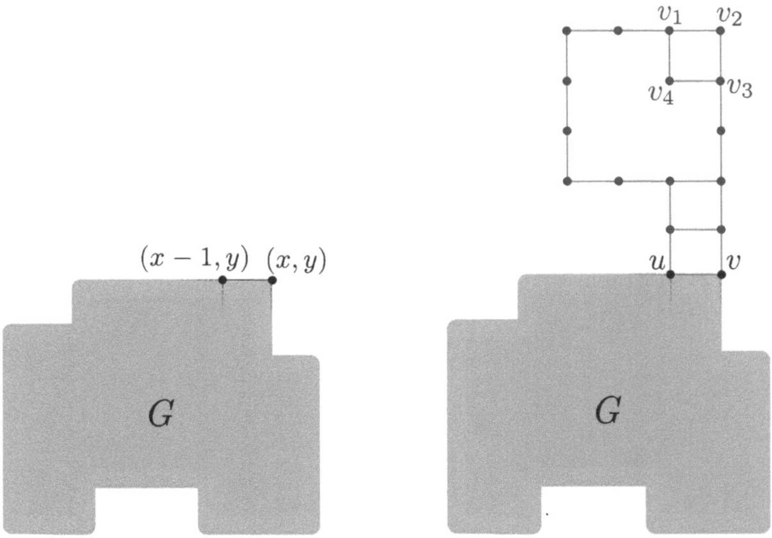

(a) The original graph G.

(b) Attaching the gadget to G to create G'.

Fig. 4. The reduction from a grid graph G to G' by using a gadget.

Corollary 3.6. *Let $G = (V, E)$ be a grid graph with $|V|$ odd. Then G is snake-winnable if and only if $\Theta(|V| - 3, 2, 2)$ is a spanning subgraph of G.*

We will use the characterization from Corollary 3.6 to show that the snake problem on grid graphs is NP-hard. To do so, we reduce from the Hamiltonian cycle problem on grid graphs, which is NP-complete [9].

Theorem 3.7. *The snake problem is* NP*-hard, even restricted to grid graphs.*

Proof (Proof sketch.). We give a general description of the reduction, further details can be found in the full version. Given some instance G of the Hamiltonian cycle problem on grid graphs, we create a new graph G' by attaching the gadget depicted in Fig. 4. By construction, G' has a $\Theta(|V| - 3, 2, 2)$ spanning subgraph if and only if G has a Hamiltonian cycle. □

We leave open whether the snake problem is in NP, as it seems unclear to us whether there is a compact and efficiently verifiable description of winning strategies.

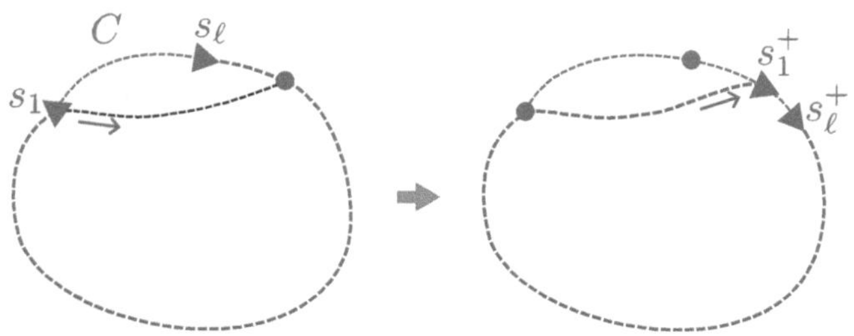

Fig. 5. The snake leaves C from s_1 and returns to C at s_1^+.

4 The Girth of Snake-Winnable Graphs

In this section, we prove that any non-Hamiltonian snake-winnable graph has a girth of at most 6. We also show this bound is tight by giving an example of a non-Hamiltonian snake-winnable graph that has a girth of 6.

Definition 4.1. *The **girth** of a graph G, denoted by $g(G)$, is the length of the shortest cycle in G.*

Let $G = (V, E)$ be a graph and C be some cycle in G. We say a cycle C *contains* the snake if the entire path formed by the snake lies on C. To prove that any non-Hamiltonian graph with a girth greater than 6 is not snake-winnable, we use the following. When the snake reaches length $|V|-3$, there are four possible scenarios: the snake can be contained in a cycle of length $|V|-3$, a cycle of length $|V|-2$, a cycle of length $|V|-1$, or in no cycle. We will show that in each of these scenarios, the apple placer has a winning strategy. Note that there can be no cycle of length $|V|$ that contains the snake since this would make the graph Hamiltonian.

The following lemma shows that, as the snake grows longer, the girth required to make certain types of moves becomes smaller.

Lemma 4.2. *Let C be a cycle in G that contains the snake and let ℓ be the current length of the snake. Suppose the head of the snake leaves C and returns to C after visiting m vertices in $\overline{C}$. Then $g(G) \le |C| - \ell + 2m + 2$.*

Proof. The idea behind the proof is as follows. The path the snake takes outside of C, together with the section of C it "skips", forms a cycle. But the snake cannot just return anywhere on C, since it must move to an unoccupied vertex or the tail. Hence, the length of the section it can skip is bounded. We will show that the length of this skipped section is at most $|C| - \ell + m + 1$. Combined with the path the snake takes through $\overline{C}$, this will result in a cycle of length at most $|C| - \ell + 2m + 2$.

We first note that if $\ell \le 2m+2$, then $|C| - \ell + 2m + 2 \ge |C|$. Since the existence of C implies $g(G) \le |C|$, the statement is trivially true. Hence, we may assume that $\ell > 2m + 2 > m$, which means that while the head moves through $\overline{C}$, the tail remains on C.

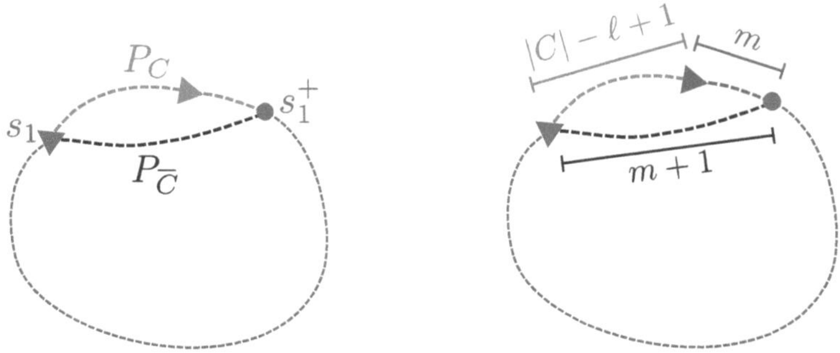

(a) The paths P_C on C and $P_{\overline{C}}$ through $\overline{C}$.

(b) The path P_C has length at most $|C| - \ell + 1 + m$ and $P_{\overline{C}}$ has length $m + 1$.

Fig. 6. If the snake returns to C after visiting m vertices, G must contain a cycle of length $|C| - \ell + 2m + 2$.

Suppose the head of the snake leaves C and returns to C after visiting m vertices in $\overline{C}$, as depicted in Fig. 5. Let $S = (s_1, \ldots, s_\ell)$ be the position of the snake right before the head leaves C. Similarly, let $S^- = (s_1^-, \ldots, s_\ell^-)$ be the position of the snake right before the head returns to C and $S^+ = (s_1^+, \ldots, s_\ell^+)$ the position right after. By P_C we denote the section of C that the snake skips, as depicted in Fig. 6(a). More precisely, P_C is the path the head would have taken had it stayed on C, with endpoints s_1 and s_1^+. When the snake is in position $S^- = (s_1^-, \ldots, s_\ell^-)$, all the unoccupied vertices on C are between s_1 (from where the snake left C) and s_ℓ^-. From s_1^-, the head will move to one of these unoccupied vertices or to s_ℓ^-. Hence, the length of P_C is maximized if the head returns to C by moving to its tail and $s_1^+ = s_\ell^-$. This is depicted in Fig. 6(b).

Since the snake makes m moves before returning to C, we know that $s_\ell^- = s_{\ell-m}$. Thus, the length of P_C is maximized if $s_1^+ = s_{\ell-m}$. In this case, P_C consists of $(s_\ell, \ldots, s_{\ell-m})$ and the path from s_1 to s_ℓ through the section of C that was unoccupied by S. The section $(s_\ell, \ldots, s_{\ell-m})$ has length m. Before the head leaves C, there are $|C| - \ell$ unoccupied vertices between s_1 and s_ℓ. This forms a path of length $|C| - \ell + 1$. Combined, we obtain that P_C has length at most $|C| - \ell + 1 + m$.

The head moves from s_1 to s_1^+ by visiting m vertices in $\overline{C}$. It follows that there must also exist some $s_1 s_1^+$-path of length at most $m+1$ with all internal vertices in $\overline{C}$, as depicted in Fig. 6(b). We will call this path $P_{\overline{C}}$, as depicted in Fig. 6(a). By combining P_C and $P_{\overline{C}}$, we obtain a cycle of length at most $|C| - \ell + 2m + 2$. It follows that $g(G) \leq |C| - \ell + 2m + 2$. □

Corollary 4.3. *Let $G = (V, E)$ be a graph with $g(G) > 2k$. Let $\ell = |V| - k$ be the length of the snake, with $k \geq 2$. Suppose G has a cycle C that contains the snake. If the head of the snake leaves C, then it must visit all vertices in $\overline{C}$ before returning to C.*

Observation 4.4. *Let $\overline{S}$ be the current set of unoccupied vertices. To win, the snake will have to visit all vertices in $\overline{S}$ at some point in the future.*

Using Corollary 4.3 and Observation 4.4, we can prove the following lemma in a similar way as Lemma 4.2.

Lemma 4.5. *Let $G = (V, E)$ be a graph with $g(G) > 2k$. Let $\ell = |V| - k$ be the length of the snake with $k \geq 2$. Suppose G has a cycle C that contains the snake with $|C| > \ell$. If the apple is on some vertex in $\overline{C}$, then the snake will lose.*

Corollary 4.6. *Let $G = (V, E)$ be a graph with $g(G) > 6$. Consider the moment the snake grows to length $|V| - 3$. If there is a cycle C of length $|V| - 1$ or $|V| - 2$ that contains the snake, then the snake will lose.*

Corollary 4.6 excludes two out of the four possible scenarios for when the snake reaches length $|V| - 3$. The two that remain are: the snake is contained in a cycle of length $|V| - 3$, and there is no cycle that contains the snake. We only state the lemmas for these two cases. Full proofs can be found in the full version.

Lemma 4.7. *Let $G = (V, E)$ be a non-Hamiltonian graph with $g(G) > 6$. Consider the moment the snake grows to length $\ell = |V| - 3$ and suppose there is no cycle in G that contains the snake. Then, the snake will lose.*

Lemma 4.8. *Let $G = (V, E)$ be a non-Hamiltonian graph with $g(G) > 6$. Consider the moment the snake grows to length $\ell = |V| - 3$ and suppose there is a cycle C of length $|V| - 3$ that contains the snake. Then, the snake will lose.*

Theorem 4.9. *Let $G = (V, E)$ be a non-Hamiltonian graph with $g(G) > 6$. Then, G is not snake-winnable.*

Proof. Consider the moment the snake grows to length $|V| - 3$. Since G is non-Hamiltonian, there cannot be any cycle of length $|V|$. Hence, the snake is either contained in a cycle of length $|V| - 1$, a cycle of length $|V| - 2$, a cycle of length $|V| - 3$, or in no cycle.

If the snake is contained in a cycle of length $|V| - 1$ or $|V| - 2$, then the snake will lose by Corollary 4.6. If the snake is contained in a cycle of length $|V| - 3$, then it will lose by Lemma 4.8. Finally, if the snake is not contained in any cycle, then it will lose by Lemma 4.7. It follows that once the snake reaches length $|V| - 3$, the apple placer can guarantee that the snake loses, and thus G is not snake-winnable. $\square$

The graph in Fig. 7 shows the bound from Theorem 4.9 is tight, as it is non-Hamiltonian, has girth 6, and is snake-winnable. The winning strategy for the snake can be found in the full version.

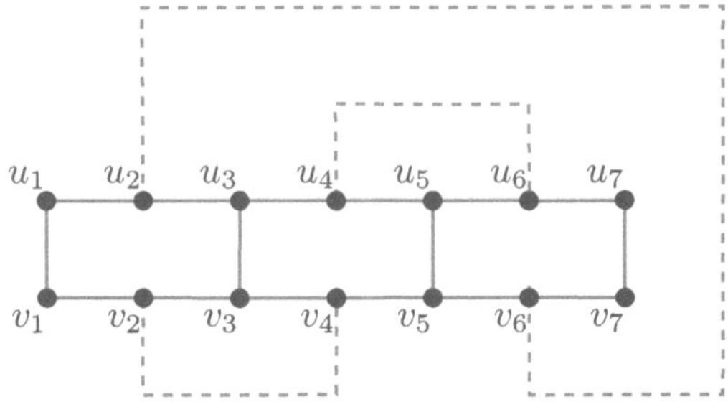

Fig. 7. A non-Hamiltonian graph with a girth of 6 that is snake-winnable.

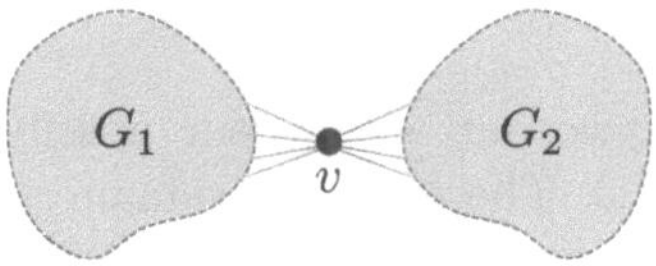

Fig. 8. A snake-winnable graph with cut vertex v. The subgraphs $G_1 + v$ and $G_2 + v$ are both complete.

5 Snake-Winnable Graphs with Vertex-Connectivity 1

One of the challenges for the snake is that its body can block access to certain parts of the graph. Intuitively, this becomes more of an issue when the graph has low vertex-connectivity. In this section, we prove that for vertex-connectivity 1, there is only one type of graph that is snake-winnable.

Definition 5.1. *Let $G = (V, E)$ be a connected simple graph. If there is a single vertex v such that $G - v$ is disconnected, then G has has* ***vertex-connectivity*** *1, which we denote as $\kappa(G) = 1$. We call v a* ***cut vertex*** *of G.*

Observation 5.2. *Let v be a cut vertex of G. If $G - v$ has at least 3 components, then G is not snake-winnable.*

Lemma 5.3. *Let $G = (V, E)$ be a graph with $\kappa(G) = 1$ and let v be a cut vertex of G. Suppose that G_1 and G_2 are the only two connected components of $G - v$ with $|V(G_1)| = |V(G_2)| = m$ and $m \geq 2$. Furthermore, let $G_1 + v$ and $G_2 + v$ both be complete. Then G is snake-winnable.*

Proof (sketch). The graph G is depicted in Fig. 8. We give a sketch of the snake's strategy, further details can be found in the full version. The snake can ensure that once it grows to length $m+1$, it occupies exactly all of the vertices in either $G_1 + v$ or $G_2 + v$, with its tail on v. Since both $G_1 + v$ and $G_2 + v$ are complete, it can move to the tail vertex, placing its head on v. The remaining unoccupied vertices are either $V(G_1)$ or $V(G_2)$, and thus, all the remaining unoccupied vertices are adjacent to v. Regardless of where the next apple is placed, the snake can directly move to it from v and the tail remains in place. The snake repeats this until all of the remaining vertices are occupied. □

It turns out the graph from Lemma 5.3 is the only type of snake-winnable graph with vertex-connectivity 1. To show this, we use the following two lemmas that show the limitations of the snake on graphs with vertex-connectivity 1.

Lemma 5.4. *Let $G = (V, E)$ be a graph with $\kappa(G) = 1$ and let v be a cut vertex of G. Let G_1 and G_2 be two different connected components of $G - v$, with $|V(G_1)| = m$. Suppose the head of the snake is in G_1 and the snake has a length of at least $m + 2$. If there is an unoccupied vertex in G_2, then the snake will lose.*

Lemma 5.5. *Let $G = (V, E)$ be a graph with $\kappa(G) = 1$ and let v be a cut vertex of G. Let G_1 and G_2 be two different connected components of $G - v$, with $|V(G_1)| = m$. Suppose the head of the snake is in G_1, the snake has a length of at least $m + 1$, and the apple is on some vertex a in G_1. If there is some unoccupied vertex in G_2, then the snake will lose.*

We now show that if the two components of $G - v$ have different sizes, then the apple placer has a winning strategy.

Lemma 5.6. *Let $G = (V, E)$ be a graph with $\kappa(G) = 1$ and let v be a cut vertex of G. Let G_1 and G_2 be two different components of $G - v$. If $|V(G_1)| \neq |V(G_2)|$ then G is not snake-winnable.*

Proof. First note that if either $|V(G_1)| = 1$ or $|V(G_2)| = 1$, then G has a vertex of degree 1 and is not snake-winnable by Observation 2.5. Hence, we may assume that both G_1 and G_2 have at least two vertices.

Let $|V(G_1)| = m_1$ and $|V(G_2)| = m_2$ and suppose $m_1 \neq m_2$. Without loss of generality, we assume that $m_1 < m_2$. When the snake grows to length $m_1 - 1$, there must be some vertex u_1 in G_1 that is unoccupied. The apple placer chooses u_1 as the next apple location. When the snake eats the apple on u_1, it grows to length m_1. Since $m_1 < m_2$, there must be some unoccupied vertex u_2 in G_2. The apple placer chooses u_2 as the next apple location.

When the snake eats the apple on u_2, it is either entirely in G_2, or both the cut vertex v and u_2 are occupied by the snake. Thus, at least two of the occupied vertices are in $G_2 + v$. Since the snake now has length $m_1 + 1$, it follows that there must be some unoccupied vertex v_1 in G_1. The apple placer chooses v_1 as the next apple location. When the snake eats the apple on v_1, it grows to length $m_1 + 2$ with its head in G_1. Furthermore, there must be an unoccupied vertex in G_2 and by Lemma 5.4, the snake will lose. □

It remains to show that if either $G_1 + v$ or $G_2 + v$ is incomplete, then G is not snake-winnable. We only give the result, the full proof can be found in the full version.

Lemma 5.7. *Let $G = (V, E)$ be a graph with $\kappa(G) = 1$ and let v be a cut vertex of G. Let G_1 and G_2 be two different connected components of $G - v$ and suppose $G_1 + v$ and $G_2 + v$ are not both complete. Then G is not snake-winnable.*

Theorem 5.8. *Let $G = (V, E)$ have vertex-connectivity 1 and let v be a cut vertex of G. Furthermore, let G_1 and G_2 be two different connected components of $G - v$. Then G is snake-winnable if and only if $G - v$ has two components, $|V(G_1)| = |V(G_2)| \geq 2$, and $G_1 + v$ and $G_2 + v$ are both complete.*

Proof. By Lemma 5.3, if $G - v$ has two components, $|V(G_1)| = |V(G_2)| \geq 2$, and $G_1 + v$ and $G_2 + v$ are both complete, then G is snake-winnable.

By Observation 5.2, G is not snake-winnable if $G - v$ has more than two components. If either $|V(G_1)| = 1$ or $|V(G_2)| = 1$, then G had a vertex of degree 1 and is not snake-winnable by Observation 2.5. By Lemma 5.6, if $|V(G_1)| \neq |V(G_2)|$, then G is not snake-winnable. By Lemma 5.7, G is also not snake-winnable if either $G_1 + v$ or $G_2 + v$ is incomplete. □

6 Open Problems

Having characterized snake-winnable graphs for odd-sized bipartite graphs, a natural question is whether a similar characterization exists for bipartite graphs with an even number of vertices, or even just for grid graphs. Similarly, given the characterization of snake-winnable graphs with vertex-connectivity 1, we wonder whether the snake-winnable graphs with vertex-connectivity 2 can also be characterized. Finally, the question of whether the snake problem is in NP also remains open.

References

1. Aigner, M., Fromme, M.: A game of cops and robbers. Discret. Appl. Math. **8**(1), 1–12 (1984). https://doi.org/10.1016/0166-218X(84)90073-8
2. Bonato, A., Golovach, P., Hahn, G., Kratochvíl, J.: The capture time of a graph. Discret. Math. **309**(18), 5588–5595 (2009). https://doi.org/10.1016/j.disc.2008.04.004. Combinatorics 2006, A Meeting in Celebration of Pavol Hell's 60th Birthday (2006)
3. Bonato, A., Nowakowski, R.: The Game of Cops and Robbers on Graphs, vol. 61. Student Mathematical Library (2011). https://doi.org/10.1090/stml/061
4. Chvátal, V., Erdős, P.: Biased positional games. In: Alspach, B., Hell, P., Miller, D. (eds.) Algorithmic Aspects of Combinatorics. Annals of Discrete Mathematics, vol. 2, pp. 221–229. Elsevier (1978). https://doi.org/10.1016/S0167-5060(08)70335-2
5. De Biasi, M., Ophelders, T.: The complexity of snake and undirected NCL variants. Theoret. Comput. Sci. **748**, 55–65 (2018). https://doi.org/10.1016/j.tcs.2017.10.031
6. Demaine, E.D., Hearn, R.A.: Playing games with algorithms: algorithmic combinatorial game theory. In: Albert, M.H., Nowakowski, R.J. (eds.) Games of No Chance 3, pp. 3–56. Cambridge University Press (2009)
7. Duchêne, E., et al.: Complexity of maker-breaker games on edge sets of graphs. Discret. Appl. Math. **361**, 502–522 (2025). https://doi.org/10.1016/j.dam.2024.11.012

8. Hefetz, D., Krivelevich, M., Stojaković, M., Szabó, T.: Maker-breaker games. In: Hefetz, D., Krivelevich, M., Stojaković, M., Szabó, T. (eds.) Positional Games. OS, vol. 44, pp. 13–25. Springer, Basel (2014). https://doi.org/10.1007/978-3-0348-0825-5_2
9. Itai, A., Papadimitriou, C.H., Szwarcfiter, J.L.: Hamilton paths in grid graphs. SIAM J. Comput. **11**(4), 676–686 (1982). https://doi.org/10.1137/0211056
10. Kendall, G., Parkes, A., Spoerer, K.: A survey of NP-complete puzzles. ICGA J. **31**(1), 13–34 (2008). https://doi.org/10.3233/ICG-2008-31103
11. Nowakowski, R., Winkler, P.: Vertex-to-vertex pursuit in a graph. Discret. Math. **43**(2), 235–239 (1983). https://doi.org/10.1016/0012-365X(83)90160-7

On Graphs Coverable by Chubby Shortest Paths

Meike Hatzel[1(✉)] and Michał Pilipczuk[2]

[1] Discrete Mathematics Group, Institute for Basic Science (IBS), Daejeon, South Korea
research@meikehatzel.com

[2] Institute of Informatics, University of Warsaw, Warsaw, Poland
michal.pilipczuk@mimuw.edu.pl

Abstract. Dumas, Foucaud, Perez, and Todinca [SIAM J. Disc. Math., 2024] proved that if the vertex set of a graph G can be covered by k shortest paths, then the pathwidth of G is bounded by $\mathcal{O}(k \cdot 3^k)$. We prove a coarse variant of this theorem: if in a graph G one can find k shortest paths such that every vertex is at distance at most ρ from one of them, then G is $(3, 12\rho)$-quasi-isometric to a graph of pathwidth $k^{\mathcal{O}(k)}$ and maximum degree $\mathcal{O}(k)$, and G admits a path-partition-decomposition whose bags are coverable by $k^{\mathcal{O}(k)}$ balls of radius at most 2ρ and vertices from non-adjacent bags are at distance larger than 2ρ. We also discuss applications of such decompositions in the context of algorithms for finding maximum distance independent sets and minimum distance dominating sets in graphs.

Keywords: coarse graph structure · geodesics · pathwidth

1 Introduction

The aim of *coarse graph theory* is to study the metric structure in graphs. Recently, Georgakopoulous and Papasoglu [11] launched a systematic investigation of the coarse counterpart of the theory of Graph Minors, with the following overarching conjecture in mind: If a graph G excludes some fixed graph as a *fat minor*, then the metric structure of G resembles that of a graph excluding a classic minor (there are subtleties regarding this statement, see the discussion in [8]). Here, fat minors are a coarse counterpart of classic minors, and the resemblance is measured through the existence of a *quasi-isometry*: a roughly bijective mapping between graphs that roughly preserves distances.[1]

The investigation of the theory of coarse minors naturally leads to studying coarse counterparts of classic graph parameters. The general principle of the coarse theory is that disjointness corresponds to farness, and intersection corresponds to closeness. Hence, in the coarse counterpart of treewidth, we would

[1] This concept was also studied before under the name of *metric minors* by Chepoi, Dragan, Newman, Rabinovich, and Vaxès [5].

H. Fernau and P. Kindermann (Eds.): WG 2025, LNCS 16124, pp. 244–257, 2026.
https://doi.org/10.1007/978-3-032-11835-6_18

consider tree decompositions where the bags are not of bounded size but can be covered by a bounded number of balls of bounded radius. Very recently, Nguyen, Scott, and Seymour [14], and independently Hickingbotham [13] proved that the existence of such a tree decomposition is equivalent to being quasi-isometric to a graph of bounded treewidth, and the natural analogue of this statement also works for pathwidth [13]. In a yet unpublished work, Abrishami, Czyżewska, Kluk, Pilipczuk, Pilipczuk, and Rzążewski [1] study the connection between such tree decompositions and the existence of balanced separators consisting of a bounded number of bounded-radius balls.

The drive permeating the recent work on coarse graph theory is to understand which results from classic structural graph theory can be lifted to the coarse setting and to what extent. See [11] for a broad overview and [2–4,6–9,13–15] for examples of individual works of this kind. In this paper, we add one more result to this growing list. Precisely, we examine the following elegant theorem proved recently by Dumas, Foucaud, Perez, and Todinca [10].

Theorem 1 ([10]). *Suppose G is a graph in which one can find a family $\mathcal{P}$ of k shortest paths so that every vertex of G belongs to some path of $\mathcal{P}$. Then G has pathwidth bounded by $\mathcal{O}(k \cdot 3^k)$.*

The main contribution of this work is the following coarse counterpart of Theorem 1. Here, for positive integers k, ρ, we call a graph G *(k, ρ)-geodesic-coverable* if in G there is a family $\mathcal{P}$ of k shortest paths so that every vertex is at distance at most ρ from some path of $\mathcal{P}$.

Theorem 2. *Suppose G is a (k, ρ)-geodesic-coverable graph, for some positive integers k, ρ. Then G is $(3, 12\rho)$-quasi-isometric to a graph of pathwidth at most $k^{\mathcal{O}(k)}$ and maximum degree less than $26k$. Furthermore, the vertices of G can be partitioned into parts $C_1, C_2, \ldots, C_\ell$ so that every part can be covered by $k^{\mathcal{O}(k)}$ balls of radius 2ρ, and every two vertices at distance at most 2ρ belong to the same part or two consecutive parts.*

The proof of Theorem 2 follows the strategy proposed by Dumas et al. [10] and lifts it naturally to the coarse setting. Namely, we fix a root vertex v and partition the graph into BFS layers starting from v. The key technical statement is the following: every BFS layer S can be covered by $k^{\mathcal{O}(k)}$ balls of radius 2ρ. The idea here, inspired by [10], is to classify the shortest paths from v to the vertices of S according to their *types*. Very roughly, every shortest path Q from v to a vertex of S can be technically massaged so that it first travels for a significant distance along one shortest path P_1 from the family $\mathcal{P}$ witnessing geodesic-coverability, then jumps through a short "connector" to another shortest path $P_2 \in \mathcal{P}$ and travels for a significant distance along it, and so on; the sequence $P_1, P_2, \ldots$ is the type of Q. There are only $k^{\mathcal{O}(k)}$ relevant types to consider, and it turns out that paths of the same type starting in the same vertex lead to vertices that are at distance at most $k^{\mathcal{O}(k)} \cdot \rho$ from each other. This allows us to cover S with $k^{\mathcal{O}(k)}$ balls of radius $k^{\mathcal{O}(k)}$, and an additional argument is needed to cover each of those balls with $k^{\mathcal{O}(k)}$ balls of radius 2ρ. Once the coverage property of the

BFS layers is established, we naturally construct the partition $C_1, C_2, \ldots, C_\ell$ so that each C_i consists of 2ρ consecutive layers. The quasi-isometry statement is obtained by combining all these observations with a generic construction. We note that the proof naturally gives a linear-time algorithm to construct a partition $C_1, C_2, \ldots, C_\ell$ with the asserted properties.

Compared to Theorem 1, Theorem 2 gives a $k^{\mathcal{O}(k)}$ bound instead of $2^{\mathcal{O}(k)}$. In [10] , Dumas et al. first gave a simpler argument giving a bound of $k^{\mathcal{O}(k)}$ on the number of relevant types, and then proceeded with a more complicated argument reducing the bound to $2^{\mathcal{O}(k)}$. In our proof, we use the simpler argument, but we also lose an additional $k^{\mathcal{O}(k)}$ factor later in the proof, when proving that paths of the same type lead to vertices not far from each other. Compared to Theorem 1, this part of the proof gets far more complicated in the coarse setting.

The partition $C_1, C_2, \ldots, C_\ell$ as described in Theorem 2 shall be called a *distance-2ρ path-partition-decomposition* of G. We see it as the natural coarse counterpart of the linear variant of *tree-partition-decompositions*, underlying the graph parameter *tree-partition width*; see an overview of this parameter e.g. in the work of Wood [16]. More generally, a *distance-ρ tree-partition-decomposition* of a graph G would be a tree T together with a partition of the vertex set of G into parts $\{C_x : x \in V(T))\}$ so that whenever $\mathsf{dist}(u, v) \leqslant \rho$, u and v must belong to the same part C_x or to two parts C_x, C_y so that x, y are adjacent in T. We observe that such a decomposition can be used to efficiently solve some computational problems of metric character in G, by means of dynamic programming. We showcase this on the examples of the computation of $\alpha_{2\rho}(G)$—the maximum number of disjoint balls of radius ρ that can be packed in G—and $\gamma_\rho(G)$—the minimum number of balls of radius ρ needed to cover all the vertices of G.

Theorem 3. *There exists an algorithm that given an n-vertex graph G and a distance-ρ tree-partition-decomposition of G of width at most k and degree at most Δ, computes $\alpha_{2\rho}(G)$ in time $n^{\mathcal{O}(\Delta k)}$.*

Theorem 4. *There exists an algorithm that given an n-vertex graph G and a distance-ρ tree-partition-decomposition of G of width at most k and degree at most Δ, computes $\gamma_\rho(G)$ in time $n^{\mathcal{O}(\Delta^2 k)}$.*

By combining Theorems 2 to 4 we obtain the following algorithmic consequence.

Corollary 1. *Suppose G is an n-vertex (k, ρ)-geodesic-coverable graph, for some positive integers k, ρ. Then $\alpha_{4\rho}(G)$ and $\gamma_{2\rho}(G)$ can be computed in time $n^{k^{\mathcal{O}(k)}}$.*

In the language of *parameterized complexity*, this places the computation of $\alpha_{4\rho}(G)$ and $\gamma_{2\rho}(G)$ in the complexity class XP (*slicewise polynomial*) when parametrised by the smallest k such that the graph is (k, ρ)-geodesic-coverable.

This extended abstract focuses on proving Theorem 2. For the proofs of Theorem 3 and Theorem 4 and additional omitted proofs (marked with ♠), we refer the reader to the full version of the paper [12].

2 Preliminaries

Paths and Walks. A *walk* P of length ℓ in a graph G is a sequence of vertices $v_0, \ldots, v_\ell$ such that $v_i v_{i+1} \in E(G)$ for all $0 \leqslant i < \ell$; and P is a *path* if the vertices $v_0, \ldots, v_\ell$ are additionally pairwise different. We sometimes identify P with the subgraph of G containing these edges and vertices. We identify v_0 as the *startvertex*, $\mathsf{start}(P)$, and v_ℓ as the *endvertex*, $\mathsf{end}(P)$, in order to distinguish the two ways of traversing a walk: *forward* from $\mathsf{start}(P)$ to $\mathsf{end}(P)$, or *backward* from $\mathsf{end}(P)$ to $\mathsf{start}(P)$. We refer to P as a v_0-v_ℓ-walk (or v_0-v_ℓ-path, if it is a path). If $x, y \in V(P)$, then we write xPy for the subwalk of P starting in x and ending in y. We write xP instead of $xP\mathsf{end}(P)$ and Px instead of $\mathsf{start}(P)Px$. For two walks P and Q with $\mathsf{end}(P) = \mathsf{start}(Q)$, we write $P \cdot Q$ for their concatenation. We also combine these two notations to write PxQ instead of $Px \cdot xQ$.

Metric Aspects. The *distance* between two vertices $u, v \in V(G)$, denoted as $\mathsf{dist}_G(u, v)$, is the length of the shortest u-v-path in G. We drop the index in case the graph is evident from the context. We extend the notation to sets X, Y of vertices in the expected way: $\mathsf{dist}_G(u, X) := \min_{v \in X} \mathsf{dist}_G(u, v)$ and $\mathsf{dist}_G(Y, X) := \min_{u \in Y} \mathsf{dist}_G(u, X)$. If P and Q are paths, we write $\mathsf{dist}(u, P)$ and $\mathsf{dist}(Q, P)$ instead of $\mathsf{dist}(u, V(P))$ and $\mathsf{dist}(V(Q), V(P))$. A *geodesic* in G is a path that is a shortest path between its endpoints. Note that every subpath of a geodesic is a geodesic as well. A path/walk P is a *γ-almost shortest path/walk* if $||P| - \mathsf{dist}(\mathsf{start}(P), \mathsf{end}(P))| \leqslant \gamma$. Note that every subpath/subwalk of a γ-almost shortest path/walk is again a γ-almost shortest path/walk.

For a vertex u and a non-negative integer ρ, the radius-ρ ball around v is

$$\mathsf{B}_\rho(v) := \{u \in V(G) \mid \mathsf{dist}(u, v) \leqslant \rho\}.$$

A collection of paths $\mathcal{P}$ is a *distance-ρ path-cover* if every vertex of G lies at distance at most ρ of some vertex of a path in $\mathcal{P}$. We say that G is *(k, ρ)-geodesic coverable* if there is a distance-ρ path-cover $\mathcal{P}$ of size k and all paths in $\mathcal{P}$ are geodesics. We then also say that $\mathcal{P}$ is a *distance-ρ geodesic-cover*, or a *(k, ρ)-geodesic-cover* of G.

A *distance-ρ independent set* in G is a set $I \subseteq V(G)$ such that $\mathsf{dist}(u, v) > \rho$ for all distinct $u, v \in I$. The maximum size of a distance-ρ independent subset of a set $A \subseteq V(G)$ is denoted by $\alpha_\rho(A, G)$, and the distance-ρ independence number of G is defined as $\alpha_\rho(G) := \alpha_\rho(V(G), G)$.

A *distance-ρ dominating set* of a set $A \subseteq V(G)$ in a graph G is a set $D \subseteq V(G)$ such that for every $u \in A$ there exists $d \in D$ satisfying $\mathsf{dist}(u, d) \leqslant \rho$; we will also say that D *distance-ρ dominates* A. The minimum size of a distance-ρ dominating set of A is denoted as $\gamma_\rho(A, G)$. We say that A is *(k, ρ)-coverable* if $\gamma_\rho(A, G) \leqslant k$. The *distance-$\rho$ domination number* of G is defined as $\gamma_\rho(G) := \gamma_\rho(V(G), G)$.

We note the following simple relation between the quantities defined above.

Lemma 1 (♠). For every graph G, vertex set $A \subseteq V(G)$, and a positive integer ρ, we have $\alpha_{2\rho}(A, G) \leqslant \gamma_\rho(A, G)$.

Decompositions. We use the following coarse counterparts of *path-partition-decompositions* and *tree-partition-decompositions*. Similarly to other notions in coarse graph theory, these are obtained by requiring the parts of the decomposition to be coverable by a small number of balls. Fix a positive integer ρ.

Definition 1. *A* distance-ρ path-partition-decomposition *of a graph G is a sequence $C_1, \ldots, C_\ell$ of subsets of $V(G)$ such that*

- *$\{C_1, C_2, \ldots, C_\ell\}$ is a partition of $V(G)$, and*
- *for all $u, v \in V(G)$ with $\mathsf{dist}(u, v) \leqslant \rho$ and $u \in C_i, v \in C_j$, we have $|i-j| \leqslant 1$.*

The width *of this path-partition-decomposition is defined as* $\max_{1 \leqslant i \leqslant n} \{\gamma_\rho(C_i, G)\}$.

Definition 2. *A* distance-ρ tree-partition-decomposition *of a graph G consists of a tree T and, for every node $x \in V(T)$, a set of vertices $C_x \subseteq V(G)$ such that*

- *$\{C_x : x \in V(T)\}$ is a partition of $V(G)$, and*
- *for all $u, v \in V(G)$ with $\mathsf{dist}(u, v) \leqslant \rho$ and $u \in C_x, v \in C_y$, the nodes x and y are either equal or adjacent in T.*

The width *of this tree-partition-decomposition is* $\max_{1 \leqslant i \leqslant n} \{\gamma_\rho(C_i, G)\}$, *and its* degree *is the maximum degree of T.*

Thus, the width of a path- or tree-partition-decomposition is at most k if and only if every part C_x is (k, ρ)-coverable. Let us also recall the classic notions of *path-decompositions* and *pathwidth*.

Definition 3. *A* path-decomposition *of a graph G is a sequence $D_1, \ldots, D_\ell$ of subsets of $V(G)$ such that*

- *for every $e \in E(G)$ there is an $i \in [\ell]$ with $e \subseteq D_i$, and*
- *for all $i \leqslant j \leqslant k$ holds $D_i \cap D_k \subseteq D_j$.*

The width *of such a decomposition is the size of its largest set, and the* pathwidth *of G is the minimal width over all path-decompositions of G.*

Quasi-isometries. Let G and H be graphs. A function $\varphi : V(G) \to V(H)$ is an *(m, a)-quasi-isometry from G to H* if for all $u, v \in V(G)$ we have

$$m^{-1} \cdot \mathsf{dist}_G(u, v) - a \leqslant \mathsf{dist}_H(\varphi(u), \varphi(v)) \leqslant m \cdot \mathsf{dist}_G(u, v) + a,$$

and for every $w \in V(H)$ there exists $u \in V(G)$ with $\mathsf{dist}_H(w, \varphi(u)) \leqslant a$. We say G is *(m, a)-quasi-isometric* to H if there is an (m, a)-quasi-isometry from G to H. Intuitively, the vertices of G are mapped into H preserving the distances to some extend and all of the vertices in H are *relevant* in the sense that they do not lie too far from vertices $V(G)$ gets mapped to.

We use the following generic construction of a quasi-isometry between a graph, and its "coarsening" induced by a maximal distance independent set. We use the terminology from [1], but the method is folklore; see e.g. [11, Observation 2.1] or [3, Theorem 4].

Let G be a graph and I be a maximal distance-ρ independent set in G. The *(I, ρ)-distance graph* of G, denoted $H_\rho(I, G)$, is defined as follows

- $V(H_\rho(I,G)):=I$, and
- for all $x, y \in I$ we have $xy \in E(H_\rho(I,G))$ if and only if $\mathsf{dist}_G(x,y) \leqslant 3\rho$.

Note that $H_\rho(I,G)$ is unweighted, so single edges in this graph roughly correspond to distances between ρ and 3ρ in G. To get a closer connection between the shortest path metrics, we also define $\widehat{H}_\rho(I,G)$ as the graph obtained from $H_\rho(I,G)$ by replacing every edge with a path of length 3ρ. With this definition, we have that G and $\widehat{H}_\rho(I,G)$ are quasi-isometric in the sense below.

Lemma 2 ([1], ♠). *Let G be a graph and I be an inclusionwise maximal distance-ρ independent set in G. For every $u \in V(G)$, let $\varphi(u)$ be a vertex of I closest to u in G. (If there are several such vertices, choose an arbitrary one.) Then φ is a $(3, 3\rho)$-quasi-isometry from G to $\widehat{H}_\rho(I,G)$.*

3 Snapwalks

In this section, we discuss the concept of *snapwalks*. Intuitively, a snapwalk in a graph G is a sequence consisting alternately of subpaths of geodesics from a fixed geodesic-cover $\mathcal{P}$ of G, and short "connectors" joining the endpoints of those subpaths in order. The idea is that every shortest path Q in G can be "snapped" to a snapwalk by (roughly) snapping every vertex on Q to the closest geodesic from $\mathcal{P}$. However, we want to avoid reusing the same geodesic of $\mathcal{P}$ several times in a snapwalk, so every segment between two consecutive "snappings" to some $P \in \mathcal{P}$ is replaced by simply sliding along P.

We proceed to a formal description. Let G be a (k, ρ)-geodesic-coverable graph and $\mathcal{P}$ a (k, ρ)-geodesic-cover of G. A *ρ-snapwalk* Q is a sequence $(R_0, Q_1, R_1, Q_2, R_2, \dots, Q_{k'}, R_{k'})$ with $k' \leqslant k$ such that

- every Q_i is a subpath of a geodesic in $\mathcal{P}$;
- for every $P \in \mathcal{P}$ there is at most one i such that Q_i is a subpath of P;
- every R_i is walk in G of length at most $2\rho + 1$, and moreover, R_0 and $R_{k'}$ are of length at most ρ; and
- $\mathsf{end}(R_{i-1}) = \mathsf{start}(Q_i)$ and $\mathsf{end}(Q_i) = \mathsf{start}(R_i)$, for all $i \in \{1, \dots, k'\}$.

We define $\mathsf{concat}(Q):=R_0 \cdot Q_1 \cdot R_1 \cdot Q_2 \cdot R_2 \cdot \dots \cdot Q_{k'} \cdot R_{k'}$.

We generalise notation for paths and walks to snapwalks by applying it to the concatenation instead wherever sensible, for example, $|Q|:=|\mathsf{concat}(Q)|$. The geodesics of $\mathcal{P}$ that the paths Q_i are subpaths of will be called *visited* by Q.

The following lemma establishes that every path in G lies "near to" a snapwalk.

Lemma 3. *Let G be a (k, ρ)-geodesic-coverable graph and $\mathcal{P}$ a (k, ρ)-geodesic-cover of G. For every shortest path P in G there exists a ρ-snapwalk Q with $\mathsf{start}(Q) = \mathsf{start}(P)$, $\mathsf{end}(Q) = \mathsf{end}(P)$, and $||P| - |Q|| \leqslant 4\rho k$.*

Proof. Let $\pi\colon V(G) \to \bigcup_{P_i \in \mathcal{P}} V(P_i)$ map every vertex of G to a vertex on a path in $\mathcal{P}$ minimising the distance to it. (If multiple vertices minimise the distance, select an arbitrary one.) Note that $\mathsf{dist}(x, \pi(x)) \leqslant \rho$, for $\mathcal{P}$ is a distance-ρ geodesic-cover. Also, let $\pi^{\mathcal{P}} : V(G) \to \mathcal{P}$ map every vertex x to a path of $\mathcal{P}$ containing $\pi(x)$, i.e. $\pi(x) \in \pi^{\mathcal{P}}(x)$. For a walk S, we write $\pi^{\mathcal{P}}(S){:=}\pi^{\mathcal{P}}(V(S)) \subseteq \mathcal{P}$.

We now prove the following statement, which implies the desired one, by induction on $k'{:=}|\pi^{\mathcal{P}}(P)|$: There is a ρ-snapwalk Q with $\mathsf{start}(Q) = \mathsf{start}(P)$, $\mathsf{end}(Q) = \mathsf{end}(P)$, and $||P| - |Q|| \leqslant 4\rho k'$ which only visits paths in $\pi^{\mathcal{P}}(P)$.

If $\pi^{\mathcal{P}}(x) = \pi^{\mathcal{P}}(y) = P_1 \in \mathcal{P}$ for all $x, y \in V(P)$, then there is a path R_0 of length at most ρ from $\mathsf{start}(P)$ to $\pi(\mathsf{start}(P))$ and a path R_1 from $\mathsf{end}(P)$ to $\pi(\mathsf{end}(P))$. The walk $R_0{\cdot}\pi(\mathsf{start}(P))P_1\pi(\mathsf{end}(P)){\cdot}R_1$ yields the desired snapwalk.

So assume $\pi^{\mathcal{P}}(P)$ contains at least two geodesics. Let $u{:=}\mathsf{start}(P)$ and $P_1{:=}\pi^{\mathcal{P}}(u)$. Let $v \in V(P)$ be the last vertex along P with $\pi^{\mathcal{P}}(v) = P_1$ and v' be its successor on P (if $v = \mathsf{end}(P)$, we proceed as in the base case). We define R_0 to be a shortest path from u to $\pi(u)$ and R_1° to be a shortest path from v to $\pi(v)$; both are of length at most ρ. Also, let $Q_1{:=}\pi(u)P_1\pi(v)$. Note that, as both Pv and Q_1 are shortest paths, we have $||Pv| - |Q_1|| \leqslant 2\rho$. As $\pi^{\mathcal{P}}(v'P)$ contains a geodesic less than $\pi^{\mathcal{P}}(P)$, by induction, there is a ρ-snapwalk $Q' = (R_0', Q_1', R_1', \ldots, Q_{k'-1}', R_{k'-1})$ starting in v' and ending in $\mathsf{end}(P)$ with $||v'P| - |Q'|| \leqslant 4\rho(k'-1)$ and only using geodesics from $\pi^{\mathcal{P}}(v'P)$, so in particular not using P_1. Next, we define

$$Q{:=}(R_0,\, Q_1,\, R_1^\circ \cdot vv' \cdot R_0',\, Q_1',\ldots,\, Q_{k'-1}',\, R_{k'-1}).$$

Note that $|R_0| \leqslant \rho$ and $|R_1^\circ \cdot vv' \cdot R_0'| \leqslant 2\rho + 1$. Further, we have

$$||P| - |Q|| \leqslant |R_0| + ||Pv| - |Q_1|| + |R_1^\circ| + ||v'P| - |Q'|| \leqslant 4\rho + 4\rho(k'-1) = 4\rho k'.$$

We conclude that Q satisfies all the requirements for the desired ρ-snapwalk. □

Next, we need to massage the snapwalks a bit in order to achieve a certain minimality property. Fix a graph G with a (k, ρ)-geodesic-cover $\mathcal{P}$. We define a sequence of values $\rho_0, \ldots, \rho_k$ as follows:

$$\rho_0{:=}2\rho + 1 \qquad \text{and} \qquad \rho_i{:=}(2k+5) \cdot \rho_{i-1} \quad \text{for } i = 1, 2, \ldots, k.$$

For $\ell \in \{0, 1, \ldots, k\}$, an *$\ell$-simplified ρ-snapwalk* Q is a sequence $(R_0, Q_1, R_1, Q_2, R_2, \ldots, Q_{k'}, R_{k'})$ such that

- every Q_i is subpath of a geodesic in $\mathcal{P}$, with every geodesic in $\mathcal{P}$ containing at most one Q_i as a subpath,
- every R_i is a walk in G of length at most ρ_ℓ, and
- every Q_i is of length larger than $(2k+3)\rho_\ell$.

Note that we allow $k' = 0$; then Q consists only of the walk R_0. Again, the k' paths of $\mathcal{P}$ that contain the paths Q_i as subpaths are called *visited* by Q. Same as for snapwalks, we generalise the notation for paths and walks to simplified snapwalks by applying it to the concatenation wherever sensible.

The intuition behind simplified snapwalks is that we allow the walks R_i to be somewhat longer, but in return we require that all paths Q_i are significantly longer than all the walks R_i. In the next lemma we show that every snapwalk can be turned into a simplified snapwalk by merging some walks in the sequence.

Lemma 4. *Let G be a graph with a (k, ρ)-geodesic-cover $\mathcal{P}$. Then for every ρ-snapwalk P, there is some $\ell \in \{0, 1, \ldots, k\}$ and an ℓ-simplified ρ-snapwalk P' such that* $\mathsf{concat}(P) = \mathsf{concat}(P')$.

Proof. Let an *ℓ-presimplified ρ-snapwalk* be defined exactly as an ℓ-simplified ρ-snapwalk, except we do not require the last item of the definition—the lower bound on the lengths of the paths Q_i. Thus, every ρ-snapwalk, in particular also P, is a 0-presimplified ρ-snapwalk. Denote $P_0{:=}P$ and perform the following iterative construction of $P_1, P_2, \ldots$, where each P_i is an i-presimplified ρ-snapwalk and $\mathsf{concat}(P) = \mathsf{concat}(P_1) = \mathsf{concat}(P_2) = \ldots$:

- If P_i is an i-simplified ρ-snapwalk, finish the construction by setting $P'{:=}P_i$.
- Otherwise, denoting $P_i = (R_0, Q_1, R_1, \ldots, R_{k'-1}, Q_{k'}, R_{k'})$, we have $|Q_j| \leqslant (2k+3)\rho_i$ for some $j \in \{1, \ldots, k'\}$. Obtain P_{i+1} from P_i by replacing the walks R_{j-1}, Q_j, R_j with their concatenation $R_{j-1} \cdot Q_j \cdot R_j$. Observe that

$$|R_{j-1} \cdot Q_j \cdot R_j| \leqslant \rho_i + (2k+3)\rho_i + \rho_i = \rho_{i+1},$$

  so P_i is indeed an $(i+1)$-presimplified ρ-snapwalk.

Each step of the construction reduces the length of the considered sequence of walks by 2. As this length is at most $2k+1$ at the beginning (for P_0), the construction makes at most k steps in total and results in finding an ℓ-simplified snapwalk P' with $\mathsf{concat}(P) = \mathsf{concat}(P')$, for some $\ell \in \{0, 1, \ldots, k\}$. □

Let $\mathcal{P}$ be a (k, ρ)-geodesic-cover of a graph G and $Q = (R_0, Q_1, R_1, \ldots, Q_{k'}, R_{k'})$ be an ℓ-simplified ρ-snapwalk visiting k' geodesics. For every Q_i, let $P_i \in \mathcal{P}$ be the geodesic Q_i is a subpath of. If Q_i follows P_i forwards, we write $\overrightarrow{P_i}$, otherwise we write $\overleftarrow{P_i}$. This yields a tuple $(\overline{P}_1, \ldots, \overline{P}_{k'}, \ell)$ with $\overline{P}_i \in \left\{\overrightarrow{P_i}, \overleftarrow{P_i}\right\}$. We call this tuple the *type* of Q, written $\mathsf{type}(Q)$.

4 Geodesic-Coverability and Coarse Pathwidth

In this section, we prove our main result, Theorem 2. We first observe that in geodesic-coverable graphs, every ball can be covered with a relatively small number of balls of radius 2ρ.

Lemma 5. *Let G be a (k, ρ)-geodesic-coverable graph. Then for all $v \in V(G)$ and all $\ell \in \mathbb{N}$, the ball $\mathsf{B}_{\ell\rho}(v)$ is $(2k(\ell+1), 2\rho)$-coverable.*

Proof. Let $\mathcal{P}$ be a (k,ρ)-geodesic-cover of G and $v \in V(G)$. For every $P \in \mathcal{P}$, let I_P be an inclusionwise maximal distance-ρ independent subset of $V(P) \cap \mathsf{B}_{(\ell+1)\rho}(v)$. Let a and b be the first and the last vertex of I_P, respectively, in the order of appearance on P. By the triangle inequality, we have $\mathsf{dist}(a,b) \leqslant \mathsf{dist}(a,v) + \mathsf{dist}(v,b) \leqslant 2(\ell+1)\rho$. Since P is a geodesic, the subpath of P between a and b has length at most $2(\ell+1)\rho$. As all the vertices of I_P lie on this subpath and I_P is distance-ρ independent, we conclude that $|I_P| \leqslant 2(\ell+1)$.

Let $D := \bigcup_{P\in\mathcal{P}} I_P$; then $|D| \leqslant |\mathcal{P}| \cdot 2(\ell+1) = 2k(\ell+1)$. Note that for each $P \in \mathcal{P}$, the maximality of I_P implies that I_P distance-ρ dominates $V(P) \cap \mathsf{B}_{(\ell+1)\rho}(v)$; and the union of the sets $V(P) \cap \mathsf{B}_{(\ell+1)\rho}(v)$ over all $P \in \mathcal{P}$ distance-ρ dominates $\mathsf{B}_{\ell\rho}(v)$, as $\mathcal{P}$ is a (k,ρ)-geodesic-cover. Hence D distance-2ρ dominates $\mathsf{B}_{\ell\rho}(v)$, and thus D witnesses that $\mathsf{B}_{\ell\rho}(v)$ is $(2k(\ell+1), 2\rho)$-coverable. □

The next technical lemma contains the core argument of the proof: two simplified snapwalks that start at the same vertex, have the same type, and have roughly the same length must lead to vertices that are not too far from each other.

Lemma 6. *Let G be a graph with a (k,ρ)-geodesic-cover $\mathcal{P}$. Suppose C and D are two ℓ-simplified ρ-snapwalks such that*

- $\mathsf{start}(C) = \mathsf{start}(D)$ *and the first walk* (R_0) *on both C and D is of length* 0,
- $\mathsf{type}(C) = \mathsf{type}(D)$,
- $||C| - |D|| \leqslant \gamma$, *for some non-negative integer γ, and*
- C, D *are δ-almost shortest paths with $\delta = 4k\rho + 2\rho_\ell + 2(k-k')\rho_\ell$ where k' is the number of geodesics in* $\mathsf{type}(C)$.

Then $\mathsf{dist}(\mathsf{end}(C), \mathsf{end}(D)) \leqslant \gamma + k \cdot (4k\rho + 2\rho_\ell + 2k\rho_\ell) + k \cdot (4\rho_\ell)$.

Proof. By reversing the geodesics of $\mathcal{P}$ if necessary, without loss of generality we may assume that $\mathsf{type}(C) = (\overrightarrow{P_1}, \overrightarrow{P_2}, \ldots, \overrightarrow{P_{k'}}) = \mathsf{type}(D)$. For $F \in \{C, D\}$, denote

$$F = R_0^F, Q_1^F, R_1^F, Q_2^F, R_2^F, \ldots, R_{k'-1}^F, Q_{k'}^F, R_{k'}^F, \qquad \text{where}$$

- $|R_0^C| = |R_0^D| = 0$ and $|R_i^C|, |R_i^D| \leqslant \rho_\ell$, for $i \in \{1, \ldots, k'\}$;
- Q_i^C and Q_i^D are subpaths of P_i, oriented the same way as P_i; and
- $|Q_i^C|, |Q_i^D| > (2k+2)\rho_\ell$, for all $i \in \{1, \ldots, k'\}$.

We prove by induction over k', that is, the number of geodesics the snapwalks C and D visit, that $\mathsf{dist}(\mathsf{end}(C), \mathsf{end}(D)) \leqslant \gamma + k'(4k\rho + 2\rho_\ell + 2k\rho_\ell) + k'(4\rho_\ell)$. We conduct the induction on slightly relaxed objects—*relaxed ℓ-simplified ρ-snapwalks*—in which we allow the subpaths Q_1^C and Q_1^D to be arbitrarily short (in other words, the assertion $|Q_i^C|, |Q_i^D| > (k+2)\rho_\ell$ only needs to hold for $i \geqslant 2$); clearly, this holds for the initial input. Let $\rho' := \rho_\ell$ and $u := \mathsf{start}(C) = \mathsf{start}(D)$. Note that $u \in V(P_1)$ and $u = \mathsf{start}(Q_1^C) = \mathsf{start}(Q_1^D)$.

In the case $k' = 0$ we have that $u = \mathsf{end}(C) = \mathsf{end}(D)$. Consider then the case $k' = 1$. Let $x := \mathsf{end}(Q_1^C)$ and $y := \mathsf{end}(Q_1^D)$. Without loss of generality, we assume that x occurs on P_1 not later than y. By assumption, we have $|R_1^C|, |R_1^D| \leqslant \rho'$. As we have $\gamma \geqslant ||D| - |C|| = |(|xP_1\,y| + |R_1^D|) - |R_1^C|| \geqslant |xP_1\,y| - \rho'$, we obtain

$$\mathsf{dist}(\mathsf{end}(C), \mathsf{end}(D)) \leqslant |R_1^C| + |xP_1\,y| + |R_1^D| \leqslant \gamma + 3\rho'.$$

We proceed to the main case $k' > 1$. Again, let $x{:=}\mathsf{end}(Q_1^C) = \mathsf{start}(R_1^C)$ and $y{:=}\mathsf{end}(Q_1^D) = \mathsf{start}(R_1^D)$ and assume, without loss of generality, that x occurs on P_1 not later than y does. Additionally, let $x'{:=}\mathsf{end}(R_1^C) = \mathsf{start}(Q_2^C)$ and $y'{:=}\mathsf{end}(R_1^D) = \mathsf{start}(Q_2^D)$; note that x' and y' lie on P_2. We distinguish three different cases: **(1)** y' lies on P_2 before x' and $z{:=}\mathsf{end}(Q_2^D)$ lies between them, **(2)** y' lies on P_2 before x' and z lies after x', and **(3)** y' occurs on P_2 at the same point or after x'.

In case **(1)**, we define $A{:=}y'D\,z = Q_2^D$, $R{:=}R_1^C \cdot x'P_2z$ and $S{:=}xP_1y$. Because $y'P_2x'$ is a shortest path and $|R_1^C|, |R_1^D| \leqslant \rho'$, we get $|R| - \rho' + |A| \leqslant |S| + 2\rho'$, and thus, $|R| \leqslant |S| + 3\rho' - |A|$. On the other hand, because D is a δ-almost shortest path, we obtain $|S| + |A| - \delta \leqslant |R|$. So we have

$$|S| + |A| - \delta \leqslant |S| + 3\rho' - |A|, \qquad \text{implying}$$
$$|Q_2^D| = |A| \leqslant \frac{3\rho' + \delta}{2} \leqslant \frac{(2k+5)\rho' + 2k\rho}{2} \leqslant (2k+3)\rho'.$$

This contradicts the assumption that D is a relaxed ℓ-simplified ρ-snapwalk. So this case simply cannot happen.

We continue with case **(2)**. Here, we define $A{:=}xD\,x'$. As D is a δ-almost shortest walk and $|R_1^C| = |xC\,x'| \leqslant \rho'$, we get $|A| \leqslant \rho' + \delta$. We define snapwalks:

- $C'{:=}x'C$, i.e., C' is the suffix of C starting from Q_2^C; and
- $D'{:=}x'D$, i.e., D' is the suffix of D starting from Q_2^D, but with Q_2^D trimmed to $x'Q_2^D$.

Note that both C' and D' are again relaxed ℓ-simplified ρ-snapwalks (here we make use of them being relaxed, as we shorten the subpaths of D on P_2, which now is the first geodesic the walks visit) and visit $k' - 1$ geodesics of $\mathcal{P}$. As C' is a subwalk of C and D' is a subwalk of D, both are again δ-almost shortest walks. Additionally, we obtain $||C'| - |D'|| \leqslant \gamma + ||A| - |R_1^C|| \leqslant \gamma + \delta + 2\rho'$. Thus, by induction, we obtain that

$$\begin{aligned}\mathsf{dist}(\mathsf{end}(C), \mathsf{end}(D)) &\leqslant \gamma + \delta + 2\rho' + (k'-1)(4k\rho + 2\rho' + 2k\rho') + (k'-1)(4\rho') \\ &\leqslant \gamma + k'(4k\rho + 2\rho' + 2k\rho') + k'(4\rho').\end{aligned}$$

Finally, we consider case **(3)**, where we define $A{:=}x'P_2y'$ and $B{:=}xD\,y'$. In this case, we define snapwalks:

- $C'{:=}x'C$, i.e., C' is the suffix of C starting from Q_2^C; and
- $D'{:=}A \cdot y'D$, i.e., D' is the suffix of D starting from Q_2^D, but with Q_2^D extended by prepending A.

Note that thus, both C' and D' are relaxed ℓ-simplified ρ-snapwalks that start at x' and visit $k'-1$ geodesics of $\mathcal{P}$. As P_2 is a geodesic, we know that A is a shortest path and hence $|A| \leqslant |B|+|R_1^C| \leqslant |B|+\rho'$. On the other hand, xBy is a shortest path as well, hence $|B| = |xBy|+|R_1^D| \leqslant |R_1^C|+|A|+2|R_1^D| \leqslant |A|+3\rho'$. Thus $||A|-|B|| \leqslant 3\rho'$, and $|(|uP_1x|+|R_1^C|+|C'|)-(|uP_1x|+|B|+|D'|-|A|)| \leqslant \gamma$, implying $||C'|-|D'|| \leqslant \gamma+4\rho'$.

As xD is δ-almost shortest, by $|A| \leqslant |B|+\rho'$ we obtain that $R_1^C \cdot D'$ is $(\delta+2\rho')$-almost shortest, which in turn implies that D' is $(\delta+2\rho')$-almost shortest as well. C' as a subpath of C is again δ-almost shortest. Note that $(\delta+2\rho') \leqslant 4k\rho+2\rho'+(k-(k'-1))2\rho'$. So we may apply induction to C' and D' to obtain that

$$\begin{aligned}\mathsf{dist}(\mathsf{end}(C),\mathsf{end}(D)) &\leqslant \gamma+4\rho'+(k'-1)(4k\rho+2\rho'+2k\rho')+(k'-1)(4\rho')\\ &\leqslant \gamma+k'(4k\rho+2\rho'+2k\rho')+k'(4\rho').\end{aligned}$$

Note that, indeed, none of the cases uses any lower bound on $|Q_1^C|$ or $|Q_1^D|$, so relaxed ℓ-simplified ρ-snapwalks suffice. □

We have the following easy consequence that concerns snapwalks with the same start vertex, but not necessarily lying on a geodesic from $\mathcal{P}$.

Corollary 2 (♠). Let G be a graph with a (k,ρ)-geodesic-cover $\mathcal{P}$. Suppose C and D are two ℓ-simplified ρ-snapwalks with

- $\mathsf{start}(C) = \mathsf{start}(D)$,
- $\mathsf{type}(C) = \mathsf{type}(D)$,
- $||C|-|D|| \leqslant \gamma$ for some non-negative integer γ, and
- C, D are δ-almost shortest paths, where $\delta = 4k\rho$.

Then $\mathsf{dist}(\mathsf{end}(C),\mathsf{end}(D)) \leqslant \gamma+k(4k\rho+2\rho_\ell+2k\rho_\ell)+(6k+4)\rho_\ell$.

In the following lemma, we use the tools we have developed to show that every BFS layer is well-coverable.

Lemma 7. *Let G be a (k,ρ)-geodesic-coverable graph. Then for every $u \in V(G)$ and every non-negative integer d, the set $\mathsf{S}_d(u) := \{v \in V(G) \mid \mathsf{dist}(u,v) = d\}$ is $(200 \cdot 14^k \cdot k^{2k+4}, 2\rho)$-coverable.*

Proof. For every $v \in \mathsf{S}_d(u)$ we fix a shortest u-v-path P_v of length d. We also fix a ρ-snapwalk Q_v° corresponding to P_v in the sense of Lemma 3. Note that we have $d \leqslant |Q_v^\circ| \leqslant d+4k\rho$. We may then apply Lemma 4 to find an ℓ-simplified snapwalk Q_v with $\mathsf{concat}(Q_v) = \mathsf{concat}(Q_v^\circ)$, for some $\ell \in \{0,1,\dots,k\}$. Note that Q_v is a $4k\rho$-almost shortest path. Let $\mathcal{Q} := \{Q_v : v \in \mathsf{S}_d(u)\}$ and $\mathcal{T} := \{\mathsf{type}(Q) : Q \in \mathcal{Q}\}$.

For every type $t \in \mathcal{T}$, we now fix an endvertex x_t of some path $Q \in \mathcal{Q}$ with $\mathsf{type}(Q) = t$. By Corollary 2 and Lemma 4, $\mathsf{type}(Q_v) = t$ entails $\mathsf{end}(P_v) = \mathsf{end}(Q_v) \in \mathsf{B}_{k(4k\rho+2k\rho_k)+(6k+4)\rho_k}(x_t)$. Thus, $\mathsf{S}_d(u)$ is coverable by $|\mathcal{T}| \leqslant (2k)^{k+1}$ balls of radius $k(4k\rho+2k\rho_k)+(6k+4)\rho_k \leqslant 16k^2\rho_k \leqslant 48k^2(2k+5)^k \cdot \rho$.

Now, by Lemma 5, every such ball is again $(2k(48k^2(2k+5)^k+1), 2\rho)$-coverable, so also $(100k^3(7k)^k, 2\rho)$-coverable. Thus, $\mathsf{S}_d(u)$ is $((2k)^{k+1} \cdot 100k^3(7k)^k, 2\rho)$-coverable, hence also $(200 \cdot 14^k \cdot k^{2k+4}, 2\rho)$-coverable □

We can now prove our main result, which we restate for convenience. Note that in the adopted terminology, the properties expected from $C_1, \dots, C_\ell$ just assert that it is a distance-2ρ path-partition-decomposition of width $k^{\mathcal{O}(k)}$.

Theorem 2. *Suppose G is a (k, ρ)-geodesic-coverable graph, for some positive integers k, ρ. Then G is $(3, 12\rho)$-quasi-isometric to a graph of pathwidth at most $k^{\mathcal{O}(k)}$ and maximum degree less than $26k$. Furthermore, the vertices of G can be partitioned into parts $C_1, C_2, \dots, C_\ell$ so that every part can be covered by $k^{\mathcal{O}(k)}$ balls of radius 2ρ, and every two vertices at distance at most 2ρ belong to the same part or two consecutive parts.*

Proof. Write $\eta := 200 \cdot 14^k \cdot k^{2k+4}$ for brevity, and note that $\eta \in k^{\mathcal{O}(k)}$. Without loss of generality, we may assume that G is connected; otherwise, we may apply the reasoning to every connected component of G separately and naturally combine the outcomes. Fix an arbitrary vertex u of G and for $i = 1, 2, 3, \dots$, define

$$C_i := \{v \in V(G) \mid (i-1) \cdot 2\rho \leqslant \mathsf{dist}(u, v) < i \cdot 2\rho\}.$$

Letting ℓ be the largest integer such that C_ℓ is non-empty, we see that $C_1, \dots, C_\ell$ is a distance-2ρ path-partition-decomposition of G. Further, for each $i \in \{1, \dots, \ell\}$ we have, by Lemma 7, that $\mathsf{S}_{(i-1)\cdot 2\rho}(u)$ is $(\eta, 2\rho)$-coverable. Since every vertex of C_i is at distance at most 2ρ from some vertex of $\mathsf{S}_{(i-1)\cdot 2\rho}(u)$, we conclude that C_i is $(\eta, 4\rho)$-coverable, hence also $(10k\eta, 2\rho)$-coverable by Lemma 5. Therefore, $C_1, \dots, C_\ell$ is a distance-2ρ path-partition-decomposition of G of width at most $10k\eta \in k^{\mathcal{O}(k)}$, as required.

To construct the quasi-isometry, let I be a maximal distance-4ρ independent set in G and let $H := H_{4\rho}(I, G)$ be the $(I, 4\rho)$-distance graph of G. Further, let $\widehat{H} := \widehat{H}_{4\rho}(I, G)$ be the graph obtained from H by replacing every edge by a path of length 12ρ. By Lemma 2, G is $(3, 12\rho)$-quasi-isometric to $\widehat{H}$, hence it remains to prove that $\mathsf{pw}(\widehat{H}) \leqslant k^{\mathcal{O}(k)}$ and $\Delta(\widehat{H}) \leqslant 26k$. Since subdividing edges does not change the pathwidth and the maximum degree (assuming they are at least 2), it suffices to show that (i) $\mathsf{pw}(H) \leqslant k^{\mathcal{O}(k)}$ and (ii) $\Delta(H) < 26k$.

For the first property, for $i \in \{1, \dots, \ell\}$ we set $D_i := (C_i \cup C_{i+1} \cup C_{i+2} \cup C_{i+3}) \cap I$. Recall that each set C_i is $(10k\eta, 2\rho)$-coverable, hence each set D_i is $(40k\eta, 2\rho)$-coverable. As D_i is also a distance-4ρ independent set, from Lemma 1 we conclude that $|D_i| \leqslant 40k\eta \in k^{\mathcal{O}(k)}$. It remains to argue that $D_1, \dots, D_\ell$ is a path decomposition of H. The only non-trivial check is that whenever vertices $u, v \in I$ are adjacent in H, there is some t such that $u, v \in D_t$. Recall that u, v being adjacent in H entails $\mathsf{dist}_G(u, v) \leqslant 12\rho$. Hence, if i and j are such that $u \in C_i$ and $v \in C_j$, then $|i - j| \leqslant 3$. It follows that $u, v \in D_t$ where $t = \min(i, j)$.

Finally, for the second property, consider any $u \in I$. By the definition of the distance graph, for every neighbour v of u in H we have $\mathsf{dist}_G(u, v) \leqslant 12\rho$. So $\{v \in I \mid \mathsf{dist}_H(u, v) \leqslant 1\}$ is a distance-4ρ independent set contained in $\mathsf{B}_{12\rho}(v)$. By Lemma 5, $\mathsf{B}_{12\rho}(v)$ is $(26k, 2\rho)$-coverable. From Lemma 1 we now infer that $|\{v \in I \mid \mathsf{dist}_H(u, v) \leqslant 1\}| \leqslant 26k$, so the degree of u in H is smaller than $26k$. □

Acknowledgments. Meike Hatzel's research was supported by the Institute for Basic Science (IBS-R029-C1). Michał Pilipczuk's research was supported by the project BOBR that has received funding from the European Research Council (ERC) under the European Union's Horizon 2020 research and innovation programme, grant agreement No. 948057.

Disclosure of Interests. The authors have no competing interests to declare that are relevant to the content of this article.

References

1. Abrishami, T., Czyżewska, J., Kluk, K., Pilipczuk, M., Pilipczuk, M., Rzążewski, P.: On coarse tree decompositions and coarse balanced separators. Preprint, arXiv:2502.20182 (2025). https://arxiv.org/abs/2502.20182
2. Ahn, J., Gollin, J.P., Huynh, T., Kwon, O.: A coarse Erdős-Pósa theorem. In: 36th Annual ACM-SIAM Symposium on Discrete Algorithms, SODA 2025, pp. 3363–3381. SIAM (2025). https://doi.org/10.1137/1.9781611978322.109
3. Albrechtsen, S., Huynh, T., Jacobs, R.W., Knappe, P., Wollan, P.: A Menger-type theorem for two induced paths. SIAM J. Discret. Math. **38**(2), 1438–1450 (2024). https://doi.org/10.1137/23M1573082
4. Berger, E., Seymour, P.D.: Bounded-diameter tree-decompositions. Combinatorica **44**(3), 659–674 (2024). https://doi.org/10.1007/S00493-024-00088-1
5. Chepoi, V., Dragan, F.F., Newman, I., Rabinovich, Y., Vaxès, Y.: Constant approximation algorithms for embedding graph metrics into trees and outerplanar graphs. Discrete Comput. Geom. **47**(1), 187–214 (2011). https://doi.org/10.1007/s00454-011-9386-0
6. Chepoi, V., Dragan, F.F., Vaxès, Y.: Core congestion is inherent in hyperbolic networks. In: Proceedings of the 28th annual ACM-SIAM Symposium on Discrete Algorithms, SODA 2017, Barcelona, Spain, January 16–19, 2017, pp. 2264–2279. Philadelphia, PA: Society for Industrial and Applied Mathematics (SIAM); New York, NY: Association for Computing Machinery (ACM) (2017). https://doi.org/10.1137/1.9781611974782.149
7. Chepoi, V., Estellon, B.: Packing and covering δ-hyperbolic spaces by balls. In: Approximation, randomization, and combinatorial optimization. Algorithms and techniques. 10th international workshop, APPROX 2007, and 11th international workshop, RANDOM 2007, Princeton, NJ, USA, August 20–22, 2007. Proceedings, pp. 59–73. Berlin: Springer (2007). https://doi.org/10.1007/978-3-540-74208-1_5
8. Davies, J., Hickingbotham, R., Illingworth, F., McCarty, R.: Fat minors cannot be thinned (by quasi-isometries). Preprint, arXiv:2405.09383 (2024). https://arxiv.org/abs/2405.09383
9. Dujmović, V., Joret, G., Micek, P., Morin, P.: Erdős-Pósa property of cycles that are far apart. Preprint, arXiv:2412.13893 (2024). https://arxiv.org/abs/2412.13893
10. Dumas, M., Foucaud, F., Perez, A., Todinca, I.: On graphs coverable by k shortest paths. SIAM J. Discret. Math. **38**(2), 1840–1862 (2024). https://doi.org/10.1137/23M1564511
11. Georgakopoulos, A., Papasoglu, P.: Graph minors and metric spaces. Preprint, arXiv:2305.07456 (2023). https://arxiv.org/abs/2305.07456

12. Hatzel, M., Pilipczuk, M.: On graphs coverable by chubby shortest paths (2025). https://arxiv.org/abs/2503.02160
13. Hickingbotham, R.: Graphs quasi-isometric to graphs with bounded treewidth. Preprint, arXiv:2501.10840 (2025). https://arxiv.org/abs/2501.10840
14. Nguyen, T., Scott, A., Seymour, P.: Coarse tree-width. Preprint, arXiv:2501.09839 (2025). https://arxiv.org/abs/2501.09839
15. Nguyen, T., Scott, A., Seymour, P.: A counterexample to the coarse Menger conjecture. J. Comb. Theory Ser. B **173**, 68–82 (2025). https://doi.org/10.1016/j.jctb.2025.01.004, https://www.sciencedirect.com/science/article/pii/S0095895625000061
16. Wood, D.R.: On tree-partition-width. Eur. J. Comb. **30**(5), 1245–1253 (2009). https://doi.org/10.1016/J.EJC.2008.11.010

Matchings with Five Directions in Hypercubes Extend to Hamilton Cycles and Paths with Prescribed Ends

Jiří Fink and Vojtěch Hotmar(✉)

Department of Theoretical Computer Science and Mathematical Logic, Faculty of Mathematics and Physics, Charles University, Prague, Czechia
fink@ktiml.mff.cuni.cz, vojtech.hotmar@pm.me

Abstract. The n-dimensional hypercube graph Q_n has vertices representing all binary strings of length n, with edges between strings differing in exactly one bit position. The Ruskey-Savage conjecture states that every matching of Q_n can be extended into a Hamilton cycle. We prove that matchings of Q_n containing edges spanning at most $d = 5$ directions can be extended into a Hamilton cycle. We also characterize when these matchings spanning at most $d = 5$ directions can be extended into a Hamilton path between two prescribed vertices. Our proofs work for arbitrary d and n where $d \leq n$, assuming certain extension properties hold in Q_d, which we verified computationally for $d = 5$.

Keywords: Hypercube · Hamilton cycle · Hamilton path · Matchings spanning limited in number of directions

1 Introduction

The *n-dimensional hypercube* Q_n is the graph whose vertices are all binary strings of length n, with edges connecting strings that differ in exactly one bit position. There is extensive literature on the structural properties of hypercubes stemming from research on topological structure and analysis of hypercubic interconnection networks [20]. For example, Gros in 1872 [11] proved that Q_n contains a *Hamilton cycle*, a cycle visiting every vertex exactly once, for every $n \geq 2$. A *Hamilton path* in the hypercube Q_n is a path visiting every vertex exactly once and is also called a Binary Gray Code in the literature. Such a path corresponds to a listing of all binary strings of length n where every two consecutive strings differ in exactly one position. Frank Gray notably exploited this connection in 1953, constructing a special Hamilton cycle in Q_n called the Reflected Binary Gray Code and using it in his U.S. Patent 2632058 for signal processing. Since then, Gray codes have found applications in diverse areas such as information retrieval, image processing, and data compression, with alternative constructions satisfying additional properties being widely studied [12,16].

This work is supported by Czech Science Foundation grant GA 22-15272S and it is an extended abstract of arXiv:2501.19029.

H. Fernau and P. Kindermann (Eds.): WG 2025, LNCS 16124, pp. 258–269, 2026.
https://doi.org/10.1007/978-3-032-11835-6_19

1.1 Extending Matchings to Hamilton Cycles

Ruskey and Savage [15] posed the following fundamental question, which remains open:

Conjecture 1 (Ruskey and Savage [15]). Every matching of the hypercube Q_n can be extended into a Hamilton cycle where $n \geq 2$.

One natural approach toward this problem considers *perfect matchings* only, where each vertex is incident to exactly one edge. Kreweras [13] conjectured that every perfect matching in the n-dimensional hypercube with $n \geq 2$ extends to a Hamilton cycle. This conjecture was popularized by Knuth [12] and proven by Fink [6]. Fink's proof actually establishes a stronger result, where $K(Q_n)$ denotes the complete graph on the vertices of Q_n:

Theorem 1 (Fink [6]). *For $n \geq 2$ and every perfect matching M of $K(Q_n)$, there exists a perfect matching N of Q_n such that $M \cup N$ is a Hamilton cycle of $K(Q_n)$.*

This result inspired several generalizations [1,9]. For instance, the authors of [1] showed that Kreweras' conjecture also holds for some sparse spanning regular subgraphs of hypercubes. Using the following theorem, Alahmadi et al. [1] proved that for every perfect matching M in Q_n, the set of edges e of Q_n such that $M \cup \{e\}$ cannot be extended into a Hamilton cycle forms a matching of Q_n.

Theorem 2 (Alahmadi et al. [1], Theorem 4). *Let $n \geq 2$ and let M be a perfect matching on Q_n. Let $e_1, e_2 \in E(Q_n) \setminus M$ incident with the same vertex. Then Q_n has a perfect matching N such that N contains one of e_1, e_2 and $M \cup N$ is a Hamilton cycle of Q_n.*

To state further results, we introduce some terminology. The *direction* of an edge in Q_n is the bit position $i \in [n] := \{1, \ldots, n\}$ where the two vertices' binary representations differ. A matching M *spans a direction* i if it contains at least one edge in direction i, and *spans at most d directions* if its edges use at most d distinct directions. A vertex is *covered* by a matching if the matching contains an edge incident to that vertex. The *parity* of a vertex is the parity of the number of ones in its binary string representation.

Shujia Wang and Fan Wang [19] presented another relaxation to the Ruskey-Savage conjecture:

Theorem 3 (Shujia Wang, Fan Wang [19]). *For $n \geq 5$, let M be a matching on the hypercube Q_n such that $|M| < 10 \cdot 2^{n-5}$ and M spans at most 5 directions. Then M can be extended to a Hamilton cycle.*

We expand upon their idea and prove that the size restriction on M is unnecessary:

Theorem 4. *For $n \geq 2$, let M be a matching on the hypercube Q_n that spans at most 5 directions. Then M can be extended to a Hamilton cycle.*

Our proof of Theorem 4 works for arbitrary numbers of directions d, assuming the following conjecture holds:

Conjecture 2. Given a matching M on the hypercube Q_n and two vertices $u, v \in V(Q_n)$ of opposite parity, where u is not covered by M, there exists a Hamilton path between u and v extending M on Q_n where $n \geq 2$.

Observation 5. *Conjecture 2 implies Conjecture 1.*

We also verify the following observation using a computer program[1] which provides us with Theorem 4:

Observation 6. *Conjecture 2 holds for $2 \leq d \leq 5$.*

For general d, we have the following conditional statement.

Theorem 7. *For $n \geq d \geq 2$, let M be a matching on the hypercube Q_n that spans at most d directions. If Conjecture 2 holds for dimension d, then M can be extended to a Hamilton cycle.*

Figure 1 illustrates the logical relationships between the conjectures and theorems discussed above and those introduced in subsequent sections.

1.2 Extending Matchings to Hamilton Paths

It is well-known that the hypercube Q_n contains a Hamilton path between any two prescribed vertices of opposite parity [17].

To state the following results, we introduce additional notation. For vertices u and v, let uv denote the edge between u and v. For a vertex u and direction i, let u^i denote the unique neighbor of u in direction i. If u is covered by a matching M, let u^M denote the other endpoint of the edge in M incident to u. The *parity of an edge* is the parity of the endpoint which has fewer ones in its binary representation. A *layer in direction* i consists of all edges in direction i. A *half-layer* is a subset of a layer containing only edges of the same parity. An *almost half-layer* is a set of edges that becomes a half-layer when one additional edge is added; it is u*-avoiding* if the missing edge is incident to vertex u.

Gregor, Novotný, and Škrekovski [10] considered the problem of extending perfect matchings of Q_n to a Hamilton paths between two prescribed endpoints of different parity. Their proof works in the more general setting of the complete bipartite graph $B(Q_n)$ which contains an edge between all pairs of vertices of Q_n with opposite parities.

[1] https://gitlab.com/jirka.fink/matching_of_five_directions.

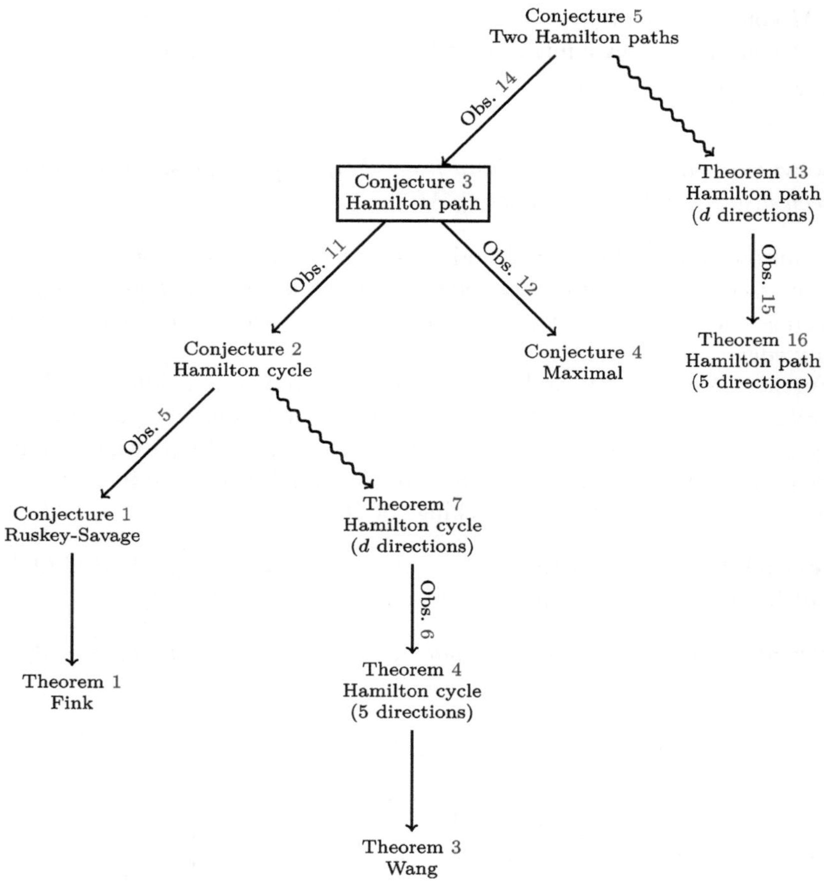

Fig. 1. Logical relationships between conjectures and theorems. Solid arrows represent direct implications via observations, while wavy arrows indicate conditional implications that hold when the source conjecture is true.

Theorem 8 (Gregor, Novotný, and Škrekovski [10], **Theorem 2).** *Let $n \geq 5$, and let u, v be vertices of opposite parity in Q_n. A perfect matching M of $B(Q_n)$ with $uv \notin M$ can be extended to a Hamilton path, by edges from Q_n, with endpoints u and v if and only if $(M \setminus \{uu^M, vv^M\}) \cup \{u^M v^M\}$ contains no half-layers.*

Motivated by this result, we characterize when arbitrary matchings of Q_n can be extended to Hamilton paths between prescribed vertices, which is the main conjecture of this paper:

Conjecture 3. Let $n \geq 5$ and $u, v \in V(Q_n)$ be vertices of opposite parity. For a matching M on Q_n, there exists a Hamilton path between u and v extending M if and only if none of these conditions hold:

C1. M contains a half-layer and u and v are both covered by it,
C2. M contains a u-avoiding almost half-layer in the direction i where $v = u^i$, $j \neq i$ and $uu^j, vv^j \in M$,
C3. $uv \in M$.

Observation 9. *If any C-condition from Conjecture 3 holds, no Hamilton path between u and v extends a matching M.*

Counterexamples for $n = 4$ and perfect matchings were presented in [10], and we discovered an additional counterexample for maximal (but not perfect) matchings shown in Fig. 2, where a matching is *maximal* if it cannot be extended by any edge.

Furthermore, Dvořák and Fink [3] constructed a matching of $B(Q_n)$ of size $\Theta(2^n/\sqrt{n})$ which cannot be extended into a Hamilton cycle for $n \geq 9$. Using a similar construction, we present a matching of $B(Q_n)$ which cannot be extended to a Hamilton path between any two vertices for $n \geq 9$, so we restrict Conjecture 3 to consider only matchings of Q_n.

Observation 10. *For every $n \geq 9$, there exists a matching M of $B(Q_n)$ that cannot be extended to a Hamilton path using only edges from Q_n.*

Observation 11. *Conjecture 3 implies Conjecture 2, for dimension $d \geq 5$.*

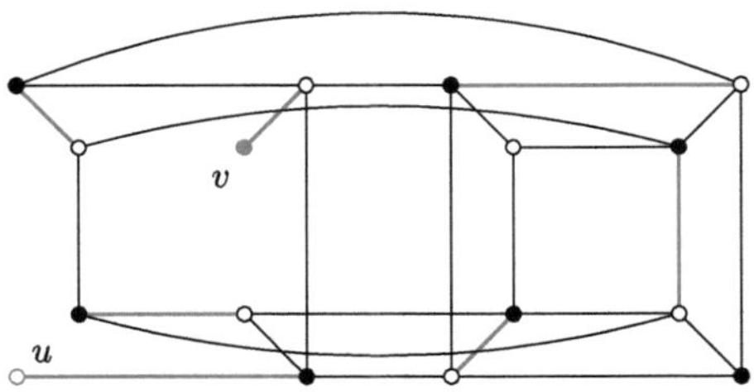

Fig. 2. The forbidden case for $n = 4$. Orange edges are the matching M and the orange vertices are the vertices u, v which cannot be together extended to a Hamilton path extending M between u and v. Black edges are the remaining usable edges in Q_4. (Color figure online)

We consider the following simplification of Conjecture 3 for maximal matchings:

Conjecture 4. Let $n \geq 5$, and let u, v be vertices of opposite parity in Q_n. A maximal matching M of Q_n with $uv \notin M$ can be extended to a Hamilton path with endpoints u and v if and only if $(M \setminus \{uu^M, vv^M\}) \cup \{u^M v^M\}$ contains no half-layers.

Observation 12. *Conjecture 3 implies Conjecture 4.*

This shows the relation between our Conjecture 3 and Theorem 8, and it implies that Conjecture 3 holds for perfect matchings and matchings extendable to a perfect one. Our main results for Hamilton paths are:

Theorem 13. *Let $n \geq d \geq 5$, and $u, v \in Q_n$ be vertices of opposite parity and M be a matching on Q_n spanning at most d directions. Assuming Conjecture 5 holds for dimension d, then M can be extended into a Hamilton path between u and v if and only if none of the C-conditions hold for M.*

Our proof technique relies on the following stronger statement:

Conjecture 5. Let $n \geq 5$, and let $u, v \in Q_n$ be vertices of opposite parity and M be a matching on Q_n. Then if and only if none of the C-conditions hold for M there exists two distinct Hamilton paths between u and v extending M.

Observation 14. *Conjecture 5 implies Conjecture 3.*

We verify the following observation through computer verification:

Observation 15. *Conjecture 5 holds for $d = 5$.*

Immediately yielding:

Theorem 16. *Let $n \geq 5$, and $u, v \in Q_n$ be vertices of opposite parity and M be a matching on Q_n spanning at most 5 directions. Then M can be extended into a Hamilton path between u and v if and only if none of the C-conditions hold for M.*

1.3 Related Results

Research on Hamilton cycles in hypercubes satisfying certain additional properties has received considerable attention; see e.g. the survey by Savage [16]. This research is motivated by applications in computer networks, where hypercubes serve as network topologies with desirable properties like small degree and diameter.

Vandenbussche and West [18] conjectured that every matching can be extended into a 2-factor, which was proven by Fink [7]. Fink and Mütze [8] proved that every matching of Q_n can be extended into a cycle visiting at least a 2/3-fraction of all vertices.

Dvořák [5] showed that any set of at most $2n - 3$ edges of Q_n, $n \geq 2$, that induces vertex-disjoint paths is contained in a Hamilton cycle. Dvořák and Gregor [4] proved that for any set E of at most $2n - 4$ edges in Q_n, $n \geq 5$, that form disjoint paths and two vertices x and y of opposite parity that are neither internal vertices of the paths nor endpoints of the same path, there is a Hamilton path with end vertices x and y that contains all of E. Dimitrov et al. [2] proved that the hypercube Q_n contains a Hamilton cycle avoiding a given matching except a forbidden configuration. Other related results are nicely surveyed in the introduction of the recently published paper [8].

1.4 Overview of Our Approach

We consider matchings M in Q_n spanning at most $d \leq n$ directions. It is natural to partition Q_n into 2^{n-d} subcubes of dimension d so that every edge of M belongs to one subcube. We find Hamilton cycles extending M by constructing cycles within each subcube that can be interconnected into a single Hamilton cycle.

A similar approach was used by Gregor [9], who proved that every perfect matching N of Q_n can be extended into a Hamilton cycle by adding only edges I inside prescribed subcubes if and only if $N \cup I$ is connected. The connectivity condition is necessary, so in our case where $M \subseteq I$, we must use some edges between subcubes in the Hamilton cycle extending M, but our approach uses only a few of them.

1.5 Organization

Section 2 presents the extension to Hamilton cycles, proving Theorem 7 for matchings spanning at most 5 directions. Section 3 establishes the characterization of Hamilton path extensions, culminating in an outline of the proof of 13. For the full version with all of the proofs, see arXiv:2501.19029.

2 Extending to a Hamilton Cycle

In this section, we show how to extend a matching M on the hypercube Q_n that spans at most d directions into a Hamilton cycle. First, we prove Theorem 7 giving the extension for $n \geq d \geq 2$, but it assumes that Conjecture 2 holds for dimension d. Then, using Observation 6 which verifies Conjecture 2 for $d = 5$, we prove Theorem 4.

First, we introduce some notation. The *Cartesian product* $G \square H$ of graphs G and H has vertex set $V(G) \times V(H)$, with edges between vertices (u, v) and (u', v') if either $u = u'$ and $vv' \in E(H)$, or $v = v'$ and $uu' \in E(G)$. Note that $Q_n = Q_{n-d} \square Q_d$. We call the copies of H in G *canonical copies*. In this paper, we consider matchings on Q_n that span at most d directions, and we can assume without loss of generality that these directions are the last d directions. Therefore, every edge uv of these matchings belongs to one canonical copy of Q_d in Q_n, and this canonical copy is uniquely identified by the first $n - d$ directions u which are the same as those in v. For a vertex $u \in V(Q_n)$, we denote by u^L and u^R its first (Left) $n - d$ and last (Right) d coordinates, respectively. Let $Q_n[u^L]$ be the subgraph of Q_n induced by vertices w such that $w^L = u^L$. In other words, $Q_n[u^L]$ is the canonical copy of Q_n containing u. Similarly for a matching M spanning only the last d directions, $M[u^L]$ denotes the edges of M in $Q_n[u^L]$.

Proof (Proof of Theorem 7). If $n = d$, the theorem is a direct consequence of Observation 5. Assuming $n > d$, we can without loss of generality consider only matchings that span the last d directions. If M is perfect, we invoke Theorem 1 for an immediate proof. Otherwise, for a non-perfect matching, there

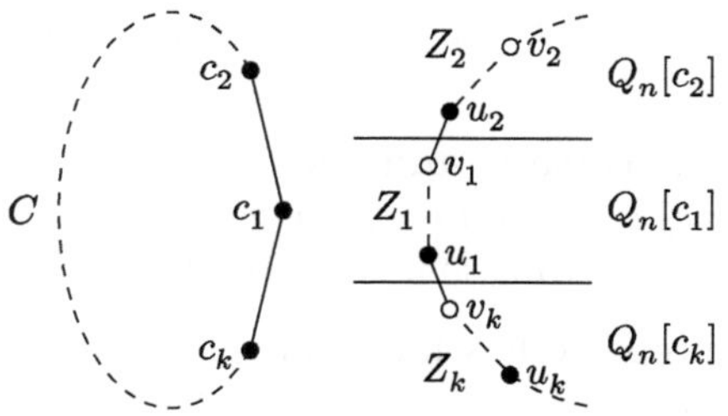

Fig. 3. Illustration of proof of Theorem 7. On the left is a Hamilton cycle $C = (c_1, \ldots, c_k)$ on Q_{n-d} and on the right the solid lines are edges between subcubes and the dashed lines are the Hamilton paths on the subcubes.

is a vertex v_1 avoided by M. Let $k = 2^{n-d}$, and consider a Hamilton cycle $C = (c_1, c_2, \ldots, c_k)$ on Q_{n-d}, such that $Q_n[c_1]$ is the subcube containing v_1.

The objective is for each $i \in [k]$ to construct a Hamilton path Z_i in $Q_n[c_i]$ extending $M[c_i]$ and connecting some vertices u_i and v_i. Also ensuring that u_i and v_{i-1} are adjacent for $i \in \{2, 3, \ldots, k\}$, as well as v_k to u_1. These paths, together with edges $u_i v_{i-1}$ for $i \in \{2, 3, \ldots, k\}$ and $v_k u_1$, complete a Hamilton cycle extending M, which implies the statement. Figure 3 shows how the proof works.

The paths Z_i are constructed inductively, starting from vertex v_1 avoided by M in $Q_n[c_1]$. For each $i \in \{2, 3, \ldots, k\}$, u_i is chosen as v_{i-1}'s sole neighbor in $Q_n[c_i]$. From Observation 5 each $Q_n[c_i]$ contains a Hamilton cycle C_i extending the matching $M[c_i]$. Selecting v_i among u_i's neighbors in C_i, such that $u_i v_i \notin M$, creates a Hamilton path $Z_i = E(C_i) \setminus \{u_i v_i\}$ between u_i and v_i extending $M[c_i]$.

Upon reaching v_k, u_1 is identified as the sole v_k's neighbor in $Q_n[c_1]$. Applying Conjecture 2 to $M[c_1]$ and vertices u_1 and v_1 yields a Hamilton path Z_1 from u_1 to v_1, extending $M[c_1]$ and completing the proof. □

Proof (Proof of Theorem 4). The theorem directly follows from Observation 6 and Theorem 7. □

3 Extending to a Hamilton Path

The extension of matchings to Hamilton paths presents additional challenges beyond the cycle case. Half-layers create rigid structures that constrain possible paths through the hypercube, making them natural obstacles to path extension.

This phenomenon was observed by Gregor, Novotný and Škrekovski [10] in their study of path extensions in bipartite graphs derived from hypercubes. Their work, presented in Theorem 8, characterizes when perfect matchings in $B(Q_n)$ can be extended to Hamilton paths. The dimension requirement $n \geq 5$ proves essential, as it provides sufficient flexibility for routing paths through the hypercube while accommodating matching constraints. Smaller dimensions lack this flexibility, as demonstrated by $n = 4$ where certain matching configurations provably [10] prevent Hamilton path extensions.

This characterization builds on a fundamental structural property of half-layers:

Lemma 1 (Gregor, Novotný, and Škrekovski [10], Lemma 4). *Let H_i and H_j be half-layers of Q_n of different directions i and j, respectively. Then, every edge of H_i is incident with some edge of H_j.*

This intersection property has important implications. If a matching contains a half-layer in some direction i, it cannot contain even a single edge of a half-layer in any other direction j, as these edges would necessarily share endpoints. Consequently, if a perfect matching contains two half-layers, they must be in the same direction with opposite parity. By [10, Lemma 6], a matching of Q_n contains an almost half-layer in at most one direction, leading to the same constraint for matchings with two almost half-layers for $n \geq 4$.

The conditions C1 and C2 have structural implications. When either holds, M must contain either a half-layer or an almost half-layer in some direction i and by previous discussion only in that direction. Furthermore, C1 requires u and v to be covered by the half-layer, while C2 requires those vertices to be covered by edges in direction j different from i. Thus, only C1 or C2 may hold at once (Fig. 4). Moreover, these conditions naturally suggest how M could be completed to a perfect matching: any extension must use only edges in direction i to complete the half-layer or almost half-layer structure. We call this completion the *unique extension of* M, as it aligns with the conditions characterized in Theorem 8. Since any Hamilton path extending the unique extension must use all its edges, such a path also extends M.

The pathological nature of half-layers is most evident when considering the C-conditions. Their necessity follows from similar arguments as in the proof of [10, Theorem 3], using counting arguments based on vertex parity and the structural constraints imposed by half-layers and almost half-layers.

We now outline our approach for proving Theorem 13. The proof divides into two main cases based on whether the endpoints u and v lie in the same subcube or different subcubes of the decomposition $Q_n = Q_{n-d} \square Q_d$.

When u and v lie in different subcubes, that is if $u^L \neq v^L$, our approach depends on the parity relationship of u^L and v^L. If these have different parity, we can construct a Hamilton path between them in Q_{n-d} and use it to guide our construction of paths through the Q_d subcubes. If they share the same parity, no Hamilton path between them exists in Q_{n-d}. Instead we identify one vertex to avoid and construct a Hamilton path in Q_{n-d} that bypasses this vertex [14, Corollary 4]. We then carefully attach the corresponding subcube to complete the construction.

When u and v occupy the same subcube, that is if $u^L = v^L$, we employ a cyclic structure in Q_{n-d} similar to our approach for Hamilton cycles. The key challenge is finding appropriate connection points for the global path within the originating subcube containing both endpoints. The approach differs based on whether C-conditions hold locally in the originating subcube.

In all cases, the final Hamilton path emerges from connecting Hamilton paths within individual Q_d subcubes according to the structure we establish in Q_{n-d}.

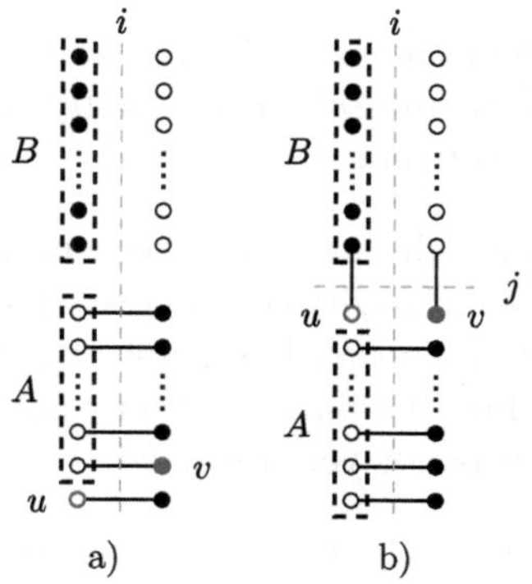

Fig. 4. Necessary conditions for the existence of Hamilton paths in a hypercube, a) condition C1, u and v covered by a half-layer b) condition C2, M contains u and v avoiding almost half-layer in direction i and $uu^j, vv^j \in M, j \neq i$.

To analyze these path constructions systematically, we need a framework for connecting Hamilton paths across multiple Q_d subcubes. When we traverse a path in Q_{n-d}, we visit a sequence of subcubes in Q_n, and we must understand how to thread a Hamilton path through this sequence while respecting the matching constraints.

This motivates the following definition. A *sequence of m cubes of dimension d* is the graph $P_m \square Q_d$, where P_m is the path on m vertices. For a path $Y = (y_1, \ldots, y_m)$ in Q_{n-d} where $m \leq 2^{n-d}$, its induced sequence in Q_n consists of subcubes $G_i = Q_n[y_i]$ and edges between them, with corresponding matchings $M_i = M[y_i]$ for $i \in [m]$. We refer to this as the *induced sequence of cubes and matchings of path Y*.

Any cube sequence embeds as a subgraph into Q_n for sufficiently large n, allowing us to extend our hypercube notation to these sequences. Vertices and edges in cube sequences exhibit a dual nature. They possess both local properties within their canonical copy of Q_d and global properties from their embedding in Q_n. For instance, in $Q_n = Q_{n-d} \square Q_d$, a vertex $u \in V(Q_n)$ with $u^L = \mathbf{1} \in V(Q_{n-d})$ and $u^R = \mathbf{0} \in V(Q_d)$ is locally even in its Q_d copy due to $u^R = \mathbf{0}$, while its global parity in Q_n is determined by $u^L = \mathbf{1}$.

A *sequence of half-layers in $P_m \square Q_d$* consists of half-layers in each canonical copy of Q_d, all sharing the same direction and global parity. These sequences exhibit an important alternation property. When traversing consecutive cubes in the sequence, half-layers of the same global parity must alternate between odd and even local parity within their respective Q_d copies. This property is inherent to the embedding structure. We now establish our first key structural result for such sequences:

Lemma 2. *Let $d \geq 5$ and $G = P_m \square Q_d$ with $m \geq 2$. For $i \in [m]$, let M_i be a matching of the i-th canonical copy G_i of Q_d, and let $M = \cup_{i \in [m]} M_i$. For vertices $u \in G_1, v \in G_m$ of opposite global parity, there exists a Hamilton path between u and v extending M if and only if no sequence of half-layers* *in G contained in M covers both u and v, assuming Conjecture 3 holds for dimension d. Moreover, when connecting consecutive canonical copies of Q_d, we can choose from 2^{d-3} different edges.*

This lemma fully handles the case of Theorem 13 where endpoints lie in different subcubes with different parity coordinates. For same-parity coordinates, we need additional flexibility in our path routing, that is represented by the lower bound of 2^{d-3} when threading the path between subcubes. We are then able to attach an extra cube that was not yet traversed:

Lemma 3. *Let $d \geq 5$ and $G = P_2 \square Q_d$ with canonical copies G_1, G_2 of Q_d containing matchings M_1, M_2 respectively. Let Z_1 and Z_2 be two distinct Hamilton paths extending M_1 in G_1, where Z_1 is between u and v_1 and Z_2 is between u and v_2. Note that v_1 and v_2 are not necessarily distinct. Assuming Conjecture 3 holds for dimension d, there exists a Hamilton path in G extending $M_1 \cup M_2$ from u to at least one of $\{v_1, v_2\}$.*

Having handled cases where endpoints u and v lie in different canonical copies of Q_d, we now address the remaining case where they occupy the same subcube.

Our strategy uses a cycle structure in Q_{n-d}. We construct Hamilton paths through all subcubes along this cycle except the originating subcube containing u and v, connecting them via two carefully chosen vertices in that originating subcube.

When no C-conditions hold locally, Conjecture 5 provides two distinct Hamilton paths between u and v within their subcube. We select an edge not part of the matching from one of these paths and use its endpoints as connection points. The availability of two distinct paths ensures such an edge exists.

When C-conditions do hold locally, no Hamilton path exists between u and v. Instead, we decompose the subcube into two disjoint paths. One path starts from u and another starts from v, together covering all vertices. The endpoints of these paths become our connection points for the global construction. The following result captures this situation:

Lemma 4. *Let $n \geq 5$ and M be a matching of Q_n. Let $u, v \in V(Q_n)$ be vertices of different parity with $uv \notin M$. If condition C1 or C2 from Conjecture 3 holds for u, v, and M, then there exist vertex-disjoint paths Z_u and Z_v where Z_u connects u to z_u, Z_v connects v to z_v, with z_u adjacent to u, z_v adjacent to v, and z_u and z_v non-adjacent to each other, such that $Z_u \cup Z_v$ covers all vertices of Q_n while extending M.*

With these lemmas established, we can construct Hamilton paths through all subcubes along the cycle in Q_{n-d} and correctly connect them to the originating subcube containing u and v. Whether C-conditions hold locally or not, we obtain appropriate connection points that enable a global Hamilton path extending the matching M. This completes our analysis of all subcases, providing the foundation for the full proof of Theorem 13.

Acknowledgments. We thank the anonymous referees for their valuable comments and suggestions that improved the quality of this paper. We are grateful to Jan Kratochvíl for his opposition of the thesis on which this paper is based, which helped strengthen our arguments and presentation. We acknowledge the use of large language models for grammar checking and style improvements.

References

1. Alahmadi, A., Aldred, R.E., Alkenani, A., Hijazi, R., Solé, P., Thomassen, C.: Extending a perfect matching to a Hamiltonian cycle. Discret. Math. Theor. Comput. Sci. (Online Edn.) **17**(1), 241–254 (2015)
2. Dimitrov, D., Dvořák, T., Gregor, P., Škrekovski, R.: Gray codes avoiding matchings. Discret. Math. Theor. Comput. Sci. **11**, 123–147 (2009)
3. Dvořák, T., Fink, J.: Gray codes extending quadratic matchings. J. Graph Theory (2018). https://doi.org/10.1002/jgt.22371
4. Dvořák, T., Gregor, P.: Hamiltonian paths with prescribed edges in hypercubes. Discret. Math. **307**(16), 1982–1998 (2007)
5. Dvořák, T.: Hamiltonian cycles with prescribed edges in hypercubes. SIAM J. Discret. Math. **19**(1), 135–144 (2005)
6. Fink, J.: Perfect matchings extend to Hamilton cycles in hypercubes. J. Comb. Theory Ser. B **97**(6), 1074–1076 (2007)
7. Fink, J.: Matchings extend into 2-factors in hypercubes. Combinatorica **39**(1), 77–84 (2018). https://doi.org/10.1007/s00493-017-3731-8
8. Fink, J., Mütze, T.: Matchings in hypercubes extend to long cycles. In: Rescigno, A.A., Vaccaro, U. (eds.) IWOCA 2024. LNCS, vol. 14764, pp. 14–27. Springer, Cham (2024). https://doi.org/10.1007/978-3-031-63021-7_2
9. Gregor, P.: Perfect matchings extending on subcubes to Hamiltonian cycles of hypercubes. Discret. Math. **309**(6), 1711–1713 (2009)
10. Gregor, P., Novotný, T., Škrekovski, R.: Extending perfect matchings to gray codes with prescribed ends. Electron. J. Comb. **25**(2), 2–56 (2018)
11. Gros, L.: Théorie du Baguenodier. Aimé Vingtrinier, Lyon (1872)
12. Knuth, D.E.: The Art of Computer Programming, Volume 4, Fascicle 2: Generating All Tuples and Permutations. Addison-Wesley Professional (2005)
13. Kreweras, G.: Matchings and Hamiltonian cycles on hypercubes. Bull. Inst. Combin. Appl. **16**, 87–91 (1996)
14. Lewinter, M., Widulski, W.: Hyper-Hamilton laceable and caterpillar-spannable product graphs. Comput. Math. Appl. **34**(11), 99–104 (1997). https://doi.org/10.1016/s0898-1221(97)00223-x
15. Ruskey, F., Savage, C.: Hamilton cycles that extend transposition matchings in cayley graphs of S_n. SIAM J. Discret. Math. **6**(1), 152–166 (1993)
16. Savage, C.: A survey of combinatorial Gray codes. SIAM Rev. **39**(4), 605–629 (1997)
17. Simmons, G.J.: Almost all n-dimensional rectangular lattices are Hamilton-laceable. In: Presented at the 9th Southeastern Conference on Combinatoric Graph Theory and Computing (1977)
18. Vandenbussche, J., West, D.B.: Extensions to 2-factors in bipartite graphs. Electron. J. Comb. **20**(3), 1–10 (2013)
19. Wang, S., Wang, F.: A kind of matchings extend to Hamiltonian cycles in hypercubes. RAIRO-Oper. Res. **58**(6), 5237–5254 (2024)
20. Xu, J.: Topological Structure and Analysis of Interconnection Networks, vol. 7. Springer, Heidelberg (2013)

Succinct Preferential-Attachment Graphs

Ziad Ismaili Alaoui[1], Namrata[1], and Sebastian Wild[1,2(✉)]

[1] University of Liverpool, Liverpool, UK
{ziad.ismaili-alaoui,n.namrata,sebastian.wild}@liverpool.ac.uk
[2] University of Marburg, Marburg, Germany
wild@informatik.uni-marburg.de

Abstract. Computing over compressed data combines the space saving of data compression with efficient support for queries directly on the compressed representation. Such data structures are widely applied in text indexing and have been successfully generalised to trees. For graphs, support for computing over compressed data remains patchy; typical results in the area of succinct data structures are restricted to a specific class of graphs and use the same, worst-case amount of space for any graph from this class.

In this work, we design a data structure whose space usage automatically improves with the compressibility of the graph at hand, while efficiently supporting navigational operations (simulating adjacency-list access). Specifically, we show that the space usage approaches the *instance-optimal* space when the graph is drawn according to the classic Barabási-Albert model of preferential-attachment graphs. Our data-structure techniques also work for arbitrary graphs, guaranteeing a size asymptotically no larger than an entropy-compressed edge list. A key technical contribution is the careful analysis of the instance-optimal space usage.

(Omitted proofs are found online in the extended arXiv version (arxiv.org/abs/2506.21436) of this paper).

1 Introduction

In this paper, we design a compressed representation for graphs generated according to the Barabási-Albert model of preferential attachment approaching the instance-optimal space usage of $\lg(1/p)$ bits, for p the probability of the stored graph, that supports efficient operations within the same space. (Here and throughout, $\lg = \log_2$).

The motivation and techniques of our work span the fields of information theory, compression algorithms, and succinct data structures. Information theory studies random processes (*sources*) for generating combinatorial objects, as a way to quantify the intrinsic information content in an object x. Compression methods are algorithms to efficiently construct representations of x that come close to this lower bound for the size of representations. Succinct data structures

This work was supported by EPSRC grant EP/X039447/1.

H. Fernau and P. Kindermann (Eds.): WG 2025, LNCS 16124, pp. 270–285, 2026.
https://doi.org/10.1007/978-3-032-11835-6_20

aim to support efficient queries for an object $x \in \mathcal{X}$ using $\lg|\mathcal{X}|(1+o(1))$ bits of space. This space usage is asymptotically optimal in the *worst case* over $\mathcal{X}$, but can, in principle, be improved to $\lg(1/\mathbb{P}[x])$ bits of space when the object is drawn randomly from $\mathcal{X}$ with probability $\mathbb{P}[x]$. *Universal (source) codes*, as studied in information theory, approximate that space usage over a whole class of random sources "automatically", i.e., *without knowing* the actual probabilities $\mathbb{P}[x]$. In some restricted cases (namely strings [13] and trees [24]), such universal codes have successfully been augmented with efficient query support. When applied to graphs, information theory, compression algorithms, and succinct data structures all are, by comparison, still in their infancy. A short survey of existing work on graphs from these angles is given in Sect. 1.2.

A classical notion of compressibility of a text T is its (zeroth-order) *empirical entropy* $H_0(t)$, (formally defined in Sect. 2). $H_0(T)$ gives a lower bound for the size of any representation that independently encodes individual characters of T, such as the well-known Huffman codes. It is also the entropy rate of the maximum-likelihood memoryless source for T, or the entropy of the character distribution obtained by Bayesian inference, i.e., starting with a uniform prior distribution, for each $T[i]$, update our prior (to a Beta distribution) to make $T[i]$ more likely to appear in the future.

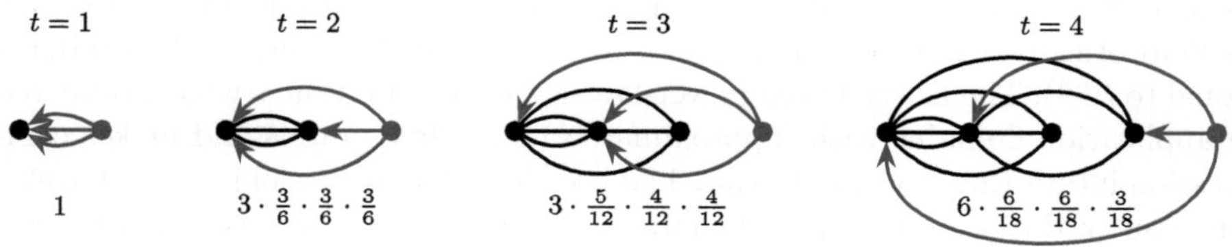

Fig. 1. The Barabási-Albert PA $(M; n)$ model of iteratively growing graphs with $M = 3$ for $n = 4$ time steps. The probability of the shown choices of M targets are given for each time step, yielding $\mathbb{P}[G_4] = 5/864$ overall. The information content is thus $\lg(1/\mathbb{P}[G_4]) \approx 7.43$ bits; the degree-entropy is $H_0^{\text{deg}}(G_4) \approx 15.37$.

Several competing notions of empirical entropies for trees have been proposed [17,18,20] (see also [28, Part II]); for graphs, the grounding in information theory is weaker still. A simple and intuitive notion of empirical entropy for (directed) graphs G can be obtained as follows: Concatenate the adjacency lists (the out-neighbours) of all vertices and consider the empirical entropy $H_0(A)$ of the resulting *adjacency string* $A = A(G)$ over the alphabet $V(G)$. Observe that each vertex $v \in V(G)$ appears exactly $d_{\text{in}}(v)$ times (for $d_{\text{in}}(v)$ the in-degree of v) in $A(G)$; hence we call this quantity $H_0^{\text{deg}}(G)$, the *degree entropy* of G. For ease of presentation, we focus on graphs where the out-degree of all vertices is the same and a canonical order of the vertices is given; then it is easy to reconstruct G from $A(G)$. Note that $H_0^{\text{deg}}(G) = H_0(A(G)) \le m \lg n$, where the latter is the number of bits used by an uncompressed adjacency list. (Here $n = |V(G)|$ is the number of vertices and $m = |E(G)|$ the number of edges).

We will show that despite its simplicity, $H_0^{\deg}(G)$ can be a meaningful benchmark for graph compression. We point out here that this is at least partially surprising since $A(G)$ fixes a specific ordering (and hence, naming) of vertices and edges. We (almost always) consider the graph unchanged when (1) the edges are listed in different order, and, in many applications, (2) when arbitrary renaming of vertices occurs. In the latter case, we only seek to preserve the *structure* of a graph (i.e., the *unlabelled graph* or the equivalence class under relabelling), e.g., to encode labels separately. In addition, *undirected* edges are given a direction by listing them in $A(G)$. So, in general $A^{-1}(G)$ is far from unique and $H_0^{\deg}(G)$ is *not* in general a lower bound for the description length for G. This complication is indicative of what makes graph compression harder than text compression.

As a challenging testbed for compressed graph data structures, we hence consider *unlabelled, undirected* graphs from a highly non-uniform distribution. Specifically, we assume the *Barabási-Albert model* [3] of random *preferential-attachment graphs*. It generates a graph iteratively as follows (Fig. 1). At each time step t, we add one new vertex v_t and draw M random neighbours for v_t, where M is a fixed parameter of the model. The M neighbours $a_{t,1}, \ldots, a_{t,M}$ are chosen, independently and with repetitions allowed, from the existing vertices $v_0, \ldots, v_{t-1}$ with probability *proportional* to the (current) *degree* of vertex v_i, $\mathbb{P}[u_{t,j} = v_i] \propto d_t(v_i)$. This rule follows the *Matthew principle* (rich vertices get richer) and produces graphs where the degree-distribution follows a power law (asymptotically, the probability of a vertex to collect total degree k is proportional to k^{-3}). The heavy-tailed power-law degree distribution is widely observed in application domains with (approximately) scale-free graphs and makes compression interesting and non-trivial. The Barabási-Albert model is also a natural analogue of Bayesian inference of character distributions: whenever we see a vertex as the target of an edge, we consider this target as more likely for future choices.

1.1 Our Results

Suppose G is a graph generated by the Barabási-Albert model, and let $\mathbb{P}[G]$ be the probability for G to arise in this process (where we assume that M is a fixed parameter and we stop the generation process after the nth vertex). Any encoding for graphs that is simultaneously optimal for all such graphs must use (close to) $\lg(1/\mathbb{P}[G])$ bits of space to encode G. Our first set of results relate the amount of information in G, $\lg(1/\mathbb{P}[G])$, to the empirical degree-entropy of G, $H_0^{\deg}(G)$: despite their seemingly unrelated origins, the two quantities *coincide* up to an error term.

Theorem 1.1 (Instance-specific lower bound). *Let G be a labelled graph generated by preferential attachment, $G \sim \mathrm{PA}(M; n)$, and further assume that apart from a fixed-size seed graph, the graph G is simple. Then*

$$\lg(1/\mathbb{P}[G]) \quad = \quad H_0^{\deg}(G) \pm O(nM \lg M).$$

Note that $O(nM \lg M)$ is a lower order term for typical graphs when M is not too big, that is $\lg M = o(\lg n)$. The analysis (Sect. 3) uses Gibbs' inequality to compare different notions of empirical entropy, a trick that might be of independent interest.

For *unlabelled* graphs G, the bound reduces by $n \lg n$ bits:

Theorem 1.2 (Unlabelled lower bound). *Let G be as in Theorem 1.1 and consider its structure $S = S(G)$, i.e., a random* unlabelled *graph generated by preferential attachment. Then*

$$\lg(1/\mathbb{P}[S]) \;\geq\; H_0^{\text{deg}}(S) - n \lg n \;-\; O(nM \lg M).$$

(Note that $H_0^{\text{deg}}(G) = H_0^{\text{deg}}(S(G))$.) It may seem obvious that the bound would reduce precisely by the $\lg(n!) \sim n \lg n$ bits needed to store n vertex labels since any unlabelled graph can correspond to at most $n!$ labelled graphs; however, we point out that (1) for general distributions over labelled graphs, it is not true that these $n!$ labelled graphs are equally likely, and (2) a strictly higher lower bound is the actual truth for certain S, e.g., graphs with linear diameter. We obtain a tight bound in Sect. 3, but its general growth with n is opaque. We do not know whether the bound in Theorem 1.2 is tight. Due to space constraints, some proofs are deferred to the extended arXiv version of this paper.

The degree entropy thus precisely captures the asymptotic information content of a preferential-attachment graph. We next show that we can represent G using close to $H_0^{\text{deg}}(G)$ bits of space and support efficient queries on the compressed representation.

Theorem 1.3 (Ultrasuccinct preferential-attachment graphs). *Let G be obtained from* PA $(M; n)$. *There are data structures that support operations finding (1) the ith in-neighbour, (2) the ith out-neighbour, (3) the degree of a vertex, and (4) testing adjacency, all in $O(\lg n)$ time, in the following total space:*

(a) Storing the labelled graph G uses $H_0^{\text{deg}}(G) + o(Mn)$ bits of space.
(b) Storing the unlabelled graph $S = S(G)$ uses at most $H_0^{\text{deg}}(S)(1 - \frac{1}{M}) + 2n + o(Mn) \leq (M-1)n \lg n + 2n + o(Mn)$ bits of space.

By Theorem 1.1, the space for (a) is asymptotically optimal. The space for (b) almost matches Theorem 1.2; in particular, it does so on average: $\frac{1}{M} H_0^{\text{deg}}(S)$ is close to $n \lg n$ unless S is very compressible. (Note that $A(S)$ has Mn characters, so $H_0^{\text{deg}}(S) = Mn \cdot H_0^{\text{pc}}(A(S))$ and $\frac{1}{M} H_0^{\text{deg}}(S) = n \cdot H_0^{\text{pc}}(A(S))$, for H_0^{pc} the *per-character entropy*; $H_0^{\text{pc}}(A(S))$ is always between 0 and $\lg n$.)

The labelled graph data structure uses an entropy-compressed wavelet tree to represent the string $A(G)$. To reduce the space for an unlabelled graph, our data structure uses one outgoing edge per vertex (except v_0) as a *parent edge* in an ordinal tree. These can be stored using a succinct tree data structure using only 2 bits per edge if we use the tree indices to identify the graph vertices. The remaining out-neighbours form string A', which replaces A and is now n entries shorter, saving up to $\lg n$ bits each. By judiciously choosing which edges to turn into tree edges, we can moreover *retain* the *compressibility* of A'; we show that our choice implies that the per-character entropy in A' is never larger than that of A (Appendix C of the extended arXiv version).

The unlabelled-graph data structure assigns new labels to the vertices of the graph, which have to be used in queries; the relabelling function can be provided during construction, but cannot be stored as part of the data structure in the stated space. A use case for the unlabelled-graph data structure is any task where we only need to identify a small subset of vertices; e.g., computing network-analysis metrics such as betweenness centrality for a subset of important vertices. We then store the names of important vertices in addition to the data structure for $S(G)$.

1.2 Related Work

The closest work to ours is the analysis of $\mathrm{PA}\,(M;n)$ by Łuczak et al. [22,23]. They compute the entropy of the distribution $G \sim \mathrm{PA}\,(M;n)$, $\mathbb{E}[\lg(1/\mathbb{P}[G])] = (M-1)n\lg n\cdot(1+o(1))$ and (only in the conference version [23]) describe a compression algorithm achieving this entropy in expectation up to a redundancy of $O(n\lg\lg n)$. Our work improves their compression method to redundancy $O(n)$ (using a succinct tree instead of their backtracking numbers), provides instance-specific bounds, and supports efficient operations on the compressed representation. (It is not clear whether their compression algorithm can be turned into an efficient compressed data structure).

Many more specialized models of random graphs have been suggested and studied. An example that also features a compressed data structure is the geometric inhomogeneous random graph (GIRG) model [8]. The expected compressed space matches the entropy up to constant factors. It is not clear if instance-optimal space is easily achievable. As stated, the compression method requires a geometric realization of the graph as input.

Another related area is applied graph compression, where methods are evaluated empirically on benchmarks from specific domains. As a representative example, we mention the webgraph framework [6], which provides space-efficient data structures tuned to large web graphs.

The idea to use wavelet trees for the edges of a graph is discussed in Navarro's survey on wavelet trees [26] (and may be folklore). This approach alone is inherently labelled and thus cannot escape the lower bound of Theorem 1.1. The closest prior work to our data structure is the GLOUDS graph representation of Fischer and Peters [14]. GLOUDS partitions the edges into tree edges and non-tree edges using a standard breadth-first search to obtain an ordinal tree; it uses a custom representation for the tree, as well. GLOUDS does not achieve optimal space for $G \sim \mathrm{PA}\,(M;n)$ with constant M. In its original form it also does not adapt to $H_0^{\mathrm{deg}}(G)$ (although entropy compressing their array H would move towards the latter).

There are many further works that are related in spirit to our work, but with a distinctly different angle. For certain specific classes of graphs, *succinct* data structures have been designed: interval graphs [2,16], chordal graphs [25], permutation graphs [29], circle graphs [1], bounded clique-width graphs [10], series-parallel graphs [9]. For representing any graph from a class of graphs with g_n graphs on vertex set $[n]$, these data structures use $\lg(g_n)(1 + o(1))$ bits of space for representing one such graph; this invariably corresponds to a uniform distribution over the class of graphs under consideration. Supported operations vary, but always include finding neighbours of a given vertex.

For trees, data structures beating the worst-case optimal space on compressible inputs have been considered. Ultrasuccinct trees [19] adapt to the degree entropy of a tree. Hypersuccinct trees [24] have been shown to simultaneously yield optimal space for a large variety of entropy measures, both empirical (such as degree entropy) and entropy rates of random tree sources. Here we lift these results to "ultrasuccinct graphs".

On the information theory of structures (unlabelled graphs), a line of work studied the entropy of random unlabelled graphs. Apart from the works of Łuczak et al. [22,23] on preferential attachment graphs, small-world graphs [21], the Erdős-Renyi graphs [11], and a vertex-copying model [30] have been studied.

Lossless compression of graphs is surveyed by Besta and Hoefler [5]; see also [7]. Typical examples consider an application and exploit specifics of graphs arising there. A statistical model of data and efficient queries are often not available.

2 Preliminaries

In this section, we collect some definitions and known results that we build on in this work. We write $[a..b]$ for $\{a, a+1, \ldots, b\}$ and abbreviate $[n] = [1..n]$; also $[a..b) = [a..b-1]$. We use standard graph terminology; specifically, for a graph G and a vertex $v \in V(G)$, we write $d(v) = d_G(v)$ for the degree of v in G, i.e., the number of (potentially parallel) edges incident at v. For directed graphs (digraphs), we distinguish in-degree and out-degree: $d(v) = d_{\text{in}}(v) + d_{\text{out}}(v)$; we also write $N_{\text{out}}(v)$ for the out-neighbourhood of v, i.e., the (multi)set of vertices w for which there is a (bundle of) edge(s) vw in $E(G)$. We call a DAG *M-out-regular* if $d_{\text{out}}(v) = M$ for all vertices $v \in V \setminus \{v_0\}$.

2.1 Empirical Entropy

For a text $T[1..n]$ over alphabet $\Sigma = [1..\sigma]$, the *zeroth-order empirical entropy* $H_0(T)$ is given by

$$H_0(T) \quad = \quad \sum_{c=1}^{\sigma} |T|_c \lg\left(\frac{n}{|T|_c}\right),$$

where $|T|_c = |\{i \in [1..n] : T[i] = c\}|$ denotes the number of occurrences of c in T. Additionally, we define $H_0^{\text{pc}}(T) = H_0(T)/|T|$ to be the *per-character zeroth-order empirical entropy* of T. As is standard, we set $0\lg(n/0) := 0$ for notational convenience.

2.2 Preferential-Attachment Graphs

We consider the classical Barabási-Albert model of generating a random (undirected) graph (with parallel edges); see Fig. 1. Given a target size n and a parameter $M \in \mathbb{N}$, we grow a graph G iteratively, where vertex v_t arrives at time t, for $t = 0, \ldots, n$. We denote by G_t the graph after this arrival at time t; the vertex set is $V(G_t) = \{v_0, \ldots, v_t\}$. G_0 is the single isolated vertex v_0; G_1 has M parallel edges between v_0 and v_1. G_t for $t \geq 2$ results from G_{t-1} by adding v_t and M edges from v_t to targets $a_{t,1}, \ldots, a_{t,M} \in V(G_{t-1}) = \{v_0, \ldots, v_{t-1}\}$, where the $a_{t,1}, \ldots, a_{t,M}$ are mutually independent and identically distributed with

$$\mathbb{P}[a_{t,1} = v_\ell] \;=\; \cdots \;=\; \mathbb{P}[a_{t,M} = v_\ell] \;=\; \frac{d_{G_{t-1}}(v_\ell)}{2|E(G_{t-1})|} \qquad (\ell = 0, \ldots, t-1).$$

Note that if we orient the edges from new to old vertices in $G_n \sim \mathrm{PA}\,(M; n)$ (in the order they arrived to the graph, i.e., from large to small t), we obtain a directed acyclic graph with uniform out-degrees: $d_{\mathrm{out}}(v_t) = M$ for $t \in [1..n]$ ($d_{\mathrm{out}}(v_0) = 0$).

We will abbreviate $d_t(v) = d_{G_t}(v)$. Note that $|E(G_t)| = tM$ and $|V(G_t)| = t+1$. We write $G_n \sim \mathrm{PA}\,(M; n)$ to indicate that G_n has been randomly chosen according to this preferential-attachment process.

For $G \sim \mathrm{PA}\,(M; n)$, we obtain $H_0^{\deg}(G)$ as $H_0^{\deg}(G) = H_0(A(G))$ where

$$A(G) = [a_{1,1}, \ldots, a_{1,M}, \;\ldots,\; a_{n,1}, \ldots, a_{n,M}].$$

Let $S(G)$ denote the *structure* of G, i.e., the class of graphs which are isomorphic to G. We also refer to these as *unlabelled graphs*. We define the unlabelled graph distribution $PA^u(M; n)$ as the probability distribution on the family of $S(G)$, where the probability of each class is the sum of the probabilities that a labelled version of $S(G)$ is $\mathrm{PA}\,(M; n)$, that is,

$$\sum_{H \in S(G)} \mathbb{P}[H = \mathrm{PA}\,(M; n)]\,.$$

2.3 Succinct Data Structures

For the reader's convenience, we collect used results on succinct data structures here. First, we cite the compressed bit vectors of Pătrașcu [27].

Lemma 2.1 (Compressed bit vector). *Let $B[1..n]$ be a bit vector of length n, containing m 1-bits. For any constant $c > 0$, there is a data structure using $\lg\binom{n}{m} + O\big(\frac{n}{\lg^c n}\big) \leq m\lg\big(\frac{n}{m}\big) + O\big(\frac{n}{\lg^c n} + m\big)$ bits of space that supports in $O(1)$ time operations (for $i \in [1..n]$):*

- $\mathsf{access}(B, i)$: *return* $B[i]$, *the bit at index* i *in* B;
- $\mathsf{rank}_\alpha(B, i)$: *return the number of bits with value* $\alpha \in \{0, 1\}$ *in* $B[1..i]$;
- $\mathsf{select}_\alpha(B, i)$: *return the index of the* i*th bit with value* $\alpha \in \{0, 1\}$.

Using wavelet trees, we can support rank and select queries on arbitrary static strings/sequences while compressing them to zeroth-order empirical entropy.

Lemma 2.2 (Wavelet tree [26]). *Let* $S[1..n]$ *be a array with entries* $S[i] \in \Sigma = [1..\sigma]$. *There is a data structure using* $H_0(S) + o(n)$ *bits of space that supports the following queries in* $O(\lg \sigma)$ *time (without access to* S *at query time):*

- $\mathsf{access}(S, i)$: *return* $S[i]$, *the symbol at index* i *in* S;
- $\mathsf{rank}_\alpha(S, i)$: *return the number of indices with value* $\alpha \in \Sigma$ *in* $S[1..i]$;
- $\mathsf{select}_\alpha(S, i)$: *return the index of the* i*th occurrence of value* $\alpha \in \Sigma$ *in* S.

Proof. Many variants of wavelet trees are possible [26]; for our result, we use the plain encoding of the characters with $\lceil \lg \sigma \rceil$ bits each (balanced wavelet tree). All levels except the last are full, so we can concatenate all bitvectors of nodes in the wavelet tree in level order (pointerless wavelet tree [26, §2.3]) and support navigation up and down the tree. Moreover, using Lemma 2.1 for this bitvector of length at most $\lceil \lg \sigma \rceil n$ yields space $H_0(S) + o(n)$ ([26, §3.1]). □

We lastly need a succinct data structure for ordinal trees. Several options exist for the basic queries we need [4,15]; (see [16, App. A] for a comprehensive review).

Lemma 2.3 (Succinct ordinal trees). *Let* T *be an ordinal tree on* n *vertices. There is a data structure using* $2n + o(n)$ *bits of space that supports the following queries in* $O(1)$ *time (where nodes are identified with their preorder index in* T*):*

- $\mathsf{parent}(T, v)$: *return the parent of* v *in* T;
- $\mathsf{degree}(T, v)$: *return the number of children of* v *in* T;
- $\mathsf{child}(T, v, i)$: *return the* i*th child of* v *in* T.

3 Space Lower Bound

In this section, we derive the instance-specific lower bounds for preferential-attachment graphs. Section 3.1 gives the main result for *labelled* graphs. Appendix A of the extended arXiv version (arxiv.org/abs/2506.21436) gives the proof for the *unlabelled* case.

Let $G \sim \mathrm{PA}(M; n)$ be a labelled preferential-attachment graph. Recall that G is drawn iteratively, where vertex v_t arrives at time t, for $t = 1, \ldots, n$. The graph after the arrival of v_t is G_t, where $V(G_t) = \{v_0, \ldots, v_t\}$. The total number of edges in $G = G_n$ is $m = Mn$.

For time $t = 1, \ldots, n$, let $N_{\text{out}}(v_t)$ denote the random multiset of size M that contains the M out-neighbours of v_t randomly sampled from the set $V(G_{t-1}) = \{v_0, v_1, \ldots, v_{t-1}\}$. The vertex v_t attaches to a vertex in $V(G_{t-1})$ with probability proportional to its degree at time $t-1$:

$$P_i^{(t)} = \mathbb{P}\left[v_i \in N_{\text{out}}(v_t) \,\middle|\, G_{t-1}\right] = \frac{d_{t-1}(v_i)}{2(t-1)M}\,. \tag{1}$$

Multinomial Distribution of $N_{\text{out}}(v_t)$. The frequencies of appearance of vertices as out-neighbours of v_t follow a multinomial distribution. The number of trials in this case is M, where each trial when v_t chooses the neighbour from the set $V(G_{t-1})$ is independent of each other. Also, each trial has t possible outcomes $v_0, v_1, \ldots v_{t-1}$, with probabilities as given by Eq. (1). Therefore, we have (conditional on G_{t-1}), that $(C_0^{(t)}, \ldots, C_{t-1}^{(t)}) \sim \text{Mult}(M; P_0^{(t)}, P_1^{(t)}, \ldots, P_{t-1}^{(t)})$, where $C_i^{(t)}$ denote the number of times the vertex v_i is chosen randomly as a neighbour of v_t at time t. Therefore,

$$\mathbb{P}\left[(C_0^{(t)}, \ldots, C_{t-1}^{(t)}) = (c_0, \ldots c_{t-1}) \,\middle|\, G_{t-1}\right] = \binom{M}{c_0, \ldots, c_{t-1}} (P_0^{(t)})^{c_0} \ldots (P_{t-1}^{(t)})^{c_{t-1}}.$$

3.1 Proof of Theorem 1.1

In this subsection, we derive a lower bound on $\lg(1/\mathbb{P}[G])$ for $\mathbb{P}[G]$ the probability that a given graph $G \sim \text{PA}\,(M;n)$ results from a preferential-attachment process $\text{PA}\,(M;n)$.

Recall that $G = G_n$. Using the product rule, we get

$$\mathbb{P}[G] = \mathbb{P}\big[N_{\text{out}}(v_n) \,\big|\, G_{n-1}\big] \cdot \mathbb{P}\big[N_{\text{out}}(v_{n-1}) \,\big|\, G_{n-2}\big] \cdots \mathbb{P}\big[N_{\text{out}}(v_2) \,\big|\, G_1\big] \cdot \mathbb{P}[G_1]\,.$$

This implies

$$\begin{aligned} \lg\left(\frac{1}{\mathbb{P}[G]}\right) &= \sum_{t=2}^{n} \lg\left(\frac{1}{\mathbb{P}\big[N_{\text{out}}(v_t) \,\big|\, G_{t-1}\big]}\right) \\ &= \sum_{t=2}^{n}\left[\sum_{i=0}^{t-1} C_i^{(t)} \lg\left(\frac{2(t-1)M}{d_{t-1}(v_i)}\right) - \lg\left(\binom{M}{C_0^{(t)}, \ldots, C_{t-1}^{(t)}}\right)\right]. \end{aligned} \tag{2}$$

Consider $-\sum_{t=2}^{n} \lg\left(\binom{M}{C_0^{(t)}, \ldots, C_{t-1}^{(t)}}\right)$. We have

$$-\lg\left(\binom{M}{C_0^{(t)}, \ldots, C_{t-1}^{(t)}}\right) = \lg\left(\frac{C_0^{(t)}! \ldots C_{t-1}^{(t)}!}{M!}\right) \geq \lg\left(\frac{1}{M!}\right) = -\lg(M!).$$

Now let us consider the first summand in (2),

$$\sum_{t=2}^{n}\sum_{i=0}^{t-1} C_i^{(t)} \lg\left(\frac{2(t-1)M}{d_{t-1}(v_i)}\right)$$
$$= \sum_{t=1}^{n-1}\left(\sum_{i=0}^{t} C_i^{(t+1)} \lg(2tM)\right) - \sum_{t=1}^{n-1}\left(\sum_{i=0}^{t} C_i^{(t+1)} \lg(d_t(v_i))\right). \quad (3)$$

Let us analyse the first summand. With $\sum_{i=0}^{t} C_i^{(t+1)} = M$, we find

$$\begin{aligned}\sum_{t=1}^{n-1}\left(\sum_{i=0}^{t} C_i^{(t+1)} \lg(2tM)\right) &= \sum_{t=1}^{n-1} \lg(2tM)\left(\sum_{i=0}^{t} C_i^{(t+1)}\right)\\ &= M(n-1)\lg(2M) + M\lg((n-1)!).\end{aligned}$$

Now consider the second summand in (3). Swapping the order of summation yields

$$-\sum_{t=1}^{n-1}\left(\sum_{i=0}^{t} C_i^{(t+1)} \lg(d_t(v_i))\right) = -\sum_{i=0}^{n-1}\left(\sum_{t=i+1}^{n-1} C_i^{(t+1)} \lg(d_t(v_i))\right).$$

Let $T_i = \{t \geq 2 : C_i^{(t)} > 0\}$ denote the set of timestamps such that the vertex v_i is selected as an out-neighbour; we have $T_i = \{t_1^{(i)}, \ldots, t_{d_{\text{in}}(v_i)}^{(i)}\}$ with $t_1^{(i)} \leq \cdots \leq t_{d_{\text{in}}(v_i)}^{(i)}$, i.e., just before time $t_k^{(i)}$, v_i's total degree was $d_{t_k^{(i)}-1}(v_i) = M + k - 1$.[1] So far, the analysis works for general graphs; for the following simplification, we assume that no parallel edges[2] are added in the graph (for $t \geq 2$). Then $C_i^{(t)} \leq 1$ and the summand becomes

$$\begin{aligned}-\sum_{i=0}^{n-1}\sum_{\substack{t=i+1\\ t\geq 2}}^{n} C_i^{(t)} \lg(d_{t-1}(v_i)) &= -\sum_{i=0}^{n-1}\sum_{k=1}^{d_{\text{in}}(v_i)} \lg(M+k-1)\\ &= -\sum_{i=0}^{n-1}\left(\sum_{k=1}^{d_{\text{in}}(v_i)+M-1} \lg(k) - \sum_{k=1}^{M-1}\lg(k)\right)\\ &= -\sum_{i=0}^{n-1} \lg((d(v_i)-1)!) + n\lg((M-1)!).\end{aligned}$$

[1] For ease of notation, we do not count the M edges from v_1 to v_0 in $d_{\text{in}}(v_0)$; then for all vertices, $d(v_i) = d_{\text{in}}(v_i) + M$. That is also why we only include times $t \geq 2$ in T_i.

[2] We note that G_M necessarily contains parallel edges, and our assumptions concern the edges chosen for $t > M$. The contribution of these parallel edges in the calculation below can be shown to be $O(M)$ overall, so we ignore their presence for legibility.

Therefore, Eq. (2) becomes

$$\lg\left(\frac{1}{\mathbb{P}[G]}\right) \geq M(n-1)\lg(2M) + M(\lg((n-1)!)) - \sum_{i=0}^{n-1}\lg((d(v_i)-1)!) + n\lg((M-1)!) - (n-1)\lg(M!) = Mn\lg(2Mn) - \sum_{i=0}^{n-1} d(v_i)\lg(d(v_i)) \pm O(Mn)$$

(using $d(v) = d_{\text{in}}(v) + M$)

$$= \underbrace{Mn\sum_{i=0}^{n-1}\frac{d_{\text{in}}(v_i)}{Mn}\lg\left(\frac{2Mn}{d(v_i)}\right)}_{(*)} - \underbrace{\sum_{i=0}^{n-1} M\lg(d(v_i))}_{(\dagger)} \pm O(Mn)$$

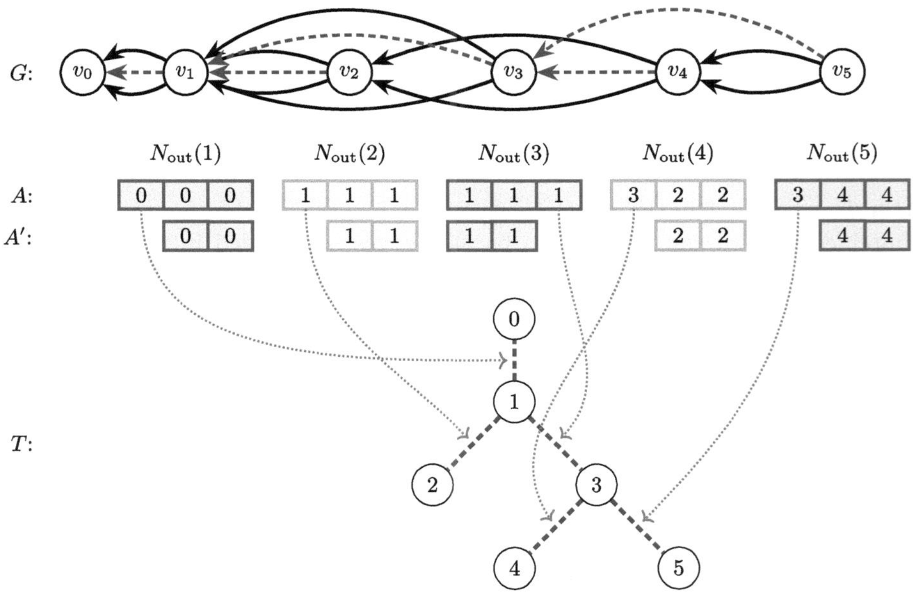

Fig. 2. Illustration of the construction of T, A and A' on a sample preferential-attachment graph where $n = 5$ and $M = 3$. The dashed red edges in G represent the edges that are represented in T. Note that the vertices are labelled by their pre-order index in T. (Color figure online)

(bounding (∗) by Gibbs' inequality, and (†) by the log-sum inequality)

$$\geq \; Mn\sum_{i=0}^{n-1}\frac{d_{\text{in}}(v_i)}{Mn}\lg\left(\frac{Mn}{d_{\text{in}}(v_i)}\right) \;\pm\; O(Mn\lg M)$$

$$= \; H_0^{\text{deg}}(G) \;\pm\; O(Mn\lg M). \tag{4}$$

This proves Theorem 1.1.

4 Data Structures

In this section, we describe data structures that can represent a directed graph generated by the preferential-attachment process in compressed form, while allowing for efficient navigational queries without decompression.

For labelled graphs, we only use the wavelet tree portion of the data structure described in the following. We omit the simple modifications here and present the data structure for unlabelled graphs in detail; (see also [26, §5.3] for the labelled case).

4.1 Construction

Let $G' \sim \text{PA}\,(M;n)$ be generated by the Barabási-Albert model. By iteratively "peeling off" degree-M vertices, we uniquely obtain the directed version $G = DAG(G')$ [22].

We now sort vertices by indegree and construct a tree T informally by choosing, for each vertex v_t in G (except for v_0), a parent among its out-neighbours with minimal (in-)degree. Formally, define a bijective *rank* function $\sigma : \{v_0,\ldots,v_n\} \to [0..n]$ such that $d_{\text{in}}(u) > d_{\text{in}}(v)$ implies $\sigma(u) > \sigma(v)$. Construct a tree T with $V(T) = V(G)$ and

$$E(T) \;=\; \bigcup_{t=1}^{n}\Big\{(v_t,\; \underset{w\in N_{\text{out}}(v_t)}{\arg\min}\;\sigma(w))\Big\}.$$

Observe that T is indeed a tree since it consists of $n+1$ vertices and n edges and is acyclic by definition. Vertex v_0 is the root of T.

We will store T using Lemma 2.3, where operations refer to nodes via their preorder index. We hence perform a preorder traversal of T from its root, and relabel $v_t \in V(G)$ by its index in the traversal, starting from 0.

For each vertex $v \neq 0$ in G, define an order on its outgoing edges such that the edge (v,u), for some u, is the first if and only if $(v,u) \in E(T)$, and all other vertices (v,w), $u \neq w$, are ordered arbitrarily. Finally, construct a static array A' such that the $(i+1)$th ($i \in [1..M)$) out-neighbour of vertex j in G is stored in $A'[(j-1)(M-1)+i]$. See Fig. 2 for an illustration of the construction on a sample graph. We store A' in a wavelet tree using Lemma 2.2.

4.2 Operations

Building upon the aforementioned data structures, we present algorithms to efficiently handle a variety of navigational queries on G. Define $\mathsf{source}(i) = \lceil \frac{i}{M-1} \rceil$ as the source of the edge whose target is stored in $A'[i]$.

Computing the ith Out-Neighbour of v. The ith *out-neighbour* of v in G can be determined as follows: if $i = 1$, it is the parent of v in T; otherwise, it is given by $A'[(v-1)(M-1)+i-1]$ (the $(i-1)$th element of $A'[(v-1)(M-1)+1..v(M-1)]$, which represents all but the first out-neighbour of v).

Computing the ith in-Neighbour of v. To determine the ith *in-neighbour* of v, we distinguish two cases:

1. If v has at least i children in T, then its ith in-neighbour in G is simply its ith child in T.
2. Otherwise, if $i > \mathsf{degree}(T, v)$, we determine the $(i-\mathsf{degree}(T, v))$th occurrence of v in A'. The vertex whose neighbourhood includes this occurrence is found by computing $\mathsf{source}(\mathsf{select}_v(S, i - \mathsf{degree}(T, v)))$.

Using these two operations, we can also iterate through the neighbourhoods or return all (in- resp. out-) neighbours.

Computing the In-Degree of v. The in-degree of a vertex v can be determined by summing the number of its children in T (computable by $\mathsf{degree}(T, v)$) and its occurrences in A' (computable by $\mathsf{rank}_v(A', n(M-1))$.

Determining if u and v are Adjacent. If u and v are adjacent, either (1) u is the parent of v in T, $\mathsf{parent}(T, v) = u$, (2) vice versa, $\mathsf{parent}(T, u) = v$, or (3) v occurs in $A'[(u-1)(M-1)+1..u(M-1)]$ or (4) vice versa, u occurs in $A'[(v-1)(M-1)+1..v(M-1)]$. (If either of them is v_0, two cases need not be checked.) Conditions (1) and (2) are directly supported on T; for (3) and (4), we can use rank: If $\mathsf{rank}_v(A', u(M-1)) - \mathsf{rank}_v(A', (u-1)(M-1)) \geq 1$, then v occurs at least once, so v and u are adjacent. (4) is similar.

To conclude the proof of Theorem 1.3, it remains to analyse the space usage of our data structure; details are given in Appendix B of the extended arXiv version (arxiv.org/abs/2506.21436).

Remark 4.1 (General graphs). We point out that the data-structure techniques above can be extended to general graphs. Using a compressed bit vector to mark where the next neighbourhoods begin, we can support the same queries with $n \lg(m/n) + O(n)$ extra space.

5 Conclusion

We designed a compressed representation for graphs generated by the Barabási-Albert model of random preferential-attachment graphs approaching the instance-optimal $\lg(1/\mathbb{P}[G])$ bits of space, where $\mathbb{P}[G]$ is the probability for the graph to arise in the model. We further related $\lg(1/\mathbb{P}[G])$ to the empirical degree-entropy of G, $H_0^{\deg}(G)$; they coincide up to an error term in both labelled and unlabelled graphs. Our compressed representation supports navigational queries in $O(\log n)$ time and can simulate access to an adjacency-list representation.

Future work might study other models of preferential attachment. For example, the model introduced by Cooper and Frieze [12], where the number of edges added follows a given distribution. One can also consider a non-linear model in which the probability of a vertex, when M edges are added, being chosen is proportional to its degree raised to some power α. (Here $\alpha = 1$).

More broadly speaking, this work merely made initial strides into data structures that adapt to information-theoretic measures of compressibility in graphs, leaving many avenues for future work open. For example, unlike in trees, a convincing notion of higher-order entropy for graphs is not yet emerging, let alone data structures approaching them.

References

1. Acan, H., Chakraborty, S., Jo, S., Nakashima, K., Sadakane, K., Rao Satti, S.: Succinct navigational oracles for families of intersection graphs on a circle. Theoret. Comput. Sci. **928**, 151–166 (2022). https://doi.org/10.1016/j.tcs.2022.06.022
2. Acan, H., Chakraborty, S., Jo, S., Satti, S.R.: Succinct encodings for families of interval graphs. Algorithmica **83**(3), 776–794 (2020). https://doi.org/10.1007/s00453-020-00710-w
3. Barabási, A.L., Albert, R.: Emergence of scaling in random networks. Science **286**(5439), 509–512 (1999). https://doi.org/10.1126/science.286.5439.509
4. Benoit, D., Demaine, E.D., Munro, J.I., Raman, R., Raman, V., Rao, S.S.: Representing trees of higher degree. Algorithmica **43**(4), 275–292 (2005). https://doi.org/10.1007/s00453-004-1146-6
5. Besta, M., Hoefler, T.: Survey and taxonomy of lossless graph compression and space-efficient graph representations (2019)
6. Boldi, P., Vigna, S.: The webgraph framework I: compression techniques. In: Proceedings of the 13th International Conference on World Wide Web, WWW 2004, pp. 595–602. ACM (2004). https://doi.org/10.1145/988672.988752
7. Bouritsas, G., Loukas, A., Karalias, N., Bronstein, M.: Partition and code: learning how to compress graphs. In: Ranzato, M., Beygelzimer, A., Dauphin, Y., Liang, P., Vaughan, J.W. (eds.) Advances in Neural Information Processing Systems, vol. 34, pp. 18603–18619. Curran Associates, Inc. (2021)
8. Bringmann, K., Keusch, R., Lengler, J.: Geometric inhomogeneous random graphs. Theoret. Comput. Sci. **760**, 35–54 (2019). https://doi.org/10.1016/j.tcs.2018.08.014
9. Chakraborty, S., Jo, S., Sadakane, K., Rao Satti, S.: Succinct data structures for SP, block-cactus and 3-leaf power graphs. Int. J. Found. Comput. Sci. 1–18 (2023). https://doi.org/10.1142/s012905412341006x
10. Chakraborty, S., Jo, S., Sadakane, K., Rao Satti, S.: Succinct data structures for bounded clique-width graphs. Discret. Appl. Math. **352**, 55–68 (2024). https://doi.org/10.1016/j.dam.2024.03.016

11. Choi, Y., Szpankowski, W.: Compression of graphical structures: fundamental limits, algorithms, and experiments. IEEE Trans. Inf. Theory **58**(2), 620–638 (2012). https://doi.org/10.1109/tit.2011.2173710
12. Cooper, C., Frieze, A.: A general model of web graphs. Random Struct. Algorithms **22**(3), 311–335 (2003)
13. Ferragina, P., Manzini, G.: Opportunistic data structures with applications. In: Annual Symposium on Foundations of Computer Science (FOCS). IEEE Computer Society (2000). https://doi.org/10.1109/sfcs.2000.892127
14. Fischer, J., Peters, D.: GLOUDS: representing tree-like graphs. J. Discrete Algorithms **36**, 39–49 (2016). https://doi.org/10.1016/j.jda.2015.10.004
15. Geary, R.F., Raman, R., Raman, V.: Succinct ordinal trees with level-ancestor queries. ACM Trans. Algorithms **2**(4), 510–534 (2006). https://doi.org/10.1145/1198513.1198516
16. He, M., Munro, J.I., Nekrich, Y., Wild, S., Wu, K.: Distance oracles for interval graphs via breadth-first rank/select in succinct trees. In: International Symposium on Algorithms and Computation (ISAAC). LIPIcs, pp. 25:1–25:18. Schloss Dagstuhl (2020)
17. Hucke, D., Lohrey, M., Seelbach Benkner, L.: Entropy bounds for grammar-based tree compressors. In: IEEE International Symposium on Information Theory (ISIT), pp. 1687–1691. IEEE (2019). https://doi.org/10.1109/isit.2019.8849372
18. Hucke, D., Lohrey, M., Benkner, L.S.: A comparison of empirical tree entropies. In: Boucher, C., Thankachan, S.V. (eds.) SPIRE 2020. LNCS, vol. 12303, pp. 232–246. Springer, Cham (2020). https://doi.org/10.1007/978-3-030-59212-7_17
19. Jansson, J., Sadakane, K., Sung, W.K.: Ultra-succinct representation of ordered trees with applications. J. Comput. Syst. Sci. **78**(2), 619–631 (2012). https://doi.org/10.1016/j.jcss.2011.09.002
20. Kieffer, J.C., Yang, E.H., Szpankowski, W.: Structural complexity of random binary trees. In: 2009 IEEE International Symposium on Information Theory. IEEE (2009). https://doi.org/10.1109/isit.2009.5205704
21. Kontoyiannis, I., Lim, Y.H., Papakonstantinopoulou, K., Szpankowski, W.: Compression and symmetry of small-world graphs and structures. Commun. Inf. Syst. **22**(2), 275–302 (2022). https://doi.org/10.4310/cis.2022.v22.n2.a5
22. Luczak, T., Magner, A., Szpankowski, W.: Asymmetry and structural information in preferential attachment graphs. Random Struct. Algorithms **55**(3), 696–718 (2019). https://doi.org/10.1002/rsa.20842
23. Luczak, T., Magner, A., Szpankowski, W.: Compression of preferential attachment graphs. In: IEEE International Symposium on Information Theory (ISIT), pp. 1697–1701. IEEE (2019). https://doi.org/10.1109/isit.2019.8849739
24. Munro, J.I., Nicholson, P.K., Seelbach Benkner, L., Wild, S.: Hypersuccinct trees – new universal tree source codes for optimal compressed tree data structures and range minima. In: Mutzel, P., Pagh, R., Herman, G. (eds.) European Symposium on Algorithms (ESA), pp. 70:1–70:18. Schloss Dagstuhl - Leibniz-Zentrum für Informatik (2021). https://doi.org/10.4230/LIPICS.ESA.2021.70
25. Munro, J.I., Wu, K.: Succinct data structures for chordal graphs. In: 29th International Symposium on Algorithms and Computation, ISAAC 2018, 16–19 December 2018, Jiaoxi, Yilan, Taiwan, pp. 67:1–67:12 (2018). https://doi.org/10.4230/LIPIcs.ISAAC.2018.67
26. Navarro, G.: Wavelet trees for all. J. Discrete Algorithms **25**, 2–20 (2014)
27. Pătraşcu, M.: Succincter. In: Symposium on Foundations of Computer Science (FOCS). IEEE (2008). https://doi.org/10.1109/focs.2008.83

28. Seelbach Benkner, L.: Combinatorial and information-theoretic aspects of tree compression. Ph.D. thesis, University of Siegen (2023). https://doi.org/10.25819/UBSI/10429
29. Tsakalidis, K., Wild, S., Zamaraev, V.: Succinct permutation graphs. Algorithmica **85**(2), 509–543 (2023). https://doi.org/10.1007/s00453-022-01039-2
30. Turowski, K., Magner, A., Szpankowski, W.: Compression of dynamic graphs generated by a duplication model. Algorithmica **82**(9), 2687–2707 (2020). https://doi.org/10.1007/s00453-020-00699-2

A Parameterized Complexity Analysis of Bounded Height Depth-First Search Trees

Lars Jaffke[1(✉)], Paloma T. de Lima[2], Wojciech Nadara[3,4], and Emmanuel Sam[5]

[1] NHH Norwegian School of Economics, Bergen, Norway
lars.jaffke@nhh.no
[2] IT University of Copenhagen, Copenhagen, Denmark
palt@itu.dk
[3] University of Warsaw, Warsaw, Poland
w.nadara@mimuw.edu.pl
[4] Technical University of Denmark, Lyngby, Denmark
[5] University of Bergen, Bergen, Norway
emmanuel.sam@uib.no

Abstract. Computing bounded depth decompositions is a bottleneck in many applications of the treedepth parameter. The fastest known algorithm, which is due to Reidl, Rossmanith, Sánchez Villaamil, and Sikdar [ICALP 2014], runs in $2^{\mathcal{O}(k^2)} \cdot n$ time and it is a big open problem whether the dependency on k can be improved to $2^{o(k^2)} \cdot n^{\mathcal{O}(1)}$. We show that the related problem of finding DFS trees of bounded height can be solved faster in $2^{\mathcal{O}(k \log k)} \cdot n$ time. As DFS trees are treedepth decompositions, this circumvents the above mentioned bottleneck for this subclass of graphs of bounded treedepth. This problem has recently found attention independently under the name MinHLT and our algorithm gives a positive answer to an open problem posed by Golovach [Dagstuhl Reports, 2023]. We complement our main result by studying the complexity of MinHLT and related problems in several other settings. First, we show that it remains NP-complete on chordal graphs, and give an FPT-algorithm on chordal graphs for the dual problem, called DualMinHLT, asking for a DFS tree of height at most $n - k$, parameterized by k. The parameterized complexity of DualMinHLT on general graphs is wide open. Lastly, we show that DualMinHLT and two other problems concerned with finding DFS trees with few or many leaves are FPT parameterized by k plus the treewidth of the input graph.

Keywords: DFS tree · Treedepth · Lineal topology · Treewidth · Chordal graphs

1 Introduction

Treedepth [23] is a graph parameter with many interesting algorithmic applications. As it is more restrictive than treewidth, there are several problems that

H. Fernau and P. Kindermann (Eds.): WG 2025, LNCS 16124, pp. 286–301, 2026.
https://doi.org/10.1007/978-3-032-11835-6_21

can be solved more efficiently on graphs of bounded treedepth than on graphs of bounded treewidth, in particular in terms of space complexity [14,16,22,24,25], and in some cases in terms of time complexity [9,18]. One downside of treedepth is that the current fastest algorithm to compute decompositions of depth k runs in $2^{\mathcal{O}(k^2)} \cdot n$ time [26], and it is a big open problem to determine whether the dependence on k can be improved to $2^{o(k^2)}$—even if we allow a higher degree of the polynomial dependence on n, or if we only want to have a constant-factor approximation instead of an optimal decomposition.

There has been extensive study on the computation of the treedepth of graphs [2,3,12,19]. Regarding the complexity of TREEDEPTH parameterized by k, Ossona de Mendez and Nešetřil showed by Courcelle's theorem that the problem is FPT parameterized by k. Later, Reidl, Rossmanith, Sanchez Villaamil, and Sikdar [26] gave an algorithm that solves this problem explicitly in time $2^{\mathcal{O}(k\cdot\mathsf{tw}(G))} \cdot n$ and $2^{\mathcal{O}(k^2)} \cdot n$ via dynamic programming on tree decompositions (which can be 2-approximated in $2^{\mathcal{O}(k)} \cdot n$ time [20]). This algorithm improves on a previous algorithm by Bodlaender [2] that runs time in $2^{\mathcal{O}(k\cdot\mathsf{tw}(G)^2)} \cdot n^2$ time. Recently Nadara et al. [21] presented an algorithm that computes treedepth in $2^{\mathcal{O}(k^2)} \cdot n^{\mathcal{O}(1)}$ time and $n^{\mathcal{O}(1)}$ space, improving on the algorithm by Reidl et al., which uses exponential space.

Bounded treedepth decompositions are rooted trees T of bounded height on the vertex set of a graph G that do not give rise to any cross-edges, i.e., for each edge $uv \in E(G)$, u and v are in an ancestral relationship in T. This property is shared with trees resulting from executing a depth-first search (DFS) on G; with the difference that in a treedepth decomposition T, the edges of T need not be edges of G. Therefore, the class of graphs that have a DFS tree of height at most k is a subclass of the class of graphs of treedepth at most k. Our first main result is that in this subclass of graphs of treedepth at most k, decompositions can be computed faster, namely in $2^{\mathcal{O}(k \log k)} \cdot n$ time, see below. We would like to point out here that the parameters treedepth and "minimum height of a DFS tree" are in fact functionally equivalent: The treedepth of a path on 2^k vertices equals $k+1$ [23], and the height of a DFS tree is at most the length of the longest path in the graph.

The problem of computing DFS trees of small height has recently been studied by Sam, Fellows, Rosamond, and Golovach [30], motivated by several algorithmic applications [1,8,10,13]. They referred to DFS trees as *lineal topologies*, and we keep the problem names introduced in their work for consistency.

MINIMUM HEIGHT LINEAL TOPOLOGY (MINHLT)

Input:	Graph G, integer k
Question:	Does G have a DFS tree (lineal topology) of height at most k?

Fellows, Friesen, and Langston [11] showed that the above problem is NP-complete and hard to approximate. Sam et al. [30] observed that MINHLT parameterized by the target height k is FPT via the following win/win-approach.

If a graph has a DFS tree of height at most k, then it has treedepth at most k, which in turn implies that the treewidth is at most k. The property of having a DFS tree with height at most k can be expressed in MSO which yields the result via Courcelle's Theorem for bounded treewidth graphs [5]. However, an explicit algorithm for MinHLT was not known, and asked for as an open problem by Golovach (see Sect. 4.4 in [15]). Our first contribution is a resolution of this problem via the following explicit algorithm.

Theorem 1. *There is an algorithm that solves* MinHLT *given an n-vertex graph G and an integer k in $2^{\mathcal{O}(k \log k)} \cdot n$ time. If (G, k) is a yes-instance, then the algorithm returns a certifying DFS tree.*

In [30], several variations of MinHLT are studied from the classical and parameterized points of view, and the coarse (parameterized) complexity of most problems is understood, apart from one notable exception:

Dual Minimum Height Lineal Topology (DualMinHLT) par. by k

Input: Graph G on n vertices, integer k

Question: Does G have a DFS tree (lineal topology) of height at most $n - k$?

Curiously, all attempts at classifying the parameterized complexity of DualMinHLT have remained fruitless. We conjecture that it is $W[1]$-hard, however there is no matching XP algorithm for that claim and to the best of our knowledge, it is not known whether it is solvable in polynomial time even for the case $k = 2$. While we do not resolve this problem, we give insight into it by considering its parameterized complexity on *chordal* graphs by giving an FPT-algorithm for it. To justify the relevance of this algorithm, we first show that the non-parameterized version of MinHLT, and hence of DualMinHLT, remains NP-complete on chordal graphs.

Theorem 2. MinHLT *(and hence* DualMinHLT*) remains NP-complete on chordal graphs. Moreover,* DualMinHLT *can be solved in $\mathcal{O}(n^2) + 2^{\mathcal{O}(k^2)}$ time on chordal graphs.*

Lastly, we consider the parameterized complexity of DualMinHLT on graphs of bounded treewidth. We observe that this property is expressible in MSO_2 and therefore FPT by k plus the treewidth of the input graph. As by-products, we observe the same for MinLLT and MaxLLT, which ask for DFS trees with at most and at least k leaves, respectively.

Theorem 3. DualMinHLT, MinLLT, *and* MaxLLT *parameterized by k plus the treewidth of the input graph are fixed-parameter tractable.*

Related Work. On the complexity of finding DFS trees with many or few leaves, Sam et al. [30] proved that the MinLLT and MaxLLT problems are NP-complete. When parameterized by k, they showed that MinLLT is para-NP-hard and MaxLLT is W[1]-hard. By an application of Courcelle's theorem, they demonstrated that the parametric duals of these problems, namely, DUALMIN-LLT and DUALMAXLLT, respectively, are both FPT parameterized by k. Subsequently, it was shown in [29] that both problems admit polynomial kernels with $\mathcal{O}(k^3)$ vertices, which implies FPT algorithms running in $k^{\mathcal{O}(k)} \cdot n^{O(1)}$ time. The authors also gave polynomial kernels for MIN-LLT and MAX-LLT parameterized by the vertex cover number.

Overview and Methods. In Sect. 3, we show that MINHLT remains NP-hard on chordal graphs via a polynomial-time reduction from the strongly NP-hard WEIGHTED TREEDEPTH problem [7] to MINHLT on chordal graphs.

In Sect. 4 we present an algorithm that, exploiting the structure provided by the clique tree of the chordal graph checks whether $\mathsf{MinHLT}(G) \leq n - k$ in $\mathcal{O}(n^2) + 2^{\mathcal{O}(k^2)}$ time and $\mathcal{O}(n^2)$ space.

In Sect. 5 we present an algorithm that, given a graph G and a positive integer k, checks if $\mathsf{MinHLT}(G) \leq k$ in $k^{\mathcal{O}(k)} \cdot n$ time. It relies on the fact that if G is a YES-instance, then $\mathsf{tw}(G) \leq \mathsf{td}(G) \leq k$. It uses a tree decomposition of width $2k$ obtained by the 2-approximation algorithm described in [20] and adapts the strategy for computing $\mathsf{td}(G)$ by Reidl, Rossmanith, Sanchez Villaamil, and Sikdar [26]. However, notably, we are able to improve over the natural adaptation of that algorithm that would lead to a $2^{\mathcal{O}(k^2)}$ time complexity.

The proof of Theorem 3 is via Courcelle's Theorem [5] and given in the full version [17], as are other statements marked with "★".

2 Preliminaries

We consider simple, finite, undirected, and connected graphs. For a graph G with the vertex set $V(G)$ and the edge set $E(G)$, we use n and m to denote the number of vertices $|V(G)|$ and the number of edges $|E(G)|$ of G, respectively, if this does not cause any confusion.

A graph G is a *subgraph* of H, written $G \subseteq H$, if $V(G) \subseteq V(H)$ and $E(G) \subseteq E(H)$. A graph G is an *induced subgraph* of H if $V(G) = S \subseteq V(H)$ and $E(G) = \{uv \mid uv \in E(G), \{u, v\} \subseteq S\}$. In this case we say that G is the *subgraph of H induced by S*, written $G = H[S]$.

Let G be a graph. A sequence $v_1, \ldots, v_r$ of vertices of G is a *walk from v_1 to v_r* if for all $i \in [r-1]$, $v_i v_{i+1} \in E(G)$. We say that $v_1, \ldots, v_r$ is a *path* if all vertices are pairwise distinct, and a *cycle* if $v_1, \ldots, v_{r-1}$ are pairwise distinct and $v_r = v_1$. A graph G is *connected* if there is a path from any vertex of G to any other vertex. A *(connected) component* of G is a maximal connected subgraph. We may denote connected components as vertex sets when no confusion is possible.

A graph is a *tree* if it is connected and does not contain a cycle. A graph is a *forest* if its connected components are trees. A tree T is a *spanning tree of G* if $T \subseteq G$ and $V(T) = V(G)$. A tree T is *rooted* if it has a unique designated vertex $r \in V(T)$, called the *root*, inducing an ancestral relationship on some pairs of vertices in T: We say that u is an *ancestor* of v if u lies on the path from r to v in T. In this case we also call v a *descendant* of u. If either u is an ancestor of

v or v is an ancestor of u, we say that u and v are in an *ancestral relationship*, or *comparable in* T. A path P in T is *vertical* if all pairs of vertices on it are comparable. If u is an ancestor of v and $uv \in E(T)$, then u is the *parent* of v and v a *child* of u. A *rooted subtree* of a rooted tree T consists of a vertex $v \in V(T)$ along with all of its descendants. A forest F where all components are rooted trees is called a *rooted forest.* We denote the set of roots of F by $R(F)$. (With slight abuse of notation, we may treat F as a regular, unrooted, forest whenever convenient).

Let T be a rooted tree. If v is a leaf, then its *height* is 1. Otherwise, the *height of* v is the maximum height of any of its children plus 1. The *height of a rooted tree* is the height of its root. The *depth* of the root is 1, and the depth of any non-root node is the depth of its parent plus 1.

DFS Trees. Let G be a connected graph and $T \subseteq G$ a spanning tree of G. We say that an edge $uv \in E(G)$ is a *cross-edge for* T if u and v are not comparable in T. We say that T is a *DFS tree of* G if G has no cross-edges for T.

Tree Decompositions.

Definition 1. *Let* G *be a graph. A* tree decomposition *of* G *is a pair* (T, β) *where* T *is a tree and* $\beta\colon V(T) \to 2^{V(G)}$ *is such that:*

1. $\bigcup_{t \in V(T)} \beta(t) = V(G)$.
2. *For all* $uv \in E(G)$, *there is some* $t \in V(T)$ *such that* $\{u, v\} \subseteq \beta(t)$.
3. *For all* $v \in V(G)$: $T[\{t \mid v \in \beta(t)\}]$ *is connected.*

For each $t \in V(T)$, *we call the set* $\beta(t)$ *the* bag at t. *The* width *of* (T, β) *is* $\max_{t \in V(T)} |\beta(t)| - 1$.

Definition 2. *Let* G *be a graph and* (T, β) *be a tree decomposition of* G. *We say that* G *is a* nice tree decomposition *if* T *is a rooted binary tree and each node* $t \in V(T)$ *is one of the following types.*

Introduce Vertex. *The node* t *either has one child* t' *and* $\beta(t) = \beta(t') \cup \{v\}$ *for some vertex* $v \notin \beta(t')$ *or* t *is a leaf and* $\beta(t) = \{v\}$. *We say that* v *is* introduced at t.

Forget Vertex. *The node* t *has one child* t' *and* $\beta(t') = \beta(t) \cup \{v\}$ *for some* $v \notin \beta(t)$. *We say that* v *and all edges incident to it are* forgotten at t.

Join. *The node* t *has two children* t_1 *and* t_2 *with* $\beta(t) = \beta(t_1) = \beta(t_2)$.

For each $t \in V(T)$, *we denote by* G_t *the subgraph of* G *induced on all vertices that were introduced at* t *or below.*

In a nice tree decomposition we may assume that each vertex is forgotten exactly once (but may be introduced multiple times), and that the number of nodes is $w^{\mathcal{O}(1)}n$, where w is its width and n the number of vertices of the underlying graph [6] (Fig. 1).

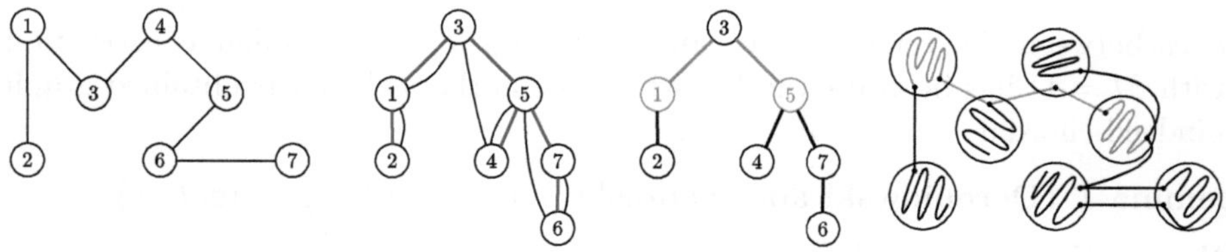

Fig. 1. Example of a graph (a path) with its (suboptimal) treedepth decomposition and a DFS tree of its blown-up version "guided" by this treedepth decomposition.

Treedepth Decompositions

Definition 3. *Let G be a graph. A treedepth decomposition of G is a rooted forest F such that $V(F) = V(G)$ and for every edge $uv \in E(G)$, we have that either u is an ancestor of v in F or v is an ancestor of u.*

We note that the definition of a treedepth decomposition is very similar to the definition of DFS tree with one notable difference—DFS trees have to additionally satisfy that edges of F are also edges of G. The treedepth of G, denoted by $\mathsf{td}(G)$, is the lowest height a treedepth decomposition of G might have. As every DFS tree of G is also a treedepth decomposition of G, we get that $\mathsf{td}(G) \leq \mathsf{MinHLT}(G)$. It is known that $\mathsf{MinHLT}(G) \leq 2^{\mathsf{td}(G)}$, as the treedepth of a path on 2^k vertices equals $k+1$ [23] and $\mathsf{MinHLT}(G)$ is at most the size of the longest path in G.

Chordal Graphs. A graph G is *chordal* if it has no cycle of length four or more as an induced subgraph. It is well-known that G is chordal if and only if it has a tree decomposition where all bags are maximal cliques.

3 MinHLT is NP-Hard on Chordal Graphs

Theorem 4. *The* MinHLT *problem is NP-hard on chordal graphs.*

We will reduce from the problem known as Weighted Treedepth on trees. In the Weighted Treedepth problem we are given a graph G with a positive weight function $w\colon V(G) \to \mathbb{N}^+$ and an integer k and we ask whether there exists a treedepth decomposition of G such that on any vertical path, the sum of the weights of the vertices on it is at most k. The lowest value of k for which such treedepth decomposition exists will be denoted by $\mathsf{td}(G, w)$.

Theorem 5 (Dereniowski and Nadolski [7]). Weighted Treedepth *with positive integral weights is strongly NP-hard on trees.*

Given an instance (T, w) of Weighted Treedepth, where T is a tree, and a positive integer M, we construct a blown-up version of T—the graph $G_M(T, w)$—as follows. For each $x \in V(T)$ we create a set V_x of size $w(x) \cdot M$ that induces a clique in $G_M(T, w)$. For each edge $xy \in E(T)$ we add all possible edges between the vertices of V_x and V_y in $G_M(T, w)$. A similar construction (with $M = 1$) has been used in [7], where the authors show the obtained graph is indeed chordal.

Lemma 1 (Dereniowski and Nadolski [7]). $\mathsf{td}(G_1(T, w)) = \mathsf{td}(T, w)$

Observation 1 (★). $\mathsf{td}(G_M(T, w)) = M \cdot \mathsf{td}(T, w)$

Lemma 2 (★). *Let (T, w) be an instance of* Weighted Treedepth *where T is a tree with n vertices and let $M \in \mathbb{Z}^+$ be such that $M \geq n$. Then*

$$M \cdot \mathsf{td}(T, w) \leq \mathsf{MinHLT}(G_M(T, w)) \leq M \cdot \mathsf{td}(T, w) + n^2.$$

Proof (Proof of Theorem 4). We give a reduction from Weighted Treedepth with positive integral weights on trees. By Theorem 5, this problem is NP-hard even if the weights are polynomial in the size of the tree. Let (T, w) be an instance of Weighted Treedepth with positive integral weights that are polynomial in $|V(T)| = n$. Let $M = n^2+1$. As M is polynomial in n, the function $w_M(v) = M \cdot w(v)$ stays polynomial in n. Consequently, the size of $G_M(T, w)$ is polynomial in n as well. Moreover, $G_M(T, w)$ is a chordal graph [7]. Based on Lemma 2, we have that

$$M \cdot \mathsf{td}(T, w) \leq \mathsf{MinHLT}(G_M(T, w)) \leq M \cdot \mathsf{td}(T, w) + n^2 < M \cdot (\mathsf{td}(T, w) + 1).$$

Therefore, we can conclude that $\mathsf{td}(T, w) = \left\lfloor \frac{\mathsf{MinHLT}(G_M(T,w))}{M} \right\rfloor$. Hence, we deduce that MinHLT is NP-hard on chordal graphs. □

4 DualMinHLT Is FPT on Chordal Graphs

This section will prove that DualMinHLT is FPT on chordal graphs. More formally, we prove the following theorem:

Theorem 6. *Let G be a chordal graph on n vertices and m edges and k be a positive integer. Then, we can check if $\mathsf{MinHLT}(G) \leq n - k$ in $\mathcal{O}(n^2) + 2^{\mathcal{O}(k^2)}$ time and $\mathcal{O}(n^2)$ space. Moreover, if this inequality holds, then we can return a certifying DFS tree in the same time and space complexities.*

Lemma 3 (★). *Let G be a graph, $S \subseteq V(G)$ be a subset of vertices such that $G[S]$ has a Hamiltonian path and $C_1, C_2, \ldots, C_c \subseteq V(G) - S$ be a partition of $V(G) - S$ into connected components. Then $\mathsf{MinHLT}(G) \leq |S| + \max(|C_1|, \ldots, |C_c|)$. Moreover, if we are given any Hamiltonian path of $G[S]$, then we can return a certifying DFS tree in $\mathcal{O}(|E(G)|)$ time.*

We need the following lemma which has been proved as Lemma 2.5 in [27]. We would like to point out that the statement of [27, Lemma 2.5] is weaker than the following in two ways. First, its statement is purely existential, while we need to be able to compute the desired set in polynomial time. Second, it only speaks about a set of vertices, while we require the set to be a subset of a bag of a tree decomposition. However, a close inspection of the proof of [27, Lemma 2.5] reveals that it proves the following, stronger, statement.

Lemma 4 (Cf. Lemma 2.5 in [27]). *Let G be an undirected graph on n vertices and (T, β) be its tree decomposition of width w. There exists $K \subseteq V(G)$, which is a subset of some bag $\beta(v)$, such that if $C_1, \ldots, C_c$ are vertex sets of connected components of $G - K$, then we have $2 \cdot \max(|C_1|, \ldots, |C_c|) \leq |C_1| + |C_2| + \ldots + |C_c|$. Moreover, K can be found in $\mathcal{O}(n \cdot w)$ time.*

We proceed to the most challenging part of proving Theorem 6, which is the following lemma:

Lemma 5 (★). *Let G be a chordal graph on n vertices, k be a positive integer and K be a clique in G such that $n - 2k < |K| \leq n - k$ given to us. Then, we can check if $\mathsf{MinHLT}(G) \leq n - k$ in $\mathcal{O}(n+m) + 2^{\mathcal{O}(k^2)}$ time and $\mathcal{O}(n+m)$ space. Moreover, if this inequality holds, then we can return a certifying DFS tree in the same time and space complexities.*

This lemma is a restriction of Theorem 6 to the hardest case of a "moderate clique". We show how to prove Theorem 6 assuming correctness of Lemma 5.

Proof (Proof of Theorem 6). We first compute a clique tree $\mathcal{T}$ of G, that is, a tree decomposition where each bag is a maximal clique, in linear time [28]. If G is a clique, then we obviously have $\mathsf{MinHLT}(G) = n$. If G is not a clique, then we can apply Lemma 4 and get a set K such that we have that $2 \cdot \max(|C_1|, \ldots, |C_c|) \leq |C_1| + |C_2| + \ldots + |C_c|$, where $C_1, \ldots, C_c$ are vertex sets of connected components of $G - K$. Since K is the subset of some bag of $\mathcal{T}$, it is a clique. Let us consider a few cases based on the size of K:

The "small clique case": $|K| \leq n - 2k$
The "moderate clique case": $n - 2k < |K| \leq n - k$
The "large clique case": $n - k < |K|$

In the large clique case we clearly have $\mathsf{MinHLT}(G) \geq |K| > n - k$, so we return *NO*. In the small clique case we apply Lemma 3 (as K is clearly Hamiltonian and we can easily get its Hamiltonian path as well) and get that $\mathsf{MinHLT}(G) \leq |K| + \max(|C_1|, \ldots, |C_c|)$ and get a certifying DFS tree as well. Note that using Lemma 4 we can obtain that $|K| + \max(|C_1|, \ldots, |C_c|) \leq \frac{2|K| + |C_1| + \ldots + |C_c|}{2} = \frac{n + |K|}{2} \leq n - k$, hence we can return *YES* and a certifying DFS tree as well. What remains is the most challenging moderate clique case, which is handled by Lemma 5. □

5 An Explicit FPT-Algorithm for MinHLT

In [30], Sam, Fellows, Rosamond, and Golovach showed that MinHLT is FPT by solution size, using Courcelle's Theorem. In this section, we give an explicit algorithm for this problem which was mentioned as an open problem by Golovach in [15].

Theorem 1. *There is an algorithm that solves* MinHLT *given an n-vertex graph G and an integer k in $2^{\mathcal{O}(k \log k)} \cdot n$ time. If (G, k) is a yes-instance, then the algorithm returns a certifying DFS tree.*

To prove Theorem 1, we follow a similar strategy as the FPT-algorithm to compute treedepth due to Reidl, Rossmanith, Sanchez Villaamil, and Sikdar [26]. Perhaps the most interesting part about our algorithm is not the fact that it can be obtained as an adaptation of the algorithm from [26], but the fact that thanks to specific properties of DFS trees, the MinHLT problem turns out to be solvable in $k^{\mathcal{O}(k)} \cdot n$ time, rather than $2^{\mathcal{O}(k^2)}$ time complexity which is required for the fastest known algorithms for computing treedepth exactly [21,26], which would also be the time complexity of the natural adaptation of [26] to MinHLT.

For this section, instead of thinking that a DFS tree F of a graph G has the same set of vertices, we prefer thinking in a way that they have different vertex sets, but we are additionally provided with a bijective mapping ρ from $V(G)$ to $V(F)$.

Definition 4. *We say that (F, ρ) is a* full solution *for G if F is a tree and ρ is a bijective mapping from $V(G)$ to $V(F)$ such that $uv \in E(G)$ implies that $\rho(u)$ and $\rho(v)$ are in an ancestral relation in F (that is, one of them is the ancestor of the other in F) and $uv \in E(F)$ implies $\rho^{-1}(u)\rho^{-1}(v) \in E(G)$.*

The starting point is this: If a graph has a DFS tree of height at most k, then this tree witnesses that G has treedepth at most k. Since for each graph G, we have that $\mathsf{tw}(G) \leq \mathsf{td}(G)$, and since we can compute a 2-approximation of the treewidth of a graph G in $2^{\mathcal{O}(\mathsf{tw}(G))} \cdot |V(G)|$ time [20], we may assume that we have a tree decomposition of width at most $2k$ of the input graph at hand. Below, we describe a dynamic programming algorithm that uses this tree decomposition to decide if G has a DFS tree of height at most k.

Definition 5. *Let T be a rooted tree and X be a subset of its vertices. The ancestral closure of X, denoted $\mathsf{Clos}_T(X)$, is the subgraph of T induced on vertices of X and all of their ancestors.*

Definition 6. *A partial solution for a graph H is a pair (D, τ), where D is a rooted tree of height at most k and $\tau\colon V(H) \to V(D)$ is an injective mapping such that for any $uv \in E(H)$ we have that $\tau(u)$ and $\tau(v)$ are in an ancestral relation in D, for any $uv \in E(D)$ such that $u, v \in \tau(V(H))$, we have that $\tau^{-1}(u)\tau^{-1}(v) \in E(H)$ and any leaf of D belongs to $\tau(V(H))$.*

Intuitively, such partial solution represents a candidate leading to a full solution for a supergraph G of H, where the vertices of D that are not in the $\tau(V(H))$ will correspond to some vertices of $V(G) \setminus V(H)$. Let us make this intuition formal through the following definition:

Definition 7. *Let H be a subgraph of G, (D, τ) be a partial solution for H and (F, ρ) be a full solution for G. We say that (D, τ)* extends to *(F, ρ) if D is a subgraph of F, they have the same root and τ and ρ agree on $V(H)$.*

Lemma 6. *If (D, τ) extends to (F, ρ), then $D = \mathsf{Clos}_F(\rho(V(H)))$ (Fig. 2).*

Proof. Note that D is a connected subgraph of F containing the root of F and the whole $\rho(V(H))$, hence $\mathsf{Clos}_F(\rho(V(H))) \subseteq D$. Let us assume now for the contradiction that $V(D) \setminus V(\mathsf{Clos}_F(\rho(V(H)))$ is nonempty and let v be any vertex in it. Let u be any leaf of D among the descendants of v (we can possibly have $u = v$). Then, it belongs to $\tau(V(H))$ based on the definition of a partial solution. Hence $u \in \rho(V(H))$, but then $v \in V(\mathsf{Clos}_F(\rho(V(H))))$, which is a contradiction. □

Let $\mathcal{T} = (T, \beta)$ be a tree decomposition of G of width at most $2k$ that we have already computed. We will define *partial solutions for nodes of* T as well.

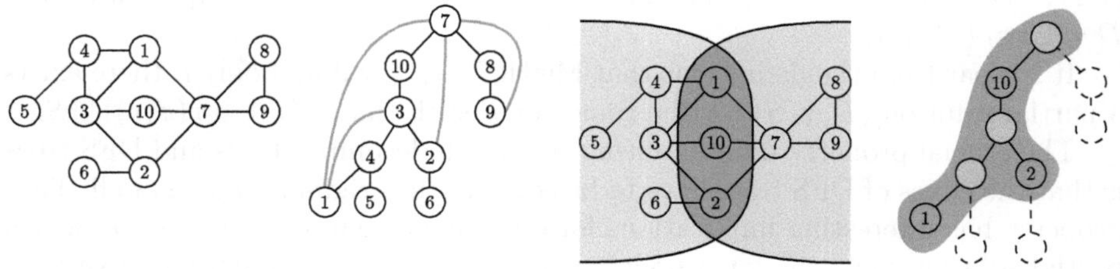

Fig. 2. The first picture represents an input graph G. The second picture shows an example of its DFS tree. The third picture depicts a separation of G, where purple vertices are separating already processed grey vertices from not yet processed blue vertices (think that purple vertices are vertices of some bag and grey vertices are already forgotten vertices in the subtree below it). The circled part of the fourth picture shows a promising partial solution for that separation that is induced by the shown DFS tree. (Color figure online)

Definition 8. *A partial solution for a node $t \in V(T)$ is a triple (D, τ, S), where D is a rooted tree, τ is an injective mapping from $\beta(t)$ to $V(D)$, and $S \subseteq V(D)$, such that there exists a partial solution (D', τ') for G_t such that*

- $D = \mathsf{Clos}_{D'}(\tau(\beta(t)))$,
- *τ and τ' agree on $\beta(t)$,*
- $\tau'(V(G_t) \setminus \beta(t)) \cap V(D) = S$ *and* $\tau'(V(G_t) \setminus \beta(t)) \setminus V(D) = V(D') \setminus V(D)$.

We additionally say that such (D, τ, S) extends to (D', τ').

Intuitively, a partial solution for a node t is a partial solution for G_t stripped off any redundant information relating to already forgotten vertices (where forgotten vertices means $V(G_t) \setminus \beta(t)$) that will never influence our future possibilities of extending such partial solution to a full solution for the whole graph G.

In this setting, the vertices $V(D')$ can be partitioned into the following parts:

1. $\tau(\beta(t))$—images of the vertices from the node t
2. S—vertices on paths from some $v \in \tau(\beta(t))$ to the root that have a vertex that is already forgotten as their preimage in τ'

3. $V(D) \setminus (\tau(\beta(t)) \cup S)$—vertices on paths from some $v \in \tau(\beta(t))$ to the root that might have a vertex that will be introduced later as their preimage in τ'
4. $V(D') \setminus V(D)$—vertices that are images in τ' of forgotten vertices and not on a path between r and any $v \in \tau(\beta(t))$

As vertices forgotten below t are not adjacent to anybody from $V(G) \setminus V(G_t)$, we do not need to remember anything about how the part outside of D looks like, and for any vertex $u \in S$, we also do not need to remember who specifically was mapped to u beside the fact that there existed a vertex like that. Consequently, we say that a partial solution (D, τ, S) for t extends to a full solution (F, ρ) if there exists a partial solution (D', τ') for G_t such that it extends to (F, ρ) and (D, τ, S) extends to (D', τ'). Similarly, if (D, τ, S) extends to (F, ρ), we have that $D = \mathsf{Clos}_F(\tau(\beta(t)))$.

It is clear from the definitions that $\mathsf{MinHLT}(G) \leq k$ if and only if there exists a partial solution (D, τ, S) for the root r of T such that $V(D) = \tau(\beta(r)) \cup S$.

The crucial property differentiating treedepth decompositions and DFS trees is that the edges of DFS trees have to be edges of the original graph as well. That property has interesting implications for our partial solutions. Let us consider a partial solution (D, τ, S) for a node t and assume that there exist two vertices $u, v \in V(D)$ such that $u \in S$, $v \notin S \cup \tau(\beta(t))$ and $uv \in E(D)$. Such partial solution cannot be extended to a full solution for G as in any extension of that partial solution to a full solution, v has to be an image of a vertex from $V(G) \setminus V(G_t)$, while u was an image of a vertex from $V(G_t) \setminus \beta(t)$. However, these two sets are not adjacent to each other in G, contradicting the fact that for an edge of a full solution, its preimages are adjacent to each other. Hence, we may restrict our computations only to partial solutions such that sets S and $V(D) \setminus (S \cup \tau(\beta(t)))$ are not adjacent to each other as other partial solutions do not extend to any full solutions. Such partial solutions will be called *promising*. In other words, each connected component of $D \setminus \tau(\beta(t))$ has to be either fully in S, or disjoint from S. We then prove the following:

Lemma 7. *For a node t and a pair (D, τ), there are only $2^{\mathcal{O}(k)}$ sets S such that (D, τ, S) is a promising partial solution for t.*

Proof. Assume that there exists S such that (D, τ, S) is a promising partial solution for t, as otherwise the claim is vacuously true. Note that, as $D = \mathsf{Clos}_{D'}(\tau'(\beta(t)))$ for some partial solution (D', τ') for G_t that it extends to, D has at most one degree-one vertex that is not within $\tau(\beta(t))$ (the root of D is the only possible vertex with such property). It implies that the number of connected components of $D \setminus \tau(\beta(t))$ is at most $|\beta(t)| \leq 2k+1$. As each of them has to be either fully in S or disjoint from S, there are at most $2^{2k+1} = 2^{\mathcal{O}(k)}$ possible values of S such that (D, τ, S) is a promising partial solution. □

Lemma 8 (★). *lem:trees-shapes For a node t, there are only $k^{\mathcal{O}(k)}$ non-isomorphic pairs (D, τ) across all partial solutions for t.*

Corollary 1. *For a node t, there are only $k^{\mathcal{O}(k)}$ promising partial solutions (D, τ, S).*

Also, as for any partial solution (D, τ, S), we have that D is a sum of at most $|\beta(t)| \leq 2k+1$ paths of length at most k from its root, we get the following observation, after which we present the algorithm.

Observation 2. For any partial solution (D, τ, S) for any node t, we have $|V(D)| = \mathcal{O}(k^2)$.

Introduce vertex. Let t be an introduce node, introducing the vertex v, with child t^*, that is, $\beta(t) = \beta(t^*) \cup \{v\}$. Let (D^*, τ^*, S^*) be a promising partial solution for t^* and let (D, τ, S) be a hypothetical promising partial solution for t such that there exists a full solution (F, ρ) such that (D, τ, S) and (D^*, τ^*, S^*) both extend to (F, ρ). As τ and ρ agree on $\beta(t)$, while τ^* and ρ agree on $\beta(t^*)$, it follows that τ and τ^* agree on $\beta(t) \cap \beta(t^*) = \beta(t^*)$. As $D^* = \mathsf{Clos}_F(\tau'(\beta(t')))$, $D = \mathsf{Clos}_F(\tau(\beta(t)))$ and $\beta(t) = \beta(t^*) \cup \{v\}$, we conclude that $D = \mathsf{Clos}_F(\tau^*(\beta(t^*)) \cup \tau(v)) = D^* \cup P_{r\tau(v)}$, where r denotes the root of F and $P_{r\tau(v)}$ is the path between r and $\tau(v)$ in F (we remind that $\tau(v) = \rho(v)$). If this path is not fully contained within D^*, there is a unique deepest vertex u of D^* such that all vertices of $V(P_{r\tau(v)}) \setminus V(D^*)$ are descendants of u, or in other words $V(P_{r\tau(v)}) \setminus V(D^*)$ induces a path in F appended at some vertex $u \in V(D^*)$. As $|V(D^*)| = \mathcal{O}(k^2)$, there are at most $\mathcal{O}(k^2)$ choices for u and there are at most k choices for the length of $P_{r\tau(v)}$, which gives rise to at most $\mathcal{O}(k^3)$ non-isomorphic possibilities for D. In transitions where there is any vertex added (that is, if $V(P_{r\tau(v)}) \setminus V(D^*) \neq \emptyset$), all added vertices that are not $\tau(v)$ cannot belong to S and cannot be images of τ, so τ and S are uniquely determined, and that in fact gives rise to at most $\mathcal{O}(k^3)$ partial solutions (D, τ, S). In transitions where $P_{r\tau(v)}$ is fully contained within D^*, $\tau(v)$ has to belong to $V(D^*)$ and the choice of it uniquely determines such partial solution (D, τ, S), which gives rise to at most $\mathcal{O}(k^2)$ promising partial solutions (note that τ has to stay injective and satisfy $\tau(v) \notin S^*$). For all of these at most $\mathcal{O}(k^3) + \mathcal{O}(k^2) = \mathcal{O}(k^3)$ candidate promising partial solutions, we check if they are indeed promising, and if yes, we add them to the set of promising partial solutions for t. Note that the procedure generating these candidates did not depend on the choice of (F, ρ).
In the special case where t is a leaf, we create promising partial solutions (D, τ, S), where D is a rooted path on at most k vertices, whose leaf is $\tau(v)$ and $S = \emptyset$.

Forget vertex. Let t be a forget vertex v node with a child t^*, let (D^*, τ^*, S^*) be a promising partial solution for t^* and let (D, τ, S) be a hypothetical partial solution for t such that there exists a full solution (F, ρ) such that both (D, τ, S) and (D^*, τ^*, S^*) extend to (F, ρ) (we will see that there is at most one such (D, τ, S) and that it does not depend on the choice of (F, ρ)). By forgetting v, we have to add $\tau(v)$ to S^* and remove v from the domain of τ, that is, $S := S^* \cup \{\tau^*(v)\}$ and $\tau = \tau^*|_{\beta(t)}$. In case $\tau^*(v)$ was a leaf of D^*, D will have a smaller vertex set than D^* given by $D = \mathsf{Clos}_{D^*}(\tau(\beta(t)))$ (and if it was not, then $D = D^*$). Note that the restriction from D^* to $\mathsf{Clos}_{D^*}(\tau(\beta(t)))$ may incur corresponding restriction of S to $S \cap V(\mathsf{Clos}_{D^*}(\tau(\beta(t))))$. However, if for D defined that way we have that $V(D^*) \setminus V(D) \not\subseteq S^* \cup \{\tau^*(v)\}$, then such transition is illegal as we remove a vertex that was never assigned its preimage. We also have to additionally check that such defined (D, τ, S) is promising. If none of these checks failed, we add (D, τ, S) to the set of promising partial solutions for t.

Join. Let t be a join node with children t_1 and t_2. We consider all pairs (D_1, τ_1, S_1) and (D_2, τ_2, S_2) of promising partial solutions for t_1 and t_2, respectively. If (D_1, τ_1) is not isomorphic to (D_2, τ_2) we skip that pair, otherwise they can be thought of having the same set of vertices. If $S_1 \cap S_2 = \emptyset$, we add $(D_1, \tau_1, S_1 \cup S_2)$ to the set of promising partial solutions for t. It is easy to see that such partial solution will always be promising.

After we compute sets of promising partial solutions for each node t according to the above rules, we check if there exists a partial solution (D, τ, S) for the root r of T such that $V(D) = \tau(\beta(r)) \cup S$. If yes, we output that $\mathsf{MinHLT}(G) \leq k$. If needed, the certifying full solution can be restored by investigating the tree of transitions leading to that particular partial solution. All of these transitions can be realized in $k^{\mathcal{O}(1)}$ time per transition. There are at most $k^{\mathcal{O}(k)}$ transitions considered per a node t, therefore, the whole algorithm takes $k^{\mathcal{O}(k)} \cdot n$ time.

6 Conclusion

In this work we considered the parameterized complexity of MinHLT, the problem of deciding if a graph has a DFS tree of height at most k. We gave an explicit FPT algorithm for the problem running in time $k^{\mathcal{O}(k)} \cdot n$. This algorithm uses a similar strategy as the one for treedepth [26], but we achieve an improvement in the running time compared to the natural adaptation of this algorithm, that would lead to a $2^{\mathcal{O}(k^2)} \cdot n$ running time. We further showed the problem remains NP-hard on chordal graphs, but its dual version DualMinHLT is FPT. Finally, we also show DualMinHLT, MaxLLT and MinLLT are FPT parameterized by k plus treewidth.

One of the most interesting open problems in this area is the question of the parameterized complexity of DualMinHLT, or even the complexity of DualMinHLT for fixed k. If a DFS of a graph G starting at any vertex of G results in a Hamiltonian path, then G is called *randomly traceable* [4]. Such graphs characterize the NO-instances of DualMinHLT for $k = 1$. They are the complete graph K_m on m vertices, the cycle C_m on $m \geq 3$ vertices, and the complete bipartite graph $K_{m,m}$ on $2m$ vertices. Since these graphs are recognizable in linear time, the DualMinHLT is linear-time solvable for $k = 1$. Perhaps generalizations of such characterizations can give an opening for the study of the DualMinHLT for fixed k. As the next step in that direction we state the following conjecture:

Conjecture 1. The problem of determining whether an n-vertex graph has a DFS-tree of height at most $n - 2$ is solvable in polynomial time.

However, we also believe that DualMinHLT is not FPT.

Conjecture 2. DualMinHLT is $W[1]$-hard.

Acknowledgement. Lima acknowledges support of the Independent Research Fund Denmark grant agreement number 2098-00012B. Sam acknowledges support from the Research Council of Norway grant "Parameterized Complexity for Practical Computing (PCPC)" (NFR, no. 274526). Nadara acknowledges support of the Independent Research Fund Denmark grant 2020–2023 (9131-00044B) "Dynamic Network Analysis" (while being employed in Denmark) and European Union's Horizon 2020 research and innovation programme, grant agreement No. 948057—BOBR (while being employed in Poland).

Disclosure of Interests. The authors have no competing interests.

References

1. Bodlaender, H.L.: On linear time minor tests with depth-first search. J. Algorithms **14**(1), 1–23 (1993)
2. Bodlaender, H.L., et al.: Rankings of graphs. SIAM J. Discret. Math. **11**(1), 168–181 (1998). https://doi.org/10.1137/S0895480195282550
3. Bodlaender, H.L., Gilbert, J.R., Hafsteinsson, H., Kloks, T.: Approximating treewidth, pathwidth, frontsize, and shortest elimination tree. J. Algorithms **18**(2), 238–255 (1995). https://doi.org/10.1006/JAGM.1995.1009
4. Chartrand, G., Kronk, H.V.: Randomly traceable graphs. SIAM J. Appl. Math. **16**(4), 696–700 (1968). https://doi.org/10.1137/0116056
5. Courcelle, B.: The monadic second-order logic of graphs. I. Recognizable sets of finite graphs. Inf. Comput. **85**(1), 12–75 (1990). https://doi.org/10.1016/0890-5401(90)90043-H
6. Cygan, M., et al.: Parameterized Algorithms. Springer, Cham (2015). https://doi.org/10.1007/978-3-319-21275-3
7. Dereniowski, D., Nadolski, A.: Vertex rankings of chordal graphs and weighted trees. Inf. Process. Lett. **98**(3), 96–100 (2006). https://doi.org/10.1016/j.ipl.2005.12.006
8. Downey, R.G., Fellows, M.R.: Fundamentals of Parameterized Complexity. Springer, Cham (2013). https://doi.org/10.1007/978-1-4471-5559-1
9. Dvořák, P., Knop, D.: Parameterized complexity of length-bounded cuts and multicuts. Algorithmica **80**(12), 3597–3617 (2018). https://doi.org/10.1007/S00453-018-0408-7
10. Fellows, M.R., Langston, M.A.: On search decision and the efficiency of polynomial-time algorithms. In: Proceedings of the Twenty-First Annual ACM Symposium on Theory of Computing, STOC 1989, pp. 501–512. Association for Computing Machinery, New York (1989). https://doi.org/10.1145/73007.73055
11. Fellows, M.R., Friesen, D.K., Langston, M.A.: On finding optimal and near-optimal lineal spanning trees. Algorithmica **3**(1–4), 549–560 (1988)
12. Fomin, F.V., Giannopoulou, A.C., Pilipczuk, M.: Computing tree-depth faster than 2^n. Algorithmica **73**(1), 202–216 (2015). https://doi.org/10.1007/S00453-014-9914-4
13. Freuder, E.C., Quinn, M.J.: Taking advantage of stable sets of variables in constraint satisfaction problems. In: Proceedings of the 9th International Joint Conference on Artificial Intelligence, IJCAI 1985, vol. 2, pp. 1076–1078. Morgan Kaufmann Publishers Inc., San Francisco (1985)

14. Fürer, M., Yu, H.: Space saving by dynamic algebraization based on tree-depth. Theory Comput. Syst. **61**(2), 283–304 (2017). https://doi.org/10.1007/S00224-017-9751-3
15. Ganian, R., Montecchiani, F., Nöllenburg, M., Zehavi, M., Khazaliya, L.: New frontiers of parameterized complexity in graph drawing (Dagstuhl seminar 23162). Dagstuhl Rep. **13**(4), 58–97 (2023). https://doi.org/10.4230/DagRep.13.4.58, https://drops.dagstuhl.de/entities/document/10.4230/DagRep.13.4.58
16. Hegerfeld, F., Kratsch, S.: Solving connectivity problems parameterized by tree-depth in single-exponential time and polynomial space. In: Paul, C., Bläser, M. (eds.) Proceedings of the 37th International Symposium on Theoretical Aspects of Computer Science (STACS 2020). LIPIcs, vol. 154, pp. 29:1–29:16. Schloss Dagstuhl - Leibniz-Zentrum für Informatik (2020). https://doi.org/10.4230/LIPICS.STACS.2020.29
17. Jaffke, L., Lima, P.T., Nadara, W., Sam, E.: A parameterized complexity analysis of bounded height depth-first search trees. CoRR abs/2502.16723 (2025). https://doi.org/10.48550/ARXIV.2502.16723
18. Kellerhals, L., Koana, T.: Parameterized complexity of geodetic set. J. Graph Algorithms Appl. **26**(4), 401–419 (2022). https://doi.org/10.7155/JGAA.00601
19. Kobayashi, Y., Tamaki, H.: Treedepth parameterized by vertex cover number. In: Guo, J., Hermelin, D. (eds.) Proceedings of the 11th International Symposium on Parameterized and Exact Computation (IPEC 2016). LIPIcs, vol. 63, pp. 18:1–18:11. Schloss Dagstuhl - Leibniz-Zentrum für Informatik (2016). https://doi.org/10.4230/LIPICS.IPEC.2016.18
20. Korhonen, T.: A single-exponential time 2-approximation algorithm for treewidth. In: Proceedings of the 62nd IEEE Annual Symposium on Foundations of Computer Science (FOCS 2021), pp. 184–192. IEEE (2021). https://doi.org/10.1109/FOCS52979.2021.00026
21. Nadara, W., Pilipczuk, M., Smulewicz, M.: Computing treedepth in polynomial space and linear FPT time. In: Chechik, S., Navarro, G., Rotenberg, E., Herman, G. (eds.) Proceedings of the 30th Annual European Symposium on Algorithms (ESA 2022). LIPIcs, vol. 244, pp. 79:1–79:14. Schloss Dagstuhl - Leibniz-Zentrum für Informatik (2022). https://doi.org/10.4230/LIPICS.ESA.2022.79
22. Nederlof, J., Pilipczuk, M., Swennenhuis, C.M.F., Wegrzycki, K.: Hamiltonian cycle parameterized by treedepth in single exponential time and polynomial space. SIAM J. Discret. Math. **37**(3), 1566–1586 (2023). https://doi.org/10.1137/22M1518943
23. Nešetřil, J., Ossona de Mendez, P.: Bounded height trees and tree-depth. In: Nešetřil, J., Ossona de Mendez, P. (eds.) Sparsity: Algorithms and Combinatorics, vol. 28, pp. 115–144. Springer, Heidelberg (2012). https://doi.org/10.1007/978-3-642-27875-4_6
24. Pilipczuk, M., Siebertz, S.: Polynomial bounds for centered colorings on proper minor-closed graph classes. J. Comb. Theory B **151**, 111–147 (2021). https://doi.org/10.1016/J.JCTB.2021.06.002
25. Pilipczuk, M., Wrochna, M.: On space efficiency of algorithms working on structural decompositions of graphs. ACM Trans. Comput. Theory **9**(4), 18:1–18:36 (2018). https://doi.org/10.1145/3154856
26. Reidl, F., Rossmanith, P., Villaamil, F.S., Sikdar, S.: A faster parameterized algorithm for treedepth. In: Esparza, J., Fraigniaud, P., Husfeldt, T., Koutsoupias, E. (eds.) ICALP 2014. LNCS, vol. 8572, pp. 931–942. Springer, Heidelberg (2014). https://doi.org/10.1007/978-3-662-43948-7_77

27. Robertson, N., Seymour, P.D.: Graph minors. II. Algorithmic aspects of tree-width. J. Algorithms **7**(3), 309–322 (1986). https://doi.org/10.1016/0196-6774(86)90023-4, https://www.sciencedirect.com/science/article/pii/0196677486900234
28. Rose, D.J., Tarjan, R.E., Lueker, G.S.: Algorithmic aspects of vertex elimination on graphs. SIAM J. Comput. **5**(2), 266–283 (1976). https://doi.org/10.1137/0205021
29. Sam, E., Bergougnoux, B., Golovach, P.A., Blaser, N.: Kernelization for finding lineal topologies (depth-first spanning trees) with many or few leaves. In: Fernau, H., Jansen, K. (eds.) FCT 2023. LNCS, vol. 14292, pp. 392–405. Springer, Cham (2023). https://doi.org/10.1007/978-3-031-43587-4_28
30. Sam, E., Fellows, M.R., Rosamond, F.A., Golovach, P.A.: On the parameterized complexity of the structure of lineal topologies (depth-first spanning trees) of finite graphs: the number of leaves. In: Mavronicolas, M. (ed.) CIAC 2023. LNCS, vol. 13898, pp. 353–367. Springer, Cham (2023). https://doi.org/10.1007/978-3-031-30448-4_25

Expanders in Models of Social Networks

Marc Kaufmann, Johannes Lengler(✉), Ulysse Schaller, and Konstantin Sturm

Department of Computer Science, ETH Zürich, Zürich, Switzerland
{kamarc,lenglerj,ulysses,kosturm}@ethz.ch

Abstract. A common model for social networks are Geometric Inhomogeneous Random Graphs (GIRGs), in which vertices draw a random position in some latent geometric space, and the probability of two vertices forming an edge depends on their geometric distance. The geometry may be modelled in two ways: either two points are defined as close if they are similar in *all* dimensions, or they are defined as close if they are similar in *some* dimensions.

The first option is mathematically more natural since it can be described by metrics. However, the second option is arguably the better model for social networks if the different dimensions represent features like profession, kinship, or interests. In such cases, nodes already form bonds if they align in some, but not all dimensions.

For the first option, it is known that the resulting networks are poor expanders. We study the second option in the form of Minimum-Component Distance GIRGs, and find that those behave the opposite way for dimension $d \geq 2$, and that they have strong expanding properties. More precisely, for a suitable constant $C > 0$, the subgraph induced by vertices of (expected) degree at least $(\log n)^C$ forms an expander. Moreover, we study how the expansion factor of the resulting subgraph depends on the choice of C, and show that this expansion factor is $\omega(1)$ except for sets that already take up a constant fraction of the vertices.

This has far-reaching consequences, since many algorithms and mixing processes are fast on expander graphs.

Keywords: Geometric Inhomogeneous Random Graphs · Social Network Models · Expander Graphs · Minimum-Component Distance

1 Introduction

Personal attributes affect whom we know and whom we befriend. Most of our acquaintances are similar to us in some dimensions, for example, our colleagues share the same employer with us, our neighbours live in (almost) the same place, many of our friends share our hobbies, and so on. A common approach for generative models of social networks is thus to define a latent geometric space which captures similarity along such dimensions, and places edges (possibly in a randomized manner) between nodes that are geometrically close to each other.

A well-studied model with such a latent space are Geometric Inhomogeneous Random Graphs (GIRGs), which capture many structural properties observed

H. Fernau and P. Kindermann (Eds.): WG 2025, LNCS 16124, pp. 302–315, 2026.
https://doi.org/10.1007/978-3-032-11835-6_22

in real networks. In particular, they are sparse (have a linear number of edges), small worlds, exhibit heavy-tailed degree distributions and have a large clustering coefficient [4]. Moreover, they are navigable [5], match betweenness centrality of real networks [7], reproduce the runtimes of various algorithms on real social networks [1], and have been used to explain differences in epidemic curves in social networks [20, 22] and the effectiveness of different epidemiological interventions [14].

GIRGs are generated as follows. Each node v independently draws a random position x_v in a d-dimensional geometric ground space, where the different dimensions may be used to represent attributes like profession, place of residence, hobbies, kinship, and so on. Moreover, each node v also independently draws a random weight w_v from a power-law distribution, which will correspond to its expected degree, up to a constant factor. Every pair of nodes then forms an edge with a probability that increases with their weights and decreases with their geometric distance (the exact connection probability can be found in Definition 2).

However, this construction leaves an important degree of freedom: how to measure the geometric distance in the ground space? Of course, an obvious possibility is to use a standard mathematical distance as induced by the Euclidean distance or the L^∞ distance. However, those natural choices come with an unwanted side effect: in the Euclidean or L^∞ distance, two points only have small distance if they are close *along all dimensions*. If they differ strongly in a single coordinate, this suffices to make their geometric distance large. In terms of social networks, this is not desirable, since it would mean that two nodes who are similar in, say, hobby, residency and kinship, but not in their profession, would have a large geometric distance and thus a low probability to form an edge. This clearly does not match the distribution of bonds in social networks.

For this reason, it was proposed in [4] to use the *Minimum-Component Distance* (MCD) instead, where the geometric distance is given by the minimum distance along all dimensions.[1] Figure 1 gives a realization of an MCD-GIRG in two dimensions. Note that the MCD is not a metric and does not obey the triangle inequality. For example, consider three points x, y, z where x and y are close (have small MCD) because they agree along the first dimension, and y and z are close because they agree along the second dimension. Then this does not imply that x and z are also close. This matches the fact that nodes in social networks may belong to very different social circles. If x and y are co-workers, and y and z are close relatives, then it makes sense to consider x and y as "close" (for example, they probably have many shared neighbours in the network due to their shared work environment), and to also consider y and z as close, but this does not imply that x and z are close as well.

[1] MCD-GIRGs and Euclidean GIRGs have been jointly generalized by BDF-GIRGs, which inherit their realistic properties but offer a much richer spectrum of edge formation mechanisms by allowing distance functions with arbitrary combinations of minima and maxima [18]. For more information on BDF-GIRGs see Sect. 1.3.

Strikingly, it was shown in [4] that many of the fundamental properties of GIRGs do not depend on the underlying geometry. In particular, for *any* geometric distance function[2], the resulting model still has the *ultra-small world* property that average graph distances (hop distances) are only of order $O(\log\log n)$, where n is the number of vertices in the graph. Moreover, the model has a power-law degree distribution, and in both the MCD case and the case of classical Euclidean norms[3] it has high clustering coefficient $\Theta(1)$. None of these properties depend on the choice of the geometric distance function. However, in this paper we show a fundamental difference between the Euclidean GIRG model and the MCD-GIRG model: unlike the Euclidean version, MCD-GIRGs have strong *expansion properties*.

1.1 Our Contribution

We show that in MCD-GIRGs, the subgraph induced by vertices of large degree forms an $\omega(1)$-expander. Intuitively, this means that, any vertex subset S of this subgraph, with a mild restriction on the cardinality of S, has external neighbourhood whose cardinality is at least a $\omega(1)$-multiple of $|S|$. In the following, τ denotes the power-law exponent of the degree distribution, i.e., the number of vertices of degree larger than x is $\Theta(nx^{1-\tau})$, where n is the total number of vertices in the graph. We assume $2 < \tau < 3$, which is arguably the most common case for social networks, see Sect. 2 for a discussion. For a graph $\mathcal{G} = (\mathcal{V}, \mathcal{E})$ and a vertex set $S \subseteq \mathcal{V}$, we write $N_{ext}(S) := \{v \in \mathcal{V} \setminus S : \exists u \in S \text{ with } uv \in \mathcal{E}\}$ for the external neighbourhood of S. Our main result can be summarized as follows.

Theorem 1 (Main theorem, simplified). *Let $\mathcal{G} = (\mathcal{V}, \mathcal{E})$ be an MCD-GIRG on n vertices with dimension $d \geq 2$, power-law exponent $\tau \in (2, 3)$ and $\alpha > 1$. Let $\gamma > \frac{1}{3-\tau}$ and consider the subgraph $\mathcal{G}' = (\mathcal{V}', \mathcal{E}')$ induced by vertices of degree at least $\log^{\gamma} n$. There exists a constant $\varepsilon > 0$ such that with high probability[4] the following holds. All subsets $S \subset \mathcal{V}'$ of size $|S| \leq \frac{1}{2}|\mathcal{V}'|$ satisfy $|N_{ext}(S) \cap \mathcal{V}'| \geq \varepsilon|S|$. Moreover, all subsets $\emptyset \neq S \subset \mathcal{V}'$ of size $|S| = o(|\mathcal{V}'|)$ satisfy $|N_{ext}(S) \cap \mathcal{V}'| = \omega(|S|)$.*

The same statement is true when $\mathcal{G}'$ is replaced by the graph $\mathcal{G}''$ induced by vertices of degree in the interval $[\log^{\gamma} n, 2\log^{\gamma} n]$.

In other words, the induced subgraph $\mathcal{G}'$ is a vertex expander with an expansion factor $\omega(1)$ for all subsets S whose size is $o(|\mathcal{V}'|)$. The exact value of this expansion factor depends on the parameter γ and thus on the subgraph that we consider. It is actually polylogarithmic if S contains at most a polylogarithmic fraction of $\mathcal{V}'$. See Theorem 2 for the full statement containing the precise expansion factor we obtain. We mention here that the case of linear-sized subsets S

[2] With some very mild technical conditions, in particular that the distance function is symmetric and that the volume of a ball of radius r grows continuously with r.

[3] Or the L^{∞} norm. In fact, the choice of the norm does not matter since all norms in d-dimensional space are equivalent.

[4] *With high probability* (whp) means that the probability converges to 1 as $n \to \infty$.

was already treated in [24], as we discuss in more detail in Sect. 2.2. The main contribution of this paper is to treat the case $|S| = o(|\mathcal{V}'|)$ and in particular to show that in this case we have an expansion factor of $\omega(1)$.

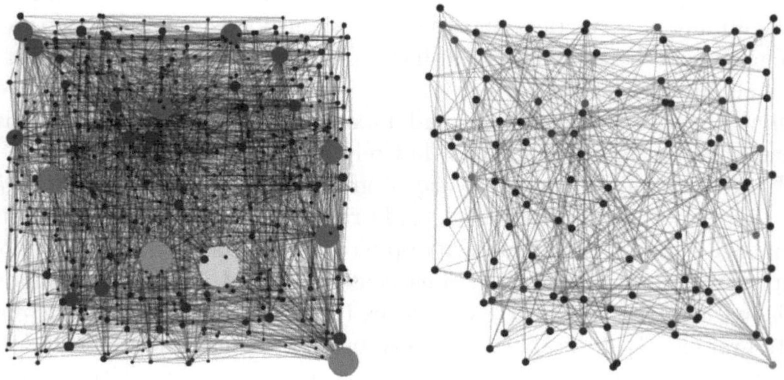

Fig. 1. MCD GIRG (left), with parameters $d = 2, n = 500, \tau = 2.1, \alpha = 1.5$, and its induced subgraph of weights larger than $\log^2 n$ (right). Larger (and lighter-coloured) disks indicate nodes of larger weight/degree. Disks are only scaled on the left illustration to increase visibility.

1.2 Implications

Expander graphs have an astonishing wealth of applications across mathematics and computer science – and consequently, MCD-GIRGs inherit a rich list of such useful and desirable properties. For an introduction to the field, we refer the reader to the survey papers [13,25]. We discuss some of the most important applications and properties of expanders in the following paragraphs.

Remark. Some of the applications described below assume that the expander graph in question is *almost regular*, by which we mean that the minimum and maximum degree in the graph only differ by a constant factor. But since our result also holds in the graph $\mathcal{G}'' = (\mathcal{V}'', \mathcal{E}'')$, and in $\mathcal{G}''$ the minimum and maximum degree (induced in $\mathcal{G}''$) are easily seen to differ by at most factor of $2+o(1)$, this subgraph is indeed regular up to a constant factor. We also mention that $\mathcal{V}''$ is large, $|\mathcal{V}''| = n/\log^{O(1)} n$, and the whole giant component of $\mathcal{G}$ is well-connected to $\mathcal{V}''$. Whp, all vertices of degree larger than $2\log^\gamma n$ have neighbours in $\mathcal{V}''$, and it was shown in the so-called *bulk lemma* in [4] that whp every vertex in $\mathcal{V}$ in the giant component has a path of at most polylogarithmic length to a vertex of degree at least $\log^\gamma n$.

– *Rapid Mixing of Random Walks:* On almost regular expanders, random walks will quickly reach a near-stationary distribution. This allows for random sampling of vertices (within the induced subgraph), even if the graph is only accessible locally, which can be used for representative polling and computation of expectations of vertex properties. A practical application of random walks is the PageRank algorithm, which is used by search engines on the web graph. If the graph has expansion properties then PageRank converges quickly. Rapid mixing may also be beneficial for property testing: sampling a small number of vertices and their neighbourhoods may demonstrate more efficiently whether a given graph has some local property – or at least has a large subgraph with this property.

- *Distributed Computing:* The rapid mixing property is very useful in distributed computing. Algorithms that rely on message passing between nodes will converge quickly on this type of graph. One example is best-of-k voting (for any $k \geq 2$), where each node randomly samples k neighbours and updates its opinion to the majority opinion among them: On certain expander graphs, with best-of-k voting, consensus is reached in $O(\log n)$ rounds [27]. This matches the asymptotic consensus time of best-of-three voting on complete graphs [11]. In graphical models, belief propagation, a message-passing algorithm for inference, often converges quickly on graphs with good expansion. Similarly, algorithms like distributed averaging and load balancing may benefit from good expansion. This has great potential for the design of distributed algorithms of this type for MCD-GIRGs, where the algorithm is first run on the induced subgraph $\mathcal{G}'$ (or the almost regular $\mathcal{G}''$), where all these theoretical performance guarantees are present.
- *Robustness to Edge or Vertex Failures:* Expanders are robust to random failures. Even after removing a random small constant fraction of edges or vertices, the remaining graph is still well-connected. Since expanders offer several short paths, low-congestion and robust routing are possible [26].
- *Information Dissemination:* Information – or a rumour or disease – can spread very quickly through the graph. We get for free that a rumour will spread to most large-degree vertices in polylogarithmically many rounds, since the expansion property ensures rapid spreading through the subgraph of vertices with degree $[\log^\gamma n, 2\log^\gamma n]$, as well as to all their neighbours. For the same reason, in an SI model the infection will spread among the large-degree vertices in $o(\log n)$ rounds.
- *Fast Algorithms for Graph Problems:* Problems such as determining connectivity or finding shortest (approximate) paths can often be solved more efficiently on expanders or graphs with expander-like components. More generally, many *NP-hard problems*, such as graph colouring, finding large independent sets or balanced separators, minimum bisection, maximum cut, as well as determining the graph bandwidth, have good approximation algorithms on expanders [6]. Some of these approximation algorithms are among the most central and widely used tools in algorithm design, especially due to their natural connections to the hierarchical divide-and-conquer paradigm [26].
- Expander graphs also appear in many other algorithms. In streaming algorithms, when processing a stream of edges representing the graph, expander properties allow to maintain certain graph characteristics such as connectivity using limited memory.

1.3 Related Work

The GIRG model with classical Euclidean geometry has been intensively studied over the last decade. A popular predecessor model are *Hyperbolic Random Graphs* (HRG) [23], which turned out to be a special case of GIRGs [3]. Although HRG can be defined for higher dimensions, they are traditionally only studied in the one-dimensional case $d = 1$.[5] Note that in this case there are no differences between Euclidean geometry and MCD. The GIRG model was invented independently, with only minor technical deviations, as *Scale-Free Percolation* (SFP) [8] and has been studied under this name as well. See [21] for a comparison of SFP and GIRG.

A particularly important result on Euclidean GIRGs is that they have small separators: if we split the geometric space into two halves via a hyperplane, then there are only $O(n^{1-\varepsilon})$ edges crossing this hyperplane, for some $\varepsilon > 0$ that depends on the model parameters [3]. This implies that it is possible to split the giant component (of either $\mathcal{G}$ or $\mathcal{G}'$) into two components of equal size by removing $O(n^{1-\varepsilon})$ edges from the graph. Those separators exist at all scales. Consider any cube $\mathcal{C}$ containing x vertices, then the expected number of edges crossing the boundary of this cube is $O(x^{1-\varepsilon})$. In particular, this yields a vertex set of size x whose exterior neighbourhood has size at most $O(x^{1-\varepsilon})$, which is asymptotically smaller than x if $x \to \infty$. Not only is this construction possible for any growing function $x = \omega(1)$, but we may also place the cube $\mathcal{C}$ at *any* position in the geometric space. This means that small separators and non-expanding sets are ubiquitous in Euclidean GIRGs.

While there is ample work on the models with Euclidean geometry, work on substantially different geometries is still sparse. The first paper to consider this was [4], where the MCD-GIRG model was introduced as an example for a non-metric distance function and it was shown that the power-law degree sequence, the $O(\log \log n)$ average hop distance, and the poly-logarithmic diameter are independent of the underlying geometry. The first substantial structural difference between MCD-GIRGs and Euclidean GIRGs was shown in [24], where it was proven that the giant component of MCD-GIRGs does not have sublinear separators, in contrast to the Euclidean case. In other words, every vertex subset of linear size of the giant component has an exterior neighbourhood of size $\Omega(n)$ in MCD-GIRGs. This result is both weaker and stronger compared to our result: it is stronger because it includes all vertices, also vertices of smaller weight, as long as they are in the giant component. But it is weaker because it only covers very large sets of size $\Theta(n)$, not smaller sets. This makes it less useful for applications, since expanders need that vertex sets at all scales are expanding. Moreover, the expansion factor was naturally only constant in [24], whereas we obtain a much larger expansion factor unless the vertex set has linear size.

The MCD was extended in [18] to general combinations of maxima and minima. The resulting distance functions were called *Boolean Distance Functions* (BDF), and the corresponding network model BDF-GIRGs, which inherits the

[5] In the HRG terminology this still involves a two-dimensional disk since the weight is encoded as one dimension. Nevertheless, it corresponds to the $d = 1$ case.

desirable properties of real-world networks captured by Euclidean GIRGs such as sparsity, small diameter and ultra-small average distances, and jointly generalizes MCD-GIRGs and Euclidean GIRGs. An example of such a BDF would be $dist := \min\{dist_{work}, \max\{dist_{hobby}, dist_{residence}\}\}$, indicating that two individuals are likely to know each other if they are either working in closely related fields, or if they share both the same hobby and live in the same place. There are two main results in [18]. The first one is a complete classification of the occurrence of sublinear separators. A sub-category, *single-coordinate outer max* (SCOM) GIRGs behave as Euclidean GIRGs in this regard and have sublinear separators of the giant component. All other BDF-GIRGs behave like MCD-GIRGs and have no such sublinear separators. This result from [18] thus extended the analysis from [24] to arbitrary BDFs. The second result in [18] is that all BDFs satisfy the following *stochastic triangle inequality*. Consider a radius r and a point x in a geometric space with some BDF distance function d. If we sample two points y, z uniformly at random from the ball of radius r around x, then $\mathbb{P}(d(y,z) \leq 2r) = \Omega(1)$, independently of r. Note that the probability is 1 when d is a metric, but recall that BDFs are in general not metrics, e.g. the MCD is not a metric. It was shown in [18] that this stochastic triangle inequality suffices to guarantee that all BDF-GIRGs exhibit strong clustering, more precisely a clustering coefficient of $\Theta(1)$.

It is natural to ask whether the same characterization also applies to separators of all scales, i.e. whether the result of this paper generalizes to all BDFs which are not SCOM. However, our proof methods do not easily generalize to the whole class of BDFs, so this remains an open problem.

Another aspect in which Euclidean and MCD-GIRGs differ is rumour spreading. While our result immediately implies that rumours need at most polylogarithmic time to spread through an MCD-GIRG (because it takes polylogarithmic time to reach the first vertex in $\mathcal{V}''$, then polylogarithmic time to reach *every* vertex in $\mathcal{V}''$, and finally polylogarithmic time to reach all other vertices from $\mathcal{V}$), more precise results were obtained in [16]. There it was shown that there are three different speeds at which a rumour can reach a constant fraction of the vertex set: *ultra-fast* in time $O(\log\log n)$; *fast* in polylogarithmic time; or *slow* in time $n^{\Omega(1)}$. The latter possibility had first been observed in a slightly different model, *Spatial Preferential Attachment* (SPA) [9]. Note that our result rules out slow rumour spreading in MCD-GIRGs. However, it has already been shown in [16] that all three regimes can occur in Euclidean GIRGs depending on the model parameters, but that rumour spreading is always ultra-fast in MCD-GIRGs (assuming the power-law exponent τ is in the interval $(2,3)$). This coincides with the non-geometric setting of Chung-Lu graphs, where rumours also spread ultra-fast when $2 < \tau < 3$ [10].

Finally, an efficient algorithm for sampling MCD-GIRGs in time $O(n)$ was recently presented in [7]. Even though several algorithms had been developed for sampling Euclidean GIRGs in linear time [2,3], those algorithms are fundamentally based on the existence of small separators in the underlying geometric space. Hence, those algorithms could not be transferred to MCD-GIRGs. The algorithm [7] closes this gap and allows to efficiently sample and study MCD-GIRG empirically.

Organization of the Paper. The remainder of the paper is organized as follows. In Sect. 2 we formally introduce MCD-GIRGs and give the formal statement of our main result. Section 3 then gives a proof overview and intuition behind the overall strategy. The formal proof of our main result can be found in the extended version of this paper [17] (in Appendix A.2).

2 Formal Definitions and Results

Notation and Terminology. We write $[n] := \{1, ..., n\}$. The ith component of a vector e is denoted by e_i. In other words, we will denote the ith component of the position of a vertex v by $x_{v,i}$ for $i \in [d]$. We denote the degree of a vertex v constrained to a set of other vertices S by $\deg_S(v)$. We denote by uv the edge connecting two vertices u and v. We say that an event $\mathcal{A}$ occurs with high probability if $\mathbb{P}[\mathcal{A}] = 1 - o(1)$ as the number of vertices n tends to ∞. We denote the natural logarithm of x by $\log x$.

2.1 (MCD-)GIRGs

As our ground space, we consider the d-dimensional torus $\mathbb{T}^d := \mathbb{R}^d/\mathbb{Z}^d$. This can be viewed as the d-dimensional unit cube $[0,1]^d$, with opposite faces identified, which has the advantage of yielding a bounded and symmetric ground space. In this topology, the absolute difference for $a, b \in [0,1]$ is defined as $|a-b|_T := \min\{|a-b|, 1-|a-b|\}$. As we will only deal with the torus topology, we will simply write $|x|$ instead of $|x|_T$.

We now define the distance function, which we will use to construct the graphs. Consider $x_u, x_v \in \mathbb{T}^d$ with $x_u = (x_{u,1}, ..., x_{u,d})$ and $x_v = (x_{v,1}, ..., x_{v,d})$. We define the Minimum-Component Distance (MCD) as

$$\|x_u - x_v\|_{\min} := \min\{|x_{u,i} - x_{v,i}|_T \mid 1 \leq i \leq d\}.$$

This distance function gives rise to the measurable volume function $V_{\min} : \mathbb{R}_0^+ \to [0,1]$ given by $V_{\min}(r) := \mathrm{Vol}\left(\{x \in \mathbb{T}^d \mid |x| \leq r\}\right)$, which is symmetric, translation-invariant and continuous. For $r \to 0$, the volume function satisfies $V_{\min}(r) = \Theta(r)$ [24]. See also Fig. 2 (left) for an illustration of "balls", which are cross-shaped in this geometry. The distance and the volume function together allow us to define Geometric Inhomogeneous Random Graphs (GIRGs).

GIRGs have originally been introduced in [3], though only for Euclidean distances, and have been generalized to arbitrary geometries in [4]. We begin by formalizing the concept of a power-law distribution.

Definition 1. *Let $\tau > 1$. A discrete random variable X is said to follow a* power-law with exponent τ *if $\mathbb{P}(X = x) = \Theta(x^{-\tau})$ for $x \in \mathbb{N}$. A continuous random variable X is said to follow a* power-law with exponent τ *if it has a density function f_X satisfying $f_X(x) = \Theta(x^{-\tau})$ for $x \geq 1$.*

Now, we are ready to formally define MCD-GIRGs. Throughout the paper, we will assume that the three parameters τ, α and d below are constants.

Definition 2 (MCD-GIRGs). *Let $\tau > 2$, $\alpha > 1$ and $d \in \mathbb{N}$ and let $\mathcal{D}$ be a power-law distribution on $[1, \infty)$ with exponent τ. A* MCD-Geometric Inhomogeneous Random Graph (MCD-GIRG) *with vertex set $\mathcal{V} = [n]$ and edge set $\mathcal{E}$ is obtained by the following three-step procedure:*

1. *Every vertex $v \in \mathcal{V}$ draws i.i.d. a* weight *$w_v \sim \mathcal{D}$.*
2. *Every vertex $v \in \mathcal{V}$ draws independently and u.a.r. a position $x_v \in \mathbb{T}^d$.*
3. *For every two distinct vertices $u, v \in \mathcal{V}$, add the edge $uv \in \mathcal{E}$ independently with probability*

$$\mathbb{P}[uv \in \mathcal{E} \mid w_u, w_v, x_u, x_v] = \Theta\Big(\min\Big\{\frac{w_u w_v}{n \cdot V_{min}(\|x_u - x_v\|_{min})}, 1\Big\}^{\alpha}\Big), \qquad (1)$$

where the hidden constants are uniform over all u, v.

With the connection probability given above, the expected degree of a vertex v is of the same order as its weight w_v (Lemma 4.3 in [4]) and its actual degree is a sum of independent random variables which converges to the Poisson distribution in the limit. The restrictions on the range of the power-law parameter τ come from the observation that the degrees in many real networks, in particular social networks, approximately follow a power-law with τ in this range [28]. In particular, such a power-law distribution has infinite second moment. Nonetheless, the resulting GIRGs are sparse, that is, have only a linear number of edges. We call an edge a *strong tie* if the minimum in (1) is taken by 1, and *weak tie* otherwise. The latter are edges that exist although they are unlikely given the weights and positions of the vertices. The parameter α quantifies the influence of the geometry in the sense that it controls the number of weak ties (increasing α reduces the number of weak ties). These weak ties are an important concept in sociology [12] and influence the spreading of rumours [16] and viral infections [14,20,22]. Strong ties, on the other hand, are edges which have at least a constant probability of existing. The geometric region around a vertex v of the radius r that satisfies $V_{\min}(r) = w_v/n$, is also referred to as the *ball of influence* of v. It is the region around v where all edges are strong ties, regardless of the weight of the other endpoint.

2.2 Main Result

In order to state our main result, we first formally introduce expander graphs. We call a set S of vertices ζ-expanding if the external neighbourhood of S is at least ζ times the size of S itself.

Definition 3 (Expanding set). *Let $\zeta > 0$. For a graph $\mathcal{G} = (\mathcal{V}, \mathcal{E})$, we call a subset $S \subset \mathcal{V}$ of vertices a ζ-*expanding set *(for $\mathcal{G}$) if the external neighbourhood $N_{ext}(S) = \{v \in \mathcal{V} \setminus S : \exists u \in S$* with *$uv \in \mathcal{E}\}$ of S is of size at least $|N_{ext}(S)| \geq \zeta \cdot |S|$. We call ζ the* expansion factor *of the expanding set.*

We are now ready to formally state our main result, which asserts that in MCD-GIRGs, the subgraph induced by vertices of sufficiently large polylogarithmic weight is an expander with high probability. Furthermore, the expansion factor is typically of polylogarithmic order (and always at least constant).

Theorem 2. *For each $d \geq 2$ there is $c_d > 1$ such that the following holds. Let $\mathcal{G} = (\mathcal{V}, \mathcal{E})$ be an MCD-GIRG on n vertices of dimension $d \geq 2$, power-law exponent $\tau \in (2,3)$ and $\alpha > 1$, let $\gamma > \frac{1}{3-\tau}$ and $c' > 0$ be constants and consider the subgraph $\mathcal{G}' = (\mathcal{V}', \mathcal{E}')$ induced by vertices of weight at least $c' \log^\gamma n$. There exists a constant $\varepsilon > 0$ such that with high probability all subsets $S \subset \mathcal{V}'$ satisfying $|S| \leq \varepsilon|\mathcal{V}'|$ are expanding sets for $\mathcal{G}'$ with expansion factor at least*

$$\varepsilon \cdot \min\left\{ \log^{\gamma(3-\tau)} n, \left(\tfrac{|\mathcal{V}'|}{|S|}\right)^{1-1/c_d} \right\}. \tag{2}$$

The same result holds if we replace $\mathcal{G}'$ by the subgraph of $\mathcal{G}$ induced by all vertices with weight in the interval $[c_1' \log^\gamma n, c_2' \log^\gamma n]$ for two arbitrary constants $c_2' > c_1' > 0$. Moreover, all statements are still true if we replace $\mathcal{G}'$ by the graph induced by all vertices of degree *(in $\mathcal{G}$) at least $c' \log^\gamma n$, or by all vertices of degree in the interval $[c_1' \log^\gamma n, c_2' \log^\gamma n]$.*

The larger the weights we consider, the better the expansion factor $\log^{\gamma(3-\tau)} n$ becomes. Eventually, when the size of S approaches $\mathcal{V}'$, this expansion factor is no longer achievable simply because we run out of vertices. In this case, we prove a smaller expansion factor, but this factor is still $\omega(1)$ whenever $|S| = o(|\mathcal{V}'|)$ and becomes constant for $|S| = \Theta(|\mathcal{V}'|)$. Note that this is optimal, simply because the number of remaining vertices in $\mathcal{V}' \setminus S$ is then $O(|S|)$.

We highlight that the assumption on $d \geq 2$ is crucial, as in one dimension, the MCD is equivalent to the Euclidean metric, where we encounter small separators at any scale. We also remark that each vertex in $\mathcal{V}$ has probability $\Theta(\log^{\gamma(1-\tau)} n)$ to have weight at least $\log^\gamma n$. Therefore, whp $|\mathcal{V}'| = \Theta(n \log^{\gamma(1-\tau)} n) = n^{1-o(1)}$.

Moreover, let us discuss where the condition $\gamma > \frac{1}{3-\tau}$ comes from. By [19, Lemma 10] the expected degree of a vertex $v \in \mathcal{V}'$ of weight w_v in $\mathcal{G}'$ is $\mathbb{E}[\deg_{\mathcal{G}'}(v)] = \Theta(w_v \log^{\gamma(2-\tau)})$. Since $w_v \geq \log^\gamma n$, this is $\Omega(\log^{\gamma(3-\tau)} n) = \omega(\log n)$ as $\gamma > \frac{1}{3-\tau}$. Hence, every vertex has expected degree $\omega(\log n)$ in $\mathcal{G}'$, and also in $\mathcal{G}''$. In particular, whp $\mathcal{G}'$ and $\mathcal{G}''$ do not have isolated vertices.

The condition on γ is also tight, for the same reason. By the same calculation as above, for $\gamma < \frac{1}{3-\tau}$ the expected degree of a vertex of weight $\Theta(\log^\gamma n)$ in $\mathcal{G}'$ or $\mathcal{G}''$ is $o(\log n)$, and it is not hard to see that then the induced subgraph contains isolated vertices whp, which implies that it is not an expander. Indeed, since the degree of a vertex v with given weight $w_v = \Theta(\log^\gamma n)$ in $\mathcal{G}'$ is binomially distributed, any vertex of such a weight - of which there are $\Theta(n \cdot \log^{\gamma(1-\tau)} n) = n^{1-o(1)}$ many - would be isolated in $\mathcal{G}'$ with probability at least $e^{-o(\log n)} = n^{-o(1)}$. So in expectation, there would be $\Theta(n^{1-o(1)} \cdot n^{-o(1)}) = n^{1-o(1)}$ isolated vertices in the induced subgraph. Those form a (large) set which is not expanding at all, as it has zero neighbours. This also holds whp using an Azuma-Hoeffding-type estimate.[6] Hence, the condition $\gamma > \frac{1}{3-\tau}$ is tight.

Finally, we note that the case $\varepsilon|\mathcal{V}'| \leq |S| \leq \frac{1}{2}|\mathcal{V}'|$ has already been treated in very similar form in [24], which is why we exclude this case from the statement. There, it was shown that any linear-sized vertex subset of the giant component of $\mathcal{G}$ has an external neighbourhood of linear size and hence constant expansion. This statement and its proof can readily be adapted to give the analogous result for the induced subgraph $\mathcal{G}'$. Since $\mathcal{G}'$ is connected, we may also drop the condition on the giant component, hence obtaining that every set $S \subseteq \mathcal{V}'$ with $\varepsilon|\mathcal{V}'| \leq |S| \leq \frac{1}{2}|\mathcal{V}'|$ has external neighbourhood of size at least $\varepsilon|\mathcal{V}'|$.

3 Proof Sketch

In this section, we provide an intuitive overview of the proof. The formal, detailed proof can be found in Appendix A.2 in [17].

We want to lower-bound the external neighbourhood of a given vertex set S. We begin by sampling an MCD-GIRG and restricting to the subgraph induced by vertices of weight at least $\log^\gamma n$. Next, we partition the ground space into *strips* of width $\ell = \Theta(\frac{\log^{2\gamma} n}{n})$ along each dimension. Figure 2 (right) illustrates this. The idea behind this strip width is that two vertices of weight $\log^\gamma n$ in the same strip have a connection probability in $\Theta(1)$.

For each (set) size s we will find a suitable value k such that whp the following claim holds:

"There is no set S of vertices with $|S| \geq s$ such that in every dimension, S intersects at most k strips".

In a second step, we show that every vertex $v \in \mathcal{V}'$ has whp $\Omega(w_v \log^{\gamma(2-\tau)} n) = \Omega(\log^{\gamma(3-\tau)} n)$ neighbours *in the same strip*, in both $\mathcal{G}'$ and $\mathcal{G}''$. Since every set S of size s intersects at least k strips in some dimension, the vertices in S have at least $k \cdot \Omega(\log^{\gamma(3-\tau)} n)$ distinct neighbours. To obtain the size of the *external* neighbourhood, we must ensure that vertices inside S itself are not counted. By simply subtracting $|S|$ from the total neighbourhood count, we obtain the desired lower bound on the size of the external neighbourhood.

Formalizing the strategy outlined above, our main result will leverage the following two key propositions, which are proven in Appendix A.2 in [17] along with the theorem itself. The first proposition provides a lower bound on the number of strips[7] which a set of at least s many vertices will intersect whp.

Proposition 1. *For each $d \geq 2$ there exists $c_d > 1$ such that the following holds. Let $\mathcal{G} = (\mathcal{V}, \mathcal{E})$ be an MCD-GIRG on n vertices of dimension $d \geq 2$ with power-law exponent $\tau \in (2, 3)$ and $\alpha > 1$, let $\gamma > \frac{1}{3-\tau}$ and $c' > 0$ be constants*

[6] See also Remark 5.13. in [15] which shows that whp there exists a vertex with weight less than $\log^{\frac{1}{3-\tau}} n$ which lies outside the giant component.

[7] For the formal definition of i-strips see Definition 4 in [17].

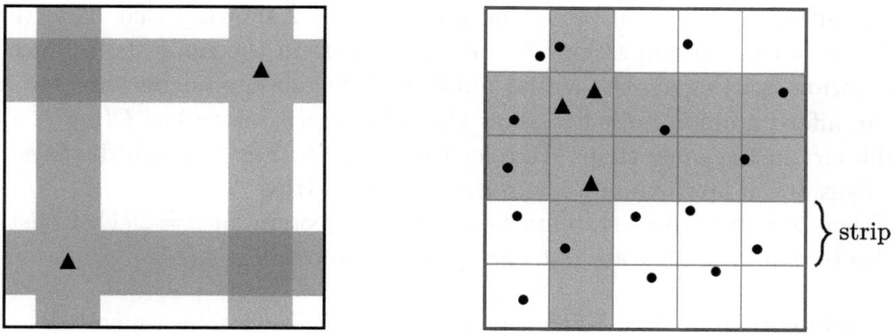

Fig. 2. Left: Balls in the Minimum-Component Distance are cross-shaped. **Right:** Illustration of the proof technique in two dimensions. The vertices marked by triangles form the adversarially chosen set S for $s = 3$. Although the adversary has managed to have S only intersect with one vertical strip, it does not achieve $k = 1$ because S intersects with two horizontal strips. In fact, for $s = 3$ and $k = 1$ the claim holds as there is no set of three vertices spanning only one strip in each dimension.

and consider the subgraph $\mathcal{G}' = (\mathcal{V}', \mathcal{E}')$ induced by vertices of weight at least $c' \log^\gamma n$. For every $s = \omega(1)$ there exists

$$k = s \cdot \Omega\Big(\min\Big\{1, \tfrac{1}{\log^{\gamma(3-\tau)} n} \cdot \Big(\tfrac{|\mathcal{V}'|}{s}\Big)^{1-1/c_d}\Big\}\Big), \tag{3}$$

such that the following holds with high probability: for every $S \subset \mathcal{V}'$ with $|S| \geq s$, there is a coordinate $1 \leq i \leq d$ such that S has non-empty intersection with at least k distinct i-strips.

The proof of Proposition 1 is a counting argument. We simply count the number of sets of k strips in each dimension and the number of sets S of s vertices, and estimate the probability that S is contained in the chosen strips. By a union bound over all sets of strips and all S we show that whp there exists no combination of k strips per dimension and set S that falls entirely into these strips.

The second proposition gives a lower bound on the number of neighbours of weight $\Theta(\log^\gamma n)$ of any set $S \subseteq \mathcal{V}'$, expressed as a function of k.

Proposition 2. *Let $\mathcal{G} = (\mathcal{V}, \mathcal{E})$ be an MCD-GIRG on n vertices of dimension $d \geq 2$ with power-law exponent $\tau \in (2, 3)$ and $\alpha > 1$, let $\gamma > \frac{1}{3-\tau}$ and $c' > 0$ $c_2' > c_1' > 0$ be constants and consider the subgraph $\mathcal{G}' = (\mathcal{V}', \mathcal{E}')$ induced by vertices of weight at least $c' \log^\gamma n$. Then with high probability the following holds: for every $S \subset \mathcal{V}'$ that has non-empty intersection with at least k distinct i-strips (for some coordinate $1 \leq i \leq d$), the number of neighbours $N_{c_1',c_2'}(S)$ with weights in the interval $[c_1' \log^\gamma n, c_2' \log^\gamma n]$ of S is at least*

$$N_{c_1',c_2'}(S) \geq \Omega(k \cdot \log^{\gamma(3-\tau)} n).$$

As outlined above, we prove the proposition by showing that each vertex in $\mathcal{V}'$ has in expectation $\Omega(\log^{\gamma(3-\tau)} n)$ neighbours in the same strip. Since the expectation is $\omega(\log n)$, the actual number of neighbours is concentrated, and we can afford a union bound to show that whp *every* vertex has $\Omega(\log^{\gamma(3-\tau)} n)$ neighbours in the same strip. We refer the reader to Fig. 1, which displays how the edges are in approximate alignment with the strips.

Taken together, these statements imply the theorem for $s = \omega(1)$. Constant-sized sets require a separate but analogous argument.

Acknowledgments. Marc Kaufmann and Ulysse Schaller were supported by the Swiss National Science Foundation [grant number 200021_192079].

References

1. Bläsius, T., Fischbeck, P.: On the external validity of average-case analyses of graph algorithms. ACM Trans. Algorithms **20**(1) (2024)
2. Bläsius, T., Friedrich, T., Katzmann, M., Meyer, U., Penschuck, M., Weyand, C.: Efficiently generating geometric inhomogeneous and hyperbolic random graphs. Netw. Sci. **10**(4), 361–380 (2022)
3. Bringmann, K., Keusch, R., Lengler, J.: Geometric inhomogeneous random graphs. Theoret. Comput. Sci. **760**, 35–54 (2019)
4. Bringmann, K., Keusch, R., Lengler, J.: Average distance in a general class of scale-free networks. Adv. Appl. Probab. 1–36 (2024)
5. Bringmann, K., Keusch, R., Lengler, J., Maus, Y., Molla, A.R.: Greedy routing and the algorithmic small-world phenomenon. J. Comput. Syst. Sci. **125**, 59–105 (2022)
6. Chuzhoy, J., Gao, Y., Li, J., Nanongkai, D., Peng, R., Saranurak, T.: A deterministic algorithm for balanced cut with applications to dynamic connectivity, flows, and beyond. In: 2020 IEEE 61st Annual Symposium on Foundations of Computer Science (FOCS), pp. 1158–1167 (2019)
7. Dayan, B., Kaufmann, M., Schaller, U.: Expressivity of geometric inhomogeneous random graphs—metric and non-metric. In: Botta, F., Macedo, M., Barbosa, H., Menezes, R. (eds) CompleNet-Live 2024. Springer Proceedings in Complexity, pp. 85–100. Springer, Cham (2024). https://doi.org/10.1007/978-3-031-57515-0_7
8. Deijfen, M., Van der Hofstad, R., Hooghiemstra, G.: Scale-free percolation. In: Annales de l'IHP Probabilités et statistiques, vol. 49, pp. 817–838 (2013)
9. Feldman, J., Janssen, J.: High degree vertices and spread of infections in spatially modelled social networks. In: Bonato, A., Chung Graham, F., Prałat, P. (eds.) WAW 2017. LNCS, vol. 10519, pp. 60–74. Springer, Cham (2017). https://doi.org/10.1007/978-3-319-67810-8_5
10. Fountoulakis, N., Panagiotou, K., Sauerwald, T.: Ultra-fast rumor spreading in social networks, pp. 1642–1660 (2012). https://doi.org/10.1137/1.9781611973099.130
11. Ghaffari, M., Lengler, J.: Nearly-tight analysis for 2-choice and 3-majority consensus dynamics. In: Proceedings of the 2018 ACM Symposium on Principles of Distributed Computing, PODC 2018, pp. 305–313. Association for Computing Machinery, New York (2018)

12. Granovetter, M.S.: The strength of weak ties. Am. J. Sociol. **78**(6), 1360–1380 (1973)
13. Hoory, S., Linial, N., Wigderson, A.: Expander graphs and their applications. Bull. Amer. Math. Soc. **43**, 439–561 (2006)
14. Jorritsma, J., Hulshof, T., Komjáthy, J.: Not all interventions are equal for the height of the second peak. Chaos Solitons Fractals **139**, 109965 (2020)
15. Jorritsma, J., Komjáthy, J., Mitsche, D.: Cluster-size decay in supercritical kernel-based spatial random graphs (2024). https://arxiv.org/abs/2303.00724
16. Kaufmann, M., Lakis, K., Lengler, J., Ravi, R.R., Schaller, U., Sturm, K.: Rumour spreading depends on the latent geometry and degree distribution in social network models. arXiv preprint arXiv:2408.01268 (2024)
17. Kaufmann, M., Lengler, J., Schaller, U., Sturm, K.: Expanders in models of social networks. arXiv e-prints arXiv:2506.19485 (2025). https://doi.org/10.48550/arXiv.2506.19485
18. Kaufmann, M., Ravi, R.R., Schaller, U.: Sublinear cuts are the exception in BDF-GIRGs. In: The 13th International Conference on Complex Networks and their Applications (2024)
19. Koch, C., Lengler, J.: Bootstrap percolation on geometric inhomogeneous random graphs. Internet Math. 05c80 (2021)
20. Komjáthy, J., Lapinskas, J., Lengler, J., Schaller, U.: Polynomial growth in degree-dependent first passage percolation on spatial random graphs. Electron. J. Probab. **29**, 1–48 (2024)
21. Komjáthy, J., Lodewijks, B.: Explosion in weighted hyperbolic random graphs and geometric inhomogeneous random graphs. Stochast. Process. Appl. **130**(3), 1309–1367 (2020)
22. Komjáthy, J., Lapinskas, J., Lengler, J., Schaller, U.: Four universal growth regimes in degree-dependent first passage percolation on spatial random graphs I (2024)
23. Krioukov, D., Papadopoulos, F., Kitsak, M., Vahdat, A., Boguná, M.: Hyperbolic geometry of complex networks. Phys. Rev. E-Stat. Nonlinear Soft Matt. Phys. **82**(3), 036106 (2010)
24. Lengler, J., Todorovic, L.: Existence of small separators depends on geometry for geometric inhomogeneous random graphs. arXiv preprint arXiv:1711.03814 (2017)
25. Lubotzky, A.: Expander graphs in pure and applied mathematics. Bull. Am. Math. Soc. **49** (2011)
26. Racke, H.: Minimizing congestion in general networks. In: 2002 Proceedings of the 43rd Annual IEEE Symposium on Foundations of Computer Science, pp. 43–52 (2002)
27. Shimizu, N., Shiraga, T.: Quasi-majority functional voting on expander graphs. Random Struct. Algorithms **65**(4), 613–643 (2024)
28. Voitalov, I., van der Hoorn, P., van der Hofstad, R., Krioukov, D.: Scale-free networks well done. Phys. Rev. Res. **1**, 033034 (2019)

Connectivity-Preserving Minimum Separator in AT-Free Graphs

Batya Kenig(✉)

Technion, Israel Institute of Technology, Haifa, Israel
batyak@technion.ac.il

Abstract. Let A and B be disjoint, non-adjacent vertex-sets in an undirected, connected graph G, whose vertices are associated with positive weights. We address the problem of identifying a minimum-weight subset of vertices $S \subseteq \mathtt{V}(G)$ that, when removed, disconnects A from B while preserving the internal connectivity of both A and B. We call such a subset of vertices a *connectivity-preserving*, or *safe* minimum A, B-separator. Deciding whether a safe A, B-separator exists is NP-hard by reduction from the 2-DISJOINT CONNECTED SUBGRAPHS problem [15], and remains NP-hard even for restricted graph classes that include planar graphs [14], and P_ℓ-free graphs if $\ell \geq 5$ [15]. In this work, we show that if G is AT-free then in polynomial time we can find a safe A, B-separator of minimum weight, or establish that no safe A, B-separator exists.

Keywords: Graph Separation · 2-disjoint-connected subgraphs · AT-Free

1 Introduction

Let G be a simple, undirected, connected graph, whose vertices are associated with positive weights, and let $A, B \subseteq \mathtt{V}(G)$ be disjoint, non-adjacent subsets of $\mathtt{V}(G)$. That is, G does not contain an edge that has an endpoint in A and an endpoint in B. An A, B-separator is a subset of vertices $S \subseteq \mathtt{V}(G) \backslash (A \cup B)$, such that in the graph G–S that results from G by removing S and its adjacent edges, A and B are disconnected; for every pair $a \in A$ and $b \in B$ there is no path in G–S between a and b. We say that S is a *minimal* A, B-separator if no strict subset of S is an A, B-separator. The objective is to compute an A, B-separator with minimum weight to disconnect A from B, while preserving the connectivity of A and B in G–S. That is, the graph G–S has two distinct, connected components A_1 and B_1 such that $A \subseteq A_1$ and $B \subseteq B_1$. For brevity, we call such a connectivity-preserving A, B-separator a *safe* A, B-separator. In this paper, we study the problem of finding a safe A, B-separator of minimum weight, if one exists, or determine that no safe A, B-separator exists. We refer to this problem as MIN SAFE SEPARATOR.

H. Fernau and P. Kindermann (Eds.): WG 2025, LNCS 16124, pp. 316–330, 2026.
https://doi.org/10.1007/978-3-032-11835-6_23

Finding minimum separators, and minimum (edge) separators, under connectivity constraints is crucial in various domains, and has been studied extensively [1,7,8,10,11]. In network security, for instance, a common challenge during denial-of-service attacks is to isolate compromised nodes from the rest of the network while preserving the connectivity among un-compromised nodes [7,10]. Connectivity preserving minimum separators are also used to model clustering problems with constraints enforcing clustering of certain objects together [11]. The connectivity preserving minimum vertex (edge) separator problem is defined as follows [11]. The input is a connected graph G with positive vertex (or edge) weights, and three vertices s, t, v. The objective is to compute a vertex (edge) separator of minimum weight to disconnect s from t, while preserving the connectivity of s and v [11]. There are two variants of this problem, one that seeks a connectivity-preserving minimum edge separator, and one that seeks a connectivity-preserving minimum vertex separator. The latter corresponds to MIN SAFE SEPARATOR where $|A| = 2$ and $|B| = 1$. It was shown in [11], that even if $|A| = 2$ and $|B| = 1$, it is NP-complete to approximate the minimum safe vertex separator within $\alpha \log(|\mathtt{V}(G)|)$, for any constant α. Interestingly, Duan and Xu [11] showed that the connectivity preserving minimum edge separator can be solved in polynomial time in planar graphs. Recently, Bentert et al. [1] extended this result and presented a randomized algorithm that finds a connectivity preserving minimum-cardinality A, B-edge separator in planar graphs, in time $2^{|A|+|B|} \cdot |\mathtt{V}(G)|$.

MIN SAFE SEPARATOR is the natural optimization variant of the intensively studied 2-DISJOINT CONNECTED SUBGRAPHS problem [9,13–15,19,21,23]. The 2-DISJOINT CONNECTED SUBGRAPHS problem receives as input an undirected graph G, together with two disjoint subsets of vertices $A, B \subseteq \mathtt{V}(G)$. The goal is to decide whether there exist two disjoint subsets $A_1, B_1 \subseteq \mathtt{V}(G)$, such that $A \subseteq A_1$, $B \subseteq B_1$ and A_1 and B_1 are connected in G. The 2-DISJOINT CONNECTED SUBGRAPHS problem remains NP-complete even in very restricted settings where $|A| = 2$, $|B|$ is unbounded, and the input graph is a line graph (a subclass of claw-free graphs) [19]. It is also NP-complete for planar graphs [14], and in many other settings [15,19]. In general graphs, 2-DISJOINT CONNECTED SUBGRAPHS is NP-Complete even if $|A| = |B| = 2$ by reduction from the INDUCED DISJOINT PATHS PROBLEM [17]. It is easy to show that deciding whether a safe A, B-separator even exists is NP-complete by reduction from 2-DISJOINT CONNECTED SUBGRAPHS, and the formal proof is deferred to the full version of thid paper [18].

We consider MIN SAFE SEPARATOR in the class of *asteroidal triple-free graphs*, also known as *AT-free* graphs. An *asteroidal triple* is a set of three mutually non-adjacent vertices, such that every pair of vertices from this triple is joined by a path that avoids the neighborhood of the third. AT-free graphs are exactly those graphs that contain no such triple. AT-free graphs are intensively studied and include as a subclass the set of cobipartite graphs, cocomparability graphs, cographs, interval graphs, permutation graphs, and trapezoid graphs [6,13]. In previous work, Golovach, Kratsch, and Paulusma [13] presented a dynamic programming algorithm for k-DISJOINT CONNECTED SUBGRAPHS in

AT-free graphs, where k is fixed. For the case of $k = 2$ their algorithm involves "guessing" at least 12 vertices and has a runtime of $O(n^{15})$ for an n-vertex, AT-free graph (cf. [13]). Our algorithm goes well beyond deciding whether a safe A, B-separator exists, and actually finds a minimum, safe A, B-separator (or decides that none exist) orders of magnitude faster. Towards this goal, we prove new properties of minimal separators in AT-free graphs that may be of independent interest, and show how questions regarding the existence of disjoint connected subgraphs can be translated to questions regarding the existence of certain minimal separators in the graph.

Theorem 1. *Let G be a simple, undirected, connected, weighted AT-free graph, and let $A, B \subseteq V(G)$ be a pair of non-empty, disjoint, vertex-sets. There is an algorithm that finds a safe A, B-separator of minimum weight, or establishes that no safe A, B-separator exists in time $O(n^4 \cdot T(n,m))$ where $n = |V(G)|, m = |E(G)|$, and $T(n,m)$ is the time to find a minimum s,t-separator in G for some pair of vertices $s,t \in V(G)$.*

The problem of finding a minimum s,t-separator in an undirected graph can be reduced, by standard techniques [12], to the problem of finding a minimum s,t-cut, or maximum flow from s to t. Following a sequence of improvements to max-flow algorithms in the past few years [4,16,20], the current best running time is $O(m^{1+o(1)})$ [3,5].

Organization. The rest of this paper is organized as follows. Following preliminaries in Sect. 2, we establish some basic results on minimal s,t-separators, and on minimal separators between vertex sets, in Sect. 3. In Sect. 4, we give an overview of the algorithm, high-level pseudo-code, and map the results that need to be proved to establish its correctness and runtime guarantee. In Sect. 5, we prove several results about minimal s,t-separators and minimal s,t-separators in AT-free graphs in particular. The main theorem behind the algorithm is proved in Sect. 6, where we also present the pseudo-code of the main component. Due to space restrictions, some of the proofs and technical details are deferred to the full version of this paper [18].

2 Preliminaries and Notation

Let G be an undirected graph with nodes $\mathtt{V}(G)$ and edges $\mathtt{E}(G)$, where $n = |\mathtt{V}(G)|$, and $m = |\mathtt{E}(G)|$. We assume a positive weight function on the vertices $w : \mathtt{V}(G) \to \mathbb{Z}^+$. We also assume, without loss of generality, that G is connected. For $A, B \subseteq \mathtt{V}(G)$, we abbreviate $AB \stackrel{\text{def}}{=} A \cup B$; for $v \in \mathtt{V}(G)$ we abbreviate $vA \stackrel{\text{def}}{=} \{v\} \cup A$. Let $v \in V$. We denote by $N_G(v) \stackrel{\text{def}}{=} \{u \in \mathtt{V}(G) : (u,v) \in \mathtt{E}(G)\}$ the neighborhood of v, and by $N_G[v] \stackrel{\text{def}}{=} N_G(v) \cup \{v\}$ the *closed* neighborhood of v. For a subset of vertices $T \subseteq \mathtt{V}(G)$, we denote by $N_G(T) \stackrel{\text{def}}{=} \bigcup_{v \in T} N_G(v) \backslash T$, and $N_G[T] \stackrel{\text{def}}{=} N_G(T) \cup T$. We denote by $G[T]$ the subgraph of G induced by T. Formally, $\mathtt{V}(G[T]) = T$, and $\mathtt{E}(G[T]) = \{(u,v) \in \mathtt{E}(G) : \{u,v\} \subseteq T\}$. For a subset $S \subseteq \mathtt{V}(G)$, we abbreviate $G-S \stackrel{\text{def}}{=} G[\mathtt{V}(G)\backslash S]$; for $v \in \mathtt{V}(G)$, we abbreviate $G-v \stackrel{\text{def}}{=} G-\{v\}$. We say that G' is a *subgraph* of G if it results from G by removing vertices and edges; formally, $\mathtt{V}(G') \subseteq \mathtt{V}(G)$ and $\mathtt{E}(G') \subseteq \mathtt{E}(G)$. In that case, we also say that G is a *supergraph* of G'.

Let $(u, v) \in \mathtt{E}(G)$. *Contracting the edge* (u, v) *to vertex* u results in a new graph G' where:

$$\mathtt{V}(G') = \mathtt{V}(G)\backslash\{v\} \qquad \text{and} \qquad \mathtt{E}(G') = \mathtt{E}(G-v) \cup \{(u, x) : x \in N_G(v)\}$$

Let $u, v \in \mathtt{V}(G)$. A *simple path* between u and v, called a u, v-path, is a finite sequence of distinct vertices $u = v_1, \ldots, v_k = v$ where, for all $i \in [1, k-1]$, $(v_i, v_{i+1}) \in \mathtt{E}(G)$, and whose ends are u and v. A u, v-path is *chordless* or *induced* if $(v_i, v_j) \notin \mathtt{E}(G)$ whenever $|i - j| > 1$.

A subset of vertices $A \subseteq \mathtt{V}(G)$ is said to be *connected* in G if $G[A]$ contains a path between every pair of vertices in A. A subset of vertices $A \subseteq \mathtt{V}(G)$ is said to be a *connected component* of G if A is connected, and A' is not connected for every subset of vertices $A \subset A' \subseteq \mathtt{V}(G)$ that properly contains A. Let $A \subseteq \mathtt{V}(G)$ and $u \in \mathtt{V}(G)\backslash A$, where $G[uA]$ is connected. *Contracting the connected vertex-set* uA *to vertex* u results in a new graph G' where $\mathtt{V}(G') = \mathtt{V}(G)\backslash A$ and $\mathtt{E}(G') = \mathtt{E}(G-A) \cup \{(u, a) : a \in N_G(A)\}$. It is easy to see that contracting a connected vertex-set uA to u is equivalent to multiple edge contractions.

Let $V_1, V_2 \subseteq \mathtt{V}(G)$ denote two disjoint vertex subsets of $\mathtt{V}(G)$. We say that V_1 and V_2 are adjacent if there is at least one pair of adjacent vertices $v_1 \in V_1$ and $v_2 \in V_2$. We say that there is a path between V_1 and V_2 if there exist vertices $v_1 \in V_1$ and $v_2 \in V_2$ such that there is a path between v_1 and v_2.

Three mutually non-adjacent vertices of a graph form an *asteroidal triple* if every two of them are connected by a path avoiding the neighborhood of the third. A graph is *AT-free* if it does not contain any asteroidal triple. By this definition, if G is AT-free, then every induced subgraph of G is AT-free.

3 Minimal Separators

Let $s, t \in \mathtt{V}(G)$. For $X \subseteq \mathtt{V}(G)$, we let $\mathcal{C}(G-X)$ denote the set of connected components of $G-X$. The vertex set X is called a *separator* of G if $|\mathcal{C}(G-X)| \geq 2$, an *$s, t$-separator* if s and t are in different connected components of $\mathcal{C}(G-X)$, and a *minimal s, t-separator* if no proper subset of X is an s, t-separator of G. For an s, t-separator X, we denote by $C_s(G-X)$ and $C_t(G-X)$ the connected components of $\mathcal{C}(G-X)$ containing s and t respectively. In other words, $C_s(G-X) = \{v \in \mathtt{V}(G) : \text{ there is a path from } s \text{ to } v \text{ in } G-X\}$.

Lemma 1 ([2]). *An s, t-separator $X \subseteq \mathtt{V}(G)$ is a minimal s, t-separator if and only if $N_G(C_s(G-X)) = N_G(C_t(G-X)) = X$.*

A subset $X \subseteq \mathtt{V}(G)$ is a *minimal separator* of G if there exist a pair of vertices $u, v \in \mathtt{V}(G)$ such that X is a minimal u, v-separator. A connected component $C \in \mathcal{C}(G-X)$ is called a *full component* of X if $N_G(C) = X$. By Lemma 1, X is a minimal u, v-separator if and only if the components $C_u(G-X)$ and $C_v(G-X)$ are full. We denote by $\mathcal{S}_{s,t}(G)$ the set of minimal s, t-separators of G, and by $\mathcal{S}(G)$ the set of minimal separators of G.

3.1 Minimal Separators Between Vertex-Sets

Let $A, B \subseteq \mathtt{V}(G)$ be disjoint and non-adjacent. A subset $S \subseteq \mathtt{V}(G)\backslash AB$ is an A, B-separator if, in the graph $G-S$, there is no path between A and B. We say that S is a minimal A, B-separator if no proper subset of S is an A, B-separator. We denote by $\mathcal{S}_{A,B}(G)$ the set of minimal A, B-separators of G. In the full version of this paper [18], we prove the following two technical lemmas that show how finding minimal separators between vertex-sets can be reduced to the problem of finding minimal separators between singleton vertices.

Lemma 2. *Let A and B be two disjoint, non-adjacent subsets of $\mathtt{V}(G)$. Then $S \in \mathcal{S}_{A,B}(G)$ if and only if S is an A, B-separator, and for every $w \in S$, there exist two connected components $C_A, C_B \in \mathcal{C}(G-S)$ such that $C_A \cap A \neq \emptyset$, $C_B \cap B \neq \emptyset$, and $w \in N_G(C_A) \cap N_G(C_B)$.*

Observe that Lemma 2 implies Lemma 1. By Lemma 2, it holds that $S \in \mathcal{S}_{s,t}(G)$ if and only if S is an s, t-separator and $S \subseteq N_G(C_s(G-S)) \cap N_G(C_t(G-S))$. By definition, $N_G(C_s(G-S)) \subseteq S$ and $N_G(C_t(G-S)) \subseteq S$, and hence $S = N_G(C_s(G-S)) \cap N_G(C_t(G-S))$, and $S = N_G(C_s(G-S)) = N_G(C_t(G-S))$.

Lemma 3. *Let $A \subseteq \mathtt{V}(G)\backslash\{s,t\}$. Let H be the graph that results from G by adding all edges between s and $N_G[A]$. That is, $\mathtt{E}(H) = \mathtt{E}(G) \cup \{(s,v) : v \in N_G[A]\}$. Then $\mathcal{S}_{sA,t}(G) = \mathcal{S}_{s,t}(H)$.*

3.2 Minimal s, t-Separators: Some Basic Properties

The following are basic results used by our algorithms. Due to space restrictions, the proofs of Lemmas 4 and 5 are deferred to the full version of this paper [18].

Lemma 4. *Let $s, t \in \mathtt{V}(G)$, and $A \subseteq \mathtt{V}(G)\backslash\{s,t\}$ such that $G[sA]$ is connected. Let H be the graph where $\mathtt{V}(H) = \mathtt{V}(G)\backslash A$ that results from G by contracting all edges in $G[sA]$. Then (1) $\mathcal{S}_{s,t}(H) = \{S \in \mathcal{S}_{s,t}(G) : A \subseteq C_s(G-S)\}$, and (2) If $S \in \mathcal{S}_{s,t}(H)$, then $C_s(G-S) = C_s(H-S) \cup A$ and $C_t(G-S) = C_t(H-S)$.*

Lemma 5. *Let $s, t \in \mathtt{V}(G)$, and let $S, T \in \mathcal{S}_{s,t}(G)$. The following holds:*

$$C_s(G-S) \subseteq C_s(G-T) \iff S \subseteq T \cup C_s(G-T) \iff T \subseteq S \cup C_t(G-S).$$

Proof. If $C_s(G-S) \subseteq C_s(G-T)$ then $N_G(C_s(G-S)) \subseteq C_s(G-T) \cup N_G(C_s(G-T))$. Since $S, T \in \mathcal{S}_{s,t}(G)$, then by Lemma 1, it holds that $S = N_G(C_s(G-S))$ and $T = N_G(C_s(G-T))$. Therefore, $S \subseteq C_s(G-T) \cup T$. Hence, $C_s(G-S) \subseteq C_s(G-T) \Longrightarrow S \subseteq C_s(G-T) \cup T$. If $S \subseteq C_s(G-T) \cup T$, then by definition, $S \cap C_t(G-T) = \emptyset$. Therefore, $C_t(G-T)$ is connected in $G-S$. By definition, this means that

$C_t(G\text{–}T) \subseteq C_t(G\text{–}S)$. Therefore, $N_G(C_t(G\text{–}T)) \subseteq C_t(G\text{–}S) \cup N_G(C_t(G\text{–}S))$. Since $S, T \in \mathcal{S}_{s,t}(G)$, then by Lemma 1, it holds that $S = N_G(C_t(G\text{–}S))$ and $T = N_G(C_t(G\text{–}T))$. Consequently, $T \subseteq S \cup C_t(G\text{–}S)$. So, we have shown that $C_s(G\text{–}S) \subseteq C_s(G\text{–}T) \implies S \subseteq T \cup C_s(G\text{–}T) \implies T \subseteq S \cup C_t(G\text{–}S)$. If $T \subseteq S \cup C_t(G\text{–}S)$, then by definition, $T \cap C_s(G\text{–}S) = \emptyset$. Therefore, $C_s(G\text{–}S)$ is connected in $G\text{–}T$. Consequently, $C_s(G\text{–}S) \subseteq C_s(G\text{–}T)$. □

Algorithm 1: MinSafeSep.

Input: Connected, weighted, AT-free graph G, and $\emptyset \subset A, B \subseteq \mathtt{V}(G)$.
Output: A minimum-weight, safe A, B-separator, or $\bot$ if none exist.

if $A \cap N_G[B] \neq \emptyset$ **then return** $\bot$;
$G \leftarrow G\text{–}(N_G(A) \cap N_G(B))$
Let $s \in A$, and $t \in B$
$\mathbb{F}_{sA}(G) \leftarrow \mathsf{CloseTo}(G, s, t, A \setminus \{s\})$
$\mathbb{F}_{tB}(G) \leftarrow \mathsf{CloseTo}(G, t, s, B \setminus \{t\})$
Initialize $R \leftarrow \bot$
forall $S_A \in \mathbb{F}_{sA}(G)$ *and* $S_B \in \mathbb{F}_{tB}(G)$ **do**
  if $S_A \subseteq S_B \cup C_s(G\text{–}S_B)$ **then**
    Let $G(S_A, S_B)$ be the graph that results from G by contracting $C_s(G\text{–}S_A)$ to s and $C_t(G\text{–}S_B)$ to t ; `// Lemma 4.`
    $T_{AB} \leftarrow \mathsf{MinSep}(G(S_A, S_B), s, t)$
    if $R = \bot$ *or* $w(R) > w(T_{AB})$ **then** $R \leftarrow T_{AB}$;
  end
end
return $R \cup (N_G(A) \cap N_G(B))$

4 Algorithm Overview

In this Section, we give an overview of the algorithm, and map the results that need to be proved to establish its correctness and runtime guarantee.

Definition 1. *Let $A \subseteq V(G)$. We say that $S \in \mathcal{S}_{s,t}(G)$ is* close to sA *if:*

1. *$A \subseteq C_s(G\text{–}S)$.*
2. *For every $T \in \mathcal{S}_{s,t}(G) \setminus \{S\}$, if $A \subseteq C_s(G\text{–}T)$ then $C_s(G\text{–}T) \not\subseteq C_s(G\text{–}S)$.*

We denote by $\mathbb{F}_{sA}(G)$ the minimal s,t-separators that are close to sA. By Definition 1, we have that if $S \in \mathcal{S}_{s,t}(G)$ where $A \subseteq C_s(G\text{–}S)$, then there exists a $T \in \mathbb{F}_{sA}(G)$, such that $A \subseteq C_s(G\text{–}T) \subseteq C_s(G\text{–}S)$. We make this formal in Lemma 6 whose proof is deferrred to the full version [18].

Lemma 6. *Let $A \subseteq V(G)$, and let $S \in \mathcal{S}_{s,t}(G)$ where $A \subseteq C_s(G\text{–}S)$. There exists a $T \in \mathbb{F}_{sA}(G)$ where $C_s(G\text{–}T) \subseteq C_s(G\text{–}S)$.*

If no restrictions are made to $A \subseteq \mathtt{V}(G)$, there may be an unbounded number of minimal s,t-separators that are close to sA. If $G[sA]$ is connected then, following a result by Takata [22], the minimal s,t-separator close to sA is unique and can be found in time $O(m)$.

Lemma 7 ([22]). *Let $A \subseteq \mathtt{V}(G)$, where $G[sA]$ is connected. If $sA \cap N_G[t] = \emptyset$ then $N_G(sA)$ contains a unique minimal s,t-separator, which can be found in $O(m)$ time.*

Corollary 1. *Let $A \subseteq \mathtt{V}(G)$ such that $G[sA]$ is connected. If $sA \cap N_G[t] = \emptyset$ there exists a unique minimal s,t-separator that is close to sA, which can be found in $O(m)$ time.*

Proof. Let $T \in \mathcal{S}_{s,t}(G)$ where $T \subseteq N_G(sA)$. By Lemma 7, T is unique. Let $S \in \mathcal{S}_{s,t}(G)$ that is close to sA. By Definition 1, $A \subseteq C_s(G{-}S)$. Therefore, $T \subseteq N_G(sA) \subseteq S \cup C_s(G{-}S)$. By Lemma 5, we have that $C_s(G{-}T) \subseteq C_s(G{-}S)$. By Definition 1, $S = T$. □

In Sect. 6, we prove that if G is AT-free then $|\mathbb{F}_{sA}(G)| \leq n^2$ ($|\mathbb{F}_{tB}(G)| \leq n^2$), and the set $\mathbb{F}_{sA}(G)$ ($\mathbb{F}_{tB}(G)$) can be computed in time $O(n^2m)$.

Theorem 2. *Let G be an AT-free graph, $s,t \in \mathtt{V}(G)$ two distinguished vertices, and $A \subseteq \mathtt{V}(G)\backslash\{s,t\}$. Let $T_s \in \mathcal{S}_{s,t}(G)$ where $T_s \subseteq N_G(s)$. If $A \subseteq C_s(G{-}T_s) \cup T_s \cup C_t(G{-}T_s)$, then there are at most n minimal s,t-separators that are close to sA, and they can be found in time $O(nm)$. Otherwise, at most n^2 minimal s,t-separators are close to sA, and they can be found in time $O(n^2m)$.*

In Sect. 6, we prove Theorem 2, and present algorithm $\mathsf{CloseTo}$ that receives as input an AT-free graph G, vertices $s,t \in \mathtt{V}(G)$ and subset $A \subseteq \mathtt{V}(G)$ ($B \subseteq \mathtt{V}(G)$), and computes $\mathbb{F}_{sA}(G)$ ($\mathbb{F}_{tB}(G)$) in time $O(n^2m)$. If $T_s \in \mathcal{S}_{s,t}(G)$ is the unique minimal s,t-separator where $T_s \subseteq N_G(s)$ (Lemma 7), and if $A \subseteq C_s(G{-}T_s) \cup T_s \cup C_t(G{-}T_s)$, then algorithm $\mathsf{CloseTo}$ computes $\mathbb{F}_{sA}(G)$ (or $\mathbb{F}_{tB}(G)$) in time $O(nm)$.

We now describe the algorithm for MIN SAFE SEP, presented in Algorithm 1, that receives as input a vertex-weighted, AT-free graph G, and a pair of vertex sets $A, B \subseteq \mathtt{V}(G)$. If $A \cap N_G[B] \neq \emptyset$, then no A,B-separator exists and the algorithm returns $\perp$ in line 1. Every A,B-separator must include $N_G(A) \cap N_G(B)$. Therefore, the algorithm processes the graph $G{-}(N_G(A) \cap N_G(B))$ (line 2). Since G is AT-free, then $G{-}(N_G(A) \cap N_G(B))$ is also AT-free. The algorithm relates minimal s,t-separators to minimal, safe A,B-separators using the following.

Lemma 8. *A subset $S \subseteq \mathtt{V}(G)$ is a safe, minimal A,B-separator if and only if for every pair of vertices $s \in A$ and $t \in B$ it holds that $S \in \mathcal{S}_{s,t}(G)$ where $A \subseteq C_s(G{-}S)$ and $B \subseteq C_t(G{-}S)$.*

Proof. Let $s \in A$, $t \in B$. If $S \in \mathcal{S}_{s,t}(G)$ where $A \subseteq C_s(G{-}S)$ and $B \subseteq C_t(G{-}S)$ then clearly S is a safe A,B-separator. By Lemma 1, it holds that $S = N_G(C_s(G{-}S)) \cap N_G(C_t(G{-}S))$. By Lemma 2, it holds that S is a minimal, safe A,B-separator.

Now, let S be a minimal, safe A, B-separator, where $C_A, C_B \in \mathcal{C}(G-S)$ contain A and B respectively. By Lemma 2, it holds that $S = N_G(C_A) \cap N_G(C_B)$. By Lemma 1, $S \in \mathcal{S}_{s,t}(G)$ for every pair of vertices $s \in A$ and $t \in B$, where $C_s(G-S) = C_A$ and $C_t(G-S) = C_B$. □

Take any $S \in \mathcal{S}_{s,t}(G)$ such that $A \subseteq C_s(G-S)$ and $B \subseteq C_t(G-S)$. By Lemma 6, there exists an $S_A \in \mathbb{F}_{sA}(G)$ such that $C_s(G-S_A) \subseteq C_s(G-S)$, and an $S_B \in \mathbb{F}_{tB}(G)$ such that $C_t(G-S_B) \subseteq C_t(G-S)$. Let $G(S_A, S_B)$ denote the graph that results from G by contracting $C_s(G-S_A)$ to vertex s and $C_t(G-S_B)$ to vertex t. By Lemma 8, and Lemma 4, every $T \in \mathcal{S}_{s,t}(G(S_A, S_B))$ is a safe, minimal A, B-separator. Consequently, by Lemma 6 and Lemma 8, we have that S is a minimal, safe, A, B-separator if and only if $S \in \mathcal{S}_{s,t}(G(S_A, S_B))$ for some pair of minimal s, t-separators $S_A \in \mathbb{F}_{sA}(G)$ and $S_B \in \mathbb{F}_{tB}(G)$. Moreover, S is a minimum, safe A, B-separator if and only if S is a minimum s, t-separator of $G(S_A, S_B)$ for some pair of minimal s, t-separators $S_A \in \mathbb{F}_{sA}(G)$ and $S_B \in \mathbb{F}_{tB}(G)$. The loop in lines 7–13 runs over all pairs $S_A \in \mathbb{F}_{sA}(G)$ and $S_B \in \mathbb{F}_{tB}(G)$, generates the graph $G(S_A, S_B)$ in line 9, and finds a minimum-weight s, t-separator of $G(S_A, S_B)$ in line 10. The algorithm returns the minimum over all combinations of $S_A \in \mathbb{F}_{sA}(G)$ and $S_B \in \mathbb{F}_{tB}(G)$ in line 14.

Theorem 3. *Given a weighted, connected, AT-free graph G, and two vertex-sets $A, B \subseteq \mathtt{V}(G)$, Algorithm* MinSafeSep *returns a minimum-weight, safe A, B-separator if one exists, or $\perp$ otherwise, and runs in time $O(|\mathbb{F}_{sA}(G)| \cdot |\mathbb{F}_{tB}(G)| \cdot T(n, m))$, where $s \in A, t \in B$, and $T(n, m)$ is the time to compute a minimum-weight s, t-separator.*

Theorem 2 establishes that in AT-free graphs, if no assumptions are made to the input vertex-sets $A, B \subseteq \mathtt{V}(G)$, then $|\mathbb{F}_{sA}(G)| \leq n^2$, $|\mathbb{F}_{tB}(G)| \leq n^2$, and that $\mathbb{F}_{sA}(G)$ and $\mathbb{F}_{tB}(G)$ can be computed in time $O(n^2 m)$. It further establishes that if there exist vertices $s \in A$, and $t \in B$ such that $A \backslash \{s\} \subseteq C_s(G-T_s) \cup T_s \cup C_t(G-T_s)$ and $B \backslash \{t\} \subseteq C_t(G-T_t) \cup T_t \cup C_s(G-T_t)$ where $T_s, T_t \in \mathcal{S}_{s,t}(G)$ are the unique minimal s, t-separators close to s and t, respectively (Lemma 7), then $|\mathbb{F}_{sA}(G)| \leq n$, $|\mathbb{F}_{tB}(G)| \leq n$, and $\mathbb{F}_{sA}(G)$ and $\mathbb{F}_{tB}(G)$ can be computed in time $O(nm)$. It immediately follows from Theorem 2 and Theorem 3:

Corollary 2. *Let G be a weighted, connected, AT-free graph, and $A, B \subseteq \mathtt{V}(G)$. If there exist vertices $s \in A$ and $t \in B$ such that $A \backslash \{s\} \subseteq C_s(G-T_s) \cup T_s \cup C_t(G-T_s)$ and $B \backslash \{t\} \subseteq C_t(G-T_t) \cup T_t \cup C_s(G-T_t)$ then Algorithm* MinSafeSep *returns a minimum-weight, safe A, B-separator if one exists, or $\perp$ otherwise, in time $O(n^2 T(n, m))$. Otherwise, the runtime is $O(n^4 T(n, m))$.*

Recall that $T(n, m) = O(m^{1+o(1)})$ is the time to compute a minimum-weight s, t-separator [5]. The rest of this paper is devoted to proving Theorem 2.

5 Essential Findings: Minimal s, t-Separators

In this Section, we prove several results about minimal s, t-separators that are crucial for proving Theorem 2. In Sect. 5.1, we establish a result concerning minimal s,t-separators close to sA, where $A \subseteq \mathtt{V}(G)$ (Definition 1). In Sect. 5.2, we establish results on minimal s,t-separators in AT-free graphs. Some of the proofs are deferred to the full version of this paper [18].

5.1 Results on Close Minimal s,t-Separators

Lemma 9. *Let $T_s \in \mathcal{S}_{s,t}(G)$ where $T_s \subseteq N_G(s)$. Let $A \subseteq V(G) \setminus (T_s \cup C_s(G-T_s) \cup N_G[t])$ such that $T_s \subseteq N_G(a)$ for every $a \in A$. There are at most $|T_s|$ minimal s,t-separators that are close to sA, which can be found in time $O(|T_s| \cdot m)$.*

To illustrate Lemma 9, consider Fig. 1 where $A = \{a_1, a_2\}$. Let $T_s = \{v_1, \dots, v_\ell\}$. By the assumption of the lemma, $T_s \subseteq N_G(a_1)$ and $T_s \subseteq N_G(a_2)$. Lemma 9 establishes that there are at most $|T_s|$ minimal s,t-separators close to sA, which can be found in time $O(|T_s| \cdot m)$. The proof of Lemma 9 establishes that $S \in \mathcal{S}_{s,t}(G)$ is close to sA if and only if S is close to sv_iA for some $v_i \in T_s$. The complete proof is available in [18].

Lemma 10. *Let $S \in \mathcal{S}_{s,t}(G)$ such that $S \subseteq N_G(t)$, and let $u \in V(G) \setminus \{s,t\}$. If $u \notin C_s(G-S)$ then for every $T \in \mathcal{S}_{s,t}(G)$, it holds that $u \notin C_s(G-T)$.*

By Lemma 7, a minimal s,t-separator $S \in \mathcal{S}_{s,t}(G)$, such that $S \subseteq N_G(t)$, is unique, and can be found in polynomial time. An immediate consequence of Lemma 10 is that we can, in polynomial time, test whether there exists a minimal s,t-separator $S \in \mathcal{S}_{s,t}(G)$ such that $u \in C_s(G-S)$, for a distinguished vertex $u \in \mathtt{V}(G)$. To do so, we find the unique $S \in \mathcal{S}_{s,t}(G)$ such that $S \subseteq N_G(t)$. If $u \in C_s(G-S)$ then the answer is clearly yes. Otherwise, by Lemma 10, it holds that $u \notin C_s(G-T)$ for any $T \in \mathcal{S}_{s,t}(G)$.

Lemma 11. *Let $T_s \in \mathcal{S}_{s,t}(G)$ where $T_s \subseteq N_G(s)$. Let $D \in \mathcal{C}(G-T_s)$ where $s \notin D$ and $t \notin D$. Define $T_D \stackrel{\text{def}}{=} T_s \cap N_G(D)$. For every $A \subseteq D$ it holds that:*

$$\{\mathcal{S}_{s,t}(G) : A \subseteq C_s(G-S)\} = \bigcup_{v \in T_D} \{S \in \mathcal{S}_{s,t}(G) : v \in C_s(G-S)\}$$

To illustrate Lemma 11, consider Fig. 2, which shows $T_s \in \mathcal{S}_{s,t}(G)$ where $T_s \subseteq N_G(s)$, $D \in \mathcal{C}(G-T_s)$, vertex $a \in A \subseteq D$, and $T_D \stackrel{\text{def}}{=} N_G(D)$. Observe that T_D is, by definition, an s,a-separator of G where $D = C_a(G-T_D)$ (see Fig. 2). Let $S \in \mathcal{S}_{s,t}(G)$. Lemma 11 establishes that $A \subseteq C_s(G-S)$ if and only if $C_s(G-S) \cap T_D \neq \emptyset$. The complete proof appears in [18].

5.2 Minimal s,t-Separators in AT-Free Graphs

In any graph G, it holds that $N_G(s) \cap N_G(t) \subseteq S$ for every $S \in \mathcal{S}_{s,t}(G)$. Therefore, finding a minimum s,t-separator in G is equivalent to finding a minimum s,t-separator in $G-(N_G(s) \cap N_G(t))$. If G is AT-free, then every induced subgraph of G is AT-free, and hence $G-(N_G(s) \cap N_G(t))$ is AT-free is well. Consequently, we make the assumption that $N_G(s) \cap N_G(t) = \emptyset$. In this Section, we prove some useful properties of minimal separators in AT-free graphs; proofs are deferred to the full version [18].

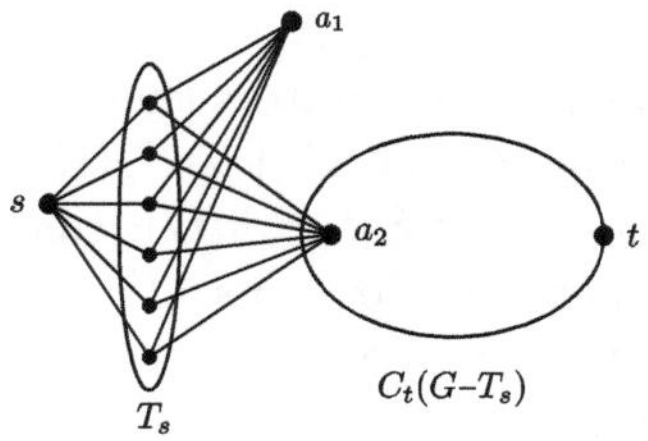

Fig. 1. Illustration–Lemma 9.

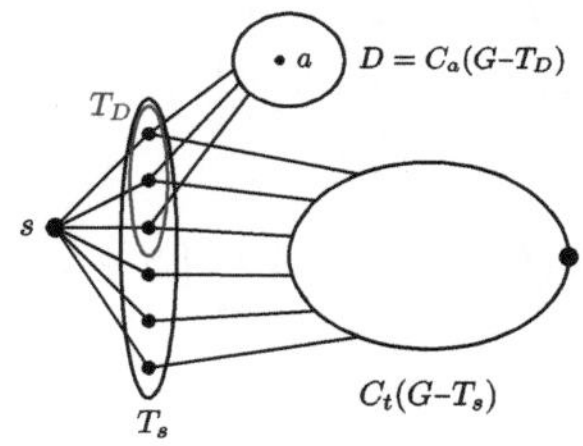

Fig. 2. Illustration–Lemma 11.

Lemma 12. *Let G be AT-free, $T_s \in \mathcal{S}_{s,t}(G)$ where $T_s \subseteq N_G(s) \backslash N_G[t]$, and $C_1, C_2 \in \mathcal{C}(G-T_s) \backslash \{C_s(G-T_s)\}$. Then $N_G(C_1) \subseteq N_G(C_2)$ (or $N_G(C_2) \subseteq N_G(C_1)$).*

Corollary 3. *Let G be AT-free, $T_s \in \mathcal{S}_{s,t}(G)$ where $T_s \subseteq N_G(s) \backslash N_G[t]$, and $\emptyset \subset A \subseteq V(G)$, such that $A \cap (C_s(G-T_s) \cup T_s \cup C_t(G-T_s)) = \emptyset$. Define $\{C_1, \ldots, C_\ell\} \stackrel{\text{def}}{=} \{C \in \mathcal{C}(G-T_s) : C \cap A \neq \emptyset\}$, and $S^* \stackrel{\text{def}}{=} \bigcap_{i=1}^{\ell} N_G(C_i)$. Then:*

$$\{S \in \mathcal{S}_{s,t}(G) : A \subseteq C_s(G-S)\} = \bigcup_{v \in S^*} \{S \in \mathcal{S}_{s,t}(G) : v \in C_s(G-S)\}$$

6 Finding All Close Minimal s, t-Separators in AT-Free Graphs

In this section, we prove Theorem 2 that forms the basis of our algorithm. The proof relies on the results established in Sect. 5.

THEOREM 2. *Let G be an AT-free graph, $s, t \in V(G)$ two distinguished vertices, and $A \subseteq V(G) \backslash \{s, t\}$. Let $T_s \in \mathcal{S}_{s,t}(G)$ where $T_s \subseteq N_G(s)$. If $A \subseteq C_s(G-T_s) \cup T_s \cup C_t(G-T_s)$, then there are at most n minimal s, t-separators that are close to sA, and they can be found in time $O(nm)$. Otherwise, at most n^2 minimal s, t-separators are close to sA, and they can be found in time $O(n^2m)$.*

The procedure CloseTo that returns the minimal s, t-separators of G that are close to sA in time $O(n^2m)$ (or in time $O(nm)$), and its detailed runtime analysis, is deferred to the full version of this paper [18]. We now prove Theorem 2.

Proof. Let $T_t, T_s \in \mathcal{S}_{s,t}(G)$, where $T_t \subseteq N_G(t)$ and $T_s \subseteq N_G(s)$. By Lemma 7, T_s and T_t are the unique minimal s,t-separators that are close s and t, respectively, and they can be found in time $O(m)$. If $A \not\subseteq C_s(G-T_t)$, then by Lemma 10, it holds that $A \not\subseteq C_s(G-S)$ for every $S \in \mathcal{S}_{s,t}(G)$. Hence, if $A \not\subseteq C_s(G-T_t)$, then there are no minimal s,t-separators close to sA (i.e., $\mathbb{F}_{sA}(G) = \emptyset$). So, we assume that $A \subseteq C_s(G-T_t)$. For every $S \in \mathcal{S}_{s,t}(G)$ where $A \subseteq C_s(G-S)$ it must hold that $N_G(sA) \cap N_G(t) \subseteq S$. Since $G-(N_G(sA) \cap N_G(t))$ is an induced subgraph of an AT-free graph G, then $G-(N_G(sA) \cap N_G(t))$ is AT-free as well. Therefore, we assume that G is AT-free, and that $N_G(sA) \cap N_G(t) = \emptyset$.

Let $A_1 \stackrel{\text{def}}{=} A \cap (C_s(G-T_s) \cup T_s \cup C_t(G-T_s))$, and $A_2 \stackrel{\text{def}}{=} A \backslash A_1$. Let $\{C_1, \dots, C_\ell\} \stackrel{\text{def}}{=} \{C \in \mathcal{C}(G-T_s) : C \cap A_2 \neq \emptyset\}$, and $S^* \stackrel{\text{def}}{=} \bigcap_{i=1}^{\ell} N_G(C_i)$. By Corollary 3:

$$\{S \in \mathcal{S}_{s,t}(G) : A_2 \subseteq C_s(G-S)\} = \bigcup_{v \in S^*} \{S \in \mathcal{S}_{s,t}(G) : v \in C_s(G-S)\} \tag{1}$$

Therefore, we have that:

$$\begin{aligned}
&\{S \in \mathcal{S}_{s,t}(G) : A \subseteq C_s(G-S)\} && (2)\\
&= \{S \in \mathcal{S}_{s,t}(G) : A_1 \subseteq C_s(G-S)\} \cap \{S \in \mathcal{S}_{s,t}(G) : A_2 \subseteq C_s(G-S)\} \\
&\underbrace{=}_{(1)} \{S \in \mathcal{S}_{s,t}(G) : A_1 \subseteq C_s(G-S)\} \cap \Big(\bigcup_{v \in S^*} \{S \in \mathcal{S}_{s,t}(G) : v \in C_s(G-S)\}\Big) \\
&= \bigcup_{v \in S^*} \{S \in \mathcal{S}_{s,t}(G) : A_1 v \in C_s(G-S)\} && (3)
\end{aligned}$$

Since $S^* \subseteq T_s$, then $A_1 v \subseteq C_s(G-T_s) \cup T_s \cup C_t(G-T_s)$. Therefore, if we show that for every $v \in S^*$, there are at most n minimal s,t-separators that are close to svA_1 (i.e., $|\mathbb{F}_{svA_1}| \leq n$), which can be found in time $O(nm)$, then we get that there are at most $|S^*| \cdot n \leq n^2$ minimal s,t-separators that are close to sA, which can be found in time $O(n^2 m)$. Overall, to prove the claim we need to show that if $A \subseteq C_s(G-S) \cup T_s \cup C_t(G-T_s)$, then there are at most n minimal s,t-separators that are close to sA, which can be found in time $O(nm)$. The rest of the proof is devoted to this setting.

Claim 1: $\{S \in \mathcal{S}_{s,t}(G) : A \subseteq C_s(G-S)\} \subseteq \mathcal{S}_{sA,t}(G)$.

Proof. Let $S \in \mathcal{S}_{s,t}(G)$ where $A \subseteq C_s(G-S)$. Then S is an sA,t-separator. By Lemma 1, $S = N_G(C_t(G-S)) = N_G(C_s(G-S))$. By Lemma 2, $S \in \mathcal{S}_{sA,t}(G)$. □

By Lemma 3, we have that $\mathcal{S}_{sA,t}(G) = \mathcal{S}_{s,t}(H)$ where $\mathtt{V}(H) = \mathtt{V}(G)$ and $\mathtt{E}(H) = \mathtt{E}(G) \cup \{(s,a) : a \in N_G[A]\}$. Let $S_1 \in \mathcal{S}_{s,t}(H)$ where $S_1 \subseteq N_H(s)$. By Lemma 7, S_1 is unique and can be found in time $O(m)$. In addition, $S_1 \in \mathcal{S}_{sA,t}(G)$. Let $C_s, C_t \in \mathcal{C}(G-S_1)$ be the connected components of $G-S_1$ that contain vertices s and t, respectively.

Claim 2: For every $S \in \mathcal{S}_{s,t}(G)$: if $A \subseteq C_s(G-S)$, then $C_s(G-S_1) \subseteq C_s(G-S)$.

Proof. Let $S \in \mathcal{S}_{s,t}(G)$ where $A \subseteq C_s(G\text{–}S)$. By Claim 1, it holds that $S \in \mathcal{S}_{sA,t}(G) = \mathcal{S}_{s,t}(H)$. Since $S_1 \in \mathcal{S}_{s,t}(H)$ where $S_1 \subseteq N_H(s)$, then $S_1 \subseteq S \cup C_s(H\text{–}S)$, and by Lemma 5, that $C_s(H\text{–}S_1) \subseteq C_s(H\text{–}S)$. Since $S \in \mathcal{S}_{s,t}(G)$ where $A \subseteq C_s(G\text{–}S)$, then $C_s(H\text{–}S) = C_s(G\text{–}S)$. Since $\mathtt{E}(G) \subseteq \mathtt{E}(H)$, then:

$$C_s(G\text{–}S_1) \subseteq C_s(H\text{–}S_1) \underbrace{\subseteq}_{\text{Lemma 5}} C_s(H\text{–}S) \underbrace{=}_{A \subseteq C_s(G\text{–}S)} C_s(G\text{–}S).$$

Claim 3: $A \cap (T_s \cup C_s(G\text{–}T_s)) \subseteq C_s(G\text{–}S_1)$.

Proof. Since $T_s \subseteq N_G(s)$, then $T_s \subseteq S_1 \cup C_s(G\text{–}S_1)$. By Lemma 5, $C_s(G\text{–}T_s) \subseteq C_s(G\text{–}S_1)$. Since $S_1 \in \mathcal{S}_{sA,t}(G)$, then $A \cap S_1 = \emptyset$. Since $A \cap T_s \subseteq N_G(s)$, then $A \cap T_s \subseteq C_s(G\text{–}S_1)$. □

Consider the graph $G\text{–}S_1$. There are two cases: $A \subseteq C_s(G\text{–}S_1)$ and $A \not\subseteq C_s(G\text{–}S_1)$. Recall that $C_s \stackrel{\text{def}}{=} C_s(G\text{–}S_1)$, where $S_1 \in \mathcal{S}_{s,t}(H)$ and $S_1 \subseteq N_H(s)$.

Case 1: $A \subseteq C_s$**.** Since $S_1 \in \mathcal{S}_{s,t}(H) = \mathcal{S}_{sA,t}(G)$ and $A \subseteq C_s$, then by Lemma 2 it holds that $S_1 = N_G(C_s) \cap N_G(C_t)$. By Lemma 1, we have that $S_1 \in \mathcal{S}_{s,t}(G)$. We claim that S_1 is the unique minimal s,t-separator that is close to sA. Let $S \in \mathcal{S}_{s,t}(G)$ such that $A \subseteq C_s(G\text{–}S)$. By Claim 2, $C_s(G\text{–}S_1) \subseteq C_s(G\text{–}S)$. Hence, for this case the Theorem is proved.

Case 2: $A \not\subseteq C_s$**.** Let $A' \stackrel{\text{def}}{=} A \backslash C_s$. By Claim 3, we have that $A' \subseteq \mathtt{V}(G) \backslash (C_s(G\text{–}T_s) \cup T_s)$. Since $A \subseteq C_s(G\text{–}T_s) \cup T_s \cup C_t(G\text{–}T_s)$, we have that $A' \subseteq C_t(G\text{–}T_s)$. Therefore, for every $a \in A'$ there is an a,t-path in G that resides entirely in $C_t(G\text{–}T_s)$, and hence avoids $N_G[s]$. By our assumption that $A' \subseteq A \subseteq C_s(G\text{–}T_t)$, there is an s,a-path in G that resides entirely in $C_s(G\text{–}T_t)$, and hence avoids $N_G[t]$.

Define $Q_s \stackrel{\text{def}}{=} N_G(C_s)$. By definition, Q_s is an s,t-separator, and $Q_s \subseteq S_1$, and hence $Q_s = N_G(C_s) \cap S_1$. Since $S_1 \in \mathcal{S}_{sA,t}(G)$, then by Lemma 2, it holds that $S_1 \subseteq N_G(C_t)$. Therefore, Q_s is an s,t-separator where $Q_s \subseteq N_G(C_s) \cap N_G(C_t)$. By Lemma 1, $Q_s \in \mathcal{S}_{s,t}(G)$; see Fig. 3 for illustration.

Claim 4: For every $a \in A'$, it holds that $Q_s \subseteq N_G(a)$.

Proof. Suppose, by way of contradiction, that $Q_s \not\subseteq N_G(a)$ for some $a \in A'$, and let $v \in Q_s \backslash N_G(a)$. By Definition, $v \in Q_s \subseteq N_G(C_s) \cap N_G(C_t)$. Therefore, there is an s,t-path in G that passes through v, denoted $P^v_{s,t}$, that resides entirely in $C_s \cup \{v\} \cup C_t$ (see Fig. 3). Since $C_s \stackrel{\text{def}}{=} C_s(G\text{–}S_1)$, where $a \notin S_1 \cup C_s$, and $C_t \stackrel{\text{def}}{=} C_t(G\text{–}S_1)$ where $a \notin C_t \cup S_1$, then $N_G[a] \cap (C_s \cup C_t) = \emptyset$. Combined with the assumption that $v \notin N_G[a]$, we get that $N_G[a] \cap (C_s \cup \{v\} \cup C_t) = \emptyset$, and hence $N_G[a] \cap \mathtt{V}(P^v_{s,t}) = \emptyset$. Therefore, there is an s,t-path in G (via v) that avoids $N_G[a]$. Since there exists an s,a-path in G that avoids $N_G[t]$ (i.e., $P_{a,s}$) and an a,t-path in G that avoids $N_G[s]$ (i.e., $P_{a,t}$), we get that s,a,t form an asteroidal triple in G (see Fig. 3). But this is a contradiction. Therefore, $Q_s \subseteq N_G(a)$. □

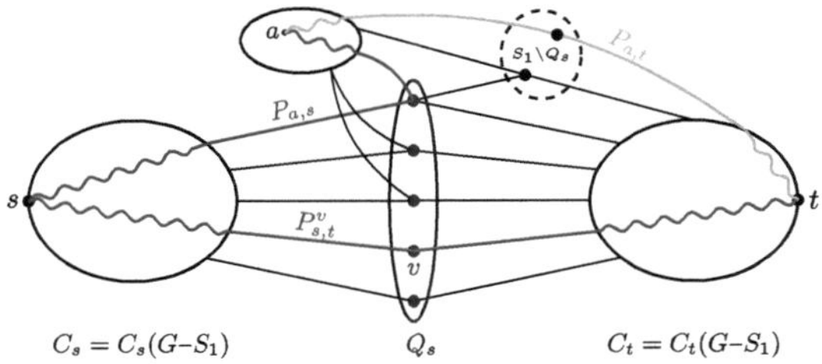

Fig. 3. Illustration for the proof of Theorem 2 (Case 2).

Let $S \in \mathcal{S}_{s,t}(G)$ where $A \subseteq C_s(G-S)$. By Claim 4, $Q_s \subseteq N_G(a)$ for every $a \in A' \subseteq A$. Since $A' \neq \emptyset$, then $Q_s \subseteq S \cup C_s(G-S)$. Since $S, Q_s \in \mathcal{S}_{s,t}(G)$, then by Lemma 5, we have that $C_s(G-Q_s) \subseteq C_s(G-S)$. Therefore, we get that:

$$\{S \in \mathcal{S}_{s,t}(G) : A \subseteq C_s(G-S)\} \subseteq \{S \in \mathcal{S}_{s,t}(G) : C_s(G-Q_s) \subseteq C_s(G-S)\} \quad (4)$$

Since $C_s(G-Q_s)$ is a connected component, then by Lemma 4, we have that:

$$\{S \in \mathcal{S}_{s,t}(G) : C_s(G-Q_s) \subseteq C_s(G-S)\} = \mathcal{S}_{s,t}(M) \quad (5)$$

where M is the graph that results from G by contracting $C_s(G-Q_s)$ to vertex s. Also, by Lemma 4, we have that $C_s(G-S) = C_s(M-S) \cup C_s(G-Q_s)$, and $C_t(G-S) = C_t(M-S)$ for every $S \in \mathcal{S}_{s,t}(M)$. Let $D \stackrel{\text{def}}{=} A \setminus C_s(G-Q_s)$.

Claim 5: $\mathbb{F}_{sD}(M) = \mathbb{F}_{sA}(G)$.

Proof: Let $S \in \mathbb{F}_{sA}(G)$. By definition, $\mathbb{F}_{sA}(G) \subseteq \{S \in \mathcal{S}_{s,t}(G) : A \subseteq C_s(G-S)\}$. From (4) and (5), we have that $S \in \mathcal{S}_{s,t}(M)$, where $C_s(G-S) = C_s(M-S) \cup C_s(G-Q_s)$. Since $A \subseteq C_s(G-S)$, then $D = A \setminus C_s(G-Q_s) \subseteq C_s(M-S)$. Suppose, by way of contradiction, that $S \notin \mathbb{F}_{sD}(M)$. Since $D \subseteq C_s(M-S)$, it means that there exists an $S' \in \mathcal{S}_{s,t}(M)$ where $D \subseteq C_s(M-S') \subset C_s(M-S)$. By (5), we have that $S' \in \mathcal{S}_{s,t}(G)$ where $C_s(G-S') = C_s(M-S') \cup C_s(G-Q_s) \subset C_s(M-S) \cup C_s(G-Q_s) = C_s(G-S)$. But then, $A \subseteq C_s(G-S') \subset C_s(G-S)$, contradicting the assumption that $S \in \mathbb{F}_{sA}(G)$. Therefore, $\mathbb{F}_{sA}(G) \subseteq \mathbb{F}_{sD}(M)$.

Now, let $S \in \mathbb{F}_{sD}(M)$. From (5), we have that $S \in \mathcal{S}_{s,t}(G)$ where $C_s(G-S) = C_s(M-S) \cup C_s(G-Q_s)$. Since $D \subseteq C_s(M-S)$, we have that $A \subseteq C_s(G-S)$. If $S \notin \mathbb{F}_{sA}(G)$, then there exists an $S' \in \mathcal{S}_{s,t}(G)$ where $A \subseteq C_s(G-S') \subset C_s(G-S)$. From (4) and (5), we have that $S' \in \mathcal{S}_{s,t}(M)$, where $C_s(G-S') = C_s(M-S') \cup C_s(G-Q_s)$. Therefore, we have that $A \subseteq C_s(G-S') = C_s(M-S') \cup C_s(G-Q_s) \subset C_s(M-S) \cup C_s(G-Q_s) = C_s(G-S)$. In particular, this means that $D \subseteq C_s(M-S')$ and $C_s(M-S') \subset C_s(M-S)$, contradicting the assumption that $S \in \mathbb{F}_{sD}(M)$. □

Since $\mathbb{F}_{sD}(M) = \mathbb{F}_{sA}(G)$, we are left to show that $|\mathbb{F}_{sD}(M)| \leq n$, and that $\mathbb{F}_{sD}(M)$ can be computed in time $O(nm)$. By definition of contraction, we have that $N_M(s) \supseteq N_G(C_s(G-Q_s)) = Q_s$, and that $C_t(M-Q_s) = C_t(G-Q_s)$. Consequently, we have that $Q_s \in \mathcal{S}_{s,t}(M)$ where $Q_s \subseteq N_M(s)$. By Lemma 7, Q_s

is the unique minimal s, t-separator of M that is close to s. For every $a \in A$, either $a \in C_s(G - S_1) \subseteq C_s(G - Q_s)$, or by Claim 4, $Q_s \subseteq N_G(a) \subseteq N_M(a)$. By Lemma 9, there are at most $|Q_s| \leq n$ minimal s, t-separators that are close to sD in M, and they can be found in time $O(|Q_s| \cdot m) = O(nm)$.

7 Conclusion

In this paper, we presented the first polynomial-time algorithm for finding a connectivity-preserving, minimum-weight A, B-separator in AT-free graphs, a general class encompassing interval, cocomparability, cobipartite, and trapezoid graphs. Our algorithm leverages key properties of minimal separators in AT-free graphs for an efficient solution. To our knowledge, this is also the first polynomial-time algorithm to find a connectivity-preserving A, B-separator when A and B are unbounded in any non-trivial, infinite graph class. Additionally, our results on minimal separators in AT-free graphs may be of independent interest, offering insights applicable to other problems.

References

1. Bentert, M., Drange, P.G., Fomin, F.V., Golovach, P.A., Korhonen, T.: Two-sets cut-uncut on planar graphs. In: Bringmann, K., Grohe, M., Puppis, G., Svensson, O. (eds.) 51st International Colloquium on Automata, Languages, and Programming, ICALP 2024, 8–12 July 2024, Tallinn, Estonia. LIPIcs, vol. 297, pp. 22:1–22:18. Schloss Dagstuhl - Leibniz-Zentrum für Informatik (2024). https://doi.org/10.4230/LIPICS.ICALP.2024.22
2. Berry, A., Bordat, J.P., Cogis, O.: Generating all the minimal separators of a graph. Int. J. Found. Comput. Sci. **11**(3), 397–403 (2000). https://doi.org/10.1142/S0129054100000211
3. van den Brand, J., et al.: A deterministic almost-linear time algorithm for minimum-cost flow. In: 64th IEEE Annual Symposium on Foundations of Computer Science, FOCS 2023, Santa Cruz, CA, USA, 6–9 November 2023, pp. 503–514. IEEE (2023). https://doi.org/10.1109/FOCS57990.2023.00037
4. van den Brand, J., et al.: Minimum cost flows, MDPs, and ℓ_1-regression in nearly linear time for dense instances. In: Khuller, S., Williams, V.V. (eds.) STOC 2021: 53rd Annual ACM SIGACT Symposium on Theory of Computing, Virtual Event, Italy, 21–25 June 2021, pp. 859–869. ACM (2021). https://doi.org/10.1145/3406325.3451108
5. Chen, L., Kyng, R., Liu, Y.P., Peng, R., Gutenberg, M.P., Sachdeva, S.: Maximum flow and minimum-cost flow in almost-linear time. In: 2022 IEEE 63rd Annual Symposium on Foundations of Computer Science (FOCS), pp. 612–623 (2022). https://doi.org/10.1109/FOCS54457.2022.00064
6. Corneil, D.G., Olariu, S., Stewart, L.: Asteroidal triple-free graphs. SIAM J. Discret. Math. **10**(3), 399–430 (1997). https://doi.org/10.1137/S0895480193250125
7. Cullenbine, C., Wood, R.K., Newman, A.M.: Theoretical and computational advances for network diversion. Networks **62**(3), 225–242 (2013). https://doi.org/10.1002/NET.21514

8. Curet, N.D.: The network diversion problem. Milit. Oper. Res. **6**(2), 35–44 (2001). http://www.jstor.org/stable/43943674
9. Cygan, M., Pilipczuk, M., Pilipczuk, M., Wojtaszczyk, J.O.: Solving the 2-disjoint connected subgraphs problem faster than 2^n. Algorithmica **70**(2), 195–207 (2014). https://doi.org/10.1007/s00453-013-9796-x
10. Duan, Q., Jafarian, J.H., Al-Shaer, E., Xu, J.: On DDoS attack related minimum cut problems. CoRR abs/1412.3359 (2014). http://arxiv.org/abs/1412.3359
11. Duan, Q., Xu, J.: On the connectivity preserving minimum cut problem. J. Comput. Syst. Sci. **80**(4), 837–848 (2014). https://doi.org/10.1016/J.JCSS.2014.01.003
12. Even, S., Even, G.: Graph Algorithms, 2nd edn. Cambridge University Press (2012)
13. Golovach, P.A., Kratsch, D., Paulusma, D.: Detecting induced minors in AT-free graphs. Theor. Comput. Sci. **482**, 20–32 (2013). https://doi.org/10.1016/J.TCS.2013.02.029
14. Gray, C., Kammer, F., Löffler, M., Silveira, R.I.: Removing local extrema from imprecise terrains. Comput. Geom. **45**(7), 334–349 (2012). https://doi.org/10.1016/j.comgeo.2012.02.002, https://www.sciencedirect.com/science/article/pii/S0925772112000533
15. van 't Hof, P., Paulusma, D., Woeginger, G.J.: Partitioning graphs into connected parts. Theor. Comput. Sci. **410**(47), 4834–4843 (2009). https://doi.org/10.1016/j.tcs.2009.06.028, https://www.sciencedirect.com/science/article/pii/S0304397509004198
16. Kathuria, T., Liu, Y.P., Sidford, A.: Unit capacity maxflow in almost $O(m^{\frac{4}{3}})$ time. In: Irani, S. (ed.) 61st IEEE Annual Symposium on Foundations of Computer Science, FOCS 2020, Durham, NC, USA, 16–19 November 2020, pp. 119–130. IEEE (2020). https://doi.org/10.1109/FOCS46700.2020.00020
17. Kawarabayashi, K., Kobayashi, Y.: The induced disjoint paths problem. In: Lodi, A., Panconesi, A., Rinaldi, G. (eds.) IPCO 2008. LNCS, vol. 5035, pp. 47–61. Springer, Heidelberg (2008). https://doi.org/10.1007/978-3-540-68891-4_4
18. Kenig, B.: Connectivity-preserving minimum separator in at-free graphs (2025). https://arxiv.org/abs/2506.03612
19. Kern, W., Martin, B., Paulusma, D., Smith, S., Leeuwen, E.J.V.: Disjoint paths and connected subgraphs for H-free graphs. Theor. Comput. Sci. **898**, 59–68 (2022). https://doi.org/10.1016/j.tcs.2021.10.019, https://www.sciencedirect.com/science/article/pii/S0304397521006344
20. Liu, Y.P., Sidford, A.: Faster energy maximization for faster maximum flow. In: Makarychev, K., Makarychev, Y., Tulsiani, M., Kamath, G., Chuzhoy, J. (eds.) Proccedings of the 52nd Annual ACM SIGACT Symposium on Theory of Computing, STOC 2020, Chicago, IL, USA, 22–26 June 2020, pp. 803–814. ACM (2020). https://doi.org/10.1145/3357713.3384247
21. Paulusma, D., van Rooij, J.M.M.: On partitioning a graph into two connected subgraphs. Theor. Comput. Sci. **412**(48), 6761–6769 (2011). https://doi.org/10.1016/j.tcs.2011.09.001, https://www.sciencedirect.com/science/article/pii/S0304397511007675
22. Takata, K.: Space-optimal, backtracking algorithms to list the minimal vertex separators of a graph. Discret. Appl. Math. **158**(15), 1660–1667 (2010). https://doi.org/10.1016/j.dam.2010.05.013
23. Telle, J.A., Villanger, Y.: Connecting terminals and 2-disjoint connected subgraphs. In: Brandstädt, A., Jansen, K., Reischuk, R. (eds.) WG 2013. LNCS, vol. 8165, pp. 418–428. Springer, Heidelberg (2013). https://doi.org/10.1007/978-3-642-45043-3_36

Polynomial-Size Enumeration Kernelizations for Long Path Enumeration

Christian Komusiewicz[1], Diptapriyo Majumdar[2], and Frank Sommer[1](✉)

[1] Institute of Computer Science, Friedrich Schiller University Jena, Jena, Germany
{c.komusiewicz,frank.sommer}@uni-jena.de

[2] Indraprastha Institute of Information Technology Delhi, New Delhi, India
diptapriyo@iiitd.ac.in

Abstract. Enumeration kernelization for parameterized enumeration problems was defined by Creignou et al. [Theory Comput. Syst. 2017] and later refined by Golovach et al. [J. Comput. Syst. Sci. 2022] to polynomial-delay enumeration kernelization. We consider ENUM LONG-PATH, the enumeration variant of the LONG-PATH problem, from the perspective of enumeration kernelization. Formally, given an undirected graph G and an integer k, the objective of ENUM LONG-PATH is to enumerate all paths of G having exactly k vertices. We consider the structural parameters vertex cover number, dissociation number, and distance to clique and provide polynomial-delay enumeration kernels of polynomial size for each of these parameters.

Keywords: Enumeration Problems · Parameterized Enumeration · Long Path · Expansion Lemma · Polynomial-Delay Enumeration Kernelization

1 Introduction

Kernelization is one of the most important contributions of parameterized complexity to the toolbox for handling hard computational problems [5,8,13]. The idea in kernelization is to develop polynomial-time algorithms that shrink any instance x of a parameterized problem L with parameterization κ to an equivalent instance x', the size of which is upper-bounded by $g(\kappa(x))$ for some function g. The crucial innovation of kernelization is that the size bound g allows for a theory of the effectiveness of polynomial-time preprocessing, a goal that seems virtually unreachable in a nonparameterized setting. Accordingly, the aim is to obtain a kernelization with g growing as slowly as possible.

Originally limited to decision problems, several extensions of kernelization adapt the notion to optimization [24], counting [19,23], and enumeration problems [4,6,14,15]. For enumeration problems, the focus of this work, a first

Research of Diptapriyo Majumdar has been supported by Science and Engineering Research Board (SERB) grant SRG/2023/001592. Research of Frank Sommer has been supported by the Alexander von Humboldt Foundation.

H. Fernau and P. Kindermann (Eds.): WG 2025, LNCS 16124, pp. 331–344, 2026.
https://doi.org/10.1007/978-3-032-11835-6_24

attempt at a suitable kernelization definition, was to ask that x' should not only be equivalent to x but furthermore that x' contains *all* solutions of x [6]. This definition has several drawbacks: First, it is restricted to subset-minimization problems. Second, the parameter is necessarily at least as large as the solution size. Finally, one may hope for kernels only in the case that the total number of solutions is bounded by a function of $\kappa(x)$. These drawbacks prompted Creignou et al. [4] to give a different definition of enumeration kernelization that asks for two algorithms: The first algorithm is the kernelization; it shrinks the input instance in polynomial time to the *kernel*, an instance whose size is bounded in the parameter. The second algorithm is the *solution-lifting* algorithm which allows to enumerate all solutions of the input instance from the solutions of the kernel. The crucial restriction for the solution-lifting algorithm is that it should be an FPT-delay algorithm, that is, the time spent between outputting consecutive solutions should be bounded by $f(\kappa(x)) \cdot |x|^{\mathcal{O}(1)}$ for some function f. This definition gives the desired property that an enumeration problem has an FPT-delay algorithm if and only if it has an enumeration kernel [4]. As discussed by Golovach et al. [14], however, any enumeration problem with an FPT-delay algorithm automatically admits an enumeration kernel of *constant size*. The crux of kernelization—providing a measure g of the effectiveness of data reduction—is undermined by this fact. This led Golovach et al. [14] to define enumeration kernels whose solution-lifting algorithms are only allowed to have *polynomial* delay. This particular running time bound has several desirable consequences [14]: First, a problem has an enumeration kernel of constant size if and only if it admits a polynomial-delay enumeration algorithm. This now directly corresponds to kernelization of decision problems, where a problem admits a kernelization of constant size if and only if it is polynomial-time solvable. Second, a parameterized enumeration problem admits an FPT-delay algorithm if and only if it admits a polynomial-delay enumeration kernel. Hence, the stricter delay requirement for the solution-lifting algorithm still suffices to capture all problems with FPT-delay. The size of this automatically implied kernel is, however, exponential in the parameter. This now makes it interesting to ask for polynomial-delay enumeration kernelization of *polynomial size* (henceforth *pd-ps kernel*). A further informal argument in favor of polynomial-delay enumeration kernels is that the limitation to polynomial delay often results in conceptually simple solution-lifting algorithms. Thus, solutions that are not in the kernel are similar to the kernel solutions and the kernel essentially captures all interesting solution types.

Summarizing, a pd-ps kernel provides (i) a compact representation of all types of solutions to the input instance, and (ii) a guarantee that for every solution of the output instance, the collection of solutions that are not "contained" in the kernel can be enumerated easily, that is, in polynomial delay. A further advantage of polynomial-delay enumeration kernels, which we discuss in more detail in Sect. 2, is that they provide a means to bound the total running time that an enumeration algorithm spends on outputs without polynomial delay. In particular, one can observe that the smaller the kernel, the better this running time bound will be.

By the above, enumeration kernels with polynomial-delay solution-lifting algorithms are a desirable type of algorithms for hard parameterized enumeration problems and obtaining a small kernel size is an interesting goal in the design of these algorithms. A priori it is unclear, however, whether there are many problems for which we can find such kernelizations: The only currently known pd-ps kernels are given for enumeration variants of the MATCHING CUT and d-CUT problem [14,22], two graph problems that have received some interest but are certainly not among the most famous ones. Thus, it is open whether there are further examples of pd-ps kernels, ideally for well-studied hard problems. In this work, we thus consider the following well-known problem as a candidate for application of the framework.

ENUM LONG-PATH
Input: An undirected graph $G = (V, E)$ and an integer k.
Goal: Enumerate all k-paths, that is, all paths having exactly k distinct vertices in G.

The corresponding decision version of the ENUM LONG-PATH problem is the NP-complete LONG-PATH problem which asks if an undirected graph G has a k-path. LONG-PATH is well-studied from the perspective of parameterized complexity [1,9,11,12,16] and kernelization [2,3,17,18]. In particular, LONG-PATH admits a polynomial kernel when parameterized by the vertex cover number, the distance to cluster, and the max-leaf number of the input graph [3]. In contrast, LONG-PATH admits no polynomial kernel when parameterized by the solution size k [2] or by the (vertex deletion) distance to outerplanar graphs [3].

Our Contributions. The two above-mentioned kernelization hardness results for LONG-PATH directly exclude pd-ps kernels for parameterization by solution size k or by distance to outerplanar graphs. We thus consider ENUM LONG-PATH parameterized by two parameters that are lower-bounded by the distance to outerplanar graphs: the vertex cover number (vc) and the dissociation number (diss) of the input graph. We also consider the distance to clique (dtc) of the input graph which is incomparable to vc and diss. Our first main result is the following.

Theorem 1. ENUM LONG-PATH *parameterized by* vc *admits an* $\mathcal{O}(n \cdot m \cdot k^2)$-*delay enumeration kernel with* $\mathcal{O}(\mathsf{vc}^2)$ *vertices.*

The quadratic bound on the number of vertices is the same as for the known kernel for the decision problem [3]. Thus, Theorem 1 generalizes the best decision kernel to the enumeration problem. To prove Theorem 1 we make use of (i) the "new expansion lemma" [10,13] and (ii) the enumeration of maximal matchings of a given cardinality with polynomial delay [20,21]. To the best of our knowledge, this is the first application of the expansion lemma and its variants [10,13,27] in the context of enumeration algorithms.

Our next result is a polynomial-delay enumeration kernel when the parameter is diss, the dissociation number of the graph. This number is the vertex deletion distance to graphs where each connected component has at most two vertices.

Theorem 2 **($\star$).** Enum Long-Path *parameterized by* diss *admits an* $\mathcal{O}(\mathsf{diss} \cdot n)$*-delay enumeration kernelization with* $\mathcal{O}(\mathsf{diss}^3)$ *vertices*[1].

Our approach even gives a kernelization for the more general r-coc parameter, which is the vertex deletion distance to graphs whose connected components have size at most r.

Corollary 1 **($\star$).** Enum Long-Path *parameterized by* r-coc *for constant* r *admits an* $\mathcal{O}(r\text{-coc} \cdot n \cdot 2^r)$*-delay enumeration kernelization with* $\mathcal{O}((r\text{-coc})^3)$ *vertices that can be computed in* $\mathcal{O}(2^r \cdot r \cdot (r\text{-coc})^2 \cdot n)$ *time.*

Ideally, we would like to provide kernels for even smaller parameters such as the vertex integrity or the treedepth of G. For these parameters, however, already decision kernels are unlikely due to the following folklore result. Here, maxcp denotes the size of a largest connected component. This parameter is at least as large as vertex integrity and treedepth.

Proposition 1 **($\star$).** Enum Long-Path *has no pd-ps kernel when parameterized by* maxcp *unless* NP $\subseteq$ coNP/poly.

To obtain Theorems 1 and 2, we exploited that the length of a longest path can be upper-bounded by $\mathcal{O}(\mathsf{vc})$ and $\mathcal{O}(\mathsf{diss})$, respectively.

To show that this is not a necessary condition for the existence of pd-ps kernels, we consider parameterization by dtc, the vertex deletion distance to a clique which can be smaller than k. We obtain the following.

Theorem 3 **($\star$).** Enum Long-Path *parameterized by* dtc *admits an* $\mathcal{O}(n \cdot \mathsf{dtc}(G))$*-delay enumeration kernelization with* $\mathcal{O}(\mathsf{dtc}^3)$ *vertices.*

2 Preliminaries

We use standard graph theoretic notations [7]. We also use the following tool.

Proposition 2 (New q-Expansion Lemma - Lemma 3.2 of [10]). *Let q be a positive integer and H be a bipartite graph with bipartition (A, B). Then, there exists a polynomial-time algorithm that computes $\widehat{A} \subseteq A$ and $\widehat{B} \subseteq B$ such that there is a q-expansion of $\widehat{A}$ onto $\widehat{B}$ such that* **(i)** $N_H(\widehat{B}) \subseteq \widehat{A}$, *and* **(ii)** $|B \setminus \widehat{B}| \leq q \cdot |A \setminus \widehat{A}|$.

[1] Due to lack of space proofs of results marked with ($\star$) can be found in the full version.

Parameterized Enumeration and Enumeration Kernelization. For details on parameterized complexity and kernelization, we refer to [5,8]. Creignou et al. [4] defined parameterized enumeration as follows. An *enumeration problem* over a finite alphabet Σ is a tuple (L, Sol) such that **(i)** $L \subseteq \Sigma^*$ is a decision problem, and **(ii)** $\mathsf{Sol} : \Sigma^* \to 2^{\Sigma^*}$ is a computable function and $\mathsf{Sol}(x) \neq \emptyset$ if and only if $x \in L$. If $x \in L$, then $\mathsf{Sol}(x)$ is the finite *set of solutions* to x. A *parameterized enumeration problem* is a triple $\Pi = (L, \mathsf{Sol}, \kappa(x))$ where (L, Sol) is an enumeration problem and $\kappa : \Sigma^* \to \mathbb{N}$ is the parameterization.

An enumeration algorithm $\mathcal{A}$ for a parameterized enumeration problem Π is a deterministic algorithm that given $x \in \Sigma^*$, outputs $\mathsf{Sol}(x)$ exactly without duplicates and terminates after a finite number of steps. For $x \in L$ and $1 \leq i \leq |\mathsf{Sol}(x)|$, the *i-th delay* of $\mathcal{A}$ is the time taken between outputting the i-th solution and the $(i+1)$-th solution of $\mathsf{Sol}(x)$. The 0*-th delay* is the *precalculation time*, that is, the time taken from the start of the algorithm $\mathcal{A}$ to outputting the first solution. The *postcalculation* time is the $|\mathsf{Sol}(x)|$-th delay, that is, the time from the last output to the termination of $\mathcal{A}$. If an enumeration algorithm $\mathcal{A}$ guarantees that every delay is bounded by $f(\kappa(x)) \cdot |x|^{\mathcal{O}(1)}$, then $\mathcal{A}$ is an *FPT-delay enumeration algorithm* (*FPT-delay* algorithm). If every delay is bounded by $|x|^{\mathcal{O}(1)}$, then $\mathcal{A}$ is a *polynomial-delay enumeration algorithm.* We will show kernelization results for the following definition of enumeration kernels.

Definition 1 ([14]). *Let $\Pi = (L, \mathsf{Sol}, \kappa)$ be a parameterized enumeration problem. A* polynomial-delay enumeration kernel *for Π is a pair of algorithms $\mathcal{A}$ and $\mathcal{B}$ with the following properties.*

(i) For every instance $x \in \Sigma^*$, the kernelization algorithm *$\mathcal{A}$ runs in time polynomial in $|x| + \kappa(x)$ and outputs an instance $y \in \Sigma^*$ such that $|y| + \kappa(y) \leq g(\kappa(x))$ for a computable function $g : \mathbb{N} \to \mathbb{N}$, and*
(ii) for every $s \in \mathsf{Sol}(y)$, the solution-lifting algorithm *$\mathcal{B}$ computes with delay polynomial in $|x| + |y| + \kappa(x) + \kappa(y)$, a nonempty set of solutions $S_s \subseteq \mathsf{Sol}(x)$ such that $\{S_s \mid s \in \mathsf{Sol}(y)\}$ is a partition of $\mathsf{Sol}(x)$.*

The function $g(\cdot)$ is the size *of the polynomial-delay enumeration kernel. If $g(\cdot)$ is a polynomial function, then Π is said to admit a polynomial-delay enumeration kernel of* polynomial size.

Observe that, by condition (ii) in the above definition, $x \in L$ if and only if $y \in L$. Observe from the definition that every polynomial-delay enumeration kernelization also corresponds to a kernel for the decision version of the corresponding problem.

Before stating our results, we show that the existence of polynomial delay enumeration kernels of a small size also entails running time guarantees concerning the non-polynomial part of the total running time. To put this bound into context, let us mention that the existence of an enumeration algorithm with FPT delay implies that there is an enumeration algorithm where the precalculation time is $f(\kappa(x))$ and all further delays and the postcalculation time are polynomial [25]. The bound for $f(\cdot)$ is, however, usually prohibitively large. For polynomial-delay enumeration kernels we obtain the following.

Proposition 3 (⋆). *Let $\Pi = (L, \mathsf{Sol}, \kappa)$ be a parameterized enumeration problem such that there is an algorithm that outputs for any given $x \in \Sigma^*$ the set* Sol(x) *in $T(|x|)$ time. If Π admits a polynomial-delay enumeration kernel of size g, then Π admits an algorithm with $T(g(\kappa(x))) + |x|^{\mathcal{O}(1)}$ precalculation time and polynomial delay.*

The above is similar to decision problems where small kernels and exact exponential-time algorithms directly give concrete running time bounds for FPT-algorithms.

3 Parameterization by the Vertex Cover Number

In this section, we describe a polynomial-delay enumeration kernel for Enum Long-Path when parameterized by vc, the size of a minimum vertex cover of the input graph. We first describe the kernelization algorithm and then the solution-lifting algorithm. We assume without loss of generality that G has no isolated vertex, as isolated vertices cannot be part of any k-path for $k > 1$ and Enum Long-Path is trivial for $k = 1$.

Marking Scheme and Kernelization Algorithm. To obtain our kernelization, we mark the vertices of G that we will keep in the kernel. Given the input graph $G = (V, E)$ and an integer k, we first invoke a linear-time algorithm [26] that outputs a vertex cover X of G such that $|X| \leq 2{\cdot}\mathsf{vc}(G)$. We mark all vertices in X. Now, let $I = V(G) \setminus X$. We consider a partition $I = I_1 \uplus I_2$ such that I_1 is the set of all vertices of I that have degree one in G. We call the vertices of I_1 *pendant.* We then construct an auxiliary bipartite graph H with bipartition $(\binom{X}{2}, I_2)$ as follows: a vertex $u \in I_2$ is adjacent to a vertex $\{x, y\} \in \binom{X}{2}$ in H if and only if $\{x, y\} \subseteq N_G(u)$. This bipartite graph construction is inspired by Bodlaender et al. [3] who also exploited connections between k-paths of G and matchings in H. Since G has no isolated vertex and I_2 excludes the pendant vertices, H has no isolated vertex. Then, we invoke Proposition 2, the new q-expansion lemma, with $q = 3$ on the graph H. That is, in polynomial time we obtain a set $\widehat{A} \subseteq \binom{X}{2}$ and a set $\widehat{B} \subseteq I_2$ such that there is a 3-expansion $M \subseteq E(H)$ of $\widehat{A}$ onto $\widehat{B}$ such that **(i)** $N_H(\widehat{B}) \subseteq \widehat{A}$ and **(ii)** $|I_2 \setminus \widehat{B}| \leq q \cdot |\binom{X}{2} \setminus \widehat{A}|$. Furthermore, there exists a set $B' \subseteq \widehat{B}$ of size $3 \cdot |\widehat{A}|$ such that each edge of M has exactly one endpoint in B', the vertices in $\widehat{B}$ saturated by M. We mark all vertices in $B' \cup (I_2 \setminus \widehat{B})$. If no 3-expansion exists, that is, if $\widehat{A} = \emptyset$ and thus also $\widehat{B} = B' = \emptyset$, then $I_2 \setminus \widehat{B} = I_2$. Consequently the kernel contains all vertices of I_2 in that case, and I_2 is small by property **(ii)** of Proposition 2. Otherwise, it is sufficient to mark the vertices $B' \subseteq \widehat{B}$ which are saturated by M. To prove this, we first define a notion of equivalence for k-paths in G and G'. Then, in Lemma 1, we show that the marked vertices are sufficient to keep for each k-path P in G an equivalent one in G'. Moreover, for every $x \in X$, we mark one arbitrary pendant vertex $u \in I_1$ such that $xu \in E(G)$. Let I_1^+ denote the set of all marked pendant vertices of I_1. It is sufficient to mark one vertex of I_1^+ per $x \in X$ since each k-path can contain at most one such vertex (in the beginning or the end). Now, G' is the subgraph induced by all marked vertices, that is, $V(G') = X \cup I_+^1 \cup (I_2 \setminus \widehat{B}) \cup B'$. Since G' is an induced subgraph of G and by properties (i) and (ii) of Proposition 2 (new q-expansion lemma), we may observe the following.

Observation 1 ($\star$). *Let G' be the graph obtained from G after invoking the above marking scheme. Then, G' has $\mathcal{O}(|X|^2)$ vertices. Moreover, any k-path of G' is also a k-path of G.*

We let $I' := I \cap V(G')$ denote the independent set in G'. Observe that I' consists of the marked pendant vertices I_1^+, the vertices B' saturated by the 3-expansion M, and the vertices of $I_2 \setminus \widehat{B}$ which have at least one neighbor in $\binom{X}{2} \setminus \widehat{A}$. Moreover, let H' be the subgraph of H induced by $V(G')$ with bipartition $(\binom{X}{2}, (I_2 \setminus \widehat{B}) \cup B')$.

Signatures and Equivalence among k-Paths. While the kernelization algorithm is sufficient to prove that (G', X, k) is equivalent to (G, X, k) for the decision version of the problem, we need to specify some relation between the k-paths of G and the k-paths of G' for the solution-lifting algorithm. Based on the description of the marking scheme, every $x \in X$ has at most one neighbor $v \in I_1^+$ in the graph G'. If $x \in X$ has a unique pendant neighbor $v \in I_1$ in G, then we call v a *1-pendant* vertex of I_1. Otherwise, if $x \in X$ has at least two pendant neighbors in I_1, then we call each of these pendant neighbors of x a *2-pendant* vertex of I_1. This distinction is important since a k-path in G may use an unmarked vertex v of I_1, but then v has to be a 2-pendant vertex.

Let $\mathcal{R} \subseteq I'$ denote the set of all vertices of G' that are in $I_2 \setminus \widehat{B}$ or 1-pendant vertices of I_1. Roughly speaking, $\mathcal{R}$ contains the vertices of I' (and also of I) which are somewhat unique; in contrast, $I' \setminus \mathcal{R}$ contains the vertices which have many similar vertices in $I \setminus I'$. Our definition of equivalence for k-paths will essentially make paths equivalent if they have the same interaction with $X \cup \mathcal{R}$, where the vertices of $\mathcal{R}$ are included because of their "uniqueness". We formalize this interaction as follows.

Definition 2. *Let $P = (v_1, \ldots, v_k)$ be a k-path of G or G'. Then, the function $\mathsf{sig}_P : V(P) \cap (X \cup \mathcal{R}) \to [k]$ with $\mathsf{sig}_P(v_i) = i$ is called the* signature *of P.*

Informally, the *signature* of P is the mapping of the vertices of $X \cup \mathcal{R}$ to their position in P; for an example see Fig. 1. In other words, the signature is the restriction of the position function pos_P, which maps each vertex to its index, to $X \cup \mathcal{R}$. Observe that one can compute the signature sig_P of a given k-path in linear time.

In order to establish a relation between the k-paths of G and the k-paths of G', we define the notion of equivalent k-paths as follows.

Definition 3. *Two k-paths P_1 and P_2 of G are said to be* equivalent *if*

(i) P_1 and P_2 have the same signature.

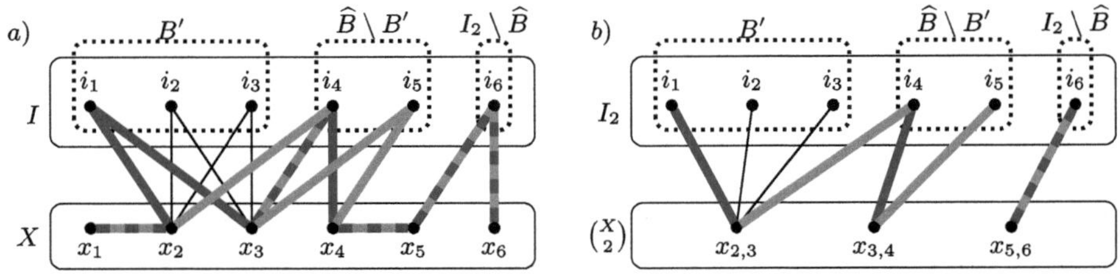

Fig. 1. Part *a)* shows a graph G and its decomposition into the modulator X and the independent set I. Here, the vertices in I_1 are not shown. The two paths $P_1 = (x_1, x_2, i_1, x_3, i_4, x_4, x_5, i_6, x_6)$ (in red) and $P_2 = (x_1, x_2, i_4, x_3, i_5, x_4, x_5, i_6, x_6)$ (in blue) have the same signature $\mathsf{sig}(x_1) = 1, \mathsf{sig}(x_2) = 2, \mathsf{sig}(x_3) = 4, \mathsf{sig}(x_4) = 6, \mathsf{sig}(x_5) = 7, \mathsf{sig}(i_6) = 8, \mathsf{sig}(x_6) = 9$. Part *b)* shows the corresponding auxiliary bipartite graph and the highlighted edges show the edges occupied (see paragraph above Lemma 2 for a definition) of both paths P_1 and P_2.

(ii) P_1 starts (respectively ends) at a 2-pendant vertex of I if and only if P_2 starts (respectively ends) at a 2-pendant vertex of I.

An example of two equivalent k-paths is shown in Fig. 1. Note that Item i of Definition 3 implies that $V(P_1) \cap (X \cup \mathcal{R}) = V(P_2) \cap (X \cup \mathcal{R})$. Also, note that in Item ii of Definition 3 it is sufficient to consider 2-pendant vertices, since all 1-pendant vertices are contained in $\mathcal{R}$.

To obtain k-paths from a signature, we use another mapping *extension* from $I \setminus \mathcal{R}$ to the indices of P such that the combination of both mappings yields a k-path in G (or G'). Its formal definition is given as follows.

Definition 4. *Let P' be a k-path in G' and let $I^* \subseteq I$. A mapping $\mathsf{exts} : I^* \to [k]$ is an* extension *of the signature $\mathsf{sig}_{P'}$ if (i) exts is injective, and (ii) there is a k-path P in G such that $\mathsf{sig}_{P'} \cup \mathsf{exts}$ is the position function pos_P.*

Note that if $I^* \subseteq I'$, then the k-path P obtained by $\mathsf{sig}_{P'} \cup \mathsf{exts}$ is a k-path in G'. The equivalence definition is given for k-paths of G but since G' is an induced subgraph of G this also implicitly defines equivalence between k-paths of G' and between k-paths of G and k-paths of G'. It is clear that the above definition is an equivalence relation for the paths of G. Hence, a k-path P is equivalent to itself. Consequently, every k-path P of G' has some equivalent k-path in G, namely P itself. We now show that the reverse is also true.

Lemma 1 ($\star$). *Let P be a k-path of G. Then, G' has a k-path P' that is equivalent to P.*

We now observe a structural property that illustrates a correspondence between the matchings in the auxiliary bipartite graph H and the k-paths of G. This allows the solution-lifting algorithm to make use of polynomial-delay algorithms for matching enumeration. If $xu, uy \in E(G)$ are two edges of P such that $x, y \in X$ and $u \in I$, then the pair $\{x, y\}$ is adjacent to u in H. In such a situation, the path P is said to *occupy* the edge $(\{x,y\},u)$ of H. We state the following lemma that also holds true for G' (for an example see Fig. 1).

Lemma 2 (⋆). *Let P be a k-path in G and let $H = (\binom{X}{2}, I_2)$ be the auxiliary bipartite graph. Then, the edges of H that are occupied by P form a matching in H. Moreover, given a k-path P of G and the auxiliary bipartite graph H, the edges of H that are occupied by P can be computed in $\mathcal{O}(n+m)$ time.*

Avoiding Duplicate Enumeration. To correctly design the solution-lifting algorithm, we need to make sure that every k-path P of G is outputted for exactly one k-path P' of G'. The natural idea is that we output P for a path P' that is equivalent. In case two distinct k-paths P_1 and P_2 of G' are equivalent to a k-path P of G, then both P_1 and P_2 have the same signature where $X' = V(P_1) \cap (X \cup \mathcal{R}) = V(P_2) \cap (X \cup \mathcal{R})$ and for every $u \in X'$, $\mathsf{sig}_{P_1}(u) = \mathsf{sig}_{P_2}(u)$. In such a case, if we enumerate all k-paths of G that are equivalent to a k-path P' of G', then the same k-path of G can be output multiple times. To circumvent this issue, we introduce a lexicographic order on equivalent paths which allows to output equivalent paths only for solution paths that are minimal with respect to this order.

We formalize the order as follows. First, we assume some fixed total order $\prec$ on the vertices of G and thus also on the vertices of G'. Now, for a path P' consider the subsequence $(v_1,\ldots,v_q)$ of P' containing only the vertices of $I \setminus \mathcal{R}$ and call this the I-sequence of P'. In other words, the I-sequence is the order of vertices of the extension of P'. We say that a path P' is *lexicographically smaller* than a path P^* when the I-sequence of P' is lexicographically smaller than the I-sequence of P^*.

Lemma 3 (⋆). *Let P be a k-path in G'. There is an $\mathcal{O}(k \cdot (m'+n'))$-time algorithm that checks if P is the lexicographically smallest k-path among all k-paths in G' with the same signature.*

Solution-Lifting Algorithm. Given a k-path P of G', let $F(H,P)$ be the set of edges from H that are occupied by P. It follows from Lemma 2 that $F(H,P)$ forms a matching in H. Without loss of generality, assume that P is the lexicographically smallest k-path for the signature $\mathsf{sig}_P : V(P) \cap (X \cup \mathcal{R}) \to [k]$. Let $\mathcal{A}$ denote the endpoints of $F(H,P)$ that are in $\binom{X}{2}$. If we are to enumerate the collection of all k-paths in G that intersect $V(G) \setminus V(G')$ and are equivalent to P, then every enumerated k-path P^* must occupy edges that also form a matching in H and satisfy the conditions (i) and (ii) of Definition 3. To avoid duplicates, we ensure that the enumerated paths contain at least one vertex that is not from G'. The following lemma summarizes the properties of the solution-lifting algorithm.

Lemma 4. *Let P be a k-path of G'. Then, there exists an algorithm with delay $\mathcal{O}(n \cdot m \cdot k^2)$ that enumerates all the k-paths P^* of G exactly once such that **(i)** P^* contains at least one vertex from G that is not in G', and **(ii)** P^* is equivalent to P.*

Proof. Let P be a k-path of G' with signature $\mathsf{sig}_P : V(P) \cap (X \cup \mathcal{R}) \to [k]$. We invoke Lemma 2 to compute the edges $F(H, P)$ of H that are occupied by P in $\mathcal{O}(k + n + m)$ time. For each position i in P that is not used by a vertex in $V(P) \cap (X \cup \mathcal{R})$ and each vertex $u \in V(G) \setminus V(G')$, we enumerate all paths P^* in G that are equivalent to P such that u is the first vertex of P^* that is not contained in $V(G')$ and $\mathsf{pos}_{P^*}(u) = i$. For this, we build an auxiliary graph H^* with bipartition A^*, B^* based on H and then enumerate all maximal matchings in H^*. The graph H^* is constructed as follows:

- First, in H^* we keep only those vertices $\{a, b\} \in \binom{X}{2}$ where a and b are connected by a vertex w outside of $X \cup \mathcal{R}$ in P. Note that $w \in B'$ for each such vertex. These vertices are added to A^*.
- If $1 < i < k$ and $\mathsf{pos}_P(a) = i - 1$ and $\mathsf{pos}_P(b) = i + 1$, then keep u as the only neighbor of $\{a, b\}$ in H^*; otherwise, discard the choice of u and i. Additionally, if $i = 1$ check if u is adjacent to the second vertex $x \in X$ of P. If this is not the case, then discard the choice of u and i. Symmetrically, if $i = k$, then check if u is adjacent to the penultimate vertex $y \in X$ of P. If this is not the case, then discard the choice of u and i.
- For each vertex $\{a, b\}$ in H^* with $\mathsf{pos}_P(a) < \mathsf{pos}_P(b) < i$ remove the edges between $\{a, b\}$ and any vertex v such that $v \notin V(G')$ or $v \in \mathcal{R}$. In other words, only keep the edges $(\{a, b\}, v)$ where $v \in B'$.
- For each vertex $\{a, b\}$ in H^* with $i < \mathsf{pos}_P(a) < \mathsf{pos}_P(b)$ remove the edges between $\{a, b\}$ and any vertex v such that $v \in \mathcal{R} \cup \{u\}$, that is, we only keep those edges where $v \in \widehat{B} \setminus \{u\}$.
- Finally, if $i \neq 1$ and the first vertex of P is not from $X \cup \mathcal{R}$, then let $x \in X$ denote the second vertex of P and add a new vertex $\{x, x^*\}$ to H and make $\{x, x^*\}$ in H^* adjacent to any vertex from $I' \setminus \mathcal{R}$ that is in G a neighbor of x. Similarly, if $i \neq k$ and the last vertex of P is not from $X \cup \mathcal{R}$, then let $y \in X$ denote the penultimate vertex of P. Add a new vertex $\{y, y^*\}$ to H^* and make $\{y, y^*\}$ adjacent to every vertex of $I' \setminus \mathcal{R}$ that is a neighbor of y. If added, $\{x, x^*\}$ and $\{y, y^*\}$ belong to A^*.

Now, for each such graph H^*, we enumerate all matchings that saturate all vertices in A^*. This can be achieved by enumerating all maximum matchings which can be done with $\mathcal{O}(n \cdot m)$-delay [21, paragraph above Theorem 13], [20, Theorem 18] on graphs with n vertices and m edges. For each enumerated matching M, we output the following path P^*, defined via pos_{P^*}.

For the vertex u, we set $\mathsf{pos}_{P^*}(u) := i$. The path P^* has the same signature as P, that is, for each vertex $v \in X \cup \mathcal{R}$, we set $\mathsf{pos}_{P^*}(v) := \mathsf{pos}_P(v)$. For each edge $\{\{a, b\}, v\} \in M$ with $a, b \in X$, assuming $\mathsf{pos}_P(a) < \mathsf{pos}_P(B)$, we set $\mathsf{pos}^*_P(v) = \mathsf{pos}_P(a) + 1$. For the edge $\{\{x, x^*\}, v\} \in M$, if it exists, we set $\mathsf{pos}_{P^*}(v) = 1$. For the edge $\{\{y, y^*\}, v\} \in M$, if it exists, we set $\mathsf{pos}_{P^*}(v) = k$. Now, pos_{P^*} indeed gives a k-path: since A^* is saturated by M and since the signatures of P^* and P agree, the codomain of pos_{P^*} is $[k]$. Moreover, by the construction of H^* and the fact that M is a matching, every vertex that is not in $X \cup \mathcal{R}$ is adjacent to its predecessor and successor (except for the first and last vertex if they are not in $X \cup \mathcal{R}$).

Thus, every outputted path is equivalent to P and by the inclusion of u it contains at least one vertex that is not from G'. The proof that no path is output twice is deferred to the full version.

It remains to show the bound on the delay. Observe that there are k choices for i and up to n choices for u. For each fixed pair i, u, we now use the algorithm of Kobayashi et al. [21, paragraph above Theorem 13], [20, Theorem 18] yielding a delay of $\mathcal{O}(n \cdot m)$. They argue that at most n times an augmenting path in $\mathcal{O}(m)$ time needs to be found. In other words, the delay can also be bounded by $\mathcal{O}(\nu(G) \cdot m)$, where $\nu(G)$ is the size of a maximum matching. In our case $\nu(G) = k$ since A^* contains at most k vertices. Thus, the overall delay is $\mathcal{O}(n \cdot m \cdot k^2)$. □

Theorem 1. Enum Long-Path *parameterized by* vc *admits an* $\mathcal{O}(n \cdot m \cdot k^2)$-*delay enumeration kernel with* $\mathcal{O}(\mathsf{vc}^2)$ *vertices.*

Proof. Our enumeration kernelization has two parts. The first part is the kernelization algorithm and the second part is the solution-lifting algorithm. Without loss of generality, let (G, X, k) be the input instance such that X is a vertex cover of G having at most $2 \cdot \mathsf{vc}(G)$ vertices. The kernelization algorithm described by the *marking scheme* outputs the instance (G', X, k). By Observation 1, G' has $\mathcal{O}(|X|^2) = \mathcal{O}(\mathsf{vc}^2)$ vertices.

Our solution-lifting algorithm works as follows. By construction, $(\widehat{B} \setminus B') \cup (I_1 \setminus I_1^+)$ are the only vertices of G that are not in G'. Let P' be a k-path of (G', X, k) such that $X' = (X \cup \mathcal{R}) \cap V(P')$. Moreover, let $\mathsf{sig}_{P'} : X' \to [k]$ be the signature of P'. Our first step is to output P' itself. Afterwards, we invoke Lemma 3 and check whether P' is the lexicographically smallest path of G' having the signature same as $\mathsf{sig}_{P'} : X' \to [k]$. If P' is the lexicographically smallest, then we invoke Lemma 4 to enumerate all the k-paths that are equivalent to P' and contain at least one vertex that is not from $V(G')$. Otherwise, if P' is not the lexicographically smallest for the signature $\mathsf{sig}_{P'} : X' \to [k]$, then the enumeration algorithm outputs only the path P'.

Since the enumeration algorithm invoked by Lemma 4 runs with a delay of $\mathcal{O}(n \cdot m \cdot k^2)$ and it never enumerates any other path $P \neq P'$ in G that is contained in G', this completes the proof that we have a polynomial-delay enumeration kernelization with $\mathcal{O}(\mathsf{vc}^2)$ vertices. □

Our techniques can also be used for Enum Long-Cycle and variants where all paths/cycles of length *at least* k need to be outputted.

Proposition 4 (⋆). Enum Long-Path At Least, Enum Long-Cycle, *and* Enum Long-Cycle At Least *parameterized by* vc *admits an* $\mathcal{O}(n \cdot m \cdot k^2)$-*delay enumeration kernel with* $\mathcal{O}(\mathsf{vc}^2)$ *vertices.*

4 Further Parameterizations

The pd-ps kernel for Enum Long-Path parameterized by the dissociation number $\mathsf{diss}(G)$ works roughly as follows. Given the instance (G, k), we first find a

3-approximation X of a minimal dissociation set using a greedy polynomial-time algorithm. Let $I = V(G) \setminus X$. By definition, $G[I]$ consists of connected components of size at most 2. Thus, the length k of the path can be upper-bounded by $3 \cdot (|X| + 1)$.

In our marking scheme we mark connected components of $G[I]$ as *rare* and *frequent* so that all possible structures of a k-path can be preserved in the kernel. Let us call a component C of $G[I]$ a neighbor component of some vertex v in X if v is adjacent to at least one vertex of C. Roughly speaking, a connected component C of $G[I]$ is now rare if there is some pair of vertices in X which has only few common neighbor components, one of which is C. Other components of $G[I]$ are frequent. We mark all vertices in rare components and, for each pair of vertices in X with many common neighbor components in $G[I]$ we mark $\mathcal{O}(|X|)$ neighbor components. This ensures that the number of marked vertices is cubic in $|X|$. The parameter k is not changed.

For the solution-lifting algorithm, we define the *signature* $\mathsf{sig}_{P'}$ of each k-path P' in G as the mapping of the rare vertices and the vertices in the modulator X of P' to their indices in P' and an *extension* of P' is another mapping from frequent vertices to the indices of P' such that the combination of both mappings yields a k-path in G which contains at least one non-kernel vertex. Based on the signature, we construct equivalence classes $\mathcal{P}_i$ of k-paths in G and equivalence classes $\mathcal{P}'_i$ of k-paths in G' and show there is a one-to-one correspondence between these equivalence classes. Afterwards, we define a *suitable* k-path in each equivalence class $\mathcal{P}'_i$ of G'. Then, if $P'_i \in \mathcal{P}'_i$ is not suitable, we only output P'_i, and if P'_i is suitable we output all k-paths in $\mathcal{P}_i$ which contain at least one non-kernel vertex, that is, all k-paths of $\mathcal{P}_i \setminus \mathcal{P}'_i$. One observation simplifies this step compared to the kernelization for the vertex cover parameterization: The only vertices that need to be replaced are frequent and there are always sufficiently many such vertices. This allows us to replace the matching enumeration by an enumeration algorithm that branches on all choices of frequent vertices in each position.

For parameterization by dtc, we need to resolve the issue that k can be much larger than dtc. To this end, our kernelization works in 2 stages: First, the clique and parameter k are reduced simultaneously. Second, we employ a marking scheme on the remaining clique similar to our marking scheme for diss. For the solution-lifting algorithm, we also need to be careful since the parameter k' for the kernel might be much smaller than k. In particular, this complicates the signature definition, where we need to introduce gaps which specify that between two modulator vertices, there can be a (possibly longer) path. The solution-lifting algorithm then has two phases. In the first phase, we speficy the first and last vertices of each gap. In the second phase, we enumerate all possibilities to insert further clique vertices in the gaps, to increase the path length to k.

5 Conclusion

We initiated a systematic study of enumeration kernels for the problems of finding paths (and cycles) with (at least) k vertices, leading to polynomial-delay enumeration kernels of polynomial size (pd-ps kernels) for the parameters vc, diss, r-coc, and dtc. It is open whether our positive results can be extended to a pd-ps kernel for the parameters distance to cluster, feedback vertex set, or neighborhood diversity (for which even a polynomial-size kernel for the decision problem seems to be open). More generally, it is open to develop pd-ps kernels for other well-studied problems.

References

1. Björklund, A., Husfeldt, T., Kaski, P., Koivisto, M.: Narrow sieves for parameterized paths and packings. J. Comput. Syst. Sci. **87**, 119–139 (2017)
2. Bodlaender, H.L., Downey, R.G., Fellows, M.R., Hermelin, D.: On problems without polynomial kernels. J. Comput. Syst. Sci. **75**(8), 423–434 (2009)
3. Bodlaender, H.L., Jansen, B.M.P., Kratsch, S.: Kernel bounds for path and cycle problems. Theor. Comput. Sci. **511**, 117–136 (2013)
4. Creignou, N., Meier, A., Müller, J., Schmidt, J., Vollmer, H.: Paradigms for parameterized enumeration. Theory Comput. Syst. **60**(4), 737–758 (2017)
5. Cygan, M., et al.: Parameterized Algorithms. Springer, Heidelberg (2015)
6. Damaschke, P.: Parameterized enumeration, transversals, and imperfect phylogeny reconstruction. Theor. Comput. Sci. **351**(3), 337–350 (2006)
7. Diestel, R.: Graph Theory. Graduate texts in mathematics, 4th edn., vol. 173. Springer, Heidelberg (2012)
8. Downey, R.G., Fellows, M.R.: Fundamentals of Parameterized Complexity. Texts in Computer Science. Springer, Heidelberg (2013)
9. Fellows, M.R., Lokshtanov, D., Misra, N., Mnich, M., Rosamond, F.A., Saurabh, S.: The complexity ecology of parameters: an illustration using bounded max leaf number. Theory Comput. Syst. **45**(4), 822–848 (2009)
10. Fomin, F.V., Le, T., Lokshtanov, D., Saurabh, S., Thomassé, S., Zehavi, M.: Subquadratic kernels for implicit 3-hitting set and 3-set packing problems. ACM Trans. Algorithms **15**(1), 13:1–13:44 (2019)
11. Fomin, F.V., Lokshtanov, D., Panolan, F., Saurabh, S.: Efficient computation of representative families with applications in parameterized and exact algorithms. J. ACM **63**(4), 29:1–29:60 (2016)
12. Fomin, F.V., Lokshtanov, D., Panolan, F., Saurabh, S.: Representative families of product families. ACM Trans. Algorithms **13**(3), 36:1–36:29 (2017)
13. Fomin, F.V., Lokshtanov, D., Saurabh, S., Zehavi, M.: Kernelization: Theory of Parameterized Preprocessing. Cambridge University Press (2019)
14. Golovach, P.A., Komusiewicz, C., Kratsch, D., Le, V.B.: Refined notions of parameterized enumeration kernels with applications to matching cut enumeration. J. Comput. Syst. Sci. **123**, 76–102 (2022)
15. Gomes, G.C.M., Juliano, E., Martins, G., dos Santos, V.F.: Matching (multi)cut: algorithms, complexity, and enumeration. In: Proceedings of the 19th International Symposium on Parameterized and Exact Computation (IPEC 2024). LIPIcs, vol. 321, pp. 25:1–25:15. Schloss Dagstuhl - Leibniz-Zentrum für Informatik (2024)
16. Goyal, P., Misra, N., Panolan, F., Zehavi, M.: Deterministic algorithms for matching and packing problems based on representative sets. SIAM J. Discret. Math. **29**(4), 1815–1836 (2015)
17. Jansen, B.M.P.: Turing kernelization for finding long paths and cycles in restricted graph classes. J. Comput. Syst. Sci. **85**, 18–37 (2017)

18. Jansen, B.M.P., Pilipczuk, M., Wrochna, M.: Turing kernelization for finding long paths in graph classes excluding a topological minor. Algorithmica **81**(10), 3936–3967 (2019)
19. Jansen, B.M.P., van der Steenhoven, B.: Kernelization for counting problems on graphs: preserving the number of minimum solutions. In: Proceedings of the 18th International Symposium on Parameterized and Exact Computation (IPEC 2023). LIPIcs, vol. 285, pp. 27:1–27:15. Schloss Dagstuhl - Leibniz-Zentrum für Informatik (2023)
20. Kobayashi, Y., Kurita, K., Wasa, K.: Polynomial-delay enumeration of large maximal matchings. CoRR abs/2105.04146 (2021). https://arxiv.org/abs/2105.04146
21. Kobayashi, Y., Kurita, K., Wasa, K.: Polynomial-delay and polynomial-space enumeration of large maximal matchings. In: Proceedings of the 48th International Workshop on Graph-Theoretic Concepts in Computer Science (WG 2022). Lecture Notes in Computer Science, vol. 13453, pp. 342–355. Springer, Cham (2022)
22. Komusiewicz, C., Majumdar, D.: Enumeration kernels of polynomial size for cuts of bounded degree. CoRR abs/2308.01286 (2023). https://doi.org/10.48550/arXiv.2308.01286
23. Lokshtanov, D., Misra, P., Saurabh, S., Zehavi, M.: Kernelization of counting problems. In: Proceedings of the 15th Innovations in Theoretical Computer Science Conference (ITCS 2024). LIPIcs, vol. 287, pp. 77:1–77:23. Schloss Dagstuhl - Leibniz-Zentrum für Informatik (2024)
24. Lokshtanov, D., Panolan, F., Ramanujan, M.S., Saurabh, S.: Lossy kernelization. In: Proceedings of the 49th Annual ACM SIGACT Symposium on Theory of Computing (STOC 2017), pp. 224–237. ACM (2017)
25. Meier, A.: Parametrised enumeration (2020). https://doi.org/10.15488/9427. Habilitation thesis, Leibniz Universität Hannover
26. Savage, C.D.: Depth-first search and the vertex cover problem. Inf. Process. Lett. **14**(5), 233–237 (1982)
27. Thomassé, S.: A $4k^2$ kernel for feedback vertex set. ACM Trans. Algorithms **6**(2), 32:1–32:8 (2010)

Efficient Parameterized Approximation

Stefan Kratsch and Pascal Kunz(✉)

Algorithm Engineering, Humboldt-Universität zu Berlin, Berlin, Germany
{stefan.kratsch,kunzpasc}@hu-berlin.de

Abstract. It is well known that NP-hard problems can be solved efficiently on instances that are sufficiently well-structured. Many admit exact parameterized algorithms with running time $f(k) \cdot n^{\mathcal{O}(1)}$ where k is some structural measure such as treewidth. For small k, this is comparable to a polynomial-time exact algorithm and outperforms exact exponential-time algorithms for a large range of k.

In this work, we are interested instead in leveraging instance structure for polynomial-time approximation algorithms. Concretely, we aim for polynomial-time algorithms that produce a solution of value at most/least $c\,\mathrm{OPT} \pm f(k)$ where c is a constant. Unlike for standard parameterized algorithms, we do not assume that any structural information is provided with the input. Ideally, we can obtain algorithms with small additive error only, i.e., $c = 1$ and $f(k)$ is polynomial or even linear in k. For small k, this is similarly comparable to a polynomial-time exact algorithm and it will beat general case approximation for a large range of k.

We study (Connected) Vertex Cover, Chromatic Number, and Triangle Packing. As parameters we consider the size of minimum modulators to graph classes on which the respective problem is tractable. For most cases we give efficient algorithms that compute a solution of size at least/most $\mathrm{OPT} \pm k$. For Vertex Cover, most of our algorithms are tight under the Unique Games Conjecture and have considerably better approximation guarantees than standard 2-approximations whenever the modulator is smaller than the optimum solution.

Keywords: Approximation algorithms · parameterized complexity · structural parameters

1 Introduction

Many important (graph) problems are NP-hard, ruling out polynomial-time exact algorithms that optimally solve every input, unless P = NP. Taking into account only the *size* of inputs, the best general (i.e., worst-case) results one can thus hope for are polynomial-time approximation algorithms and super-polynomial-time (usually exponential-time) exact algorithms. It is also well known that hard problems usually admit efficient algorithms on sufficiently well-structured inputs. How much can the *structure* of inputs help us to improve upon the best general-case approximation or exact algorithms?

H. Fernau and P. Kindermann (Eds.): WG 2025, LNCS 16124, pp. 345–359, 2026.
https://doi.org/10.1007/978-3-032-11835-6_25

Regarding the influence of structure on the *time complexity* of (mainly) exact algorithms, this is the main focus of the field of *parameterized complexity*. Therein, structural properties are quantified by *parameters* (such as solution size, similarity to a tractable special case, or treewidth) and one seeks algorithms that run in time $f(k) \cdot |x|^{\mathcal{O}(1)}$ for inputs (x, k) with parameter value k, called *fixed-parameter tractable (FPT)* algorithms. For small values of the parameter k this is almost as good as an efficient algorithm and for a certain range of k it will outperform the best known exact exponential-time algorithm. Such algorithms are known for a wide variety of problems with an (equal) variety of parameters.

There has been far less work on leveraging structure to design polynomial-time approximation algorithms.[1] The natural objective here is to study the influence of structure on the *approximation ratio* of polynomial-time algorithms. Mimicking the benefits of FPT-algorithms, i.e., closeness to a polynomial-time exact algorithm for small parameter value, leads us to *additive approximation* as the main goal: In polynomial time an *efficient parameterized algorithm* with parameter k should return a solution of cost at most $\mathrm{OPT} + f(k)$ (resp., value at least $\mathrm{OPT} - f(k)$) for a minimization (resp. maximization) problem; for small values of k this is close to an exact solution. Clearly, it is desirable for $f(k)$ to be no more than polynomial in k, or even linear in k, such as to beat the best unparameterized (i.e., standard) approximation algorithm for a large range of k.

Intuitively, for many standard problems we should not expect an additive error of at most $\mathrm{poly}(k)$ for parameters that do not grow under disjoint union of instances, such as treewidth: As a toy example consider VERTEX COVER with parameter k being the largest size of a connected component. By taking a disjoint union of $\mathrm{poly}(n) + 1$ copies of an instance with n vertices, the overall solution with error at most $\mathrm{poly}(k) \leq \mathrm{poly}(n)$ would have to contain an optimal solution for at least one copy (yielding P = NP).[2] At the same time, cost at most $\mathrm{OPT} + 2^k$ is straightforward to get for this example: It holds trivially when $n \leq 2^k$; and for $n > 2^k$ we can exactly solve each connected component and obtain an optimal solution in polynomial time. This is clearly reminiscent of what is known about the existence of *polynomial kernelizations* (a parameterized form of efficient preprocessing), which mostly works only for parameters like solution size or distance to a tractable case (i.e., size of a suitable modulator). It also emphasizes once more that we should mainly be concerned with polynomial or even linear additive error (and will hence focus on modulator-based parameters).

Among the few works on efficient parameterized approximation algorithms, the *structural rounding* framework of Demaine et al. [12] is closest to our present research interest. They also focus on modulator-based parameters, i.e., on the distance of inputs to some known tractable case of the problem, and show a

[1] The subfield of *parameterized approximation algorithms* studies approximation algorithms that run in FPT-time $f(k) \cdot |x|^{\mathcal{O}(1)}$, especially for problems with no exact FPT-algorithm, but not polynomial-time approximation.

[2] CHROMATIC NUMBER is a notable exception to this observation. Indeed, it is well known that any graph can in polynomial time be colored with at most $k + 1$ colors where k is its degeneracy, which subsumes e.g. treewidth at most k.

general approach for getting better than general $\alpha \cdot \mathrm{OPT}$ approximation results for instances that are sufficiently close to a specific tractable case (i.e., with sufficiently small parameter value). This involves a sufficiently good approximation of the distance, efficiently solving the remaining tractable case, and lifting back to an approximate solution of the initial input (with cost/value change proportional to the approximate distance). Here the final step implicitly produces an additive approximation but it is traded for $\alpha \cdot \mathrm{OPT}$ using the assumption that the distance is not too large.[3] Lavallee et al. [26] evaluated the performance of their approach applied to VERTEX COVER on practical inputs and found that, empirically, it can outperform the standard 2-approximation algorithms. In the field of exponential-time parameterized approximation, the work of Inamdar et al. [20] is most similar to ours. It, too, focuses on "distance-to-triviality" parameters, i.e., parameters that measure a graph's distance to a graph class on which a given problem can be solved efficiently.

Our Results. Like Demaine et al. [12], we focus on graph problems parameterized by the vertex-deletion distance to a known tractable graph class $\mathcal{C}$ for the problem in question, i.e., the size of a *modulator* X such that $G - X \in \mathcal{C}$. We similarly do not assume that such a modulator is given with the input, though this is commonly assumed to be given for FPT-algorithms and kernelization. In this setting, however, we focus mainly on getting cost at most $\mathrm{OPT} + \alpha k$ (resp. value at least $\mathrm{OPT} - \alpha k$) for $\alpha > 0$ as low as possible, rather than aiming for $c \cdot \mathrm{OPT}$ (resp. $\frac{1}{c} \cdot \mathrm{OPT}$) for $c > 1$ better than general-case approximation. In many cases, we improve over a straightforward application of structural rounding by not computing an (approximate) modulator but instead seeking a direct argument for why the generated solution fits the claimed error bound.

We will explore the relationship between efficient parameterized approximation on the one hand and FPT algorithms and structural rounding on the other in Sect. 3. We then present efficient parameterized approximation algorithms for concrete problems. We focus mainly on VERTEX COVER, CONNECTED VERTEX COVER, CHROMATIC NUMBER, and TRIANGLE PACKING and in Sects. 4 and 5 we obtain the results listed in Table 1. See Sect. 2 for definitions of the parameters. Proofs and some result statements (including all results on CONNECTED VERTEX COVER and TRIANGLE PACKING) are deferred to the full version [24] as indicated by the symbol $\star$.

Related Work. There a few other works that also make use of structure in the input for polynomial-time approximation. Chalermsook et al. [8] give an $f(\mathrm{tw})$-factor approximation for INDEPENDENT SET. There have also been several works on EDGE DISJOINT PATHS [9,14,16] and, in the realm of temporal graph problems, for TEMPORAL EXPLORATION [15], each parameterized by treewidth. In contrast to our results, this approximation guarantee is multiplicative. In the spirit of "above guarantee" parameterization, Mishra et al. [29]

[3] Their framework applies as well for distance to a (better than general) α-approximable case and will then allow lift to a slightly worse than that $\alpha \cdot \mathrm{OPT}$ (but better than general) approximation.

Table 1. An overview of our results: The third column lists the solution quality guaranteed by the algorithm we give. Here, k refers to the parameter value and c to an arbitrary constant. Throughout, the input consists of just a graph with no further information about its structure. For WEIGHTED VERTEX COVER, where the input is a weighted graph, the parameter value is the minimum weight of a modulator to the respective graph class whereas for the other problems it is the minimum size of a modulator. Results marked with $\star$ are entirely deferred to the full version.

Problem	Parameter is modulator to	Guarantee	Ref.
WEIGHTED VERTEX COVER	cograph	$\mathrm{OPT_{VC}} + 2k$	Corollary 1
	interval	$\mathrm{OPT_{VC}} + 4k$	Corollary 1
	cluster	$\mathrm{OPT_{VC}} + k$	Theorem 5
	cocluster	$\mathrm{OPT_{VC}} + k$	Corollary 2
	forest	$\mathrm{OPT_{VC}} + k$	Theorem 6
	chordal	$\frac{3}{2}\,\mathrm{OPT_{VC}} + k$	$\star$
VERTEX COVER	bipartite	no $(2-\varepsilon)\,\mathrm{OPT_{VC}} + ck$	Corollary 3
	split	$\mathrm{OPT_{VC}} + k$	$\star$
CONNECTED VERTEX COVER	split	$\mathrm{OPT_{VC}} + k$	$\star$
CHROMATIC NUMBER	bipartite	$2 + k$	Corollary 4
	planar	$4 + k$	Corollary 4
	chordal	$\mathrm{OPT_{COL}} + k$	Theorem 9
	cograph	$\mathrm{OPT_{COL}} + k$	Theorem 10
	cochordal	$2\,\mathrm{OPT_{COL}} + k - 1$	Theorem 11
	$(P_3 + K_1)$-free	$\mathrm{OPT_{COL}} + k$	$\star$
TRIANGLE PACKING	cluster	$\mathrm{OPT_{TP}} - k$	$\star$
	cocluster	$\mathrm{OPT_{TP}} - k$	$\star$

showed that one can in polynomial time output a vertex cover of size at most $\mathrm{OPT_{MM}} + \mathcal{O}(\log n \log\log n)(\mathrm{OPT_{VC}} - \mathrm{OPT_{MM}})$ where $\mathrm{OPT_{MM}}$ denotes the size of a maximum matching. LONGEST CYCLE. Kashaev and Schäfer [22] investigate VERTEX COVER approximation on graphs that are close to bipartite in that they have a small odd girth. Several standard approximation algorithms could also be interpreted as efficient parameterized approximations. For example, a standard 2-approximation for VERTEX COVER actually computes a vertex cover of size at most $\mathrm{OPT_{VC}} + \mathrm{OPT_{MM}}$.

2 Preliminaries

We use standard graph notation (see the full version or [13]). We will use the following parameters in this work: If $\mathcal{C}$ is a graph class and $G = (V, E)$ a graph, then $M \subseteq V$ is a *modulator* to $\mathcal{C}$ if $G - M \in \mathcal{C}$. It is a *minimum modulator* to $\mathcal{C}$ if $|M'| \geq |M|$ for all modulators M'. If $w \colon V \to \mathbb{Q}_{\geq 0}$ is a weight on G, then M is a *minimum-weight modulator* to $\mathcal{C}$ if $w(M') \geq w(M)$ for all modulators M'. Most parameters we will consider will be the minimum size or weight of a modulator to a certain graph class. We will now define the relevant graph classes. A modulator to an edgeless/acyclic/bipartite graph is a *vertex cover/feedback vertex set/odd cycle transversal.* A graph $= (V, E)$ is a *split graph* if $V = I \uplus C$ where $G[I]$ is edgeless and $G[C]$ is complete. It is *chordal* if it does not contain an induced cycle on at least four vertices. It is *cochordal* if $\overline{G}$ is chordal. It is a *cluster graph* if it does not contain an induced path on three vertices. It is a *cocluster* if $\overline{G}$ is a cluster graph. It is a *cograph* if it does not contain an induced path on four vertices. It is an *interval graph* if there is a set $\mathcal{I}$ of intervals on the real line and a bijection $\tau \colon V \to \mathcal{I}$ such that $\{u, v\} \in E$ if and only if $\tau(u)$ and $\tau(v)$ overlap. A graph is $(P_3 + K_1)$-free if it does not contain an induced $P_3 + K_1$ (the graph consisting of a path on three vertices and an isolated vertex).

3 Efficient Parameterized Approximation

An *optimization problem* over an alphabet Σ is defined by a function $\mathcal{P} \colon \Sigma^* \times \Sigma^* \to \mathbb{R} \cup \{\pm\infty\}$. An optimization problem can be a minimization or a maximization problem. Intuitively, $\mathcal{P}(I, X)$ is the value of a solution X in an instance I. For each $I \in \Sigma^*$, the number of $X \in \Sigma^*$ such that $\mathcal{P}(I, X) \notin \{\pm\infty\}$ should be finite. Then, we define the optimum value of $\mathcal{P}$ on I as $\mathrm{OPT}_{\mathcal{P}}(I) := \min_{X \in \Sigma^*} \mathcal{P}(I, X)$ if $\mathcal{P}$ is a minimization problem or as $\mathrm{OPT}_{\mathcal{P}}(I) := \max_{X \in \Sigma^*} \mathcal{P}(I, X)$ if it is a maximization problem. A *constant-factor approximation* with ratio $\alpha \geq 1$ for $\mathcal{P}$ is a polynomial-time algorithm that, given an instance $I \in \Sigma^*$, outputs a solution $X \in \Sigma^*$ such that $\mathcal{P}(I, X) \leq \alpha \, \mathrm{OPT}_{\mathcal{P}}(I)$ if $\mathcal{P}$ is a minimization problem or $\mathcal{P}(I, X) \geq \frac{1}{\alpha} \mathrm{OPT}_{\mathcal{P}}(I)$ if it is a maximization problem.

A *parameterized optimization problem* is a pair consisting of an optimization problem $\mathcal{P}$ and a function $\kappa \colon \Sigma^* \to \mathbb{N}$. Let $\alpha \geq 1$ and $f \colon \mathbb{N} \to \mathbb{R}_{\geq 0}$. An *efficient parameterized approximation* (EPA) with additive error of f and ratio α is a polynomial-time algorithm that given an instance $I \in \Sigma^*$ outputs a solution $X \in \Sigma^*$ with $\mathcal{P}(I, X) \leq \alpha \, \mathrm{OPT}_{\mathcal{P}} + f(\kappa(I))$ if $\mathcal{P}$ is a minimization problem or $\mathcal{P}(I, X) \geq \frac{1}{\alpha} \mathrm{OPT}_{\mathcal{P}} - f(\kappa(I))$ if it is a maximization problem. An EPA is not given access to the parameter value $\kappa(I)$ or any structure witnessing the parameter value, as designing EPAs is often trivial when such a structure is given. While in parameterized complexity it is fairly common to consider the solution size as a parameter, this is not interesting in the context of efficient parameterized approximation, since this is equivalent to normal polynomial-time approximation. For instance an EPA for a minimization problem with ratio α and additive error $f(k)$ where the parameter k is the optimum solution size is simply a polynomial-time approximation with factor $\alpha + f(\mathrm{OPT})/\mathrm{OPT}$. Graph parameters that do not increase when a graph is copied, while popular in parameterized algorithms, are also of little use if one wants an EPA with polynomial additive error for most common graph problems, as we noted in the introduction. This is why we focus almost exclusively on modulator parameters.

On the other hand, there are trivial EPAs with superpolynomial additive error for such parameterizations. We will say that $(\mathcal{P}, \kappa)$ is *fixed-parameter tractable*, if there is a computable function $f \colon \mathbb{N} \to \mathbb{N}$ and an algorithm with running time bounded by $f(\kappa(I)) \cdot |I|^{\mathcal{O}(1)}$ that given an instance $I \in \Sigma^*$ outputs an optimum solution, that is, a solution $X \in \Sigma^*$ with $\mathcal{P}(I, X) = \mathrm{OPT}_{\mathcal{P}}(I)$. This is a somewhat non-standard definition in parameterized complexity, as in this field one generally assumes that an FPT algorithm is given the parameter value and sometimes even a structure witnessing this parameter value as input and also because when dealing with exact parameterized algorithms one usually considers only decision problems. In practice, neither of these deviations from the normal definitions should make much of a difference for typical problems, parameters, or algorithms.

We will show that, if a parameterized optimization problem satisfying certain mild conditions admits an FPT algorithm, then it also admits an EPA with ratio 1 and additive error which depends on the superpolynomial part of the running time of the FPT algorithm. This is similar to the relationship between FPT algorithms and kernelization where it is known (see [11], for instance) that an FPT algorithm implies a kernelization, albeit one with superpolynomial size. We need the additional assumption that for some constant c there is a polynomial-time algorithm that outputs a solution X with $\mathcal{P}(I, X) \leq \mathrm{OPT}_{\mathcal{P}}(I) + |I|^c$ if $\mathcal{P}$ is a minimization problem or $\mathcal{P}(I, X) \geq \mathrm{OPT}_{\mathcal{P}}(I) - |I|^c$ if it is a maximization problem. This condition is trivial for most typical optimization problems (for instance, for VERTEX COVER, one may output the entire graph).

Theorem 1 ($\star$). *Let $(\mathcal{P}, \kappa)$ be a parameterized minimization (resp. maximization) problem such that there is a polynomial-time algorithm that outputs a solution X with $\mathcal{P}(I, X) \leq \mathrm{OPT}_{\mathcal{P}}(I) + |I|^c$ (resp. $\mathcal{P}(I, X) \geq \mathrm{OPT}_{\mathcal{P}}(I) - |I|^c$) for a constant c. If $(\mathcal{P}, \kappa)$ is FPT with a running time of $f(k) \cdot |I|^{\mathcal{O}(1)}$, then $(\mathcal{P}, \kappa)$ admits an EPA with ratio* 1 *and additive error f.*

Unlike in the analogous statement for kernelization, the converse of Theorem 1 is not true. Consider, for example, EDGE COLORING. Testing whether a graph can be edge colored with three colors is NP-hard [19], so, unless P = NP, there is no FPT algorithm for EDGE COLORING parameterized by the number of colors. There is, however, an EPA with ratio and additive error 1 because Vizing's algorithm [30] only uses OPT +1 colors. On the other hand, there are problems that admit a polynomial kernel, but do not admit an EPA with ratio 1 and polynomial additive error. Consider the TREE DELETION SET problem, where the task is to find a minimum modulator to a tree, parameterized by the solution size. This problem admits a polynomial kernel, but no $\mathrm{OPT}^{\mathcal{O}(1)}$-factor approximation, unless P = NP [18]. As we noted above, an EPA with polynomial additive error would be equivalent to such an approximation.

As we noted in the introduction, the present work is closely related to the "structural rounding" framework introduced by Demaine et al. [12]. Let Ψ be a class of edit operations on graphs. Such a class may include vertex deletions, edge deletions, edge additions, etc. A graph G' is *γ-editable* from a graph G under Ψ, if there is a sequence of edits $\psi_1, \ldots, \psi_k \in \Psi$ with $k \leq \gamma$ such that $G' = \psi_k(\psi_{k-1}(\cdots \psi_1(G) \cdots))$. The graph G is *γ-close* to the graph class $\mathcal{C}$, if there is a graph $G' \in \mathcal{C}$ such that G' is γ-editable from G under Ψ. A graph minimization (resp. maximization) problem $\mathcal{P}$ is *stable* under Ψ with constant c' if $\text{OPT}_{\mathcal{P}}(G') \leq \text{OPT}_{\mathcal{P}}(G) + c'\gamma$ (resp. $\text{OPT}_{\mathcal{P}}(G') \leq \text{OPT}_{\mathcal{P}}(G) - c'\gamma$) for any graph G' that is γ-editable from G under Ψ. A minimization (resp. maximization) problem $\mathcal{P}$ can be *structurally lifted* with respect to Ψ with constant c if there is a polynomial-time algorithm that given graphs G, G', an edit sequence such that $G' = \psi_k(\psi_{k-1}(\cdots\psi_1(G)\cdots))$, and a solution S' for G' and outputs a solution S for G such that $\mathcal{P}(G, S) \leq \mathcal{P}(G', S') + ck$ (resp. $\mathcal{P}(G, S) \geq \mathcal{P}(G', S') - ck$). Let $\mathcal{Q}(\mathcal{C}, \Psi)$ denote the problem of finding a minimum number of edits from Ψ in an input graph to obtain a graph in $\mathcal{C}$. The main theorem of Demaine et al. [12] can be restated as guaranteeing the existence of an EPA. For completeness and because our statement of the theorem differs from that of Demaine et al., we include a proof in the full version.

Theorem 2 (Demaine et al. [12], $\star$**).** *Let $\mathcal{P}$ be a minimization problem (resp. maximization problem) that is stable under Ψ with constant c' and can be structurally lifted with respect to Ψ with constant c. Suppose that $\mathcal{P}$ admits an α-approximation algorithm on the graph class $\mathcal{C}$ and that $\mathcal{Q}(\mathcal{C}, \Psi)$ admits a β-approximation algorithm on general graphs. Then, $\mathcal{P}$ parameterized by the closeness k of a graph to $\mathcal{C}$ under Ψ admits an EPA with ratio α and additive error $(\alpha\beta c' + c\beta)k$ (resp. $(\frac{1}{\alpha}\beta c' + c\beta)k$).*

Consider VERTEX COVER parameterized by the size of a minimum feedback vertex set. In other words, Ψ contains vertex deletions and $\mathcal{C}$ is the class of acyclic graphs. It is easy to see that VERTEX COVER is stable under Ψ with constant $c' = 0$ and that it can be structurally lifted with constant $c = 1$. Moreover, there is a 1-approximation for VERTEX COVER on acyclic graphs and $\mathcal{Q}(\mathcal{C}, \Psi)$, i.e. the FEEDBACK VERTEX SET problem, has a 2-approximation [3,6]. Hence, Theorem 2 implies an EPA with ratio 1 and additive error of $2k$. In Sect. 4.2, we will improve on this with an EPA with ratio 1 and additive error k.

4 Vertex Cover

In this section, we will consider the problems VERTEX COVER and WEIGHTED VERTEX COVER. We will give several efficient parameterized approximations for these two problems parameterized by the size of a minimum modulator to certain graph classes on which they can be solved in polynomial time. Most of these EPAs have a ratio of 1. We briefly observe that such an EPA is at the same time an EPA for the INDEPENDENT SET problem with the same parameter.

Observation 1 ($\star$) *If* (WEIGHTED) VERTEX COVER *with parameter κ has an EPA with ratio* 1 *and additive error f, then* (WEIGHTED) INDEPENDENT SET *with parameter κ has an EPA with the same ratio and the same additive error.*

Several algorithms will use the well-known local ratio method. It is based on the following adaptation of what is known as the *Local Ratio Theorem* (cf. e.g. [5]). For simplicity, we only state and prove the lemma for VERTEX COVER.

Lemma 2 (⋆). *Let $\mathcal{C}$ be a graph class and $\mathcal{P}$ be the problem of finding a minimum-weight modulator to $\mathcal{C}$. Let $\alpha, \beta \geq 1$. Let $G = (V, E)$ be a graph and $w, w_1, w_2 \colon V \to \mathbb{Q}_{\geq 0}$ be weight functions with $w = w_1 + w_2$. Let X be a vertex cover in G such that $w_i(X) \leq \alpha \operatorname{OPT}_{\mathrm{WVC}}(G, w_i) + \beta \operatorname{OPT}_{\mathcal{P}}(G, w_i)$ for $i \in \{1, 2\}$. Then, $w(X) \leq \alpha \operatorname{OPT}_{\mathrm{WVC}}(G, w) + \beta \operatorname{OPT}_{\mathcal{P}}(G, w)$.*

We use $\alpha(H)$ to denote the independence number of H. If $\mathcal{F}$ is a graph class, then a graph G is *$\mathcal{F}$-free* if G does not contain any graph in $\mathcal{F}$ as an induced subgraph. By eliminating copies of graphs in $\mathcal{F}$ using Lemma 2 and then solving Weighted Vertex Cover exactly on $\mathcal{F}$-free graphs, we get the following:

Theorem 3 (⋆). *Let $\mathcal{F}$ be a graph class such that*

(i) there is a polynomial-time algorithm that, given a graph $G = (V, E)$, either correctly decides that G is $\mathcal{F}$-free or outputs $S \subseteq V$ such that $G[S] \in \mathcal{F}$,
(ii) $\alpha^* := \max_{H \in \mathcal{F}} \alpha(H)$ is bounded, and
(iii) Weighted Vertex Cover can be solved exactly in polynomial time on $\mathcal{F}$-free graphs.

Let $\mathcal{P}$ denote the problem of finding a minimum-weight modulator to the class of $\mathcal{F}$-graphs. Then, there is a polynomial-time algorithm that, given a weighted graph (G, w), finds a vertex cover of weight at most $\operatorname{OPT}_{\mathrm{WVC}}(G, w) + \alpha^* \operatorname{OPT}_{\mathcal{P}}(G, w)$.

This directly implies that there is an algorithm that computes a vertex cover of weight at most $\operatorname{OPT}_{\mathrm{WVC}} + 2k$ where k is the minimum weight of a modulator to a cograph ($\operatorname{OPT}_{\mathrm{CGVD}}$). With a little more work (which is necessary because the forbidden subgraphs for interval graphs have unbounded independence number) based on Cao's algorithm for Interval Vertex Deletion [7], it also follows that one can compute a vertex cover of weight at most $\operatorname{OPT}_{\mathrm{WVC}} + 4k$ where k is the minimum weight of a modulator to an interval graph ($\operatorname{OPT}_{\mathrm{IVD}}$).

Corollary 1. *There is a polynomial-time algorithm that, given a weighted graph (G, w), computes a vertex cover of weight at most $\operatorname{OPT}_{\mathrm{WVC}}(G, w) + 2 \operatorname{OPT}_{\mathrm{CGVD}}(G, w)$ or at most $\operatorname{OPT}_{\mathrm{WVC}}(G, w) + 4 \operatorname{OPT}_{\mathrm{IVD}}(G, w)$*

4.1 Cluster and Co-cluster Vertex Deletion

We now consider the problem of computing vertex covers in graphs that are close to being cluster graphs. We will show that there is a polynomial-time algorithm that outputs a vertex cover of weight at most $\operatorname{OPT}_{\mathrm{WVC}} + \operatorname{OPT}_{\mathrm{CVD}}$ where $\operatorname{OPT}_{\mathrm{CVD}}$ is the weight of a minimum modulator to a cluster graph.

Let G be a graph and $\alpha \geq 1$. Let $H = (W, F)$ be an induced subgraph of G and $w_H \colon W \to \mathbb{Q}_{\geq 0}$ a weight function on H that is not necessarily related to any weight function on G. We will say that (H, w_H) is *strongly α-good* in G if w_H is not uniformly 0 and $w_H(W) \leq \alpha \cdot \operatorname{OPT}_{\mathrm{CVD}}(H, w_H)$. We will call (H, w_H) *centrally α-good* in G with respect to $v_0 \in W$ if (i) $N_G[v_0] \subseteq W$, (ii) $w_H(v) \geq 1$ for all vertices $v \in N_G[v_0]$, and (iii) $w_H(W) \leq \alpha \cdot \operatorname{OPT}_{\mathrm{CVD}}(H, w_H) + 1$.

Theorem 4 (Aprile et al. [2]). *There is a polynomial-time algorithm that given a true twin-free graph G outputs an induced subgraph $H = (W, F)$ of G and a weight function $w_H : W \to \mathbb{Q}_{\geq 0}$ on H such that (H, w_H) is strongly 2-good in G or centrally 2-good in G.*

Our EPA for Weighted Vertex Cover parameterized by the size of a minimum modulator to a cluster graph is very similar to the 2-approximation for CVD given by Aprile et al. [2]. It is based on eliminating true twins using a simple reduction rule, finding a strongly or centrally 2-good subgraph using Theorem 4, and eliminating it using Lemma 2.

Theorem 5 (⋆). *There is a polynomial-time algorithm that, given a weighted graph (G, w), computes a vertex cover of weight at most* $\mathrm{OPT}_{\mathrm{WVC}}(G, w) + \mathrm{OPT}_{\mathrm{CVD}}(G, w)$.

Theorem 5 can easily be adapted to the parameter $\mathrm{OPT}_{\mathrm{CCVD}}$, the weight of a minimum modulator to a cocluster.

Corollary 2 (⋆). *There is a polynomial-time algorithm that, given a weighted graph (G, w), computes a vertex cover of weight at most* $\mathrm{OPT}_{\mathrm{WVC}}(G, w) + \mathrm{OPT}_{\mathrm{CCVD}}(G, w)$.

The results in Theorem 5 and Corollary 2 are tight in the following sense: Let $k \in \{\mathrm{OPT}_{\mathrm{CVD}}, \mathrm{OPT}_{\mathrm{CCVD}}\}$. A vertex cover of weight at most $(1-\varepsilon)\,\mathrm{OPT}_{\mathrm{VC}} + k$ or $\mathrm{OPT}_{\mathrm{VC}} + (1-\varepsilon)k$ would be a $(2-\varepsilon)$-approximation for Vertex Cover, contradicting the Unique Games Conjecture [23] Both of these results also improve on the the solution obtained by applying the structural rounding approach and Theorem 2. This approach only yields a solution of weight at most $\mathrm{OPT}_{\mathrm{VC}} + 2\,\mathrm{OPT}_{\mathrm{CVD}}$ resp. $\mathrm{OPT}_{\mathrm{VC}} + 2\,\mathrm{OPT}_{\mathrm{CCVD}}$.

4.2 Feedback Vertex Number

Now, we consider the parameterization of Weighted Vertex Cover (VC) by the weight of a minimum feedback vertex set ($\mathrm{OPT}_{\mathrm{FVS}}$). We will show that Vertex Cover has a polynomial-time approximation that computes a vertex cover of weight at most $\mathrm{OPT}_{\mathrm{WVC}}(G) + \mathrm{OPT}_{\mathrm{FVS}}(G)$ in any weighted graph G. We employ an argument reminiscent of that used by Lokshtanov et al. [27] to obtain an FPT algorithm for Vertex Cover parameterized by the size of a Kőnig deletion set. Our algorithm applies a known reduction rule for Vertex Cover due to Nemhauser and Trotter [31,32], then computes a 2-approximate feedback vertex set, which it adds to the solution in its entirety, and then computes a minimum vertex cover in the remaining forest. We show that this yields a vertex cover of the requisite weight.

Theorem 6 (⋆). *There is a polynomial-time algorithm that, given a weighted graph (G, w), computes a vertex cover of weight at most* $\mathrm{OPT}_{\mathrm{WVC}}(G, w) + \mathrm{OPT}_{\mathrm{FVS}}(G, w)$.

4.3 A Lower Bound

We will now show that there is probably no EPA for VERTEX COVER parameterized by the size of an odd cycle transversal ($\mathrm{OPT}_{\mathrm{OCT}}$) with ratio less than 2 and linear additive error. This is a consequence of the following known approximation lower bound:

Theorem 7 (Bansal and Khot [4]). *Assuming the Unique Games Conjecture (UGC) and that* $\mathrm{P} \neq \mathrm{NP}$, *then for any constant* $\varepsilon, \delta > 0$ *there is no polynomial-time algorithm that, given an* n*-vertex graph containing two disjoint independent each of size* $(\frac{1}{2} - \varepsilon)n$, *finds an independent set of size at* δn.

Corollary 3 (⋆). *Let* $\alpha < 2$ *and* β *be given constants. Assuming the UGC and that* $\mathrm{P} \neq \mathrm{NP}$, *there is no polynomial-time algorithm that, given a graph* G, *computes a vertex cover of size at most* $\alpha\,\mathrm{OPT}_{\mathrm{VC}}(G) + \beta\,\mathrm{OPT}_{\mathrm{OCT}}(G)$.

The best known approximation algorithm for ODD CYCLE TRANSVERSAL is due to Agarwal et al. [1] and achieves a ratio of $\mathcal{O}(\sqrt{\log n})$. As observed by Kratsch and Wahlström [25], this can be combined with an FPT algorithm to compute an odd cycle transversal of size at most $\mathrm{OPT}^{3/2}$. With a straightforward application of the structural rounding approach, this gives a polynomial-time algorithm that computes a vertex cover of size at most $\mathrm{OPT}_{\mathrm{VC}} + \mathrm{OPT}_{\mathrm{OCT}}^{3/2}$.

5 Coloring

In this section, we will give several efficient parameterized algorithms for the CHROMATIC NUMBER problem. All of these algorithms compute a coloring that uses at most $\alpha\,\mathrm{OPT}_{\mathrm{COL}}(G - M) + \beta|M|$ colors where M is a modulator in the graph G to a graph class on which CHROMATIC NUMBER can be solved efficiently. This is a slightly stronger result than that required in our definition of an EPA.

5.1 Bounded Chromatic Number

We will start by considering as parameter the size of a modulator to any hereditary graph class $\mathcal{C}$ in which all graphs have a chromatic number bounded by a constant c and on which a c-coloring can be found in polynomial time. The idea of the algorithm is as follows. Starting from the empty graph, we add the vertices one by one and compute a coloring of the vertices that have already been added. For the analysis, we fix an unknown minimum modulator M and show that after each step the number of colors used is at most $c + k$, where k is the number of vertices from M already added. Each time we add a vertex we attempt to color it without using a new color, while re-coloring some previously added vertices. We attempt this by "guessing" c colors (using the fact that c is a constant) that do not contain a vertex from M and then apply the coloring algorithm for $\mathcal{C}$ to the subgraph induced by the c colors along with the new vertex. If this fails for every possible choice of the c colors, then the new vertex is assigned a new color.

If the vertex being added is not a modulator vertex and we have not previously used fewer than $c + k$ colors, then there must be c colors that are not assigned to any vertex M. Then, the new vertex along with the vertices that are assigned those c colors induce a graph in $\mathcal{C}$ and the recoloring attempt will succeed. If the vertex being added is from M or we have previously used fewer than the allowed number of colors, then there is no harm in assigning a new color.

Theorem 8 ($\star$). *Let $\mathcal{C}$ be a graph class and $c \in \mathbb{N}$ be a constant such that every graph in $\mathcal{C}$ is c-colorable and there is a polynomial-time algorithm that c-colors graphs in $\mathcal{C}$. Then, there is a polynomial-time algorithm that colors any graph with $c + k$ colors where k is the size of a minimum modulator to $\mathcal{C}$.*

Let $\mathrm{OPT}_{\mathrm{OCT}}$ and $\mathrm{OPT}_{\mathrm{PVD}}$ be the size of a minimum modulator to a bipartite and to a planar graph, respectively. Since bipartite graphs can be 2-colored and planar graphs can be 4-colored in polynomial time [33], Theorem 8 implies:

Corollary 4. *There are polynomial-time algorithms that color any graph with at most $2 + \mathrm{OPT}_{\mathrm{OCT}}$ colors or $4 + \mathrm{OPT}_{\mathrm{PVD}}$ colors.*

The results in Corollary 4 are tight in the sense that there are graphs that require $2 + \mathrm{OPT}_{\mathrm{OCT}}$ or $4 + \mathrm{OPT}_{\mathrm{PVD}}$ colors. Like in the discussion in Sect. 4.3, we can get a coloring with at most $2 + \mathrm{OPT}_{\mathrm{OCT}}^{3/2}$ using structural rounding. The best known approximation for a modulator to a planar graph achieves a ratio of $\mathcal{O}(\mathrm{OPT}^{\varepsilon})$ for any $\varepsilon > 0$ [21]. Hence, the structural rounding approach yields a coloring with $4 + \mathrm{OPT}_{\mathrm{PVD}}^{1+\varepsilon}$ colors. It is not possible to achieve $\mathrm{OPT}_{\mathrm{COL}} + f(\mathrm{OPT}_{\mathrm{PVD}})$ colors, unless $\mathrm{P} = \mathrm{NP}$, because coloring a planar graph optimally is NP-hard [17].

5.2 Cograph, Chordal, and Cochordal Deletion

In the following, we will explore the power of greedy algorithms for EPAs for Chromatic Number. We will show that with two simple greedy strategies, we can get EPAs for this problem parameterized by the size of a modulator to a cograph, a chordal graph, or a cochordal graph. It is well-known that any graph G can be colored with $\mathrm{degen}(G) + 1$ colors in polynomial time [28] where $\mathrm{degen}(G)$ is the degeneracy of G. By showing that $\mathrm{degen}(G) \leq \mathrm{OPT}_{\mathrm{COL}}(G - M) + |M| - 1$, we get the following:

Theorem 9 ($\star$). *There is a polynomial-time algorithm that colors any graph G with at most $\mathrm{OPT}_{\mathrm{COL}}(G - M) + |M|$ colors where M is a minimum modulator to a chordal graph in G.*

Using the fact that in a cograph every maximal clique intersects every maximal independent set [10], we obtain the following by greedily choosing a maximal independent set and assigning all vertices in it the same color.

Theorem 10 ($\star$). *There is a polynomial-time algorithm that colors any graph G with at most $\mathrm{OPT}_{\mathrm{COL}}(G - M) + |M|$ colors where M is a minimum modulator to a cograph in G.*

The same algorithm is also an EPA when the parameter is the size of a modulator to cochordal graph.

Theorem 11 (⋆). *There is a polynomial-time algorithm that colors any graph* G *with at most* $2\,\mathrm{OPT}_{\mathrm{COL}}(G - M) + |M| - 1$ *colors where* M *is a minimum modulator to a cochordal graph in* G.

6 Conclusion

We have initiated a dedicated study of efficient parameterized approximation algorithms with small additive error depending on some structural parameter. This is motivated by the goal of leveraging structure for improved polynomial-time approximation results, similarly to how fixed-parameter tractable algorithms improve upon exact exponential-time algorithms within certain regimes of input structure. As a byproduct of the structural rounding framework of Demaine et al. [12], such error bounds can be obtained for a variety of hard graph problems relative to modulator-based parameters. By focusing directly on the question of additive errors and, as far as possible, seeking to avoid an explicit (approximate) computation of the relevant modulators, we obtain better error bounds than by straightforward application of the framework. In many cases, further improvement of our bounds would contradict known lower bounds for approximation, or the bounds are tight for certain inputs. Arguably, this kind of efficient parameterized approximation algorithm is a sound alternative to computing exact solutions in time exponential in the parameter via FPT-algorithms.

For future work, more examples of efficient parameterized approximation algorithms with error bound better than via the structural rounding framework are of great interest. Ideally, the (approximate) computation of the parameter/modulator can be avoided. Let us name some specific problems related to our results: Vertex Cover relative to the size of a modulator to chordal graphs, graphs of treewidth at most two, or claw-free graphs. Weighted Vertex Cover and Triangle Packing with modulator to split graphs. Chromatic Number with modulator to weakly chordal, comparability, or perfect graphs. For problems that resist polynomial/linear error bounds one may also study parameterization by the size of an *edge* modulator (deletion/addition/editing) to the class in question. As an example, is there a linear additive error for Vertex Cover with edge modulator to bipartite graphs?

Certainly lower bounds are of great interest as well. Note that these are challenging because one needs to construct inputs with a small but *unknown modulator* to some target class, while being hard to find a good approximation for the target problem. As an example, with a given optimal modulator to bipartite graphs it would clearly be no challenge to beat the lower bound of $(2-\varepsilon)\,\mathrm{OPT} + c \cdot \mathrm{OPT}_{\mathrm{OCT}}$ for Vertex Cover.

References

1. Agarwal, A., Charikar, M., Makarychev, K., Makarychev, Y.: $O(\sqrt{\log n})$ approximation algorithms for Min UnCut, Min 2CNF deletion, and directed cut problems. In: Proceedings of the Thirty-Seventh Annual ACM Symposium on Theory of Computing, pp. 573–581 (2005). https://doi.org/10.1145/1060590.1060675
2. Aprile, M., Drescher, M., Fiorini, S., Huynh, T.: A tight approximation algorithm for the cluster vertex deletion problem. Math. Program. 1–23 (2021). https://doi.org/10.1007/s10107-021-01744-w
3. Bafna, V., Berman, P., Fujito, T.: A 2-approximation algorithm for the undirected feedback vertex set problem. SIAM J. Discret. Math. **12**(3), 289–297 (1999). https://doi.org/10.1137/S0895480196305124
4. Bansal, N., Khot, S.: Optimal long code test with one free bit. In: Proceedings of the 50th Annual Symposium on Foundations of Computer Science (FOCS), pp. 453–462 (2009). https://doi.org/10.1109/FOCS.2009.23
5. Bar-Yehuda, R., Bendel, K., Freund, A., Rawitz, D.: Local ratio: a unified framework for approximation algorithms. ACM Comput. Surv. **36**(4), 422–463 (2004). https://doi.org/10.1145/1041680.1041683
6. Becker, A., Geiger, D.: Optimization of Pearl's method of conditioning and greedy-like approximation algorithms for the vertex feedback set problem. Artif. Intell. **83**(1), 167–188 (1996). https://doi.org/10.1016/0004-3702(95)00004-6
7. Cao, Y.: Linear recognition of almost interval graphs. In: Proceedings of the 2016 Annual ACM-SIAM Symposium on Discrete Algorithms (SODA), pp. 1096–1115 (2016). https://doi.org/10.1137/1.9781611974331.ch77
8. Chalermsook, P., Fomin, F., Hamm, T., Korhonen, T., Nederlof, J., Orgo, L.: Polynomial-time approximation of Independent Set parameterized by treewidth. In: Proceedings of the 31st Annual European Symposium on Algorithms (ESA 2023), pp. 33:1–33:13 (2023). https://doi.org/10.4230/LIPIcs.ESA.2023.33
9. Chekuri, C., Naves, G., Shepherd, F.B.: Maximum edge-disjoint paths in k-sums of graphs. In: Fomin, F.V., Freivalds, R., Kwiatkowska, M., Peleg, D. (eds.) ICALP 2013. LNCS, vol. 7965, pp. 328–339. Springer, Heidelberg (2013). https://doi.org/10.1007/978-3-642-39206-1_28
10. Corneil, D., Lerchs, H., Burlingham, L.S.: Complement reducible graphs. Discret. Appl. Math. **3**(3), 163–174 (1981). https://doi.org/10.1016/0166-218X(81)90013-5
11. Cygan, M., et al.: Parameterized Algorithms. Springer, Cham (2015). https://doi.org/10.1007/978-3-319-21275-3
12. Demaine, E.D., et al.: Structural rounding: Approximation algorithms for graphs near an algorithmically tractable class. In: Proceedings of the 27th Annual European Symposium on Algorithms (ESA), pp. 37:1–37:15 (2019). https://doi.org/10.4230/LIPIcs.ESA.2019.37
13. Diestel, R.: Graph theory. Springer (2024)
14. Ene, A., Mnich, M., Pilipczuk, M., Risteski, A.: On routing disjoint paths in bounded treewidth graphs. In: Proceedings of the 15th Scandinavian Symposium and Workshops on Algorithm Theory (SWAT 2016), pp. 15:1–15:15 (2016). https://doi.org/10.4230/LIPIcs.SWAT.2016.15
15. Erlebach, T., Hoffmann, M., Kammer, F.: On temporal graph exploration. J. Comput. Syst. Sci. **119**, 1–18 (2021). https://doi.org/10.1016/j.jcss.2021.01.005

16. Fleszar, K., Mnich, M., Spoerhase, J.: New algorithms for maximum disjoint paths based on tree-likeness. In: Proceedings of the 24th Annual European Symposium on Algorithms (ESA 2016), pp. 42:1–42:17 (2016). https://doi.org/10.4230/LIPIcs.ESA.2016.42
17. Garey, M., Johnson, D., Stockmeyer, L.: Some simplified NP-complete graph problems. Theoret. Comput. Sci. **1**(3), 237–267 (1976). https://doi.org/10.1016/0304-3975(76)90059-1
18. Giannopoulou, A.C., Lokshtanov, D., Saurabh, S., Suchý, O.: Tree deletion set has a polynomial kernel (but no $\mathrm{OPT}^{O(1)}$ approximation). SIAM J. Discret. Math. **30**(3), 1371–1384 (2016). https://doi.org/10.1137/15M1038876
19. Holyer, I.: The NP-completeness of edge-coloring. SIAM J. Comput. **10**(4), 718–720 (1981). https://doi.org/10.1137/0210055
20. Inamdar, T., Kanesh, L., Kundu, M., Ramanujan, M.S., Saurabh, S.: FPT approximations for packing and covering problems parameterized by elimination distance and even less. In: Proceedings of the 43rd IARCS Annual Conference on Foundations of Software Technology and Theoretical Computer Science (FSTTCS 2023), pp. 28:1–28:16 (2023). https://doi.org/10.4230/LIPIcs.FSTTCS.2023.28
21. Jansen, B.M.P., Włodarczyk, M.: Lossy planarization: a constant-factor approximate kernelization for planar vertex deletion. In: Proceedings of the 54th Annual ACM SIGACT Symposium on Theory of Computing (STOC 2022), p. 900–913 (2022). https://doi.org/10.1145/3519935.3520021
22. Kashaev, D., Schäfer, G.: Round and bipartize for vertex cover approximation. In: Approximation, Randomization, and Combinatorial Optimization. Algorithms and Techniques (APPROX/RANDOM 2023), pp. 20:1–20:20 (2023). https://doi.org/10.4230/LIPIcs.APPROX/RANDOM.2023.20
23. Khot, S., Regev, O.: Vertex cover might be hard to approximate to within $2 - \epsilon$. J. Comput. Syst. Sci. **74**(3), 335–349 (2008). https://doi.org/10.1016/j.jcss.2007.06.019
24. Kratsch, S., Kunz, P.: Efficient parameterized approximation (2025), https://arxiv.org/abs/2501.14461
25. Kratsch, S., Wahlström, M.: Compression via matroids: a randomized polynomial kernel for odd cycle transversal. ACM Trans. Algorithms **10**(4) (2014). https://doi.org/10.1145/2635810
26. Lavallee, B., Russell, H., Sullivan, B.D., van der Poel, A.: Approximating vertex cover using structural rounding. In: Proceedings of the 22nd Workshop on Algorithm Engineering and Experiments (ALENEX), pp. 70–80 (2020). https://doi.org/10.1137/1.9781611976007.6
27. Lokshtanov, D., Narayanaswamy, N.S., Raman, V., Ramanujan, M.S., Saurabh, S.: Faster parameterized algorithms using linear programming. ACM Trans. Algorithms **11**(2) (2014). https://doi.org/10.1145/2566616
28. Matula, D.W.: A min-max theorem for graphs with application to graph coloring. SIAM Rev. **10**, 481–482 (1968). https://doi.org/10.1137/1010115
29. Mishra, S., Raman, V., Saurabh, S., Sikdar, S., Subramanian, C.R.: The complexity of König subgraph problems and above-guarantee vertex cover. Algorithmica **61**(4), 857–881 (2010). https://doi.org/10.1007/s00453-010-9412-2
30. Misra, J., Gries, D.: A constructive proof of Vizing's theorem. Inf. Process. Lett. **41**(3), 131–133 (1992). https://doi.org/10.1016/0020-0190(92)90041-S
31. Nemhauser, G.L., Trotter, L.E., Jr.: Properties of vertex packing and independence system polyhedra. Math. Program. **6**(1), 48–61 (1974). https://doi.org/10.1007/BF01580222

32. Nemhauser, G.L., Trotter, L.E., Jr.: Vertex packings: structural properties and algorithms. Math. Program. **8**(1), 232–248 (1975). https://doi.org/10.1007/BF01580444
33. Robertson, N., Sanders, D.P., Seymour, P., Thomas, R.: Efficiently four-coloring planar graphs. In: Proceedings of the Twenty-Eighth Annual ACM Symposium on Theory of Computing (STOC 1996), pp. 571–575 (1996). https://doi.org/10.1145/237814.238005

Structural Parameterizations for Induced and Acyclic Matching

Michael Lampis and Manolis Vasilakis(✉)

Université Paris-Dauphine, PSL University, CNRS, LAMSADE, Paris, France
michail.lampis@dauphine.fr, emmanouil.vasilakis@dauphine.eu

Abstract. We revisit the (structurally) parameterized complexity of Induced Matching and Acyclic Matching, two problems where we seek to find a maximum independent set of edges whose endpoints induce, respectively, a matching and a forest. Chaudhary and Zehavi [WG '23, SIDMA '25] recently studied these problems parameterized by treewidth, denoted by tw. We resolve several of the problems left open in their work and extend their results as follows: (i) for Acyclic Matching, Chaudhary and Zehavi gave an algorithm of running time $6^{\mathrm{tw}} n^{\mathcal{O}(1)}$ and a lower bound of $(3-\varepsilon)^{\mathrm{tw}} n^{\mathcal{O}(1)}$ (under the SETH); we close this gap by, on the one hand giving a more careful analysis of their algorithm showing that its complexity is actually $5^{\mathrm{tw}} n^{\mathcal{O}(1)}$, and on the other giving a pw-SETH-based lower bound showing that this running time cannot be improved (even for pathwidth), (ii) for Induced Matching we show that their $3^{\mathrm{tw}} n^{\mathcal{O}(1)}$ algorithm is optimal under the pw-SETH (in fact improving over this for pathwidth is *equivalent* to falsifying the pw-SETH) by adapting a recent reduction for Bounded Degree Vertex Deletion, (iii) for both problems we give FPT algorithms with single-exponential dependence when parameterized by clique-width and in particular for Induced Matching our algorithm has running time $3^{\mathrm{cw}} n^{\mathcal{O}(1)}$, which is optimal under the pw-SETH from our previous result.

Keywords: Acyclic Matching · Clique-width · Induced Matching · Parameterized Complexity · SETH · Treewidth

1 Introduction

Given a graph G, a *matching* $M \subseteq E(G)$ of G is a subset of its edges such that every vertex of G is incident with at most one edge in M. Matchings are central in Computer Science [54], and problems involving them find numerous practical applications such as matching kidney donors to recipients and clients to server clusters to name a few [56]. Even though finding a matching of maximum cardinality in general graphs is well-known to be polynomial-time solvable [57], further demanding that the graph induced by the vertices of the matching adheres to

This work is partially supported by ANR project ANR-21-CE48-0022 (S-EX-AP-PE-AL).

H. Fernau and P. Kindermann (Eds.): WG 2025, LNCS 16124, pp. 360–376, 2026.
https://doi.org/10.1007/978-3-032-11835-6_26

a specific constraint oftentimes renders the problem NP-hard. Recently, Chaudhary and Zehavi [17,18] studied a plethora of such NP-hard problems under the lens of parameterized complexity, providing a variety of algorithms and posing some open questions, some of which we answer in this work.

We first define the problems in question. Let $G = (V, E)$ be a graph and $M \subseteq E$ a matching of G, where $V_M \subseteq V$ denotes the set of vertices of G incident with edges in M. We say that matching M is *induced* (resp., *acyclic*) if $G[V_M]$ is 1-regular (resp., acyclic). In that case, given a graph G and an integer ℓ, Induced Matching (resp., Acyclic Matching) asks whether G has an induced (resp., acyclic) matching of size at least ℓ. Originally introduced as the "risk-free" marriage problem [66], Induced Matching has been extensively studied due to its numerous applications (see [37]). This problem and its variations also have connections to various problems such as Maximum Feasible Subsystem [13,32] and storylines extraction [48]. As for Acyclic Matching, it was introduced by Goddard et al. [36] along with some other variations of Maximum Matching. Both problems are known to be NP-hard [11,36,66], thus in this paper we examine them under the perspective of parameterized complexity.[1] Most relevant to us are the results concerning their complexity under structural parameterizations. Both problems are known to be FPT by the treewidth tw of the input graph [18,40,59], with the state-of-the-art being due to Chaudhary and Zehavi [18] who proposed algorithms of running time[2] $\mathcal{O}^\star(3^{\text{tw}})$ and $\mathcal{O}^\star(6^{\text{tw}})$ for Induced Matching and Acyclic Matching respectively. Furthermore, the authors complement their algorithms with $\mathcal{O}^\star((\sqrt{6} - \varepsilon)^{\text{pw}})$ and $\mathcal{O}^\star((3 - \varepsilon)^{\text{pw}})$ SETH-based lower bounds respectively, where pw denotes the pathwidth of the input graph. Lastly, as mentioned in [46], Induced Matching is expressible in MSO_1 logic, thus it is FPT parameterized by clique-width due to standard metatheorems [23].

Our Contribution. We start with the Induced Matching problem. Our first result is to show the optimality of the $\mathcal{O}^\star(3^{\text{tw}})$ algorithm of Chaudhary and Zehavi [18]. As a matter of fact, for the parameterization by the pathwidth pw of the input graph, we show that for all $\varepsilon > 0$, obtaining an $\mathcal{O}^\star((3 - \varepsilon)^{\text{pw}})$ algorithm is *equivalent* to falsifying the pw-SETH, a weakening of the SETH that was recently postulated by Lampis [51]. Note that this implies that an $\mathcal{O}^\star((3 - \varepsilon)^{\text{pw}})$ algorithm would falsify not just the standard form of the SETH (for k-CNF formulas), but (among others) also the SETH for circuits of depth εn [50] and the Set Cover Conjecture [25]. Our reduction follows along the lines of the one for Bounded Degree Vertex Deletion by Lampis and Vasilakis [52], though adapted to the machinery of the pw-SETH. Moving on, we consider the parameterization by the clique-width cw and develop an (optimal due to our previous result) $\mathcal{O}^\star(3^{\text{cw}})$ algorithm based on dynamic programming. Although obtaining an algorithm of running time $\mathcal{O}^\star(6^{\text{cw}})$ is quite straightforward, improving this to

[1] We assume the reader is familiar with the basics of parameterized complexity, as given e.g. in [26].

[2] Standard $\mathcal{O}^\star$ notation is used to suppress polynomial factors.

our obtained $\mathcal{O}^\star(3^{\text{cw}})$ running time requires some extra machinery. In particular, to bypass the bottleneck of the computation time in the union nodes, we employ ideas from [6,65] and consider *merged states* while solving the counting version of the problem. Doing so allows us to subsequently use the FFT technique introduced by Cygan and Pilipczuk [27] and populate the whole table in quasilinear time. Next we move towards the ACYCLIC MATCHING problem. We first provide a more careful analysis of the recent algorithm by Chaudhary and Zehavi [18] and show that in fact it runs in time $\mathcal{O}^\star(5^{\text{tw}})$. Our main result however is to show that improving over this, even for the parameterization by pathwidth, results in the pw-SETH failing. Finally, we consider ACYCLIC MATCHING parameterized by clique-width. Bergougnoux and Kanté [4] developed a framework based on the rank-based approach of Bodlaender, Cygan, Kratsch, and Nederlof [5] to deal with connectivity problems in this setting and developed algorithms with single-exponential dependence on the parameter for various problems, including FEEDBACK VERTEX SET. Adapting the latter algorithm allows us to obtain an $2^{\mathcal{O}(\text{cw})}n^{\mathcal{O}(1)}$ algorithm for ACYCLIC MATCHING, which is optimal under the ETH.

Related Work. Both INDUCED MATCHING and ACYCLIC MATCHING have been extensively studied with respect to their polynomial-time solvability in specific graph classes. We mention in passing that both problems remain NP-hard even in very restricted graph classes; for the former, such results are known for bipartite graphs of maximum degree 3 [55] and planar graphs of maximum degree 4 [44], while for the latter for planar bipartite graphs of maximum degree 3 [40]. For further results in specific graph classes we refer to [9–12,15,29,37,39,43,44,46, 47,55,69] for INDUCED MATCHING and to [3,40,62,63] for ACYCLIC MATCHING. Exact exponential-time algorithms for INDUCED MATCHING have been proposed in the literature [16,68], while both problems have also been studied with respect to their (in)approximability; see [1,13,14,20,21,31,32,34,53,61,64] and [2,3,35, 62] respectively.

From the viewpoint of Parameterized Complexity, both INDUCED MATCHING and ACYCLIC MATCHING are W[1]-hard parameterized by the natural parameter ℓ even in bipartite graphs, yet become FPT for various graph classes, including planar graphs [28,40,59]; as a matter of fact, in case of INDUCED MATCHING in planar graphs, various kernelization algorithms have been developed [33,42]. Recently, Chaudhary and Zehavi [19] also provided parameterized inapproximability results. Below guarantee parameterizations have also been studied [19,45,60,67].

Regarding structural parameterizations, both problems are FPT parameterized by the treewidth of the input graph, denoted by tw. For INDUCED MATCHING, Moser and Sikdar [59] developed a $\mathcal{O}^\star(4^{\text{tw}})$-time DP algorithm which was subsequently improved to $\mathcal{O}^\star(3^{\text{tw}})$ by Chaudhary and Zehavi [18] who also showed a $\mathcal{O}^\star((\sqrt{6}-\varepsilon)^{\text{pw}})$ lower bound under the SETH (pw denotes the pathwidth). As for ACYCLIC MATCHING, Hajebi and Javadi [40] noticed that the problem is expressible in MSO_2, thus it is FPT parameterized by treewidth due to Courcelle's theorem [22]; Chaudhary and Zehavi [18] later provided an explicit algorithm of complexity $\mathcal{O}^\star(6^{\mathrm{tw}})$ as well as a $\mathcal{O}^\star((3-\varepsilon)^{\mathrm{pw}})$ lower bound under the SETH. For structural parameters apart from treewidth, it is known that Induced Matching is expressible in MSO_1 logic [46], thus FPT parameterized by clique-width due to standard metatheorems [23], and Acyclic Matching is FPT parameterized by modular-width [40], while neither problem admits (under standard assumptions) a polynomial kernel parameterized by either the vertex cover number or the distance to clique of the input graph [19,38].

2 Preliminaries

Throughout the paper we use standard graph notations [30], and we assume familiarity with the basic notions of parameterized complexity [26]. All graphs considered are undirected without loops. Let $G=(V,E)$ be a graph. Then, $\mathtt{cc}(G)$ denotes the set of its connected components. Given a subset of its vertices $S \subseteq V$, $G[S]$ denotes the subgraph induced by S while $G-S$ denotes $G[V \setminus S]$. For a weight function w on the vertices of G and a subset $S \subseteq V$, we denote by $\mathrm{w}(S)$ the sum of the weights of all vertices of S. Given a subset of its edges $M \subseteq E$, V_M denotes the vertices incident with the edges in M while $G[V_M]$ denotes the graph induced by those vertices. A *matching* M of G is a subset of its edges such that every vertex of G is incident with at most one edge in M.

For $x, y \in \mathbb{Z}$, let $[x,y]=\{z \in \mathbb{Z} \mid x \leq z \leq y\}$ while $[x]=[1,x]$. Given a (partial) function $f\colon A \to B$, for every $b \in B$ we denote by $f^{-1}(b)$ the set of preimages of b under f, that is, $f^{-1}(b)=\{a \in A \mid f(a)=b\}$. Notice that in the case of partial functions it is not necessary for f to be defined over the whole set A. For a (partial) function f we denote by $f[v \mapsto \alpha]$ the (partial) function $(f \setminus \{(v,f(v))\}) \cup \{(v,\alpha)\}$, viewing f as a set. Standard $\mathcal{O}^\star$ notation is used to suppress polynomial factors. Proofs of statements marked with ($\star$) can be found in the full version of this article.

Clique-Width. A graph of clique-width k can be constructed through a sequence of the following operations on vertices that are labeled with at most k different labels. We can use (1) introducing a single vertex v of an arbitrary label i, denoted $i(v)$, (2) disjoint union of two labeled graphs, denoted $H_1 \oplus H_2$, (3) introducing edges between *all* pairs of vertices of two distinct labels i and j in a labeled graph H, denoted $\eta_{i,j}(H)$, and (4) changing the label of *all* vertices of a given label i in a labeled graph H to a different label j, denoted $\rho_{i \to j}(H)$. An expression describes a graph G if G is the final graph given by the expression (after we remove all the labels). The *width* of an expression is the number of different labels it uses. The clique-width of a graph is the minimum width of an expression describing it [24]. For a labeled graph H and $v \in V(H)$, let $\mathrm{lab}_H(v)$ denote the label of v in H, while $\mathrm{lab}_H^{-1}(i)=\{v \in V(H) \mid \mathrm{lab}_H(v)=i\}$ denotes the set of vertices of H of label i. For $S \subseteq V(H)$, let $H[S]$ denote the labeled subgraph of H induced by S. A clique-width expression is *irredundant* if whenever the operation $\eta_{i,j}$ is applied on a graph G, there is no edge between an i-vertex and a j-vertex in G, and we will use such expressions to simplify our algorithms. We remark that any clique-width expression can be transformed in linear time into an irredundant one of the same width [24].

pw-SETH. The *primal pathwidth SETH* (pw-SETH) states that, for all $\varepsilon > 0$, 3-SAT requires time at least $(2-\varepsilon)^{\text{pw}} n^{\mathcal{O}(1)}$, where pw is the pathwidth of the primal graph of the input formula.[3] In other words, the pw-SETH posits that the simple DP algorithm which solves 3-SAT in time $\mathcal{O}^\star(2^{\text{pw}})$ is best possible. Beating this algorithm seems to encapsulate the difficulty of improving upon simple DP algorithms in general and indeed the pw-SETH is *equivalent* to many tight lower bounds for problems parameterized by linear structure widths (pathwidth or linear clique-width) [51]. The pw-SETH also has the advantage of being implied by several other standard assumptions, such as the Set Cover Conjecture and the SETH for circuits of depth εn [50], making lower bound results based on the pw-SETH seem more believable.

In this paper we present two such lower bound results based on the pw-SETH, in one case (for INDUCED MATCHING parameterized by pathwidth) showing that breaking our bound is *equivalent* to the pw-SETH. For this, we recall some notions and results from [51] which we will make use of later in our reductions, and in particular in Lemma 3 and Theorem 9. Informally, reductions in this context start from a constraint satisfaction problem CSP with alphabet size equal to the desired base of the lower bound, similarly to SETH-based reductions [49]. However, rather than reducing from a CSP instance parameterized by the number of variables we reduce from an instance parameterized by the pathwidth of its primal graph. To facilitate the reduction, we use the fact that it is hard (under pw-SETH) to distinguish between CSP instances which are satisfiable and instances which are unsatisfiable even if for the majority of variables we are allowed to select multiple assignments, albeit while only modifying values in a monotone way along the given path decomposition.

Definition 1 (**[51, Definition 3.2]**). *Suppose we have a* CSP *instance* ψ *over an alphabet* $[B]$ *of size* $B \geq 2$*, with variable set* V*, a path decomposition of its primal graph* $B_1, \ldots, B_t$*, and an injective function* b *mapping each constraint to the index of a bag that contains all its variables. A* multi-assignment *is a function* σ *that takes as input a variable* $x \in V$ *and an index* $j \in [t]$ *such that* $v \in B_j$ *and returns a value in* $[B]$*. We will say that:*

1. *A multi-assignment* σ *is* satisfying *for* ψ *if for each constraint* c*, the assignment* $\sigma_c(x) = \sigma(x, b(c))$*, that is, the restriction of the multi-assignment to* $b(c)$*, satisfies the constraint.*
2. *A multi-assignment* σ *is* monotone *if for all* $x \in V$ *and* $j_1 < j_2$ *with* $x \in B_{j_1} \cap B_{j_2}$ *we have* $\sigma(x, j_1) \leq \sigma(x, j_2)$*.*
3. *A multi-assignment* σ *is* consistent *for* $x \in V$ *if for all* $j_1, j_2 \in [t]$ *such that* $x \in B_{j_1} \cap B_{j_2}$ *we have* $\sigma(x, j_1) = \sigma(x, j_2)$*.*

[3] The *primal graph* of a formula contains a vertex for each variable and an edge when two variables appear in a common constraint.

Corollary 1 ([51, **Corollary 3.1**]). *For all $\varepsilon > 0, B \geq 2$ we have the following. Suppose there is an algorithm with the following properties:*

- *it takes as input a* 4-CSP *instance ψ over an alphabet of size B, a partition of its variables into two sets V_1, V_2, a path decomposition of its primal graph of width p where each bag contains at most $\mathcal{O}(B \log p)$ variables of V_2, and an injective function b mapping each constraint to a bag that contains its variables;*
- *it decides if there exists a monotone satisfying multi-assignment σ, which is consistent for the variables of V_2.*

If the supposed algorithm runs in time $\mathcal{O}((B-\varepsilon)^p|\psi|^{\mathcal{O}(1)})$, then the pw-SETH is false.

3 Induced Matching

In this section we consider the Induced Matching problem. Chaudhary and Zehavi [18] developed an algorithm of running time $\mathcal{O}^\star(3^{\mathrm{tw}})$, where tw denotes the treewidth of the input graph, and left as an open question whether this is optimal. As our first result, in Theorem 1 we show that this is indeed the case, as obtaining an algorithm of running time $\mathcal{O}^\star((3-\varepsilon)^{\mathrm{pw}})$ is equivalent to falsifying the pw-SETH. Next, we develop a DP algorithm of running time $\mathcal{O}^\star(3^{\mathrm{cw}})$ for the problem, where cw denotes the clique-width of the input graph. Since $\mathrm{cw}(G) \leq \mathrm{pw}(G) + 2$ for any graph G, our algorithm is optimal under the pw-SETH due to Theorem 1.

3.1 Lower Bound

We first state the main theorem of this section, which we then prove in Lemmas 2 and 3.

Theorem 1. *For all $\varepsilon > 0$, we have the following: there exists an algorithm deciding* Induced Matching *in time $\mathcal{O}^\star((3-\varepsilon)^{\mathrm{pw}})$, where pw denotes the pathwidth of the input graph, if and only if the pw-SETH is false.*

Lemma 2 ($\star$). *If the pw-SETH is false, then there exists $\varepsilon > 0$ and an algorithm that solves* Induced Matching *in time $\mathcal{O}^\star((3-\varepsilon)^{\mathrm{pw}})$, where pw denotes the pathwidth of the input graph.*

For the other direction, we reuse the gadgets of the reduction presented by Lampis and Vasilakis [52] to prove that Bounded Degree Vertex Deletion when $\Delta = 1$ cannot be solved in $\mathcal{O}^\star((3-\varepsilon)^{\mathrm{pw}})$ under the SETH; the latter problem asks, given a graph G, $\Delta \geq 0$, and $k > 0$, whether there exists a set $S \subseteq V(G)$ of size at most k such that the maximum degree of $G - S$ is at most Δ.

Lemma 3 ($\star$). *If there exists $\varepsilon > 0$ and an algorithm that solves* Induced Matching *in time $\mathcal{O}^\star((3-\varepsilon)^{\mathrm{pw}})$, where pw denotes the pathwidth of the input graph, then the pw-SETH is false.*

3.2 Algorithm

In this section we prove the following theorem.

Theorem 4. *There is an algorithm that, given a graph G and an irredundant clique-width expression ψ of G of width* cw, *counts the number of induced matchings of G of each size $\ell \in [0, n/2]$ in time $\mathcal{O}^\star(3^{\text{cw}})$.*

Proof. Before describing our algorithm we start with some definitions and notation. Given a tuple σ, we refer to its w-th entry by $\sigma(w)$. Let H be a graph and $S \subseteq V(H)$ a subset of its vertices. We say that a vertex $v \in S$ is *unsaturated* in $H[S]$ if its degree is $\deg_{H[S]}(v) = 0$, else if $\deg_{H[S]}(v) = 1$ it is *saturated* in $H[S]$ instead. We say that S is a *partial matching* of H if the maximum degree of $H[S]$ is at most 1, in which case all of its vertices are either saturated or unsaturated in $H[S]$. The *size of a partial matching* S of H is equal to the number of its vertices, that is, $|S|$. Finally, we say that a partial matching S of H is *extensible* if there exists an induced matching M of G such that $V_M \cap V(H) = S$.

Let $\mathcal{H}$ denote the set of all cw-labeled graphs generated by subexpressions of ψ. For $H \in \mathcal{H}$ and $i \in [\text{cw}]$, we define the partial function $\text{sgn}_{H,i} \colon V(H) \to \{0, 1, 2\}$ for $S \subseteq V(H)$ as follows:

$$\text{sgn}_{H,i}(S) = \begin{cases} 0 & \text{if } S \cap \text{lab}_H^{-1}(i) = \varnothing, \\ 1 & \text{if } S \cap \text{lab}_H^{-1}(i) \neq \varnothing \text{ and } \deg_{H[S]}(v) = 1 \text{ for all } v \in S \cap \text{lab}_H^{-1}(i), \\ 2 & \text{if } S \cap \text{lab}_H^{-1}(i) = \{v\} \text{ and } \deg_{H[S]}(v) = 0. \end{cases}$$

We further define the tuple $\text{sgn}_H(S) = \big(\text{sgn}_{H,1}(S), \ldots, \text{sgn}_{H,\text{cw}}(S)\big)$ to be the *signature* of S in H if $\text{sgn}_{H,i}(S)$ is well-defined for all $i \in [\text{cw}]$. Notice that $\text{sgn}_H(S)$ is only defined for a subset of the partial matchings of H, while there are at most 3^{cw} different signatures. For improved readability, we sometimes refer to tuples $\sigma \in \{0, 1, 2\}^{\text{cw}}$ as signatures. We first argue that for any induced matching M of G, the signature of $V_M \cap V(H)$ is well-defined for all $H \in \mathcal{H}$.

Claim. Let M be an induced matching of G. For all $H \in \mathcal{H}$ it holds that $\text{sgn}_H(V_M \cap V(H))$ is well-defined.

Proof of the claim. Notice that for all $H \in \mathcal{H}$ it holds that $H[V_M \cap V(H)]$ is a subgraph of $G[V_M]$, therefore the maximum degree of $H[V_M \cap V(H)]$ is at most 1. Consequently, $V_M \cap V(H)$ is a partial matching of H. Fix $H \in \mathcal{H}$ and let $S = V_M \cap V(H)$. It suffices to prove that $\text{sgn}_{H,i}(S)$ is well-defined for all $i \in [\text{cw}]$.

For the sake of contradiction, assume that $\text{sgn}_{H,i}(S)$ is not well-defined for some $i \in [\text{cw}]$. Since S is a partial matching of H, that can only be the case when $S \cap \text{lab}_H^{-1}(i)$ contains at least two vertices, say v, v', and at least one of them, say v, is unsaturated in $H[S]$, that is, $\deg_{H[S]}(v) = 0$. Let $u \in V_M$ such that $\{v, u\} \in M$. Since $\deg_{H[S]}(v) = 0$ it holds that either $u \notin V(H)$ or $\{v, u\} \notin E(H)$. This means that at some point further in the clique-expression there exists a join operation with $H_1, H_2 \in \mathcal{H}$ and $H_2 = \eta_{i',j'}(H_1)$ that adds all edges between vertices of labels i' and j', with v, v' having label i' and u having label j'. In that case however, it holds that the degree of $u \in V_M \cap V(H_2)$ in $H_2[V_M \cap V(H_2)]$ is at least 2 as it is adjacent to both v and v', a contradiction.$\diamond$

Our algorithm proceeds by dynamic programming along ψ in a bottom-up fashion. In particular, for every $H \in \mathcal{H}$ it stores a table $\mathrm{DP}_H[\cdot,\cdot]$ where, on a high-level, for $k \in [0,n]$ and $\sigma \in \{0,1,2\}^{\mathrm{cw}}$, $\mathrm{DP}_H[k,\sigma]$ contains the number of partial matchings S of H of size k and signature σ, that is, with $|S| = k$ and $\mathrm{sgn}_H(S) = \sigma$. We now proceed to describe how to populate the DP tables, and in Lemmas 5 to 7 we prove that they indeed contain the number of partial matchings of appropriate size and signature.

Singleton $H = i(v)$. For $H = i(v)$, notice that $\mathrm{lab}_H^{-1}(i) = \{v\}$ and $\mathrm{lab}_H^{-1}(w) = \varnothing$ for all $w \in [\mathrm{cw}] \setminus \{i\}$. Consequently, for $k \in [0,n]$ and $\sigma \in \{0,1,2\}^{\mathrm{cw}}$ we set

$$\mathrm{DP}_H[k,\sigma] = \begin{cases} 1 & \text{if } k = 0 \text{ and } \sigma(w) = 0 \text{ for all } w \in [\mathrm{cw}], \\ 1 & \text{if } k = 1 \text{ and } \sigma(w) = 0 \text{ for all } w \in [\mathrm{cw}] \setminus \{i\} \text{ and } \sigma(i) = 2, \\ 0 & \text{otherwise.} \end{cases} \tag{1}$$

Notice that all but two entries in $\mathrm{DP}_H[\cdot,\cdot]$ are 0, and by Eq. (1) we can fill the table $\mathrm{DP}_H[\cdot,\cdot]$ in time $\mathcal{O}^\star(3^{\mathrm{cw}})$.

Joining labels with edges, $H = \eta_{i,j}(H')$. For each pair of distinct $i, j \in [\mathrm{cw}]$, we define a partial function $f_{i,j} \colon \{0,1,2\}^{\mathrm{cw}} \to \{0,1,2\}^{\mathrm{cw}}$. In particular, for $\sigma' \in \{0,1,2\}^{\mathrm{cw}}$ we set $f_{i,j}(\sigma') = \sigma$ such that $\sigma(w) = \sigma'(w)$ for all $w \in [\mathrm{cw}] \setminus \{i,j\}$ and

$$(\sigma(i), \sigma(j)) = \begin{cases} (\sigma'(i), \sigma'(j)) & \text{if } 0 \in \{\sigma'(i), \sigma'(j)\}, \\ (1,1) & \text{if } \sigma'(i) = \sigma'(j) = 2. \end{cases}$$

For a fixed size $k \in [0,n]$ and signature $\sigma \in \{0,1,2\}^{\mathrm{cw}}$ we set the value of $\mathrm{DP}_H[k,\sigma]$ to be

$$\mathrm{DP}_H[k,\sigma] = \sum_{\sigma' \in f_{i,j}^{-1}(\sigma)} \mathrm{DP}_{H'}[k,\sigma']. \tag{2}$$

If $f_{i,j}^{-1}(\sigma) = \varnothing$ we set $\mathrm{DP}_H[k,\sigma] = 0$. Notice that computing $f_{i,j}^{-1}(\sigma)$ requires $\mathcal{O}(\mathrm{cw})$ time, thus by Eq. (2) we can fill the table $\mathrm{DP}_H[\cdot,\cdot]$ in time $\mathcal{O}^\star(3^{\mathrm{cw}})$.

Lemma 5 ($\star$). *Let $H = \eta_{i,j}(H')$. Assume that for all $k \in [0,n]$ and $\sigma \in \{0,1,2\}^{\mathrm{cw}}$, $\mathrm{DP}_{H'}[k,\sigma]$ is equal to the number of partial matchings S of H' such that $|S| = k$ and $\mathrm{sgn}_{H'}(S) = \sigma$. Then, $\mathrm{DP}_H[k,\sigma]$ is equal to the number of partial matchings S of H such that $|S| = k$ and $\mathrm{sgn}_H(S) = \sigma$.*

Relabeling, $H = \rho_{i\to j}(H')$. For each pair of distinct $i, j \in [\mathrm{cw}]$, we define a partial function $g_{i\to j} \colon \{0,1,2\}^{\mathrm{cw}} \to \{0,1,2\}^{\mathrm{cw}}$. In particular, for $\sigma' \in \{0,1,2\}^{\mathrm{cw}}$ we set $g_{i\to j}(\sigma') = \sigma$ such that $\sigma(w) = \sigma'(w)$ for all $w \in [\mathrm{cw}] \setminus \{i,j\}$, $\sigma(i) = 0$, and

$$\sigma(j) = \begin{cases} \max\{\sigma'(i), \sigma'(j)\} & \text{if } 0 \in \{\sigma'(i), \sigma'(j)\}, \\ 1 & \text{if } \sigma'(i) = \sigma'(j) = 1. \end{cases}$$

For a fixed size $k \in [0,n]$ and signature $\sigma \in \{0,1,2\}^{\mathrm{cw}}$ we set the value of $\mathrm{DP}_H[k,\sigma]$ to be

$$\mathrm{DP}_H[k,\sigma] = \sum_{\sigma' \in g_{i\to j}^{-1}(\sigma)} \mathrm{DP}_{H'}[k,\sigma']. \tag{3}$$

If $g_{i\to j}^{-1}(\sigma) = \varnothing$ we set $\mathrm{DP}_H[k,\sigma] = 0$. Notice that computing $g_{i\to j}^{-1}(\sigma)$ requires $\mathcal{O}(\mathrm{cw})$ time, thus by Eq. (3) we can fill the table $\mathrm{DP}_H[\cdot,\cdot]$ in time $\mathcal{O}^\star(3^{\mathrm{cw}})$.

Lemma 6 (⋆). *Let $H = \rho_{i\to j}(H')$. Assume that for all $k \in [0,n]$ and $\sigma \in \{0,1,2\}^{\mathrm{cw}}$, $\mathrm{DP}_{H'}[k,\sigma]$ is equal to the number of partial matchings S of H' such that $|S| = k$ and $\mathrm{sgn}_{H'}(S) = \sigma$. Then, $\mathrm{DP}_H[k,\sigma]$ is equal to the number of partial matchings S of H such that $|S| = k$ and $\mathrm{sgn}_H(S) = \sigma$.*

Disjoint Union, $H = H_1 \oplus H_2$. For the Union nodes, an approach similar to the previous cases would yield a total running time of $\mathcal{O}^\star(9^{\mathrm{cw}})$ to populate the table $\mathrm{DP}_H[\cdot,\cdot]$, which can be further improved to $\mathcal{O}^\star(6^{\mathrm{cw}})$ by some slightly more careful analysis. In order to bring this down to $\mathcal{O}^\star(3^{\mathrm{cw}})$ we proceed in a different way by using the techniques introduced in [6,27,65].

We start with some definitions. Let $\tilde{\sigma} \in \{0,\tilde{1},2\}^{\mathrm{cw}}$ be a tuple with cw entries, each taking values over the set $\{0,\tilde{1},2\}$. We call any such tuple a *pseudosignature* and we say that the signature $\sigma \in \{0,1,2\}^{\mathrm{cw}}$ is *compatible* with the pseudosignature $\tilde{\sigma}$ if for all $w \in [\mathrm{cw}]$ it holds that (i) if $\tilde{\sigma}(w) = 0$ then $\sigma(w) = 0$, (ii) if $\tilde{\sigma}(w) = \tilde{1}$ then $\sigma(w) \in \{0,1\}$, and (iii) if $\tilde{\sigma}(w) = 2$ then $\sigma(w) = 2$. Let for $H' \in \mathcal{H}$, $\widetilde{\mathrm{DP}}_{H'}[\cdot,\cdot]$ be the table which in entry $\widetilde{\mathrm{DP}}_{H'}[k,\tilde{\sigma}]$ contains the number of partial matchings in H' of size k whose signatures are compatible with $\tilde{\sigma}$. Our main idea is to first compute the tables $\widetilde{\mathrm{DP}}_{H_1}[\cdot,\cdot]$ and $\widetilde{\mathrm{DP}}_{H_2}[\cdot,\cdot]$, then from those compute the table $\widetilde{\mathrm{DP}}_H[\cdot,\cdot]$, and lastly use the latter to populate the table $\mathrm{DP}_H[\cdot,\cdot]$.

Step 1. Let $H_i \in \{H_1, H_2\}$. For $\lambda \in [0,\mathrm{cw}]$ we say that $\tilde{\sigma}_{\mathsf{m}} \in \{0,\tilde{1},2\}^\lambda \times \{0,1,2\}^{\mathrm{cw}-\lambda}$ is a λ-*mixed signature*. We say that the signature $\sigma \in \{0,1,2\}^{\mathrm{cw}}$ is *compatible* with the λ-mixed signature $\tilde{\sigma}_{\mathsf{m}}$ if the following hold:

- For all $w \in [\lambda]$ it holds that (i) if $\tilde{\sigma}_{\mathsf{m}}(w) = 0$ then $\sigma(w) = 0$, (ii) if $\tilde{\sigma}_{\mathsf{m}}(w) = \tilde{1}$ then $\sigma(w) \in \{0,1\}$, and (iii) if $\tilde{\sigma}_{\mathsf{m}}(w) = 2$ then $\sigma(w) = 2$.
- For all $w \in [\lambda+1,\mathrm{cw}]$, $\sigma(w) = \tilde{\sigma}_{\mathsf{m}}(w)$.

We proceed inductively. For $\lambda \in [0,\mathrm{cw}]$ let $\widetilde{\mathrm{DP}}^{\lambda}_{H_i}[\cdot,\cdot]$ denote a table taking values over $[0,n] \times \{0,\tilde{1},2\}^\lambda \times \{0,1,2\}^{\mathrm{cw}-\lambda}$, where intuitively $\widetilde{\mathrm{DP}}^{\lambda}_{H_i}[k,\tilde{\sigma}_{\mathsf{m}}]$ is equal to the number of partial matchings of H_i of size k whose signature is compatible with $\tilde{\sigma}_{\mathsf{m}}$. For $\lambda = 0$, this is precisely the table $\mathrm{DP}_{H_i}[\cdot,\cdot]$, while for $\lambda = \mathrm{cw}$ this will be the table $\widetilde{\mathrm{DP}}_{H_i}[\cdot,\cdot]$.

It suffices to explain how to compute $\widetilde{\mathrm{DP}}^{\lambda+1}_{H_i}[\cdot,\cdot]$ from the table $\widetilde{\mathrm{DP}}^{\lambda}_{H_i}[\cdot,\cdot]$ in time $\mathcal{O}^\star(3^{\mathrm{cw}})$, and repeat the whole procedure cw times. Let $k \in [0,n]$ and $\tilde{\sigma}_{\mathsf{m}} \in \{0,\tilde{1},2\}^{\lambda+1} \times \{0,1,2\}^{\mathrm{cw}-(\lambda+1)}$ be a $(\lambda+1)$-mixed signature. Consider two cases. If $\tilde{\sigma}_{\mathsf{m}}(\lambda+1) \in \{0,2\}$, then set $\widetilde{\mathrm{DP}}^{\lambda+1}_{H_i}[k,\tilde{\sigma}_{\mathsf{m}}] = \widetilde{\mathrm{DP}}^{\lambda}_{H_i}[k,\tilde{\sigma}_{\mathsf{m}}]$. Otherwise, it holds that $\tilde{\sigma}_{\mathsf{m}}(\lambda+1) = \tilde{1}$, and let $\tilde{\sigma}'_{\mathsf{m}}, \tilde{\sigma}''_{\mathsf{m}} \in \{0,\tilde{1},2\}^\lambda \times \{0,1,2\}^{\mathrm{cw}-\lambda}$ denote the λ-mixed signatures obtained from $\tilde{\sigma}_{\mathsf{m}}$ by substituting the $(\lambda+1)$-th entry with 1 and 0 respectively, that is, $\tilde{\sigma}'_{\mathsf{m}} = \tilde{\sigma}_{\mathsf{m}}[(\lambda+1) \mapsto 0]$ and $\tilde{\sigma}''_{\mathsf{m}} = \tilde{\sigma}_{\mathsf{m}}[(\lambda+1) \mapsto 1]$. In that case, set $\widetilde{\mathrm{DP}}^{\lambda+1}_{H_i}[k,\tilde{\sigma}_{\mathsf{m}}] = \widetilde{\mathrm{DP}}^{\lambda}_{H_i}[k,\tilde{\sigma}'_{\mathsf{m}}] + \widetilde{\mathrm{DP}}^{\lambda}_{H_i}[k,\tilde{\sigma}''_{\mathsf{m}}]$.

Step 2. Utilizing the tables $\widetilde{\mathrm{DP}}_{H_1}[\cdot,\cdot]$ and $\widetilde{\mathrm{DP}}_{H_2}[\cdot,\cdot]$ constructed in Step 1, along with making use of the FFT technique introduced by Cygan and Pilipczuk [27], we compute in time $\mathcal{O}^\star(3^{\mathrm{cw}})$ the table $\widetilde{\mathrm{DP}}_H[\cdot,\cdot]$, also indexed over $[0,n] \times \{0,\tilde{1},2\}^{\mathrm{cw}}$. Initially we set every entry of said table to 0.

In the following, fix $k \in [0,n]$ and a subset $A \subseteq [\mathrm{cw}]$. Let $\mathcal{S}_A \subseteq \{0,\tilde{1},2\}^{\mathrm{cw}}$ denote the set of pseudosignatures such that $\tilde{\sigma} \in \mathcal{S}_A$ iff for all $w \in [\mathrm{cw}]$, $\tilde{\sigma}(w) = \tilde{1} \iff w \in A$. We show how to compute all entries $\widetilde{\mathrm{DP}}_H[k,\tilde{\sigma}]$ with $\tilde{\sigma} \in \mathcal{S}_A$ in time $\mathcal{O}^\star(2^{\mathrm{cw}-|A|})$. Since $\sum_{A\subseteq[\mathrm{cw}]} 2^{\mathrm{cw}-|A|} = 3^{\mathrm{cw}}$ and k takes $n+1$ different values, this suffices for the stated time bound. Let $w_1,\ldots,w_{\mathrm{cw}-|A|}$ denote in increasing order the non $\tilde{1}$-valued coordinates of a pseudosignature in $\mathcal{S}_A$, that is, for all $j \in [\mathrm{cw}-|A|]$ and $\tilde{\sigma} \in \mathcal{S}_A$, we have that $\tilde{\sigma}(w_j) \neq \tilde{1}$.

We first fix $k_1, k_2 \in [0,k]$ such that $k_1 + k_2 = k$. Notice that there are $\mathcal{O}(n)$ such pairs of k_1, k_2. Intuitively, k_i represents the number of vertices of the partial matching of H_i that account towards the total size of the partial matching in H. Next we define a *binary encoding* $\mathrm{bin}_A \colon \mathcal{S}_A \to [0, 2^{(\mathrm{cw}-|A|)} - 1]$ of every pseudosignature in $\mathcal{S}_A$ such that for all $\tilde{\sigma} \in \mathcal{S}_A$ the j-th bit of the binary representation of $\mathrm{bin}_A(\tilde{\sigma})$, denoted by $\mathrm{bin}_{A,j}(\tilde{\sigma})$, is equal to

$$\mathrm{bin}_{A,j}(\tilde{\sigma}) = \begin{cases} 0 & \text{if } \tilde{\sigma}(w_j) = 0, \\ 1 & \text{if } \tilde{\sigma}(w_j) = 2. \end{cases}$$

For every entry $\widetilde{\mathrm{DP}}_{H_i}[k_i,\tilde{\sigma}]$ with $\tilde{\sigma} \in \mathcal{S}_A$ we introduce a monomial $\widetilde{\mathrm{DP}}_{H_i}[k_i,\tilde{\sigma}] \cdot x^{\mathrm{bin}_A(\tilde{\sigma})}$, and we set P_{A,H_i,k_i} to denote the polynomial comprised of all such monomials, that is,

$$P_{A,H_i,k_i} = \sum_{\tilde{\sigma}\in\mathcal{S}_A} \widetilde{\mathrm{DP}}_{H_i}[k_i,\tilde{\sigma}] \cdot x^{\mathrm{bin}_A(\tilde{\sigma})}.$$

Notice that the degree of P_{A,H_i,k_i} is at most $2^{\mathrm{cw}-|A|}$, while all coefficients (i.e., the number of partial solutions) are upper-bounded by 2^n. Let $P^{r_i}_{A,H_i,k_i}$ denote the restriction of P_{A,H_i,k_i} to monomials whose degree has *exactly* r_i out of the $\mathrm{cw}-|A|$ bits being 1. Fix $r_1, r_2 \in [0, \mathrm{cw}-|A|]$ and multiply polynomials $P^{r_1}_{A,H_1,k_1}$ and $P^{r_2}_{A,H_2,k_2}$ in $\mathcal{O}^\star(2^{\mathrm{cw}-|A|})$ time using FFT.[4] We iterate over all $\tilde{\sigma} \in \mathcal{S}_A$ having exactly $r_1 + r_2$ bits of $\mathrm{bin}_A(\tilde{\sigma})$ being 1 and check what coefficient, say s, stands in front of $x^{\mathrm{bin}_A(\tilde{\sigma})}$ in the resulting polynomial. We update the value of $\widetilde{\mathrm{DP}}_H[k,\tilde{\sigma}]$ by adding to its previous value $+s$. We repeat over all $r_1, r_2 \in [0, \mathrm{cw}-|A|]$, $k_1, k_2 \in [0,k]$, and $k \in [0,n]$, needing in total $\mathcal{O}^\star(2^{\mathrm{cw}-|A|})$ time. Finally, we fill the whole table $\widetilde{\mathrm{DP}}_H[\cdot,\cdot]$ in time $\mathcal{O}^\star(3^{\mathrm{cw}})$ by repeating over all $A \subseteq [\mathrm{cw}]$.

Step 3. Given $\widetilde{\mathrm{DP}}_H[\cdot,\cdot]$, we populate the values of $\mathrm{DP}_H[\cdot,\cdot]$ by repeating a procedure analogous to the one described in Step 1. The total running time is once again $\mathcal{O}^\star(3^{\mathrm{cw}})$.

---

[4] Recall that using FFT we can multiply two polynomials of degree n whose coefficients are upper-bounded by $2^{\mathcal{O}(n)}$ in $\mathcal{O}(n \log n)$ time [58].

Lemma 7 (⋆). *Let $H = H_1 \oplus H_2$. Assume that for all $i \in \{1,2\}$, $k \in [0,n]$, and $\sigma \in \{0,1,2\}^{\mathrm{cw}}$, $\mathrm{DP}_{H_i}[k,\sigma]$ is equal to the number of partial matchings S of H_i such that $|S| = k$ and $\mathrm{sgn}_{H_i}(S) = \sigma$, Then, $\mathrm{DP}_H[k,\sigma]$ is equal to the number of partial matchings S of H such that $|S| = k$ and $\mathrm{sgn}_H(S) = \sigma$.*

Wrapping Up. We argue about the correctness of the algorithm by a bottom-up induction on the clique-width cw-expression of G. We claim that for all $H \in \mathcal{H}$, $k \in [0,k]$, and $\sigma \in \{0,1,2\}^{\mathrm{cw}}$, $\mathrm{DP}_H[k,\sigma]$ is equal to the number of partial matchings in H of size k and signature σ. Let $H = i(v)$ be a singleton graph for some label $i \in [\mathrm{cw}]$. There are only two possible partial matchings in H, the first being $\{v\}$ and the second being $\varnothing$. Notice that then by Eq. (1), the statement holds for singleton graphs. For the Join, Relabel, and Union nodes the statement holds due to Lemmas 5 to 7. As for the algorithm's running time, notice that for any cw-labeled graph $H \in \mathcal{H}$ arising in ψ, the table $\mathrm{DP}_H[\cdot,\cdot]$ can be filled in time $\mathcal{O}^\star(3^{\mathrm{cw}})$. As the number of such graphs is $\mathcal{O}(|\psi|)$, this implies that our algorithm runs in time $\mathcal{O}^\star(3^{\mathrm{cw}})$. Finally, notice that for all $H \in \mathcal{H}$, the number of induced matchings of H of size ℓ is equal to the sum of $\mathrm{DP}_H[2\ell,\sigma]$ over all $\sigma \in \{0,1\}^{\mathrm{cw}}$. This concludes the proof of Theorem 4.

4 Acyclic Matching

In this section we consider the Acyclic Matching problem. Chaudhary and Zehavi [18] recently gave an $\mathcal{O}^\star(6^{\mathrm{tw}})$ algorithm for this problem, where tw denotes the treewidth of the input graph. We first notice that, with some slightly more careful analysis of its running time, we can show that the algorithm in fact runs in time $\mathcal{O}^\star(5^{\mathrm{tw}})$. Our main contribution however is to show in Sect. 4.1 that this is optimal under the pw-SETH, even for the more restricted parameterization by pathwidth. Lastly, in Sect. 4.2 we employ the rank-based approach introduced by Bodlaender, Cygan, Kratsch, and Nederlof [5] and adapted to the parameterization by clique-width by Bergougnoux and Kanté [4] to develop a $2^{\mathcal{O}(\mathrm{cw})}n^{\mathcal{O}(1)}$ algorithm for Acyclic Matching parameterized by the clique-width cw of the input graph.

Theorem 8 (⋆). *Given a graph G along with a nice tree decomposition of G of width tw, there is an algorithm that computes the size of the maximum acyclic matching in G in time $\mathcal{O}^\star(5^{\mathrm{tw}})$. The algorithm is randomized, cannot give false positives, and may give false negatives with probability at most 1/3.*

4.1 Lower Bound

Here we state the following lower bound for Acyclic Matching.

Theorem 9 (⋆). *For any constant $\varepsilon > 0$, there is no $\mathcal{O}^\star((5-\varepsilon)^{\mathrm{pw}})$ algorithm deciding* Acyclic Matching, *where* pw *denotes the pathwidth of the input graph, unless the pw-SETH is false.*

4.2 Parameterization by Clique-Width

Before describing our algorithm, we first define some necessary notions.

Partitions. A *partition* p of a set L is a collection of non-empty subsets of L that are pairwise non-intersecting and such that $\bigcup_{p_i \in p} p_i = L$; each set in p is called a *block* of p. The set of partitions of a finite set L is denoted by $\Pi(L)$, and $(\Pi(L), \sqsubseteq)$ forms a lattice where $p \sqsubseteq q$ if for each block p_i of p there is a block q_j of q with $p_i \subseteq q_j$. The join operation of this lattice is denoted by $\sqcup$. For example, we have $\{\{1,2\},\{3,4\},\{5\}\} \sqcup \{\{1\},\{2,3\},\{4\},\{5\}\} = \{\{1,2,3,4\},\{5\}\}$. Let $\#\,\mathrm{block}(p)$ denote the number of blocks of a partition p. Observe that $\varnothing$ is the only partition of the empty set. A *weighted partition* is an element of $\Pi(L) \times \mathbb{N}$ for some finite set L. For $p \in \Pi(L)$ and $X \subseteq L$, let $p_{\downarrow X} \in \Pi(X)$ be the partition $\{p_i \cap X \mid p_i \in p\} \setminus \{\varnothing\}$, and for a set Y, let $p_{\uparrow Y} \in \Pi(L \cup Y)$ be the partition $p \cup \left(\bigcup_{y \in Y \setminus L} \{\{y\}\}\right)$.

Theorem 10 (⋆). *There is an algorithm that, given an integer ℓ as well as a graph G along with an irredundant clique-width expression ψ of G of width* cw, *determines whether G has an acyclic matching of size at least ℓ in time $2^{\mathcal{O}(\mathrm{cw})} n^{\mathcal{O}(1)}$.*

Proof (sketch). The proof follows along the lines of the $2^{\mathcal{O}(\mathrm{cw})} n^{\mathcal{O}(1)}$ algorithm for Feedback Vertex Set by Bergougnoux and Kanté [4]. There, the authors develop a general framework, expanding upon the one of Bodlaender et al. [5], in order to cope with various problems with connectivity constraints on graphs of bounded clique-width. Using their framework, we can perform DP over the clique-width expression and, starting from a partial solution, eventually build an optimal one.

Instead of Acyclic Matching, we solve the equivalent problem of asking, given a graph G and an integer k, whether there exists $S \subseteq V(G)$ of size $|S| \geq k$ such that $G[S]$ is a forest that contains a perfect matching. Notice that G has an acyclic matching of size ℓ if and only if there exists such a set S of size $|S| \geq 2\ell$. In the following let $\mathcal{H}$ denote the set of all cw-labeled graphs generated by subexpressions of the given clique-width expression ψ of the input graph G.

In order to deal with Feedback Vertex Set, in [4] the authors solve the Maximum Induced Tree problem and introduce a *signature* on the partial solutions, that encodes (i) the number of vertices that a partial solution contains from each label class (0, 1, or at least 2), and (ii) in case a solution contains at least 2 vertices of the same label class, whether this label class will partake in a join operation with a non-empty set in the future; as a matter of fact, the number of such join operations can be at most 1, as otherwise a cycle is formed. For our problem, we have that a partial solution of $H \in \mathcal{H}$ is a pair (S, M) such that M is a matching of H with $V_M \subseteq S \subseteq V(H)$. In that case, the signature of a partial solution is, for each label $i \in [\mathrm{cw}]$, whether $S \cap \mathrm{lab}_H^{-1}(i)$ is empty, a singleton, or contains at least 2 vertices, and whether $(S \setminus V_M) \cap \mathrm{lab}_H^{-1}(i)$ is empty or not. We notice that in fact $|(S \setminus V_M) \cap \mathrm{lab}_H^{-1}(i)| \leq 1$, thus leading to 6 different states $\Gamma = \{\gamma_0, \gamma_{\tilde{1}}, \gamma_1, \gamma_{\tilde{2}}, \gamma_2, \gamma_{-2}\}$ per label class, with the index denoting the number of vertices in $S \cap \text{lab}_H^{-1}(i)$, the tilde denoting that $|(S \setminus V_M) \cap \text{lab}_H^{-1}(i)| = 1$, and γ_2 (resp., γ_{-2}) denoting that the label class will (resp., will not) partake in a join operation in the future. This, along with the machinery of the framework of [4], allows us to detect any possible cycles as well as forcing the final solution to be a forest containing a perfect matching.

5 Conclusion

In this paper we studied the Induced Matching and Acyclic Matching problems under the perspective of fine-grained structural parameterized complexity. In particular, we presented tight lower bounds under the pw-SETH for the pathwidth parameterization of both problems (equivalence even in the case of Induced Matching), rendering known treewidth algorithms optimal. Furthermore, we considered the parameterization by clique-width and developed a tight $\mathcal{O}^\star(3^{\text{cw}})$ algorithm for Induced Matching, as well as an $2^{\mathcal{O}(\text{cw})}n^{\mathcal{O}(1)}$ algorithm for Acyclic Matching, which is optimal under the ETH. As a direction for future work, it would be interesting to develop an algorithm for Acyclic Matching parameterized by clique-width that is optimal under the (pw-)SETH. Although recently there has been a flurry of results regarding optimal algorithms for connectivity problems parameterized by clique-width [7,8,41], the case of Feedback Vertex Set, which is closely related to Acyclic Matching, remains open.

References

1. Baste, J., Fürst, M., Rautenbach, D.: Linear programming based approximation for unweighted induced matchings - breaking the Δ barrier. Discret. Optim. **38**, 100593 (2020). https://doi.org/10.1016/J.DISOPT.2020.100593
2. Baste, J., Fürst, M., Rautenbach, D.: Acyclic matchings in graphs of bounded maximum degree. Discret. Math. **345**(7), 112885 (2022). https://doi.org/10.1016/J.DISC.2022.112885
3. Baste, J., Rautenbach, D.: Degenerate matchings and edge colorings. Discret. Appl. Math. **239**, 38–44 (2018). https://doi.org/10.1016/J.DAM.2018.01.002
4. Bergougnoux, B., Kanté, M.M.: Fast exact algorithms for some connectivity problems parameterized by clique-width. Theor. Comput. Sci. **782**, 30–53 (2019). https://doi.org/10.1016/J.TCS.2019.02.030
5. Bodlaender, H.L., Cygan, M., Kratsch, S., Nederlof, J.: Deterministic single exponential time algorithms for connectivity problems parameterized by treewidth. Inf. Comput. **243**, 86–111 (2015). https://doi.org/10.1016/J.IC.2014.12.008
6. Bodlaender, H.L., van Leeuwen, E.J., van Rooij, J.M.M., Vatshelle, M.: Faster algorithms on branch and clique decompositions. In: Hliněný, P., Kučera, A. (eds.) MFCS 2010. LNCS, vol. 6281, pp. 174–185. Springer, Heidelberg (2010). https://doi.org/10.1007/978-3-642-15155-2_17
7. Bojikian, N., Kratsch, S.: Tight algorithm for connected odd cycle transversal parameterized by clique-width (2024), https://arxiv.org/abs/2402.08046

8. Bojikian, N., Kratsch, S.: A tight monte-carlo algorithm for steiner tree parameterized by clique-width. In: 51st International Colloquium on Automata, Languages, and Programming, ICALP 2024. LIPIcs, vol. 297, pp. 29:1–29:18. Schloss Dagstuhl - Leibniz-Zentrum für Informatik (2024). https://doi.org/10.4230/LIPICS.ICALP.2024.29
9. Brandstädt, A., Eschen, E.M., Sritharan, R.: The induced matching and chain subgraph cover problems for convex bipartite graphs. Theor. Comput. Sci. **381**(1–3), 260–265 (2007). https://doi.org/10.1016/J.TCS.2007.04.006
10. Brandstädt, A., Hoàng, C.T.: Maximum induced matchings for chordal graphs in linear time. Algorithmica **52**(4), 440–447 (2008). https://doi.org/10.1007/S00453-007-9045-2
11. Cameron, K.: Induced matchings. Discret. Appl. Math. **24**(1–3), 97–102 (1989). https://doi.org/10.1016/0166-218X(92)90275-F
12. Cameron, K., Sritharan, R., Tang, Y.: Finding a maximum induced matching in weakly chordal graphs. Discret. Math. **266**(1–3), 133–142 (2003). https://doi.org/10.1016/S0012-365X(02)00803-8
13. Chalermsook, P., Laekhanukit, B., Nanongkai, D.: Graph products revisited: tight approximation hardness of induced matching, poset dimension and more. In: Proceedings of the Twenty-Fourth Annual ACM-SIAM Symposium on Discrete Algorithms, SODA 2013, pp. 1557–1576. SIAM (2013). https://doi.org/10.1137/1.9781611973105.112
14. Chalermsook, P., Laekhanukit, B., Nanongkai, D.: Independent set, induced matching, and pricing: Connections and tight (subexponential time) approximation hardnesses. In: 54th Annual IEEE Symposium on Foundations of Computer Science, FOCS 2013, pp. 370–379. IEEE Computer Society (2013). https://doi.org/10.1109/FOCS.2013.47
15. Chang, J.: Induced matchings in asteroidal triple-free graphs. Discret. Appl. Math. **132**(1–3), 67–78 (2003). https://doi.org/10.1016/S0166-218X(03)00390-1
16. Chang, M., Chen, L., Hung, L.: Moderately exponential time algorithms for the maximum induced matching problem. Optim. Lett. **9**(5), 981–998 (2015). https://doi.org/10.1007/S11590-014-0813-Z
17. Chaudhary, J., Zehavi, M.: *P*-matchings parameterized by treewidth. In: Graph-Theoretic Concepts in Computer Science - 49th International Workshop, WG 2023, LNS, vol. 14093, pp. 217–231. Springer (2023). https://doi.org/10.1007/978-3-031-43380-1_16
18. Chaudhary, J., Zehavi, M.: P-matchings parameterized by treewidth. SIAM J. Discret. Math. **39**(2), 1280–1311 (2025). https://doi.org/10.1137/23M160013X
19. Chaudhary, J., Zehavi, M.: Parameterized results on acyclic matchings with implications for related problems. J. Comput. Syst. Sci. **148**, 103599 (2025). https://doi.org/10.1016/J.JCSS.2024.103599
20. Chlebík, M., Chlebíková, J.: Complexity of approximating bounded variants of optimization problems. Theor. Comput. Sci. **354**(3), 320–338 (2006). https://doi.org/10.1016/J.TCS.2005.11.029
21. Chlebík, M., Chlebíková, J.: Approximation hardness of dominating set problems in bounded degree graphs. Inf. Comput. **206**(11), 1264–1275 (2008). https://doi.org/10.1016/J.IC.2008.07.003
22. Courcelle, B.: The monadic second-order logic of graphs. I. Recognizable sets of finite graphs. Inf. Comput. **85**(1), 12–75 (1990). https://doi.org/10.1016/0890-5401(90)90043-H

23. Courcelle, B., Makowsky, J.A., Rotics, U.: Linear time solvable optimization problems on graphs of bounded clique-width. Theory Comput. Syst. **33**(2), 125–150 (2000). https://doi.org/10.1007/S002249910009
24. Courcelle, B., Olariu, S.: Upper bounds to the clique width of graphs. Discret. Appl. Math. **101**(1–3), 77–114 (2000). https://doi.org/10.1016/S0166-218X(99)00184-5
25. Cygan, M., et al.: On problems as hard as CNF-SAT. ACM Trans. Algorithms **12**(3), 41:1–41:24 (2016). https://doi.org/10.1145/2925416
26. Cygan, M., et al.: Parameterized Algorithms. Springer (2015). https://doi.org/10.1007/978-3-319-21275-3
27. Cygan, M., Pilipczuk, M.: Exact and approximate bandwidth. Theor. Comput. Sci. **411**(40–42), 3701–3713 (2010). https://doi.org/10.1016/j.tcs.2010.06.018
28. Dabrowski, K.K., Demange, M., Lozin, V.V.: New results on maximum induced matchings in bipartite graphs and beyond. Theor. Comput. Sci. **478**, 33–40 (2013). https://doi.org/10.1016/J.TCS.2013.01.027
29. Demange, M., Ekim, T.: A note on the NP-hardness of two matching problems in induced subgrids. Discret. Math. Theor. Comput. Sci. **15**(2), 233–242 (2013). https://doi.org/10.46298/DMTCS.606
30. Diestel, R.: Graph theory, graduate texts in mathematics, vol. 173. Springer (2017). https://doi.org/10.1007/978-3-662-53622-3
31. Duckworth, W., Manlove, D.F., Zito, M.: On the approximability of the maximum induced matching problem. J. Discrete Algorithms **3**(1), 79–91 (2005). https://doi.org/10.1016/J.JDA.2004.05.001
32. Elbassioni, K.M., Raman, R., Ray, S., Sitters, R.: On the approximability of the maximum feasible subsystem problem with 0/1-coefficients. In: Proceedings of the Twentieth Annual ACM-SIAM Symposium on Discrete Algorithms, SODA 2009, pp. 1210–1219. SIAM (2009). https://doi.org/10.1137/1.9781611973068.131
33. Erman, R., Kowalik, L., Krnc, M., Walen, T.: Improved induced matchings in sparse graphs. Discret. Appl. Math. **158**(18), 1994–2003 (2010). https://doi.org/10.1016/J.DAM.2010.08.026
34. Fürst, M., Leichter, M., Rautenbach, D.: Locally searching for large induced matchings. Theor. Comput. Sci. **720**, 64–72 (2018). https://doi.org/10.1016/J.TCS.2018.02.006
35. Fürst, M., Rautenbach, D.: On some hard and some tractable cases of the maximum acyclic matching problem. Ann. Oper. Res. **279**(1–2), 291–300 (2019). https://doi.org/10.1007/S10479-019-03311-1
36. Goddard, W., Hedetniemi, S.M., Hedetniemi, S.T., Laskar, R.C.: Generalized subgraph-restricted matchings in graphs. Discret. Math. **293**(1–3), 129–138 (2005). https://doi.org/10.1016/J.DISC.2004.08.027
37. Golumbic, M.C., Lewenstein, M.: New results on induced matchings. Discret. Appl. Math. **101**(1–3), 157–165 (2000). https://doi.org/10.1016/S0166-218X(99)00194-8
38. Gomes, G.C.M., Masquio, B.P., Pinto, P.E.D., dos Santos, V.F., Szwarcfiter, J.L.: Disconnected matchings. Theor. Comput. Sci. **956**, 113821 (2023). https://doi.org/10.1016/J.TCS.2023.113821
39. Habib, M., Mouatadid, L.: Maximum induced matching algorithms via vertex ordering characterizations. Algorithmica **82**(2), 260–278 (2020). https://doi.org/10.1007/S00453-018-00538-5
40. Hajebi, S., Javadi, R.: On the parameterized complexity of the acyclic matching problem. Theor. Comput. Sci. **958**, 113862 (2023). https://doi.org/10.1016/J.TCS.2023.113862

41. Hegerfeld, F., Kratsch, S.: Tight algorithms for connectivity problems parameterized by clique-width. In: 31st Annual European Symposium on Algorithms, ESA 2023. LIPIcs, vol. 274, pp. 59:1–59:19. Schloss Dagstuhl - Leibniz-Zentrum für Informatik (2023). https://doi.org/10.4230/LIPICS.ESA.2023.59
42. Kanj, I.A., Pelsmajer, M.J., Schaefer, M., Xia, G.: On the induced matching problem. J. Comput. Syst. Sci. **77**(6), 1058–1070 (2011). https://doi.org/10.1016/J.JCSS.2010.09.001
43. Klemz, B., Rote, G.: Linear-time algorithms for maximum-weight induced matchings and minimum chain covers in convex bipartite graphs. Algorithmica **84**(4), 1064–1080 (2022). https://doi.org/10.1007/S00453-021-00904-W
44. Ko, C.W., Shepherd, F.B.: Bipartite domination and simultaneous matroid covers. SIAM J. Discret. Math. **16**(4), 517–523 (2003). https://doi.org/10.1137/S089548019828371X
45. Koana, T.: Induced matching below guarantees: average paves the way for fixed-parameter tractability. In: 40th International Symposium on Theoretical Aspects of Computer Science, STACS 2023. LIPIcs, vol. 254, pp. 39:1–39:21. Schloss Dagstuhl - Leibniz-Zentrum für Informatik (2023). https://doi.org/10.4230/LIPICS.STACS.2023.39
46. Kobler, D., Rotics, U.: Finding maximum induced matchings in subclasses of claw-free and P_5-free graphs, and in graphs with matching and induced matching of equal maximum size. Algorithmica **37**(4), 327–346 (2003). https://doi.org/10.1007/S00453-003-1035-4
47. Krishnamurthy, C.M., Sritharan, R.: Maximum induced matching problem on hhd-free graphs. Discret. Appl. Math. **160**(3), 224–230 (2012). https://doi.org/10.1016/J.DAM.2011.08.027
48. Kumar, R., Mahadevan, U., Sivakumar, D.: A graph-theoretic approach to extract storylines from search results. In: Proceedings of the Tenth ACM SIGKDD International Conference on Knowledge Discovery and Data Mining, 2004, pp. 216–225. ACM (2004). https://doi.org/10.1145/1014052.1014078
49. Lampis, M.: Finer tight bounds for coloring on clique-width. SIAM J. Discret. Math. **34**(3), 1538–1558 (2020). https://doi.org/10.1137/19M1280326
50. Lampis, M.: Circuits and backdoors: Five shades of the SETH (2024), https://arxiv.org/abs/2407.09683
51. Lampis, M.: The primal pathwidth SETH. In: Proceedings of the 2025 Annual ACM-SIAM Symposium on Discrete Algorithms, SODA 2025, pp. 1494–1564. SIAM (2025). https://doi.org/10.1137/1.9781611978322.47
52. Lampis, M., Vasilakis, M.: Structural parameterizations for two bounded degree problems revisited. ACM Trans. Comput. Theory **16**(3), 17:1–17:51 (2024). https://doi.org/10.1145/3665156
53. Lin, M.C., Mestre, J., Vasiliev, S.: Approximating weighted induced matchings. Discret. Appl. Math. **243**, 304–310 (2018). https://doi.org/10.1016/J.DAM.2018.01.009
54. Lovász, L., Plummer, M.D.: Matching theory, vol. 367. American Mathematical Soc. (2009)
55. Lozin, V.V.: On maximum induced matchings in bipartite graphs. Inf. Process. Lett. **81**(1), 7–11 (2002). https://doi.org/10.1016/S0020-0190(01)00185-5
56. Manlove, D.F.: Algorithmics of matching under preferences, series on theoretical computer science, vol. 2. WorldScientific (2013). https://doi.org/10.1142/8591
57. Micali, S., Vazirani, V.V.: An $O(\sqrt{|V|} \cdot |E|)$ algorithm for finding maximum matching in general graphs. In: 21st Annual Symposium on Foundations of Computer

Science, pp. 17–27. IEEE Computer Society (1980). https://doi.org/10.1109/SFCS.1980.12
58. Moenck, R.T.: Practical fast polynomial multiplication. In: Proceedings of the third ACM Symposium on Symbolic and Algebraic Manipulation, SYMSAC 1976, pp. 136–148. ACM (1976). https://doi.org/10.1145/800205.806332
59. Moser, H., Sikdar, S.: The parameterized complexity of the induced matching problem. Discret. Appl. Math. **157**(4), 715–727 (2009). https://doi.org/10.1016/J.DAM.2008.07.011
60. Moser, H., Thilikos, D.M.: Parameterized complexity of finding regular induced subgraphs. J. Discrete Algorithms **7**(2), 181–190 (2009). https://doi.org/10.1016/J.JDA.2008.09.005
61. Orlovich, Y.L., Finke, G., Gordon, V.S., Zverovich, I.E.: Approximability results for the maximum and minimum maximal induced matching problems. Discret. Optim. **5**(3), 584–593 (2008). https://doi.org/10.1016/J.DISOPT.2007.11.010
62. Panda, B.S., Chaudhary, J.: Acyclic matching in some subclasses of graphs. Theor. Comput. Sci. **943**, 36–49 (2023). https://doi.org/10.1016/J.TCS.2022.12.008
63. Panda, B.S., Pradhan, D.: Acyclic matchings in subclasses of bipartite graphs. Discret. Math. Algorithms Appl. **4**(4) (2012). https://doi.org/10.1142/S1793830912500504
64. Rautenbach, D.: Two greedy consequences for maximum induced matchings. Theor. Comput. Sci. **602**, 32–38 (2015). https://doi.org/10.1016/J.TCS.2015.08.002
65. van Rooij, J.M.M., Bodlaender, H.L., Rossmanith, P.: Dynamic programming on tree decompositions using generalised fast subset convolution. In: Fiat, A., Sanders, P. (eds.) ESA 2009. LNCS, vol. 5757, pp. 566–577. Springer, Heidelberg (2009). https://doi.org/10.1007/978-3-642-04128-0_51
66. Stockmeyer, L.J., Vazirani, V.V.: NP-completeness of some generalizations of the maximum matching problem. Inf. Process. Lett. **15**(1), 14–19 (1982). https://doi.org/10.1016/0020-0190(82)90077-1
67. Xiao, M., Kou, S.: Parameterized algorithms and kernels for almost induced matching. Theor. Comput. Sci. **846**, 103–113 (2020). https://doi.org/10.1016/J.TCS.2020.09.026
68. Xiao, M., Tan, H.: Exact algorithms for maximum induced matching. Inf. Comput. **256**, 196–211 (2017). https://doi.org/10.1016/J.IC.2017.07.006
69. Zito, M.: Induced matchings in regular graphs and trees. In: Widmayer, P., Neyer, G., Eidenbenz, S. (eds.) WG 1999. LNCS, vol. 1665, pp. 89–101. Springer, Heidelberg (1999). https://doi.org/10.1007/3-540-46784-X_10

A Structural Description of Zykov and Blanche Descartes Graphs

Malory Marin(✉), Stéphan Thomassé, Nicolas Trotignon, and Rémi Watrigant

ENS de Lyon, CNRS, Université Claude Bernard Lyon 1, LIP, UMR 5668, Lyon 69342, Cedex 07, France
malory.marin@ens-lyon.fr

Abstract. Around 1950, Zykov and Blanche Descartes proposed the two first explicit constructions of triangle-free graphs with arbitrarily large chromatic number. We define a Zykov (resp. Blanche Descartes) graph as any induced subgraph of a graph created using Zykov's construction (resp. Blanche Descartes's). We give a structural characterization of Zykov graphs based on a specific type of stable set, that we call splitting stable set. It implies that recognizing this class is NP-complete, while being FPT in the treewidth of the input graph. We provide similar results for Blanche Descartes graphs. We also observe that any Blanche Descartes graph is a Zykov graph.

1 Introduction

A class of graphs is *hereditary* if it is closed under taking induced subgraphs. A hereditary class of graphs $\mathcal{C}$ is *χ-bounded* if there exists a function f such that every graph G in $\mathcal{C}$ satisfies $\chi(G) \leq f(\omega(G))$. Some classes are χ-bounded and some are not. There has been much research about this topic. Some questions remain open, such as the Gyárfás-Sumner conjecture, see [16].

It turns out that some conjectures about χ-boundedness have been disproved by studying carefully some constructions of triangle-free graphs of arbitrarily large chromatic number defined much before the conjectures were first stated. In particular, the construction of Burling [2], defined in 1965, turned out to be a counter-example to a conjecture of Scott (see [10]), to a conjecture of Trotignon (see [5,12–14]) and to a conjecture of Thomassé, Trotignon and Vušković (see [15]). Also the question of solving several algorithmic problems in non-χ bounded classes attracted some attention (see [15]).

More insight about other classical constructions giving non-χ-bounded classes is therefore of interest. Here, we study the two first constructions of triangle-free graphs of arbitrarily large chromatic number, defined respectively by Zykov [18] in 1949 and Blanche Descartes [6,7] in 1954. We view them as hereditary classes, defined by taking all induced subgraphs of graphs occuring in the constructions. Our main result is a structural characterization of both Zykov and Blanche Descartes graphs (see Theorems 1 and 7). They imply that recognizing both classes is NP-complete (see Theorems 3 and 9), but tractable in polynomial

H. Fernau and P. Kindermann (Eds.): WG 2025, LNCS 16124, pp. 377–389, 2026.
https://doi.org/10.1007/978-3-032-11835-6_27

time when parameterized by the treewidth of the input graph (see Theorems 2 and 8). They also imply that some classical optimization problems are NP-hard for Zykov or Blanche Descartes graphs (see Theorems 10 and 11). We should stress that our motivation is not really algorithmic. We rather view our results as meaningful regarding the structure of Zykov and Blanche Descartes graphs. For instance, we observe that every Blanche Descartes graph is a Zykov graph (see Lemma 9). Our results also allow easy constructions of non-Zykov and non-Blanche Descartes graphs, which is not obvious. In particular we show that there exists non-Zykov graphs of arbitrarily large girth (see Theorem 6). However, NP-completeness makes the existence of a nice description of minimal non-Zykov and minimal non-Blanche Descartes graphs unlikely.

Before giving the formal definitions, let us briefly survey some other famous constructions of triangle-free graphs of high chromatic number. For the famous Mycielski's construction [9], the situation is simple: an easy induction shows that any triangle-free graph is an induced subgraph of some Mycielski graph. Thus, the hereditary class of induced subgraphs of Mycielski graphs is exactly the class of triangle-free graphs, which is obviously recognizable in polynomial time. For Burling graphs, several characterizations are given by Pournajafi and Trotignon [12], and recently, Rzążewski and Walczak [15] proved that recognizing Burling graphs and computing a maximum stable set in a Burling graph are polynomial-time tractable. A subclass of Zykov graphs, the so-called twincut graphs [1], is known to contain graphs of arbitrarily large chromatic number, while being recognizable in polynomial time. The structure of Blanche Descartes graphs has been already investigated by Kostochka and Nešetřil [8], but their results are of a completely different flavour.

Let us now give all the formal definitions.

Zykov Graphs. Zykov's construction is inductive. The graph Z_1 is the graph on one vertex. Suppose that $k \geq 1$ is an integer and that $Z_1, \dots, Z_k$ are all defined. Then Z_{k+1} is obtained as follows:

- Take the disjoint union of $Z_1, \dots, Z_k$.
- For each k-tuple of vertices $v_1 \in V(Z_1), \dots, v_k \in V(Z_k)$, add a vertex v adjacent to $v_1, \dots, v_k$.

For the proofs in the next sections, we denote by S_{k+1} the set of vertices added in the second item of the definition above.

For the sake of completeness, let us recall the main properties of Zykov's construction. It is a routine matter to prove by induction that for all $k \geq 1$, Z_k is triangle-free and k-colorable. To prove that at least k colors are needed to color Z_k, the proof is also by induction. Suppose it is true for $1, \dots, k$ and suppose for a contradiction that Z_{k+1} is k-colorable. In a k-coloring of Z_{k+1}, color 1 must be used in Z_1, color 2 in Z_2, and so on (up to a permutation of colors). Hence, there exists a k-tuple of vertices $v_1 \in V(Z_1), \dots, v_k \in V(Z_k)$ that uses k distinct colors. One more color is needed for the vertex adjacent to this k-tuple. We proved that Z_k is a triangle-free graph of chromatic number k.

A *Zykov graph* is any graph isomorphic to an induced subgraph of Z_k for some integer $k \geq 1$. In Sect. 2, we prove our results about Zykov graphs.

Blanche Descartes Graphs. Blanche Descartes's construction is also inductive. The graph D_1 is the graph on one vertex. Suppose that $k \geq 1$ is an integer and that D_k is defined and has n vertices. Then D_{k+1} is obtained as follows:

- Take a stable set S on $k(n-1)+1$ new vertices.
- For each n-tuple of vertices T in S, add a copy of D_k and a matching between T and D_k.

Let us recall the main properties of Blanche Descartes's construction. It is a routine matter to prove by induction that for all $k \geq 1$, D_k is (C_3, C_4, C_5)-free and k-colorable. To prove that at least k colors are needed to color D_k, the proof is also by induction. Suppose it is true for $1, \dots, k$ and suppose for a contradiction that D_{k+1} is k-colorable. By the pigeonhole principle, in a k-coloring of D_{k+1}, at least one n-tuple of vertices of S must be monochromatic. Hence, the copy of D_k that is matched to this n-tuple is colored with $k-1$ colors, a contradiction to the induction hypothesis. We proved that D_k has chromatic number k.

Observe that as presented here, the construction is not deterministic because different matchings can lead to non-isomorphic graphs. More formally, D_k should be defined as the set of graphs which can be constructed using this method after k steps. To avoid heavy notation, along the paper we consider that D_k is one of the Blanche Descartes graphs that can be constructed after k steps. We leave as an open question whether the characterization of Blanche Descartes graphs that we obtain is independent of the chosen matchings.

A *Blanche Descartes graph* is any graph isomorphic to an induced subgraph of any D_k for some integer $k \geq 1$. In Sect. 3, we prove our results about Blanche Descartes graphs.

Due to space constraints, some proofs (those marked with a *) were omitted and are available in the extended version of the paper.

2 Zykov Graphs

In this section, we start by providing a characterization of Zykov graphs based on certain special stable sets. Using this characterization, we derive the main result: the proof that recognizing Zykov graphs is NP-complete.

2.1 Characterization of Zykov Graphs

A stable set A in some graph G is *splitting* if every vertex of A has at most one neighbor in each connected component of $G \setminus A$. Note that every graph contains a splitting stable set: the empty set. Note also that G contains a non-empty splitting stable set if and only if at least one of its connected component does.

Lemma 1. *For all graphs G, the following conditions are equivalent:*

(a) G is a Zykov graph.
(b) Every induced subgraph H of G contains a non-empty splitting stable set A such that $H \setminus A$ is a Zykov graph.
(c) G contains a non-empty splitting stable set A such that $G \setminus A$ is a Zykov graph.

Proof. To prove that (a) implies (b), it is enough to prove that every connected induced subgraph H of some graph Z_k contains a non-empty splitting stable set (its removal yields a Zykov graph because Zykov graphs form a hereditary class by definition). So, let H be a connected subgraph of some graph Z_k and suppose that k is minimal with respect to this property. By the minimality of k, H contains some vertices of the set S_k used to construct Z_k from $Z_1, \ldots, Z_{k-1}$ as in the definition. Hence, $V(H) \cap S_k$ is a non-empty splitting stable set of H, so Condition (b) holds.

Clearly, (b) implies (c).

To prove that (c) implies (a), suppose that G contains a non-empty splitting stable set $A = \{a_1, \ldots, a_n\}$ such that $G \setminus A$ is a Zykov graph. Our goal is to find a large enough integer m such that G can be viewed as an induced subgraph of Z_m.

Let $C_1, \ldots, C_\ell$ be the connected components of $G \setminus A$. Let k be an integer large enough to satisfy the following properties: $|V(Z_k)| \geq |A|$ and each component of $G \setminus A$ is an induced subgraph of Z_k (this is possible since $G \setminus A$ is a Zykov graph by assumption). We claim that G is isomorphic to an induced subgraph of $Z_{k+\ell+1}$. Recall that $Z_{k+\ell+1}$ is obtained from the disjoint union of $Z_1, \ldots, Z_{k+\ell}$. Since $|V(Z_k)| \geq |A|$, we may consider distinct vertices $b_1, \ldots, b_n$ in Z_k. We view each component C_i as an induced subgraph of Z_{k+i}. For each vertex a_i, we define a $(k+\ell)$-tuple $T_i = (c_{i,1}, \ldots, c_{i,k+\ell})$ of vertices with $c_{i,j} \in V(Z_j)$ for $j = 1, \ldots, k+\ell$. For $j = 1, \ldots, k-1$, we choose $c_{i,j}$ to be an arbitraty vertex in Z_j. For $j = k$, we choose $c_{i,j} = b_i$. For $j = k+1, \ldots, k+\ell$, if a_i has a neighbor $a_{i,j}$ in C_j, we set $c_{i,j} = a'_{i,j}$ and if a_i has no neighbor in C_j, we choose $c_{i,j}$ to be any vertex from $Z_j \setminus C_j$ (note that $Z_j \setminus C_j \neq \emptyset$ since $j > k$). We now view a_i as the vertex associated to the $(k+\ell)$-tuple T_i. Note that the vertices $c_{i,k}$ for $i = 1, \ldots, n$ guarantee that all T_i's are distinct. This proves that G is an induced subgraph of $Z_{k+\ell+1}$. □

Theorem 1. *A graph G is a Zykov graph if and only if all induced subgraphs of G contain a non-empty splitting stable set.*

Proof. If G is a Zykov graphs, the conclusion holds by Condition (b) of Lemma 1. Conversely, suppose that G is such that all induced subgraphs of G contain a non-empty splitting stable set. Let us prove by induction on $|V(G)|$ that G is a Zykov graph. If $|V(G)| = 1$, $G = Z_1$ is a Zykov graph. If $|V(G)| > 1$, then by assumption G contains a non-empty splitting stable set A. By the induction hypothesis, $G \setminus A$ is a Zykov graph. Hence, by Condition (c) of Lemma 1, G is a Zykov graph. □

Some Properties of Zykov Graphs. We state now several simple observations, needed in the sequel or possibly useful for a further study of the structure of Zykov graphs.

Lemma 2. *The class of Zykov graphs is closed under subdividing edges.*

Proof. Let H be obtained from some Zykov graph G by subdividing edges. Let $Y \subseteq V(H)$, and notice that if $H[Y]$ contains a vertex of degree at most 1, then this vertex forms a splitting stable set of $H[Y]$. Then, if there is no such vertex, $H[Y]$ is the subdivision of an induced subgraph $G[X]$ of G, where $X \subseteq V(G)$. Notice that if A is a splitting stable set of $G[X]$, then it is also a splitting stable set of $H[Y]$. It follows from Theorem 1 that all induced subgraphs of H have a non-empty splitting stable set. □

Lemma 3. *The class of Zykov graphs is closed under taking subgraphs.*

Proof. This follows from Lemma 2 and the fact that a subgraph of some graph G can be obtained from G by a sequence of subdivision of edges and deletion of vertices. □

We now give tools to prove that some graphs are not Zykov graphs.

Lemma 4. *If C is a cycle of some graph G, then every splitting stable set A of G contains either no vertex of C, or at least two vertices of C.*

Proof. If A contains a unique vertex v of C, then v has at least two neighbors in the connected component of $G \setminus A$ that contains $C \setminus \{v\}$, a contradiction to the definition of a splitting stable set. □

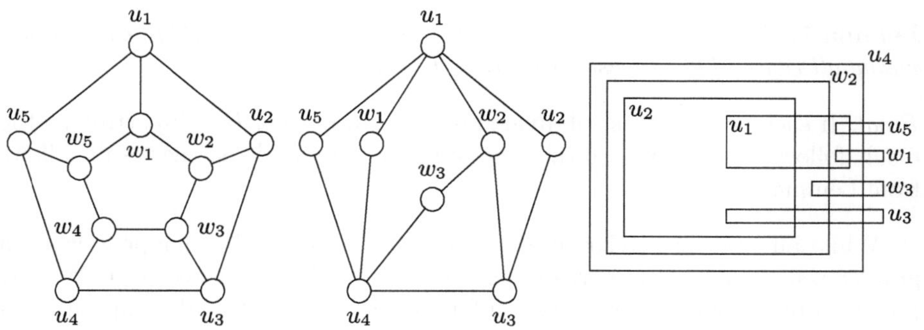

Fig. 1. Graphs F and F', and a frame representation of F', proving that F' is indeed a Burling graph [12].

Lemma 4 gives a simple criterion to prove that some graphs are not Zykov. The graph F represented in Fig. 1 is not Zykov. Indeed, if it were Zykov, it would contain by Theorem 1 a non-empty splitting stable set A. Up to symmetry, A

must contain u_1, and therefore by a sequence of applications of Lemma 4, A must contain w_2, u_3, w_4 and u_5, a contradiction. By a similar argument one can easily check that the graph F' from the same figure is not Zykov. It is also easy to check that F' is the smallest non-Zykov triangle-free graph. Indeed, such a graph must be non-bipartite (as will be shown in Lemma 6), and must therefore contain an odd cycle of length at least 5. A brute-force check of all possible ways to attach two vertices to a cycle of length 5 or to add chords to a cycle of length 7 only yields graphs with a splitting stable set whose removal yields a forest, so all triangle-free graphs on at most seven vertices are Zykov graphs. Notice that the graph F' is a Burling graph (by the frame representation of Fig. 1), and thus Burling graphs are not contained in the class of Zykov graphs.

On Specific Graph Classes. The well-known Gyárfás-Sumner conjecture states that any hereditary class of graphs that forbids a tree is χ-bounded. It is already established that Zykov graphs are not counterexamples to this conjecture, implying that every tree appears in the Zykov construction. Theorem 1 provides a direct proof of this fact, and we can further strengthen the result by showing that any bipartite graph is a Zykov graph.

Lemma 5. *All forests are Zykov graphs.*

Proof. Forests are Zykov graphs by Theorem 1 because a vertex of degree at most 1 in any graph forms a splitting stable set. □

Lemma 6. *All bipartite graphs are Zykov graphs.*

Proof. Any side of the bipartition of a bipartite graph forms a splitting stable set whose removal yields a forest. Hence, the conclusion holds by Condition (c) of Lemma 1 and Lemma 5. □

Lemma 7. *Let G be any graph (possibly not Zykov). If H is obtained by subdividing all edges of G at least once, then H is a Zykov graph.*

Proof. If each edge of G is subdivided exactly once, then H is bipartite, so the result follows from Lemma 6. If there are more subdivisions, the result follows from Lemma 2. □

While all trees are Zykov graphs, one may wonder what happens when a graph resembles a tree, or more formally, when it has bounded treewidth. We first provide a brief definition of treewidth for completeness. A tree-decomposition of a graph G is a pair (T, μ) where T is a tree and μ is a map from $V(T)$ to $2^{V(G)}$ such that:

- for every $v \in V(G)$, the set $\{t \in V(T) : v \in \mu(t)\}$ induces a non-empty subtree of T, and
- for every $uv \in E(G)$, there exists $t \in V(T)$ such that $\{u, v\} \subseteq \mu(t)$.

The width of (T, μ) is $\max_{t \in V(T)} |\mu(t)| - 1$, and the treewidth of G, denoted by $\mathrm{tw}(G)$, is the minimum width of a tree-decomposition of G. Moreover, there exists an algorithm that returns a tree-decomposition of width $O(\mathrm{tw}(G))$ in time $2^{O(\mathrm{tw}(G))} \cdot n$. For more details, see Cygan et al. [4].

A straightforward generalization of the graph F (illustrated in Fig. 1) with arbitrarily large odd cycles provides an example of a non-Zykov graph that is triangle-free and has treewidth 3. At first glance, it seems non-trivial to decide whether a graph is a Zykov graph, even if it has bounded treewidth. However, using Theorem 1, we show that there exists a linear-time algorithm to determine whether the input graph is a Zykov graph when it has bounded *treewidth.*

A Fixed Parameter Tractable (FPT) algorithm is an algorithm that decides whether an instance of a decision problem is positive in time $f(k) \cdot n^c$, where c is some fixed constant, f is a computable function, n is the size of the instance, and k is a chosen parameter depending on the problem.

Theorem 2. *There exists an algorithm that, given an input graph G, decides if G is a Zykov graph in time $f(tw(G)) \cdot n$, where n is the number of vertices of G, $tw(G)$ its treewidth and f a computable function.*

Proof. By a classical theorem of Courcelle [3], it suffices to construct a monadic second-order logic formula, denoted by Φ, such that a graph G satisfies Φ if and only if it is a Zykov graph. We provide a high-level overview of the formula Φ, based on the characterization from Theorem 1.

A graph $G = (V, E)$ is a Zykov graph if and only if, for any subset of vertices $V' \subseteq V$, there exists a subset $S \subseteq V'$ such that S is an splitting stable set for the induced subgraph $G[V']$. The property of being a splitting set can be expressed as follows: for any subset $C \subseteq V' \setminus S$, if $G[C]$ is connected, then each vertex $v \in S$ is adjacent to at most one vertex in C. Furthermore, the properties of a subset of vertices inducing either a stable set or a connected subgraph can also be expressed in monadic second-order logic. □

Notice that in Fig. 1, the graph F' has treewidth 3 and is not a Zykov graph. In Fig. 2, we give a non-Zykov triangle-free graph of treewidth 2.

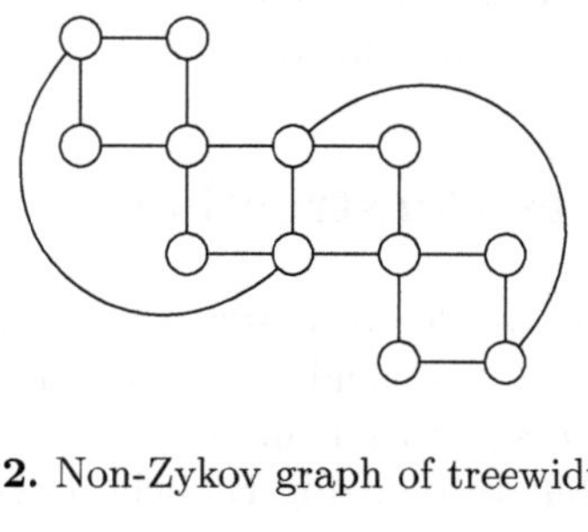

Fig. 2. Non-Zykov graph of treewidth 2.

While we have established that recognizing Zykov graphs is tractable for graphs with bounded treewidth, the situation changes drastically for arbitrary graphs. In the following, we prove that the problem becomes NP-complete when the restriction on treewidth is lifted.

Theorem 3 (*). *Recognizing Zykov graphs is NP-complete.*

Without the characterization of Zykov graphs using splitting stable sets, even the existence of an NP certificate would be far from trivial. Indeed, the most natural certificate for this problem is, given a graph G, a Zykov graph Z_k (for some $k \geq 1$) such that G is an induced subgraph of Z_k. However, there is no guarantee that Z_k has a size polynomial in the size of G. In contrast, Theorem 1 allows us to certify that a graph is a Zykov graph by providing a sequence of non-empty splitting stable sets, which can be described in polynomial size.

The following results are byproducts of our construction. We omit the proofs that are similar to the proof of Theorem 3.

Theorem 4. *It is NP-complete to decide whether an input graph contains a non-empty splitting stable set.*

Theorem 5. *It is NP-complete to decide whether an input graph can be (vertex-wise) partitioned into a stable set and a forest in such way that every vertex of the stable set has at most one neighbor in each component of the forest.*

2.2 Augmenting the Girth

In the same way that Mycielski graphs contain all triangle-free graphs, that is, graphs of girth at least 4, one might conjecture that Zykov graphs could contain all graphs of girth at least g, for some constant g. However, Theorem 6 disproves this possibility by showing that non-Zykov graphs can have arbitrarily large girth. The proof, provided in the long version of the paper, relies on the construction of expanders with large girth.

Theorem 6 (*). *There exists non-Zykov (and therefore non-Blanche Descartes) graphs of arbitrarily large girth.*

Theorems 3, 4 and 5 may still be true under the assumption that the input graph has high girth, but we leave this open.

3 Blanche Descartes Construction

Similarly to Zykov graphs, we derive a structural characterization of Blanche Descartes graphs, also based on special stable sets. In particular, as a corollary of this characterization, we show that all Blanche Descartes graphs are Zykov graphs. Then, we prove that recognizing Blanche Descartes graphs is also NP-complete. Finally, we show that both Maximum Independent Set and 3-Coloring remain NP-complete on Blanche Descartes, and therefore on Zykov graphs as well.

3.1 Characterization and Properties of Blanche Descartes Graphs

A stable set S in some graph G is *strongly splitting* if every vertex of S has at most one neighbor in each connected component of $G \setminus S$ and each vertex of $G \setminus S$ has at most one neighbor in S. In the same way as splitting sets, a graph contains a strongly splitting stable set if and only if at least one of its connected components does.

Lemma 8. *For all graphs G, the following conditions are equivalent:*

(a) G is a Blanche Descartes graph.
(b) Every induced subgraph H of G contains a non-empty strong splitting stable set S such that $H \setminus S$ is a Blanche Descartes graph.
(c) G contains a non-empty strong splitting stable set S such that $G \setminus S$ is a Blanche Descartes graph.

Proof. To prove that (a) implies (b), it is enough to prove that every connected induced subgraph H of some Blanche Descartes graph G contains a non-empty strong splitting stable set. So, let H be a connected subgraph of some graph D_k and suppose that k is minimal with respect to this property. By the minimality of k and connectedness of H, H contains some vertices of the set S_k used to construct D_k from the copies of D_{k-1} as in the definition. Hence, $V(H) \cap S_k$ is a non-empty splitting stable set of H, so Condition (b) holds.

Clearly, (b) implies (c).

To prove that (c) implies (a), suppose that G has a non-empty strong splitting stable set S such that $G \setminus S$ is a Blanche Descartes graph. Let $C_1, \dots, C_\ell$ be the connected components of $G \setminus S$, and since $G \setminus S$ is a Blanche Descartes graph, each of them is an induced subgraph of some Blanche Descartes graph, say $D_{k_1}, \dots, D_{k_\ell}$ respectively. Let k be an integer large enough such that $k \geq \max_{1 \leq i \leq \ell} k_i$ and $k > |S|$, and let $n = |V(D_{k+\ell})|$. We claim that G is isomorphic to an induced subgraph of $D_{k+\ell+1}$. Consider the connected components $C_1, \dots, C_\ell$ as induced subgraphs of ℓ disjoint copies $C'_1, \dots, C'_\ell$ of $D_{k+\ell}$. Then, for each $i = 1, ..., \ell$, let S_i be the set of vertices of S adjacent to some vertex of C_i, and add $n - n_i$ new vertices $\{c_{i,1}, ..., c_{i,n-n_i}\} := S'_i$, where $n_i = |S_i|$ to G, and add $n - n_i$ edges to obtain a matching between $S_i \cup S'_i$ and C'_i. Notice that $S \cup S'_1 \cup ... \cup S'_\ell$ has size at most $|S| + n\ell \leq k + n\ell$. Thus, add enough vertices to complete $S \cup S'_1 \cup ... \cup S'_\ell$ into a stable set S' of size $(k+\ell)n+1$. Finally, for each n-tuple of S' which is not matched with a copy of $D_{k+\ell}$, add such a copy and a matching to obtain $D_{k+\ell+1}$. □

Theorem 7. *A graph G is a Blanche Descartes graph if and only if all induced subgraphs of G contain a non-empty strong splitting stable set.*

Proof. If G is a Blanche Descartes graphs, the conclusion holds by Condition (b) of Lemma 8. Conversely, suppose that G is such that all induced subgraphs of G contain a non-empty strong splitting stable set. Let us prove by induction on $|V(G)|$ that G is a Blanche Descartes graph. If $|V(G)| = 1$, $G = D_1$ is a Blanche Descartes graph. If $|V(G)| > 1$, then by assumption G contains a non-empty strong splitting stable set A. By the induction hypothesis, $G \setminus A$ is a Blanche Descartes graph. Hence, by Condition (c) of Lemma 8, G is a Blanche Descartes graph. □

Lemma 9. *Every Blanche Descartes graph is a Zykov graph.*

Proof. A strong splitting stable set is also by definition a splitting stable set. From Theorem 1 and Theorem 7, the result holds. □

Using a similar proof technique as in Lemma 7, it follows that Blanche Descartes graphs are closed under edge subdivision and taking subgraphs.

Lemma 10. *The class of Blanche Descartes graphs is closed under subdividing edges and under taking subgraphs.*

We now give some tools to prove that some graphs are not Blanche Descartes graphs. We omit the proofs which are very similar to their equivalent with Zykov graphs.

Lemma 11. *If S is a strong splitting set of some graph G, then there is no vertex v in G adjacent to two distinct vertices of S.*

Lemma 12. *If C is a cycle of some graph G, then every strong splitting stable set S of G contains either no vertex of C, or at least two vertices of C at distance at least 3 in C.*

Proof. Let S be a strong splitting stable set of G. By Lemma 4, S contains either no vertex of C, or at least two of them. If S contains two of them at distance 2, then there exists a vertex of $G \setminus S$ adjacent to two vertices of S, which contradicts Lemma 11. □

Notice that any vertex of degree at most 1 is also a strong splitting stable set. Since any forest has such a vertex, they are all Blanche Descartes graphs.

Lemma 13. *All forests are Blanche Descartes graphs.*

Lemma 14. *Let G be any graph (possibly not Blanche Descartes). If H is obtained by subdividing all edges of G at least twice, then H is a Blanche Descartes graph.*

Proof. If each edge of G is subdivided exactly twice, then H is Blanche Descartes because original vertices of G form a strong splitting stable set whose removal yields a forest of edges, so the result follows from Lemmas 8 and 13. If there are more subdivisions, then the result follows from Lemma 10. □

Observe that subdividing twice in Lemma 14 is best possible, since for instance subdividing each edge of K_4 once yields a non-Blanche Descartes graph. Similarly to Zykov graphs, Theorem 7 can be used to describe Blanche Descartes using a monadic second order formula. Hence the following holds.

Theorem 8. *There exists an algorithm that, given an input graph G, decides if G is a Blanche Descartes graph in time $f(tw(G)) \cdot n$, where n is the number of vertices of G, $tw(G)$ its treewidth and f a computable function.*

Similarly to Zykov graphs, recognizing Blanche Descartes graphs is NP-complete.

Theorem 9 (*). *Recognizing Blanche Descartes graphs is NP-complete.*

3.2 NP-Hard Problems on Blanche Descartes Graphs

Given a graph G and an integer k, the MAXIMUM INDEPENDENT SET problem asks whether G contains a stable set of size at least k, and the k-COLORING problem asks whether G has chromatic number at most k. In [17], it is asked whether all hereditary classes with a polynomial time algorithm for MAXIMUM INDEPENDENT SET are χ-bounded. This stayed open, until recently Rzążewski and Walczak proved that MAXIMUM INDEPENDENT SET is polynomially tractable in Burling graphs. However, it remains an open question whether any hereditary graph class in which the k-COLORING problem is in P is necessarily χ-bounded.

Here, we show that both problems remain NP-complete on Blanche Descartes graphs, and, as a corollary, on Zykov graphs as well by Lemma 9.

Theorem 10. *MAXIMUM INDEPENDENT SET is NP-complete in Blanche Descartes Graphs.*

Proof. We reduce from *MAXIMUM INDEPENDENT SET* on arbitrary graphs. From an instance (G, k), we construct the graph G^+ from G by subdividing each edge twice, meaning that we replace each edge $uv \in E(G)$ by a path uw_1w_2v in G^+. It follows from [11] that G contains an independent set of size k if and only if G^+ contains an independent set of size $k + m$, where $m = |E(G)|$. In addition, G^+ is a Blanche Descartes graph by Lemma 14. □

Theorem 11. *3-COLORING is NP-complete in Blanche Descartes Graphs.*

Proof. Before the reduction, let us construct a Blanche Descartes graph H with two special vertices a and b, such that $\chi(H) = 3$ and in any 3-coloring of H, a and b have different colors. The construction is as follows:

- Take a stable set $S = \{s_1, \ldots, s_{19}\}$ on 19 vertices.
- For each subset T of S of 7 vertices, add a copy of C_7, the cycle with 7 vertices, and a matching between T and C_7.
- Remove the edge between s_1 and the copy of C_7 associated with $\{s_1, \cdots, s_7\}$.

We first prove that H has a proper 3-coloring. Start by coloring the vertices of S such that the only monochromatic subset of size 7 is $\{s_1, \cdots, s_7\}$. It is possible for instance by giving color 2 to $s_1, \cdots, s_7$, and then color $i \mod 2$ to s_i for $8 \leq i \leq 19$. Then, notice that the copy of C_7 associated with $\{s_1, ..., s_7\}$ can be properly colored using the color 2 for the only vertex which was connected to s_1 before the edge was removed, and completing the coloring using colors 0 and 1. All the others C_7s can be colored using 3 colors since the subsets of S matched to them are not monochromatic. Indeed, let $v_1, \ldots, v_7$ be the vertices of a cycle C_7, each matched with a vertex $u_1, \ldots, u_7$ in the set S. Without loss of generality, assume that u_1 and u_2 have different colors, say 0 and 1, respectively. Then, assign color 0 to v_2. For each $i = 3, \ldots, 7$, assign to v_i a color different from the colors of both v_{i-1} and u_i. Finally, consider v_1, which has three neighbors: u_1, v_2, and v_7. Since u_1 and v_2 share the same color, there remains at least one free color for v_1.

Then, we prove that s_1 and s_8 have different colors in any 3-coloring of H. Note that in any 3-coloring of H, one subset of 7 vertices of S has to be monochromatic, by the pigeonhole principle. This subset must be $\{s_1, ..., s_7\}$, otherwise there is perfect matching between a C_7, which has no 2-coloring, and a monochromatic set of 7 vertices. Then, s_8 has to have a different color from s_1, otherwise $\{s_2, ..., s_8\}$ would also be monochromatic.

We are now ready for the reduction, which is from 3-COLORING on arbitrary graphs. Let G be an instance of this problem. We construct a new graph G' as follows : start from a copy of G, and for each $uv \in E(G)$, remove this edge, create one copy of H and identify s_1 and s_8 with respectively u and v.

A direct consequence of the properties of H is that G has a proper 3-coloring if and only if G' has one. In addition, G' is a Blanche Descartes graph. Indeed, let I be the union of all stable sets S of all copies of H in G'. Notice that $G' - I$ is a disjoint union of C_7, which is a Blanche Descartes graph, and each vertex from I is adjacent to at most one vertex from each C_7 and reciprocally. □

4 Open Problems

We conclude with some questions/problems which remain open, and seem interesting to us.

Question 1 In the definition of Blanche Descartes we used, the matching chosen at each step is arbitrary. Does the definition of Blanche Descartes graphs depend on the type of matching used? Specifically, does considering arbitrary matchings or fixed matchings result in the same class of graphs?

Question 2 Another class of triangle-free graphs of arbitrarily large chromatic number is the class of *shift graphs*, which can be defined as the underlying undirected graphs of a line graph of a tournament. Is it possible to recognize shift graphs in polynomial time?

Question 3 We say that a hereditary class $\mathcal{G}$ of triangle-free graphs of arbitrarily large chromatic number is *minimal* if for any fixed graph $H \in \mathcal{G}$, any graph $G \in \mathcal{G}$ excluding H as an induced subgraph has chromatic number bounded by a constant (which depends on H). Zykov and Blanche Descartes do not satisfy this property, as there exist Blanche Descartes graphs of arbitrarily large girth [8].
This raises the following question: does there exist a minimal class contained within the class of Zykov graphs (respectively, Blanche Descartes graphs)?

Acknowledgments. The authors are partially supported by the French National Research Agency under research grant ANR Twin-width ANR-21-CE48-0014-01.

References

1. Bonnet, É., Bourneuf, R., Duron, J., Geniet, C., Thomassé, S., Trotignon, N.: A tamed family of triangle-free graphs with unbounded chromatic number (2023), arXiv:2304.04296

2. Burling, J.P.: On Coloring Problems of Families of Polytopes. University of Colorado at Boulder (1965)
3. Courcelle, B., Engelfriet, J.: Graph Structure and Monadic Second-Order Logic: a Language-Theoretic Approach, vol. 138. Cambridge University Press (2012)
4. Cygan, M., et al.: Parameterized Algorithms, vol. 5. Springer (2015)
5. Davies, J.: Triangle-free graphs with large chromatic number and no induced wheel. J. Graph Theory **103**(1), 112–118 (2023)
6. Descartes, B.: A three colour problem. Eureka **9**(21), 24–25 (1947)
7. Descartes, B.: Solution to advanced problem no. 4526. Amer. Math. Monthly **61**(352), 216 (1954)
8. Kostochka, A.V., Nešetřil, J.: Properties of Descartes' construction of triangle-free graphs with high chromatic number. Comb. Probab. Comput. **8**(5), 467–472 (1999)
9. Mycielski, J.: Sur le coloriage des graphs. In: Colloquium Mathematicae, vol. 3, pp. 161–162 (1955)
10. Pawlik, A., Kozik, J., Krawczyk, T., Lasoń, M., Micek, P., Trotter, W.T., Walczak, B.: Triangle-free geometric intersection graphs with large chromatic number. Discrete Compututational Geom. **50**(3), 714–726 (2013)
11. Poljak, S.: A note on the stable sets and coloring of graphs. Comment. Math. Univ. Carol. **15**, 307–309 (1974)
12. Pournajafi, P., Trotignon, N.: Burling graphs revisited, part I: new characterizations. Eur. J. Comb. **110**, 103686 (2023)
13. Pournajafi, P., Trotignon, N.: Burling graphs revisited, part II: structure. Eur. J. Comb. **116**, 103849 (2024)
14. Pournajafi, P., Trotignon, N.: Burling graphs revisited, part III: applications to χ-boundedness. Eur. J. Comb. **116**, 103850 (2024)
15. Rzążewski, P., Walczak, B.: Polynomial-time recognition and maximum independent set in Burling graphs. arXiv preprint arXiv:2407.16666 (2024)
16. Scott, A., Seymour, P.: A survey of χ-boundedness. J. Graph Theory **95**(3), 473–504 (2020)
17. Thomassé, S., Trotignon, N., Vušković, K.: A polynomial turing-kernel for weighted independent set in bull-free graphs. Algorithmica **77**, 619–641 (2017)
18. Zykov, A.A.: On some properties of linear complexes. Matematicheskii sbornik **66**(2), 163–188 (1949)

Realizing Temporal Transportation Trees

George B. Mertzios[1(✉)], Hendrik Molter[2], Nils Morawietz[3,4], and Paul G. Spirakis[5]

[1] Department of Computer Science, Durham University, Durham, UK
george.mertzios@durham.ac.uk
[2] Department of Computer Science, Ben-Gurion University of the Negev, Beer-Sheva, Israel
molterh@post.bgu.ac.il
[3] Institute of Computer Science, Friedrich Schiller University Jena, Jena, Germany
nils.morawietz@uni-jena.de
[4] LaBRI, Université de Bordeaux, Bordeaux, France
[5] Department of Computer Science, University of Liverpool, Liverpool, UK
p.spirakis@liverpool.ac.uk

Abstract. In this paper, we study the complexity of the *periodic temporal graph realization* problem with respect to *upper bounds* on the fastest path durations among its vertices. This constraint with respect to upper bounds appears naturally in *transportation network design* applications where, for example, a transportation network is given, and the goal is to appropriately schedule periodic travel routes, while not exceeding some desired upper bounds on the travel times. In our work, we focus only on underlying *tree topologies*, which are fundamental in many transportation network applications.

As it turns out, the periodic upper-bounded temporal tree realization problem (TTR) has a very different computational complexity behavior than both (i) the classic graph realization problem with respect to shortest path distances in static graphs and (ii) the periodic temporal graph realization problem with *exact* given fastest travel times (which was recently introduced). First, we prove that, surprisingly, TTR is NP-hard, even for a constant period Δ and when the input tree G satisfies at least one of the following conditions: (a) G is a star, or (b) G has constant maximum degree. Second, we prove that TTR is fixed-parameter tractable (FPT) with respect to the number of leaves in the input tree G, via a novel combination of techniques for totally unimodular matrices and mixed integer linear programming.

G. B. Mertzios—Supported by the EPSRC grant EP/P020372/1.

H. Molter—Supported by the ISF, grant nr. 1470/24, by the European Union's Horizon Europe research and innovation programme under grant agreement 949707, and by the European Research Council, grant nr. 101039913 (PARAPATH).

N. Morawietz—Partially supported by the French ANR, project ANR-22-CE48-0001 (TEMPOGRAL).

P. G. Spirakis—Supported by the EPSRC grant EP/P02002X/1.

H. Fernau and P. Kindermann (Eds.): WG 2025, LNCS 16124, pp. 390–404, 2026.
https://doi.org/10.1007/978-3-032-11835-6_28

Keywords: Temporal graph · periodic temporal labeling · fastest temporal path · graph realization · Mixed Integer Linear Programming

1 Introduction

A temporal (or dynamic) graph is a graph whose topology is subject to discrete changes over time. This paradigm reflects the structure and operation of a great variety of modern networks; social networks, wired or wireless networks whose links change dynamically, transportation networks, and several physical systems are only a few examples of networks that change over time [33,41,44]. Inspired by the foundational work of Kempe et al. [34], we adopt here a simple model for temporal graphs, in which the vertex set remains unchanged while each edge is equipped with a set of integer time labels.

Definition 1 (temporal graph [34]). *A* temporal graph *is a pair* (G, λ), *where* $G = (V, E)$ *is an underlying (static) graph and* $\lambda : E \to 2^{\mathbb{N}}$ *is a* labeling *function which assigns to every edge of* G *a set of discrete time labels.*

Here, whenever $t \in \lambda(e)$, we say that the edge e is *active* or *available* at time t. In the context of temporal graphs, where the notion of vertex adjacency is time-dependent, the notions of path and distance also need to be redefined. The most natural temporal analogue of a path is that of a *temporal* (or *time-dependent*) path, which is a path of the underlying static graph whose time labels strictly increase along the edges of the path. The *duration* of a temporal path is the number of discrete time steps needed to traverse it. Finally, a temporal path from vertex u to vertex v is *fastest* if it has the smallest duration among all temporal paths from u to v, see Fig. 1 for an illustration.

The *graph realization* problem with respect to some graph property $\mathcal{P}$ is to compute a graph that satisfies $\mathcal{P}$, or to decide that no such graph exists. The main motivation for graph realization problems stems both from network design and from "verification" applications in engineering. In *network design* applications, the goal is to design network topologies having a desired property [2,29]. On the other hand, in *verification* applications, given the outcomes of some exact experimental measurements, the aim is to (re)construct an input network that complies with them. If such a reconstruction is not possible, this proves that the measurements are incorrect or implausible (resp. that the algorithm making the computations is incorrectly implemented).

One example of a graph realization problem is the recognition of probe interval graphs, in the context of the physical mapping of DNA, see [38,39] and [27, Chapter 4]. Analyzing the computational complexity of the graph realization problems for various natural and fundamental graph properties $\mathcal{P}$ requires a deep understanding of these properties. Among the most studied parameters for graph realization are constraints on the distances between vertices [6,7,10,15,16,31,45,46,48], on the vertex degrees [5,18,26,28,30], on the eccentricities [4,9,32,37], and on connectivity [14,22–25,28], among others. Although the majority of studies of graph realization problems concern *static* graphs, the *temporal* (periodic) graph realization problem with respect to given (exact) delays of the fastest temporal paths among vertices has been recently studied in [20,35], motivated by *verification* applications where exact measurements are given.

In this paper, we consider the (periodic) temporal graph realization problem with respect to the durations of the fastest temporal paths from a *network design* perspective, i. e., where only *upper bounds* on the durations of fastest temporal paths are given. One important application domain of network design problems is that of *transportation network* design where, for example, a road network is given, and the goal is to appropriately schedule periodic travel routes while not exceeding some desired upper bounds on travel times. Our work is motivated by the fact that in many transportation network applications the underlying graph has a *tree* structure [3,8,11]. For example, many airlines or railway transportation networks are even star graphs (having a big hub at the center of the network, e.g. in the capital city). The formal definition of our problem is as follows (see Sect. 2 for formal definitions of all used terminology).

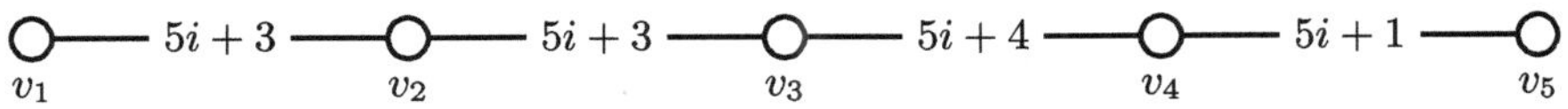

Fig. 1. Visualization of a Δ-periodic temporal graph (G, λ, Δ) with $\Delta = 5$ and the following Δ-periodic labeling $\lambda : E \to \{1, 2, \dots, 5\}$: $\lambda(\{v_1, v_2\}) = \lambda(\{v_2, v_3\}) = 3$, $\lambda(\{v_3, v_4\}) = 4$, and $\lambda(\{v_4, v_5\}) = 1$. A fastest temporal path from v_1 to v_5 first traverses $\{v_1, v_2\}$ at time 3, then $\{v_2, v_3\}$ a time 8, then $\{v_3, v_4\}$ at time 9, and then $\{v_4, v_5\}$ at time 11, and has duration 9.

PERIODIC UPPER-BOUNDED TEMPORAL TREE REALIZATION (TTR)

Input: A tree $G = (V, E)$ with $V = \{v_1, v_2, \dots, v_n\}$, an $n \times n$ matrix D of positive integers, and a positive integer Δ.

Question: Does there exist a Δ-periodic labeling $\lambda : E \to \{1, 2, \dots, \Delta\}$ such that, for every i, j, the duration of the fastest temporal path from v_i to v_j in the Δ-periodic temporal graph (G, λ, Δ) is *at most* $D_{i,j}$.

Many natural and technological systems exhibit a periodic temporal behavior. This is true even more on transportation networks [1], where the goal is to build periodic schedules of transportation units (e.g. trains, buses, or airplanes). The most natural constraint, while designing such periodic transportation schedules, is that the fastest travel time (i. e., the duration of the fastest temporal path) between a specific pair of vertices does not exceed a specific desired upper bound. That is, if the travel time between u and v in the resulting schedule is shorter than this upper bound, the schedule is even better (and thus, still feasible). Periodic temporal graphs have also been studied in various different contexts (see [12, Class 8] and [1,19,21,42,43]).

We focus on the most fundamental case of periodic temporal graphs, where every edge has the same period Δ and it appears exactly once within each period. As it turns out, TTR has a very different computational complexity behavior than both (i) the classic graph realization problem with respect to shortest path distances in static graphs [31] and (ii) the periodic temporal graph realization problem with *exact* given fastest travel times [35], both of which are polynomial-time solvable on trees. We remark that recently, the case where every edge can appear multiple times in each Δ period has been studied [20]. We defer the discussion of additional related work to the full version [40].

Our Contribution. Our results in this paper are given in two main directions. First, we prove that TTR is NP-hard, even for a constant period Δ and when the input tree G satisfies at least one of the following conditions: (a) G is a star (for any $\Delta \geq 3$), or (b) G has diameter at most 6 and constant pathwidth (even for $\Delta = 2$), or (c) G has maximum degree at most 8 and constant pathwidth (even for $\Delta = 2$). Note that, for $\Delta = 1$, TTR becomes trivial (as in this case fastest paths coincide with shortest paths in the underlying tree). On the one hand, our hardness results come in wide contrast to the classic (static) graph realization problem with respect to shortest path distances. Indeed, this static analogue is easily solvable in polynomial time in two steps [31]. On the other hand, the complexity of TTR is also surprisingly different than the periodic temporal graph realization problem with *exact* given fastest travel times. More specifically, when the input integer matrix D gives the exact fastest travel times, the problem becomes solvable in polynomial time on trees [35]. Note that our hardness results rule out the existence of FPT-algorithms for TTR for all reasonable parameters on trees besides the number of leaves (assuming P $\neq$ NP).

Second, we prove that TTR is fixed-parameter tractable (FPT) with respect to the number of leaves in the input tree G. That is, long chains of vertices of degree 2 do not affect the complexity of the problem. To provide our FPT algorithm, we reduce TTR to a number MIXED INTEGER LINEAR PROGRAM (MILP) instances that is upper-bounded by a function of the number of leaves in the input tree. Furthermore, the number of integer variables in each MILP instance is also upper-bounded by a function of the number of leaves in the input tree. This allows us to use a known FPT algorithm for MILP parameterized by the number of integer variables [17,36] to solve all MILP instances in FPT-time with respect to the number of leaves of the input tree. We prove that the TTR instance is a yes-instance if and only if at least one of the MILP instances admits a feasible solution. To this end, we use a novel combination of techniques for totally unimodular matrices and mixed integer linear programming to design the MILP instances in a way that if they admit a feasible solution, then we can assume that all variables are set to integer values.

Proofs of results marked with $\star$ are (partially) deferred to the full version of the paper [40].

2 Preliminaries and Notation

An undirected graph $G = (V, E)$ consists of a set V of vertices and a set $E \subseteq \binom{V}{2}$ of edges. We denote by $V(G)$ and $E(G)$ the vertex and edge set of G, respectively.

Let $G = (V, E)$ and $\Delta \in \mathbb{N}$, and let $\lambda : E \to \{1, 2, \ldots, \Delta\}$ be an edge-labeling function that assigns to every edge of G exactly one of the labels from $\{1, \ldots, \Delta\}$. Then we denote by (G, λ, Δ) the *Δ-periodic temporal graph* (G, L), where for every edge $e \in E$ we have $L(e) = \{\lambda(e) + i\Delta \mid i \geq 0\}$. In this case, we call λ a *Δ-periodic labeling* of G. When it is clear from the context, we drop Δ and denote the (Δ-periodic) temporal graph by (G, λ).

A *temporal* (s, z)*-walk* (or *temporal walk*) of length k from vertex $s = v_0$ to vertex $z = v_k$ in a Δ-periodic temporal graph (G, L) is a sequence $P = ((v_{i-1}, v_i, t_i))_{i=1}^{k}$ of triples (where, for every i, $\{v_{i-1}, v_i\}$ is an edge of G) which we call *transitions*, such that for all $i \in [k]$ we have that $t_i \in L(\{v_{i-1}, v_i\})$ and for all $i \in [k-1]$ we have that $t_i < t_{i+1}$. Moreover, we call P a *temporal* (s, z)*-path* (or *temporal path*) of length k if $v_i \neq v_j$ for all $i, j \in \{0, \ldots, k\}$ with $i \neq j$. Given a temporal path $P = ((v_{i-1}, v_i, t_i))_{i=1}^{k}$, we denote the set of vertices of P by $V(P) = \{v_0, v_1, \ldots, v_k\}$. A temporal (s, z)-path $P = ((v_{i-1}, v_i, t_i))_{i=1}^{k}$ is *fastest* if for all temporal (s, z)-path $P' = \left((v'_{i-1}, v'_i, t'_i)\right)_{i=1}^{k'}$ we have that $t_k - t_0 \leq t'_{k'} - t'_0$. We say that the *duration* of P is $d(P) = t_k - t_0 + 1$. Furthermore, the concept of *travel delays* is very important for our proofs.

Definition 2 (travel delays). *Let (G, λ) be a Δ-periodic temporal graph. Let $e_1 = \{u, v\}$ and $e_2 = \{v, w\}$ be two incident edges in G with $e_1 \cap e_2 = \{v\}$. We define the* travel delay *from u to w at vertex v, denoted with $\tau_v^{u,w}$, as follows.*

$$\tau_v^{u,w} = \begin{cases} \lambda(e_2) - \lambda(e_1), & \text{if } \lambda(e_2) > \lambda(e_1), \\ \lambda(e_2) - \lambda(e_1) + \Delta, & \text{otherwise.} \end{cases}$$

From Definition 2 we immediately get the following observation.

Observation 1. *Let (G, λ) be a Δ-periodic temporal graph. Let $e_1 = \{u, v\}$ and $e_2 = \{v, w\}$ be two incident edges in G with $e_1 \cap e_2 = \{v\}$. Then we have that $\tau_v^{u,w} = \Delta - \tau_v^{w,u}$ if $\lambda(e_1) \neq \lambda(e_2)$, and $\tau_v^{u,w} = \tau_v^{w,u} = \Delta$ if $\lambda(e_1) = \lambda(e_2)$.*

Intuitively, the value of $\tau_v^{u,w}$ quantifies how long a fastest temporal path waits at vertex v when first traversing edge $\{u, v\}$ and then edge $\{v, w\}$. Formally, we have the following.

Lemma 1 ($\star$). *Let $P = ((v_{i-1}, v_i, t_i))_{i=1}^{k}$ be a fastest temporal (s, z)-path. Then we have $d(P) = 1 + \sum_{i \in [k-1]} \tau_{v_i}^{v_{i-1}, v_{i+1}}$.*

3 The Classical Complexity on Restricted Instances

In this section, we prove that TTR is NP-hard even in quite restrictive settings: when the input tree is a star and $\Delta \geq 3$, or when the input tree has constant diameter or constant maximum degree, and in addition $\Delta = 2$. We remark that the diameter can be seen as a measure of how "star-like" a tree is. In particular, on trees, the diameter upper bounds the treedepth.

Theorem 1 ($\star$). *TTR is NP-hard even if the input matrix is symmetric and*

- $\Delta \geq 3$ *and* G *is a star,*
- $\Delta = 2$ *and* G *has constant diameter and a constant pathwidth, or*
- $\Delta = 2$ *and* G *has constant maximum degree and a constant pathwidth.*

As we now show, the hardness reduction on stars for $\Delta \geq 3$ is tight in the sense that for $\Delta = 2$, the problem is polynomial-time solvable on stars. More generally, we show the following.

Theorem 2 ($\star$). *For* $\Delta = 2$, *TTR can be solved in* $4^{\text{vc}} \cdot n^{O(1)}$ *time, where* vc *denotes the vertex cover number of the input tree.*

4 Main FPT-Algorithm for TTR Based on MILPs

Our hardness results from the previous section exclude FPT-algorithms for general Δ for nearly all reasonable parameters on trees like maximum degree or vertex cover number. In this section, we consider the essentially only remaining parameter on trees: the number of leaves of the tree. This parameter is larger than the maximum degree and incomparable to the vertex cover number.

Theorem 3. *TTR is fixed-parameter tractable when parameterized by the number* ℓ *of leaves of the input tree* G.

To show Theorem 3 we present a Turing-reduction from TTR to Mixed Integer Linear Program (MILP).

Mixed Integer Linear Program (MILP)

Input: A vector x of n variables of which some are considered integer variables, a constraint matrix $A \in \mathbb{R}^{m \times n}$, and two vectors $b \in \mathbb{R}^m$, $c \in \mathbb{R}^n$.

Task: Compute an assignment to the variables (if one exists) such that all integer variables are set to integer values, $Ax \leq b$, $x \geq 0$, and $c^\intercal x$ is maximized.

Given an instance (G, D, Δ) of TTR, we will produce several MILP instances. We will prove that (G, D, Δ) is a yes-instance if and only if at least one of the MILP instances admits a feasible solution. The number of produced instances is upper-bounded by a function of the number ℓ of leaves in G. Each of the produced MILP instances will have a small number of integer variables. More precisely, the number of integer variables will be upper-bounded by a function of the number ℓ of leaves in G. This will allow us to upper-bound the running time necessary to solve the MILP instances using the following known result.

Theorem 4 ([17,36]). *MILP is fixed-parameter tractable when parameterized by the number of integer variables.*

Furthermore, we build our MILP formulations in a specific way that ensures that there always exist optimal solutions where *all* variables are set to integer values. Informally, we ensure that the constraint matrix for the rational variables is totally unimodular. This allows us to use the following result.

Lemma 2 ([13]). *Let $A_{frac} \in \mathbb{R}^{m \times n_2}$ be totally unimodular. Then for any $A_{int} \in \mathbb{R}^{m \times n_1}$, $b \in \mathbb{R}^m$, and $c \in \mathbb{R}^{n_1+n_2}$, the MILP $\max c^\intercal x$ subject to $(A_{int}\ A_{frac})x \leq b, x \geq 0$, where $x = (x_{int}\ x_{frac})^\intercal$ with the first n_1 variables (i.e., x_{int}) being the integer variables, has an optimal solution where all variables are integer.*

The main idea for the reduction is the following. Given an instance (G, D, Δ) of TTR, we create variables for the travel delays at the vertices of G. A labeling can easily be computed from the travel delays. We use the constraints to ensure that the durations of the fastest paths respect the upper bounds given in D.

Before describing the reduction, we make a straightforward observation.

Observation 2. *Let G be a tree with ℓ leaves. Let $V_{>2}$ be the set of vertices in G with degree larger than two. We have that $|V_{>2}| \leq \ell$. Furthermore, we have for every $v \in V_{>2}$ that the degree of v is at most ℓ.*

Furthermore, we show that there always exist solutions to yes-instances of TTR, where the two edges incident with degree-2 vertices have different labels.

Lemma 3 ($\star$). *Let (G, D, Δ) be a yes-instance of TTR. Let V_2 be the set of vertices in G with degree two. Then there exists a solution λ for (G, D, Δ) such that for all $v \in V_2$ we have the following. Let u, w denote the two neighbors of v in G, then $\lambda(\{u, v\}) \neq \lambda(\{v, w\})$.*

From now on assume that we are given an instance (G, D, Δ) of TTR. Assume the vertices in G are ordered in an arbitrary but fixed way. Each MILP instance we create will use the same set of variables. For each vertex v in G, we create the following variables:

- If v has degree two, we create a *fractional* variable x_v. This variable will correspond to the travel delay $\tau_v^{u,w}$, where u, w are the two neighbors of v in G and u is ordered before w.
- If v has degree larger than two, then for every pair u, w of neighbors of v in G, we create an *integer* variable $y_v^{u,w}$. This variable will correspond to the travel delay $\tau_v^{u,w}$.
  Furthermore, we create an *integer* variable z_e for every edge e that is incident with v. This variable will correspond to the label $\lambda(e)$ of e.

With Observation 2 we can upper-bound the number of created integer variables.

Observation 3. *Each MILP instance has $O(\ell^3)$ integer variables.*

Now we consider all possibilities of how the labels of edges incident with vertices of degree larger than two can relate to each other. Formally, let v be a vertex in G that has degree larger than two and let E_v be the set of edges incident with v. By Observation 2 we have that $|E_v| \leq \ell$. Any labeling λ partitions E_v into at most $\min(\ell, \Delta)$ sets of edges with equal labels, and the values of the labels define an ordering of those sets. We call a partitioning of E_v into at most $\min(\ell, \Delta)$ sets together with the ordering for those sets a *label configuration* for v. A set of label configurations for all vertices v in G that have degree larger than two is called a *global label configuration*. By Observation 2 we get the following.

Observation 4. *There are $O(\ell^{\ell^2})$ global label configurations.*

For each global label configuration, we create an MILP instance. From now on fix a global label configuration σ. We describe how to construct an MILP instance I_σ. Consider a vertex v in G that has degree larger than two. Let u, w be a pair of neighbors of v in G. Let $e = \{u, v\}$ and $e' = \{v, w\}$. If according to the global label configuration, we have that e and e' are in different parts of E_v and the part of e is ordered before the part of e' (that is, e has a smaller label than e'), then we add the following constraint.

$$y_v^{u,w} = z_{e'} - z_e \tag{1}$$

If the part of e is ordered after the part of e' (that is, e has a larger label than e'), then we add the following constraint.

$$y_v^{u,w} = z_{e'} - z_e + \Delta \tag{2}$$

In both above cases, we additionally add the following constraint.

$$1 \leq y_v^{u,w} \leq \Delta - 1 \tag{3}$$

If according to the global label configuration, we have that e and e' are in the same part of E_v (that is, they have the same label), then we add the following constraints.

$$z_e = z_{e'} \quad (4) \qquad y_v^{u,w} = \Delta \quad (5) \qquad 1 \leq z_e \leq \Delta \quad (6) \qquad 1 \leq z_{e'} \leq \Delta \quad (7)$$

Now we consider all vertex pairs s, t in G that have distance at least two in G, that is, the (unique) path from s to t in G has at least one internal vertex. Let $P_{s,t}$ denote the path from s to t in G. Then $V_{\text{int}}^{(s,t)} = V(P_{s,t}) \setminus \{s, t\}$ is the set of internal vertices of $P_{s,t}$. We partition $V_{\text{int}}^{(s,t)}$ into two sets. Let $V_{\text{int,deg2}}^{(s,t)} \subseteq V_{\text{int}}^{(s,t)}$ be the internal vertices of $P_{s,t}$ with degree two and let $V_{\text{int,deg>2}}^{(s,t)} \subseteq V_{\text{int}}^{(s,t)}$ be the internal vertices with degree larger than two. Now we further partition the two sets of vertices.

We define $V_{\text{int,deg2,f}}^{(s,t)} \subseteq V_{\text{int,deg2}}^{(s,t)}$ such that for all $v \in V_{\text{int,deg2,f}}^{(s,t)}$, the neighbor of v that is closer to s is ordered before the neighbor of v that is closer to t.

Analogously, we define $V^{(s,t)}_{\text{int,deg2,b}} \subseteq V^{(s,t)}_{\text{int,deg2}}$ such that for all $v \in V^{(s,t)}_{\text{int,deg2,b}}$, the neighbor of v that is closer to t is ordered before the neighbor of v that is closer to s.

For the vertices in $V^{(s,t)}_{\text{int,deg>2}}$ we additionally define the following. For $v \in V^{(s,t)}_{\text{int,deg>2}}$ denote by $u^{(s,t)}(v)$ the neighbor of v that is on the path from v to s, and denote by $w^{(s,t)}(v)$ the neighbor of v that is on the path from v to t.

We add the following constraint for the vertex pair s, t. Recall here that $D_{s,t}$ comes from the input matrix, and thus acts as a non-variable in the MILP.

$$\sum_{v \in V^{(s,t)}_{\text{int,deg2,f}}} x_v + \sum_{v \in V^{(s,t)}_{\text{int,deg2,b}}} (\Delta - x_v) + \sum_{v \in V^{(s,t)}_{\text{int,deg>2}}} y_v^{u^{(s,t)}(v), w^{(s,t)}(v)} \leq D_{s,t} - 1 \quad (8)$$

Finally, for all fractional variables x_v we add the following constraint.

$$1 \leq x_v \leq \Delta - 1 \quad (9)$$

This finishes the description of the constraints. Since we only want to check whether a feasible solution exists, we arbitrarily choose the vector c for the objective function to be the all-one vector. This finishes the description of the MILP instance I_σ. Clearly, for a given global label configuration, the MILP instance can be constructed in polynomial time.

Observation 5. *Given a global label configuration σ, the MILP instance I_σ can be computed in polynomial time.*

In the following, we prove the correctness of the algorithm. We first show that if the given instance of TTR is a yes-instance, then at least one of the created MILP instances has a feasible solution.

Lemma 4 (⋆). *Given a yes-instance (G, D, Δ) of TTR, there exists a global label configuration σ such that the MILP instance I_σ admits a feasible solution.*

Next, we show that if one of the produced MILP instances admits a feasible solution, then we are facing a yes-instance of TTR. To this end, we first show that the constraint matrix for the fractional variables of the produced MILP instances is totally unimodular. Then by Lemma 2 we have that if one of the MILP instances admits a feasible solution, then there also is a feasible solution for that instance where all fractional variables are set to integer values. To do this, we use a sufficient condition for matrices to be totally unimodular. Precisely, we use that every so-called *network matrix* is totally unimodular.

Definition 3 (network matrix). *Let $T = (V, A)$ be a directed tree, that is, a tree where each arc has an arbitrary orientation and let A' be a set of directed arcs on V, the same vertex set as T. Let M be a matrix with $|A|$ rows and $|A'|$ columns. Then M is a* network matrix *if the following holds. For all i, j let $e_i = (a, b) \in A$ and $e'_j = (s, t) \in A'$.*

- *$M_{i,j} = 1$ if arc (a, b) appears forwards in the path in T from s to t.*

- $M_{i,j} = -1$ *if arc* (a, b) *appears backwards in the path in* T *from* s *to* t.
- $M_{i,j} = 0$ *if arc* (a, b) *does not appear in the path in* T *from* s *to* t.

Lemma 5 ([47]). *Network matrices are totally unimodular.*

We show that the produced MILP instances have the following property.

Lemma 6. *Let* σ *be a global label configuration and let* I_σ *be the corresponding MILP instance. Let* M *be the constraint matrix for the fractional variables, that is, the constraint matrix obtained by treating the integer variables* $y_v^{u,w}$ *and* z_e *in all constraints of* I_σ *as arbitrary constants, deleting all constraints that do not involve any fractional variables, and deleting all duplicate rows. Then* M *is a network matrix.*

Proof. First, notice that Constraints (1), (2), (3), (4), (5), (6), and (7) do not involve fractional variables. Hence, from now on we only consider Constraints (8) and (9), where in Constraint (8) we treat the integer variables as constants. Let M be the resulting constraint matrix where also all duplicate rows are deleted. We show that M is a network matrix.

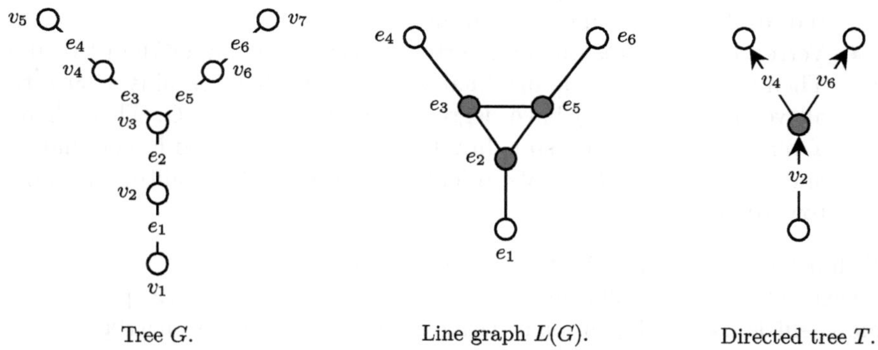

Fig. 2. Illustration of the construction of the directed tree T in Lemma 6. On the left an example tree G, where the vertices are ordered by their indices. In the middle the line graph $L(G)$ of G, where the gray vertices form a maximal clique with more than two vertices. On the right the constructed directed tree T, where the vertices of the maximal clique are merged into one vertex (gray) and the directed arcs correspond to the degree-2 vertices of G.

To this end, we define a directed tree $T = (V, A)$ (see Definition 3) as follows. We give an illustration in Fig. 2. Let $L(G)$ be the line graph of the input tree G of the TTR instance. Now we exhaustively merge all maximal cliques in $L(G)$ with more than two vertices to a single vertex. Here, the new vertex is adjacent to all remaining neighbors of the originally merged vertices. Let this graph be called G'. It is easy to see that G' is a tree, and every edge in G' corresponds to a degree-2 vertex in G, and vice versa. Furthermore, we have that every vertex v

in G' that has degree larger than two corresponds to a subtree G_v (which may be the degenerate tree consisting only of one vertex) in G such that all vertices in $V(G_v)$ have degree larger than two in G. Now we transform G' into T by replacing each edge of G' with an oriented arc. Consider a degree-2 vertex v of G and the corresponding edge e_v in G'. Let u and w be the two neighbors of v in G, where u is ordered before w. We distinguish four cases.

- Vertex u has degree two in G. Let e_u be the edge in G' corresponding to u. Then we orient e_v such that the resulting arc points away from the common endpoint of e_v and e_u.
- The degree of vertex u does not equal two and w has degree two in G. Let e_w be the edge in G' corresponding to w. Then we orient e_v such that the resulting arc points towards the common endpoint of e_v and e_w.
- Neither u nor w have degree two. Then we can assume that at least one of u and w has degree larger than two, otherwise G only has three vertices and we can solve the instance in constant time.
  - Vertex u has degree larger than two in G. Then let G_u denote the maximal subtree of G that contains u such that all vertices in $V(G_u)$ have degree larger than two in G. Then by construction of G', there is a vertex u' in G' corresponding to G_u that is incident with e_v. We orient e_v such that the resulting arc points away from u'.
  - Vertex u has degree one and vertex w has degree larger than two in G. Then let G_w denote the maximal subtree of G that contains w such that all vertices in $V(G_w)$ have degree larger than two in G. Then there is a vertex w' in G' corresponding to G_w that has degree larger than two and is incident with e_v. We orient e_v such that the resulting arc points towards w'.

This finishes the description of the directed tree T.

Next, we make the following observation: Each Constraint (9) produces the same coefficients for the rational variables in the constraint matrix as one of Constraint (8). To see this, note that each Constraint (9) has exactly one non-zero coefficient for one rational variable x_v, which is either 1 or -1. Variable x_v corresponds to a degree-2 vertex v in G. Let u and w be the two neighbors of v in G, where u is ordered before w. Then, since u and w have distance at least two in G, we have a Constraint (8) corresponding to the pair of vertices s, t with $s = u$ and $t = w$. This constraint has exactly one non-zero coefficient for a rational variable, which is x_v and the coefficient is 1. In the case where $s = w$ and $t = u$ we have the same situation, except that the coefficient is -1. Hence, since M has no duplicate rows, we can assume that each row of M corresponds to a Constraint (8).

Now consider a row of M corresponding to Constraint (8), which in turn corresponds to vertex pair s, t in G, where s and t have distance at least two in G. Let $P_{s,t}$ denote the (unique) path from s to t in G. Let $V^{(s,t)}_{\text{int,deg2,f}} \subseteq V(P_{s,t})$ and $V^{(s,t)}_{\text{int,deg2,b}} \subseteq V(P_{s,t})$ be defined as in the description of Constraint (8). Then we have that $V^{(s,t)}_{\text{int,deg2,f}} \cup V^{(s,t)}_{\text{int,deg2,b}}$ are all internal vertices of $P_{s,t}$ that have degree two in G. Furthermore, in Constraint (8) corresponding to vertex pair s, t we have that the fractional variable x_v has a non-zero coefficient if and only if $v \in V^{(s,t)}_{\text{int,deg2,f}} \cup V^{(s,t)}_{\text{int,deg2,b}}$. More specifically, the coefficient of x_v is 1 if $v \in V^{(s,t)}_{\text{int,deg2,f}}$ and the coefficient of x_v is -1 if $v \in V^{(s,t)}_{\text{int,deg2,b}}$.

Furthermore, by construction we have that each vertex in $V^{(s,t)}_{\text{int,deg2,f}} \cup V^{(s,t)}_{\text{int,deg2,b}}$ corresponds to an arc of T. It remains to show that there is a path P' in T that visits exactly the arcs corresponding to the vertices in $V^{(s,t)}_{\text{int,deg2,f}} \cup V^{(s,t)}_{\text{int,deg2,b}}$ such that the arcs corresponding to vertices in $V^{(s,t)}_{\text{int,deg2,f}}$ appear forwards in the path and the arcs corresponding to the vertices in $V^{(s,t)}_{\text{int,deg2,b}}$ appear backwards in the path. To this end, we order the vertices in $V^{(s,t)}_{\text{int,deg2,f}} \cup V^{(s,t)}_{\text{int,deg2,b}}$ in the order in which they are visited by $P_{s,t}$, that is, let $V^{(s,t)}_{\text{int,deg2,f}} \cup V^{(s,t)}_{\text{int,deg2,b}} = \{v_1, v_2, \ldots, v_k\}$. Now we show by induction on k that P' exists and that its last arc corresponds to v_k. If $k = 1$ then P' only consists of the arc e_{v_1} corresponding to v_1. If $v_1 \in V^{(s,t)}_{\text{int,deg2,f}}$, then we consider P' to be the path in which e_{v_1} appears forwards. Otherwise, we consider P' to be the path in which e_{v_1} appears backwards. Now assume $k > 1$. Let P'' be the path in T that visits the arcs corresponding to vertices $\{v_1, \ldots, v_{k-1}\}$ such that for all $i \in [k-1]$ the arc corresponding to v_i appears forwards if $v_i \in V^{(s,t)}_{\text{int,deg2,f}}$ and backwards otherwise. Furthermore, we have that P'' ends with the arc corresponding to v_{k-1}. Now we show that we can append the arc corresponding to v_k to P''. We distinguish two cases.

- In the first case, v_k is visited directly after v_{k-1} by $P_{s,t}$. Then the arcs $e_{v_{k-1}}$ and e_{v_k} in T share a common endpoint which is a degree-2 vertex. It follows that we can append the arc e_{v_k} to P'' to create a new path that ends in e_{v_k}. We still need to show that e_{v_k} is appended in the correct direction. Consider the two neighbors of v_k in G. One of the neighbors is v_{k-1}. Let the second neighbor be w. Now assume that $v_k \in V^{(s,t)}_{\text{int,deg2,f}}$. This implies that v_{k-1} is ordered before w. In this case the arc e_{v_k} in T points away from the common endpoint of $e_{v_{k-1}}$ and e_{v_k}, and hence it appears forwards in the P'' appended with e_{v_k}. The case where $v_k \in V^{(s,t)}_{\text{int,deg2,b}}$ is analogous.
- In the second case, there is a non-empty set of vertices $\widehat{V}$ with degree larger than two, that are visited by $P_{s,t}$ between v_{k-1} and v_k. Let $\widehat{G}$ denote the maximal tree in G that only contains vertices that have degree larger than two in G and that contains $\widehat{V}$. By construction, there is vertex $\widehat{v}$ in T that corresponds to $\widehat{G}$ and both $e_{v_{k-1}}$ and e_{v_k} in T have $\widehat{v}$ as one of their endpoints. First, we argue that we can append e_{v_k} to P'' and thereby create a path in T that ends with e_{v_k}. To this end, notice that the vertex visited by $P_{s,t}$ before v_{k-1} cannot be part of $\widehat{G}$, otherwise G would contain a cycle. It follows that the common endpoint of $e_{v_{k-2}}$ and $e_{v_{k-1}}$ is not $\widehat{v}$ and appending e_{v_k} to P'' indeed produces a path in T that ends with e_{v_k}. It remains to show that e_{v_k} is appended in the correct direction. Consider the two neighbors u and w of v_k in G. We have that one neighbor is contained in $\widehat{G}$, assume w.l.o.g. that

it is u. Then u is visited by $P_{s,t}$ before v_k. Now assume that $v_k \in V^{(s,t)}_{\text{int,deg2,f}}$. This implies that u is ordered before w. In this case the arc e_{v_k} in T points away from u, and hence it appears forwards in the path P'' appended with e_{v_k}. The case where $v_k \in V^{(s,t)}_{\text{int,deg2,b}}$ is analogous.

It follows that for every row in M, there is a path in T such that the conditions in Definition 3 are met. We can conclude that M is a network matrix. □

Now we are ready to show that if one of the produced MILP instances admits a feasible solution, then we are facing a yes-instance of TTR.

Lemma 7 (⋆). *Given an instance (G, D, Δ) of TTR, if there exists a global label configuration σ such that the MILP instance I_σ admits a feasible solution, then (G, D, Δ) is a yes-instance of TTR.*

Theorem 3 follows from the shown results. A formal proof is given in the full version [40].

5 Conclusion

We have initiated the investigation of the natural temporal tree realization problem TTR and shown that it is NP-hard in quite restrictive cases. On the positive side, we provided an FPT-algorithm for the number of leaves in the input tree as a parameter, essentially the only reasonable parameter on trees that is not a constant in at least one of our hardness reductions. A canonical future research direction is to investigate TTR on general graphs. For example, can our FPT-algorithm for number of leaves be transferred to general graphs with respect to the parameter max-leaf number? Further interesting parameters for general graphs include the distance to a clique (or distance to other dense graphs), maximum independent set, clique cover, or even various parameters of the input matrix D.

References

1. Arrighi, E., Grüttemeier, N., Morawietz, N., Sommer, F., Wolf, P.: Multi-parameter analysis of finding minors and subgraphs in edge-periodic temporal graphs. In: Proceedings of the 48th International Conference on Current Trends in Theory and Practice of Computer Science (SOFSEM), pp. 283–297 (2023)
2. Augustine, J., Choudhary, K., Cohen, A., Peleg, D., Sivasubramaniam, S., Sourav, S.: Distributed graph realizations. IEEE Trans. Parallel Distrib. Syst. **33**(6), 1321–1337 (2022)
3. Banavar, J.R., Colaiori, F., Flammini, A., Maritan, A., Rinaldo, A.: Topology of the fittest transportation network. Phys. Rev. Lett. **84**, 4745–4748 (2000)
4. Bar-Noy, A., Choudhary, K., Peleg, D., Rawitz, D.: Efficiently realizing interval sequences. SIAM J. Discret. Math. **34**(4), 2318–2337 (2020)
5. Bar-Noy, A., Choudhary, K., Peleg, D., Rawitz, D.: Graph realizations: maximum degree in vertex neighborhoods. Discret. Math. **346**(9), 113483 (2023)

6. Bar-Noy, A., Peleg, D., Perry, M., Rawitz, D.: Composed degree-distance realizations of graphs. Algorithmica **85**(3), 665–687 (2023)
7. Bar-Noy, A., Peleg, D., Perry, M., Rawitz, D.: Graph realization of distance sets. Theor. Comput. Sci. **1019**, 114810 (2024)
8. Barthélemy, M., Flammini, A.: Optimal traffic networks. J. Stat. Mech. Theory Exp. **2006**(07), L07002 (2006)
9. Behzad, M., Simpson, J.E.: Eccentric sequences and eccentric sets in graphs. Discrete Math. **16**(3), 187–193 (1976)
10. Bixby, R.E., Wagner, D.K.: An almost linear-time algorithm for graph realization. Math. Oper. Res. **13**(1), 99–123 (1988)
11. Bohn, S., Magnasco, M.O.: Structure, scaling, and phase transition in the optimal transport network. Phys. Rev. Lett. **98**, 088702 (2007)
12. Casteigts, A., Flocchini, P., Quattrociocchi, W., Santoro, N.: Time-varying graphs and dynamic networks. Int. J. Parallel Emergent Distrib. Syst. **27**(5), 387–408 (2012)
13. Chaudhary, J., Molter, H., Zehavi, M.: Parameterized analysis of bribery in challenge the champ tournaments. In: Proceedings of the 33rd International Joint Conference on Artificial Intelligence (IJCAI), pp. 2704–2712 (2024)
14. Chen, W.-K.: On the realization of a (p, s)-digraph with prescribed degrees. J. Franklin Inst. **281**(5), 406–422 (1966)
15. Chung, F., Garrett, M., Graham, R., Shallcross, D.: Distance realization problems with applications to internet tomography. J. Comput. Syst. Sci. **63**(3), 432–448 (2001)
16. Culberson, J.C., Rudnicki, P.: A fast algorithm for constructing trees from distance matrices. Inf. Process. Lett. **30**(4), 215–220 (1989)
17. Dadush, D., Peikert, C., Vempala, S.: Enumerative lattice algorithms in any norm via M-ellipsoid coverings. In: Proceedings of the 52nd IEEE Annual Symposium on Foundations of Computer Science (FOCS), pp. 580–589. IEEE (2011)
18. Erdös, P., Gallai, T.: Graphs with prescribed degrees of vertices. Mat. Lapok **11**, 264–274 (1960)
19. Erlebach, T., Morawietz, N., Spooner, J.T., Wolf, P.: A cop and robber game on edge-periodic temporal graphs. J. Comput. Syst. Sci. **144**, 103534 (2024)
20. Erlebach, T., Morawietz, N., Wolf, P.: Parameterized algorithms for multi-label periodic temporal graph realization. In: Proceedings of the 3rd Symposium on Algorithmic Foundations of Dynamic Networks (SAND), pp. 12:1–12:16 (2024)
21. Erlebach, T., Spooner, J.T.: A game of cops and robbers on graphs with periodic edge-connectivity. In: Proceedings of the 46th International Conference on Current Trends in Theory and Practice of Informatics (SOFSEM), pp. 64–75 (2020)
22. Frank, A.: Augmenting graphs to meet edge-connectivity requirements. SIAM J. Discret. Math. **5**(1), 25–53 (1992)
23. Frank, A.: Connectivity augmentation problems in network design. In: Mathematical Programming: State of the Art 1994 (1994)
24. Frank, H., Chou, W.: Connectivity considerations in the design of survivable networks. IEEE Trans. Circ. Theor. **17**(4), 486–490 (1970)
25. Fulkerson, D.R.: Zero-one matrices with zero trace. Pac. J. Math. **10**(3), 831–836 (1960)
26. Golovach, P.A., Mertzios, G.B.: Graph editing to a given degree sequence. Theoret. Comput. Sci. **665**, 1–12 (2017)
27. Golumbic, M.C., Trenk, A.N.: Tolerance Graphs. Cambridge Studies in Advanced Mathematics. Cambridge University Press (2004)

28. Gomory, R.E., Hu, T.C.: Multi-terminal network flows. J. Soc. Ind. Appl. Math. **9**(4), 551–570 (1961)
29. Grötschel, M., Monma, C.L., Stoer, M.: Design of survivable networks. Handbooks Oper. Res. Manage. Sci. **7**, 617–672 (1995)
30. Hakimi, S.L.: On realizability of a set of integers as degrees of the vertices of a linear graph. I. J. Soc. Ind. Appl. Math. **10**(3), 496–506 (1962)
31. Hakimi, S.L., Yau, S.S.: Distance matrix of a graph and its realizability. Quart. Appl. Math. **22**(4), 305–317 (1965)
32. Hell, P., Kirkpatrick, D.: Linear-time certifying algorithms for near-graphical sequences. Discret. Math. **309**(18), 5703–5713 (2009)
33. Holme, P., Saramäki, J.: Temporal Network Theory, vol. 2. Springer (2019)
34. Kempe, D., Kleinberg, J.M., Kumar, A.: Connectivity and inference problems for temporal networks. J. Comput. Syst. Sci. **64**(4), 820–842 (2002)
35. Klobas, N., Mertzios, G.B., Molter, H., Spirakis, P.G.: Temporal graph realization from fastest paths. In: Proceedings of the 3rd Symposium on Algorithmic Foundations of Dynamic Networks (SAND), pp. 16:1–16:18 (2024)
36. Lenstra, H.W.: Integer programming with a fixed number of variables. Math. Oper. Res. **8**, 538–548 (1983)
37. Lesniak, L.: Eccentric sequences in graphs. Period. Math. Hung. **6**, 287–293 (1975)
38. McConnell, R.M., Spinrad, J.P.: Construction of probe interval models. In: Proceedings of the 13th Annual ACM-SIAM Symposium on Discrete Algorithms (SODA), pp. 866–875 (2002)
39. McMorris, F.R., Wang, C., Zhang, P.: On probe interval graphs. Discrete Appl. Math. **88**(1), 315–324 (1998). Computational Molecular Biology DAM - CMB Series
40. Mertzios, G.B., Molter, H., Morawietz, N., Spirakis, P.G.: Realizing temporal transportation trees. *CoRR*, abs/2403.18513, (2025). https://arxiv.org/abs/2403.18513
41. Michail, O., Spirakis, P.G.: Elements of the theory of dynamic networks. Commun. ACM **61**(2), 72–72 (2018)
42. Morawietz, N., Rehs, C., Weller, M.: A timecop's work is harder than you think. In: Proceedings of the 45th International Symposium on Mathematical Foundations of Computer Science (MFCS), vol. 170, pp. 71:1–71:14 (2020)
43. Morawietz, N., Wolf, P.: A timecop's chase around the table. In: Proceedings of the 46th International Symposium on Mathematical Foundations of Computer Science (MFCS) (2021)
44. Nicosia, V., Tang, J., Mascolo, C., Musolesi, M., Russo, G., Latora, V.: Graph metrics for temporal networks. In: Temporal Networks, Springer (2013)
45. Patrinos, A.N., Louis Hakimi, S.: The distance matrix of a graph and its tree realization. Quart. Appl. Math. **30**, 255–269 (1972)
46. Rubei, E.: Weighted graphs with distances in given ranges. J. Classif. **33**, 282–297 (2016)
47. Schrijver, A.: Combinatorial Optimization: Polyhedra and Efficiency, vol. 24. Springer (2003)
48. Tamura, H., Sengoku, M., Shinoda, S., Abe, T.: Realization of a network from the upper and lower bounds of the distances (or capacities) between vertices. In: Proceedings of the 1993 IEEE International Symposium on Circuits and Systems (ISCAS), pp. 2545–2548 (1993)

Interval k-graphs : Recognition and Forbidden Obstructions

Haiko Müller[1] and Arash Rafiey[2(✉)]

[1] School of Computer Science, University of Leeds, Leeds, UK
h.muller@leeds.ac.uk
[2] Indiana State University, Terre Haute, IN, USA
arash.rafiey@indstate.edu

Abstract. We introduce the class of *interval k-graphs*, which is the generalization of interval graphs, particularly interval bigraphs. For a fixed k, we say that an input graph G with given partition $V_1, \ldots, V_k$ of its vertices into independent sets is an interval k-graph if each vertex $v \in G$ can be represented by an interval I_v from a real line so that $u \in V_i$ and $v \in V_j$, $i \neq j$ are adjacent if and only if the intervals I_u and I_v intersect.

We study the ordering characterization and forbidden obstructions of interval k-graphs and present a polynomial-time recognition algorithm for them. Special cases of interval k-graphs, particularly comparability interval k-graphs, were previously studied in [2], where the complexity of interval k-graph recognition was posed as an open problem.

1 Introduction and Problem Definition

The vertex set of a graph G is denoted by $V(G)$, and the edge set of G is denoted by $E(G)$. A graph G is called an *interval graph*, if there exists a family I_v, $v \in V(G)$, of intervals (from the real line) such that, for all different $x, y \in V(G)$ the vertices x and y are adjacent in G if and only if I_x and I_y intersect. A bigraph G is a bipartite graph with a fixed bipartition into *black* and *white* vertices. We sometimes denote these sets as B and W, and view the vertex set of G as partitioned into (B, W). A bigraph G is called an *interval bigraph* if there exists a family I_v, $v \in B \cup W$, of intervals (from the real line) such that, for all $x \in B$ and $y \in W$, the vertices x and y are adjacent in G if and only if I_x and I_y intersect. This family of intervals is called an *interval representation* of the bigraph G.

Interval bigraphs were introduced in [13] and have been studied in [5,14,19]. They are closely related to interval digraphs introduced by Sen *et al.* [22]. Interval bigraphs and interval digraphs have become of interest in such new areas as graph homomorphisms, *e.g.* [11].

Supported by Bailey Faculty Fellowship.

H. Fernau and P. Kindermann (Eds.): WG 2025, LNCS 16124, pp. 405–417, 2026.
https://doi.org/10.1007/978-3-032-11835-6_29

A *co-circular arc bigraph* is a bipartite graph whose complement is a circular arc graph (see [4] for the definitions of graph classes not introduced here). The class of interval bigraphs is a subclass of co-circular arc bigraphs. Indeed, the former class consists exactly of those bigraphs whose complement is the intersection of a family of circular arcs no two of which cover the circle [14]. There is a linear-time recognition algorithm for co-circular arc bigraphs [18]. On the other hand, the class of interval bigraphs is a super-class of proper interval bigraphs (also known as bipartite permutation graphs [23] or monotone graphs [10]), for which there is also a linear-time recognition algorithm [14,23].

Interval bigraphs can be recognized in polynomial time using the algorithm developed by Müller in [19]. This algorithm runs in time $\mathcal{O}(nm^6(n+m)\log n)$ where m is the number of edges and n is the number of vertices in the input bigraph G. A faster algorithm was developed in [21]; with a running time $\mathcal{O}(mn)$. By contrast, several linear time algorithms for recognition of *interval graphs* are known, *e.g.*, [3,6,7,17].

We use the ordering characterization of interval bigraphs in [14]. A bigraph G is an interval bigraph if and only if its vertices admit a linear ordering $<$ without any of the forbidden patterns in Fig. 1. Hence, we will rely on the existence of a linear ordering $<$ such that if $v_a < v_b < v_c$ (not necessarily consecutively) and v_a, v_b have the same color and opposite to the color of v_c then $v_a v_c \in E(G)$ implies that $v_b v_c \in E(G)$.

Fig. 1. Forbidden patterns for interval bigraphs

We call such an ordering $<$ a *bi-interval ordering* for G. There are several graph classes that can be characterized by the existence of an ordering without a number of forbidden patterns. One such class is the class of interval graphs. A graph G is an interval graph if and only if there exists an ordering $<$ of $V(G)$ such that none of the following patterns appears [8,9].

- $v_a < v_b < v_c$, $v_a v_c, v_b v_c \in E(G)$ and $v_a v_b \notin E(G)$
- $v_a < v_b < v_c$, $v_a v_c \in E(G)$ and $v_b v_c, v_a v_b \notin E(G)$

Some of the other classes of graphs that have ordering characterizations without forbidden patterns are proper interval graphs, comparability graphs, co-comparability graphs, chordal graphs, convex bipartite graphs, co-circular arc bigraphs, permutation bigraphs, and proper interval bigraphs [16]. We, in particular, mention the ordering characterization of permutation bigraphs and co-circular arc bigraphs without forbidden patterns.

A bigraph $G = (A, B, E)$ is a co-circular arc bigraph if there is a linear ordering $a_1 < \cdots < a_p < b_1 < b_2 < \cdots < b_q$ (with $A = \{a_1, \ldots, a_p\}, B =$

$\{b_1, \ldots, b_q\}$) so that if $a_i b_j, a_{i'} b_{j'} \in E$ with $i < i'$ and $j' < j$ then $a_i b_{j'} \in E$. Such an ordering $<$ is called a *min ordering* of G [15]. The forbidden patterns corresponding to a min ordering are given in Fig. 2.

Fig. 2. Forbidden Patterns for co-circular arc bigraphs

Likewise, a bigraph $G = (A, B, E)$ is a permutation bigraph if there is a linear ordering $a_1 < \cdots < a_p < b_1 < b_2 < \cdots < b_q$ (with $A = \{a_1, \ldots, a_p\}, B = \{b_1, \ldots, b_q\}$) so that if $a_i b_j, a_{i'} b_{j'} \in E$ with $i < i'$ and $j' < j$ then $a_i b'_j, a'_i b_j \in E$. Such an ordering $<$ is called a *min-max ordering* of G [12]. The forbidden patterns corresponding to a min-max ordering are given in Fig. 3.

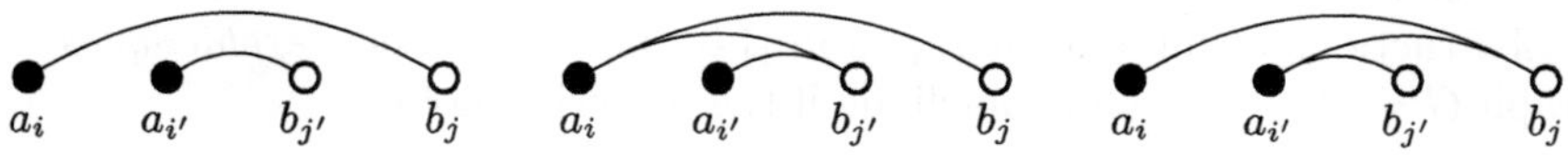

Fig. 3. Forbidden Patterns for bipartite permutation graphs (proper interval bigraphs)

Let $G = (A, B, E)$ be an interval bigraph. Let $x_1 \prec \cdots \prec x_n$ be a linear ordering of vertices of G without the forbidden patterns in Fig. 1. Let $<$ be the reverse of $\prec$; that is, let $x_n < \cdots < x_1$. Let $a_1 < a_2 < \cdots < a_p$ be the vertices in A under ordering $<$, and $b_1 < b_2 < \cdots < b_q$ be the vertices in B under ordering $<$. Then, it is easy to see that $a_1 < \cdots < a_p < b_1 < b_2 < \cdots < b_q$ provides a min ordering for G. Thus, interval bigraphs are a subclass of co-circular arc bigraphs.

Similarly, let $G = (A, B, E)$ be a permutation bigraph, and let $a_1 < \cdots < a_p < b_1 < \cdots < b_q$ be a min-max ordering of G (with $A = \{a_1, \ldots, a_p\}$, $B = \{b_1, b_2, \ldots, b_q\}$). Then, $<$ is a bi-interval ordering for G. Thus, permutation bigraphs are a subclass of interval bigraphs.

Following this line of research, we introduce a class of interval k-graphs where we still have a total ordering of the vertices of the input graph G, but there are partitions of the vertices, and the forbidden patterns are defined with colored vertices. We consider the graph G, which admits a k-coloring with the given coloring (given partitions), and we say G is an interval k-graph according to the following definition.

Definition 1 (interval k-graphs). *Let G be a k-partite graph ($k \geq 2$) with the given partite sets $V_1, V_2, \ldots, V_k$. We say that G is an* interval k-graph *if there is a family of real line intervals I_v, $v \in V(G)$, so that for all $u, v \in V(G)$ from different partite sets, $uv \in E(G)$ if and only if I_u, I_v intersect.*

Herein, we identify $V_1, \ldots, V_k$ with a k-coloring of G and simply say that two vertices have different colors whenever they belong to different partite sets. The authors of [2] studied special cases of interval k-graphs, namely cocomparability interval k-graphs.

Notice that an interval graph G is an interval k-graph where $k = |G|$. Thus, interval k-graphs generalize interval graphs. In Theorem 4 show that G is an interval k-graph if and only if G admits an ordering $u_1 < u_2 < \cdots < u_n$ of its vertices without the forbidden patterns depicted in Fig. 4.

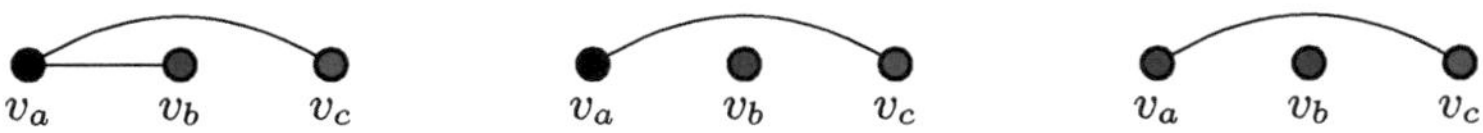

Fig. 4. The three forbidden patterns for interval k-graphs. In each pattern the three vertices have different colors.

Some Obstructions for Interval k-graphs. We have seen some of the forbidden obstructions of interval bigraphs in [14]. They are called *exobicliques*. The bigraph $G = (B, W)$ is an exobiclique if the following holds.

- B contains a nonempty part B_1 and W contains a nonempty part W_1 such that $B_1 \cup W_1$ induces a biclique in G;
- $B \setminus B_1$ contains three vertices with incomparable neighborhood in W_1 and $W \setminus W_1$ contains three vertices with incomparable neighborhoods in B_1 (Fig. 5).

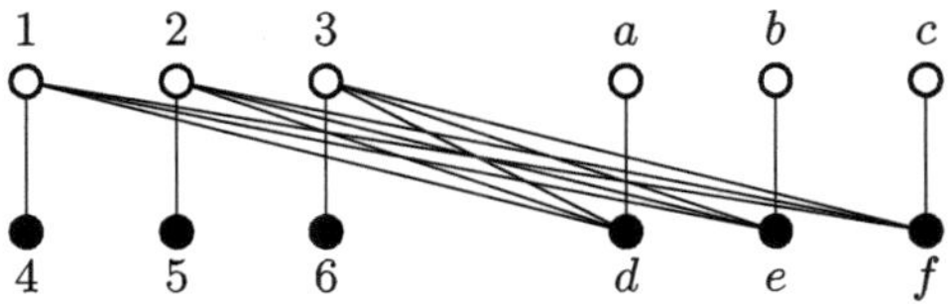

Fig. 5. Exobicliques: Here, $B = \{4, 5, 6, d, e, f\}$, $W = \{1, 2, 3, a, b, c\}$ and $B_1 = \{d, e, f\}$, $W_1 = \{1, 2, 3\}$ and $B \setminus B_1 = \{4, 5, 6\}$, $W \setminus W_1 = \{a, b, c\}$.

However, the obstruction for interval bigraphs (interval 2-graphs) are not limited to exobiclique, there is a family of these obstruction considered in [21]. Figure 6 is an obstruction of the interval bigraph that is not exobiclique. There are some new obstructions for interval k-graphs $k > 2$, depicted in Fig. 7.

1.1 Our Results and Future Work

Our primary contribution is the development of a recognition algorithm for interval k-graphs.

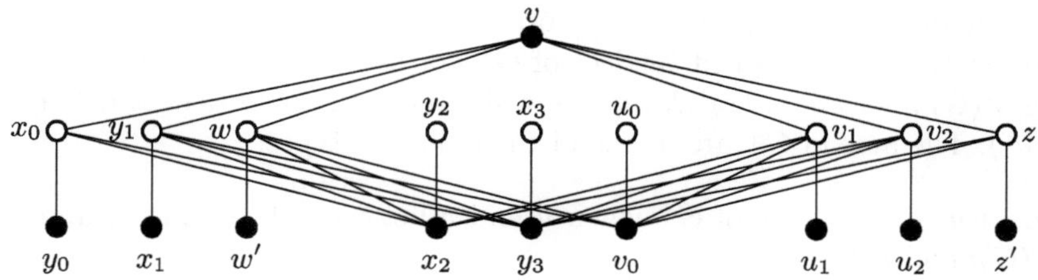

Fig. 6. An obstruction for interval bigraph which is not exobiclique

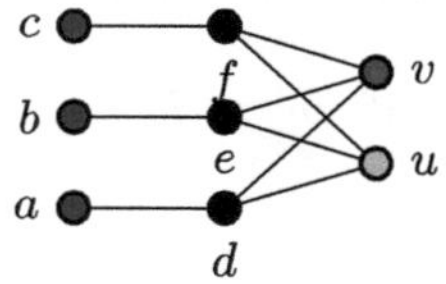

Fig. 7. Obstruction for interval 4-graphs; vertices a, b and c share one color, d, e and f share a second color, and u and v have colors three and four.

Theorem 1. *Let G be a graph with a given partition of its vertices into k partite sets $V_0, V_1, \ldots, V_{k-1}$. Then, it can be determined in $O(|V(G)||E(G)|)$ time whether G is an interval k-graph.*

An interesting challenge arises when the k-coloring of G is not provided, yet we still seek to determine whether G is an interval k-graph. However, we do not have a conclusive indication that this problem becomes NP-complete when the k-coloring (a k-partition of its vertices) is not given. The authors of [2] investigated this problem for specific cases where $k = 2, 3$. This leads us to propose the following open problem:

Open Problem 2. Can one determine in polynomial time whether a given graph G admits a k-partition $V_1, V_2, \ldots, V_k$ such that G, along with this partitioning, forms an interval k-graph?

Identifying forbidden obstructions for interval k-graphs remains a significant challenge, even for $k = 2$.

Open Problem 3. What are the forbidden obstructions for interval k-graphs?

2 Basic Definitions and Some Preliminary Results

Note that for $k > 1$, a k-partite graph G is an interval k-graph if and only if each connected component of it is an interval k-graph. In the remainder of this paper, we assume that $G = (V, E)$ is a connected k-partite graph with a fixed partition $V_1, V_2, \ldots, V_k$. By set of edges in the complete k-partite graph with partite sets $V_1, \ldots, V_k$ that are not present in G we denote by $\bar{E} = \{uv \mid u \in V_i, v \in V_j, 1 \leq i < j \leq k\} \setminus E$.

We define *pair-digraph* G^+ of G corresponding to the forbidden patterns in Fig. 4, as follows. The set of vertices of G^+ consists of all pairs (u, v) such that $u, v \in V(G)$ and $u \neq v$. For clarity, we will often refer to vertices of G^+ as *pairs* (in G^+). The arcs in G^+ are of one of the following types:

- $(u, v)(u', v)$ is an arc of G^+ when u and v belong to the same V_i and $uu' \in E(G)$, and $vu' \notin E(G)$.
- $(u, v)(u', v)$ is an arc of G^+ if $uu' \in E(G)$ and $u'v \in \bar{E}(G)$ and u, v, v' all belong to different V_i.
- $(u, v)(u, v')$ is an arc of G^+ when u and v' belong to the same V_i with $vv' \in E(G)$, and $uv \notin E(G)$.
- $(u, v)(u, v')$ is an arc of G^+ if $vv' \in E(G)$ and $uv \in \bar{E}(G)$ and u, v, v' all belong to different V_i.

Observe that if $(u, v)(u', v')$ is an arc, then both uv and $u'v'$ are non-edges of G. For two pairs $(x, y), (x', y') \in V(G^+)$ we say (x, y) *dominates* (x', y') (or (x', y') is *dominated by* (x, y)) and write $(x, y) \to (x', y')$ if there exists an arc (directed edge) from (x, y) to (x', y') in G^+. One should note that if $(x, y) \to (x', y')$ in G^+ then $(y', x') \to (y, x)$, to which property we will refer to as *skew-symmetry*.

Remark. This idea of a pair-digraph can also be applied to the forbidden patterns in Fig. 2. With $M = (A, B, E)$ being a connected bigraph, we define the *pair-digraph* M^* of M corresponding to the forbidden pattern in Fig. 2 as follows. We set $V(M^*) = \{(u, v) \mid u, v \in A \text{ or } u, v \in B\}$ and $A(M^*) = \{(u, v)(u', v') \mid uu', vv' \in E,\ uv' \notin E\}$. Notice that if $(u, v)(u', v') \in A(M^*)$ then $(u', v')(u, v'), (u, v')(u, v) \in A(M^+)$. Then, it is easy to see that all vertices of each strong component of M^* belong to the same strong component of M^+.

Lemma 1. *Let $<$ be an ordering of G without the forbidden patterns in Fig. 4, and let $(u, v) \to (u', v')$ with $u < v$. Then, $u' < v'$.*

We shall generally refer to a strong component of G^+ simply as a *component* of G^+. We shall also identify a component by its vertex (pair) set. A component in G^+ is called *non-trivial* if it contains more than one pair. For any component S of G^+, we define its *couple component*, denoted S', to be $S' = \{(u, v) :\ (v, u) \in S\}$.

The skew-symmetry property of G^+ implies the following fact.

Lemma 2. *If S is a component of G^+ then so is S'.*

In light of Lemma 2, for each component S of G^+, S and S' are couple components of each other and we shall collectively refer to them as *coupled components*. It can be easily shown that the coupled components S and S' are disjoint or equal. In the latter case, we say component S is *self-coupled*.

Definition 2 (circuit). *For $n \geq 1$, a sequence $(x_0, x_1), (x_1, x_2), \ldots, (x_{n-1}, x_n)$, (x_n, x_0) of pairs in a set $D \subseteq V(G^+)$ is called a* circuit *in D.*

Lemma 3. *If a component of G^+ contains a circuit then G is not an interval k-graph.*

If G^+ contains a self-coupled component then G is not an interval k-graph. This is because a self-coupled component of G^+ contains two such pairs as (u, v) and (v, u), which comprise a circuit of length 2 (corresponding to $n = 1$ in the definition of a circuit). However, we will show that these are not the only obstructions to interval k-graphs (and in particular interval bigraphs), that is, there are bigraphs G which are not interval bigraphs, despite G^+ not having any self-coupled component. In contrast, the obstructions to co-circular arc bigraphs are precisely the components of G^* containing both pairs $(x, y), (y, x)$ [15].

Lemma 4. *Suppose that G^+ contains no self-coupled components, and let D be any subset of $V(G^+)$ containing exactly one component from each pair of coupled components. Then, D is the set of arcs of a tournament on $V(G)$. Moreover, such a D can be chosen to be a transitive tournament if and only if G is an interval k-graph for any $k \geq 2$.*

In what follows, by a *component* we mean a non-trivial (strong) component unless we specify otherwise. For simplicity, we shall use a set S of pairs in G^+ to also denote the sub-digraph of G^+ induced by S, when no confusion arises.

We shall say two edges ab and cd of G are *independent* if the subgraph of G induced by the vertices a, b, c, and d has just the two edges ab and cd. We shall say that two disjoint induced subgraphs G_1 and G_2 of G are *independent* if there is no edge of G with one endpoint in G_1 and another endpoint in G_2.

First, to describe the algorithm, we introduce some technical definitions.

Definition 3 (reachability closure). *Let R be a subset of the pairs of G^+. Let $N^+[R]$ denote the set of all pairs in G^+ that are reachable (via a directed path in G^+) from a pair in R. (Notice that $N^+[R]$ contains R.) We call $N^+[R]$ the* reachability closure *of R. We say that a pair (u, v) is* implied *by R if $(u, v) \in N^+[R] \setminus R$. If $R = N^+[R]$, we say that R is* closed *under reachability.*

Definition 4 (envelope). *Let R be a set of pairs of G^+. The* envelope *of R, denoted $N^*[R]$, is the smallest set of pairs that contains R and is closed under both reachability and transitivity (if $(u, v), (v, w) \in N^*[R]$ then $(u, w) \in N^*[R]$).*

Lemma 5. *Let S, S' be the coupled components in G^+, so that both $N^*[S]$ and $N^*[S']$ contain a circuit. Then G is not an interval k-graph.*

The following theorem gives an ordering characterization of interval k-graphs based on forbidden patterns.

Theorem 4. *For a fixed $k \geq 2$, let G be a k-partite graph with the given partite sets $V_1, V_2, \ldots, V_k$. G is an interval k-graph if and only if G admits an ordering $u_1 < u_2 < \cdots < u_n$ of its vertices without the forbidden patterns in Fig. 4.*

Proof. We denote the right and left end-points of an interval I by $r(I)$ and $\ell(I)$, respectively.

First, suppose there is an interval representation I_v, $v \in G$ of G. Now, consider the total ordering $v_1 < v_2 < \cdots < v_n$ of G where for all i,j we have $v_i < v_j$ when either $r(I_{v_i}) < r(I_{v_j})$, or $r(I_{v_i}) = r(I_{v_j})$ and $\ell(I_{v_i}) \leq \ell(I_{v_j})$. In other words, we have $v_i < v_j$ whenever $(r(I_{v_i}), \ell(I_{v_i})) < (r(I_{v_j}), \ell(I_{v_j}))$ in the lexicographic ordering of pairs of real numbers.

Now consider three indexes $a < b < c$. Assume $v_a v_c$ is an edge of G, and vertices v_b and v_c have different colors. Since $v_a v_c \in E(G)$, I_{v_a} and I_{v_c} intersect and, hence, $\ell(I_{v_c}) < r(I_{v_a})$. Moreover, since $v_a < v_b$, we have $r(I_{v_a}) \leq r(I_{v_b})$. Therefore, $\ell(I_{v_c}) < r(I_{v_b})$; i.e., I_{v_b} and I_{v_c} intersect. Thus, $v_b v_c$ is an edge of G, implying that none of the forbidden patterns in Fig. 4 occurs.

In contrast, let $v_1 < v_2 < \cdots < v_n$ be an ordering of the vertices of G without the forbidden patterns in Fig. 4. For each i set $r(J_{v_i}) = i$ and $\ell(J_{v_i}) = \min(\{i\} \cup \{j : v_j < v_i, v_i v_j \in E(G)\})$. One can easily see that J_v, $v \in G$, is an interval representation for G. □

One can observe that if u and v are two vertices of G so that they have in-comparable neighborhoods then (u,v) is in a component of G^+. Indeed, with $uu', vv' \in E(G)$ and $uv', u'v \notin E(G)$ we get $(u,v) \rightarrow (u',v) \rightarrow (u',v') \rightarrow (u,v') \rightarrow (u,v)$. Using the same reasoning, if a pair (x,y) outside component S is dominated by a vertex in S, then $N(x) \subseteq N(y)$.

3 Structural Properties of Strong Components in G^+

The following Lemma follows from the definition of G^+.

Lemma 6. *If uu' and vv' are independent edges in G then the pairs (u,v), (u',v), (u',v'), and (u,v') form a directed four-cycle of G^+ in the given order (resp. in reverse order). In particular, (u,v), (u',v), (u',v'), and (u,v') belong to the same component of G^+.*

Lemma 7. *Let S be a component of G^+ containing a pair (u,v) then one of the following occurs.*

1. *$uv \in E(G)$ or u and v have the same colors and there exist u',v' where uu', vv' are edges of G, $uv', vu' \notin E(G)$. Furthermore, the four pairs $(u,v), (u,v')$, (u',v), and (u',v') are contained in S.*
2. *u and v have different colors and $uv \notin E(G)$ and there exists u' such that $uu' \in E(G)$, $vu' \notin E(G)$ and u,u',v all have different colors.*

The structure of the components of G^+ is quite special, and the trivial components interact with them in simple ways. A trivial component will be called a *source* if its unique pair has in-degree zero, and a *sink* if its unique pair has out-degree zero. We further explore these properties by establishing several lemmas. We need the following definition of the reachability of pairs in G^+.

Definition 5 (reachability closure) *Let R be a subset of the pairs of G^+. Let $N^+[R]$ denote the set of all pairs in G^+ that are reachable (via a directed path in G^+) from a pair in R. (Notice that $N^+[R]$ contains R.) We call $N^+[R]$ the* reachability closure *of R. We say that a pair (u, v) is* implied *by R if $(u, v) \in N^+[R] \setminus R$. If $R = N^+[R]$, we say that R is* closed *under reachability.*

Lemma 8. *A pair (a, c) is in $N^+[S] \setminus S$ (implied by S) for some component S of G^+ if one of the following occurs.*

1. *a and c have the same color. $N(a) \subseteq N(c)$, and there exist $bd, dc \in E(G)$ so that a, b, d all have different colors and $ab, ad \notin E(G)$, and $(a, b), (a, d)$ are in S.*
2. *a and c have the same color. $N(a) \subseteq N(c)$ and G contains path a, b, c, d, e, such that and $ad, be \notin E(G)$, and $(a, d), (a, e), (b, d), (b, e)$ lie in S.*
3. *$ac \in E(G)$ and G contains a path b, a, c, d such that $N(a) \setminus \{c\} \subseteq N(c)$ and $ad, bd \notin E(G)$, $(a, d), (b, d)$ lie in S.*
4. *$ac \in E(G)$ and G contains path b, d, c such that $N(a) \setminus \{c\} \subseteq N(c)$, and $ab, ad \notin E(G)$ and a, d have different colors and a, b have different colors. Furthermore, $(a, d), (a, b) \in S$.*

We emphasize that ab and de from Lemma 8 are independent edges. Inclusion $N(a) \subseteq N(c)$ implies the following corollary.

Corollary 6. *If there is an arc from a component S of G^+ to a pair $(x, y) \notin S$ then (x, y) forms a trivial component of S that is a sink component. If there is an arc to a component S of G^+ from a pair $(x, y) \notin S$ then (x, y) forms a trivial component of G^+ that is a source. In particular, if there is a directed path in G^+ from component S_1 to component S_2, then $S_1 = S_2$.*

4 Interval k-Graph Recognition

In this section, we present our algorithm for the recognition of interval bigraphs. Firstly, to describe the algorithm, we introduce some technical definitions.

Definition 7 *(envelope, $N^*[D]$). Let R be a set of pairs of G^+. The* envelope *of R, denoted $N^*[R]$, is the smallest set of pairs that contains R and is closed under both reachability and transitivity (if $(u, v), (v, w) \in N^*[R]$ then $(u, w) \in N^*[R]$).*

Remark and $N_l^*[D]$ Definition. For the purposes of the proofs, we visualize taking the envelope of R as divided into consecutive *levels*, where in the zero-th level we just replace R by its reachability closure, and in each subsequent level we replace R by the rechability closure of its transitive closure. The pairs in the envelope of R can be thought of as forming the arc of a digraph on $V(G)$, and each pair can be thought of as having a label corresponding to its level. The pairs (arcs of the digraph) in R, and those implied by R have label 0, arcs obtained by transitivity from the arcs labeled 0, as well as all arcs implied by them have label 1, and so on. More precisely, $N^*[R]$ is the disjoint union of

$R^0, R^1, \ldots, R^k$, where $R^0 = N^+[R]$ (level zero), and each R^i (level $i \geq 1$) consists of every pair (u, v) such that either (u, v) is obtainable by transitivity in R^{i-1} (meaning that there is some sequence $(u, u_1), (u_1, u_2), \ldots, (u_{r-1}, u_r), (u_r, v)$ in R^{i-1}), or (u, v) is dominated by a pair (u', v') obtainable by transitivity in R^{i-1}. Let $N_l^*[D] = \bigcup_{i=0}^{i=l} R^i$. Note that $R \subseteq N^+[R] \subseteq N^*[R]$.

Definition 8 (dictator component). *Let $\mathcal{R} = \{R_1, R_2, \ldots, R_k, S\}$ be a set of components of G^+ such that $N^*[\bigcup_{A \in \mathcal{R}} A]$ contains a circuit. We say S is a* dictator *if for every subset W of $\mathcal{R} \setminus \{S\}$, there exist a circuit in the envelope of $(\bigcup_{A \in W'} A) \cup (\bigcup_{B \in \mathcal{R} \setminus W} B)$, where $W' = \{R_i' \mid R_i \in W\}$. In other words, S is a dictator if by replacing some of the R_is with R_i's in $\mathcal{R}$ and taking the envelope of the union of elements we still get a circuit.*

Definition 9 (complete set). *A set $D_1 \subseteq V(G^+)$ is called* complete *if for every pair of coupled components R, R' of G^+, exactly one of $R \subseteq D_1$ and $R' \subseteq D_1$ holds.*

A component S is a dictator if and only if the envelope of every complete set D_1 containing S has a circuit.

Definition 10 (simple pair, complex pair). *A pair $(x, y) \in G^+$ is* simple *if it belongs to $N^+[S]$ for some component S, otherwise we call it* complex.

Definition 11. *Let D be a complete set and let C be a circuit in $N^*[D]$. We say C is a* minimal circuit *if first, the latest pair in C is created as early as possible (the smallest possible level) during the execution of $N^*[D]$; second, C has the minimum length; third, no pair in C is by transitivity.*

Definition of Color $c(u)$. For every vertex u of G we denote by $c(u)$ index of the partite class containing u. This is the color of u; $c(u) = k$ if $u \in V_k$.

High-Level Overview of the Algorithm. The algorithm begins by constructing G^+. If any component of G^+ is self-coupled, then G is not an interval k-graph. Next, we initialize an empty set D (which will store selected pairs). If we add a pair (x, y) into D, it means that in the final ordering, x must appear before y.

The core of the algorithm involves selecting components of G^+ based on the following principles: (1) If $(x, y) \in D$, then (y, x) must be discarded, (2) If $(x, y) \in D$ and $(x, y)(x', y') \in A(G^+)$, then (x', y') must also be in D, (3) If $(x, y), (y, z) \in D$, then (x, z) must also be in D.

Thus, once a component (x, y) is added to D, all pairs within its corresponding component $S_{x,y}$ are also included in D.

Stage 1: Selecting Components: We first compute an ordering $v_1, v_2, \ldots, v_n$ of the vertices of G such that if $v_i < v_j$, then the color of v_i is the same as or smaller than the color of v_j. The algorithm proceeds in steps $i = 1, 2, \ldots$, where at each step, we consider components S_{v_i, v_j} (with $i < j$) for inclusion in D. The selection follows these rules: If adding S_{v_i, v_j} to D (along with its outgoing neighbors $N^+[S_{v_i, v_j}]$) does not create a *circuit*, then we include $N^+[S_{v_i, v_j}]$ in D

and discard S_{v_j,v_i}, Otherwise, we try to add $N^+[S_{v_j,v_i}]$ to D. If this results in a *circuit*, then G is not an interval k-graph. If neither S_{v_i,v_j} nor S_{v_j,v_i} can be added to D, then G contains an *exobiclique* or one of the forbidden structures in Fig. 7.

Algorithm 1 Algorithm for recognition of interval k-graphs

function INTERVAL-K-GRAPH(G)
  Input: A connected k-partite graph G
  Output: An ordering of $V(G)$ without patterns in Figure 4 or return false
  Construct the pair-digraph G^+ of G, and compute its components; if any component is self-coupled report false
  Stage1 : Adding (non-trivial strong) components
  Initialize D to be the empty set
  let $v_1, v_2, \ldots, v_n$ be an ordering of $V(G)$ such that $i < j$ implies $c(v_i) \leq c(v_j)$
  while $\exists S_{v_i,v_j}$ and S_{v_j,v_i}, $i < j$ components in $G^+ \setminus D$ **do**
    if $D \cup N^+[S_{v_i,v_j}]$ does not have a circuit **then** add $N^+[S_{v_i,v_j}]$ to D,
      remove $N^+[S_{v_j,v_i}]$ from further consideration in this step
        ▷ add X to D means add all the pairs of X into D
      for all $(x, y) \in N^+[S_{v_i,v_j}]$ **do** $Dic(x, y) = S_{v_i,v_j}$
    else
      if $D \cup N^+[S_{v_j,v_i}]$ does not have a circuit **then** add $N^+[S_{v_j,v_i}]$ into D,
        delete $N^+[S_{v_i,v_j}]$ from further consideration here
        for all $(x, y) \in N^+[S_{v_j,v_i}]$ **do** set $Dic(x, y) = S_{v_j,v_i}$
      else report that G is not an interval k-graph
    increase i by one
  Stage2 : Computing $N^*[D]$ and detecting dictator components
  Set $En = N^*[D]$, and $\mathcal{DT} = \varnothing$ ▷ $\mathcal{DT}$ is a set of components
  while $\exists (x, y) \in En \setminus D$ **do** ▷ we consider the pairs in En level by level
    Move (x, y) into D and set $Dic(x, y) =$ DICTATOR(x, y, D)
    if $D \cup \{(x, y)\}$ contains a circuit **then** add $Dic(x, y)$ into $\mathcal{DT}$
        ▷ (x, y) is a complex pair
  Stage3 : Adding dual of dictator components
  Let $D_1 = \varnothing$
  for all components $S \in \mathcal{DT}$ **do** add $N^+[S']$ into D_1
  for all components $R \in D \setminus \mathcal{DT}$ **do** add $N^+[R]$ into D_1
  Set $D = N^*[D_1]$
  if there is a circuit in D **then** report G is not an interval k-graph
  Stage4 : Adding remaining trivial components, returning an ordering
  while $\exists$ trivial component S outside $D \cup D'$, and S is a sink component **do**
    Add S into D and remove S' from further consideration
  Output the final ordering
  for all $(u, v) \in D$ **do** set $u \prec v$
      ▷ yielding an ordering of $V(G)$ without the patterns from Figure 1
  Return the ordering $v_1 \prec v_2 < \cdots \prec v_n$ of $V(G)$ where $(v_l, v_t) \in D$ and $l < s$

```
function DICTATOR(x, y, D)
    if (x, y) ∈ N+[S] for some component S in D then return S
    if x, y have different colors and (u, y) ∈ D dominates (x, y) then
        return DICTATOR(u, y, D)                  ▷ we mean the earliest pair (u, y)
    if x, y have the same color and (x, w) ∈ D dominates (x, y) then
        return DICTATOR(x, w, D)
    if x, y have the same color and (x, y) is by transitivity on
        (x, w), (w, y) ∈ D then return DICTATOR(w, y, D)
    if x, y have different colors and (x, y) is by transitivity on
        (x, w), (w, y) ∈ D then return DICTATOR(x, w, D)
```

Stages 2 and 3: Closure and Dictator Components: Next, we compute the *closure* of D, ensuring that for each pair of components S and S', exactly one of them is in D. If a circuit C is found in $N^*[D]$, we identify a dictator component S, which cannot be included in any complete set. Consequently, we remove all these dictator components $S_1, S_2, \ldots, S_t$ from D and define a new set D_1, which includes their dual components $S'_1, S'_2, \ldots, S'_t$ along with all other elements of $D \setminus (S_1 \cup S_2 \cup \cdots \cup S_t)$. If $N^*[D_1]$ contains a *circuit*, then G is not an interval k-graph; otherwise, we update D as D_1 and set $D = N^*[D]$.
Stage 4: Handling Trivial Components Finally, we add any remaining trivial components from $G^+ \setminus (D \cup D')$ (where D' is the *dual* of D), starting with the *sink components*. This step does not introduce *circuits*, so we do not need to check for conflicts. At the end of the algorithm, we derive a final order $\prec$ setting $u \prec v$ whenever $(u, v) \in D$.

5 Generalization

For a graph H and a coloring $c\colon V(G) \to V(H)$, the graph G equipped with intervals I_v for each $v \in V(G)$ is an *interval H-graph* if, for different $u, v \in V(G)$, there is an edge $uv \in E(G)$ if and only if $I_u \cap I_v \neq \emptyset$ and $c(u)c(v) \in E(H)$. Since interval k-graphs are interval K_k-graphs, the concept of interval H-graphs generalizes interval k-graphs.

In order to decide if G is an interval H-graph we construct the auxiliary digraph G^+, as follows. The vertex set of G^+ consists of ordered pairs (u, v) where $u \neq v \in V(G)$. The arc set of G^+ is defined as follows:

- $(u, v)(u, v')$ is an arc if $uv \notin E(G)$, $c(u) \neq c(v)$, $c(v)c(v') \in E(H)$, and $vv' \in E(G)$.
- $(u, v)(u', v)$ is an arc if $c(u)c(u') \in E(H)$, $uu' \in E(G)$, and $vu' \notin E(G)$.

Clearly, if a component of G^+ contains a circuit, then G is not an interval H-graph. Assuming that no strong component of G^+ contains a circuit, the key challenge is to select, from each pair of components S and S', one component to be included in the set D, ensuring that D remains closed under reachability and transitivity. We believe that our approach for interval k-graphs could be extended to this setting. However, we may end up having a new set of obstructions.

References

1. Benzer, S.: On the topology of the genetic fine structure. Proc. Natl. Acad. Sci. U.S.A. **45**, 1607–1620 (1959)
2. Brown, D.E., Flesch, B.M., Langley, L.J.: Interval k-graphs and orders. Order **35**, 495–514 (2018)
3. Booth, K.S., Lueker, G.S.: Testing for the consecutive ones property, interval graphs, and graph planarity using PQ-tree algorithms. J. Comput. Syst. Sci. **13**(3), 335–379 (1976)
4. Brandstädt, A., Le, V.B., Spinrad, J.P.: Graph classes: a survey SIAM Monographs Discrete Math and Applications, Philadelphia (1999)
5. Brown, D.E., Lundgren, J.R., Flink, S.C.: Characterizations of interval bigraphs and unit interval bigraphs. Congressus Num. **157**, 79–93 (2002)
6. Corneil, D.G., Oleariu, S., Stewart, L.: The ultimate interval graph recognition algorithm. SODA, pp. 175–180 (1998)
7. Corneil, D.G., Olariu, S., Stewart, L.: The LBFS structure and recognition of interval graphs. SIAM J. Discrete Math. **23**(4), 1905–1953 (2009)
8. Damaschke, P.: Forbidden ordered subgraphs. Topics in Combinatorics and Graph Theory, pp. 219–229 (1990)
9. Duffus, D., Ginn, M., Rödl, V.: On the computational complexity of ordered subgraph recognition. Random Struct. Algorithms **7**(3), 223–268 (1995)
10. Dyer, M.E., Müller, H.: Quasimonotone graphs. Discret. Appl. Math. **271**, 25–48 (2019)
11. Feder, T., Hell, P., Huang, J., Rafiey, A.: Interval graphs, adjusted interval digraphs, and reflexive list homomorphisms. Discrete Appl. Math. **160**(6), 697–707 (2012)
12. Gutin, G., Hell, P., Rafiey, A., Yeo, A.: A dichotomy for minimum cost graph homomorphisms. Eur. J. Comb. **29**(4), 900–911 (2008)
13. Harary, F., Kabell, J.A., McMorris, F.R.: Bipartite intersection graphs. Comment. Math. Universitatis Carolinae **23**, 739–745 (1982)
14. Hell, P., Huang, J.: Interval bigraphs and circular arc graphs. J. Graph Theory **46**, 313–327 (2003)
15. Hell, P., Matrolilli, M., Nevisi, M., Rafiey, A.: Approximation of minimum cost homomorphisms. In: ESA, pp. 587–598 (2012)
16. Hell, P., Mohar, B., Rafiey, A.: Ordering without forbidden patterns. In: ESA, pp. 554–565 (2014)
17. Korte, N., Möhring, R.H.: An incremental linear-time algorithm for recognizing interval graphs. SIAM J. Comput. **18**(1), 68–81 (1989)
18. McConnell, R.M.: Linear time recognition of circular-arc graphs. FOCS **42**, 386–394 (2001)
19. Müller, H.: Recognizing interval digraphs and interval bigraphs in polynomial time. Discrete Appl. Math. **78**, 189–205 (1997)
20. Müller, H., Rafiey, A.: Interval H-graphs: recognition and forbidden obstructions (2025), arXiv:2503.00672, https://doi.org/10.48550/arXiv.2503.00672
21. Rafiey, A.: Recognizing interval bigraphs by forbidden patterns. J. Graph Theory **100**(3), 504–529 (2022)
22. Sen, M., Das, S., Roy, B., West, D.B.: Interval digraphs: an analogue of interval graphs. J. Graph Theory **13**, 189–202 (1989)
23. Spinrad, J.P., Brandstädt, A., Stewart, L.: Bipartite permutation graphs. Discrete Appl. Math. **18**, 279–292 (1987)

Faster Algorithms for Graph Monopolarity

Geevarghese Philip[1] and Shrinidhi Teganahally Sridhara[2(✉)]

[1] Chennai Mathematical Institute and UMI ReLaX, Chennai, India
gphilip@cmi.ac.in

[2] Univ. Bordeaux, CNRS, Bordeaux INP, LaBRI, UMR 5800, 33400 Talence, France
shrinidhi.teganahally-sridhara@u-bordeaux.fr
https://www.cmi.ac.in/ gphilip

Abstract. A graph $G = (V, E)$ is *monopolar* if its vertex set admits a partition $V = (C \uplus I)$ where $G[C]$ is a *cluster graph* and I is an *independent set* in G; this is a *monopolar partition* of G. The MONOPOLAR RECOGNITION problem—deciding whether an input graph is monopolar—is known to be NP-hard even in very restricted graph classes.

We develop fast exact exponential-time and parameterized algorithms for MONOPOLAR RECOGNITION. Our exact algorithm solves MONOPOLAR RECOGNITION in $\mathcal{O}^{\star}(1.3734^n)$ time on input graphs with n vertices, where the $\mathcal{O}^{\star}()$ notation hides polynomial factors. In fact, we develop algorithms that solve the more general problems MONOPOLAR EXTENSION and LIST MONOPOLAR PARTITION in the same running time. These are the first improvements over the trivial $\mathcal{O}^{\star}(2^n)$-time algorithms for all these problems. These problems cannot be solved in $\mathcal{O}^{\star}(2^{o(n)})$ time, under ETH.

Our FPT algorithms solve MONOPOLAR RECOGNITION in $\mathcal{O}^{\star}(3.076^{k_v})$ and $\mathcal{O}^{\star}(2.253^{k_e})$ time where k_v and k_e are, respectively, the sizes of the smallest vertex and edge modulators of the input graph to claw-free graphs. These results are a significant addition to the small number of FPT algorithms currently known for MONOPOLAR RECOGNITION.

Keywords: Graph Monopolarity · Fixed-parameter tractability · Exponential-time algorithms

1 Introduction

In this work we derive fast exponential-time and fixed-parameter tractable (FPT) algorithms for recognizing monopolar graphs. All our graphs are finite, undirected, and simple. Graph H is a *cluster graph* if each connected component of H is a complete graph. A partition of the vertex set of a graph $G = (V, E)$ into two sets C and I is a *monopolar partition* of G if the subgraph $G[C]$ induced by set C is a cluster graph and the set I is an independent set in G. A graph G

H. Fernau and P. Kindermann (Eds.): WG 2025, LNCS 16124, pp. 418–430, 2026.
https://doi.org/10.1007/978-3-032-11835-6_30

is *monopolar* if it has a monopolar partition. The primary focus of this work is on the algorithmic problem of recognizing monopolar graphs:

> MONOPOLAR RECOGNITION
> **Input:** An undirected graph $G = (V, E)$ on n vertices.
> **Question:** If G is monopolar, then output Yes. Otherwise, output No.

Monopolar graphs generalize both bipartite graphs (C is an independent set) and split graphs (C is a single clique). While both split graphs and bipartite graphs can be recognized in polynomial time, MONOPOLAR RECOGNITION is NP-hard even in very restricted graph classes such as triangle-free planar graphs of maximum degree 3 [12]. We propose fast exponential-time and FPT algorithms that solve MONOPOLAR RECOGNITION in general graphs. Each of our algorithms also outputs a monopolar partition, if the input graph is monopolar. Our first result is a fast exact exponential-time algorithm for MONOPOLAR RECOGNITION:

Theorem 1. MONOPOLAR RECOGNITION *can be solved in* $\mathcal{O}^\star(1.3734^n)$ *time: There is an algorithm which takes a graph* G *on* n *vertices as input, runs in* $\mathcal{O}^\star(1.3734^n)$ *time, and correctly decides if* G *is monopolar. If* G *is monopolar, then this algorithm also outputs one monopolar partition of the vertex set of* G.

Here the $\mathcal{O}^\star()$ notation hides polynomial factors. As far as we know, this is the first improvement over the trivial exact algorithm for this problem. We in fact obtain a similarly fast algorithm for the more general MONOPOLAR EXTENSION problem introduced by Le and Nevries [13]. Let C', I' be two disjoint subsets of the vertex set of a graph $G = (V, E)$. A monopolar partition $V = C \uplus I$ of G with $C' \subseteq C$ and $I' \subseteq I$ is called a *monopolar extension* of the pair (C', I'). Graph G is (C', I')*-monopolar extendable* if it admits a monopolar partition extending (C', I').

> MONOPOLAR EXTENSION
> **Input:** An undirected graph $G = (V, E)$ on n vertices, and subsets $C' \subseteq V$ and $I' \subseteq V$.
> **Question:** If G has a monopolar partition which extends (C', I'), then output YES. Otherwise, output NO.

Churchley and Huang [2] defined LIST MONOPOLAR PARTITION as another generalization of MONOPOLAR RECOGNITION. Given a graph $G = (V, E)$ and lists $\{\emptyset \neq L(v) \subseteq \{\hat{C}, \hat{I}\}\ ;\ v \in V\}$, a *monopolar partition of* G *that respects the list function* L is a mapping $f : V \to \{\hat{C}, \hat{I}\}$ such that (i) $f(v) \in L(v)$ holds for all $v \in V$, (ii) $f^{-1}(\hat{I})$ induces an independent set in G, and (iii) $f^{-1}(\hat{C})$ induces a cluster graph in G. The LIST MONOPOLAR PARTITION problem asks if there is a monopolar partition of a graph that respects a given list function:

> LIST MONOPOLAR PARTITION
> **Input:** An undirected graph $G = (V, E)$ on n vertices and a list function L defined on the vertices of G such that $\emptyset \neq L(v) \subseteq \{\hat{C}, \hat{I}\}$ for all $v \in V$.
> **Question:** If there exists a monopolar partition of G that respects the list function L, then output YES. Otherwise, output NO.

The problems Monopolar Extension and List Monopolar Partition are computationally equivalent modulo polynomial time (see Lemma 4), and our exponential speed-up in solving Monopolar Extension directly transfers to such a speed-up for List Monopolar Partition as well:

Theorem 2. *Let G be a graph on n vertices.*

1. *There is an algorithm that takes an instance (G, C', I') of* Monopolar Extension *as input, runs in $\mathcal{O}^\star(1.3734^n)$ time, and correctly decides if G admits a monopolar partition that is an extension of (C', I'). If such a monopolar extension exists then this algorithm also outputs one such monopolar partition of G.*
2. *There is an algorithm that takes an instance (G, L) of* List Monopolar Partition *as input, runs in $\mathcal{O}^\star(1.3734^n)$ time, and correctly decides if G admits a monopolar partition that respects the list function L. If G does admit such a monopolar partition, then this algorithm also outputs one such monopolar partition of G.*

Assuming the Exponential Time Hypothesis, there is no algorithm that can solve Monopolar Recognition—and hence, Monopolar Extension or List Monopolar Partition—in $2^{o(n)}$ time [10, Prop. 9.1]. Any improvement over the running times of Theorem 1 or Theorem 2 can therefore only be in the form of a smaller constant base for the exponential term.

We present two FPT algorithms for Monopolar Recognition parameterized by measures of *distance from triviality*. Churchley and Huang showed that Monopolar Recognition can be solved in polynomial time on claw-free graphs [3]. We show that Monopolar Recognition can be solved in FPT time for the parameter being two natural *deletion distances* to claw-free graphs.

Theorem 3. [$\star$] *There is an algorithm that takes a graph G on n vertices as input, runs in $\mathcal{O}^\star(3.076^{k_v})$ time, and correctly decides if G is monopolar. If G is monopolar, then this algorithm also outputs one monopolar partition of the vertex set of G. Here k_v is the smallest number of vertices that need to be deleted from G to obtain a claw-free graph.*

Theorem 4. [$\star$] *There is an algorithm that takes a graph G on n vertices as input, runs in $\mathcal{O}^\star(2.253^{k_e})$ time, and correctly decides if G is monopolar. If G is monopolar, then this algorithm also outputs one monopolar partition of the vertex set of G. Here k_e is the smallest number of edges that need to be deleted from G to obtain a claw-free graph.*

As far as we know, there are only two previously known results on the parameterized complexity of Monopolar Recognition (summarized below). Theorem 3 and Theorem 4 are thus a significant addition to the set of known FPT results for the problem.

An $\mathcal{O}^\star(2^n)$-time algorithm for Monopolar Recognition follows more or less directly from the definition of the problem. This can easily be improved to

$\mathcal{O}^\star(3^{\frac{n}{3}}) \approx \mathcal{O}^\star(1.4423^n)$ time by observing that we need only consider inclusion-maximal independent sets as candidates for the part I in a monopolar partition. To reduce this further to the $\mathcal{O}^\star(1.3734^n)$ bound of Theorem 1 we exploit structural properties of monopolar graphs, building in particular upon the work of Le and Nevries who showed that MONOPOLAR EXTENSION can be solved in polynomial time on *chair-free* graphs[1] [13, Corollary 3]. We significantly strengthen this result of Le and Nevries:

Theorem 5. [$\star$] *There is a polynomial-time algorithm that solves* MONOPOLAR EXTENSION *for instances $(G, (C', I'))$ where C' is a vertex modulator of G to chair-free graphs. If graph G has a monopolar partition that is an extension of (C', I'), then this algorithm also outputs one such partition.*

We may, in fact, assume C' to be an *inclusion-minimal* vertex modulator:

Lemma 1. [$\star$] *Let $G = (V, E)$ be a monopolar graph. There exists a subset $C' \subseteq V$ of its vertices such that (i) C' is a vertex modulator of G to chair-free graphs; (ii) no proper subset of C' is a vertex modulator of G to chair-free graphs, and (iii) graph G has a monopolar partition $V = (C \uplus I)$ where $C' \subseteq C$.*

It is *not* true that *every* vertex modulator to chair-free graphs can be made to "live inside" the cluster part of some monopolar partition of G. Indeed, consider a *chair-free* monopolar graph $\hat{G}$. It is not the case that arbitrary subsets of its vertices—all of which are, trivially, modulators to chair-free graphs—belong to the cluster part in some monopolar partition of $\hat{G}$. What Lemma 1 says is that there is *at least one such*—inclusion-minimal—modulator in *every* monopolar graph.

MONOPOLAR RECOGNITION thus reduces to the problem of finding a minimal vertex modulator to chair-free graphs that belongs to the "cluster part" of *some* (unknown) monopolar partition of the input graph, or ruling out that such a modulator exists. Guessing the vertices of each induced chair[2] that belongs to this unknown modulator results in a 31-way branching algorithm where the number of undecided vertices drops by 5 in each branch, yielding an $\mathcal{O}^\star(31^{\frac{n}{5}}) \approx \mathcal{O}^\star(1.987^n)$-time algorithm. Instead of looking for a "good" minimal modulator to chair-free graphs, we repeatedly find a "fresh" chair, with all undecided vertices. We then carefully branch on its vertices, assigning each vertex either to the "independent set part" or the "cluster part" of a putative monopolar partition. We stop the branching when *no* induced chair has *all* its vertices undecided; we then apply Theorem 5 to solve the remaining instance. The careful branching and early stopping lead to the considerable speed-up from the trivial $\mathcal{O}^\star(2^n)$ to the $\mathcal{O}^\star(1.3734^n)$ of Theorem 1 and Theorem 2. We get our fast FPT algorithms by "branching towards" Theorem 5, as well.

MONOPOLAR RECOGNITION has received a considerable amount of attention starting around the year 2008. Linear-time algorithms for MONOPOLAR RECOGNITION are known for *cographs* [8], *chordal graphs* [7], and *line graphs* [1]. The

[1] See Sect. 2 for a definition.
[2] See Fig. 1(f).

problem can be solved in polynomial time on P_5*-free graphs*, *chair-free graphs*, *hole-free graphs*, *maximal planar graphs*, *claw-free graphs*, *permutation graphs*, and *co-comparability graphs* [2–4,12,13]. In contrast, the problem is known to be NP-hard on *triangle-free planar graphs of maximum degree* 3 [12].

We are aware of only two papers that take up the parameterized complexity of MONOPOLAR RECOGNITION. These are both by the same set of authors, namely, Kanj, Komusiewicz, Sorge and van Leeuwen [10,11]. In the first paper [10], they show—*inter alia*—that MONOPOLAR RECOGNITION is FPT parameterized by the *number of cliques* on the "cluster side" $G[C]$ of a(n unknown) monopolar partition $V(G) = C \uplus I$ of the input graph G. They derive an algorithm that, given a graph G and a positive integer k, decides in $\mathcal{O}^\star(2^k)$ time whether G has a monopolar partition $V(G) = C \uplus I$ where the "cluster side" $G[C]$ is a disjoint union of at most k cliques. They show also that under ETH, any algorithm that solves MONOPOLAR RECOGNITION must take $2^{\Omega(n+m)}$ time on graphs with n vertices and m edges. In the second work, [11], they show that MONOPOLAR RECOGNITION has a kernel with $\mathcal{O}(k^4)$ vertices, for the same parameter k. Note that this parameter k is not comparable with either of our parameters k_v, k_e; it is easy to see that there are graphs in which each of k_v or k_e is a small constant while k is unbounded, and *vice versa*.

Organization of the Rest of the Paper. In the next section we list our notation and terminology, and prove some preliminary results. We describe our exact exponential algorithms for monopolarity and prove Lemma 1, Theorem 1, Theorem 2 and Theorem 5 in Sect. 3. We derive our FPT algorithms—and prove Theorem 3 and Theorem 4—in Sect. 4. We summarize our results and list some open problems in Sect. 5. *Proofs of results marked with a* [$\star$] *had to be excluded from this Extended Abstract due to page constraints.*

2 Preliminaries

All our graphs are finite, simple, and undirected. We follow the graph notation and terminology of Diestel [5]. Graph G is H*-free* for a graph H if H does not appear as an *induced subgraph* of G. A graph G is a *cluster graph* if each connected component of G is a complete graph; equivalently, if and only if G is P_3-free (see Fig. 1(b)).

Graph G is *claw-free* if it does not have a *claw*—Fig. 1(a)—as an induced subgraph. A subset X of the vertex set of graph G is said to be a *vertex modulator of G to claw-free graphs* if the graph $G - X$ is claw-free. A subset Y of the edge set of graph G is said to be an *edge modulator of G to claw-free graphs* if the graph $G - Y$ is claw-free. Graph G is *chair-free* if it does not have a *chair*—Fig. 1(f)—as an induced subgraph. A subset X of the vertex set of graph G is said to be a *vertex modulator of G to chair-free graphs* if the graph $G - X$ is chair-free. *We use the short forms "vertex modulator" and "edge modulator", when the fuller forms are clear from the context.*

We use P_k to denote a path with k vertices, and C_k for a cycle with k vertices. Some other small graphs that we need to refer by name are listed in Fig. 1. We use

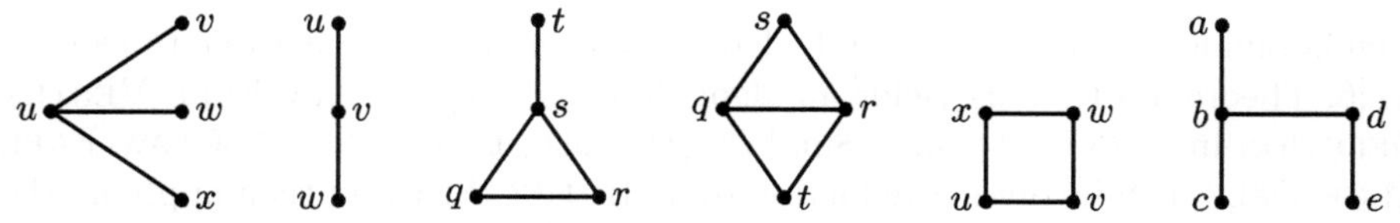

(a) Claw (b) P_3 (c) Paw $P(s,t)$ (d) Diamond $D(s,t)$ (e) C_4 (f) Chair $Ch(a,b,c,d,e)$

Fig. 1. Some small named graphs.

$Ch(a,b,c,d,e)$ to denote a *chair* with $\deg(b)=3, \deg(d)=2, \deg(a)=\deg(c)=\deg(e)=1$ as in Fig. 1(f) (a chair graph is also defined as $S_{2,1,1}$, which is a claw with edges subdivided once, zero, and zero), $P(s,t)$ to denote a *paw* with $\deg(s)=3, \deg(t)=1$, and $D(s,t)$ for a *diamond* with $\deg(s)=\deg(t)=2$.

We need the following simple properties of monopolar extensions:

Lemma 2. [$\star$] *Let C', I' be two (disjoint) subsets of the vertex set V of graph G, and let $x \notin (C' \cup I')$ be a vertex of degree 1 in G. Then G is (C', I')-monopolar extendable if and only if $G - \{x\}$ is (C', I')-monopolar extendable.*

Lemma 3. [$\star$] *A graph G is (C', I')-monopolar extendable if and only if I' is an independent set in G and the induced subgraph $G - I'$ is $((C' \cup N(I')), \emptyset)$-monopolar extendable.*

Lemma 4. [$\star$] MONOPOLAR EXTENSION *and* LIST MONOPOLAR PARTITION *are computationally equivalent modulo polynomial time.*

Le and Nevries [13] prove the following result for claw-free graphs:

Theorem 6. MONOPOLAR EXTENSION *can be solved in $O(n^4)$ time on n-vertex claw-free graphs.*

Our algorithms that prove Theorem 3 and Theorem 7 employ algorithms to solve the following parameterized deletion problems, as subroutines:

CLAW-FREE VERTEX DELETION **Parameter:** k
Input: An undirected graph $G=(V,E)$, and an integer k.
Question: If there is a set $X \subseteq V$; $|X| \leq k$ such that deleting $G - X$ is a claw-free graph, then output one such set X. Otherwise, output NO.

CLAW-FREE EDGE DELETION **Parameter:** k
Input: An undirected graph $G=(V,E)$, and an integer k.
Question: If there is a set $X \subseteq E$; $|X| \leq k$ such that deleting X from G results in a claw-free graph, then output one such set X. Otherwise, output NO.

To the best of our knowledge, the fastest way to solve CLAW-FREE VERTEX DELETION is by phrasing it as an instance of a 4-HITTING SET problem and then applying the best known FPT algorithm for 4-Hitting Set due to Dom et al. [6, Theorem 3.1]. This yields an algorithm that solves Claw-Free Vertex Deletion in $\mathcal{O}^\star(3.076^k)$ time. Similarly, the fastest algorithm for Claw-Free Edge Deletion involves a reduction to 3-Hitting Set and then applying the argument underlying Dom et al.'s algorithm for this problem [6, Proposition 3.1] which uses an algorithm for Vertex Cover as black box. Using the current best running time of $\mathcal{O}^\star(1.253^k)$ for Vertex Cover [9] here, we get an algorithm that solves Claw-Free Edge Deletion in $\mathcal{O}^\star(2.253^k)$ time.

These two fast algorithms for modulators to claw-free graphs are thus direct applications of existing results to these problems. We capture these as a theorem for later reference:

Theorem 7. Claw-Free Vertex Deletion *can be solved in* $\mathcal{O}^\star(3.076^k)$ *time, and* Claw-Free Edge Deletion *can be solved in* $\mathcal{O}^\star(2.253^k)$ *time.*

3 Exact Algorithms for Monopolarity

In this section we outline the proofs of Theorem 1 and Theorem 2. We get these faster running times by exploiting various structural properties of monopolar graphs to lift a theorem of Le and Nevries [13] for chair-free graphs, so that it works for all graphs; see Theorem 5. Le and Nevries [13] showed that Monopolar Extension, and consequently Monopolar Recognition, can be solved in polynomial time on chair-free graphs. They achieved this result by reducing Monopolar Extension on (a superclass of) chair-free graphs to 2-SAT; our algorithms make extensive use of this reduction.

If the given graph $G = (V, E)$ does not have an induced P_3 then G is a cluster graph. In this case, for any vertex $v \in V$, the partition $V = (V \setminus \{v\}) \uplus \{v\}$ is a monopolar partition of G, and there is nothing more to do. If the graph *does* contain at least one induced P_3, then we make use of the concept of a *C'-good induced P_3*, a notion introduced by Le and Nevries [13].

Definition 1. *[13] Let G be a graph, and let C' be a subset of the vertex set of G. We say that an induced P_3 in G of the form uvw is* C'-good, *if it satisfies* at least one *of the following conditions:*

1. *At least one of the vertices in the set $(N[v] \cup N[u] \cup N[w])$ is in the set C'.*
2. *At least one of the vertices u, v, w is a part of a triangle in G.*
3. *At least one of the edges of the P_3 uvw is part of an induced C_4 in G.*

Otherwise, we say that the P_3 uvw is C'-bad. *Finally, we say that graph G is* C'-good *if every induced P_3 in G is* C'-good.

Le and Nevries [13] show how, given an instance $(G, (C', \emptyset))$ of Monopolar Extension *where G is C'-good*, we can construct in polynomial time a 2-SAT formula $F(G, C')$ where (i) the variables represent vertices of G, (ii) the clauses are constructed based on the properties of small induced graphs from Fig. 1, and

(iii) the formula $F(G, C')$ is satisfiable if and only if $(G, (C', \emptyset))$ is a YES-instance of MONOPOLAR EXTENSION.

Given an instance $(G, (\tilde{C}, \tilde{I}))$ of MONOPOLAR EXTENSION, first, we use Lemma 3 to transform it into an equivalent instance $(G', (C', \emptyset))$ where $C' = \tilde{C} \cup N(\tilde{I})$ and $G' = G - \tilde{I}$. We then construct the corresponding 2-SAT formula $F(G', C')$ as described below, following Le and Nevries [13]. Intuitively, the formula is set up in such a way that the variable corresponding to a vertex v gets the value 1 in a satisfying assignment of the formula if and only if v belongs to the *independent set* part in some valid monopolar extension.

1. There is a Boolean variable v for each vertex v of G.
2. For each vertex $u \in C'$, add $(\neg u)$ to the formula (a *forced-cluster-clause*).
3. For each edge uv in G, add $(\neg u \vee \neg v)$ to the formula (an *edge-clause*).
4. For each induced paw $P(s, t)$ and each induced diamond $D(s, t)$ in G, add $(s \vee t)$ to the formula (a *paw-diamond-clause*).
5. For each induced 4-cycle $C_4 : uvwx$ in G, add the clauses $(u \vee v)$, $(v \vee w)$, $(w \vee x)$, and $(x \vee u)$ to the formula (C_4-*clauses*).
6. For each vertex $x \in C'$ and each induced P_3 of the form xvw or vxw, add the clause $(v \vee w)$ to the formula (a P_3-*clause*).

The formula $F(G, C')$ is the conjunction of all forced-cluster-clauses, edge-clauses, paw-diamond-clauses, C_4-clauses, and P_3-clauses. Le and Nevries show that *for a C'-good graph G*, the graph G is monopolar extendable with respect to $(C', \emptyset)$ if and only if the formula $F(G, C')$ is satisfiable:

Lemma 5. *[13, Lemma 1 and its proof] Let $G = (V, E)$ be a graph and C' be a subset of vertices such that G is C'-good. Then in polynomial time we can construct an instance $F(G, C')$ of 2-*SAT *with the following properties. First, G is $(C', \emptyset)$-monopolar extendable if and only if $F(G, C')$ is satisfiable. Second, if $F(G, C')$ is satisfiable then, the set I of vertices corresponding to variables set to* True *and the set C of vertices corresponding to variables set to* False *in a satisfying assignment to $F(G, C')$ together form a monopolar partition of G i.e., $V = (C \uplus I)$ is a monopolar partition that extends $(C', \emptyset)$.*

To solve MONOPOLAR RECOGNITION on an input graph G using Lemma 5 we need to find a vertex subset C' of G such that graph G is C'-good. But why should such a vertex set C' exist for *every* monopolar graph G? And even if it did, how could we find one such set? Le and Nevries [13, Corollary 3] show that *if the input graph G is chair-free* then they can use Lemma 5 as a tool to solve an instance $(G, (C', \emptyset))$ of MONOPOLAR EXTENSION in polynomial time *even if* graph G is *not* C'-good. We generalize this further; we show that in fact we can solve an instance $(G, (C', \emptyset))$ of MONOPOLAR EXTENSION in an *arbitrary graph G* in polynomial time if the set C' is a *vertex modulator of G to chair-free graphs*, even if graph G is *not* C'-good. We thus reduce MONOPOLAR RECOGNITION to the problem of finding such a vertex modulator C' of G, and then solving MONOPOLAR EXTENSION on the instance $(G, (C', \emptyset))$.

Lemma 6. [$\star$] *If a vertex subset* C' *is a vertex modulator of graph* $G = (V, E)$ *to chair-free graphs, then the instance* $(G, (C', \emptyset))$ *of* MONOPOLAR EXTENSION *can be solved in polynomial time. Furthermore if graph* G *is* $(C', \emptyset)$*-monopolar extendable, then a monopolar partition of* G *that extends* $(C', \emptyset)$ *can be obtained in polynomial time as well.*

As LIST MONOPOLAR PARTITION and MONOPOLAR EXTENSION are polynomially equivalent as shown in Lemma 4, we immediately get a similar result for LIST MONOPOLAR PARTITION when the set of vertices $\{v : L(v) = \{\hat{C}\}\}$ forms a vertex modulator of the input graph:

Corollary 1. *Let* (G, L) *be an instance of* LIST MONOPOLAR PARTITION *where the set* $C' = \{v : L(v) = \{\hat{C}\}\}$ *of vertices that are assigned to the cluster side, forms a vertex modulator of graph* G *to chair-free graphs. Such an instance can be solved in polynomial time.*

We now have all the ingredients for proving our generalization of Le and Nevries' polynomial-time algorithm for MONOPOLAR EXTENSION on chair-free graphs, namely, Theorem 5.

Let $(G = (V, E), C', I')$ be an instance of MONOPOLAR EXTENSION. If either the graph G is chair-free or the set C' is a vertex modulator, then we can solve this instance in polynomial time. Such instances form the base cases of MONOPOLAR EXTENSION that are easy to solve. We start with a simple observation:

Lemma 7. [$\star$] *Let* $(G = (V, E), C', I')$ *be an instance of* MONOPOLAR EXTENSION. *In polynomial time we can either solve the instance or derive an* equivalent *instance* $(\tilde{G}, \tilde{C}, \tilde{I})$ *of* MONOPOLAR EXTENSION *with the following property: there exists at least one induced chair—say* $Ch(a, b, c, d, e)$*—in* $\tilde{G}$ *such that* $\{a, b, c, d, e\} \cap (\tilde{C} \cup \tilde{I}) = \emptyset$.

Let $Ch(a, b, c, d, e)$ be an induced chair in $G = (V, E)$ such that $\{a, b, c, d, e\} \cap (\tilde{C} \cup \tilde{I}) = \emptyset$ holds. Let $V = (C \uplus I)$ be a(n unknown) valid monopolar partition of G. Our algorithm for MONOPOLAR EXTENSION does an exhaustive branching on the vertices b and e—see Fig. 1(f) for the notation—being in C or I:

$b \in C, e \in C$: Since $G[C]$ is a cluster graph and the edge be is not present in graph G, vertex d cannot belong to the set C. So we have $d \in I$. We cannot conclude anything about vertices a and c except that they cannot be together in set C.

$b \in I, e \in C$: Since $G[I]$ is edge-less and the edges ab, bc and bd are present in graph G, vertices a, c and d cannot belong to the set I. So we have $a \in C$, $c \in C$ and $d \in C$.

$b \in C, e \in I$: Vertex d cannot belong to the set I as de is an edge and $e \in I$. Thus $d \in C$. Also, vertex a cannot belong to the set C as the edge ad is not present and vertex c cannot belong to the set C as the edge cd is not present. So we have $d \in C$, $a \in I$ and $c \in I$.

$b \in I, e \in I$: Here, a, c and d cannot belong to the set I as ab, bc and bd are edges and $b \in I$ where $G[I]$ is needed to be an independent set. So we have $a \in C$, $c \in C$ and $d \in C$.

Let ME-ALGORITHM be a procedure (pseudocode not included in the Extended Abstract due to page constraints) which uses the above branching strategy to solve MONOPOLAR EXTENSION and returns a monopolar partition if the input is a YES-instance. The correctness of procedure ME-ALGORITHM follows from the fact that it does exhaustive branching on valid partitions, and from Corollary 1, Theorem 6, and Lemma 6.

Lemma 8. *The call* ME-ALGORITHM(G, C', I') *terminates in* $\mathcal{O}^\star(1.3734^n)$ *time where* n *is the number of vertices in graph* G.

Proof. Let $\mu = \mu(G, C', I') = |V(G) \setminus (C' \cup I')|$ be a measure function that denotes the number of vertices in graph G which are not yet in the sets C', I'. Note that the condition $\mu \leq n$ holds before the procedure ME-ALGORITHM makes any recursive call. The procedure makes recursive calls only when it finds a chair $\mathcal{C}$, all of whose five vertices contribute to the measure μ. Three of these recursive calls cause μ to drop by 5 each, since all five vertices of $\mathcal{C}$ are moved to $C' \cup I'$ in these calls. The remaining recursive call results in μ dropping by 3, since only three vertices—b, e, d—are moved to $C' \cup I'$ by this call.

The procedure stops recursing only when for each chair in graph G, at least one vertex in the chair is in $C' \cup I'$. If a vertex x in a chair $\mathcal{C}$ is moved to the set I', then the procedure moves every neighbour of x to the set C'. It follows that when the recursion stops, the set C' forms a vertex modulator of graph G. The procedure solves this instance in polynomial time using an implementation of Corollary 1.

Let $T(\mu)$ denote the number of nodes in the recursion tree of the procedure, when invoked with an instance with measure μ. The recurrence for $T(\mu)$ is then $T(\mu) = 3T(\mu - 5) + T(\mu - 3)$. This solves to $T(\mu) \leq (1.3734)^\mu$, and the lemma follows. □

We can convert an instance $G = (V, E)$ of MONOPOLAR RECOGNITION to an instance $(G = (V, E), C', I')$ of MONOPOLAR EXTENSION by setting $C' = \emptyset, I' = \emptyset$. This proves Theorem 1.

Combining Lemma 4 and Lemma 8 we get Theorem 2.

4 FPT Algorithms for Monopolarity

In this section we outline the arguments that prove Theorem 3 and Theorem 4; the proofs could not be included in this Extended Abstract due to page constraints.

We have seen in Theorem 7 that we can find a vertex modulator of graph G to claw-free graphs in time $\mathcal{O}^\star(3.076^{k_v})$ where k_v is the size of a smallest such modulator of G. It also says that we can find an *edge* modulator of G to claw-free graphs in time $\mathcal{O}^\star(2.253^{k_e})$ where k_e is the size of a smallest such modulator. Once we have such sets, we show that we can solve monopolarity in arbitrary graphs by making use of Lemma 6[3]. The idea is to find such sets and then guess the partition the elements of these sets are assigned to in any valid monopolar partition and then try to solve monopolar extension to decide monopolarity. Once fleshed out, these arguments prove Theorem 3 and Theorem 4.

5 Conclusion

We derive fast exponential-time and FPT algorithms for Monopolar Recognition. Our exact algorithm for Monopolar Recognition solves the problem in $\mathcal{O}^\star(1.3734^n)$ time on graphs with n vertices, significantly improving the trivial $\mathcal{O}^\star(2^n)$-time algorithm and the more involved $\mathcal{O}^\star(3^{\frac{n}{3}})$-time algorithm. We also show how to solve the more general problems List Monopolar Partition and Monopolar Extension in $\mathcal{O}^\star(1.3734^n)$ time. These are the fastest known exact algorithms for these three problems. We derive two FPT algorithms for Monopolar Recognition with two notions of distance from triviality as the parameters. We show that Monopolar Recognition can be solved in $\mathcal{O}^\star(3.076^{k_v})$ and $\mathcal{O}^\star(2.253^{k_e})$ time, where k_v and k_e are, respectively, the sizes of the smallest vertex and edge modulators of the input graph to claw-free graphs. These results are a significant addition to the small number of FPT algorithms known for Monopolar Recognition.

Le and Nevries [13] showed that if a graph G is chair-free then the instance $(G, (C', \emptyset))$ of Monopolar Extension can be solved in polynomial time for any vertex subset C' of graph G. We significantly generalize this result: we show that the instance $(G, (C', \emptyset))$ of Monopolar Extension can be solved in polynomial time for *any* graph G if C' is a vertex modulator of G to chair-free graphs. This generalization of Le and Nevries' result forms the basis of our fast algorithms.

Open Problems. Le and Nevries [13] show that Monopolar Recognition can be solved in polynomial time for a strict superclass of claw-free graphs, namely the class of $\{F_1, F_2, F_3\}$-free graphs for certain small graphs F_1, F_2, F_3. This class includes chair-free graphs. A natural question to ask is whether we can obtain a faster exact algorithm by proving the equivalent of Lemma 6 for $\{F_1, F_2, F_3\}$-free graphs, and then branching on induced subgraphs F_1, F_2, F_3. Recall that algorithms that solve these problem in time which is sub-exponential in n, are ruled out by ETH. It would therefore be interesting to establish a lower bound for the base of the exponential term, perhaps based on SETH.

We can ask analogous questions in the parameterized setting as well: whether we can use the sizes of vertex or edge modulators to $\{F_1, F_2, F_3\}$-free graphs as the parameters, to get faster FPT algorithms. This latter question is made interesting by the fact that each of the graphs $\{F_1, F_2, F_3\}$ has more vertices and edges than a claw. Using the known d-Hitting Set FPT algorithms as

[3] Observe that a *claw-free* graph is also a *chair-free* graph.

a black box to solve MONOPOLAR RECOGNITION, similar to our algorithms for Theorem 3 and Theorem 4, would therefore incur a significantly higher running time in terms of the respective hitting set sizes. We will need clever ideas to get around this slowdown. Since our FPT algorithms depend on algorithms to find claw-free modulators, an improvement in these algorithms would improve our FPT algorithms as well.

Acknowledgements. We extend our heartfelt gratitude to the anonymous reviewers of an earlier version of this manuscript. Their detailed and insightful feedback has helped us greatly improve the presentation.

References

1. Churchley, R., Huang, J.: Line-polar graphs: characterization and recognition. SIAM J. Discret. Math. **25**(3), 1269–1284 (2011)
2. Churchley, R., Huang, J.: List monopolar partitions of claw-free graphs. Discret. Math. **312**(17), 2545–2549 (2012)
3. Churchley, R., Huang, J.: On the polarity and monopolarity of graphs. J. Graph Theory **76**(2), 138–148 (2014)
4. Churchley, R., Huang, J.: Solving partition problems with colour-bipartitions. Graphs Comb. **30**, 353–364 (2014)
5. Diestel, R.: Graph Theory: 5th edition. Springer Graduate Texts in Mathematics. Springer (2017). https://books.google.co.in/books?id=zIxRDwAAQBAJ
6. Dom, M., Guo, J., Hüffner, F., Niedermeier, R., Truß, A.: Fixed-parameter tractability results for feedback set problems in tournaments. J. Discrete Algorithms **8**(1), 76–86 (2010)
7. Ekim, T., Hell, P., Stacho, J., de Werra, D.: Polarity of chordal graphs. Discret. Appl. Math. **156**(13), 2469–2479 (2008)
8. Ekim, T., Mahadev, N.V., de Werra, D.: Polar cographs. Discret. Appl. Math. **156**(10), 1652–1660 (2008)
9. Harris, D.G., Narayanaswamy, N.S.: A faster algorithm for vertex cover parameterized by solution size. In: Beyersdorff, O., Kanté, M.M., Kupferman, O., Lokshtanov, D. (eds.) 41st International Symposium on Theoretical Aspects of Computer Science (STACS 2024). Leibniz International Proceedings in Informatics (LIPIcs), vol. 289, pp. 40:1–40:18. Schloss Dagstuhl – Leibniz-Zentrum für Informatik, Dagstuhl, Germany (2024). https://doi.org/10.4230/LIPIcs.STACS.2024.40. https://drops.dagstuhl.de/entities/document/10.4230/LIPIcs.STACS.2024.40
10. Kanj, I., Komusiewicz, C., Sorge, M., Van Leeuwen, E.J.: Parameterized algorithms for recognizing monopolar and 2-subcolorable graphs. J. Comput. Syst. Sci. **92**, 22–47 (2018)
11. Kanj, I., Komusiewicz, C., Sorge, M., Van Leeuwen, E.J.: Solving partition problems almost always requires pushing many vertices around. SIAM J. Discret. Math. **34**(1), 640–681 (2020)
12. Le, V.B., Nevries, R.: Recognizing polar planar graphs using new results for monopolarity. In: Asano, T., Nakano, S., Okamoto, Y., Watanabe, O. (eds.) ISAAC 2011. LNCS, vol. 7074, pp. 120–129. Springer, Heidelberg (2011). https://doi.org/10.1007/978-3-642-25591-5_14

13. Le, V.B., Nevries, R.: Complexity and algorithms for recognizing polar and monopolar graphs. Theor. Comput. Sci. **528**, 1–11 (2014). https://doi.org/10.1016/j.tcs.2014.01.032. https://www.sciencedirect.com/science/article/pii/S0304397514000814

On Plane Cycles in Geometric Multipartite Graphs

Marco Ricci, Jonathan Rollin(✉), André Schulz, and Alexandra Weinberger

FernUniversität in Hagen, Hagen, Germany
{marco.ricci,jonathan.rollin,andre.schulz,
alexandra.weinberger}@fernuni-hagen.de

Abstract. A geometric graph is a drawing of a graph in the plane where the vertices are drawn as points in general position and the edges as straight-line segments connecting their endpoints. It is plane if it contains no crossing edges. We study plane cycles in geometric complete multipartite graphs. We prove that if a geometric complete multipartite graph contains a plane cycle of length t, with $t \geq 6$, it also contains a smaller plane cycle of length at least $\lceil t/2 \rceil + 1$. We further give a characterization of geometric complete multipartite graphs that contain plane cycles with a color class appearing at least twice. For geometric drawings of $K_{n,n}$, we give a sufficient condition under which they have, for each $s \leq n$, a plane cycle of length $2s$. We also provide an algorithm to decide whether a given geometric drawing of $K_{n,n}$ contains a plane Hamiltonian cycle in time $O(n \log n + nk^2) + O(k^{5k})$, where k is the number of vertices inside the convex hull of all vertices. Finally, we prove that it is NP-complete to decide if a subset of edges of a geometric complete bipartite graph H is contained in a plane Hamiltonian cycle in H.

1 Introduction

A *geometric graph* is a drawing of a graph G in the plane such that its vertices in $V(G)$ are drawn as disjoint points in general position (i.e., no three points lie on a line) and its edges in $E(G)$ are drawn as straight-line segments connecting their respective endpoints. We associate the vertices of the graph with their corresponding points in the plane and treat them interchangeable. A geometric graph is called *plane* if it contains no crossing edges. We refer to a geometric graph whose abstract graph is G as a *drawing of* G. A geometric (host) graph H *contains* a plane graph G if there is a subdrawing of H without edge crossings, whose (abstract) graph is isomorphic to G.

Past research has mostly focused on questions where the host graph is a geometric complete graph. Cabello [8] showed that for a given graph G it is NP-hard to decide whether a geometric complete graph contains a plane G. However, if G is outerplanar, this can be decided in polynomial time; see for example Bose [7]. It is easy to see that each geometric complete graph H contains plane paths and cycles with up to $|V(H)|$ vertices. This is no longer true for other

H. Fernau and P. Kindermann (Eds.): WG 2025, LNCS 16124, pp. 431–444, 2026.
https://doi.org/10.1007/978-3-032-11835-6_31

geometric host graphs. Figure 1 shows geometric complete multipartite graphs that do not contain plane cycles of certain lengths.

In this paper we consider geometric complete multipartite graphs as host graphs, with an emphasis on the bipartite case. We study whether these host graphs contain a plane cycle of a given length. We will usually refer to the vertex partition by its associated vertex coloring, where the color classes correspond to the partition classes. Throughout the paper we depict vertices from the same partition class by a common symbol/color in our figures. Moreover, we omit the edges of the host graph from our figures for a better readability, since they can be easily recovered from the vertex symbols and locations.

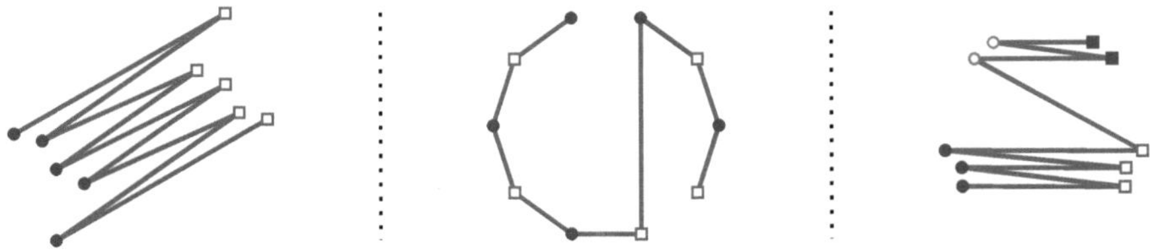

Fig. 1. Plane paths in geometric complete multipartite graphs (additional edges of the host graphs are omitted). Each graph contains a plane path on 10 vertices, but no plane 10-cycle. In fact, no plane cycle is contained in the graph on the left, the graph in the middle contains plane cycles of each even length at most 8, and the graph on the right contains exactly those plane cycles that visit at most one vertex from each color (called rainbow cycles).

There has been much interest in related questions over the years, and collections of many of the obtained results are given in the surveys by Kaneko and Kano [13] and by Kano and Urrutia [16]. Particular attention has been paid to long plane paths and cycles. For geometric complete bipartite host graphs, Kaneko and Kano [14] showed that if one color class is sufficiently larger than the other, then any drawing contains a plane path containing all vertices of the smaller color class. Many known results focus on the special case of geometric complete bipartite host graphs with n vertices of each color where all vertices are in convex position. Mulzer and Valtr [20] showed that each such host graph contains a plane path of length at least $n + \epsilon n$, for some $\epsilon > 0$, improving previous results (cf. [17,19]). For an upper bound in this case, constructions of such drawings where each plane path has length at most $\frac{4}{3}n + O(\sqrt{n})$ were given by Abellanas et al. [1] and Kynčl, Pach, and Tóth [17]. Other special (with respect to vertex placement) geometric graphs were investigated with regard to plane Hamiltonian paths and cycles by Cibulka et al. [9] and Soukup [24]. For geometric complete multipartite host graphs with an arbitrary number of color classes, Merino, Salazar and Urrutia [18] gave a lower bound on the length of the longest plane path contained in any such geometric graph and showed it is tight for odd numbers of colors. Weakening the noncrossing requirement, 1-plane Hamiltonian paths and cycles (i.e. paths and cycles of length $2n$) have been studied by

Claverol et al. [10], and Hamiltonian paths and cycles with the minimal number of crossings have been studied by Kaneko et al. [15].

The algorithmic complexity of finding plane paths or cycles is open. However, Akitaya and Urrutia [3] and Abellanas et al. [1] described algorithms that decide whether a geometric complete bipartite host graph with all vertices in convex position has a plane Hamiltonian path in $O(n^2)$ time. Bandyapadhyay et al. [4, 5] gave linear time algorithms to find plane Hamiltonian paths in drawings of complete bipartite graphs, where the vertices are mapped to the real line and the edges are drawn as circular arcs above and below the line.

We point out that a geometric drawing is uniquely determined by the vertex locations in the plane. Thus, the question whether a given geometric complete multipartite host graph H contains a plane G is equivalent to the question of whether the vertices of G can be mapped to colored points in the plane, given by $V(H)$, such that the endpoints of each edge in G are mapped to points of different colors and the resulting drawing of G is plane. Both perspectives are used in previous work. From another perspective, plane cycles can also be interpreted as polygons and there is a significant amount of research on so-called polygonizations (cf. [2]).

In this paper we study the existence and non-existence of plane cycles of specific lengths in geometric complete multipartite graphs. In Sect. 2 we consider two extremal classes of such drawings: drawings without any plane cycles and drawings with plane Hamiltonian cycles. We also investigate a possible "monotonicity" property of plane cycles, as we think that a natural candidate for a drawing with a plane cycle of a specific length is a drawing containing a longer plane cycle. We prove, among others, that if a drawing of a complete multipartite graph contains a plane cycle of length t, with $t \geq 6$, then it also contains a shorter plane cycle of length at least $\lceil \frac{t}{2} \rceil + 1$. In particular, it contains a plane cycle of length four or five. In Sect. 3 we give an FPT algorithm to decide whether a geometric bipartite graph H contains a plane Hamiltonian cycle, where the parameter is the number of vertices of H in the interior of the convex hull of all vertices. In Sect. 4 we consider the complexity of a restricted version of the same problem, where some given edges of the bipartite host graph have to be included in the plane Hamiltonian cycle. We show that the problem is NP-complete. Note that this problem is known to be NP-complete for (uncolored) complete graphs due to Akiyama et al. [2] and Jiang, Jiang, and Jiang [12].

Full proofs omitted in the following sections are given in the full version of this paper [22].

2 Monotonicity Results for Plane Cycles

In this section, we are interested in the following "monotonicity" question: Does the existence of a plane cycle of length t in a geometric complete multipartite graph H imply the existence of a plane cycle of length t' in H for each t' with $3 \leq t' \leq t$? If the given plane cycle is one where all vertices have different colors (called a *rainbow cycle*), then this can be trivially answered with yes. The question becomes interesting if at least two vertices of the given cycle have the same color; we call a cycle with this property *non-rainbow*. As complete bipartite graphs do not contain rainbow cycles, this is the only sensible case for the case of two colors (for which the question should only be asked for all even t'). There are geometric complete multipartite graphs that do not contain any plane non-rainbow cycle, including bipartite ones without any plane cycles; see Fig. 1 (left and right) for examples. Our first result characterizes those drawings. The characterization is based on four forbidden configurations $\mathcal{C}_1$–$\mathcal{C}_4$, which are depicted in Fig. 2. A formal definition is given in the full version of this paper [22]. We say that a geometric graph H *contains* configuration $\mathcal{C}_i$, if there are four or five vertices in H satisfying the constraints described by $\mathcal{C}_i$.

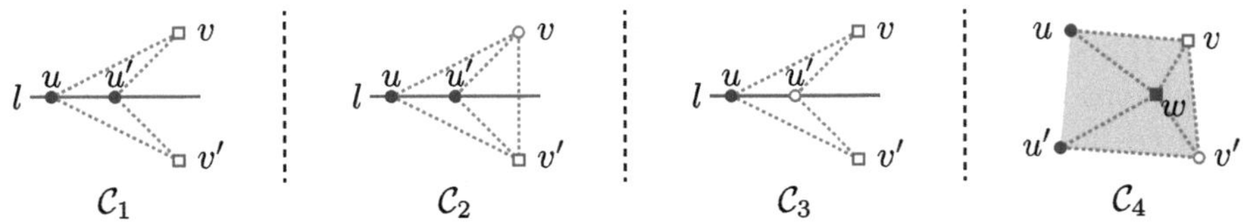

Fig. 2. The configurations $\mathcal{C}_1$–$\mathcal{C}_4$ which guarantee the existence of non-rainbow plane cycles. $\mathcal{C}_1$–$\mathcal{C}_3$: Vertices v and v' lie on different sides of the straight line l through u and u'. $\mathcal{C}_4$: Vertices $\{u, u', v, v'\}$ form a convex quadrilateral with w in its interior.

Theorem 1. *A geometric complete multipartite graph H contains a non-rainbow plane cycle if and only if it contains one of the configurations $\mathcal{C}_1$–$\mathcal{C}_4$.*

There is an algorithm that checks whether H has a non-rainbow plane cycle in time $O(|V(H)|^5)$.

Proof (sketch). A geometric complete multipartite graph which contains one of the configurations $\mathcal{C}_1 - \mathcal{C}_4$ has a non-rainbow plane cycle, as each of configurations admits such a cycle.

We now sketch the other direction of the proof in the bipartite case. Consider a geometric complete bipartite graph H which does not contain the configuration $\mathcal{C}_1$. Within the convex hull of the vertices of color ϕ_1, there is no vertex of the other color, due to the absence of configuration $\mathcal{C}_1$. This shows that there is a line ℓ separating all vertices of color ϕ_1 from all vertices of color ϕ_2. Without loss of generality assume that ℓ is horizontal. Note that each edge of H is crossing ℓ. Consider any cycle C in H and let uv denote the edge in C whose intersection with ℓ is to the left of all other intersections between C and ℓ. Let uw and vw' denote the edges in C incident to uv. Then $w \neq w'$ as H is bipartite. A brief case analysis shows that either u, v, w, and w' form configuration $\mathcal{C}_1$ (a contradiction) or uw and vw' cross each other. Hence, C is not plane and H does not contain any plane cycles. □

As $\mathcal{C}_1$ is the only forbidden configuration with only two colors, it follows that any geometric complete bipartite graph H contains a plane cycle if and only if it contains a plane cycle of length 4. With the help of Theorem 1 we can further easily check that the geometric graphs in Fig. 1 (left and right) do not have non-rainbow plane cycles. Another consequence of Theorem 1 is that any geometric complete multipartite graph that has some non-rainbow plane cycle also has such a cycle of length 4 or 5. The next result gives a partial answer to the monotonicity question.

Theorem 2. *Let H be a geometric complete multipartite graph which contains a non-rainbow plane cycle C of length t, where $t \geq 6$. Then H contains a shorter non-rainbow plane cycle of length at least $\lceil \frac{t}{2} \rceil + 1$.*

Proof (sketch). Assume, without loss of generality, $V(H) = V(C)$. We show that H contains an edge which is a "shortcut" for C. To find a suitable edge we distinguish two cases. If there is an edge in H that is not in C and not crossing C (H contains a so-called chord or diagonal of C), then this edge splits C into two shorter plane cycles. One of them contains at least $\lceil \frac{t}{2} \rceil + 1$ edges. In the full version of this paper [22], we show that this cycle is also non-rainbow.

In the second case, each edge in H that is not in C crosses C. In this case we find an edge that can be used to shorten C by exactly two edges. To find such an edge we extend C to a triangulation T of $V(H)$. By assumption, the only edges of T that are also in H are the edges of C. We consider another triangulation T' which contains an edge from H that is not in C and a sequence of flips transforming T into T', where a flip means switching the diagonal of a convex quadrilateral formed by two triangles. At some point we flip an edge that is in C. In the full version of this paper [22], we show that the first time where this happens shortens C to a non-rainbow plane cycle of length $t-2$ (see Fig. 3). □

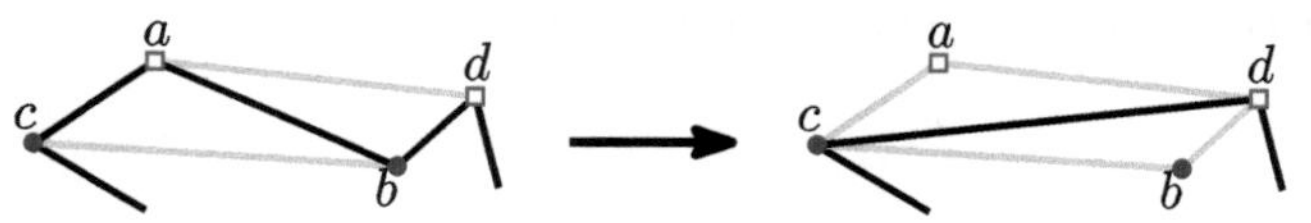

Fig. 3. Using a flip to shorten a plane cycle (black edges) by two edges.

We remark that the two cases in the proof of Theorem 2 are necessary; see Fig. 4. In particular, it is possible that in every triangulation of a plane cycle C, only the edges of C are in H. It is also possible that for every triangulation of C, no edge of C is flippable. Further note that it is not always possible to shorten an existing plane cycle by replacing a path with three edges with a single edge, as it is done in the second case of the proof of Theorem 2; an example with a plane 22-cycle is given in the full version of this paper [22]. (The geometric graph given there admits *other* plane 20-cycles).

We conclude this section with a monotonicity result for a restricted type of drawing. In the following, we associate the two partition classes of a complete

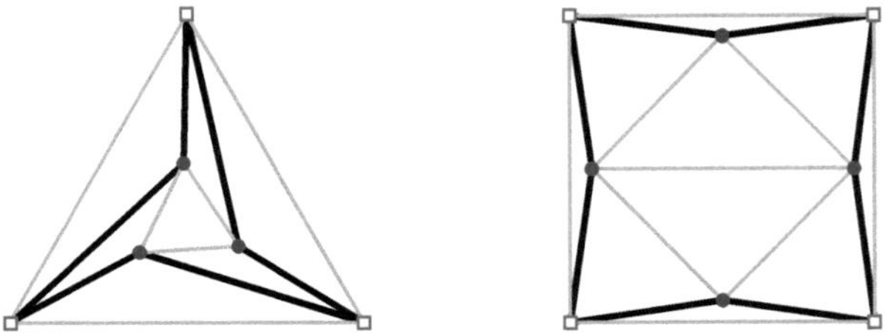

Fig. 4. Triangulations T of some plane cycles C (black edges, thick). **Left:** For any T only the edges of C are in H. **Right:** For any T no edge of C is flippable.

bipartite graph with two color classes, which we refer to as red and blue, and color the vertices accordingly. Our result generalizes the following result of Soukup (rephrased in terms of drawings): If a geometric complete bipartite graph H with n red vertices and n blue vertices has a set of blue vertices that contains in the interior of its convex hull all red vertices but no blue vertex, then H contains a plane Hamiltonian cycle [24, Remark after Theorem 1].

Theorem 3. *Let H be a geometric complete bipartite graph with n vertices of each color. If there exists a set of blue vertices that contains in the interior of its convex hull all red vertices but no blue vertex, then for each $t \in \{2, \ldots, n\}$, H contains a plane cycle of length $2t$.*

Proof (sketch). The idea is to reduce H to a subgraph from one of the following categories: Either we are left with t vertices of each color satisfying the assumptions of Soukup's result, which then gives a plane cycle of length $2t$. Or we reach one of the geometric complete bipartite graphs illustrated in Fig. 5. In the latter case we prune that graph to t vertices of each color in a systematic way and show that it contains a plane cycle of length $2t$. □

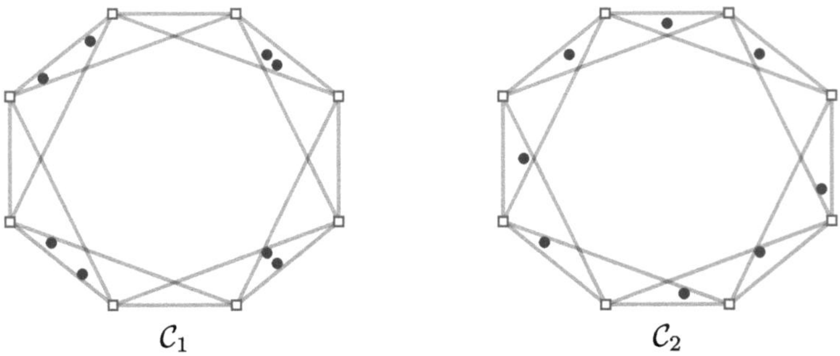

Fig. 5. Two possible drawings in the proof of Theorem 3.

3 An FPT Algorithm with Respect to the Number of Interior Vertices

In this section we consider geometric complete bipartite host graphs H with n red vertices and n blue vertices. The vertices of H that lie in the interior of the convex hull of $V(H)$ are called the *interior vertices* and the remaining vertices are called *boundary vertices.* We describe an algorithm that checks whether H contains a plane Hamiltonian cycle in time $O(n \log n + nk^2) + O(k^{5k})$, where k is the number of interior vertices.

Let I denote the set of interior vertices and let $B = V(H) \setminus I$ denote the set of boundary vertices of H. Let $w_1, \ldots, w_m$ denote the boundary vertices in counterclockwise order where w_1 is chosen arbitrarily and indices are used modulo m (in particular, $w_0 = w_m$). The case $k \leq 1$ needs to be handled separately and is covered by the following observation.

Observation 1. *If H has at most one interior vertex, then H contains a plane Hamiltonian cycle if and only if there is at most one pair of vertices w_i, w_{i+1} of the same color.*

This can be easily checked by computing the convex hull of $V(H)$ and checking the boundary vertices in counterclockwise order in $O(n \log n)$ time [6].

We continue with the case $k \geq 2$. The main idea is to split B into $O(k^2)$ parts, called critical arcs, such that it is sufficient to consider only few vertices within each part. A vertex $w_i \in B$ is *critical* if w_i and w_{i-1} are of the same color ("first kind") or there is a line through two vertices from I that separates w_i from w_{i-1} ("second kind"). Figure 6 shows an example. Let $S = \{w_{a_1}, \ldots, w_{a_s}\}$ denote the set of critical vertices where $a_1 < a_2 < \cdots < a_s$. For each critical vertex w_{a_i} the *critical arc at a_i* is the set $\{w_{a_i}, \ldots, w_{a_{i+1}-1}\}$ of boundary vertices between w_{a_i} and the next critical vertex (including the first and excluding the latter critical vertex). Observe that the critical arcs are pairwise disjoint and form a partition of B. The following lemma describes the properties of critical arcs needed later. For given vertices u and v we denote by uv the line segment between u and v, regardless of whether uv is an edge of H or not.

Lemma 1. *Let u, $v \in I$, let A and $\bar{A}$ be distinct critical arcs, and let w, $w' \in A$.*

(A1) Each critical arc is alternately colored along B.
(A2) The vertices w and w' lie on the same side of the line through uv.
(A3) Either all line segments between u and A cross all line segments between v and $\bar{A}$, or none of these segments cross.
(A4) If w comes before w' in a counterclockwise traversal of A, then uw and vw' do not cross if and only if A lies to the right of the line through u in direction of v.

We also use the following properties of plane Hamiltonian cycles which are straightforward to check.

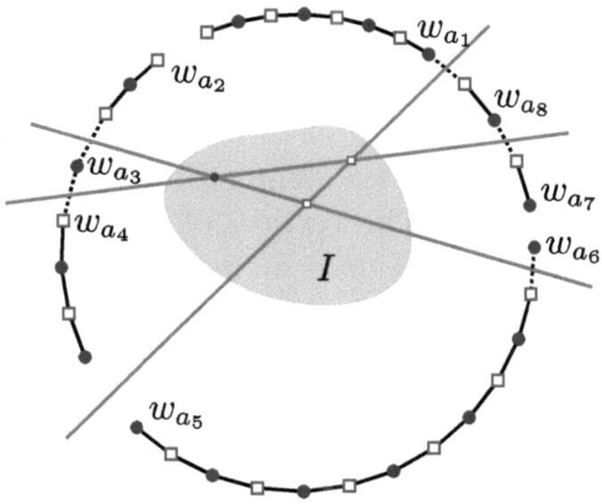

Fig. 6. The vertices w_{a_2}, w_{a_5}, and w_{a_7} are critical vertices of the first kind, w_{a_1}, w_{a_3}, w_{a_4}, w_{a_6}, and w_{a_8} are critical vertices of the second kind. Critical arcs are shown by solid black edges.

Observation 2. *Let C be a plane Hamiltonian cycle in H.*

(P1) There are at most k critical vertices of the first kind.
(P2) The cycle C visits the boundary vertices in a cyclic order along B.

We now describe the desired algorithm. The idea is to iterate over all spanning cycles of the interior vertices I (there are $O(k!)$ many) and check if one of them can be extended to the whole graph H. Observe that the assumption $k \geq 2$ ensures that there are at least two critical arcs.

An *initial cycle* F of I is a not-necessarily plane, geometric directed spanning cycle with vertex set I (that may use oriented edges from H as well as edges not in H), together with a nonempty subset G_F of the edges of F, called *gaps*, such that the edges of F not in G_F (called *fixed edges*) are in H and do not cross each other. The gaps represent those edges in F that should be replaced with longer paths using boundary vertices from B. In particular, every gap uv will be replaced by a path $u, w_i, w_{i+1}, \ldots, w_{i+j}, v$, where w_i, $w_{i+1}, \ldots, w_{i+j} \in B$. Our definition includes that in case $k = 2$, F is the directed cycle with just two edges between the two vertices in I. For technical reasons, if two gaps are adjacent in F, we insert into F a fixed zero-length dummy edge at the common endpoint and color the new vertices with the color of the replaced vertex. Note that dummy edges have both endpoints of the same color. We hence can assume for the rest of the section that each endpoint of a gap is incident to only this one gap. We call the endpoints of gaps in G_F *gap vertices*.

To extend F to B, for every gap vertex u we select a critical arc $A(u)$, and try to place a (new) neighbor for u in $A(u)$. The crucial observation is that we do not need to check all individual vertices from the selected critical arcs to decide whether F can be extended to B in this way. Instead, we find necessary and sufficient conditions for the selection of arcs alone to be "feasible". This gives a running time per selection that is independent of n as needed.

Assume we have selected a critical arc $A(u)$ for each gap vertex u, where each arc might be selected multiple times. We call the selection of critical arcs *feasible* if a labeling $u_1, \dots, u_g$ of the gap vertices along F exists, such that the following conditions hold (where $A_i = A(u_i)$):

(C1) The selected arcs $A_1, \dots, A_g$ appear counterclockwise along B in this order. Each repeatedly selected arc forms an interval in this ordering. If $A_i = A_{i+1}$ for some i (and $u_i u_{i+1}$ is not a dummy edge pair), then A_i lies to the right of the line through u_i in direction u_{i+1}.
(C2) For each i, the line segments between u_i and A_i do not cross fixed edges from F.
(C3) If $A_i \neq A_j$, then the line segments between u_i and A_i do not cross line segments between u_j and A_j.
(C4) If $u_i u_{i+1}$ is a gap, then there is no critical vertex of the first kind between A_i and A_{i+1}, that is, the union of all critical arcs from A_i, counterclockwise along B up to A_{i+1} (including the first and the latter) is alternately colored.
(C5) If $u_i u_{i+1}$ is not a gap and both vertices are of the same color, then $A_i \neq A_{i+1}$.
(C6) If $u_i u_{i+1}$ is not a gap and $A_i \neq A_{i+1}$, then A_i and A_{i+1} are adjacent along B, the (counterclockwise) last vertex in A_i is of opposite color to u_i, and the first vertex in A_{i+1} is of opposite color to u_{i+1}.
(C7) For each selected critical arc A we have $|A| \geq g(A) - \mathrm{mc}(A) + \varepsilon(A)$, where $g(A)$ is the number of gap vertices u with $A(u) = A$, $\mathrm{mc}(A)$ is the number of gaps uv with $uv \notin H$ and $A(u) = A(v) = A$, and $\varepsilon(A) = 1$ if the (counterclockwise) first vertex in A is of the same color as the (counterclockwise) first gap vertex with $A(u) = u$ and $\varepsilon(A) = 0$ otherwise.

Lemma 2 below shows that these conditions are sufficient and necessary. We briefly explain the intention behind the feasibility conditions. Feasibility is based on an ordering of the gap vertices which corresponds to their order in a counterclockwise traversal of the (given or demanded) plane Hamiltonian cycle C.

The first three conditions are responsible for planarity constraints. Condition (C1) stands for the cyclic traversal of the boundary by C, as stated by Property (P2). It also captures that edges in C between gap vertices and the same critical arc do not cross each other (cf. Property (A4)). The latter also holds for edges in C between gap vertices and distinct critical arcs, justifying (C3). Condition (C2) says that the edges in C between gap vertices and B do not cross the fixed edges of F between vertices in I. As the fixed edges and the edges along the boundary do not involve crossings by definition, this gives planarity of C. An important feature of these conditions is that their statements concern not single edges but the entire set of edges between a gap vertex and its selected critical arc. This is possible due to the properties (A2)–(A4) described in Lemma 1.

Condition (C4) (based on (A1)) says that the paths traveled by C along B use only edges from H. Conditions (C5) and (C6) describe that these paths in B neatly fit together (as C is Hamiltonian). The last condition gives a tight lower bound on the size of selected critical arcs, so that they are large enough to accommodate all necessary neighbors of gap vertices in C.

We say that F and the selected arcs are *Hamiltonian* if there is a plane Hamiltonian cycle C in H such that a counterclockwise traversal of C visits the vertices in I in the order given by F and for each gap vertex u there is an edge uw in C with $w \in A(u)$.

Lemma 2. *An initial cycle and a selection of critical arcs are Hamiltonian if and only if the selection is feasible.*

Lemma 2 implies that H contains a plane Hamiltonian cycle if and only if there is an initial cycle of I with a feasible selection of critical arcs.

The final algorithm for the case $k \geq 2$ works as follows, with missing algorithmic details explained in the full version of this paper [22]. We compute the critical arcs in $O(n \log n + nk^2)$ time. We check whether there are more than k critical vertices of the first kind (cf. (P1) from Observation 2). If not, we generate all $O(k!)$ directed spanning cycles of I. We then iterate, for each cycle, over all possible sets of gaps ($O(2^k)$ many) and check if this forms an initial cycle (fixed edges are noncrossing and in H) in $O(k^2)$ time for each. For each initial cycle, we separate adjacent gaps by dummy edges in $O(k)$ time, iterate over all $O(k^{4k})$ selections of critical arcs, and check feasibility for each selection in $O(k^2)$ time. Altogether, this needs $O(n \log n + nk^2) + O(k^{5k})$ time, as desired. In summary, the following theorem holds.

Theorem 4. *There is an algorithm to check whether a geometric complete bipartite graph with k interior vertices contains a plane Hamiltonian cycle in $O(n \log n + nk^2) + O(k^{5k})$ time.*

4 Completing Disjoint Bicolored Line Segments to a Plane Hamiltonian Cycle Is NP-Complete

In 1989, Rappaport [21] proved that completing a set of nonintersecting simple polylines to a plane simple Hamiltonian cycle is NP-complete, by reduction from the Hamiltonian path problem on planar cubic graphs [11]. In 2019, Akitaya et al. [2] and independently in 2021, Jiang, Jiang, and Jiang [12] showed that this holds for disjoint line segments as well. We show that both results generalize for a bicolored version of the problem.

Theorem 5. *It is NP-complete to decide if a set of edges of a geometric complete bipartite graph H can be completed to a plane Hamiltonian cycle in H.*

Proof. We show that the gadgets of Rappaport and Jiang, Jiang, and Jiang still work if they are embedded in a geometric complete bipartite graph. In particular, we show how to color the vertices in the gadgets such that yes-instances of the uncolored version will become yes-instances of the bipartite graph version (no-instances will remain no-instances in any case). For our arguments we briefly review the main ideas of the aforementioned reductions.

We begin with Rappaport's [21] reduction from the Hamiltonian path problem on planar cubic graphs. The input graph is drawn such that vertices and

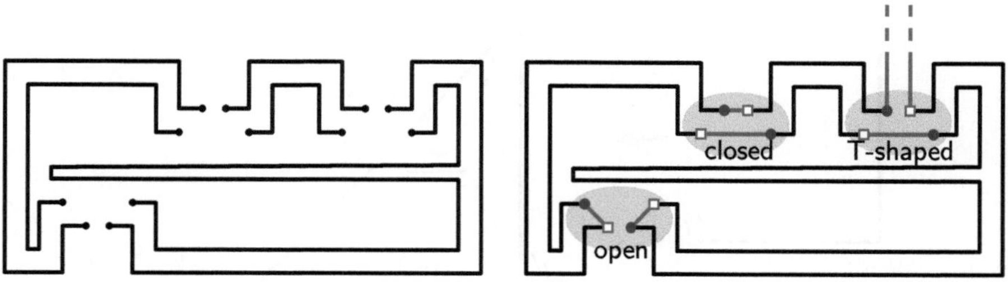

Fig. 7. Rappaport's modules. **Left:** A schematic drawing of the module. **Right:** The three possible completions for a door and the supporting coloring.

edges are represented by noncrossing horizontal and vertical line segments, respectively, such that each vertical line segment connects two horizontal line segments corresponding to adjacent vertices. Such a drawing exists for every planar cubic graph [23]. Next, the horizontal segments are replaced by so-called *modules.* Each module is based on a pair of nested polygons consisting solely of axis-parallel segments, see Fig. 7 (left). Small gaps are cut into both polygons at the three "attachment points", where the original horizontal segment connects to vertical segments (recall that the input graph is cubic). This splits all polygons into polylines. Each pair of gaps belonging to the same attachment point defines a *door*; see Fig. 7 (right). Finally, all vertical segments are removed.

Rappaport [21] proved that the final set of segments can be completed to a plane Hamiltonian cycle if and only if the input graph has a Hamiltonian path. Also, for any such completion the endpoints at every door can be connected in only three different ways (closed, open, T-shaped); see Fig. 7 (right).

In our setting each endpoint of a polyline becomes a vertex of the bipartite host graph H (we will add more vertices to the host later). We now introduce a coloring for the polyline endpoints, processing each module and each polyline in the module independently. For polylines of the outer polygon, color the counterclockwise first vertex of the polyline red and the last vertex blue. For polylines of the inner polygon, color the counterclockwise first vertex blue and the last vertex red; see Fig. 7 (right). We claim that any segment (connecting two polyline endpoints) that might be used by a completion to a plane Hamiltonian cycle is an edge in H. By construction, it is clear that closed and open doors use only edges from H. For each pair of doors from distinct modules one door comes from the upper part of its module and the other from the lower part of its module. In this case, the colorings of the two doors are inverted to each other, as if they were rotated by 180°. So, doors connected via a T-shaped completion use edges from H (joining two modules) as well. So we showed that the endpoints of the polylines in resulting yes-instances of Rappaport's reduction can be colored such that any completion to the desired cycle adds only edges from H.

Jiang, Jiang, and Jiang [12] give a reduction from Rappaport's polyline version of the problem to a version that consists only of disjoint line segments. The reduction replaces each bend of every polyline with a (suitably rotated) gadget

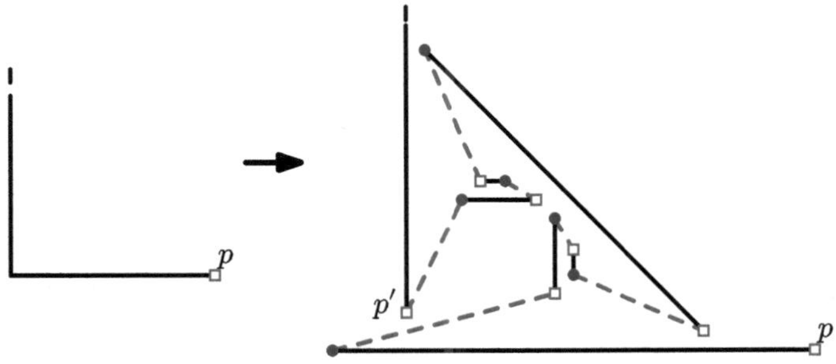

Fig. 8. Jiang, Jiang and Jiang's gadget with a suitable coloring. In each such coloring, the gadget always passes the color of p on to p'.

consisting of five disjoint line segments; see Fig. 8. Then the authors show that each completion to a noncrossing Hamiltonian cycle adds only segments within the gadgets, or segments already used in the original polyline version. Again, we reuse the gadget of their reduction. We add the endpoints of these segments as further vertices to our host graph H. The segments in the gadgets will form the set of edges from our host H that should be completed to a plane Hamiltonian cycle in H. We show that we can color the endpoints of the segments such that yes-instances lead to yes-instances in the bipartite graph version. We start with a coloring of the polyline endpoints as described above. The replacements are carried out one by one for every polyline, and the bends in every polyline are processed from one end to the other. Thus, for each bend that is currently processed, we have one incident segment whose other endpoint p is already colored. To color the vertices from the gadget, we traverse the segments of the gadget as in Fig. 8 and color the vertices alternating. Note that the vertex last colored in the gadget (called p' in Fig. 8) receives the same color as p. Hence, the last edge of the polyline(-replacement) is colored correctly, i.e. is from H. This shows that the derived edge set in H can be completed to a plane Hamiltonian cycle in H if and only if the polyline version can be completed to a plane Hamiltonian cycle. This finishes the proof. □

5 Conclusion and Open Problems

We have shown that geometric complete multipartite graphs which contain a plane cycle C also admit a shorter plane cycle of at least half the size of C. For some restricted geometric complete bipartite graphs, we have shown that they contain plane cycles of each possible even length. This prompts the following question (distinguishing between two or more colors).

Problem 1. Let H be a geometric complete multipartite graph that contains a plane cycle of length t. Does H contain a plane cycle of length t' for any (even) t' with $3 \leq t' \leq t$?

We have also given an FPT algorithm to decide whether a geometric complete bipartite graph has a plane Hamiltonian cycle. It is open whether this problem is NP-hard, which is related to an open question stated by Claverol et al. [10].

Problem 2. Is it NP-complete to decide if a given geometric complete multipartite graph contains a plane Hamiltonian cycle?

References

1. Abellanas, M., García, A., Hurtado, F., Tejel, J.: Caminos alternantes. In: Proceedings of the X Encuentros de Geometría Computacional (EGC X), pp. 7–12 (2003). https://dccg.upc.edu/people/ferran/alternatingPaths.pdf
2. Akitaya, H.A., Korman, M., Rudoy, M., Souvaine, D.L., Tóth, C.D.: Circumscribing polygons and polygonizations for disjoint line segments. In: Barequet, G., Wang, Y. (eds.) 35th International Symposium on Computational Geometry (SoCG 2019), pp. 9:1–9:17 (2019). https://doi.org/10.4230/LIPICS.SOCG.2019.9
3. Akiyama, J., Urrutia, J.: Simple alternating path problem. Discret. Math. **84**(1), 101–103 (1990). https://doi.org/10.1016/0012-365X(90)90276-N
4. Bandyapadhyay, S., Banik, A., Bhore, S., Nöllenburg, M.: Geometric planar networks on bichromatic points. In: Changat, M., Das, S. (eds.) CALDAM 2020. LNCS, vol. 12016, pp. 79–91. Springer, Cham (2020). https://doi.org/10.1007/978-3-030-39219-2_7
5. Bandyapadhyay, S., Banik, A., Bhore, S., Nöllenburg, M.: Geometric planar networks on bichromatic collinear points. Theor. Comput. Sci. **895**, 124–136 (2021). https://doi.org/10.1016/J.TCS.2021.09.035
6. de Berg, M., Cheong, O., van Kreveld, M.J., Overmars, M.H.: Computational Geometry: Algorithms and Applications, 3rd edn. Springer, Cham (2008). https://doi.org/10.1007/978-3-540-77974-2
7. Bose, P.: On embedding an outer-planar graph in a point set. Comput. Geom. **23**(3), 303–312 (2002). https://doi.org/10.1016/S0925-7721(01)00069-4
8. Cabello, S.: Planar embeddability of the vertices of a graph using a fixed point set is np-hard. J. Graph Algorithms Appl. **10**(2), 353–363 (2006). https://doi.org/10.7155/JGAA.00132
9. Cibulka, J., Kynčl, J., Mészáros, V., Stolař, R., Valtr, P.: Hamiltonian alternating paths on bicolored double-chains. In: Tollis, I.G., Patrignani, M. (eds.) GD 2008. LNCS, vol. 5417, pp. 181–192. Springer, Heidelberg (2009). https://doi.org/10.1007/978-3-642-00219-9_18
10. Claverol, M., Olaverri, A.G., Garijo, D., Seara, C., Tejel, J.: On Hamiltonian alternating cycles and paths. Comput. Geom. **68**, 146–166 (2018). https://doi.org/10.1016/J.COMGEO.2017.05.009
11. Garey, M.R., Johnson, D.S., Tarjan, R.E.: The planar Hamiltonian circuit problem is NP-complete. SIAM J. Comput. **5**(4), 704–714 (1976). https://doi.org/10.1137/0205049
12. Jiang, R., Jiang, K., Jiang, M.: Linking disjoint segments into a simple polygon is hard (2021). http://arxiv.org/abs/2108.12812. Preprint
13. Kaneko, A., Kano, M.: Discrete geometry on red and blue points in the plane—a survey—. In: Aronov, B., Basu, S., Pach, J., Sharir, M. (eds.) Discrete and computational geometry: The Goodman-Pollack Festschrift. Algorithms and Combinatorics, vol 25, pp. 551–570. Springer, Heidelberg (2003). https://doi.org/10.1007/978-3-642-55566-4_25

14. Kaneko, A., Kano, M.: On paths in a complete bipartite geometric graph. In: Akiyama, J., Kano, M., Urabe, M. (eds.) JCDCG 2000. LNCS, vol. 2098, pp. 187–191. Springer, Heidelberg (2001). https://doi.org/10.1007/3-540-47738-1_17
15. Kaneko, A., Kano, M., Yoshimoto, K.: Alternating Hamilton cycles with minimum number of crossings in the plane. Int. J. Comput. Geom. Appl. **10**(1), 73–78 (2000). https://doi.org/10.1142/S021819590000005X
16. Kano, M., Urrutia, J.: Discrete geometry on colored point sets in the plane–a survey. Graphs Combin. **37**(1), 1–53 (2021). https://doi.org/10.1007/S00373-020-02210-8
17. Kynčl, J., Pach, J., Tóth, G.: Long alternating paths in bicolored point sets. Discret. Math. **308**(19), 4315–4321 (2008). https://doi.org/10.1016/j.disc.2007.08.013
18. Merino, C., Salazar, G., Urrutia, J.: On the length of longest alternating paths for multicoloured point sets in convex position. Discret. Math. **306**(15), 1791–1797 (2006). https://doi.org/10.1016/J.DISC.2006.03.035
19. Mészáros, V.: Extremal problems on planar point sets. Ph.D. thesis, University of Szeged, Bolyai Institute (2011)
20. Mulzer, W., Valtr, P.: Long alternating paths exist. In: Cabello, S., Chen, D.Z. (eds.) 36th International Symposium on Computational Geometry (SoCG 2020), pp. 57:1–57:16 (2020). https://doi.org/10.4230/LIPIcs.SoCG.2020.57
21. Rappaport, D.: Computing simple circuits from a set of line segments is NP-complete. SIAM J. Comput. **18**(6), 1128–1139 (1989). https://doi.org/10.1137/0218075
22. Ricci, M., Rollin, J., Schulz, A., Weinberger, A.: On plane cycles in geometric multipartite graphs (2025). http://arxiv.org/abs/2506.20421. Preprint
23. Rosenstiehl, P., Tarjan, R.E.: Rectilinear planar layouts and bipolar orientations of planar graphs. Discret. Comput. Geom. **1**, 343–353 (1986). https://doi.org/10.1007/BF02187706
24. Soukup, J.: Bicolored point sets admitting non-crossing alternating Hamiltonian paths (2024). http://arxiv.org/abs/2404.06105. Preprint

Polynomial-Time Recognition and Maximum Independent Set in Burling Graphs

Paweł Rzążewski[1,2](✉) and Bartosz Walczak[3]

[1] Faculty of Mathematics and Information Science, Warsaw University of Technology, Warsaw, Poland
pawel.rzazewski@pw.edu.pl
[2] Institute of Informatics, University of Warsaw, Warsaw, Poland
[3] Department of Theoretical Computer Science, Faculty of Mathematics and Computer Science, Jagiellonian University, Kraków, Poland
bartosz.walczak@uj.edu.pl

Abstract. A Burling graph is an induced subgraph of some graph in Burling's construction of triangle-free high-chromatic graphs. Equivalently, a Burling graph is a graph that admits a so-called strict frame representation. We provide a polynomial-time algorithm to decide whether a given graph is a Burling graph and if it is, to construct its strict frame representation. The representation then enables a polynomial-time algorithm for the maximum independent set problem in Burling graphs. As a consequence, we establish Burling graphs as the first known hereditary class of graphs that admits such an algorithm while not being χ-bounded.

Keywords: Burling graph · Recognition · Independent set

1 Introduction

A class of graphs is χ-*bounded* if the chromatic number of the graphs in the class is bounded by some function of the clique number. In 1965, Burling [7] constructed an infinite sequence $B_1, B_2, \ldots$ of triangle-free graphs with $\chi(B_k) = k$ that admit an intersection representation by axis-parallel boxes in $\mathbb{R}^3$, showing that the class of graphs with such representations is not χ-bounded. This answered a question by Asplund and Grünbaum [4], who proved in 1960 that the class of graphs with intersection representations by axis-parallel rectangles in $\mathbb{R}^2$ is χ-bounded. The importance of Burling's construction has been realized only in the last decade, since Pawlik et al. [35,36] devised its intersection representations by various geometric shapes in the plane, such as straight-line segments or axis-parallel rectangular frames. Such representations imply that Burling's construction excludes induced subdivisions of 1-subdivided non-planar graphs. As a consequence, they disprove a conjecture of Scott [42] that every class of graphs excluding induced subdivisions of a fixed graph is χ-bounded. This motivated

H. Fernau and P. Kindermann (Eds.): WG 2025, LNCS 16124, pp. 445–460, 2026.
https://doi.org/10.1007/978-3-032-11835-6_32

further study of what graphs occur as induced subgraphs in Burling's construction [9,14,38,39]. Following Pournajafi and Trotignon [37–39], we call a graph a *Burling graph* if it is an induced subgraph of B_k for some k.

Burling graphs can be characterized in terms of so-called strict frame representations. A *frame representation* of a graph G assigns an axis-parallel rectangular frame to every vertex of G so that the edges of G comprise the pairs of vertices whose frames intersect. Note that a frame is an empty rectangle (excluding the interior), so nested frames do not intersect. A frame representation is *strict* if the left side of every frame intersects no other frame and lies inside no two intersecting frames (see Fig. 1). It follows from the construction in [35] that every Burling graph has a strict frame representation, and Pournajafi and Trotignon [37] proved that a graph is a Burling graph if and only if it admits a strict frame representation. Various aspects of graphs with frame representations (a.k.a. rectangle overlap graphs) were also studied in [9,27–29].

Pournajafi and Trotignon [38] asked about the complexity of recognizing Burling graphs. We answer their question by proving that the problem belongs to the complexity class P.

Theorem 1. *There is a polynomial-time algorithm which decides whether a given graph is a Burling graph and if it is, produces a strict frame representation of it.*

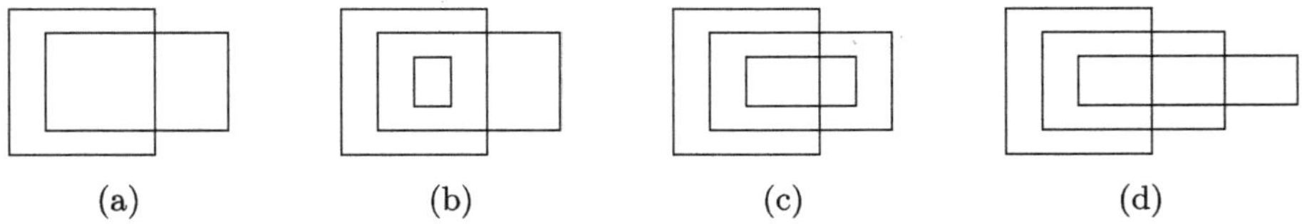

Fig. 1. Strict family of frames: (a) required configuration of every intersecting pair of frames; (b)–(d) disallowed configurations.

The maximum independent set problem is to find, given an input graph G, an independent set in G of maximum size (or maximum weight, when the vertices of G are equipped with weights). Although the problem is NP-hard in general, it might become tractable when the input graphs are "well structured", i.e., restricted to a graph class with some strong structural properties. Indeed, we know many examples of graph classes in which the maximum independent set problem can be solved in polynomial time. It turns out that these classes are also χ-bounded. A prominent example of a hereditary graph class with both discussed properties is the class of perfect graphs [21]. Numerous other examples can be found among classes that are defined by geometric representations [19,23], by bounded width parameters [1,11,12,16,44], or by excluded induced subgraphs [3, 22,24,30,31,33,41].

Motivated by such examples, Thomassé, Trotignon, and Vušković [43] asked whether *every* hereditary class of graphs that admits a polynomial-time algorithm for the maximum independent set problem is χ-bounded. Being able to construct a strict frame representation, we can solve the maximum independent set problem (with weights) in Burling graphs in polynomial time via a known dynamic programming scheme due to Gavril [19,20], which was first used in the context of so-called circle graphs. Therefore, since the class of Burling graphs is hereditary and not χ-bounded, we answer the aforesaid question in [43] in the negative.

Theorem 2. *There is a polynomial-time algorithm for the maximum independent set problem in Burling graphs with weights.*

The rest of this paper is organized as follows. In Sect. 2, we formally define strict frame representations and Burling sets—relational structures, first defined in [37], which provide a convenient abstraction over strict frame representations. In Sect. 3, we present a polynomial-time algorithm that decides whether a given graph is a Burling graph and if it is, constructs a representation by a Burling set. In Sect. 4, we describe a polynomial-time algorithm that turns a Burling set into a strict frame representation. In Sect. 5, we present a polynomial-time algorithm to find a maximum independent set given a Burling graph (with weights) and its representation by a Burling set. We conclude in Sect. 6 with various comments on related research and open problems. Proofs omitted due to space constraints can be found in the full version of this paper [40].

2 Frames and Burling Sets

The definition of the Burling sequence $B_1, B_2, \ldots$ in modern terms can be found in [9,37]. We will not need it in this paper. Instead, we will use equivalent characterizations of Burling graphs due to Pournajafi and Trotignon [37]. We start with necessary definitions.

A *frame* is the boundary of an axis-parallel rectangle in the plane, that is, a set of the form $F(\ell,r,b,t) = (\{\ell,r\} \times [b,t]) \cup ([\ell,r] \times \{b,t\})$, where $\ell,r,b,t \in \mathbb{R}$, $\ell < r$, and $b < t$. A *frame representation* of a graph G is a family of frames $\{F_v\}_{v \in V(G)}$, where $F_v = F(\ell_v, r_v, b_v, t_v)$ for all vertices v of G, such that the following conditions are satisfied for every pair of distinct vertices u and v of G:

- none of the corners (ℓ_u, b_u), (ℓ_u, t_u), (r_u, b_u), (r_u, t_u) of F_u lies on F_v;
- uv is an edge of G if and only if the frames F_u and F_v intersect.

Rephrasing [37, Definition 6.2], we call an indexed family of frames $\{F_i\}_{i \in I}$, where $F_i = F(\ell_i, r_i, b_i, t_i)$ for all $i \in I$, *strict* if it satisfies the following conditions:

- for every pair of distinct indices $i, j \in I$, if the frames F_i and F_j intersect and $\ell_i < \ell_j$, then $\ell_i < \ell_j < r_i < r_j$ and $b_i < b_j < t_j < t_i$;

- there is no triple of indices $i, j, k \in I$ such that $\ell_i < \ell_j < \ell_k < r_i < r_j$ and $b_i < b_j < b_k < t_k < t_j < t_i$.

The former means that every intersecting pair of frames in the family looks like in Fig. 1 (a), and the latter means that the family avoids the configurations in Fig. 1 (b)–(d). A graph is a *strict frame graph* if it admits a strict frame representation.[1] Observe that the class of strict frame graphs is hereditary (that is, an induced subgraph of a strict frame graph is a strict frame graph) and triangle-free (that is, no strict frame graph contains three pairwise adjacent vertices).

A *relation* on a set S is a set $R \subseteq S \times S$. We write $x \; R \; y$ to denote that $(x, y) \in R$, and we write $R|_U$ for the relation $R \cap (U \times U)$ on a subset U of S. A *cycle* in a relation R on S is a non-empty set $C \subseteq S$ whose elements can be ordered as $x_1, \ldots, x_{|C|}$ so that $x_1 \; R \cdots R \; x_{|C|} \; R \; x_1$ (in particular, $x_1 \; R \; x_1$ when $|C| = 1$). A relation R on S is *irreflexive* if there is no $x \in S$ with $x \; R \; x$, *acyclic* if there is no cycle in R (which implies irreflexivity), and *transitive* if $x \; R \; y$ and $y \; R \; z$ imply $x \; R \; z$ for all $x, y, z \in S$. A *strict partial order* is an irreflexive and transitive (and hence acyclic) relation.

Following [37, Definition 5.1], we call a triple $(S, \prec, \curvearrowright)$ a *Burling set* if S is a non-empty finite set, $\prec$ is a strict partial order on S, $\curvearrowright$ is an acyclic relation on S, and the following conditions, called *axioms*, hold for all $x, y, z \in S$:

(i) if $x \prec y$, and $x \prec z$, and $y \neq z$, then $y \prec z$ or $z \prec y$;
(ii) if $x \curvearrowright y$, and $x \curvearrowright z$, and $y \neq z$, then $y \prec z$ or $z \prec y$;
(iii) if $x \curvearrowright y$ and $x \prec z$, then $y \prec z$;
(iv) if $x \curvearrowright y$ and $y \prec z$, then $x \curvearrowright z$ or $x \prec z$.

Axiom (iii) and irreflexivity of $\prec$ imply that the relations $\prec$ and $\curvearrowright$ are disjoint. It easily follows from axiom (iv) that the relation $\prec \cup \curvearrowright$ is acyclic [37, Lemma 5.3].

Every acyclic relation R on a non-empty finite set S gives rise to an undirected graph (S, E_R) with vertex set S and edge set $E_R = \{xy \colon x, y \in S \text{ and } x \; R \; y\}$.

Theorem 3 ([37, Theorems 5.7 and 6.6]). *For every graph G, the following are equivalent:*

(1) *G is a Burling graph;*
(2) *G is a strict frame graph;*
(3) *$G = (S, E_\curvearrowright)$ for some Burling set $(S, \prec, \curvearrowright)$.*

A strict family of frames $\{F_x\}_{x \in S}$, where $F_x = F(\ell_x, r_x, b_x, t_x)$ for $x \in S$, is a *strict frame representation* of a Burling set $(S, \prec, \curvearrowright)$ if the following conditions hold for every pair of distinct elements $x, y \in S$ (see Fig. 2):

$$x \prec y \quad \text{if and only if} \quad \ell_y < \ell_x < r_x < r_y \text{ and } b_y < b_x < t_x < t_y;$$
$$x \curvearrowright y \quad \text{if and only if} \quad \ell_y < \ell_x < r_y < r_x \text{ and } b_y < b_x < t_x < t_y.$$

[1] We note that allowing the configuration in Fig. 1 (b) in the definition of strict frame representation would lead to the same class of graphs, because the vertices represented inside two intersecting frames must be disconnected from the rest of the graph.

Fig. 2. Correspondence between the relations $\prec$ and $\curvearrowright$ in a Burling set and configurations of pairs of frames in its strict frame representation.

Every strict frame representation of a Burling set $(S, \prec, \curvearrowright)$ is a strict frame representation of the graph $(S, E_{\curvearrowright})$, for which only the latter condition is required.

It is straightforward to verify that the conditions above turn every strict family of frames $\{F_x\}_{x \in S}$ into a Burling set $(S, \prec, \curvearrowright)$; see [37, Lemma 6.4]. Conversely, it is not difficult to construct a strict frame representation for any given Burling set—we provide an algorithm for this task in Sect. 4. Burling sets with their axioms (i)–(iv) thus provide an abstraction of key structural properties of strict families of frames. In particular, the core difficulty in constructing a strict frame representation for a given graph lies in constructing the corresponding Burling set—we provide an algorithm for this task in the next section.

We complete this section by defining two distinguished types of elements in a Burling set $(S, \prec, \curvearrowright)$. An element $s \in S$ is a *root* if there is no $x \in S$ with $s \prec x$ or $s \curvearrowright x$. Since the relation $\prec \cup \curvearrowright$ is acyclic, every Burling set has at least one root. Intuitively, thinking of a strict frame representation of a Burling set, a frame $F(\ell_s, r_s, b_s, t_s)$ with minimum value of ℓ_s represents a root. A *probe* is an element $p \in S$ such that there is no $x \in S$ with $p \prec x$, or $x \prec p$, or $x \curvearrowright p$. A Burling set may have no probes. Intuitively, in a strict frame representation, the frame $F(\ell_p, r_p, b_p, t_p)$ representing a probe can be extended arbitrarily far to the right, i.e., r_p can be increased arbitrarily without changing the intersection graph.

Lemma 4. *A Burling set* $(S, \prec, \curvearrowright)$ *for which the graph* $(S, E_{\curvearrowright})$ *is connected has a unique root.*

3 Recognition

For this entire section, we fix a graph G for which we want to decide whether it is a Burling graph and if it is, to construct a Burling set $(V, \prec, \curvearrowright)$ such that $G = (V, E_{\curvearrowright})$. All graph-theoretic terms will refer to the fixed graph G. In particular, a *vertex* or an *edge* will always mean a vertex or an edge of G. We will be assuming that G is triangle-free, because a graph containing a triangle is never a Burling graph, and containing a triangle can be trivially decided in polynomial time.

We will use the following notation and terminology. Let V be the set of vertices (of G). A set $S \subseteq V$ is *connected* if $S \neq \emptyset$ and the induced subgraph $G[S]$ is connected. A *component* of a set $S \subseteq V$ is an inclusion-maximal connected subset of S. A *neighbor* of a vertex v is a vertex with an edge to v. For a vertex v, let $N(v)$ denote the set of neighbors of v, and let $N[v] = N(v) \cup \{v\}$. A *neighbor* of a set $S \subseteq V$ is a vertex in $V - S$ with an edge to at least one vertex in S. For a set $S \subseteq V$, let $N(S)$ denote the set of neighbors of S, and let $N[S] = S \cup N(S)$.

A set $S' \subseteq V$ is *homogeneous* for a set $S \subseteq V$ if $S' \subseteq N(x)$ or $S' \cap N(x) = \emptyset$ for every $x \in S$. In other words, S' is homogeneous for S if all vertices in S' have the same neighbors in S. A family $\mathcal{C}$ of pairwise disjoint subsets of V is *nested* if any two distinct $C_1, C_2 \in \mathcal{C}$ satisfy at least one of the following conditions:

- $N(C_1) \subseteq N(C_2)$ and the set $N(C_1)$ is homogeneous for C_2;
- $N(C_2) \subseteq N(C_1)$ and the set $N(C_2)$ is homogeneous for C_1;
- $N(C_1) \cap N(C_2) = \emptyset$.

A *nesting order* on a nested family $\mathcal{C}$ is a strict partial order $<$ on $\mathcal{C}$ with the following properties for all distinct $C_1, C_2 \in \mathcal{C}$:

- if $C_1 < C_2$, then $N(C_1) \subseteq N(C_2)$ and the set $N(C_1)$ is homogeneous for C_2;
- if neither $C_1 < C_2$ nor $C_2 < C_1$, then $N(C_1) \cap N(C_2) = \emptyset$.

Clearly, every nested family has a nesting order. The definition directly implies a polynomial-time algorithm to test whether a family $\mathcal{C}$ is nested and if it is, to construct a nesting order on $\mathcal{C}$.

A *Burling structure* is a Burling set $(U, \prec, \curvearrowright)$ such that $U \subseteq V$ and $(U, E_{\curvearrowright})$ is an induced subgraph of G, that is, for $x, y \in U$, xy is an edge of G if and only if $x \curvearrowright y$ or $y \curvearrowright x$. Such a Burling structure is *around* a connected set S if $U = N[S]$.

Our Burling graph recognition algorithm applies the dynamic programming technique, solving the following *subproblems* on the way:

- an *unrooted subproblem* $\mathfrak{B}(X, S)$ asks for a Burling structure around S with $N(S)$ a set of probes, where $X = \emptyset$ or $X = N[v]$ for some vertex v, and S is a component of $V - X$;
- a *rooted subproblem* $\mathfrak{R}(X, r, S)$ asks for a Burling structure around S with r a root and $N(S) - \{r\}$ a set of probes, where $X = \emptyset$ or $X = N[v]$ for some vertex v, $r \in V - X$, and S is a component of $V - (X \cup \{r\})$ containing at least one neighbor of r.

The *size* of such a subproblem is the size of S. A *solution* to a subproblem is a Burling structure being asked for; if it does not exist, then the subproblem has no solution. It is clear from the description above that the number of subproblems is polynomial in the number of vertices. Careful analysis shows that if G has n vertices, then there are only $\mathcal{O}(n^2)$ subproblems—we omit the details.

Below, we describe algorithms to solve a subproblem $\mathfrak{B}(X, S)$ or $\mathfrak{R}(X, r, S)$ assuming that all smaller subproblems have been already solved. In these descriptions, *failing* means that the algorithm terminates computations and reports no solution to the subproblem.

We start by explaining the intuition behind the algorithm for an unrooted problem $\mathfrak{B}(X, S)$, describing it in terms of a strict frame representation. Recall that $X = N[v]$ for some vertex v or $X = \emptyset$. We focus on the former setting. Suppose that we have already fixed the frame F_v for v and decided that the component S of $G-X$ should be represented inside F_v. Recall that in $\mathfrak{B}(X,S)$ we ask not only for a representation of $G[S]$, but of $G[N[S]]$, requiring additionally that the vertices in $N(S)$ are probes. The frames representing such probes can be easily extended to the right in order to intersect the frame of v, which provides a clean intersection between the representation of $N[S]$ we are building in the current call and the representation of the rest of the graph that has been built so far. Our goal is to find a vertex r in S that can be the root of $G[N[S]]$. An obvious condition that the root must satisfy is as follows: for each component C of $S-\{r\}$, the graph $G[C]$ has a representation with r a root and $N(C)-\{r\}$ a set of probes; this can be verified by solving the rooted subproblem $\mathfrak{R}(X,r,C)$. Another, more subtle condition is that no two frames representing vertices from different components of $S-\{r\}$ are nested, which implies that no probe in $N(S)$ can have neighbors in two distinct components of $S-\{r\}$; consult Fig. 3. If these conditions are satisfied, it is not difficult to assemble representations of $G[N[C]]$ for all components C of $S-\{r\}$ into a desired representation of $G[N[S]]$.

Now, we turn this intuition into a formal description of the algorithm.

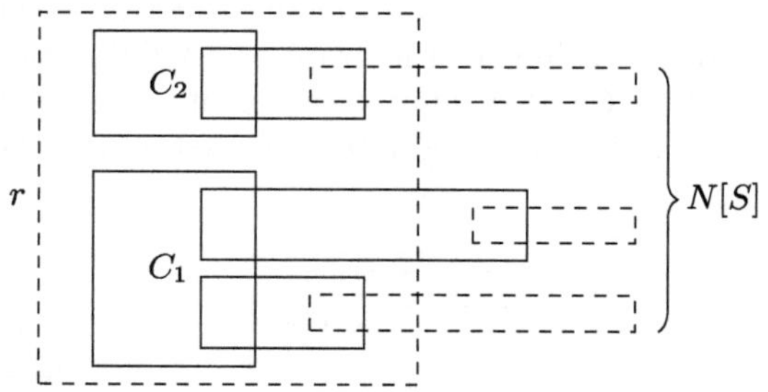

Fig. 3. The second condition in the algorithm for $\mathfrak{B}(X,S)$ with C_1 and C_2 the components of $S-\{r\}$.

Algorithm to solve $\mathfrak{B}(X,S)$.

1. Find a vertex $r \in S$ that satisfies the following two conditions, or fail if there is no such vertex:
   - for every component C of $S-\{r\}$, the subproblem $\mathfrak{R}(X,r,C)$ has a solution;
   - for every $p \in N(S)$, the set $N(p) \cap N[S]$ is contained in $C \cup \{r\}$ for some component C of $S-\{r\}$ or is equal to $\{r\}$.
2. Let $(N[C], \prec_C, \curvearrowright_C)$ be a solution to $\mathfrak{R}(X,r,C)$ for each component C of $S-\{r\}$. Let $\prec$ and $\curvearrowright$ be the unions of the relations $\prec_C$ and $\curvearrowright_C$ (respectively) over all components C of $S-\{r\}$ with an additional related pair $p \curvearrowright r$ for every $p \in N(S)$ such that $N(p) \cap N[S] = \{r\}$. Report $(N[S], \prec, \curvearrowright)$ as a solution to $\mathfrak{B}(X,S)$.

Lemma 5. *The algorithm to solve $\mathfrak{B}(X,S)$ is correct assuming that the smaller subproblems have been correctly solved.*

Now, we proceed with the description of the algorithm for a rooted subproblem $\mathfrak{R}(X, r, S)$. Again, we start with some intuition described in terms of strict frame representations; see Fig. 4. We look for a representation of $G[N[S]]$ with r a root and $N(S) - \{r\}$ a set of probes. Consider any such a representation, and let F_r be the frame representing r. Each component C of $S - N(r)$ is represented entirely inside or entirely outside F_r. Call C an *inner component* in the former and an *outer component* in the latter case. Each inner component C satisfies $N(C) \subseteq N(r)$, because the representation of any path from C to a vertex from $G - N[C]$ must cross F_r. In the representation restricted to $N[C]$, the vertices in $N(S)$ are represented as probes, so that they can intersect F_r Moreover, the way how inner components attach to their neighborhoods is very restricted—they form a nested family. Each outer component C has exactly one neighbor in $N(r)$, call it q_C. In the representation restricted to $N[C]$, q_C is a (unique) root, and the vertices in $N(C) - \{q_C\}$, which belong to $N(S) - \{r\}$, are probes, as required by the considered representation of $G[N[S]]$. There is one more condition that must be satisfied by the vertices in $N(S) - \{r\}$, like in the unrooted problem. These necessary conditions on inner and outer components and their neighborhoods can be verified looking only at the graph G, without the representation. Moreover, representations of the graphs $G[N[C]]$ obtained from solving the subproblems $\mathfrak{B}(N[r], C)$ (for inner components C) and $\mathfrak{R}(X, q_C, C)$ (for outer components C) can be combined into a desired representation of $G[N[S]]$.

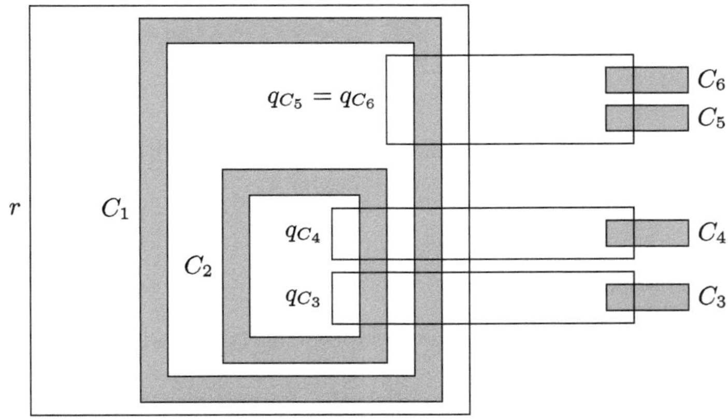

Fig. 4. A solution to a subproblem $\mathfrak{R}(X, r, S)$. The graph induced by $S - N(r)$ has six components $C_1, \ldots, C_6$, of which C_1 and C_2 are inner while $C_3, \ldots, C_6$ are outer. Gray areas are where particular components are represented.

Algorithm to solve $\mathfrak{R}(X, r, S)$.

1. Classify each component C of $S - N(r)$ as
   - an *inner component* if $N(C) \subseteq N(r)$ and the subproblem $\mathfrak{B}(N[r], C)$ has a solution;

- an *outer component* if $|N(C) \cap N(r)| = 1$ and the subproblem $\mathfrak{R}(X, q_C, C)$ has a solution, where q_C is the vertex such that $N(C) \cap N(r) = \{q_C\}$.

  Fail if some component of $S - N(r)$ cannot be classified either way. For components that can be classified both ways, choose arbitrarily.
2. Check the following condition and fail if it does not hold: for every $p \in N(S) - \{r\}$, the set $N(p) \cap N[S]$ is contained in the union of $\{r\}$ and all inner components, or is contained in $C \cup \{q_C\}$ for some outer component C, or is equal to $\{q\}$ for some $q \in N(r) \cap S$.
3. Let $\mathcal{C}$ be the family of all inner components. Check whether $\mathcal{C}$ is nested and fail if not. Compute a nesting order $<$ on $\mathcal{C}$.
4. Let $(N[C], \prec_C, \curvearrowright_C)$ be a solution to $\mathfrak{B}(N[r], C)$ for every inner component C or to $\mathfrak{R}(X, q_C, C)$ for every outer component C. Let $\prec$ and $\curvearrowright$ be the unions of the relations $\prec_C$ and $\curvearrowright_C$ (respectively) over all components C with the following additional related pairs:
   - $x \prec r$ for all $x \in C$, for all $C \in \mathcal{C}$;
   - $x \prec y$ for all $x \in C_1$ and all neighbors y of $N(C_1)$ in C_2, for all $C_1, C_2 \in \mathcal{C}$ with $C_1 < C_2$;
   - $q \curvearrowright r$ for all $q \in N(r) \cap N[S]$;
   - $p \curvearrowright q$ for all $q \in N(r) \cap S$ and all $p \in N(S) - \{r\}$ with $N(p) \cap N[S] = \{q\}$.

   Report $(N[S], \prec, \curvearrowright)$ as a solution to $\mathfrak{R}(X, r, S)$.

Lemma 6. *The algorithm to solve $\mathfrak{R}(X, r, S)$ is correct assuming that the smaller subproblems have been correctly solved.*

The graph G is a Burling graph if and only if the subproblem $\mathfrak{B}(\emptyset, C)$, which asks for a Burling structure $(C, \prec_C, \curvearrowright_C)$, has a solution for every component C of V. From Lemmas 5 and 6, by induction on the size of the subproblems, we conclude that the algorithm solves the subproblem $\mathfrak{B}(\emptyset, C)$ correctly for each component C of V. If it finds a solution $(C, \prec_C, \curvearrowright_C)$ for each component C of V (rather than reporting no solution), a Burling set $(V, \prec, \curvearrowright)$ with $G = (V, E_\curvearrowright)$ is obtained by defining $\prec$ and $\curvearrowright$ as the unions of $\prec_C$ and $\curvearrowright_C$ (respectively) over all components C of V.

It is clear that the algorithm to solve a subproblem of the form $\mathfrak{B}(X, S)$ or $\mathfrak{R}(X, r, S)$ runs in polynomial time assuming that solutions to smaller subproblems have been computed beforehand. Since the number of subproblems is polynomial, the total running time to solve $\mathfrak{B}(\emptyset, C)$ on all components C of V is polynomial. Since checking whether G is triangle-free can also be done in polynomial time, we arrive at the following conclusion.

Theorem 7. *There is a polynomial-time algorithm which, given a graph G, either declares G a Burling graph and produces a Burling set $(V, \prec, \curvearrowright)$ such that $G = (V, E_\curvearrowright)$, or reports that G is not a Burling graph.*

4 Strict Frame Representation

In this section, we provide a polynomial-time algorithm to turn a Burling set into a strict frame representation of it. The existence of such a representation, claimed in Theorem 3, was proved by Pournajafi and Trotignon [37], but that proof does not produce the representation explicitly. Instead, it relies on the existence of a strict frame representation (constructed in [35]) of a graph B_k containing a given Burling graph G as an induced subgraph, and the size of that graph B_k may be double exponential in the size of G.

We fix a Burling set $(S, \prec, \curvearrowright)$ for which we want to construct a strict frame representation $\{F_x\}_{x\in S}$ with $F_x = F(\ell_x, r_x, b_x, t_x)$ for all $x \in S$. Let $R = \prec \cup \curvearrowright$, which is an acyclic relation on S. Let R^* be the transitive closure of R, that is, for all $x, y \in S$, we have $x\ R^*\ y$ if and only if $x = x_0\ R \cdots R\ x_k = y$ for some $x_0, \dots, x_k \in S$ with $k \geqslant 1$. Since R is acyclic, R^* is also acyclic, so it is a strict partial order.

We first show how to find appropriate values ℓ_x and r_x for all $x \in S$. Consider the following conditions on a set of $2|S|$ distinct symbols ℓ_x and r_x with $x \in S$:

(1) $\ell_x < r_x$ for all $x \in S$;
(2) $\ell_x < \ell_y$ for all $x, y \in S$ with $y\ R\ x$;
(3) $r_x < r_y$ for all $x, y \in S$ with $x \prec y$ or $y \curvearrowright x$;
(4) $r_x < \ell_y$ for all $x, y \in S$ such that there is $z \in S$ with $y\ R\ z \curvearrowright x$.

Lemma 8. *The minimal binary relation $<$ defined by conditions (1)–(4) on the set $\{\ell_x, r_x\}_{x\in S}$ is acyclic.*

By Lemma 8, the symbols ℓ_x and r_x with $x \in S$ can be assigned values $1, \dots, 2|S|$ so that conditions (1)–(4) hold for the assigned values.

Now, we describe how to find appropriate values b_x and t_x for all $x \in S$. A *parent* of an element $x \in S$ is an element $z \in S$ such that $x\ R\ z$ and there is no $y \in S$ with $x\ R\ y\ R\ z$. By axioms (i)–(iii), every element of S that is not a root of $(S, \prec, \curvearrowright)$ has a unique parent. Therefore, the set S along with the parent relation forms a (directed) forest such that if $x\ R^*\ y$, then y is an ancestor of x. To determine b_x and t_x for all $x \in S$, perform a depth-first search on this forest starting from each of the roots (in some order), and record, for each $x \in S$, the discovery time b_x and the finishing time t_x. This leads to an assignment of values $1, \dots, 2|S|$ satisfying the following conditions:

(5) $b_x < t_x$ for all $x \in S$;
(6) $b_y < b_x < t_x < t_y$ for all $x, y \in S$ with $x\ R^*\ y$;
(7) $t_x < b_y$ or $t_y < b_x$ for all $x, y \in S$ with neither $x\ R^*\ y$ nor $y\ R^*\ x$.

Lemma 9. *The family $\{F(\ell_x, r_x, b_x, t_x)\}_{x\in S}$ is a strict frame representation of $(S, \prec, \curvearrowright)$.*

Proof. Conditions (1) and (5) ensure that $F(\ell_x, r_x, b_x, t_x)$ is a frame for every $x \in S$. To prove that the family $\{F(\ell_x, r_x, b_x, t_x)\}_{x\in S}$ is a strict frame representation of $(S, \prec, \curvearrowright)$, we verify the following three conditions for all $x, y \in S$:

if $x \prec y$,	then $\ell_y < \ell_x < r_x < r_y$ and $b_y < b_x < t_x < t_y$;
if $x \curvearrowright y$,	then $\ell_y < \ell_x < r_y < r_x$ and $b_y < b_x < t_x < t_y$;
if neither $x\ R\ y$ nor $y\ R\ x$,	then $r_x < \ell_y$, or $r_y < \ell_x$, or $t_x < b_y$, or $t_y < b_x$.

The first two conditions describe the configurations in Fig. 2, while the last one describes two frames that are disjoint and not nested.

If $x \prec y$, then $\ell_y < \ell_x < r_x < r_y$ by (2) and (3), and $b_y < b_x < t_x < t_y$ by (6). Likewise, if $x \curvearrowright y$, then $\ell_y < \ell_x < r_y < r_x$ by (2) and (3), and $b_y < b_x < t_x < t_y$ by (6). If neither $x\ R^*\ y$ nor $y\ R^*\ x$, then $t_x < b_y$ or $t_y < b_x$ by (7). Finally, suppose $x\ R^*\ y$ but not $x\ R\ y$. Let $x_0, \ldots, x_k \in S$ be a smallest tuple with $x = x_0\ R \cdots R\ x_k = y$ witnessing $x\ R^*\ y$. It follows that $k \geqslant 2$. If $x_i \prec x_{i+1}$ for some $i \in \{1, \ldots, k-1\}$, then $x_{i-1} \prec x_{i+1}$ by transitivity of $\prec$ and by axiom (iv), so $x_0, \ldots, x_{i-1}, x_{i+1}, \ldots, x_k$ is a smaller tuple witnessing $x\ R^*\ y$, contradicting the assumption that $x_1, \ldots, x_k$ is smallest. Therefore, we have $x\ R\ x_1 \curvearrowright \cdots \curvearrowright x_k = y$, whence it follows that $r_y = r_{x_k} < \cdots < r_{x_2} < \ell_x$. □

The following is a direct corollary to the construction above and Lemma 9.

Theorem 10. *There is a polynomial-time algorithm which constructs a strict frame representation for a given Burling set.*

Theorem 1 follows directly from Theorems 7 and 10.

5 Maximum Independent Set

In this section, we present an algorithm for the maximum (weight) independent set problem in Burling graphs. The core of the algorithm is a dynamic programming scheme developed by Gavril [19,20] (see also [8]) which reduces, for any class of families of geometric objects $\mathcal{F}$, the maximum (weight) independent set problem in the *overlap graphs* of families in $\mathcal{F}$ to the maximum weight independent set problem in the *intersection graphs* of families in $\mathcal{F}$. In our case, $\mathcal{F}$ is the class of strict families of axis-parallel rectangles (defined just like strict families of frames except that rectangles include their interiors while frames do not). Strict frame graphs are the overlap graphs of families in $\mathcal{F}$, while the intersection graphs of families in $\mathcal{F}$ turn out to be chordal.

Say that a relation R on a set S is *chordal* when it is acyclic and the following holds for all $x, y, z \in S$: if $x\ R\ y$, and $x\ R\ z$, and $y \neq z$, then $y\ R\ z$ or $z\ R\ y$. This term relates to well-known properties of chordal graphs. In particular, (S, E_R) is a chordal graph when R is a chordal relation on S, and vice versa—for every chordal graph G, orienting every edge according to a so-called perfect elimination order gives rise to a chordal relation on the vertex set. It is well known that a maximum independent set in a chordal graph with vertex weights can be computed in polynomial time [18].

Let $(V, \prec, \curvearrowright)$ be a Burling set, and let $G = (V, E_{\curvearrowright})$. Let $\mathsf{w}\colon V \to \mathbb{R}$ be a weight assignment according to which we want to compute a maximum independent set in G. For a set $S \subseteq V$, let $\mathsf{w}(S) = \sum_{x \in S} \mathsf{w}(x)$. Let $R = {\prec} \cup {\curvearrowright}$, which is an acyclic relation. This and axioms (i)–(iii) imply that the relation R is chordal. Let $V_u = \{x \in V\colon x \prec u\}$ for all $u \in V$. Note that if $x \in V_u$, then $V_x \subset V_u$. The algorithm applies the dynamic programming technique on subproblems of the form $\mathfrak{I}(S)$ defined for $S = V$ and $S = V_u$ for all $u \in V$, where the subproblem

$\mathfrak{I}(S)$ asks for a maximum independent set in the graph $(S, E_{\curvearrowright|S})$ with respect to the weight assignment w. The *size* of such a subproblem is the size of S. A *solution* to a subproblem is a maximum independent set being asked for.

Algorithm to solve $\mathfrak{I}(S)$.

1. If $S = \emptyset$, then stop and report the empty set as a solution, otherwise continue.
2. For every $u \in S$, let I_u be a solution to the subproblem $\mathfrak{I}(V_u)$, and let $\mathsf{w}^*(u) = \mathsf{w}(u) + \mathsf{w}(I_u)$.
3. Compute a maximum independent set I^* in the graph $(S, E_{R|S})$ with weight assignment w^*, using the algorithm for chordal graphs.
4. Report the set $I^* \cup \bigcup_{u \in I^*} I_u$ as a solution to $\mathfrak{I}(S)$.

Lemma 11 (cf. [20, Lemma 2]). *The algorithm to solve $\mathfrak{I}(S)$ is correct assuming that the smaller subproblems have been correctly solved.*

The subproblem $\mathfrak{I}(V)$ asks for a maximum independent set in the graph $(V, E_{\curvearrowright})$. It is clear that the algorithm to solve a subproblem of the form $\mathfrak{I}(S)$ runs in polynomial time assuming that solutions to smaller subproblems have been computed beforehand. Since there are $|V|+1$ subproblems, the total running time spent to solve $\mathfrak{I}(V)$ is polynomial. Hence, we arrive at the following conclusion.

Theorem 12. *There is a polynomial-time algorithm which, given a Burling set $(V, \prec, \curvearrowright)$ and a weight assignment $\mathsf{w}\colon V \to \mathbb{R}$, computes a maximum independent set in the graph $(V, E_{\curvearrowright})$ with respect to the weight assignment w.*

Theorem 2 follows directly from Theorems 7 and 12.

6 Concluding Remarks

Intersection graphs of axis-parallel rectangles in the plane are NP-hard to recognize [25]. An easy adaptation of the proof in [25] shows that frame graphs, i.e., graphs admitting some (not necessarily strict) frame representation, are also NP-hard to recognize. The maximum independent set problem is NP-hard for intersection graphs of axis-parallel unit squares [17] and for intersection graphs of horizontal and vertical segments [26], either of which directly implies NP-hardness for frame graphs. The situation is different for *directed frame graphs*, i.e., graphs with frame representations in which every intersecting pair of frames looks like in Fig. 1 (a) (which is the first condition from the definition of *strict*). For directed frame graphs, the relation $\curvearrowright$ defined as in Fig. 2 is chordal, and therefore the maximum independent set problem can be solved in polynomial time in the same way as described in Sect. 5 provided that the corresponding relations $\prec$ and $\curvearrowright$ are given as part of the input. This leaves the following question.

Problem 13. What is the complexity of recognizing directed frame graphs?

Burling graphs are induced subgraphs of Burling's construction of triangle-free high-chromatic graphs $(B_1, B_2, \ldots)$. What is the complexity of recognition and of computing a maximum independent set for other known constructions of triangle-free high-chromatic graphs? This is not interesting for Mycielski's construction [34], which contains all triangle-free graphs as induced subgraphs [13]. For induced subgraphs of Zykov's construction [45] and Blanche Descartes' construction [15], both problems are NP-hard [32]. For twin-cut graphs, i.e., induced subgraphs of the construction in [5], recognition is in P while computing a maximum independent set is NP-hard [6].

We will show in a follow-up work that, by contrast to the maximum independent set problem, the k-coloring problem remains NP-hard for Burling graphs for every $k \geqslant 3$, as it is for the other above-mentioned classes [6,32]. This leads to the following analog of the Thomassé–Trotignon–Vušković question from [43].

Problem 14. Is every hereditary class of graphs that admits a polynomial-time 3-coloring algorithm χ-bounded (assuming $\mathsf{P} \neq \mathsf{NP}$)?

As we have mentioned in the introduction, the recent flurry of research on Burling graphs has been mostly motivated by the conjecture of Scott [42] that the class $\mathcal{F}(H)$ of graphs excluding induced subdivisions of H is χ-bounded for every graph H. Clearly, if no subdivision of H is a Burling graph, then Scott's conjecture fails for H, because $\mathcal{F}(H)$ contains all graphs $B_1, B_2, \ldots$ from Burling's construction. Chudnovsky, Scott, and Seymour [10] conjectured that for every H, the chromatic number of graphs $G \in \mathcal{F}(H)$ is bounded by a function of the clique number of G and the maximum k such that G contains an induced subgraph isomorphic to B_k. This conjecture, if true, implies that if some subdivision of H is a Burling graph, then Scott's conjecture holds for H. Our recognition algorithm for Burling graphs can be adapted to a polynomial-time recognition algorithm of graphs H such that some subdivision of H is a Burling graph. We omit the details, because we do not find this interesting as long as the Chudnovsky–Scott–Seymour conjecture is open. However, very recently, the special case of the conjecture for graphs G that are string graphs has been proved [2]. In particular, we now know that Burling graphs form a minimal hereditary class of graphs with unbounded chromatic number—the only known class of graphs with that property other than the (trivial) class of complete graphs.

Acknowledgments. We thank Martin Milanič for bringing the question of Thomassé, Trotignon, and Vušković to our attention. We thank Jonathan Rollin and the anonymous reviewers for helpful comments.

The project was initiated at the Structural Graph Theory workshop STWOR in September 2023, which was a part of STRUG: Structural Graph Theory Bootcamp, funded by the "Excellence initiative – research university (2020–2026)" of University of Warsaw. The main part of work was done during the workshop Homonolo 2023. We thank the participants and the organizers of both events for a productive and inspiring atmosphere.

Bartosz Walczak was partially supported by the National Science Center of Poland grant 2019/34/E/ST6/00443. Paweł Rzążewski was partially supported by the National Science Center of Poland grant 2024/54/E/ST6/00094.

Disclosure of Interests. The authors have no competing interests.

References

1. Abrishami, T., et al.: Excluding a clique or a biclique in graphs of bounded induced matching treewidth. SIAM J. Discret. Math. **39**(2), 1189–1200 (2025). https://doi.org/10.1137/24M1659960
2. Abrishami, T., et al.: Burling graphs in graphs with large chromatic number (2025). https://doi.org/10.48550/arXiv.2510.19650 to appear at SODA 2026
3. Abrishami, T., Chudnovsky, M., Pilipczuk, M., Rzążewski, P.: Max weight independent set in sparse graphs with no long claws. In: Beyersdorff, O., Kanté, M.M., Kupferman, O., Lokshtanov, D. (eds.) 41st International Symposium on Theoretical Aspects of Computer Science, STACS 2024, March 12–14, 2024, Clermont-Ferrand, France, Leibniz International Proceedings in Informatics, vol. 289, pp. 4:1–15. Schloss Dagstuhl – Leibniz-Zentrum für Informatik (2024). https://doi.org/10.4230/LIPIcs.STACS.2024.4
4. Asplund, E., Grünbaum, B.: On a coloring problem. Mathematica Scandinavica **8**, 181–188 (1960). https://doi.org/10.7146/math.scand.a-10607
5. Bonnet, É., Bourneuf, R., Duron, J., Geniet, C., Thomassé, S., Trotignon, N.: A tamed family of triangle-free graphs with unbounded chromatic number. arXiv:2304.04296 (2023). https://doi.org/10.48550/arXiv.2304.04296
6. Bonnet, É., Bourneuf, R., Duron, J., Geniet, C., Thomassé, S., Trotignon, N.: Unpublished results announced in [5]
7. Burling, J.P.: On coloring problems of families of polytopes. Ph.D. thesis, University of Colorado, Boulder (1965)
8. Čenek, E., Stewart, L.: Maximum independent set and maximum clique algorithms for overlap graphs. Discret. Appl. Math. **131**(1), 77–91 (2003). https://doi.org/10.1016/S0166-218X(02)00418-3
9. Chalopin, J., Esperet, L., Li, Z., Ossona de Mendez, P.: Restricted frame graphs and a conjecture of Scott. Electron. J. Comb. **23**(1), Article P1.30 (2016). https://doi.org/10.37236/4424
10. Chudnovsky, M., Scott, A., Seymour, P.: Induced subgraphs of graphs with large chromatic number. V. Chandeliers and strings. J. Comb. Theory, Ser. B **150**, 195–243 (2021). https://doi.org/10.1016/j.jctb.2021.05.001
11. Courcelle, B., Makowsky, J.A., Rotics, U.: Linear time solvable optimization problems on graphs of bounded clique-width. In: Hromkovič, J., Sýkora, O. (eds.) Graph-Theoretic Concepts in Computer Science, 24th International Workshop, WG 1998, Smolenice Castle, Slovak Republic, 18–20 June 1998, Proceedings, Lecture Notes in Computer Science, vol. 1517, pp. 1–16. Springer (1998). https://doi.org/10.1007/10692760_1
12. Courcelle, B., Makowsky, J.A., Rotics, U.: Linear time solvable optimization problems on graphs of bounded clique-width. Theory Comput. Syst. **33**(2), 125–150 (2000). https://doi.org/10.1007/S002249910009
13. Cropper, M., Gyárfás, A., Lehel, J.: Hall ratio of the Mycielski graphs. Discret. Math. **306**(16), 1988–1990 (2006). https://doi.org/10.1016/j.disc.2005.09.020
14. Davies, J.: Triangle-free graphs with large chromatic number and no induced wheel. J. Graph Theory **103**(1), 112–118 (2023). https://doi.org/10.1002/jgt.22906
15. Descartes, B.: Solution to advanced problem no. 4526. Am. Math. Monthly **61**(5), 352–353 (1954). https://doi.org/10.2307/2307489

16. Dvořák, Z., Král', D.: Classes of graphs with small rank decompositions are χ-bounded. Eur. J. Comb. **33**(4), 679–683 (2012). https://doi.org/10.1016/j.ejc.2011.12.005
17. Fowler, R.J., Paterson, M.S., Tanimoto, S.L.: Optimal packing and covering in the plane are NP-complete. Inf. Process. Lett. **12**(3), 133–137 (1981). https://doi.org/10.1016/0020-0190(81)90111-3
18. Frank, A.: Some polynomial algorithms for certain graphs and hypergraphs. In: Nash-Williams, C., Sheehan, J. (eds.) Proceedings of the Fifth British Combinatorial Conference, Congressus Numerantium, vol. XV, pp. 211–226. Utilitas Mathematica (1976)
19. Gavril, F.: Algorithms for a maximum clique and a maximum independent set of a circle graph. Networks **3**(3), 261–273 (1973). https://doi.org/10.1002/net.3230030305
20. Gavril, F.: Maximum weight independent sets and cliques in intersection graphs of filaments. Inf. Process. Lett. **73**(5–6), 181–188 (2000). https://doi.org/10.1016/S0020-0190(00)00025-9
21. Grötschel, M., Lovász, L., Schrijver, A.: Polynomial algorithms for perfect graphs. In: Berge, C., Chvátal, V. (eds.) Topics on Perfect Graphs, North-Holland Mathematics Studies, vol. 88, pp. 325–356. North-Holland (1984). https://doi.org/10.1016/S0304-0208(08)72943-8
22. Grzesik, A., Klimošová, T., Pilipczuk, M., Pilipczuk, M.: Polynomial-time algorithm for maximum weight independent Set on P_6-free graphs. ACM Trans. Algorithms **18**(1), Article 4 (2022). https://doi.org/10.1145/3414473
23. Gyárfás, A.: On the chromatic number of multiple interval graphs and overlap graphs. Discret. Math. **55**(2), 161–166 (1985). https://doi.org/10.1016/0012-365X(85)90044-5
24. Gyárfás, A.: Problems from the world surrounding perfect graphs. Applicationes Mathematicae **19**(3–4), 413–441 (1987), https://eudml.org/doc/265405
25. Kratochvíl, J.: A special planar satisfiability problem and a consequence of its NP-completeness. Discret. Appl. Math. **52**(3), 233–252 (1994). https://doi.org/10.1016/0166-218X(94)90143-0
26. Kratochvíl, J., Nešetřil, J.: Independent set and clique problems in intersection-defined classes of graphs. Commentationes Mathematicae Universitatis Carolinae **31**(1), 85–93 (1990), https://eudml.org/doc/17810
27. Krawczyk, T., Pawlik, A., Walczak, B.: Coloring triangle-free rectangular frame intersection graphs with O(loglog$n$) colors. In: Brandstädt, A., Jansen, K., Reischuk, R. (eds.) WG 2013. LNCS, vol. 8165, pp. 333–344. Springer, Heidelberg (2013). https://doi.org/10.1007/978-3-642-45043-3_29
28. Krawczyk, T., Pawlik, A., Walczak, B.: Coloring triangle-free rectangle overlap graphs with $O(\log\log n)$ colors. Discrete Comput. Geom. **53**(1), 199–220 (2015). https://doi.org/10.1007/s00454-014-9640-3
29. Krawczyk, T., Walczak, B.: On-line approach to off-line coloring problems on graphs with geometric representations. Combinatorica **37**(6), 1139–1179 (2017). https://doi.org/10.1007/s00493-016-3414-x
30. Lokshtanov, D., Vatshelle, M., Villanger, Y.: Independent set in P_5-free graphs in polynomial time. In: Chekuri, C. (ed.) Proceedings of the Twenty-Fifth Annual ACM-SIAM Symposium on Discrete Algorithms, SODA 2014, Portland, Oregon, USA, 5–7 January 2014, pp. 570–581. SIAM (2014). https://doi.org/10.1137/1.9781611973402.43

31. Lozin, V.V., Milanič, M.: A polynomial algorithm to find an independent set of maximum weight in a fork-free graph. J. Discrete Algorithms **6**(4), 595–604 (2008). https://doi.org/10.1016/j.jda.2008.04.001
32. Marin, M., Thomassé, S., Trotignon, N., Watrigant, R.: A structural description of Zykov and Blanche Descartes graphs. In: Fernau, H., Kindermann, P. (eds.) Graph-Theoretic Concepts in Computer Science, 51st International Workshop, WG 2025, Otzenhausen, Germany, 11–13 June 2025, pp. 377–389, Revised Papers. LNCS, Springer (2025)
33. Minty, G.J.: On maximal independent sets of vertices in claw-free graphs. J. Comb. Theory, Ser. B **28**(3), 284–304 (1980). https://doi.org/10.1016/0095-8956(80)90074-X
34. Mycielski, J.: Sur le coloriage des graphs. Colloquium Mathematicae **3**(2), 161–162 (1955), https://eudml.org/doc/210000
35. Pawlik, A., et al.: Triangle-free geometric intersection graphs with large chromatic number. Discret. Comput. Geom. **50**(3), 714–726 (2013). https://doi.org/10.1007/s00454-013-9534-9
36. Pawlik, A., et al.: Triangle-free intersection graphs of line segments with large chromatic number. J. Comb. Theory, Ser. B **105**, 6–10 (2014). https://doi.org/10.1016/j.jctb.2013.11.001
37. Pournajafi, P., Trotignon, N.: Burling graphs revisited, part I: new characterizations. Eur. J. Comb. **110**, Article 103686 (2023). https://doi.org/10.1016/j.ejc.2023.103686
38. Pournajafi, P., Trotignon, N.: Burling graphs revisited, part II: structure. Eur. J. Comb. **116**, Article 103849 (2024). https://doi.org/10.1016/j.ejc.2023.103849
39. Pournajafi, P., Trotignon, N.: Burling graphs revisited, part III: applications to χ-boundedness. Eur. J. Comb. **116**, Article 103850 (2024). https://doi.org/10.1016/j.ejc.2023.103850
40. Rzążewski, P., Walczak, B.: Polynomial-time recognition and maximum independent set in Burling graphs. arXiv:2407.16666 (2024). https://doi.org/10.48550/arXiv.2407.16666
41. Sbihi, N.: Algorithme de recherche d'un stable de cardinalité maximum dans un graphe sans étoile. Discret. Math. **29**(1), 53–76 (1980). https://doi.org/10.1016/0012-365X(90)90287-R
42. Scott, A.D.: Induced trees in graphs of large chromatic number. J. Graph Theory **24**(4), 297–311 (1997). https://doi.org/10.1002/(SICI)1097-0118(199704)24:4<297::AID-JGT2>3.0.CO;2-J
43. Thomassé, S., Trotignon, N., Vušković, K.: A polynomial Turing-kernel for weighted independent set in bull-free graphs. Algorithmica **77**(3), 619–641 (2017). https://doi.org/10.1007/s00453-015-0083-x
44. Yolov, N.: Minor-matching hypertree width. In: Czumaj, A. (ed.) Proceedings of the Twenty-Ninth Annual ACM-SIAM Symposium on Discrete Algorithms, SODA 2018, New Orleans, LA, USA, 7–10 January 2018, pp. 219–233. SIAM (2018). https://doi.org/10.1137/1.9781611975031.16
45. Zykov, A.A.: O nekotorykh svoystvakh lineynykh kompleksov. Matematicheskii Sbornik, Novaya Seriya **24**(66)(2), 163–188 (1949), https://www.mathnet.ru/eng/sm5974

Author Index

H. Fernau and P. Kindermann (Eds.): WG 2025, LNCS 16124, pp. 461–462, 2026.
https://doi.org/10.1007/978-3-032-11835-6

MIX
Papier aus verantwortungsvollen Quellen
Paper from responsible sources
FSC® C105338

If you have any concerns about our products,
you can contact us on
ProductSafety@springernature.com

In case Publisher is established outside the EU,
the EU authorized representative is:
Springer Nature Customer Service Center GmbH
Europaplatz 3, 69115 Heidelberg, Germany

Printed by Libri Plureos GmbH
in Hamburg, Germany